IMPORTANT FORMULAS

- **Distance** between $P_1(x_1, y_1)$ and $P_2(x_2, y_2)$:

$$d(P_1, P_2) = \sqrt{(x_2 - x_1)^2 + (y_2 - y_1)^2}$$

- **Equation of a circle** with center $C(h,k)$ and radius r:

$$(x - h)^2 + (y - k)^2 = r^2$$

- **Slope** m of the line through $P_1(x_1, y_1)$ and $P_2(x_2, y_2)$:

$$m = \frac{y_2 - y_1}{x_2 - x_1}$$

- **Point-slope form** for a line of slope m through $P_1(x_1, y_1)$:

$$y - y_1 = m(x - x_1)$$

- **Slope-intercept form** for a line of slope m and y-intercept b:

$$y = mx + b$$

- **Linear function** f: $f(x) = ax + b$

- **Quadratic function** f: $f(x) = ax^2 + bx + c$

- **Polynomial function** f:

$$f(x) = a_n x^n + a_{n-1}x^{n-1} + \cdots + a_1 x + a_0$$

- **Rational function** f:

$$f(x) = \frac{h(x)}{g(x)}$$

where h and g are polynomial functions

- **Pythagorean Theorem**

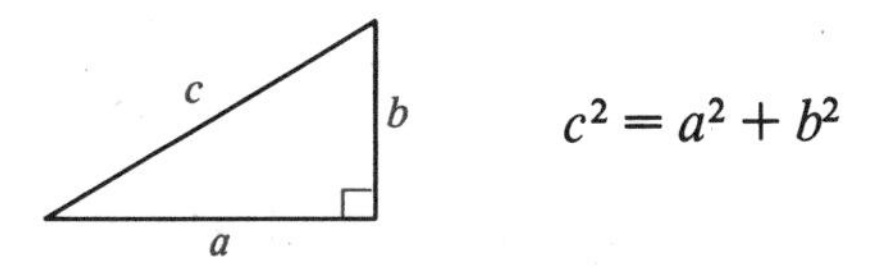

$$c^2 = a^2 + b^2$$

Formulas for area A, circumference C, volume V, and curved surface area S:

TRIANGLE

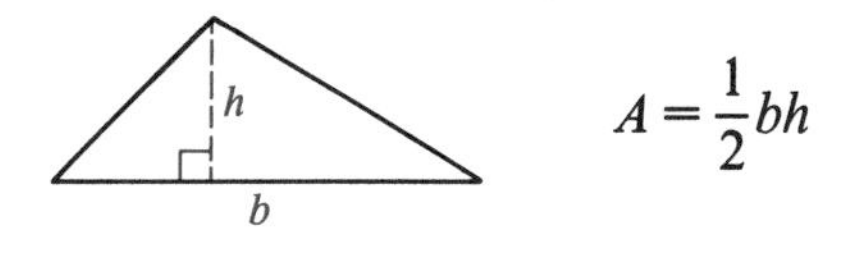

$$A = \frac{1}{2}bh$$

CIRCLE

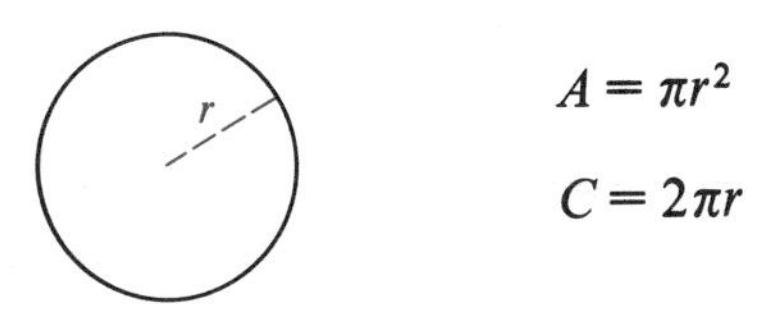

$$A = \pi r^2$$

$$C = 2\pi r$$

RIGHT CIRCULAR CYLINDER

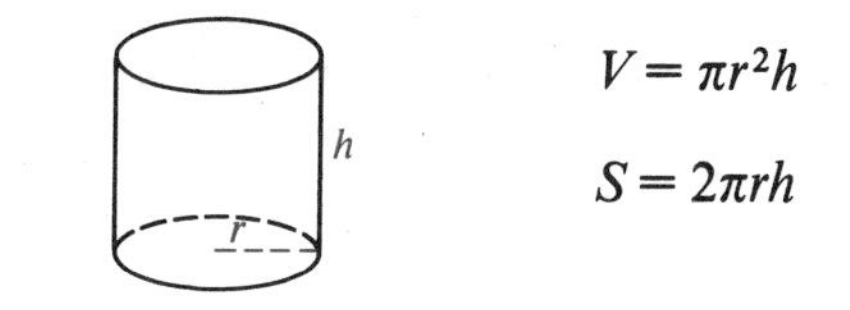

$$V = \pi r^2 h$$

$$S = 2\pi rh$$

PARALLELOGRAM

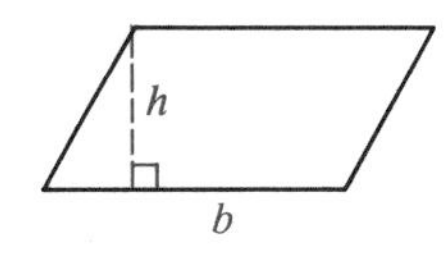

$$A = bh$$

SPHERE

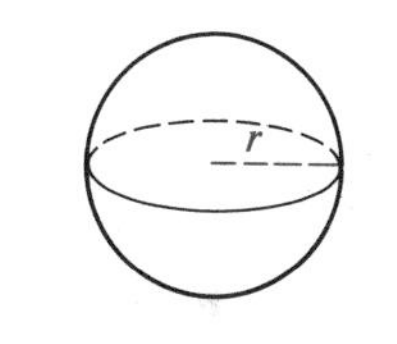

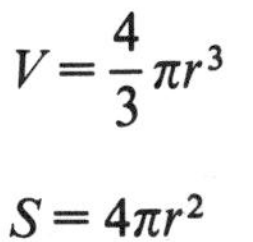

$$V = \frac{4}{3}\pi r^3$$

$$S = 4\pi r^2$$

RIGHT CIRCULAR CONE

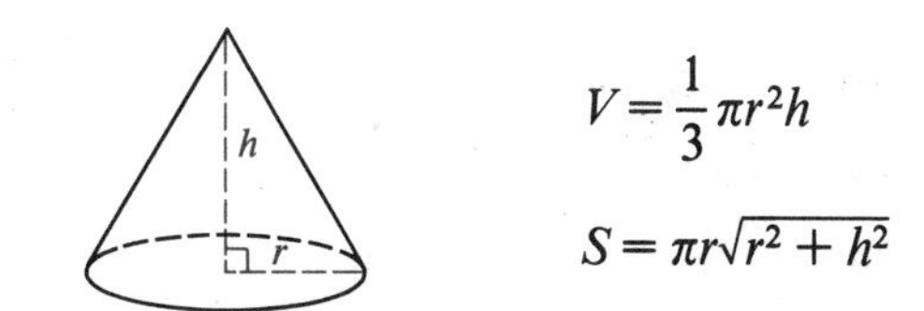

$$V = \frac{1}{3}\pi r^2 h$$

$$S = \pi r\sqrt{r^2 + h^2}$$

PRECALCULUS:
Functions and Graphs

PRECALCULUS:
Functions and Graphs

Fifth Edition

Earl W. Swokowski
Marquette University

Prindle, Weber & Schmidt ▪ Boston

PWS PUBLISHERS

Prindle, Weber & Schmidt • Duxbury Press • PWS Engineering • Breton Publishers •
20 Park Plaza • Boston, Massachusetts 02116

The previous edition of this text was published as *Functions and Graphs,* Fourth Edition.

PWS Publishers is a division of Wadsworth, Inc.

90 89 88 87 – 10 9 8 7 6 5 4 3 2 1

Printed in the United States of America

COVER IMAGE: *Four Courts, Dublin,* a dyptich by Jan Dibbets.

Photo collages with lithographs reproduced by courtesy of the publisher, Multiples, Inc., New York, and the owner, Sonesta International Hotels Corporation, Boston.

LIBRARY OF CONGRESS CATALOGING IN PUBLICATION DATA

Swokowski, Earl William.
Precalculus: functions and graphs.

Rev. ed. of: Functions and graphs. 4th ed. c1984.
Includes index.
1. Functions. 2. Algebra—Graphic methods. 3. Trigonometry.
I. Swokowski, Earl William. Functions and graphs. II. Title.
QA331.3.S95 1987 512′.1 86-25344
ISBN 0-87150-060-4

Production and design by Kathi Townes
Technical artwork by J & R Services
Text composition by Syntax International Pte. Ltd.
Text printed and bound by Arcata/Hawkins
Cover printed by John P. Pow Company, Inc.

Preface

This book is a major revision of the previous four editions of *Functions and Graphs*. One of my goals was to maintain the mathematical soundness of earlier editions, while making discussions somewhat less formal by rewriting, placing more emphasis on graphs, and by adding new examples and figures. Another objective was to stress the usefulness of the subject matter through a great variety of applied problems from many different disciplines. Finally, suggestions for improvements from users of previous editions led me to change the order in which certain topics are presented. The comments that follow highlight some of the changes and features of this edition.

CHANGES FOR THE FIFTH EDITION

- The review material in Chapter 1 is streamlined. The concepts that formerly appeared in five sections are now presented in three.
- The work on rectangular coordinate systems and lines was moved to Chapter 1.
- Chapter 2 now begins with the definition of function, and greater emphasis is given to the graphical interpretations of domain and range.
- Section 2.4, *Operations on Functions,* has not appeared in previous editions.
- The discussion of inverse functions in Section 2.5 is simplified and integrated with the concept of a one-to-one function.
- In Chapter 3, the material on division of polynomials, synthetic division, and zeros of polynomials has been reorganized.
- Chapter 4 has been completely rewritten, with much more attention given to the natural exponential and logarithmic functions and their applications.
- Logarithmic and trigonometric tables have been deemphasized, since calculators are much more efficient and accurate. Teachers who feel that students should be instructed on the use of tables and linear interpolation will find suitable material in Appendix I.
- The subject matter of trigonometry is divided into three chapters instead of two.
- Section 6.6 on the inverse trigonometric functions has been rewritten, with better motivation being given for domains and graphs.
- Applications of oblique triangles and vectors are given greater emphasis in Chapter 7.

- In Chapter 8 the method of finding solutions of systems of linear equations by means of the echelon form of a matrix is stressed.
- Partial fractions are introduced in Section 8.4 as an application of systems of linear equations.

FEATURES

Applications Previous editions contained applied problems, but most of them were in the fields of engineering, physics, chemistry, or biology. In this revision other subjects are also considered, such as physiology, medicine, sociology, ecology, oceanography, marine biology, business, and economics.

Examples Each section contains carefully chosen examples to help students understand and assimilate new concepts. Whenever feasible, applications are included to demonstrate the usefulness of the subject matter.

Exercises Exercise sets begin with routine drill problems and gradually progress to more difficult types. As a rule, applied problems appear near the end of the set, to allow students to gain confidence in manipulations and new ideas before attempting questions that require an analysis of practical situations.

There is a review section at the end of each chapter, consisting of a list of important topics and pertinent exercises.

Answers to the odd-numbered exercises are given at the end of the text. Instructors may obtain an answer booklet for the even-numbered exercises from the publisher.

Calculators Calculators are given more emphasis in this edition. It is possible to work most of the exercises without the aid of a calculator; however, instructors may wish to encourage their use to shorten numerical computations. Some sections contain problems labeled *Calculator Exercises,* for which a calculator should definitely be employed.

Text Design A new use of a second color for figures and statements of important facts should make it easier to follow discussions and remember major ideas. Graphs are usually labeled and color-coded to clarify complex figures. Many figures have been added to exercise sets to help visualize important aspects of applied problems.

Flexibility Syllabi from schools that used previous editions attest to the flexibility of the text. Sections and chapters can be rearranged in many different ways, depending on the objectives and the length of the course.

SUPPLEMENTS

Users of this text may obtain the following supplements from the publisher.

- *Complete, worked-out solutions* for instructors.
- *Students Solutions Manual* containing worked-out solutions for approximately one-third of the exercises, authored by Stephen Rodi of Austin Community College.
- *Transparency Masters* containing key definitions, theorems, and figures from the text.
- *Printed Test Items* containing three multiple-choice tests per chapter.
- *Computerized Tests* are available on disk. A computerized test generator, called **PWStest,** will produce unlimited variations of problems that test to many of the learning objectives in the book. It may be used to print final examinations, quizzes, or practice material for students having difficulties with particular concepts. Directions for using the software is included.

PWStest is provided free to schools that order *Precalculus: Functions and Graphs*, Fifth Edition from PWS Publishers.

ACKNOWLEDGMENTS

I wish to thank Michael Cullen of Loyola Marymount University for supplying all the new exercises dealing with applications. This large assortment of problems provides strong motivation for the mathematical concepts introduced in the text. Because of his significant input on exercise sets, Michael should be considered as a coauthor of this edition.

This revision has benefited from the comments of many instructors who have used my texts. In particular, I wish to thank the following individuals, who reviewed the manuscript either for this edition of *Precalculus: Functions and Graphs* or for the sixth editions of my precalculus series.

James Arnold, University of Wisconsin–Milwaukee
Thomas A. Atchison, Stephen F. Austin State University
Jeffery A. Cole, Anoka-Ramsey Community College
William E. Coppage, Wright State University
Franklin Demana, The Ohio State University
Phillip Eastman, Boise State University
Carol Edwards, St. Louis Community College
Christopher Ennis, University of Minnesota
Leonard E. Fuller, Kansas State University
James E. Hall, Westminster College
E. John Hornsby, Jr., University of New Orleans
Anne Hudson, Armstrong State College
E. Glenn Kindle, Front Range Community College
Robert H. Lohman, Kent State University
Laurence Maher, North Texas State University
Mike Mears, Manatee Community College
Richard Randell, University of Iowa
William H. Robinson, Ventura College
Albert R. Siegrist, University College of the University of Cincinnati
Christine A. Spengel, California State University–Northridge
George L. Szoke, University of Akron
Michael D. Taylor, University of Central Florida

I am grateful for the excellent cooperation of the staff of Prindle, Weber & Schmidt. Two people in the company deserve special mention. They are Senior Editor David Pallai and my production editor Kathi Townes. The present form of the book was greatly influenced by their efforts, and I owe them both a debt of gratitude. Moreover, their personal friendship has often been a source of comfort during the years we have worked together.

In addition to all of the persons named here, I express my sincere appreciation to the many unnamed students and teachers who have helped shape my views on mathematics education.

EARL W. SWOKOWSKI

Contents

CHAPTER 1

Real Numbers and Graphs

The material in this chapter is basic to the study of precalculus mathematics. ■ After reviewing concepts involving real numbers, we turn our attention to exponents and radicals, and how they may be used to simplify algebraic expressions. ■ Solutions of inequalities are considered next. ■ The chapter concludes with a discussion of rectangular coordinate systems and lines.

1.1 Real Numbers

Real numbers are employed in all phases of mathematics, and you are undoubtedly acquainted with symbols used to represent them such as

$$73, \quad -5, \quad \tfrac{49}{12}, \quad \sqrt{2}, \quad \sqrt[3]{-85}, \quad -8.674, \quad 0.33333\ldots, \quad 596.25.$$

We shall assume familiarity with the fundamental properties of addition, subtraction, multiplication, and division. Throughout this chapter, unless otherwise specified, lowercase letters $a, b, c, \ldots$ will denote real numbers.

The **positive integers** $1, 2, 3, 4, \ldots$ may be obtained by adding the real number 1 successively to itself. The **integers** consist of all positive and negative integers together with the real number 0. A **rational number** is a real number that can be expressed as a quotient a/b, where a and b are integers and $b \neq 0$. Real numbers that are not rational are called **irrational.** The ratio of the circumference of a circle to its diameter is irrational. This real number is denoted by π and the notation $\pi \approx 3.1416$ is used to indicate that π is *approximately equal* to 3.1416. Another familiar example of an irrational number is $\sqrt{2}$.

Real numbers are often expressed in terms of decimals. For rational numbers the decimals are either terminating or repeating, such as $\frac{5}{4} = 1.25$ or $\frac{177}{55} = 3.2181818\ldots$, where the digits 1 and 8 repeat indefinitely. Decimal representations for irrational numbers are always nonterminating and nonrepeating.

Real numbers may be represented geometrically by points on a line l in such a way that for each real number a there corresponds one and only one point, and conversely, to each point P on l there corresponds precisely one real number. Such an association is referred to as a **one-to-one correspondence.** We first choose an arbitrary point O, called the **origin,** and associate with it the real number 0. Points associated with the integers are then determined by laying off successive line segments of equal length on either side of O as illustrated in Figure 1.1. The points corresponding to rational numbers such as $\frac{23}{5}$ and $-\frac{1}{2}$ are obtained by subdividing the equal line segments. Points associated with irrational numbers such as π can be approximated to within any degree of accuracy by locating succes-

FIGURE 1.1

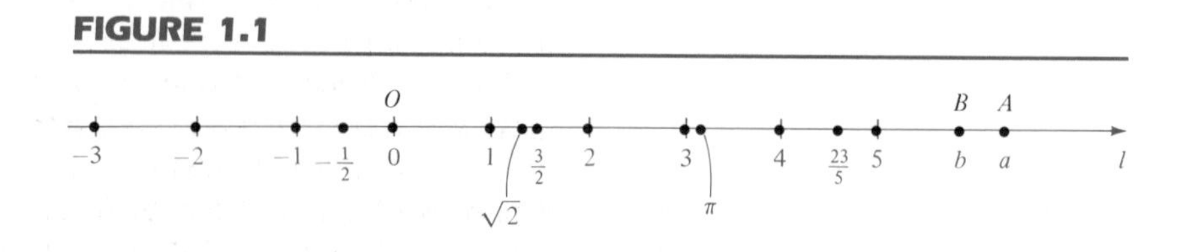

sively the points corresponding to 3, 3.1, 3.14, 3.141, 3.1415, 3.14159, and so on.

The number a that is associated with a point A on l is called the **coordinate** of A. An assignment of coordinates to points on l is called a **coordinate system** for l, and l is called a **coordinate line,** or a **real line.** A direction can be assigned to l by taking the **positive direction** along l to the right and the **negative direction** to the left. The positive direction is noted by placing an arrowhead on l as shown in Figure 1.1. Numbers that correspond to points to the right of O in Figure 1.1 are called **positive real numbers,** whereas those that correspond to points to the left of O are **negative real numbers.** The real number 0 is neither positive nor negative.

At times it is convenient to use the notation and terminology of sets. A **set** may be thought of as a collection of objects of some type. The objects are called **elements** of the set. Throughout our work $\mathbb{R}$ will denote the set of real numbers and $\mathbb{Z}$ the set of integers. If every element of a set S is also an element of a set T, then S is called a **subset** of T. For example, $\mathbb{Z}$ is a subset of $\mathbb{R}$. Two sets S and T are said to be **equal,** written $S = T$, if S and T contain precisely the same elements. The notation $S \neq T$ means that S and T are not equal.

We frequently use symbols to represent arbitrary elements of a set. For example, we may use x to denote a real number, although no *particular* real number is specified. A letter that is used to represent any element of a given set is sometimes called a **variable.** A symbol that represents a *specific* element is called a **constant.** In most of our work, letters near the end of the alphabet, such as x, y, and z, will be used for variables, whereas letters such as a, b, and c will denote constants. Throughout this text, unless otherwise specified, variables represent real numbers. The **domain of a variable** is the set of real numbers represented by the variable. To illustrate, $\sqrt{x}$ is a real number if and only if $x \geq 0$, and hence the domain of x is the set of nonnegative real numbers. Similarly, given the expression $1/(x - 2)$, we must exclude $x = 2$ in order to avoid division by zero; consequently, in this case the domain is the set of all real numbers different from 2.

If the elements of a set S have a certain property, we sometimes write $S = \{x: \quad\}$, where the property describing the variable x is stated in the space after the colon. For example, $\{x: x \text{ is an even integer}\}$ denotes the set of all even integers. Finite sets are sometimes identified by listing all the elements within braces. Thus, if the set T consists of the first five positive integers, we may write $T = \{1, 2, 3, 4, 5\}$.

If a and b are real numbers and $a - b$ is positive, we say that ***a* is greater than *b*** and write $a > b$. An equivalent statement is ***b* is less than *a*,** written $b < a$. The symbols $>$ and $<$ are called **inequality signs,** and expressions such as $a > b$ or $b < a$ are called **inequalities.** Referring to Figure 1.1, we see that if A and B are points on l with coordinates a and

b, respectively, then $a > b$ (or $b < a$) *if and only if A lies to the right of B.* As illustrations,

$$5 > 3, \qquad -6 < -2, \qquad -\sqrt{2} < 1, \qquad 2 > 0, \qquad -5 < 0.$$

Note in general that $a > 0$ if and only if a is positive, and $a < 0$ if and only if a is negative. We sometimes refer to the **sign** of a real number as being positive or negative if the number is positive or negative, respectively.

The notation $a \geq b$, which is read ***a* is greater than or equal to *b*,** means that either $a > b$ or $a = b$ (but not both). The symbol $a \leq b$ is read ***a* is less than or equal to *b*** and means that either $a < b$ or $a = b$. The expression $a < b < c$ means that both $a < b$ and $b < c$, in which case we say that ***b* is between *a* and *c*.** We may also write $c > b > a$. For instance,

$$1 < 5 < \tfrac{11}{2}, \qquad -4 < \tfrac{2}{3} < \sqrt{2}, \qquad 3 > -6 > -10.$$

Other variations of the inequality notation are used. For example, $a < b \leq c$ means both $a < b$ and $b \leq c$. Similarly, $a \leq b < c$ means both $a \leq b$ and $b < c$. Finally, $a \leq b \leq c$ means both $a \leq b$ and $b \leq c$.

If $a < b$, the symbol (a, b) is sometimes used to denote all real numbers between a and b. This set is called an **open interval.**

DEFINITION OF OPEN INTERVAL

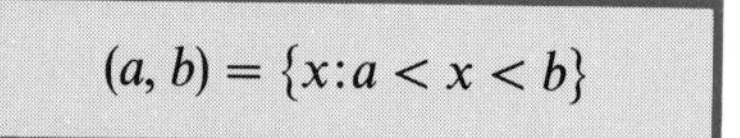

$$(a, b) = \{x : a < x < b\}$$

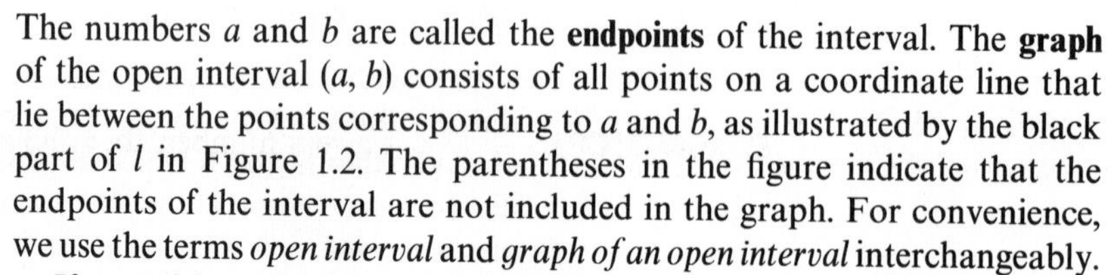

The numbers a and b are called the **endpoints** of the interval. The **graph** of the open interval (a, b) consists of all points on a coordinate line that lie between the points corresponding to a and b, as illustrated by the black part of l in Figure 1.2. The parentheses in the figure indicate that the endpoints of the interval are not included in the graph. For convenience, we use the terms *open interval* and *graph of an open interval* interchangeably.

FIGURE 1.2

Open intervals (a, b), $(-1, 3)$, and $(2, 4)$.

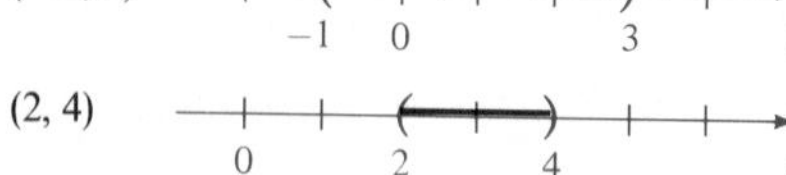

If we wish to include an endpoint of an interval, a bracket is used instead of a parenthesis. If $a < b$, then **closed intervals,** denoted by $[a, b]$, and **half-open intervals,** denoted by $[a, b)$ or $(a, b]$, are defined as follows.

$$[a, b] = \{x : a \leq x \leq b\}$$

$$[a, b) = \{x : a \leq x < b\}$$

$$(a, b] = \{x : a < x \leq b\}$$

FIGURE 1.3

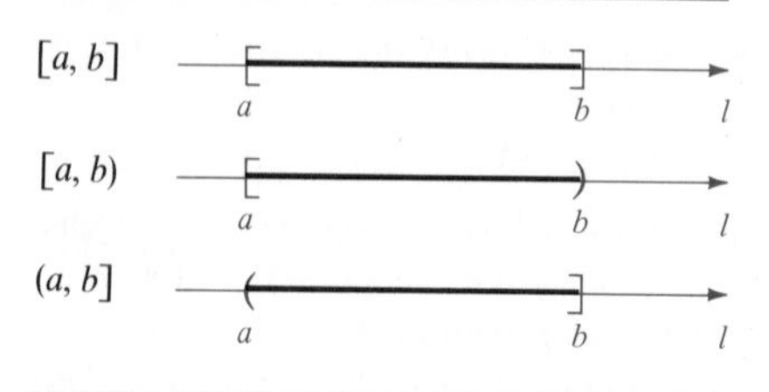

Typical graphs are sketched in Figure 1.3. A bracket indicates that the corresponding endpoint is part of the graph.

We shall sometimes employ certain **infinite intervals.** Specifically, if a is a real number, we defined

$$(-\infty, a) = \{x : x < a\}.$$

Thus, $(-\infty, a)$ denotes the set of all real numbers less than a. The symbol ∞ (**infinity**) is merely a notational device and does not represent a real number.

If we wish to include the point corresponding to a we write

$$(-\infty, a] = \{x : x \leq a\}.$$

FIGURE 1.4

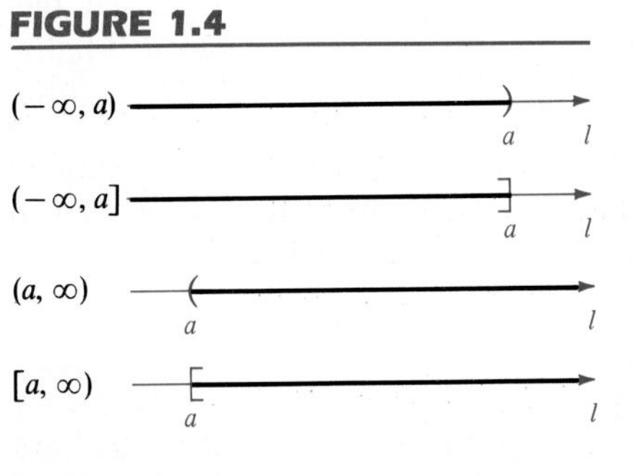

Other types of infinite intervals are defined by

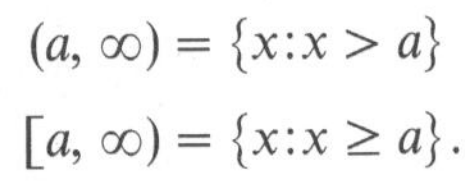

$$(a, \infty) = \{x : x > a\}$$
$$[a, \infty) = \{x : x \geq a\}.$$

The set $\mathbb{R}$ of real numbers is sometimes denoted by $(-\infty, \infty)$.

Graphs of infinite intervals for an arbitrary real number a are sketched in Figure 1.4. The absence of a parentheses or bracket, on the left for the graphs of $(-\infty, a)$ and $(-\infty, a]$, and on the right for (a, ∞) and $[a, \infty)$, indicates that the black portions extend indefinitely.

FIGURE 1.5

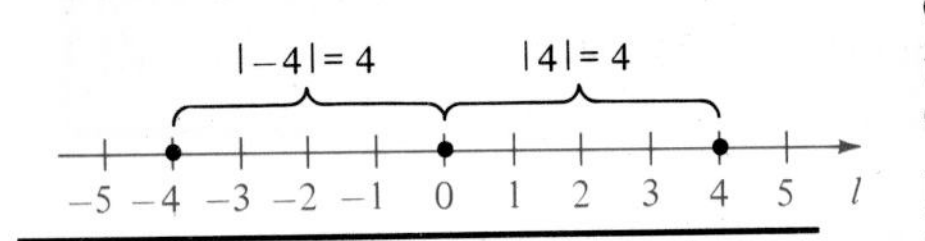

If a is a real number, then it is the coordinate of some point A on a coordinate line l, and the symbol $|a|$ is used to denote the number of units (or distance) between A and the origin, without regard to direction. The nonnegative number $|a|$ is called the *absolute value* of a. Referring to Figure 1.5, we see that for the point with coordinate -4, we have $|-4| = 4$. Similarly, $|4| = 4$. In general, if a is negative we change its sign to find $|a|$, whereas if a is nonnegative, then $|a| = a$. The next definition summarizes this discussion.

DEFINITION

If a is a real number, then the **absolute value** of a, denoted by $|a|$, is

$$|a| = \begin{cases} a & \text{if } a \geq 0. \\ -a & \text{if } a < 0. \end{cases}$$

The use of this definition is illustrated in the following example.

EXAMPLE 1 Find $|3|$, $|-3|$, $|0|$, $|\sqrt{2} - 2|$, and $|2 - \sqrt{2}|$.

SOLUTION Since 3, $2 - \sqrt{2}$, and 0 are nonnegative,

$$|3| = 3, \quad |2 - \sqrt{2}| = 2 - \sqrt{2}, \quad \text{and} \quad |0| = 0.$$

Since -3 and $\sqrt{2} - 2$ are negative, we use the formula $|a| = -a$ to obtain

$$|-3| = -(-3) = 3 \quad \text{and} \quad |\sqrt{2} - 2| = -(\sqrt{2} - 2) = 2 - \sqrt{2}.$$ ∎

Note that in Example 1, $|-3| = |3|$ and $|2 - \sqrt{2}| = |\sqrt{2} - 2|$. It can be shown that

$$|a| = |-a| \quad \text{for every real number } a.$$

We shall use the concept of absolute value to define the distance between any two points on a coordinate line. Let us begin by noting that the distance between the points with coordinates 2 and 7, shown in Figure 1.6, equals 5 units on l. This distance is the difference, $7 - 2$, obtained by subtracting the smaller coordinate from the larger. If we employ absolute values, then since $|7 - 2| = |2 - 7|$, it is unnecessary to be concerned about the order of subtraction. We shall use this as our motivation for the next definition.

FIGURE 1.6

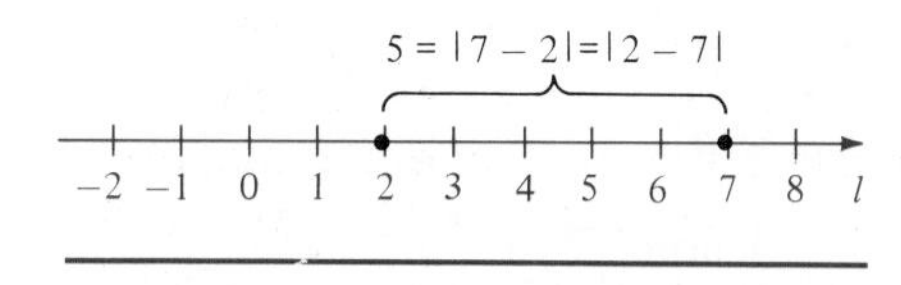

DEFINITION

> Let a and b be the coordinates of two points A and B, respectively, on a coordinate line l. The **distance between A and B,** denoted by $d(A, B)$, is
>
> $$d(A, B) = |b - a|.$$

The number $d(A, B)$ is also called the **length of the line segment AB.** Observe that since $d(B, A) = |a - b|$ and $|b - a| = |a - b|$,

$$d(A, B) = d(B, A).$$

Also note that the distance between the origin O and the point A is

$$d(O, A) = |a - 0| = |a|.$$

EXAMPLE 2 Let A, B, C, and D have coordinates -5, -3, 1, and 6, respectively, on a coordinate line l (see Figure 1.7). Find $d(A, B)$, $d(C, B)$, $d(O, A)$, and $d(C, D)$.

SOLUTION Using the definition of the distance between points,

$$d(A, B) = |-3 - (-5)| = |-3 + 5| = |2| = 2$$
$$d(C, B) = |-3 - 1| = |-4| = 4$$
$$d(O, A) = |-5 - 0| = |-5| = 5$$
$$d(C, D) = |6 - 1| = |5| = 5$$

These answers can be checked by referring to Figure 1.7. ■

FIGURE 1.7

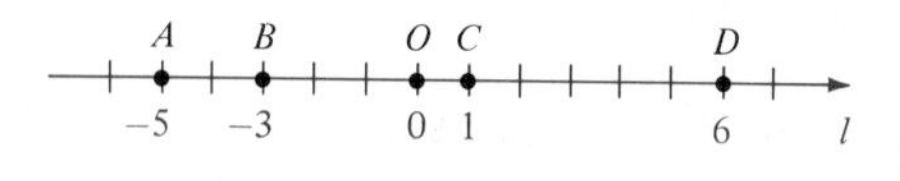

The concept of absolute value has uses other than that of finding distances between points. Generally, it is employed whenever we are interested in the "magnitude" or "numerical value" of a real number without regard to its sign.

If x is a variable, then expressions of the form

$$x + 3 = 0, \quad x^2 - 5 = 4x, \quad \text{or} \quad (x^2 - 9)\sqrt[3]{x + 1} = 0$$

are called **equations** in x. A number a is a **solution** or **root of an equation** if a true statement is obtained when a is substituted for x. We also say that a **satisfies** the equation. For example, 5 is a solution of the equation $x^2 - 5 = 4x$ since substitution gives us $(5)^2 - 5 = 4(5)$, or $20 = 20$, which is a true statement. To **solve** an equation means to find all the solutions.

An equation is called an **identity** if every number in the domain of the variable is a solution. An example of an identity is

$$\frac{1}{x^2 - 4} = \frac{1}{(x + 2)(x - 2)}$$

since this equation is true for every number in the domain of x. An equation is called a **conditional equation** if there are real numbers in the domain of the variable that are *not* solutions.

Two equations are **equivalent** if they have exactly the same solutions. For example, the equations

$$x = 3, \quad x - 1 = 2, \quad 5x = 15, \quad \text{and} \quad 2x + 1 = 7$$

are all equivalent.

We shall assume that the reader has had experience in finding solutions of equations in one variable. In particular, recall that the solutions of a **quadratic equation** $ax^2 + bx + c = 0$, for $a \neq 0$, may be obtained as follows.

QUADRATIC FORMULA

The solutions of the equation $ax^2 + bx + c = 0$, for $a \neq 0$, are given by

$$x = \frac{-b \pm \sqrt{b^2 - 4ac}}{2a}.$$

The solutions are real if $b^2 - 4ac$ is nonnegative. The case $b^2 - 4ac < 0$ is discussed in Chapter 3.

EXAMPLE 3 Find the solutions of $2x^2 + 7x - 15 = 0$.

SOLUTION Using the Quadratic Formula with $a = 2$, $b = 7$, and $c = -15$ gives us

$$x = \frac{-7 \pm \sqrt{49 + 120}}{4} = \frac{-7 \pm \sqrt{169}}{4} = \frac{-7 \pm 13}{4}.$$

Hence, $$x = \frac{-7 + 13}{4} = \frac{3}{2} \quad \text{or} \quad x = \frac{-7 - 13}{4} = -5.$$

Consequently, the solutions are $\frac{3}{2}$ and -5.

The given equation can also be solved by writing

$$2x^2 + 7x - 15 = (2x - 3)(x + 5) = 0.$$

Since a product of two real numbers can equal zero only if one of the factors is zero, we obtain

$$2x - 3 = 0 \quad \text{or} \quad x + 5 = 0.$$

This leads to the same solutions, $\frac{3}{2}$ and -5. ■

FIGURE 1.8

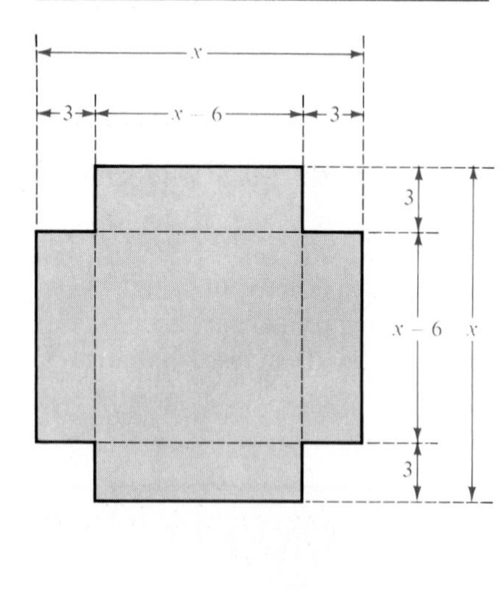

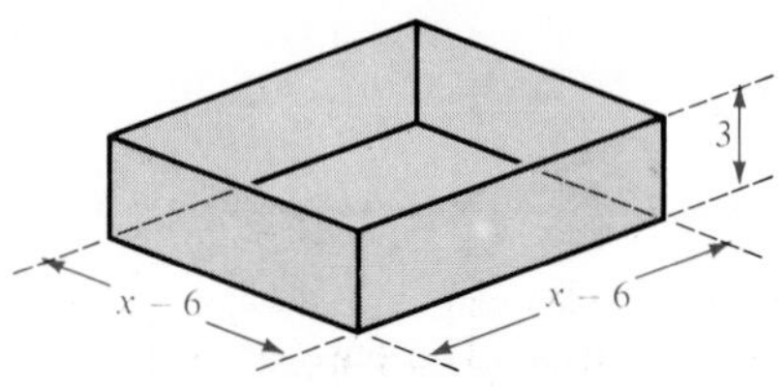

There are many applied problems that lead to quadratic equations. One is illustrated in the following example.

EXAMPLE 4 A box with a square base and no top is to be made from a square piece of tin by cutting out 3-inch squares from each corner and folding up the sides. If the box is to hold 48 cubic inches, what size piece of tin should be used?

SOLUTION We begin by drawing the picture in Figure 1.8, where x denotes the length of the side of the piece of tin.

Since the area of the base is $(x - 6)^2$ and the height is 3, we obtain the following fact:

$$\text{Volume of box} = 3(x - 6)^2.$$

Since the box is to hold 48 cubic inches,

$$3(x - 6)^2 = 48.$$

We may solve for x as follows:

$$(x-6)^2 = 16$$
$$x - 6 = \pm 4$$
$$x = 6 \pm 4.$$

Consequently, either $x = 10$ or $x = 2$.

Let us now check each of these numbers. Referring to Figure 1.8, we see that 2 is unacceptable, since no box is possible in this case. (Why?) However, if we begin with a 10-inch square of tin, cut out 3-inch corners, and fold, we obtain a box having dimensions 4 inches, 4 inches, and 3 inches. The box has the desired volume of 48 cubic inches. Thus, 10 inches is the answer to the problem. ■

As illustrated in Example 4, even though an equation is formulated correctly, it is possible, owing to the physical nature of a given problem, to arrive at meaningless solutions. These solutions should be discarded. For example, we would not accept the answer -7 years for the age of an individual, nor $\sqrt{50}$ for the number of automobiles in a parking lot.

Exercises 1.1

In Exercises 1–4 replace the symbol □ with either $<$, $>$, or $=$.

1 (a) $-7 \square -4$
(b) $3 \square -1$
(c) $1 + 3 \square 6 - 2$

2 (a) $-3 \square -5$
(b) $-6 \square 2$
(c) $\frac{1}{4} \square 0.25$

3 (a) $\frac{1}{3} \square 0.33$
(b) $\frac{125}{57} \square 2.193$
(c) $\frac{22}{7} \square \pi$

4 (a) $\frac{1}{7} \square 0.143$
(b) $\frac{3}{4} + \frac{2}{3} \square \frac{19}{12}$
(c) $\sqrt{2} \square 1.4$

Express the statements in Exercises 5–12 in terms of inequalities.

5 -8 is less than -5.

6 2 is greater than 1.9.

7 0 is greater than -1.

8 $\sqrt{2}$ is less than π.

9 x is negative.

10 y is nonnegative.

11 a is between 5 and 3.

12 b is between -2 and -6.

Express the intervals in Exercises 13–22 in terms of inequalities.

13 $(-2, 1)$

14 $[2, 6)$

15 $(3, 5]$

16 $[-3, -2]$

17 $[0, 2\pi]$

18 $(-\pi/2, \pi/2)$

19 $(-\infty, 2)$

20 $(-2, \infty)$

21 $[1, \infty)$

22 $(-\infty, 1]$

In Exercises 23–28, express the given inequality in interval notation and sketch a graph of the interval.

23 $2 < x < 5$

24 $-1 < x \leq 3$

25 $4 \geq x \geq 1$

26 $0 < x \leq 4$

27 $x > -1$

28 $x \leq 2$

29 The point on a coordinate line corresponding to $\sqrt{2}$ may be constructed by forming a right triangle with sides of length 1 as shown in the figure. Construct the points that correspond to $\sqrt{3}$ and $\sqrt{5}$, respectively. (*Hint:* Use the Pythagorean Theorem.)

FIGURE FOR EXERCISE 29

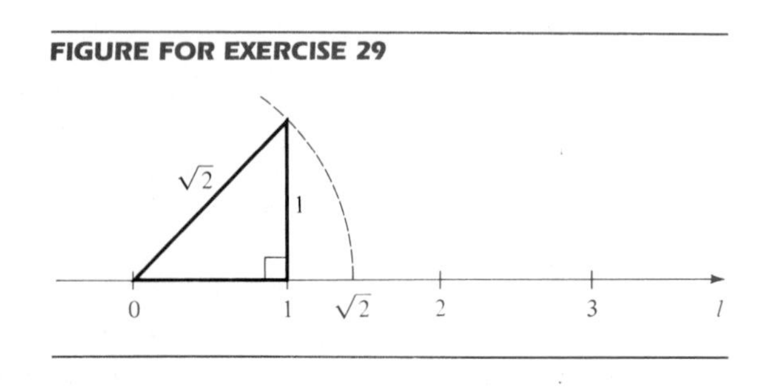

30 A circle of radius 1 rolls along a coordinate line in the positive direction. (See figure.) If point P is initially at the origin, what is the coordinate of P after (a) one complete rotation? (b) two complete rotations? (c) ten complete rotations?

FIGURE FOR EXERCISE 30

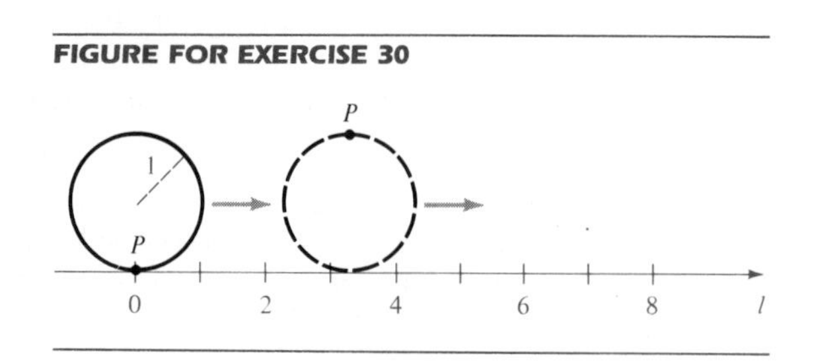

Rewrite the numbers in Exercises 31–34 without using symbols for absolute value.

31 (a) $|4 - 9|$
(b) $|-4| - |-9|$
(c) $|4| + |-9|$

32 (a) $|3 - 6|$
(b) $|0.2 - \frac{1}{5}|$
(c) $|-3| - |-4|$

33 (a) $3 - |-3|$
(b) $|\pi - 4|$
(c) $(-3)/|-3|$

34 (a) $|8 - 5|$
(b) $-5 + |-7|$
(c) $(-2)|-2|$

In Exercises 35–36 the given numbers are coordinates of three points A, B, and C (in that order) on a coordinate line l. For each, find:

(a) $d(A, B)$ (b) $d(B, C)$ (c) $d(C, B)$ and (d) $d(A, C)$.

35 $-6, -2, 4$

36 $3, 7, -5$

Use the Quadratic Formula to solve the equations in Exercises 37–44.

37 $2x^2 - x - 3 = 0$

38 $u^2 + 2u - 6 = 0$

39 $3x^2 - 2x - 8 = 0$

40 $v^2 + 3v - 5 = 0$

41 $2x^2 - 4x - 5 = 0$

42 $3x^2 - 6x + 2 = 0$

43 $4y^2 - 20y + 25 = 0$

44 $9t^2 + 6t + 1 = 0$

45 A large grain silo is to be constructed in the shape of a circular cylinder with a hemisphere attached to the top (see figure). The diameter of the silo is to be 30 ft, but the height is yet to be determined. Find the total height of the silo that will result in a capacity of $11{,}250\pi$ ft^3.

FIGURE FOR EXERCISE 45

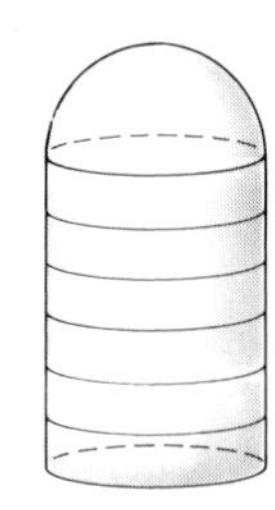

46 A farmer plans to use 180 feet of fencing to enclose a rectangular region, using part of a straight river bank instead of fencing as one side of the rectangle (see figure). Find the area of the region if the length of the side parallel to the river bank is

(a) twice the length of an adjacent side;

FIGURE FOR EXERCISE 46

(b) one-half the length of an adjacent side;

(c) the same as the length of an adjacent side.

47 An airplane flying north at 200 mph passed over a point on the ground at 2:00 P.M. Another airplane at the same altitude passed over the point at 2:30 P.M., flying east at 400 mph (see figure).

(a) If t denotes the time (in hours) after 2:30 P.M., express the distance d between the planes in terms of t.

(b) At what time were the airplanes 500 miles apart?

FIGURE FOR EXERCISE 47

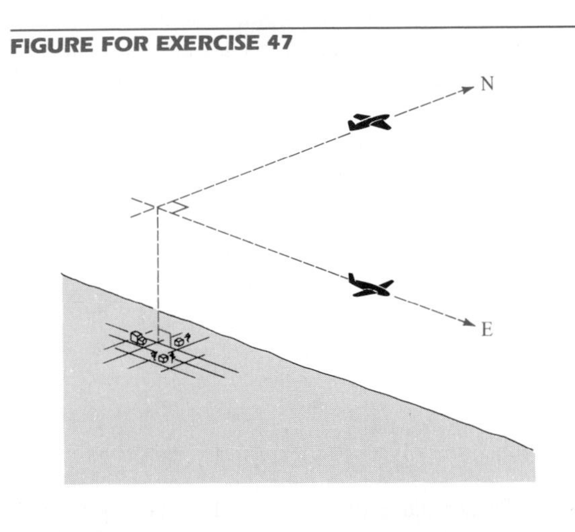

48 A piece of wire 100 inches long is cut into two pieces, and then each piece is bent into the shape of a square. If the sum of the enclosed areas is 397 in^2, find the lengths of each piece of wire.

49 The distance that a car travels between the time the driver makes the decision to hit the brakes and the time the car actually stops is called the *braking distance.* If a car is traveling v mph, the braking distance d (in feet) is approximated by $d = v + (v^2/20)$.

(a) Find the braking distance when v is 55 mph.

(b) A large cow stands peacefully in the middle of the road. If you begin to react when you are 120 feet from the cow, how fast can you be going and still avoid hitting the cow?

50 The temperature T (in °C) at which water boils is related to the elevation h (in meters above sea level) by the formula

$$h = 1000(100 - T) + 580(100 - T)^2.$$

(a) At what elevation does water boil at a temperature of 98 °C?

(b) The elevation of Mt. Everest is 29,000 feet (or 8840 meters). At what temperature will water boil at the top of this mountain? (*Hint:* Let $x = 100 - T$ and use the Quadratic Formula.)

1.2 Algebraic Expressions

If x is a variable, the symbol x^2 denotes $x \cdot x$. Similarly, $x^3 = x \cdot x \cdot x$. In general, if n is any positive integer,

$$x^n = x \cdot x \cdot \cdots \cdot x$$

where x appears n times on the right-hand side of the equal sign. The positive integer n is called the **exponent** of x in the expression x^n, and x^n is read "x to the nth **power**" or simply "x to the n."

It is important to remember that if n is a positive integer, then an expression such as $3x^n$ means $3(x^n)$ and not $(3x)^n$. The real number 3 is called the **coefficient** of x^n in the expression $3x^n$. Similarly, $-3x^n$ means $(-3)x^n$, not $(-3x)^n$. For example, we observe that

$$5 \cdot 2^4 = 5 \cdot 16 = 80 \quad \text{and} \quad -5 \cdot 2^4 = -5 \cdot 16 = -80.$$

If $x \neq 0$, the extension to nonpositive exponents is made by defining

$$x^0 = 1 \quad \text{and} \quad x^{-n} = \frac{1}{x^n}.$$

We shall assume familiarity with the following laws.

LAWS OF EXPONENTS

If x and y are real numbers and m and n are integers, then

(i) $x^m x^n = x^{m+n}$; (ii) $(x^m)^n = x^{mn}$; (iii) $(xy)^n = x^n y^n$;

(iv) $\left(\frac{x}{y}\right)^n = \frac{x^n}{y^n}$ if $y \neq 0$; (v) If $x \neq 0$, then $\frac{x^m}{x^n} = x^{m-n} = \frac{1}{x^{n-m}}$.

The Laws of Exponents can be extended to rules such as $(xyz)^n = x^n y^n z^n$ and $x^m x^n x^p = x^{m+n+p}$. To simplify statements in all problems involving exponents, we shall assume that symbols that appear in denominators represent *nonzero* real numbers.

Several illustrations of the Laws of Exponents are:

$$(3x^3y^4)(4xy^5) = (3)(4)x^3xy^4y^5 = 12x^4y^9$$

$$(2a^2b^3c)^4 = 2^4(a^2)^4(b^3)^4c^4 = 16a^8b^{12}c^4$$

$$\left(\frac{2r^3}{s^2}\right)^2(sr^{-2}) = \left(\frac{2^2r^6}{s^4}\right)\left(\frac{s}{r^2}\right) = 2^2\left(\frac{r^6}{r^2}\right)\left(\frac{s}{s^4}\right) = \frac{4r^4}{s^3}$$

$$(u^{-2}v^3)^{-3} = u^6v^{-9} = \frac{u^6}{v^9}$$

Roots of real numbers are defined by the statement

$$y = \sqrt[n]{x} \quad \text{if and only if} \quad y^n = x$$

provided that both x and y are nonnegative real numbers and n is a positive integer, or that both x and y are negative and n is an odd positive integer. The number $\sqrt[n]{x}$ is called the **principal *n*th root** of x. If $n = 2$, it is customary to write $\sqrt{x}$ instead of $\sqrt[2]{x}$ and to call $\sqrt{x}$ the (principal) **square root** of x. The number $\sqrt[3]{x}$ is referred to as the (principal) **cube root** of x. Some examples of principal roots are:

$$\sqrt{16} = 4, \qquad \sqrt[3]{-8} = -2, \qquad \sqrt[5]{\tfrac{1}{32}} = \tfrac{1}{2}.$$

Complex numbers (see Section 3.3) are needed to define $\sqrt[n]{x}$ if $x < 0$ and n is an *even* positive integer, since for all real numbers y, $y^n \geq 0$ whenever n is even.

It is important to observe that if $\sqrt[n]{x}$ exists, it is a *unique* real number. More generally, if $y^n = x$ for a positive integer n, then y is called *an* nth *root* of x. For example, both 4 and -4 are square roots of 16, since $4^2 = 16$ and also $(-4)^2 = 16$. However, the *principal* square root of 16 is 4, which we write as $\sqrt{16} = 4$.

To complete our terminology, the expression $\sqrt[n]{x}$ is called a **radical,** the number x is called the **radicand,** and n is called the **index** of the radical. The symbol $\sqrt{}$ is called a **radical sign.**

If n is any positive integer and $\sqrt[n]{x}$ exists, then

$$(\sqrt[n]{x})^n = x.$$

If $x > 0$, or if $x < 0$ and n is an odd positive integer, then

$$\sqrt[n]{x^n} = x.$$

For example,

$$(\sqrt[3]{5})^3 = 5, \qquad \sqrt{5^2} = 5, \qquad \sqrt[3]{(-2)^3} = -2, \qquad \sqrt[4]{3^4} = 3.$$

A common algebraic error is to replace $\sqrt{x^2}$ by x for *every* x; however, this is not true if x is negative. For example,

$$\sqrt{(-3)^2} = \sqrt{9} = 3 = |-3|.$$

In general, we may write

$$\sqrt{x^2} = |x| \quad \text{for every real number } x.$$

Furthermore, if n is any even positive integer, then $\sqrt[n]{x^n} = |x|$. Note that $\sqrt{x^2} = x$ if and only if $x \geq 0$.

The following laws can be proved for positive integers m and n, where it is assumed that the indicated roots exist.

LAWS OF RADICALS

(i) $\sqrt[n]{xy} = \sqrt[n]{x}\,\sqrt[n]{y}$ (ii) $\sqrt[n]{\dfrac{x}{y}} = \dfrac{\sqrt[n]{x}}{\sqrt[n]{y}}$ (iii) $\sqrt[m]{\sqrt[n]{x}} = \sqrt[mn]{x}$

Several illustrations of these laws are

$$\sqrt{50} = \sqrt{25 \cdot 2} = \sqrt{25}\sqrt{2} = 5\sqrt{2}$$

$$\sqrt[3]{-108} = \sqrt[3]{(-27)(4)} = \sqrt[3]{-27}\,\sqrt[3]{4} = -3\sqrt[3]{4}$$

$$\sqrt[3]{\sqrt{64}} = \sqrt[6]{64} = 2$$

$$\begin{aligned}\sqrt[3]{16x^3y^8} &= \sqrt[3]{(2^3x^3y^6)(2y^2)} = \sqrt[3]{(2xy^2)^3(2y^2)} \\ &= \sqrt[3]{(2xy^2)^3}\,\sqrt[3]{2y^2} = 2xy^2\sqrt[3]{2y^2}\end{aligned}$$

An important type of simplification involving radicals is **rationalizing a denominator.** The process involves beginning with a quotient that contains a radical in the denominator and then multiplying numerator and denominator by some expression so that the resulting denominator contains no radicals. Two illustrations are:

$$\frac{1}{\sqrt{5}} = \frac{1}{\sqrt{5}} \cdot \frac{\sqrt{5}}{\sqrt{5}} = \frac{\sqrt{5}}{(\sqrt{5})^2} = \frac{\sqrt{5}}{5}$$

$$\sqrt[3]{\frac{x}{y}} = \sqrt[3]{\frac{x}{y} \cdot \frac{y^2}{y^2}} = \frac{\sqrt[3]{xy^2}}{\sqrt[3]{y^3}} = \frac{\sqrt[3]{xy^2}}{y}$$

If a calculator is used to find decimal approximations of radicals, there is no advantage in writing $1/\sqrt{5} = \sqrt{5}/5$, as in the preceding illustration. However, for *algebraic* simplifications, changing expressions to such forms is often desirable.

Radicals may be used to define **rational exponents.** First, if n is a positive integer and x is a real number, we define

$$x^{1/n} = \sqrt[n]{x}$$

provided that $\sqrt[n]{x}$ is a real number. Next, if m/n is a rational number and n is a positive integer, and if x is a real number such that $\sqrt[n]{x}$ exists, then

$$x^{m/n} = (\sqrt[n]{x})^m = \sqrt[n]{x^m}.$$

Thus,

$$x^{2/3} = (\sqrt[3]{x})^2 = \sqrt[3]{x^2}.$$

It can be shown that the Laws of Exponents are true for rational exponents. For example,

$$\left(\frac{2x^{2/3}}{y^{1/2}}\right)^2 \left(\frac{3x^{-5/6}}{y^{1/3}}\right) = \left(\frac{4x^{4/3}}{y}\right)\left(\frac{3x^{-5/6}}{y^{1/3}}\right) = \frac{12x^{1/2}}{y^{4/3}}$$

If we begin with any collection of variables and real numbers, then an **algebraic expression** is the result obtained by applying additions, subtractions, multiplications, divisions, or the taking of roots. The following are examples of algebraic expressions:

$$x^3 - 2x + \frac{35}{\sqrt{2x}}, \qquad \frac{2xy + 3x}{y - 1}, \qquad \frac{4yz^{-2} + \left(\dfrac{-7}{x + w}\right)^5}{\sqrt[3]{y^2 + 5z}},$$

where x, y, z, and w are variables. If specific numbers are substituted for the variables in an algebraic expression, the resulting real number is called the **value** of the expression for these numbers. To illustrate, the value of the second expression for $x = -2$ and $y = 3$ is

$$\frac{2(-2)(3) + 3(-2)}{3 - 1} = \frac{-12 - 6}{2} = -9.$$

When working with algebraic expressions, we will assume that domains are chosen so that variables do not represent numbers that make the expressions meaningless. Thus, we assume that denominators are not zero, roots always exist, etc.

Certain algebraic expressions are given special names. If x is a variable, then a **monomial** in x is an expression of the form ax^n, where the coefficient a is a real number and n is a nonnegative integer. A *polynomial in x* is a sum of monomials in x. Another way of stating this is as follows:

DEFINITION

A **polynomial in *x*** is an expression of the form

$$a_n x^n + a_{n-1} x^{n-1} + \cdots + a_1 x + a_0,$$

where n is a nonnegative integer and each coefficient a_k is a real number.

Each of the expressions $a_k x^k$ in the sum is called a **term** of the polynomial. If a coefficient a_k is zero, we usually delete the term $a_k x^k$. The coefficient a_n of the highest power of x is the **leading coefficient** of the polynomial and, if $a_n \neq 0$, we say that the polynomial has **degree *n*.** Two polynomials are **equal** if and only if they have the same degree and corresponding coefficients are equal. If all the coefficients of a polynomial are zero, it is called the **zero polynomial** and is denoted by 0. It is customary not to assign a degree to the zero polynomial.

If some coefficients are negative, we often use minus signs between appropriate terms. To illustrate, instead of $3x^2 + (-5)x + (-7)$, we write $3x^2 - 5x - 7$ for this polynomial of degree 2.

According to the definition of degree, if c is a nonzero real number, then c is a polynomial of degree 0. Such polynomials (together with the zero polynomial) are called **constant polynomials.**

A polynomial in x may be thought of as an algebraic expression obtained by employing only additions, subtractions, and multiplications involving x. In particular, the expressions

$$\frac{1}{x} + 3x, \qquad \frac{x - 5}{x^2 + 2}, \qquad 3x^2 + \sqrt{x} - 2$$

are not polynomials since they involve divisions by variables or contain roots of variables.

EXAMPLE 1 Simplify the following:

(a) $(x^3 + 2x^2 - 5x + 7) + (4x^3 - 5x^2 + 3)$

(b) $(x^3 + 2x^2 - 5x + 7) - (4x^3 - 5x^2 + 3)$

(c) $(x^2 + 5x - 4)(2x^3 + 3x - 1)$

SOLUTION

(a) Rearranging terms and using properties of real numbers gives us

$$\begin{aligned}(x^3 + 2x^2 - 5x + 7) + (4x^3 - 5x^2 + 3) \\ &= x^3 + 4x^3 + 2x^2 - 5x^2 - 5x + 7 + 3 \\ &= (1 + 4)x^3 + (2 - 5)x^2 + (-5)x + (7 + 3) \\ &= 5x^3 - 3x^2 - 5x + 10\end{aligned}$$

This illustrates the fact that the sum of any two polynomials in x can be obtained by adding coefficients of like powers of x. The intermediate steps were used for completeness. You may omit them after you become proficient with such manipulations.

(b) $$\begin{aligned}(x^3 + 2x^2 - 5x + 7) - (4x^3 - 5x^2 + 3) \\ &= x^3 - 4x^3 + 2x^2 + 5x^2 - 5x + 7 - 3 \\ &= -3x^3 + 7x^2 - 5x + 4\end{aligned}$$

(c) $$\begin{aligned}(x^2 + 5x - 4)(2x^3 + 3x - 1) \\ &= x^2(2x^3 + 3x - 1) + 5x(2x^3 + 3x - 1) - 4(2x^3 + 3x - 1) \\ &= 2x^5 + 3x^3 - x^2 + 10x^4 + 15x^2 - 5x - 8x^3 - 12x + 4 \\ &= 2x^5 + 10x^4 + (3 - 8)x^3 + (-1 + 15)x^2 + (-5 - 12)x + 4 \\ &= 2x^5 + 10x^4 - 5x^3 + 14x^2 - 17x + 4\end{aligned}$$ ■

We may consider polynomials in more than one variable. For example, a polynomial in *two* variables, x and y, is a sum of terms, each of the form ax^my^k for some real number a and nonnegative integers m and k. An example is

$$3x^4y + 2x^3y^5 + 7x^2 - 4xy + 8y - 5.$$

Polynomials in three variables, x, y, z, or for that matter, in *any* number of variables may also be considered.

Certain products occur so frequently that they deserve special attention. We list some of these next. The validity of each formula can be checked by actually carrying out the multiplications.

PRODUCT FORMULAS

(i) $(x + y)(x - y) = x^2 - y^2$
(ii) $(ax + b)(cx + d) = acx^2 + (ad + bc)x + bd$
(iii) $(x + y)^2 = x^2 + 2xy + y^2$
(iv) $(x - y)^2 = x^2 - 2xy + y^2$
(v) $(x + y)^3 = x^3 + 3x^2y + 3xy^2 + y^3$
(vi) $(x - y)^3 = x^3 - 3x^2y + 3xy^2 - y^3$

If a polynomial is written as a product of other polynomials, then each polynomial in the product is called a **factor** of the original polynomial. The process of expressing a polynomial as a product is called **factoring.** For example, since $x^2 - 9 = (x + 3)(x - 3)$, we see that $x + 3$ and $x - 3$ are factors of $x^2 - 9$.

Factoring plays an important role in mathematics, since it may be used to reduce the study of a complicated expression to the study of several simpler expressions. For example, properties of the polynomial $x^2 - 9$ can be determined by examining the factors $x + 3$ and $x - 3$.

We shall be interested primarily in **nontrivial factors** of polynomials; that is, factors that contain polynomials of degree greater than zero. An exception to this rule is that if the coefficients are restricted to *integers*, then we usually remove a common integral factor from each term of the polynomial as follows:

$$4x^2y + 8z^3 = 4(x^2y + 2z^3).$$

An integer $a > 1$ is prime if it cannot be written as a product of two positive integers greater than 1. In similar fashion, if S denotes a set of numbers, then a polynomial with coefficients in S is said to be **prime,** or **irreducible** over S, if it cannot be written as a product of two polynomials of positive degree with coefficients in S. A polynomial may be irreducible over one set S but not over another. For example, $x^2 - 2$ is irreducible over the rational numbers, since it cannot be expressed as a product of two polynomials of positive degree that have *rational* coefficients. If we allow the factors to have *real* coefficients, then $x^2 - 2$ is not prime, since

$$x^2 - 2 = (x + \sqrt{2})(x - \sqrt{2}).$$

Similarly, $x^2 + 1$ is irreducible over the real numbers but, as we shall see in Section 3.3, not over the complex numbers. It can be shown that every polynomial $ax + b$ of degree 1 is irreducible.

It is usually difficult to factor polynomials of degree greater than 2. In simple cases the following formulas may be useful. Each can be verified by multiplication.

FACTORING FORMULAS

$x^2 - y^2 = (x + y)(x - y)$	(Difference of two squares)
$x^3 - y^3 = (x - y)(x^2 + xy + y^2)$	(Difference of two cubes)
$x^3 + y^3 = (x + y)(x^2 - xy + y^2)$	(Sum of two cubes)

EXAMPLE 2 Express each of the following as a product of polynomials that are irreducible over the set of integers:

(a) $16x^4 - 81$ (b) $x^3 - 8$ (c) $3x^3 + 2x^2 - 12x - 8$

SOLUTION

(a) We apply the difference of two squares formula twice, as follows:

$$\begin{aligned} 16x^4 - 81 &= (4x^2)^2 - 9^2 = (4x^2 + 9)(4x^2 - 9) \\ &= (4x^2 + 9)[(2x)^2 - 3^2] = (4x^2 + 9)(2x + 3)(2x - 3). \end{aligned}$$

(b) Using the difference of two cubes formula, with $y = 2$,

$$\begin{aligned} x^3 - 8 &= x^3 - 2^3 \\ &= (x - 2)(x^2 + 2x + 4). \end{aligned}$$

By trying all possibilities we can show that $x^2 + 2x + 4$ cannot be expressed as a product of two first-degree polynomials with integer coefficients.

(c) We may proceed as follows:

$$\begin{aligned} 3x^3 + 2x^2 - 12x - 8 &= (3x^3 + 2x^2) - (12x + 8) \\ &= x^2(3x + 2) - 4(3x + 2) \\ &= (x^2 - 4)(3x + 2) \\ &= (x + 2)(x - 2)(3x + 2). \end{aligned}$$

■

The technique used in Example 2(c) is called **factorization by grouping.**

Quotients of algebraic expressions are called **fractional expressions.** As a special case, a quotient of two polynomials is called a **rational expression.**

Sometimes it is necessary to combine fractional expressions and then simplify the result. Of particular importance is the following, where a, b, and d represent real numbers:

$$\frac{ad}{bd} = \frac{a}{b} \cdot \frac{d}{d} = \frac{a}{b} \cdot 1 = \frac{a}{b}; \quad \text{that is,} \quad \frac{ad}{bd} = \frac{a}{b}.$$

This is sometimes phrased "A common factor in the numerator and denominator may be *canceled* from a quotient," or "The numerator and denominator may be *divided* by the same nonzero number d." To use this technique in problems involving rational expressions, we factor both the numerator and denominator into prime factors and then cancel common factors that occur in the numerator and denominator. We refer to the resulting expression as being *simplified*, or *reduced to lowest terms.*

EXAMPLE 3 Simplify:

$$\frac{3x^2 - 5x - 2}{x^2 - 4}$$

SOLUTION Factoring the numerator and denominator and canceling common factors gives us

$$\frac{3x^2 - 5x - 2}{x^2 - 4} = \frac{(3x + 1)(x - 2)}{(x + 2)(x - 2)} = \frac{3x + 1}{x + 2}.$$

■

In the preceding example we canceled the common factor $x - 2$; that is, we divided numerator and denominator by $x - 2$. This simplification is valid only if $x - 2 \neq 0$, that is, $x \neq 2$. However, 2 is not in the domain of x, since it leads to a zero denominator when substituted in the original expression. Hence, our manipulations are valid. We shall always assume such restrictions when simplifying rational expressions.

EXAMPLE 4 Simplify:

$$\frac{2 - x - 3x^2}{6x^2 - x - 2}$$

SOLUTION

$$\frac{2 - x - 3x^2}{6x^2 - x - 2} = \frac{(1 + x)(2 - 3x)}{(2x + 1)(3x - 2)} = \frac{-(1 + x)}{2x + 1}$$

The fact that $(2 - 3x) = -(3x - 2)$ accounts for the minus sign in the final answer, since $(2 - 3x)/(3x - 2) = -(3x - 2)/(3x - 2) = -1$. Another method is to change the form of the numerator as follows:

$$\frac{2 - x - 3x^2}{6x^2 - x - 2} = \frac{-(3x^2 + x - 2)}{6x^2 - x - 2}$$

$$= -\frac{(3x - 2)(x + 1)}{(3x - 2)(2x + 1)} = -\frac{x + 1}{2x + 1}$$ ■

EXAMPLE 5 Perform the indicated operations and simplify:

(a) $\dfrac{x^2 - 6x + 9}{x^2 - 1} \cdot \dfrac{2x - 2}{x - 3}$ (b) $\dfrac{x + 2}{2x - 3} \div \dfrac{x^2 - 4}{2x^2 - 3x}$

SOLUTION

(a) $$\frac{x^2 - 6x + 9}{x^2 - 1} \cdot \frac{2x - 2}{x - 3} = \frac{(x - 3)^2}{(x + 1)(x - 1)} \cdot \frac{2(x - 1)}{x - 3}$$

$$= \frac{2(x - 3)^2(x - 1)}{(x + 1)(x - 1)(x - 3)} = \frac{2(x - 3)}{x + 1}$$

(b) $$\frac{x + 2}{2x - 3} \div \frac{x^2 - 4}{2x^2 - 3x} = \frac{x + 2}{2x - 3} \cdot \frac{2x^2 - 3x}{x^2 - 4}$$

$$= \frac{(x + 2)x(2x - 3)}{(2x - 3)(x + 2)(x - 2)} = \frac{x}{x - 2}$$ ■

When adding or subtracting two rational expressions, we usually find a common denominator and use the following properties of real numbers:

$$\frac{a}{d} + \frac{c}{d} = \frac{a + c}{d}; \qquad \frac{a}{d} - \frac{c}{d} = \frac{a - c}{d}.$$

If the denominators are not the same, a common denominator may be introduced by multiplying numerator and denominator of each of the fractions by suitable polynomials. It is usually desirable to use the **least common denominator (lcd)** of the two fractions. To find the lcd we factor each denominator into prime polynomials and then form the product of the different prime factors, using the *greatest* exponent that appears with each prime factor, as in the following examples.

EXAMPLE 6 Change to a rational expression in lowest terms:

$$\frac{6}{x(3x-2)}+\frac{5}{3x-2}-\frac{2}{x^2}$$

SOLUTION The denominators are already in factored form. Evidently, the lcd is $x^2(3x-2)$. In order to obtain three fractions having that denominator, we multiply numerator and denominator of the first fraction by x, of the second by x^2, and of the third by $3x-2$. This gives us

$$\frac{6}{x(3x-2)}+\frac{5}{3x-2}-\frac{2}{x^2}=\frac{6x}{x^2(3x-2)}+\frac{5x^2}{x^2(3x-2)}-\frac{2(3x-2)}{x^2(3x-2)}$$

$$=\frac{6x+5x^2-(6x-4)}{x^2(3x-2)}=\frac{5x^2+4}{x^2(3x-2)}.$$ ■

EXAMPLE 7 Change to a rational expression in lowest terms:

$$\frac{2x+5}{x^2+6x+9}+\frac{x}{x^2-9}+\frac{1}{x-3}$$

SOLUTION We begin by factoring denominators:

$$\frac{2x+5}{x^2+6x+9}+\frac{x}{x^2-9}+\frac{1}{x-3}=\frac{2x+5}{(x+3)^2}+\frac{x}{(x+3)(x-3)}+\frac{1}{x-3}$$

Since the lcd is $(x+3)^2(x-3)$, we multiply numerator and denominator of the first fraction by $x-3$, of the second by $x+3$, and of the third by $(x+3)^2$ and then add, as follows:

$$\frac{(2x+5)(x-3)}{(x+3)^2(x-3)}+\frac{x(x+3)}{(x+3)^2(x-3)}+\frac{(x+3)^2}{(x+3)^2(x-3)}$$

$$=\frac{(2x^2-x-15)+(x^2+3x)+(x^2+6x+9)}{(x+3)^2(x-3)}$$

$$=\frac{4x^2+8x-6}{(x+3)^2(x-3)}=\frac{2(2x^2+4x-3)}{(x+3)^2(x-3)}$$ ■

The denominators of certain fractional expressions contain sums or differences that involve radicals. Such denominators can be rationalized, as shown in the next example.

EXAMPLE 8 Rationalize the denominator of the fraction

$$\frac{1}{\sqrt{x}+\sqrt{y}}.$$

SOLUTION Multiplying numerator and denominator by $\sqrt{x}-\sqrt{y}$, we obtain

$$\begin{aligned}\frac{1}{\sqrt{x}+\sqrt{y}} &= \frac{1}{\sqrt{x}+\sqrt{y}}\cdot\frac{\sqrt{x}-\sqrt{y}}{\sqrt{x}-\sqrt{y}}\\ &= \frac{\sqrt{x}-\sqrt{y}}{(\sqrt{x})^2+\sqrt{y}\sqrt{x}-\sqrt{x}\sqrt{y}-(\sqrt{y})^2}\\ &= \frac{\sqrt{x}-\sqrt{y}}{x-y}\end{aligned}$$

■

Some problems in calculus require rationalizing a *numerator*, as in the following example.

EXAMPLE 9 Rationalize the numerator of the fraction

$$\frac{\sqrt{x+h}-\sqrt{x}}{h}, \quad \text{where } x>0 \text{ and } h>0.$$

SOLUTION We multiply numerator and denominator by the expression $\sqrt{x+h}+\sqrt{x}$ and proceed as follows:

$$\begin{aligned}\frac{\sqrt{x+h}-\sqrt{x}}{h} &= \frac{\sqrt{x+h}-\sqrt{x}}{h}\cdot\frac{\sqrt{x+h}+\sqrt{x}}{\sqrt{x+h}+\sqrt{x}}\\ &= \frac{(\sqrt{x+h})^2-\sqrt{x}\sqrt{x+h}+\sqrt{x+h}\sqrt{x}-(\sqrt{x})^2}{h(\sqrt{x+h}+\sqrt{x})}\\ &= \frac{(x+h)-x}{h(\sqrt{x+h}+\sqrt{x})}\\ &= \frac{h}{h(\sqrt{x+h}+\sqrt{x})}\\ &= \frac{1}{\sqrt{x+h}+\sqrt{x}}\end{aligned}$$

■

For certain problems in calculus it is necessary to simplify expressions of the types given in the next two examples. (Also see Exercises 83–94.)

EXAMPLE 10 Simplify the fraction

$$\frac{\dfrac{1}{(x+h)^2} - \dfrac{1}{x^2}}{h}$$

SOLUTION

$$\frac{\dfrac{1}{(x+h)^2} - \dfrac{1}{x^2}}{h} = \frac{\dfrac{x^2-(x+h)^2}{(x+h)^2x^2}}{h}$$

$$= \frac{x^2-(x^2+2xh+h^2)}{(x+h)^2x^2h}$$

$$= \frac{x^2-x^2-2xh-h^2}{(x+h)^2x^2h}$$

$$= \frac{-h(2x+h)}{(x+h)^2x^2h}$$

$$= -\frac{2x+h}{(x+h)^2x^2}$$

■

EXAMPLE 11 Simplify:

$$\frac{3x^2(2x+5)^{1/2} - x^3(\frac{1}{2})(2x+5)^{-1/2}(2)}{[(2x+5)^{1/2}]^2}$$

SOLUTION

$$\frac{3x^2(2x+5)^{1/2} - x^3(\frac{1}{2})(2x+5)^{-1/2}(2)}{[(2x+5)^{1/2}]^2} = \frac{3x^2(2x+5)^{1/2} - \dfrac{x^3}{(2x+5)^{1/2}}}{2x+5}$$

$$= \frac{\dfrac{3x^2(2x+5)-x^3}{(2x+5)^{1/2}}}{2x+5}$$

$$= \frac{6x^3+15x^2-x^3}{(2x+5)^{1/2}(2x+5)}$$

$$= \frac{5x^3+15x^2}{(2x+5)^{3/2}} = \frac{5x^2(x+3)}{(2x+5)^{3/2}}$$

An alternative solution is to eliminate the negative power $(2x+5)^{-1/2}$ in the given expression first, by multiplying numerator and denominator by

$(2x+5)^{1/2}$ as follows:

$$\frac{3x^2(2x+5)^{1/2} - x^3(\frac{1}{2})(2x+5)^{-1/2}(2)}{[(2x+5)^{1/2}]^2} \cdot \frac{(2x+5)^{1/2}}{(2x+5)^{1/2}} = \frac{3x^2(2x+5) - x^3(\frac{1}{2})(2)}{(2x+5)(2x+5)^{1/2}}$$

$$= \frac{6x^3 + 15x^2 - x^3}{(2x+5)^{3/2}}$$

The remainder of the simplification is the same. ■

Exercises 1.2

Simplify the expressions in Exercises 1–16.

1 $(3u^7v^3)(4u^4v^{-5})$

2 $(x^2yz^3)(-2xz^2)(x^3y^{-2})$

3 $(8x^4y^{-3})(\frac{1}{2}x^{-5}y^2)$

4 $\left(\frac{4a^2b}{a^3b^2}\right)\left(\frac{5a^2b}{2b^4}\right)$

5 $(3y^3)^4(4y^2)^{-3}$

6 $(-2r^2s)^5(3r^{-1}s^3)^2$

7 $\left(\frac{3x^5y^4}{x^0y^{-3}}\right)^2$

8 $(4a^2b)^4\left(\frac{-a^3}{2b}\right)^2$

9 $(27a^6)^{-2/3}$

10 $(25z^4)^{-3/2}$

11 $\left(\frac{-8x^3}{y^{-6}}\right)^{2/3}$

12 $\left(\frac{-y^{3/2}}{y^{-1/3}}\right)^3$

13 $\sqrt[3]{8a^6b^{-3}}$

14 $\sqrt[4]{81r^5s^8}$

15 $\sqrt{\frac{1}{3u^3v}}$

16 $\sqrt[3]{\frac{1}{4x^5y^2}}$

Rewrite the expressions in Exercises 17–20 in terms of rational exponents.

17 $\sqrt[4]{x^3}$

18 $\sqrt[3]{x^5}$

19 $\sqrt[3]{(a+b)^2}$

20 $\sqrt{x^2+y^2}$

Rewrite the expressions in Exercises 21–22 in terms of radicals.

21 (a) $4x^{3/2}$
(b) $(4x)^{3/2}$

22 (a) $4+x^{3/2}$
(b) $(4+x)^{3/2}$

In Exercises 23–26 replace the symbol □ with either = or ≠ and give reasons for your answers.

23 $(a^r)^2 \;\square\; a^{(r^2)}$

24 $(a^2+1)^{1/2} \;\square\; a+1$

25 $a^xb^y \;\square\; (ab)^{xy}$

26 $\sqrt{a^r} \;\square\; (\sqrt{a})^r$

In Exercises 27–32 perform the indicated operations and find the degree of the resulting polynomial.

27 $(2x^3+4x^2+3x+7)+(x^4-5x^3+x-2)$

28 $(2x^5-3x^2+2)+(-2x^4+x^2-4x+5)$

29 $(x^3+5x^2-7x+2)-(x^3+5x^2+x+2)$

30 $(2x^4-3x^2+1)-(3x^4+x+1)$

31 $(2x^3-x+5)(x^2+x+2)$

32 $(7x^4+x^2-1)(7x^4-x^3+4x)$

Find the products in Exercises 33–38.

33 $(5x^2+2y)(3x^2-7y)$

34 $(x+9y^2)(3x-4y^2)$

35 $(4x^2-5y^2)^2$

36 $(10p^2+7q^2)^2$

37 $(x-2y)^3$

38 $(u^2-3v)^3$

In Exercises 39–50, express each polynomial as a product of polynomials that are irreducible over the set of integers.

39 $3x^2+10x-8$

40 $10x^2+29x-21$

41 $6x^4-5x^3-6x^2$

42 $49-36x^2$

43 $25x^2-9$

44 $64x^3+125$

45 x^3+x^2+x+1

46 $2x^3-x^2-2x+1$

47 x^8-10x^4+9

48 x^6-64

49 x^6-1

50 $4x^3-8x^2-9x+18$

Simplify the expressions in Exercises 51–94.

51 $\frac{6x^2+7x-10}{6x^2+13x-15}$

52 $\frac{10x^2+29x-21}{5x^2-23x+12}$

53 $\frac{12y^2+3y}{20y^2+9y+1}$

54 $\frac{4z^2+12z+9}{2z^2+3z}$

55 $\frac{6-7a-5a^2}{10a^2-a-3}$

56 $\frac{6y-5y^2}{25y^2-36}$

57 $\frac{4x^3-9x}{10x^4+11x^3-6x^2}$

58 $\frac{16x^4+8x^3+x^2}{4x^3+25x^2+6x}$

59 $\frac{6r}{3r-1} - \frac{4r}{2r+5}$

60 $\frac{3s}{s^2+1} - \frac{6}{2s-1}$

61 $\frac{2x+1}{2x-1} - \frac{x-1}{x+1}$

62 $\frac{3u+2}{u-4} + \frac{4u+1}{5u+2}$

63 $\frac{9t-6}{8t^3-27} \cdot \frac{4t^2-9}{12t^2+10t-12}$

64 $\frac{a^2+4a+3}{3a^2+a-2} \cdot \frac{3a^2-2a}{2a^2+13a+21}$

65 $\frac{5a^2+12a+4}{a^4-16} \div \frac{25a^2+20a+4}{a^2-2a}$

66 $\frac{x^3-8}{x^2-4} \div \frac{x}{x^3+8}$

67 $\frac{2}{3x+1} - \frac{9}{(3x+1)^2}$

68 $\frac{4}{(5x-2)^2} + \frac{x}{5x-2}$

69 $\frac{1}{c} - \frac{c+2}{c^2} + \frac{3}{c^3}$

70 $\frac{6}{3t} + \frac{t+5}{t^3} + \frac{1-2t^2}{t^4}$

71 $\frac{5}{x-1} + \frac{8}{(x-1)^2} - \frac{3}{(x-1)^3}$

72 $\frac{8}{x} + \frac{3}{2x-4} + \frac{7x}{x^2-4}$

73 $\frac{2}{x} + \frac{7}{x^2} + \frac{5}{2x-3} + \frac{1}{(2x-3)^2}$

74 $\frac{4}{x} + \frac{3x^2+5}{x^3} - \frac{6}{2x+1}$

75 $\frac{5}{7x-3} - \frac{2}{2x+1} + \frac{4x}{14x^2+x-3}$

76 $2 + \frac{3}{x} + \frac{7x}{3x+10}$

77 $\dfrac{\frac{a}{b} - \frac{b}{a}}{\frac{1}{a} + \frac{1}{b}}$

78 $\dfrac{\frac{1}{x+1} - 5}{\frac{1}{x} - x}$

79 $\dfrac{\frac{x}{y^2} - \frac{y}{x^2}}{\frac{1}{y^2} - \frac{1}{x^2}}$

80 $\dfrac{\frac{r}{s} + \frac{s}{r}}{\frac{r^2}{s^2} - \frac{s^2}{r^2}}$

81 $\dfrac{\frac{5}{x+1} + \frac{2x}{x+3}}{\frac{x}{x+1} + \frac{7}{x+3}}$

82 $\dfrac{\frac{3}{w} - \frac{6}{2w+1}}{\frac{5}{w} + \frac{8}{2w+1}}$

83 $\frac{(x+h)^2 - 7(x+h) - (x^2 - 7x)}{h}$

84 $\frac{(x+h)^3 + 4(x+h) - (x^3 + 4x)}{h}$

85 $\dfrac{\frac{1}{(x+h)^3} - \frac{1}{x^3}}{h}$

86 $\dfrac{\frac{1}{x+h} - \frac{1}{x}}{h}$

87 $(2x^2 - 3x + 1)(4)(3x+2)^3(3) + (3x+2)^4(4x-3)$

88 $(6x-5)^3(2)(x^2+4)(2x) + (x^2+4)^2(3)(6x-5)^2(6)$

89 $(x^2-4)^{1/2}(3)(2x+1)^2(2) + (2x+1)^3(\frac{1}{2})(x^2-4)^{-1/2}(2x)$

90 $(3x+2)^{1/3}(2)(4x-5)(4) + (4x-5)^2(\frac{1}{3})(3x+2)^{-2/3}(3)$

91 $\frac{(6x+1)^3(27x^2+2) - (9x^3+2x)(3)(6x+1)^2(6)}{(6x+1)^6}$

92 $\frac{(x^2-1)^4(2x) - x^2(4)(x^2-1)^3(2x)}{(x^2-1)^8}$

93 $\frac{(x^2+4)^{1/3}(3) - (3x)(\frac{1}{3})(x^2+4)^{-2/3}(2x)}{[(x^2+4)^{1/3}]^2}$

94 $\frac{(1-x^2)^{1/2}(2x) - x^2(\frac{1}{2})(1-x^2)^{-1/2}(-2x)}{[(1-x^2)^{1/2}]^2}$

Rationalize the denominators in Exercises 95–96.

95 $\frac{\sqrt{t}-4}{\sqrt{t}+4}$

96 $\frac{4}{3-\sqrt{w}}$

Rationalize the numerators in Exercises 97–100.

97 $\frac{\sqrt{a}-\sqrt{b}}{c}$

98 $\frac{b-\sqrt{c}}{d}$

99 $\frac{\sqrt{2(x+h)+1} - \sqrt{2x+1}}{h}$

100 $\frac{\sqrt{1-x-h} - \sqrt{1-x}}{h}$

1.3 Inequalities

Let us consider the inequality

$$x^2 - 3 < 2x + 4$$

where x is a variable. If certain numbers such as 4 or 5 are substituted for x, we obtain the false statements $13 < 12$ or $22 < 14$, respectively. Other numbers, such as 1 or 2, produce the true statements $-2 < 6$ or $1 < 8$. If a true statement is obtained when x is replaced by a real number a, then a is called a **solution** of the inequality. Thus 1 and 2 are solutions of the inequality $x^2 - 3 < 2x + 4$, whereas 4 and 5 are not solutions. To **solve** an inequality means to find all solutions. We say that two inequalities are **equivalent** if they have exactly the same solutions.

One method for solving an inequality is to replace it with a list of equivalent inequalities, ending with an inequality for which the solutions are obvious. The following properties are often useful, where a, b, and c denote real numbers.

PROPERTIES OF INEQUALITIES

(i) If $a > b$ and $b > c$, then $a > c$.
(ii) If $a > b$, then $a + c > b + c$.
(iii) If $a > b$, then $a - c > b - c$.
(iv) If $a > b$ and $c > 0$, then $ac > bc$.
(v) If $a > b$ and $c < 0$, then $ac < bc$.

PROOF We will use the fact that both the sum and product of any two positive real numbers are positive. To prove (i) we first note that if $a > b$ and $b > c$, then $a - b$ and $b - c$ are both positive. Consequently, the sum $(a - b) + (b - c)$ is positive. Since the sum reduces to $a - c$, we see that $a - c$ is positive, which means that $a > c$.

To establish (ii) we again note that if $a > b$, then $a - b$ is positive. Since $(a + c) - (b + c) = a - b$, it follows that $(a + c) - (b + c)$ is positive; that is, $a + c > b + c$.

If $a > b$, then by (ii), $a + (-c) > b + (-c)$, or equivalently, $a - c > b - c$. This proves (iii).

To prove (iv) observe that if $a > b$ and $c > 0$, then $a - b$ and c are both positive and hence, so is the product $(a - b)c$. Consequently, $ac - bc$ is positive; that is, $ac > bc$.

Finally, to prove (v) we first note that if $c < 0$, then $0 - c$, or $-c$, is positive. In addition, if $a > b$, then $a - b$ is positive and hence, the product $(a - b)(-c)$ is positive. However, $(a - b)(-c) = -ac + bc$ and therefore $bc - ac$ is positive. This means that $bc > ac$, or $ac < bc$. □

Similar results are true for the symbol $<$. Thus, if $a < b$, then we have $a + c < b + c$, and so on.

EXAMPLE 1 Solve the inequality $4x - 3 < 2x + 5$ and represent the solutions graphically.

SOLUTION The following inequalities are equivalent (supply reasons):

$$4x - 3 < 2x + 5$$

$$(4x - 3) + 3 < (2x + 5) + 3$$

$$4x < 2x + 8$$

$$4x - 2x < (2x + 8) - 2x$$

$$2x < 8$$

$$\tfrac{1}{2}(2x) < \tfrac{1}{2}(8)$$

$$x < 4$$

FIGURE 1.9

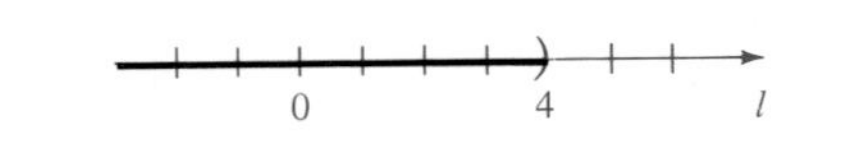

Hence, the solutions of the given inequality consist of all real numbers x such that $x < 4$; that is, all numbers in the open interval $(-\infty, 4)$. The graph is sketched in Figure 1.9. ■

EXAMPLE 2 Solve the inequality $-6 < 2x - 4 < 2$ and represent the solutions graphically.

SOLUTION A real number x is a solution of the given inequality if and only if it is a solution of *both* of the inequalities

$$-6 < 2x - 4 \quad \text{and} \quad 2x - 4 < 2.$$

The first inequality is equivalent to each of the following:

$$-6 < 2x - 4$$

$$-6 + 4 < (2x - 4) + 4$$

$$-2 < 2x$$

$$\tfrac{1}{2}(-2) < \tfrac{1}{2}(2x)$$

$$-1 < x$$

The second inequality is equivalent to each of the following:

$$2x - 4 < 2$$

$$2x < 6$$

$$x < 3$$

Thus, x is a solution of the given inequality if and only if *both*

$$-1 < x \quad \text{and} \quad x < 3,$$

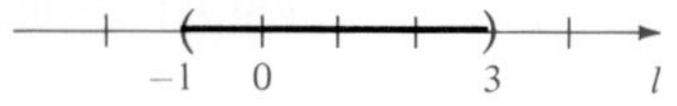

that is,

$$-1 < x < 3.$$

FIGURE 1.10

Hence, the solutions are all numbers in the open interval $(-1, 3)$. The graph is sketched in Figure 1.10.

An alternative (and shorter) method is to work with both inequalities simultaneously, as follows:

$$-6 < 2x - 4 < 2$$

$$-6 + 4 < 2x < 2 + 4$$

$$-2 < 2x < 6$$

$$-1 < x < 3$$

■

EXAMPLE 3 Solve the inequality

$$-5 \leq \frac{4 - 3x}{2} < 1$$

and sketch the graph corresponding to the solutions.

SOLUTION A number x is a solution if and only if it satisfies both of the inequalities

$$-5 \leq \frac{4 - 3x}{2} \quad \text{and} \quad \frac{4 - 3x}{2} < 1.$$

We can either work with each inequality separately or proceed as in the alternative solution of Example 2, as follows:

$$-5 \leq \frac{4 - 3x}{2} < 1$$

$$-10 \leq 4 - 3x < 2$$

$$-10 - 4 \leq -3x < 2 - 4$$

$$-14 \leq -3x < -2$$

$$(-\tfrac{1}{3})(-14) \geq (-\tfrac{1}{3})(-3x) > (-\tfrac{1}{3})(-2)$$

$$\tfrac{14}{3} \geq x > \tfrac{2}{3}$$

$$\tfrac{2}{3} < x \leq \tfrac{14}{3}.$$

FIGURE 1.11

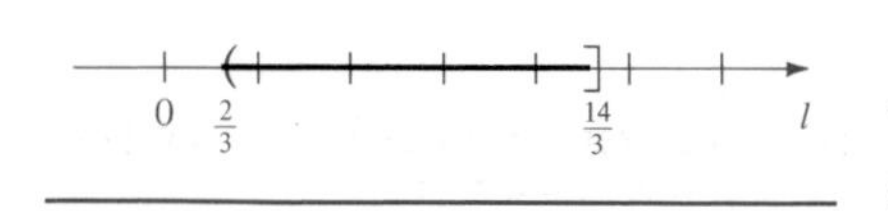

Thus, the solutions of the inequality consist of all numbers in the interval $(\frac{2}{3}, \frac{14}{3}]$. The graph is sketched in Figure 1.11. ■

The inequality $|x| < 1$ is equivalent to $-1 < x < 1$. This, in turn, means that x is in the open interval $(-1, 1)$. In general, if a is any positive real number, the following can be proved.

PROPERTIES OF ABSOLUTE VALUES ($a > 0$)

(i) $|x| < a$ if and only if $-a < x < a$.

(ii) $|x| > a$ if and only if either $x > a$ or $x < -a$.

(iii) $|x| = a$ if and only if $x = a$ or $x = -a$.

FIGURE 1.12

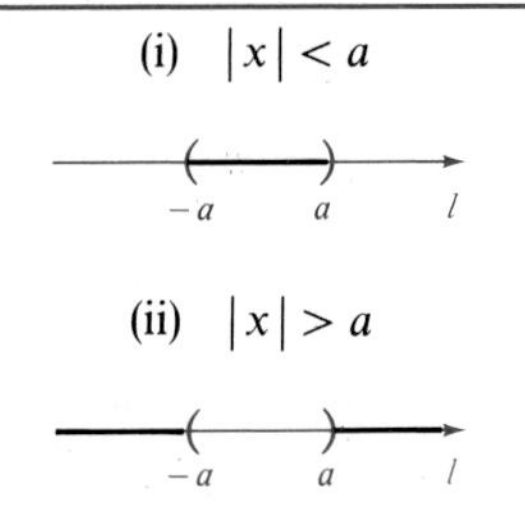

Graphs corresponding to the solutions of $|x| < a$ and $|x| > a$ are sketched in Figure 1.12. Properties (i) and (ii) are also true if the symbols $<$ and $>$ are replaced by $\leq$ and $\geq$, respectively.

EXAMPLE 4 Solve the inequality $|x - 3| < 0.1$.

SOLUTION By property (i), with $a = x - 3$ and $b = 0.1$, the inequality is equivalent to

$$-0.1 < x - 3 < 0.1$$

and hence to $$-0.1 + 3 < (x - 3) + 3 < 0.1 + 3$$

or to $$2.9 < x < 3.1.$$

Consequently, the solutions are the real numbers in the interval $(2.9, 3.1)$. ■

In the study of calculus it is necessary to consider inequalities of the type given in the next example.

EXAMPLE 5 If a and δ denote real numbers and $\delta > 0$, solve the inequality

$$|x - a| < \delta$$

and represent the solutions graphically.

SOLUTION We may proceed as in Example 4. Thus,

$$-\delta < x - a < \delta$$
$$a - \delta < a + (x - a) < a + \delta$$
$$a - \delta < x < a + \delta.$$

FIGURE 1.13

$|x - a| < \delta$

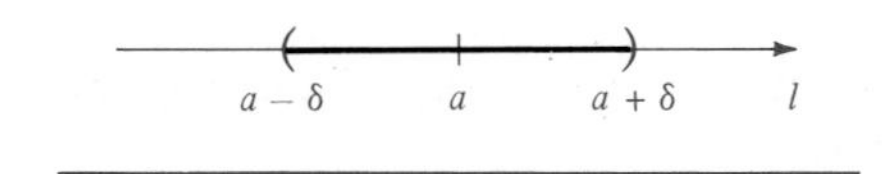

The solutions of the inequality consist of all real numbers in the interval $(a - \delta, a + \delta)$. A typical graph is sketched in Figure 1.13. ■

Note that Example 4 is the special case of Example 5 with $a = 3$ and $\delta = 0.1$.

EXAMPLE 6 Solve $|2x + 3| > 9$ and illustrate the solutions graphically.

SOLUTION By property (ii) of absolute values with $a = 2x + 3$ and $b = 9$, the solutions of the inequality are the solutions of the following *two* inequalities:

$$\text{(i)}\quad 2x + 3 > 9 \qquad \text{(ii)}\quad 2x + 3 < -9$$

Inequality (i) is equivalent to $2x > 6$, or $x > 3$. This gives us the infinite interval $(3, \infty)$. Inequality (ii) is equivalent to $2x < -12$, or $x < -6$, which leads to the interval $(-\infty, -6)$. Consequently, the solutions of the inequality $|2x + 3| > 9$ consist of the real numbers in the two infinite intervals $(-\infty, -6)$ and $(3, \infty)$. The graph is sketched in Figure 1.14. ■

FIGURE 1.14

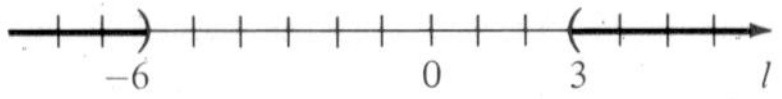

-6 0 3 l

For the solutions obtained in Example 6, we may use the **union symbol,** $\cup$, and write

$$(-\infty, -6) \cup (3, \infty)$$

to denote all real numbers that are either in $(-\infty, -6)$ or in $(3, \infty)$.

To solve inequalities involving polynomials of degree greater than 1, we shall use the following theorem, which will be discussed further in Section 3.1. In the statement of the theorem, the phrase *successive solutions* c and d means that there are no other solutions between c and d.

THEOREM

Let $a_nx^n + \cdots + a_1x + a_0$ be a polynomial. If the real numbers c and d are successive solutions of the equation

$$a_nx^n + \cdots + a_1x + a_0 = 0,$$

then when x is in the open interval (c, d) either all values of the polynomial are positive or all values are negative.

This theorem implies that if we choose *any* number k, such that $c < k < d$, and if the value of the polynomial is positive for $x = k$, then the polynomial is positive for *every* x in (c, d). Similarly, if the polynomial is negative for $x = k$, then it is negative throughout (c, d). We shall call the value of the polynomial at $x = k$ a **test value** for the interval (c, d). Test values may also be used on infinite intervals of the form $(-\infty, a)$ or (a, ∞), provided the polynomial equation has no solutions on these intervals. The use of test values is demonstrated in the following examples.

EXAMPLE 7 Solve $2x^2 - x < 3$.

SOLUTION To use the method that will be described here, *it is essential to have all nonzero terms on one side* of the inequality sign. Thus, we begin by writing

$$2x^2 - x - 3 < 0.$$

Factoring gives us

$$(x + 1)(2x - 3) < 0.$$

We see from the factored form that the equation $2x^2 - x - 3 = 0$ has solutions -1 and $\frac{3}{2}$. For reference, let us plot the corresponding points on a real axis, as in Figure 1.15. These points divide the axis into three parts and determine the following intervals:

FIGURE 1.15

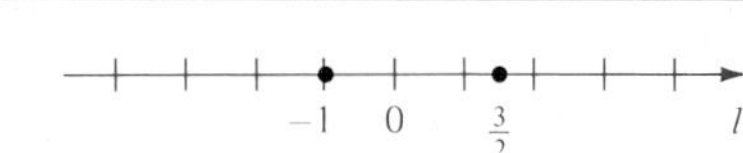

$$(-\infty, -1), \qquad (-1, \tfrac{3}{2}), \qquad (\tfrac{3}{2}, \infty).$$

The sign of the polynomial $2x^2 - x - 3$ in each interval can be found using a suitable test value.

If we choose -2 in $(-\infty, -1)$, then the polynomial $2x^2 - x - 3$ has the value

$$2(-2)^2 - (-2) - 3 = 8 + 2 - 3 = 7.$$

Since 7 is positive, it follows from the preceding theorem that the polynomial $2x^2 - x - 3$ is positive for every x in $(-\infty, -1)$.

If we choose 0 in $(-1, \frac{3}{2})$, then the polynomial has the value

$$2(0)^2 - (0) - 3 = -3.$$

Since -3 is negative, $2x^2 - x - 3 < 0$ for every x in $(-1, \frac{3}{2})$.

Finally, choosing 2 in $(\frac{3}{2}, \infty)$, we obtain

$$2(2)^2 - 2 - 3 = 8 - 2 - 3 = 3,$$

and since 3 is positive, $2x^2 - x - 3 > 0$ throughout $(\frac{3}{2}, \infty)$. It is convenient to tabulate these facts as follows:

Interval	$(-\infty, -1)$	$(-1, \frac{3}{2})$	$(\frac{3}{2}, \infty)$
k	-2	0	2
Test value of $2x^2 - x - 3$ at k	7	-3	3
Sign of $2x^2 - x - 3$ in interval	+	−	+

FIGURE 1.16

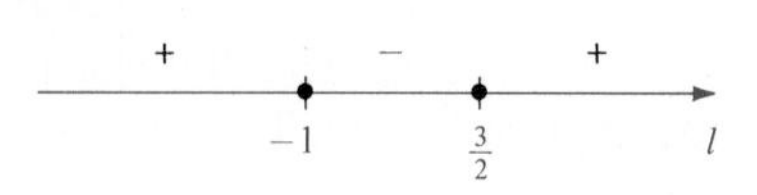

Figure 1.16 graphically illustrates the intervals in which $2x^2 - x - 3$ is positive or negative. Thus, the solutions $2x^2 - x - 3 < 0$, or equivalently $2x^2 - x < 3$, are the real numbers in the interval $(-1, \frac{3}{2})$. ■

EXAMPLE 8 Solve $x^2 > 7x - 10$ and represent the solutions graphically.

SOLUTION As in Example 7, we bring all terms to one side of the inequality sign, obtaining

$$x^2 - 7x + 10 > 0.$$

Factoring gives us $\quad (x - 2)(x - 5) > 0.$

Points corresponding to the solutions 2 and 5 of $x^2 - 7x + 10 = 0$ are plotted in Figure 1.17. Referring to the figure, we obtain the following intervals:

FIGURE 1.17

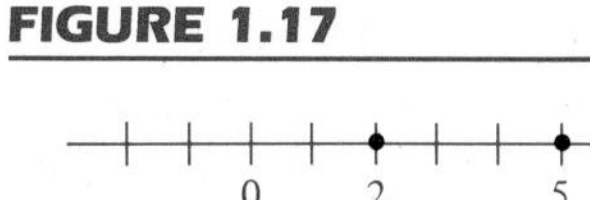

$$(-\infty, 2), \qquad (2, 5), \qquad (5, \infty).$$

We next use test values to determine the sign of $x^2 - 7x + 10$ in each interval. The following table summarizes results.

Interval	$(-\infty, 2)$	$(2, 5)$	$(5, \infty)$
k	0	3	6
Test value of $x^2 - 7x + 10$ at k	$0^2 - 7(0) + 10 = 10$	$3^2 - 7(3) + 10 = -2$	$6^2 - 7(6) + 10 = 4$
Sign of $x^2 - 7x + 10$ in interval	+	−	+

FIGURE 1.18

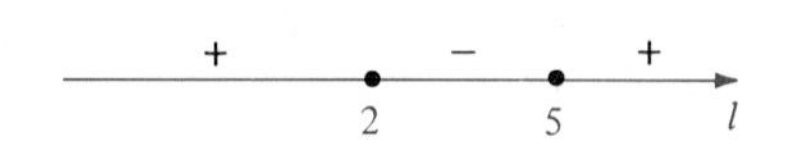

Figure 1.18 graphically illustrates where $x^2 - 7x + 10$ is positive or negative. Thus, $x^2 - 7x + 10 > 0$ if x is in either $(-\infty, 2)$ or $(5, \infty)$. The solutions of the inequality are given by $(-\infty, 2) \cup (5, \infty)$. The graph is sketched in Figure 1.19. ■

FIGURE 1.19

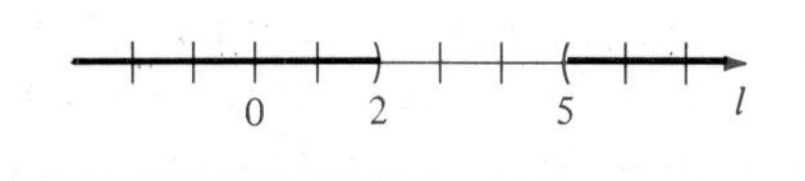

EXAMPLE 9

Solve the inequality $\dfrac{x+1}{x+3} \le 2$ and represent the solutions graphically.

SOLUTION We first take all nonzero terms to one side of the inequality symbol and then proceed as follows:

$$\frac{x+1}{x+3} \leq 2$$

$$\frac{x+1}{x+3} - 2 \leq 0$$

$$\frac{x+1-2x-6}{x+3} \leq 0$$

$$\frac{-x-5}{x+3} \leq 0$$

$$\frac{x+5}{x+3} \geq 0$$

FIGURE 1.20

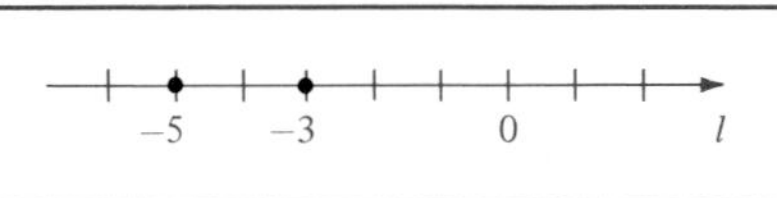

The numerator and denominator of $(x+5)/(x+3)$ equal zero at $x = -5$ and $x = -3$, respectively. For reference, we plot these points as in Figure 1.20. Note that -5 is a solution of $(x+5)/(x+3) \geq 0$, but -3 is *not* a solution since a zero denominator occurs if -3 is substituted for x. The points in the figure determine the following intervals:

$$(-\infty, -5), \qquad (-5, -3), \qquad (-3, \infty)$$

Since $(x+5)/(x+3)$ is a quotient of two polynomials, it is always positive or always negative throughout each interval. (The sign of this quotient is the same as the sign of the product $(x+5)(x+3)$.) As before, we may use test values to determine the sign in each interval. The following table summarizes results:

Interval	$(-\infty, -5)$	$(-5, -3)$	$(-3, \infty)$
k	-6	-4	0
Test value of $(x+5)/(x+3)$ at k	$\frac{1}{3}$	-1	$\frac{5}{3}$
Sign of $(x+5)/(x+3)$ in interval	$+$	$-$	$+$

FIGURE 1.21

FIGURE 1.22

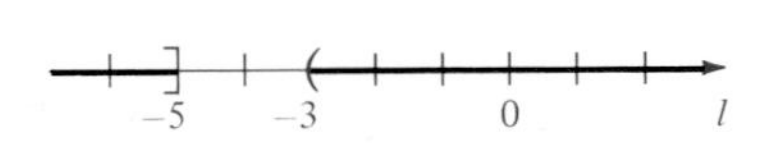

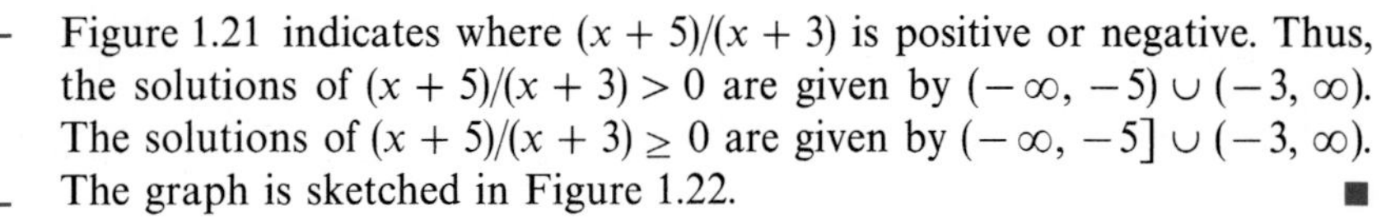

Figure 1.21 indicates where $(x+5)/(x+3)$ is positive or negative. Thus, the solutions of $(x+5)/(x+3) > 0$ are given by $(-\infty, -5) \cup (-3, \infty)$. The solutions of $(x+5)/(x+3) \geq 0$ are given by $(-\infty, -5] \cup (-3, \infty)$. The graph is sketched in Figure 1.22. ■

FIGURE 1.23

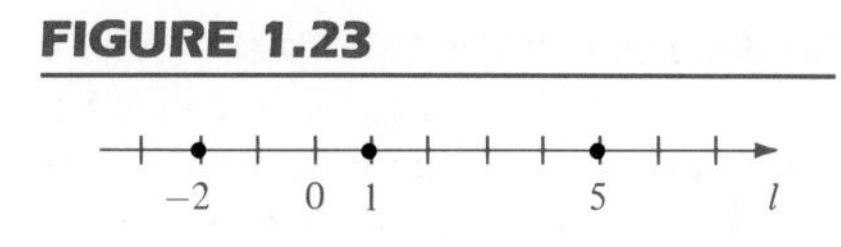

EXAMPLE 10 Solve $(x + 2)(x - 1)(x - 5) > 0$ and represent the solutions graphically.

SOLUTION The expression $(x + 2)(x - 1)(x - 5)$ is zero at -2, 1, and 5. The corresponding points are plotted on a real axis in Figure 1.23. These points determine the four intervals:

$$(-\infty, -2), \qquad (-2, 1), \qquad (1, 5), \qquad (5, \infty)$$

We next use test values, as shown in the following table.

Interval	$(-\infty, -2)$	$(-2, 1)$	$(1, 5)$	$(5, \infty)$
k	-3	0	2	6
Test value of $(x + 2)(x - 1)(x - 5)$	$(-1)(-4)(-8)$ $= -32$	$(2)(-1)(-5)$ $= 10$	$(4)(1)(-3)$ $= -12$	$(8)(5)(1)$ $= 40$
Sign of $(x + 2)(x - 1)(x - 5)$ in interval	$-$	$+$	$-$	$+$

FIGURE 1.24

Thus, the solutions of $(x + 2)(x - 1)(x - 5) > 0$ may be expressed as the union $(-2, 1) \cup (5, \infty)$. The graph is sketched in Figure 1.24. ■

Exercises 1.3

Solve the inequalities in Exercises 1–30 and express the solutions in terms of intervals.

1 $5x - 6 > 11$

2 $3x - 5 < 10$

3 $2 - 7x \leq 16$

4 $7 - 2x \geq -3$

5 $3x + 1 < 5x - 4$

6 $6x - 5 > 9x + 1$

7 $-4 < 3x + 5 < 8$

8 $6 > 2x - 6 > 4$

9 $3 \geq \dfrac{7 - x}{2} \geq 1$

10 $-2 \leq \dfrac{5 - 3x}{4} \leq \dfrac{1}{2}$

11 $0 < 2 - \frac{3}{4}x \leq \frac{1}{2}$

12 $-3 < \frac{1}{2}x - 4 \leq 0$

13 $|x - 10| < 0.05$

14 $|x - 5| < 0.001$

15 $|5 - 3x| < 7$

16 $|7x + 4| < 10$

17 $|2x + 4| > 8$

18 $|6x - 7| \geq 4$

19 $x^2 - x - 6 < 0$

20 $x^2 + 6x + 5 < 0$

21 $(2x + 1)(10 - 3x) < 0$

22 $4x^2 \geq x$

23 $\dfrac{7x}{x^2 - 16} \geq 0$

24 $\dfrac{x - 2}{x^2 - 9} < 0$

25 $\dfrac{x + 4}{2x - 1} < 3$

26 $\dfrac{1}{x - 2} > \dfrac{3}{x + 1}$

27 $x^3 - x^2 - 4x + 4 \geq 0$

28 $(x^2 - 4x + 4)(3x - 7) < 0$

29 $\dfrac{x^2 - x - 2}{x^2 - 4x + 3} \geq 0$

30 $\dfrac{x^2 + 2x - 3}{x^2 - 4x} \leq 0$

31 The relationship between the Fahrenheit and Celsius temperature scales is given by $C = \frac{5}{9}(F - 32)$. If $60 \leq F \leq 80$, express the corresponding range for C in terms of an inequality.

32 In the study of electricity, Ohm's Law states that if R denotes the resistance of an object (in ohms), E the potential difference across the object (in volts), and I the current that flows through it (in amperes), then $R = E/I$ (see figure). If the voltage is 110, what values of the resistance will result in a current that does not exceed 10 amperes?

FIGURE FOR EXERCISE 32

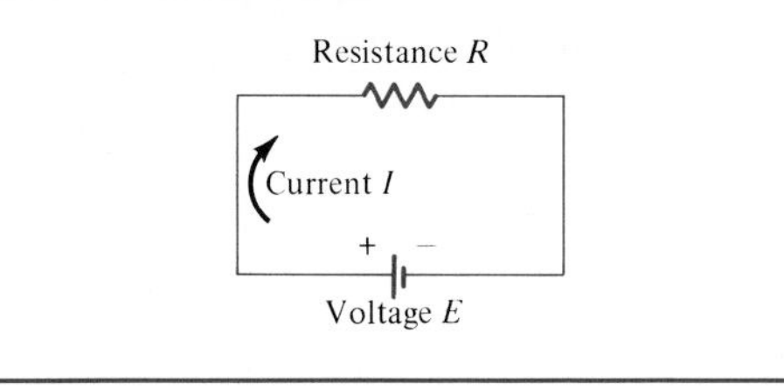

33 According to Hooke's Law, the force F (in pounds) required to stretch a certain spring x inches beyond its natural length is given by the formula $F = (4.5)x$ (see figure). If $10 \leq F \leq 18$, what are the corresponding values for x?

FIGURE FOR EXERCISE 33

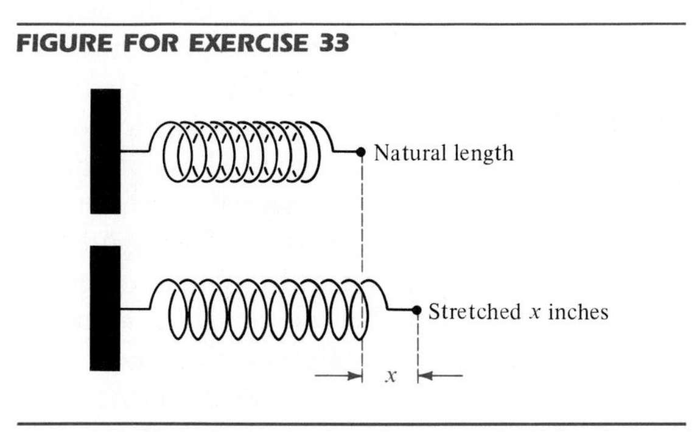

34 If two resistors R_1 and R_2 are connected in parallel in an electrical circuit, the net resistance R is given by $1/R = (1/R_1) + (1/R_2)$. If $R_1 = 10$ ohms, what values of R_2 will result in a net resistance of less than 5 ohms?

35 A convex lens has focal length $f = 5$ cm. If an object is placed at a distance of p cm from the lens, the distance q of the image from the lens is related to p and f by the formula $(1/p) + (1/q) = 1/f$ (see figure). How close must the object be to the lens for the image to be more than 12 cm from the lens?

FIGURE FOR EXERCISE 35

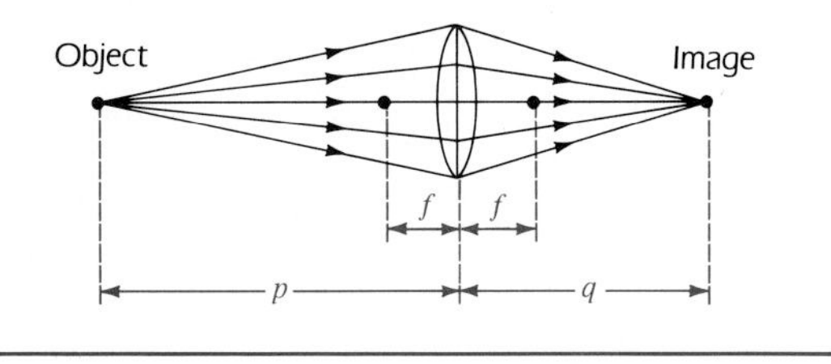

36 A construction firm is trying to decide which of two models of a type of crane to purchase. Model A costs \$50,000 and requires \$4000 per year to maintain. The corresponding figures for model B are \$40,000 initial cost and maintenance costs of \$5500 per year. For how many years must model A be used before it becomes more economical than B?

37 The braking distance d (in feet) of a car traveling v mph is approximated by $d = v + (v^2/20)$. Determine those velocities that result in braking distances of less than 75 feet.

38 For a satellite to maintain an orbit of height h km, its velocity (in km/sec) must equal $626.4/\sqrt{h + R}$ where $R = 6372$ km is the radius of the earth. What velocities will result in orbits of height more than 100 km from the earth's surface?

39 For a drug to have a beneficial effect, its concentration in the bloodstream must exceed a certain value called the *minimum therapeutic level*. Suppose that the concentration c of a drug t hours after it is taken orally is given by $c = 20t/(t^2 + 4)$ mg/liter. If the minimum therapeutic level is 4 mg/liter, determine when this level is exceeded.

40 As a jar containing 10 moles of gas A is heated, the velocity of the gas molecules increases and a second gas, B, is formed. When two molecules of gas A collide, two molecules of gas B are formed. If the number of moles of gas B after t minutes is $10t/(t + 4)$, when will there be more of gas B than gas A?

1.4 Coordinate Systems in Two Dimensions

In Section 1.1 we discussed a method of assigning coordinates to points on a line. Coordinate systems can also be introduced in planes by means of *ordered pairs.* The term **ordered pair** refers to two real numbers, where one is designated as the "first" number and the other as the "second." The symbol (a, b) is used to denote the ordered pair consisting of the real numbers a and b, where a is first and b is second. There are many uses for ordered pairs. We used them previously to denote open intervals. In this section they will represent points in a plane. Although ordered pairs are employed in different situations, there is little chance that we will confuse them, since it should always be clear from our discussion whether the symbol (a, b) represents an interval, a point, or some other mathematical object. We consider two ordered pairs (a, b) and (c, d) equal, and write

$$(a, b) = (c, d) \quad \text{if and only if} \quad a = c \quad \text{and} \quad b = d.$$

This implies, in particular, that $(a, b) \neq (b, a)$ if $a \neq b$.

A **rectangular coordinate system** may be introduced in a plane by considering two perpendicular coordinate lines in the plane that intersect in the origin O on each line. Unless specified otherwise, the same unit of length is chosen on each line. Usually one of the lines is horizontal with positive direction to the right, and the other line is vertical with positive direction upward, as indicated by the arrowheads in Figure 1.25. The two lines are called **coordinate axes,** and the point O is called the **origin.** The horizontal line is usually referred to as the ***x*-axis** and the vertical line as the ***y*-axis,** and they are labeled x and y, respectively. The plane is then called a **coordinate plane,** or an ***xy*-plane.** In certain applications different labels such as d or t are used for the coordinate lines. The coordinate axes divide the plane into four parts called the **first, second, third,** and **fourth quadrants** and labeled I, II, III, and IV, respectively. (See Figure 1.25(i).)

Each point P in an xy-plane may be assigned a unique ordered pair (a, b) as shown in Figure 1.25(ii). The number a is called the ***x*-coordinate** (or **abscissa**) of P, and b is called the ***y*-coordinate** (or **ordinate**). We say that *P has coordinates* (a, b). Conversely, every ordered pair (a, b) determines a point P in the xy-plane with coordinates a and b. We often refer to the *point* (a, b), or $P(a, b)$, meaning the point P with x-coordinate a and y-coordinate b. To **plot a point** $P(a, b)$ means to locate P in a coordinate plane and represent it by a dot, as illustrated by some points in Figure 1.25(iii).

FIGURE 1.25

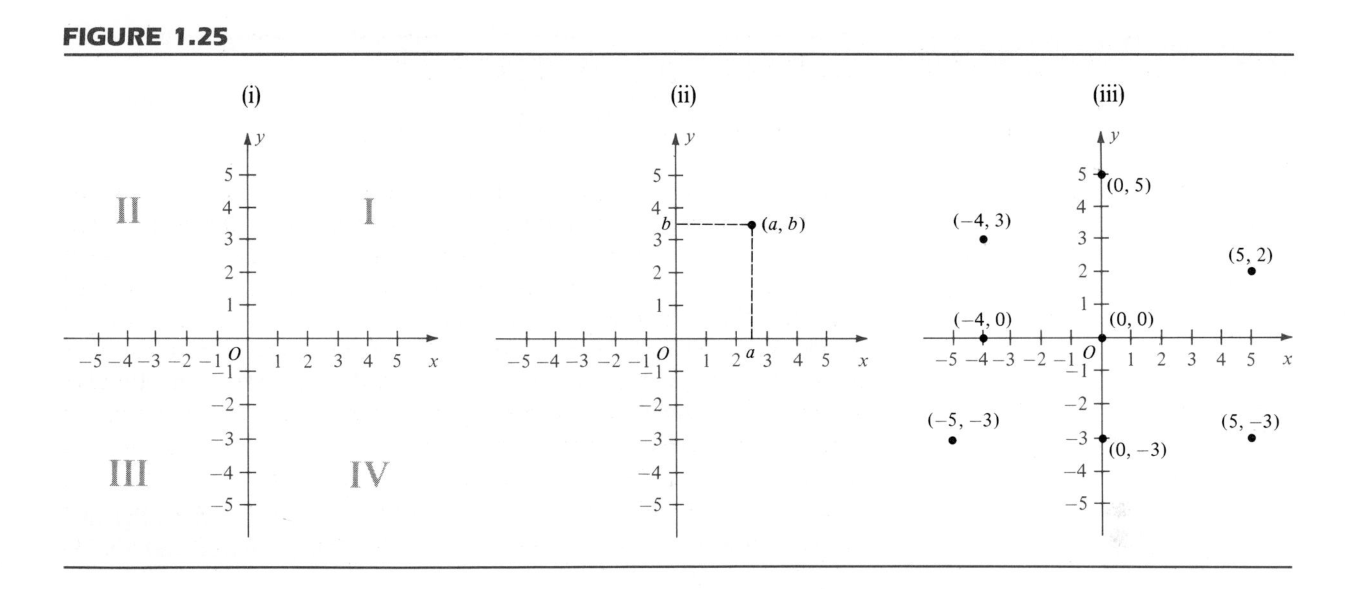

The next statement provides a formula for finding the distance between two points in a coordinate plane.

DISTANCE FORMULA

> The distance $d(P_1, P_2)$ between any two points $P_1(x_1, y_1)$ and $P_2(x_2, y_2)$ in a coordinate plane is
>
> $$d(P_1, P_2) = \sqrt{(x_2 - x_1)^2 + (y_2 - y_1)^2}.$$

PROOF If $x_1 \neq x_2$ and $y_1 \neq y_2$, then, as illustrated in Figure 1.26, the points P_1, P_2, and $P_3(x_2, y_1)$ are vertices of a right triangle. By the Pythagorean Theorem,

FIGURE 1.26

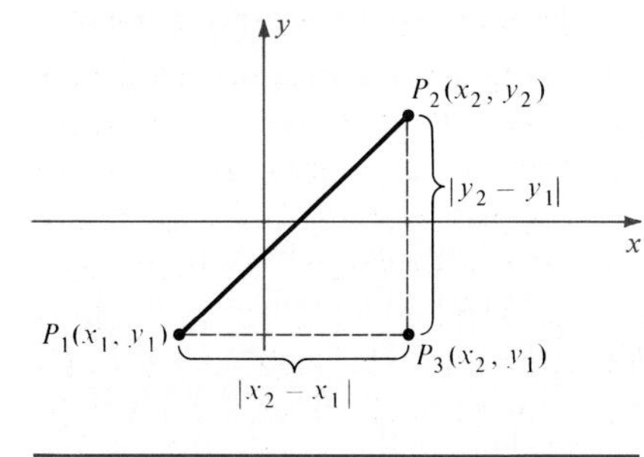

$$[d(P_1, P_2)]^2 = [d(P_1, P_3)]^2 + [d(P_3, P_2)]^2.$$

We see from the figure that

$$d(P_1, P_3) = |x_2 - x_1| \quad \text{and} \quad d(P_3, P_2) = |y_2 - y_1|.$$

Since $|a|^2 = a^2$ for every real number a, we may write

$$[d(P_1, P_2)]^2 = (x_2 - x_1)^2 + (y_2 - y_1)^2.$$

Taking the square root of each side of the last equation gives us the desired formula.

If $y_1 = y_2$, the points P_1 and P_2 lie on the same horizontal line and

$$d(P_1, P_2) = |x_2 - x_1| = \sqrt{(x_2 - x_1)^2}.$$

Similarly, if $x_1 = x_2$, the points are on the same vertical line and

$$d(P_1, P_2) = |y_2 - y_1| = \sqrt{(y_2 - y_1)^2}.$$

These are special cases of the Distance Formula.

Although we referred to Figure 1.26, the argument used in our proof is independent of the positions of the points P_1 and P_2. □

In applying the Distance Formula, note that $d(P_1, P_2) = d(P_2, P_1)$ and, hence, the order in which we subtract the x-coordinates and the y-coordinates of the points is immaterial.

EXAMPLE 1 Plot the points $A(-1, -3)$, $B(6, 1)$, and $C(2, -5)$. Prove that the triangle with vertices A, B, and C is a right triangle and find its area.

FIGURE 1.27

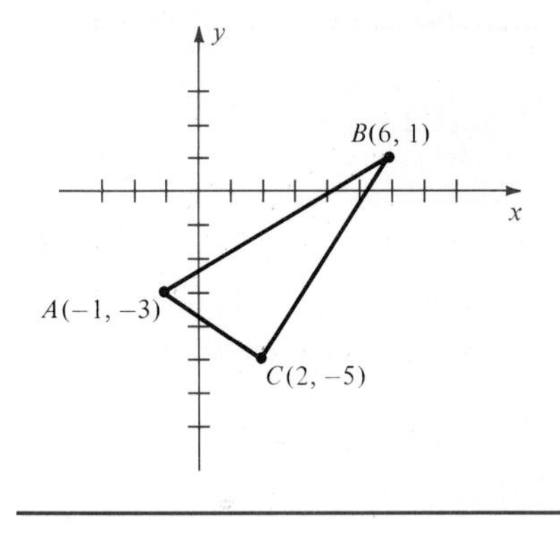

SOLUTION The points and the triangle are shown in Figure 1.27. From plane geometry, a triangle is a right triangle if and only if the sum of the squares of two of its sides is equal to the square of the remaining side. Using the Distance Formula,

$$d(A, B) = \sqrt{(-1-6)^2 + (-3-1)^2} = \sqrt{49+16} = \sqrt{65}$$
$$d(B, C) = \sqrt{(6-2)^2 + (1+5)^2} = \sqrt{16+36} = \sqrt{52}$$
$$d(A, C) = \sqrt{(-1-2)^2 + (-3+5)^2} = \sqrt{9+4} = \sqrt{13}.$$

Since $[d(A, B)]^2 = [d(B, C)]^2 + [d(A, C)]^2$, the triangle is a right triangle with hypotenuse AB. The area is $\frac{1}{2}\sqrt{52}\sqrt{13} = 13$ square units. ■

It is easy to obtain a formula for the midpoint of a line segment. Let $P_1(x_1, y_1)$ and $P_2(x_2, y_2)$ be two points in a coordinate plane and let M be the midpoint of the segment P_1P_2. The lines through P_1 and P_2 parallel to the y-axis intersect the x-axis at $A_1(x_1, 0)$ and $A_2(x_2, 0)$ and, from plane geometry, the line through M parallel to the y-axis bisects the

FIGURE 1.28

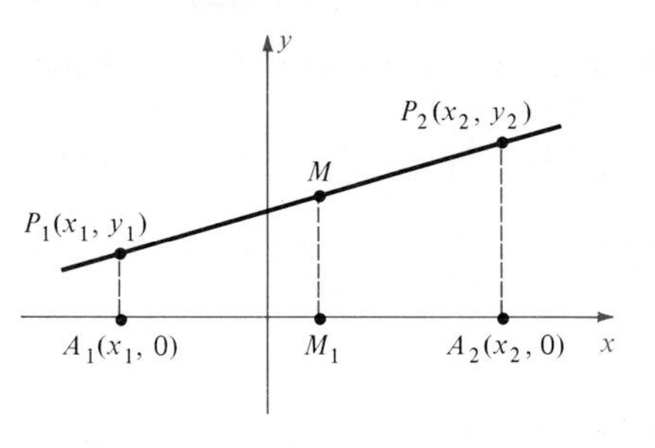

segment A_1A_2 (see Figure 1.28). If $x_1 < x_2$, then $x_2 - x_1 > 0$, and hence $d(A_1, A_2) = x_2 - x_1$. Since M_1 is halfway from A_1 to A_2, the x-coordinate of M_1 is

$$\begin{aligned} x_1 + \tfrac{1}{2}(x_2 - x_1) &= x_1 + \tfrac{1}{2}x_2 - \tfrac{1}{2}x_1 \\ &= \tfrac{1}{2}x_1 + \tfrac{1}{2}x_2 \\ &= \frac{x_1 + x_2}{2}. \end{aligned}$$

It follows that the x-coordinate of M is also $(x_1 + x_2)/2$. Similarly, the y-coordinate of M is $(y_1 + y_2)/2$. Moreover, these formulas hold for all positions of P_1 and P_2. This gives us the following result.

MIDPOINT FORMULA

> The midpoint of the line segment from $P_1(x_1, y_1)$ to $P_2(x_2, y_2)$ is
>
> $$\left(\frac{x_1 + x_2}{2}, \frac{y_1 + y_2}{2}\right).$$

EXAMPLE 2 Find the midpoint M of the line segment from $P_1(-2, 3)$ to $P_2(4, -2)$. Plot P_1, P_2, and M and verify that $d(P_1, M) = d(P_2, M)$.

SOLUTION By the Midpoint Formula, the coordinates of M are

$$\left(\frac{-2 + 4}{2}, \frac{3 + (-2)}{2}\right) \quad \text{or} \quad \left(1, \frac{1}{2}\right).$$

FIGURE 1.29

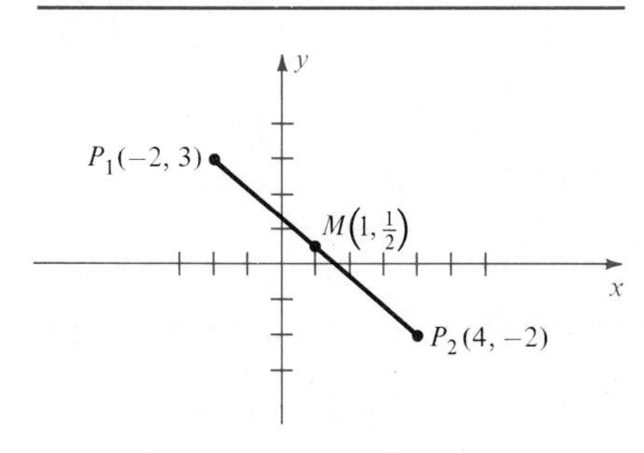

The three points P_1, P_2, and M are plotted in Figure 1.29. Using the Distance Formula,

$$d(P_1, M) = \sqrt{(-2 - 1)^2 + (3 - \tfrac{1}{2})^2} = \sqrt{9 + \tfrac{25}{4}}$$
$$d(P_2, M) = \sqrt{(4 - 1)^2 + (-2 - \tfrac{1}{2})^2} = \sqrt{9 + \tfrac{25}{4}}.$$

Hence, $d(P_1, M) = d(P_2, M)$. ■

If W is a set of ordered pairs, we may consider the point $P(x, y)$ in a coordinate plane that corresponds to the ordered pair (x, y) in W. The **graph** of W is the set of all such points. The phrase "sketch the graph of W" means to illustrate the significant features of the graph geometrically on a coordinate plane.

FIGURE 1.30

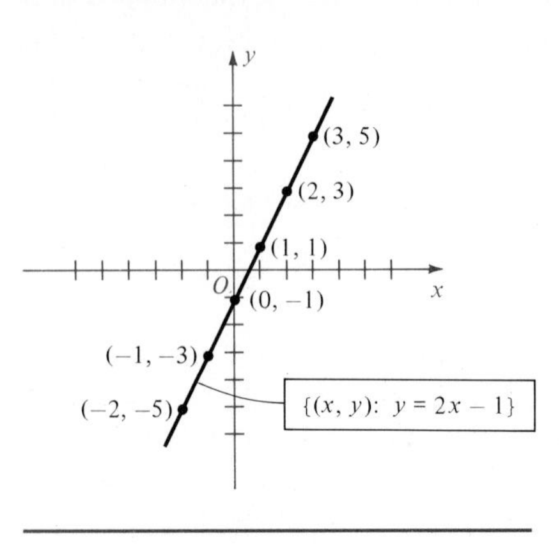

EXAMPLE 3 Sketch the graph of $W = \{(x, y) : y = 2x - 1\}$.

SOLUTION We wish to find the points (x, y), where the ordered pair (x, y) is in W. It is convenient to list coordinates of several such points in the following tabular form, where for each x, the value for y is obtained from $y = 2x - 1$.

x	-2	-1	0	1	2	3
y	-5	-3	-1	1	3	5

After we plot the points with these coordinates, it appears that they lie on a line and we sketch the graph (see Figure 1.30). Ordinarily, the few points we have plotted would not be enough to illustrate the graph; however, in this elementary case we can be reasonably sure that the graph is a line. We will prove this fact in the next section. ■

The x-coordinates of points at which a graph intersects the x-axis are called the **x-intercepts** of the graph. The y-coordinates of points at which a graph intersects the y-axis are called the **y-intercepts.** The graph in Figure 1.30 has one x-intercept, $\frac{1}{2}$, and one y-intercept, -1.

It is impossible to sketch the entire graph in Example 3, since x may be assigned values that are numerically as large as desired. Nevertheless, we call the drawing in Figure 1.30 *the graph of W* or *a sketch of the graph.* It is understood that the drawing is only a device for visualizing the actual graph and the line does not terminate as shown in the figure. In general, the sketch of a graph should illustrate enough of the graph so that the remaining parts are evident.

Given an equation in x and y, we say that an ordered pair (a, b) is a **solution** of the equation if equality is obtained when a is substituted for x and b for y. For example, $(2, 3)$ is a solution of $y = 2x - 1$, since substitution of 2 for x and 3 for y leads to $3 = 4 - 1$, or $3 = 3$. Two equations in x and y are **equivalent** if they have exactly the same solutions. The solutions of an equation in x and y determine a set W of ordered pairs, and we define the **graph of the equation** as the graph of W. Note that the graph of the equation $y = 2x - 1$ is the same as the graph of the set W in Example 3 (see Figure 1.30).

To sketch the graph of an equation we may plot points until a pattern emerges, and then sketch the graph accordingly. This is obviously a crude (and often inaccurate) way to arrive at the graph; however, it is a method often employed in elementary courses. As we progress through this text, techniques will be introduced that will enable us to sketch accurate graphs without plotting many points. To sketch graphs when complicated expressions are involved, it is usually necessary to employ methods introduced in calculus.

FIGURE 1.31

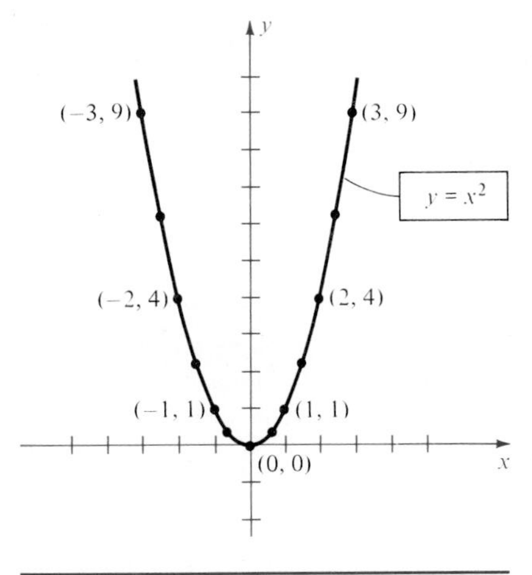

EXAMPLE 4 Sketch the graph of the equation $y = x^2$.

SOLUTION To sketch the graph, we must plot more points than in the previous example. Increasing successive x-coordinates by $\frac{1}{2}$, we obtain the following table.

x	-3	$-\frac{5}{2}$	-2	$-\frac{3}{2}$	-1	$-\frac{1}{2}$	0	$\frac{1}{2}$	1	$\frac{3}{2}$	2	$\frac{5}{2}$	3
y	9	$\frac{25}{4}$	4	$\frac{9}{4}$	1	$\frac{1}{4}$	0	$\frac{1}{4}$	1	$\frac{9}{4}$	4	$\frac{25}{4}$	9

Larger values of x produce larger values of y. For example, the points (4, 16), (5, 25), and (6, 36) are on the graph, as are (−4, 16), (−5, 25), and (−6, 36). Plotting the points given by the table and drawing a smooth curve through these points gives us the sketch in Figure 1.31, in which several points are labeled. ■

The graph in Example 4 is a **parabola.** The y-axis is called the **axis of the parabola.** The lowest point (0, 0) is the **vertex** of the parabola and we say that the parabola **opens upward.** If the graph were inverted, as would be the case for $y = -x^2$, then the parabola **opens downward** and the vertex (0, 0) is the highest point on the graph. In general, the graph of *any* equation of the form $y = ax^2$ for $a \neq 0$ is a parabola with vertex (0, 0). Parabolas may also open to the right or to the left (see Example 5). A definition of parabola and a detailed discussion of its properties may be found in Chapter 10.

If the coordinate plane in Figure 1.31 is folded along the y-axis, then the graph that lies in the left half of the plane coincides with that in the right half. We say that **the graph is symmetric with respect to the y-axis.** As in Figure 1.32(i), a graph is symmetric with respect to the y-axis provided that the point $(-x, y)$ is on the graph whenever (x, y) is on the graph.

FIGURE 1.32

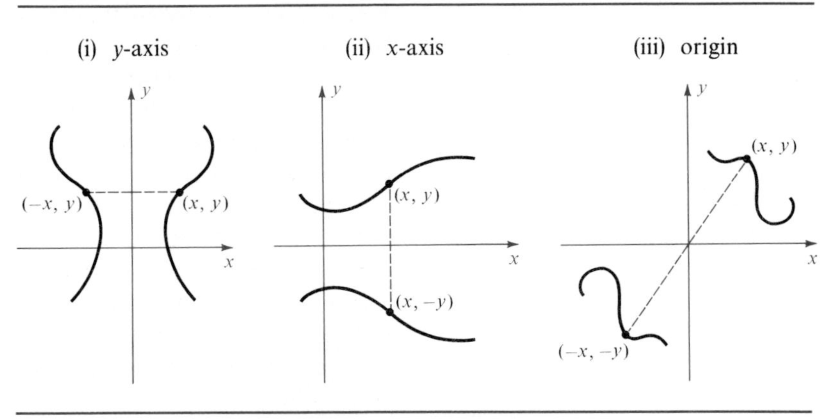

As in Figure 1.32(ii), **a graph is symmetric with respect to the x-axis** if whenever a point (x, y) is on the graph, then $(x, -y)$ is also on the graph. Certain graphs possess a type of symmetry, called **symmetry with respect to the origin.** In this situation, whenever a point (x, y) is on the graph, then $(-x, -y)$ is also on the graph, as illustrated in Figure 1.32(iii).

The following tests are useful for investigating these three types of symmetry for graphs of equations in x and y.

TESTS FOR SYMMETRY

(i) The graph of an equation is symmetric with respect to the y-axis if substitution of $-x$ for x leads to an equivalent equation.

(ii) The graph of an equation is symmetric with respect to the x-axis if substitution of $-y$ for y leads to an equivalent equation.

(iii) The graph of an equation is symmetric with respect to the origin if the simultaneous substitution of $-x$ for x and $-y$ for y leads to an equivalent equation.

If, in the equation of Example 4, we substitute $-x$ for x, we obtain $y = (-x)^2$, which is equivalent to $y = x^2$. Hence, by Symmetry Test (i), the graph is symmetric with respect to the y-axis.

If symmetry with respect to an axis exists, it is sufficient to determine the graph in half of the coordinate plane, since the remainder of the graph may be sketched by taking a mirror image, or reflection, of that half.

EXAMPLE 5 Sketch the graph of the equation $y^2 = x$.

FIGURE·1.33

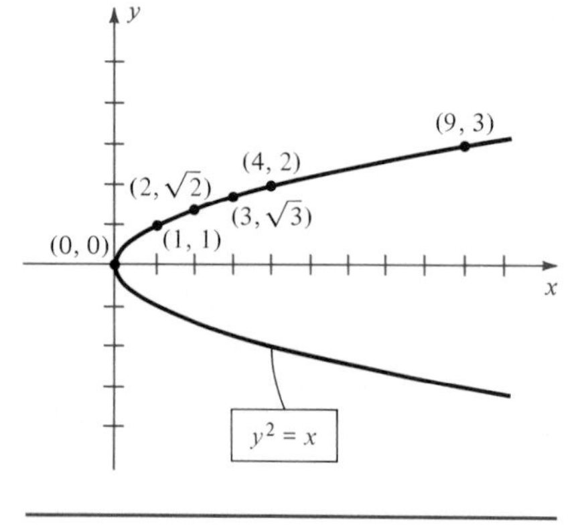

SOLUTION Since substitution of $-y$ for y does not change the equation, the graph is symmetric with respect to the x-axis. (See Symmetry Test (ii).) Thus, it is sufficient to plot points with nonnegative y-coordinates and then reflect through the x-axis. Since $y^2 = x$, the y-coordinates of points above the x-axis are given by $y = \sqrt{x}$. Coordinates of some points on the graph are listed in the following table.

x	0	1	2	3	4	9
y	0	1	$\sqrt{2} \approx 1.4$	$\sqrt{3} \approx 1.7$	2	3

A portion of the graph is sketched in Figure 1.33. The graph is a parabola that opens to the right, with its vertex at the origin. In this case the x-axis is the axis of the parabola. ■

EXAMPLE 6 Sketch the graph of the equation $4y = x^3$.

SOLUTION If we substitute $-x$ for x and $-y$ for y, then

$$4(-y) = (-x)^3 \quad \text{or} \quad -4y = -x^3.$$

FIGURE 1.34

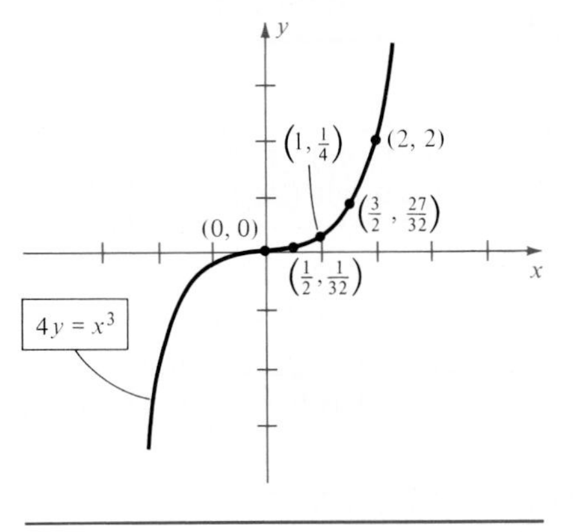

Multiplying both sides by -1, we see that the last equation has the same solutions as the equation $4y = x^3$. Hence, from Symmetry Test (iii), the graph is symmetric with respect to the origin. The following table lists some points on the graph.

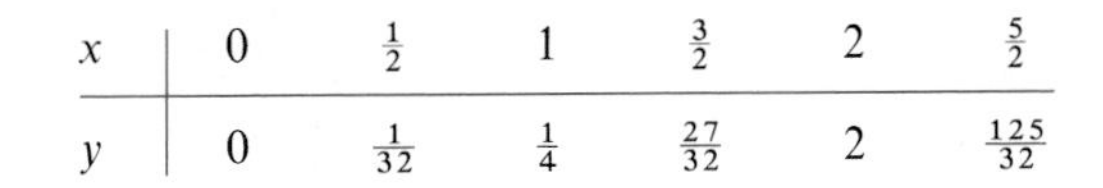

x	0	$\frac{1}{2}$	1	$\frac{3}{2}$	2	$\frac{5}{2}$
y	0	$\frac{1}{32}$	$\frac{1}{4}$	$\frac{27}{32}$	2	$\frac{125}{32}$

By symmetry (or substitution) we see that the points $(-1, -\frac{1}{4})$, $(-2, -2)$, and so on, are on the graph. Plotting points leads to the sketch in Figure 1.34. ■

If $C(h, k)$ is a point in a coordinate plane, then a circle with center C and radius $r > 0$ consists of all points in the plane that are r units from C. As shown in Figure 1.35(i), a point $P(x, y)$ is on the circle if and only if $d(C, P) = r$ or, by the Distance Formula, if and only if

$$\sqrt{(x - h)^2 + (y - k)^2} = r.$$

FIGURE 1.35

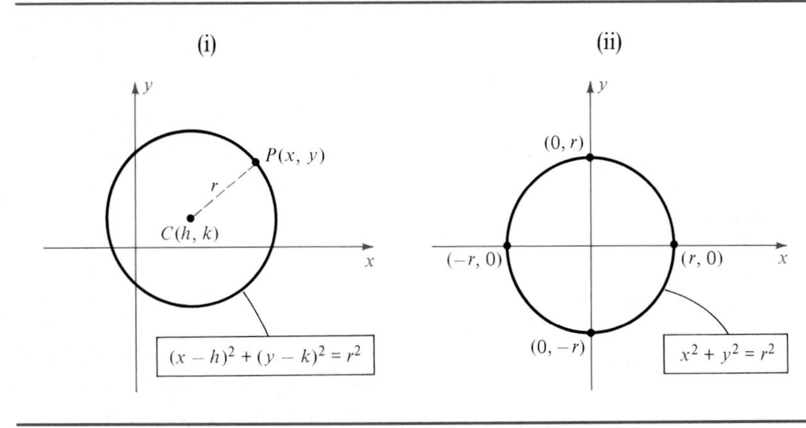

The following equivalent equation is called the **equation of a circle of radius r and center (h, k).**

EQUATION OF A CIRCLE

$$(x - h)^2 + (y - k)^2 = r^2, \qquad r > 0$$

If $h = 0$ and $k = 0$, this equation reduces to $x^2 + y^2 = r^2$, which is an equation of a circle of radius r with center at the origin (see Figure 1.35(ii)). If $r = 1$, the graph is called a **unit circle.**

FIGURE 1.36

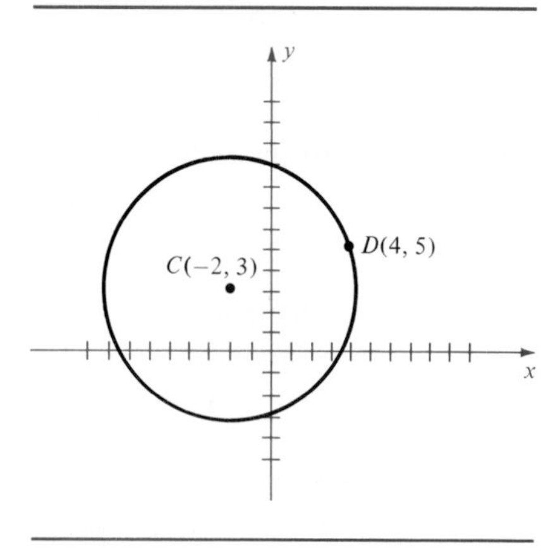

EXAMPLE 7 Find an equation of the circle that has center $C(-2, 3)$ and contains the point $D(4, 5)$.

SOLUTION The circle is illustrated in Figure 1.36. Since D is on the circle, the radius r is $d(C, D)$. By the Distance Formula,

$$r = \sqrt{(-2 - 4)^2 + (3 - 5)^2} = \sqrt{36 + 4} = \sqrt{40}.$$

Using the equation of a circle with $h = -2$, $k = 3$, and $r = \sqrt{40}$, we obtain

$$(x + 2)^2 + (y - 3)^2 = 40$$

or

$$x^2 + y^2 + 4x - 6y - 27 = 0.$$

■

Squaring terms of $(x - h)^2 + (y - k)^2 = r^2$ and simplifying leads to an equation of the form

$$x^2 + y^2 + ax + by + c = 0$$

for some real numbers a, b, and c. Conversely, if we begin with the last equation, it is always possible, by *completing squares*, to obtain an equation of the form

$$(x - h)^2 + (y - k)^2 = d.$$

The method will be illustrated in Example 8. If $d > 0$ the graph is a circle with center (h, k) and radius $r = \sqrt{d}$. If $d = 0$ the graph consists of only one point (h, k). Finally, if $d < 0$ the equation has no real solutions and hence there is no graph.

EXAMPLE 8 Find the center and radius of the circle with equation

$$x^2 + y^2 - 4x + 6y - 3 = 0.$$

SOLUTION We begin by arranging the equation as follows:

$$(x^2 - 4x) + (y^2 + 6y) = 3.$$

Next we complete the squares by adding appropriate numbers within the parentheses. Of course, to obtain equivalent equations, we must add the numbers to *both* sides of the equation. To complete the square for an expression of the form $x^2 + ax$, we add the square of half the coefficient of x, that is, $(a/2)^2$, to both sides of the equation. Similarly, for $y^2 + by$, we add $(b/2)^2$ to both sides. In this example $a = -4, b = 6, (a/2)^2 = (-2)^2 = 4$, and $(b/2)^2 = 3^2 = 9$. This leads to

$$(x^2 - 4x + 4) + (y^2 + 6y + 9) = 3 + 4 + 9$$

or

$$(x - 2)^2 + (y + 3)^2 = 16.$$

It follows that the center is $(2, -3)$ and the radius is 4. ■

Exercises 1.4

1 Plot the following points on a rectangular coordinate system: $A(5, -2)$, $B(-5, -2)$, $C(5, 2)$, $D(-5, 2)$, $E(3, 0)$, $F(0, 3)$.

2 Plot the points $A(-3, 1)$, $B(3, 1)$, $C(-2, -3)$, $D(0, 3)$, and $E(2, -3)$ on a rectangular coordinate system and then draw the line segments AB, BC, CD, DE, and EA.

3 Plot $A(0, 0)$, $B(1, 1)$, $C(3, 3)$, $D(-1, -1)$, and $E(-2, -2)$. Describe the set of all points of the form (x, x) where x is a real number.

4 Plot $A(0, 0)$, $B(1, -1)$, $C(2, -2)$, $D(-1, 1)$, and $E(-3, 3)$. Describe the set of all points of the form $(x, -x)$ where x is a real number.

5 Describe the set of all points $P(x, y)$ in a coordinate plane such that:

(a) $x = 3$ (b) $y = -1$ (c) $x \geq 0$
(d) $xy > 0$ (e) $y < 0$

6 Describe the set of all points $P(x, y)$ in a coordinate plane such that:

(a) $y = 0$ (b) $x = -5$ (c) $x/y < 0$
(d) $xy = 0$ (e) $y > 1$

In Exercises 7–12 find (a) the distance $d(A, B)$ between the given points A and B; (b) the midpoint of the segment AB.

7 $A(4, -3)$, $B(6, 2)$ **8** $A(-2, -5)$, $B(4, 6)$

9 $A(-5, 0)$, $B(-2, -2)$ **10** $A(6, 2)$, $B(6, -2)$

11 $A(7, -3)$, $B(3, -3)$ **12** $A(-4, 7)$, $B(0, -8)$

In Exercises 13 and 14 prove that the triangle with the indicated vertices is a right triangle and find its area.

13 $A(8, 5)$, $B(1, -2)$, $C(-3, 2)$

14 $A(-6, 3)$, $B(3, -5)$, $C(-1, 5)$

In Exercises 15–36 sketch the graph of the equation, and test for symmetry.

15 $y = 3x + 1$

16 $y = 4x - 3$

17 $y + 2x - 3 = 0$

18 $3x + y - 2 = 0$

19 $y = 2x^2 - 1$

20 $y = -x^2 + 2$

21 $4y = x^2$

22 $3y + x^2 = 0$

23 $y = -\frac{1}{2}x^3$

24 $y = \frac{1}{2}x^3$

25 $y = x^3 - 2$

26 $y = 2 - x^3$

27 $y = \sqrt{x}$

28 $y = \sqrt{x} - 1$

29 $y = \sqrt{-x}$

30 $y = \sqrt{x - 1}$

31 $x^2 + y^2 = 16$

32 $4x^2 + 4y^2 = 25$

33 $y = \sqrt{9 - x^2}$

34 $y = -\sqrt{4 - x^2}$

35 $x = -\sqrt{9 - y^2}$

36 $x = \sqrt{4 - y^2}$

In Exercises 37–44 find an equation of a circle that satisfies the stated conditions.

37 Center $C(3, -2)$, radius 4

38 Center $C(-5, 2)$, radius 5

39 Center at the origin, passing through $P(-3, 5)$

40 Center $C(-4, 6)$, passing through $P(1, 2)$

41 Center $C(-4, 2)$, tangent to the x-axis

42 Center $C(3, -5)$, tangent to the y-axis

43 Endpoints of a diameter $A(4, -3)$ and $B(-2, 7)$

44 Tangent to both axes, center in the first quadrant, radius 2

In Exercises 45–50 find the center and radius of the circle with the given equation.

45 $x^2 + y^2 + 4x - 6y + 4 = 0$

46 $x^2 + y^2 - 10x + 2y + 22 = 0$

47 $x^2 + y^2 + 6x = 0$

48 $x^2 + y^2 + x + y - 1 = 0$

49 $2x^2 + 2y^2 - x + y - 3 = 0$

50 $9x^2 + 9y^2 - 6x + 12y - 31 = 0$

1.5 Lines

Lines are among the simplest and most important of all geometric figures. The following concept is fundamental to the study of lines in a coordinate plane.

DEFINITION

Let l be a line that is not parallel to the y-axis, and let $P_1(x_1, y_1)$ and $P_2(x_2, y_2)$ be distinct points on l. The **slope** m of l is

$$m = \frac{y_2 - y_1}{x_2 - x_1}.$$

If l is parallel to the y-axis, then the slope is not defined.

Typical points P_1 and P_2 on a line l are shown in Figure 1.37. The numerator $y_2 - y_1$ in the formula for m is the vertical change in direction

FIGURE 1.37

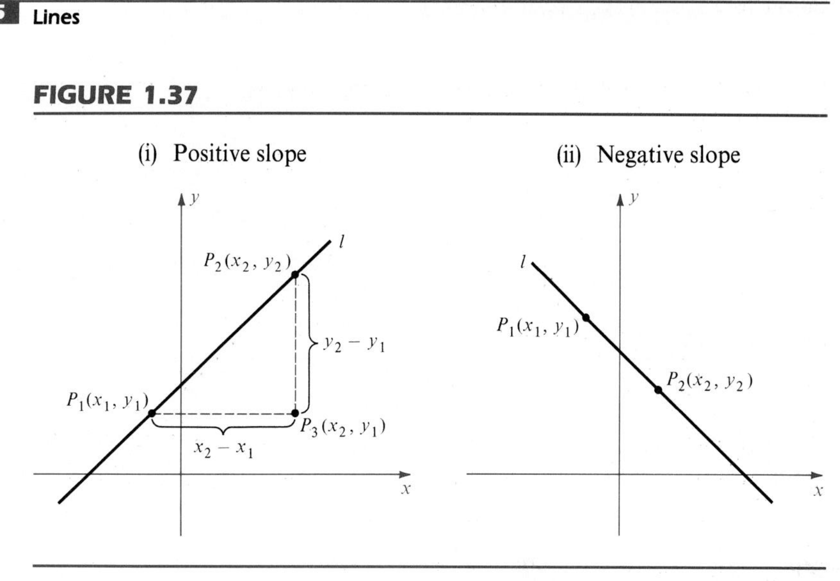

in proceeding from P_1 to P_2 and may be positive, negative, or zero. The denominator $x_2 - x_1$ is the amount of horizontal change in going from P_1 to P_2, and it may be positive or negative, but never zero, because l is not parallel to the y-axis.

In finding the slope of a line, it is immaterial which point is labeled P_1 and which is labeled P_2, since

$$\frac{y_2 - y_1}{x_2 - x_1} = \frac{y_1 - y_2}{x_1 - x_2}.$$

Consequently, we may assume that the points are labeled so that $x_1 < x_2$, as in Figure 1.37. In this situation $x_2 - x_1 > 0$, and hence the slope is positive, negative, or zero, depending on whether $y_2 > y_1$, $y_2 < y_1$, or $y_2 = y_1$. The slope of the line shown in Figure 1.37(i) is positive, whereas the slope of the line shown in (ii) of the figure is negative.

FIGURE 1.38

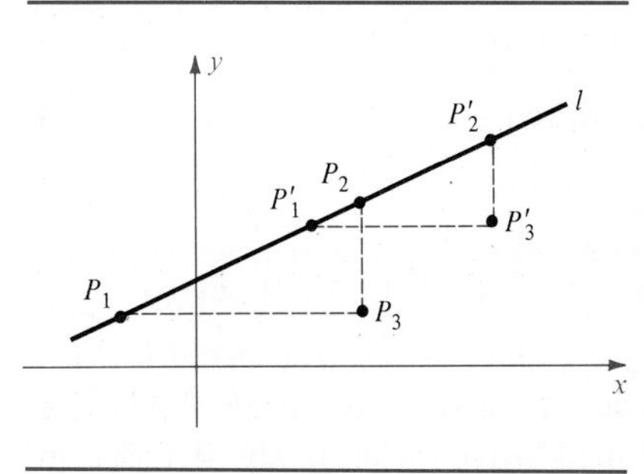

A **horizontal line** is a line that is parallel to the x-axis. A line is horizontal if and only if its slope is 0. A **vertical line** is a line that is parallel to the y-axis. The slope of a vertical line is undefined.

It is important to note that the definition of slope is independent of the two points that are chosen on l, for if other points $P'_1(x'_1, y'_1)$ and $P'_2(x'_2, y'_2)$ are used, then as in Figure 1.38, the triangle with vertices P'_1, P'_2, and $P'_3(x'_2, y'_1)$ is similar to the triangle with vertices P_1, P_2, and $P_3(x_2, y_1)$. Since the ratios of corresponding sides are equal, it follows that

$$m = \frac{y_2 - y_1}{x_2 - x_1} = \frac{y'_2 - y'_1}{x'_2 - x'_1}.$$

EXAMPLE 1 Sketch the lines through the following pairs of points and find their slopes.

(a) $A(-1, 4)$ and $B(3, 2)$ (b) $A(2, 5)$ and $B(-2, -1)$

(c) $A(4, 3)$ and $B(-2, 3)$ (d) $A(4, -1)$ and $B(4, 4)$

SOLUTION The lines are sketched in Figure 1.39.

FIGURE 1.39

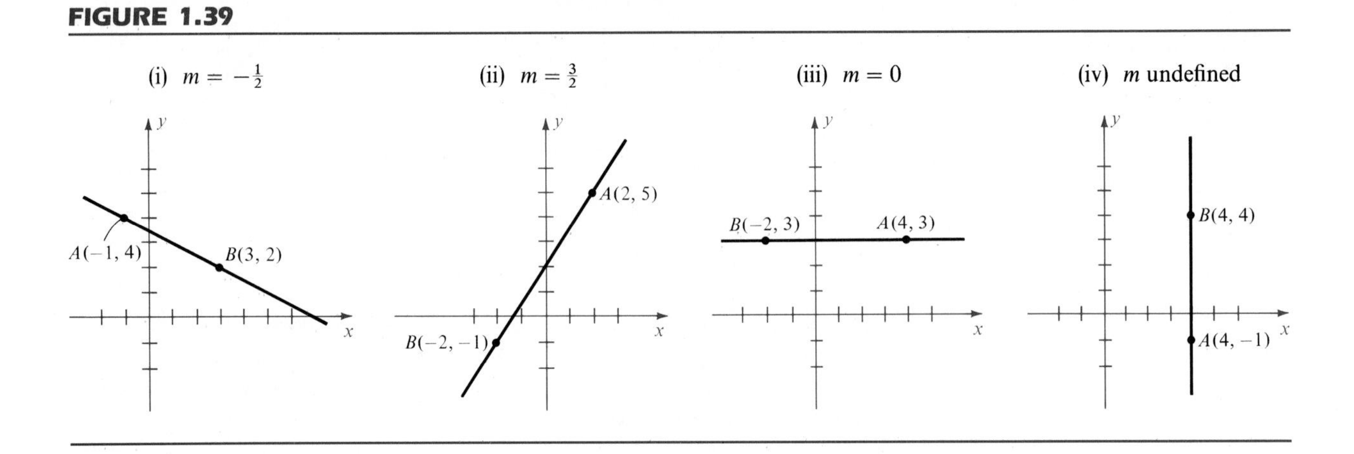

Using the definition of slope gives us

(a) $$m = \frac{2-4}{3-(-1)} = \frac{-2}{4} = -\frac{1}{2}$$

(b) $$m = \frac{5-(-1)}{2-(-2)} = \frac{6}{4} = \frac{3}{2}$$

(c) $$m = \frac{3-3}{-2-4} = \frac{0}{-6} = 0$$

(d) The slope is undefined since the line is vertical. This is also seen by noting that, if the formula for m is used, the denominator is zero. ■

EXAMPLE 2 Construct a line through $P(2, 1)$ that has slope $\frac{5}{3}$; $-\frac{5}{3}$.

SOLUTION If the slope of a line is a/b and b is positive, then for every change of b units in the horizontal direction, the line rises or falls $|a|$ units, depending on whether a is positive or negative, respectively. If $P(2, 1)$ is on the line and $m = \frac{5}{3}$, we can obtain another point on the line by starting at P and moving 3 units to the right and 5 units upward. This gives us the point $Q(5, 6)$, and the line is determined (see Figure 1.40(i)). Similarly, if $m = -\frac{5}{3}$, we move 3 units to the right and 5 units downward, obtaining $Q(5, -4)$ as in Figure 1.40(ii).

FIGURE 1.40

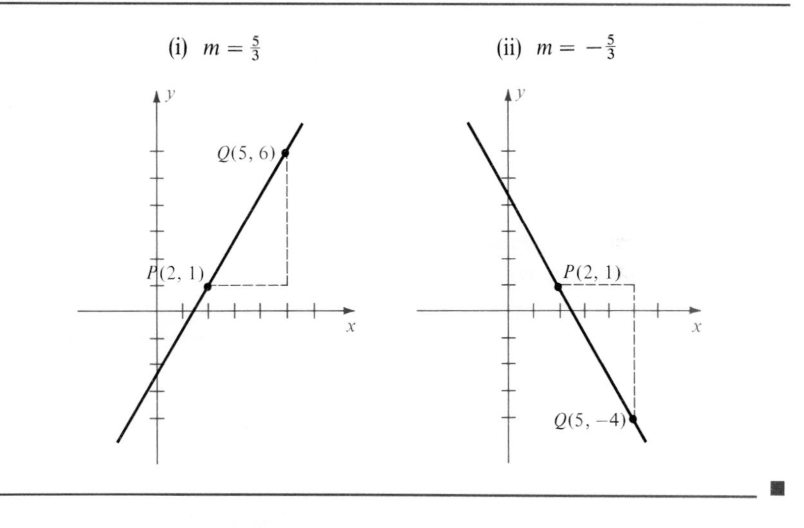

THEOREM

> (i) The graph of the equation $x = a$ is a vertical line that has x-intercept a.
>
> (ii) The graph of the equation $y = b$ is a horizontal line that has y-intercept b.

FIGURE 1.41

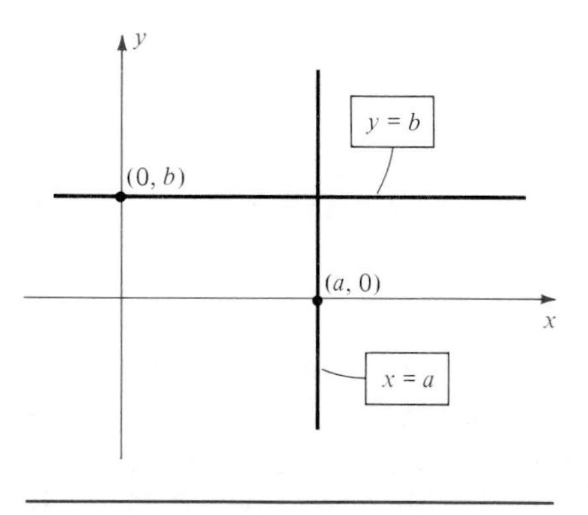

PROOF The equation $x = a$ may be written in the form $x + (0)y = a$. Some typical solutions of this equation are $(a, -2)$, $(a, 1)$, and $(a, 3)$. Evidently, every solution has the form (a, y), where y may have any value and a is fixed. It follows that the graph of $x = a$ is a line parallel to the y-axis with x-intercept a, as illustrated in Figure 1.41. This proves (i). Part (ii) is proved in similar fashion. □

Let us next find an equation of the line l through a point $P_1(x_1, y_1)$ with slope m (only one such line exists). If $P(x, y)$ is any point with $x \neq x_1$

FIGURE 1.42

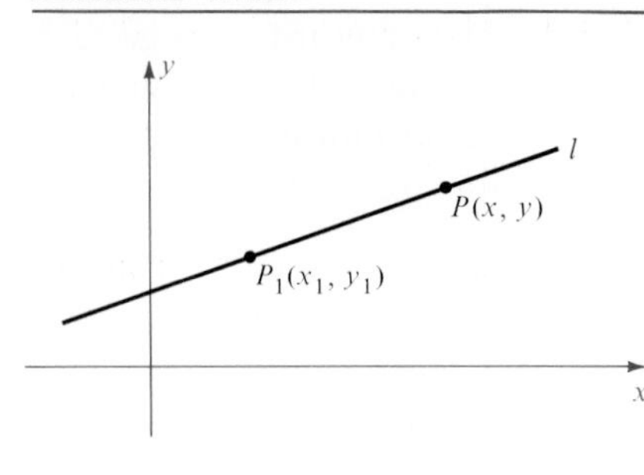

(see Figure 1.42), then P is on l if and only if the slope of the line through P_1 and P is m; that is,

$$\frac{y - y_1}{x - x_1} = m.$$

This equation may be written in the form

$$y - y_1 = m(x - x_1).$$

Note that (x_1, y_1) is also a solution of the last equation and hence the points on l are precisely the points that correspond to the solutions. This equation for l is referred to as the **Point-Slope Form.**

POINT-SLOPE FORM FOR THE EQUATION OF A LINE

An equation of the line through the point (x_1, y_1) with slope m is

$$y - y_1 = m(x - x_1).$$

EXAMPLE 3 Find an equation of the line through the points $A(1, 7)$ and $B(-3, 2)$.

SOLUTION The slope m of the line is

$$m = \frac{7 - 2}{1 - (-3)} = \frac{5}{4}.$$

We may use the coordinates of either A or B for (x_1, y_1) in the Point-Slope Form. Using $A(1, 7)$ gives us

$$y - 7 = \tfrac{5}{4}(x - 1),$$

which is equivalent to

$$4y - 28 = 5x - 5, \quad \text{or} \quad 5x - 4y + 23 = 0. \qquad ■$$

The Point-Slope Form may be rewritten as $y = mx - mx_1 + y_1$, which is of the form

$$y = mx + b$$

with $b = -mx_1 + y_1$. The real number b is the y-intercept of the graph, as may be seen by setting $x = 0$. Since the equation $y = mx + b$ displays the slope m and y-intercept b of l, it is called the **Slope-Intercept Form** for the equation of a line. Conversely, if we start with $y = mx + b$, we may write

$$y - b = m(x - 0).$$

Comparing this last equation with the Point-Slope Form, we see that the graph is a line that has slope m and passes through the point $(0, b)$. This gives us the next result.

SLOPE-INTERCEPT FORM FOR THE EQUATION OF A LINE

> The graph of the equation $y = mx + b$ is a line having slope m and y-intercept b.

We have shown that every line is the graph of an equation of the form

$$ax + by + c = 0$$

where a, b, and c are real numbers, and a and b are not both zero. We call such an equation a **linear equation** in x and y. Let us show, conversely, that the graph of $ax + by + c = 0$ is always a line provided a and b are not both zero. On the one hand, if $b \neq 0$, we may solve for y, obtaining

$$y = \left(-\frac{a}{b}\right)x + \left(-\frac{c}{b}\right),$$

which, by the Slope-Intercept Form, is an equation of a line with slope $-a/b$ and y-intercept $-c/b$. On the other hand, if $b = 0$ but $a \neq 0$, then we may solve $ax + by + c = 0$ for x, obtaining $x = -c/a$, which is the equation of a vertical line with x-intercept $-c/a$. This establishes the following important theorem.

THEOREM

> The graph of a linear equation $ax + by + c = 0$ is a line and, conversely, every line is the graph of a linear equation.

For simplicity, we shall use the terminology *the line* $ax + by + c = 0$ instead of the more accurate phrase *the line with equation* $ax + by + c = 0$.

FIGURE 1.43

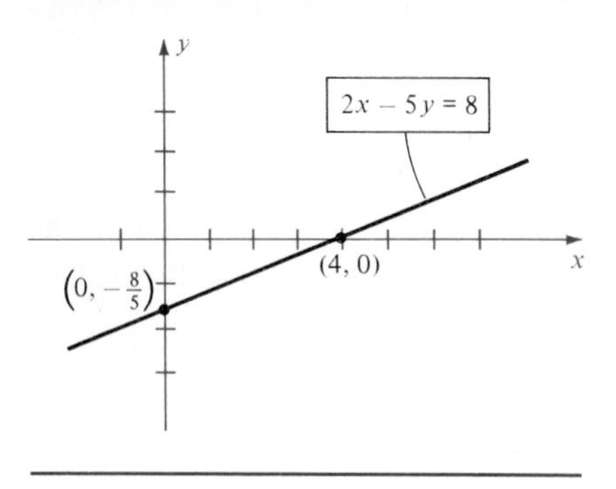

EXAMPLE 4 Sketch the graph of $2x - 5y = 8$.

SOLUTION From the preceding theorem we know that the graph is a line, and hence it is sufficient to find two points on the graph. Let us find the x- and y-intercepts. Substituting $y = 0$ in the given equation, we obtain the x-intercept 4. Substituting $x = 0$, we see that the y-intercept is $-\frac{8}{5}$. This leads to the graph in Figure 1.43.

Another method of solution is to express the given equation in Slope-Intercept Form. To do this we begin by isolating the term involving y on one side of the equal sign, obtaining

$$5y = 2x - 8.$$

Next, dividing both sides by 5 gives us

$$y = \frac{2}{5}x + \left(\frac{-8}{5}\right),$$

which is in the form $y = mx + b$. Hence, the slope is $m = \frac{2}{5}$, and the y-intercept is $b = -\frac{8}{5}$. We may then sketch a line through the point $(0, -\frac{8}{5})$ with slope $\frac{2}{5}$. ■

The following theorem can be proved.

THEOREM

Two nonvertical lines are parallel if and only if they have the same slope.

We shall use this fact in the next example.

EXAMPLE 5 Find an equation of the line that passes through the point $(5, -7)$ and is parallel to the line $6x + 3y - 4 = 0$.

SOLUTION Let us express the given equation in Slope-Intercept Form. We begin by writing

$$3y = -6x + 4$$

and then divide both sides by 3, obtaining

$$y = -2x + \frac{4}{3}.$$

The last equation is in Slope-Intercept Form with $m = -2$, and hence the slope is -2. Since parallel lines have the same slope, the required line also has slope -2. Applying the Point-Slope Form gives us

$$y + 7 = -2(x - 5).$$

This is equivalent to

$$y + 7 = -2x + 10 \quad \text{or} \quad 2x + y - 3 = 0. \qquad \blacksquare$$

The next theorem specifies conditions for perpendicular lines.

THEOREM

> Two lines with slopes m_1 and m_2 are perpendicular if and only if $m_1 m_2 = -1$.

FIGURE 1.44

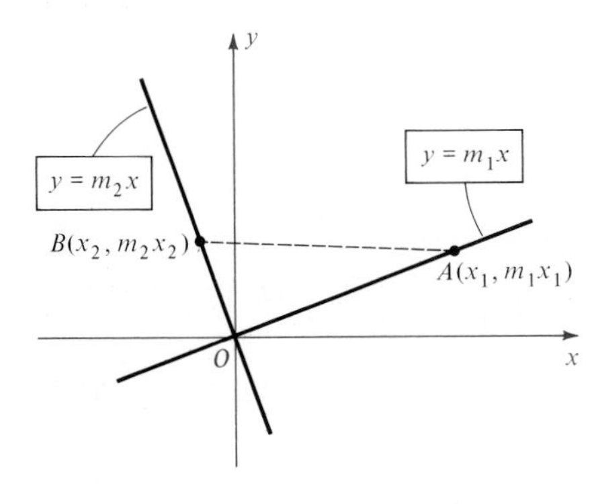

PROOF For simplicity, let us consider the special case of two lines that intersect at the origin O, as illustrated in Figure 1.44. In this case equations of the lines are $y = m_1 x$ and $y = m_2 x$. If, as in the figure, we choose points $A(x_1, m_1 x_1)$ and $B(x_2, m_2 x_2)$ different from O on the lines, then the lines are perpendicular if and only if angle AOB is a right angle. By the Pythagorean Theorem, angle AOB is a right angle if and only if

$$[d(A, B)]^2 = [d(O, B)]^2 + [d(O, A)]^2$$

or, by the Distance Formula,

$$(m_2 x_2 - m_1 x_1)^2 + (x_2 - x_1)^2 = (m_2 x_2)^2 + x_2^2 + (m_1 x_1)^2 + x_1^2.$$

Squaring the indicated terms and simplifying gives us

$$-2m_1 m_2 x_1 x_2 - 2x_1 x_2 = 0.$$

Dividing both sides by $-2x_1 x_2$, we see that the lines are perpendicular if and only if $m_1 m_2 + 1 = 0$, or $m_1 m_2 = -1$.

The same type of proof may be given if the lines intersect at *any* point (a, b). □

A convenient way to remember the conditions for perpendicularity is to note that m_1 and m_2 must be *negative reciprocals* of one another, that is, $m_1 = -1/m_2$ and $m_2 = -1/m_1$.

EXAMPLE 6 Find an equation of the line that passes through the point $(5, -7)$ and is perpendicular to the line $6x + 3y - 4 = 0$.

SOLUTION The line $6x + 3y - 4 = 0$ was considered in Example 5, and we found that its slope is -2. Hence, the slope of the required line is the negative reciprocal $-[1/(-2)]$, or $\frac{1}{2}$. Applying the Point-Slope Form gives us

$$y + 7 = \tfrac{1}{2}(x - 5),$$

which is equivalent to

$$2y + 14 = x - 5, \quad \text{or} \quad x - 2y - 19 = 0.$$

■

EXAMPLE 7 Find an equation for the perpendicular bisector of the line segment from $A(1, 7)$ to $B(-3, 2)$.

SOLUTION By the Midpoint Formula, the midpoint M of the segment AB is $(-1, \frac{9}{2})$. Since the slope of AB is $\frac{5}{4}$ (see Example 3), it follows from the preceding theorem that the slope of the perpendicular bisector is $-\frac{4}{5}$. Applying the Point-Slope Form,

$$y - \tfrac{9}{2} = -\tfrac{4}{5}(x + 1).$$

Multiplying both sides by 10 and simplifying leads to $8x + 10y - 37 = 0$.

■

Two variables x and y are said to be **linearly related** if $y = ax + b$ for some constants a and b. Linear relationships between variables occur frequently in applications. The following example gives one illustration. For other applications see Exercises 31–40.

EXAMPLE 8 The relationship between the air temperature T (in °F) and the altitude h (in feet above sea level) is approximately linear. When the temperature at sea level is 60°, an increase of 5000 feet in altitude lowers the air temperature about 18°.

(a) Express T as a function of h.

(b) What is the air temperature at an altitude of 15,000 feet?

SOLUTION

(a) If T is a linear function of h, then

$$T = ah + b$$

for some constants a and b. Since $T = 60$ when $h = 0$, we see that

$$60 = a(0) + b \quad \text{or} \quad b = 60.$$

Thus,
$$T = ah + 60.$$

In addition, we are given that if $h = 5000$, then $T = 60 - 18 = 42$. Substituting these values into the formula $T = ah + 60$, we obtain

$$42 = a(5000) + 60 \quad \text{or} \quad 5000a = -18.$$

Hence,
$$a = -\frac{18}{5000} = -\frac{9}{2500},$$

and the (approximate) formula for T is

$$T = -\frac{9}{2500}h + 60.$$

(b) Using the formula for T obtained in part (a), the (approximate) temperature when $h = 15{,}000$ is

$$T = -\frac{9}{2500}(15{,}000) + 60 = -54 + 60 = 6\,^{\circ}\text{F}.$$ ■

Exercises 1.5

In Exercises 1–4 plot the points A and B and find the slope of the line through A and B.

1 $A(-4, 6)$, $B(-1, 18)$

2 $A(6, -2)$, $B(-3, 5)$

3 $A(-1, -3)$, $B(-1, 2)$

4 $A(-3, 4)$, $B(2, 4)$

5 Show that $A(-3, 1)$, $B(5, 3)$, $C(3, 0)$, and $D(-5, -2)$ are vertices of a parallelogram.

6 Show that $A(2, 3)$, $B(5, -1)$, $C(0, -6)$, and $D(-6, 2)$ are vertices of a trapezoid.

7 Prove that the points $A(6, 15)$, $B(11, 12)$, $C(-1, -8)$, and $D(-6, -5)$ are vertices of a rectangle.

8 Prove that the points $A(1, 4)$, $B(6, -4)$, and $C(-15, -6)$ are vertices of a right triangle.

9 If three consecutive vertices of a parallelogram are $A(-1, -3)$, $B(4, 2)$, and $C(-7, 5)$, find the fourth vertex.

10 Let $A(x_1, y_1)$, $B(x_2, y_2)$, $C(x_3, y_3)$, and $D(x_4, y_4)$ denote the vertices of an arbitrary quadrilateral. Prove that the line segments joining midpoints of adjacent sides form a parallelogram.

In Exercises 11–20 find an equation of the line satisfying the given conditions.

11 Through $A(2, -6)$, slope $\frac{1}{2}$

12 Slope -3, y-intercept 5

13 Through $A(-5, -7)$, $B(3, -4)$

14 x-intercept -4, y-intercept 8

15 Through $A(8, -2)$, y-intercept -3

16 Slope 6, x-intercept -2

17 Through $A(10, -6)$, parallel to (a) the y-axis; (b) the x-axis

18 Through $A(-5, 1)$, perpendicular to (a) the y-axis; (b) the x-axis

19 Through $A(7, -3)$, perpendicular to the line with equation $2x - 5y = 8$

20 Through $(-\frac{3}{4}, -\frac{1}{2})$, parallel to the line with equation $x + 3y = 1$

21 Given $A(3, -1)$ and $B(-2, 6)$, find an equation for the perpendicular bisector of the line segment AB.

22 Find an equation for the line that bisects the second and fourth quadrants.

In Exercises 23–30 use the Slope-Intercept Form to find the slope and y-intercept of the line with the given equation, and sketch the graph of each equation.

23 $3x - 4y + 8 = 0$

24 $2y - 5x = 1$

25 $x + 2y = 0$

26 $8x = 1 - 4y$

27 $y = 4$

28 $x + 2 = \frac{1}{2}y$

29 $5x + 4y = 20$

30 $y = 0$

31 Six years ago a house was purchased for \$59,000. This year it was appraised at \$95,000. Assuming that the value of the house is linearly related to time, find a formula that specifies the value at any time after the purchase date. When was the house worth \$73,000?

32 Charles' Law for gases states that if the pressure remains constant, then the relationship between the volume V that a gas occupies and its temperature T in degrees Celsius is given by $V = V_0(1 + \frac{1}{273}T)$.

(a) What is the significance of V_0?

(b) What increase in temperature is needed to increase the volume from V_0 to $2V_0$?

(c) Sketch the graph of the equation on a TV-plane for the case $V_0 = 100$ and $T \geq 273$.

33 The electrical resistance R (in ohms) for a pure metal wire is linearly related to its temperature T (in °C) by the formula

$$R = R_0(1 + aT)$$

for some constants a and $R_0 > 0$.

(a) What is the significance of R_0?

(b) At absolute zero ($T = -273$ °C), $R = 0$. Find a.

(c) At 0 °C, silver wire has a resistance of 1.25 ohms. At what temperature is the resistance doubled?

34 The freezing point of water is 0 °C or 32 °F. The boiling point is 100 °C or 212 °F. Using this information, find a linear relationship between the temperature in °F and the temperature in °C. What temperature increase in °F corresponds to an increase in temperature of 1 °C?

35 The expected weight W (in long tons) of an adult humpback whale is related to its length L (in feet) by the linear equation

$$W = 1.70L - 42.8.$$

(a) A 30-foot humpback has been spotted by marine researchers. Estimate its weight.

(b) If the error in estimating the length could be as large as 2 feet, what is the corresponding error for the weight estimate?

36 Newborn blue whales measure approximately 24 feet and weigh 3 tons. Young whales are nursed for 7 months and, after weaning, measure an amazing 53 feet and weigh 23 tons. Let L and W denote the length (in feet) and the weight (in tons), respectively, of a whale that is t months of age.

(a) If L and t are linearly related, what is the daily increase in length? (Use 1 month = 30 days.)

(b) If W and t are linearly related, what is the daily increase in weight?

37 The owner of an ice-cream franchise must pay the parent company \$500 per month plus 10% of the profits. In addition, there are fixed costs of \$400 per month for items such as utilities and labor. Suppose that 50% of the monthly revenue is pure profit.

(a) If the monthly take-home pay M of the owner and the monthly revenue R are linearly related, express M in terms of R.

(b) How much business must be done to merely break even?

38 Several rules of thumb have been suggested for modifying adult drug dosage levels for young children. Let a denote the adult dose (in mg) and let t be the age of the child (in years). Some typical rules are

$$y = \frac{t+1}{24}a \qquad \text{(Cowling's Rule)}$$

and

$$y = \frac{2}{25}ta \qquad \text{(Friend's Rule)}$$

(a) Graph these two linear equations on the same set of axes for $0 \le t \le 12$.

(b) At approximately what age do the two rules give the same recommendation?

39 In a simple video game for young children, airplanes fly from left to right along the curve $y = 1 + (1/x)$ and can shoot their bullets in the tangent direction at creatures placed along the x-axis at $x = 1, 2, 3, 4$, and 5. Using calculus it can be shown that the slope of the tangent line at $P(1, 2)$ is $m = -1$, while the tangent line at $Q(\frac{3}{2}, \frac{5}{3})$ has slope $m = -\frac{4}{9}$. If a player shoots when the plane is at P, will a target be hit? What if the player shoots at Q?

FIGURE FOR EXERCISE 39

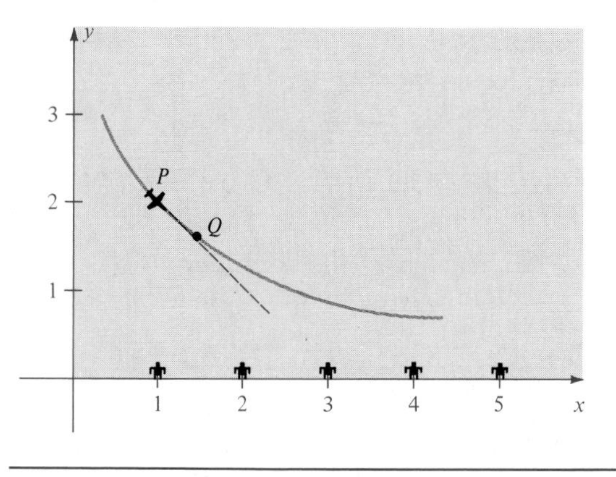

40 A hammer thrower is working on his form in a small practice area. The hammer spins, generating a circle with a radius of 5 feet, and when released, hits a tall screen that is 50 feet from the center of the throwing area. Let coordinate axes be introduced as shown in the figure (not to scale).

(a) If the hammer is released at $(-4, -3)$ and travels in the tangent direction, where will it hit the screen?

(b) If the hammer is to hit at $(0, -50)$, where on the circle should it be released?

FIGURE FOR EXERCISE 40

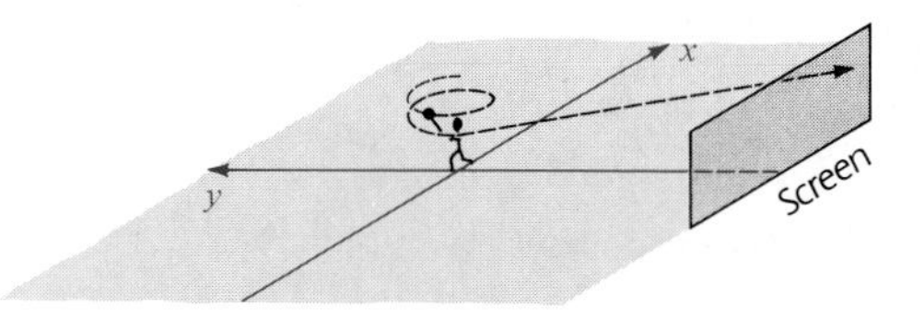

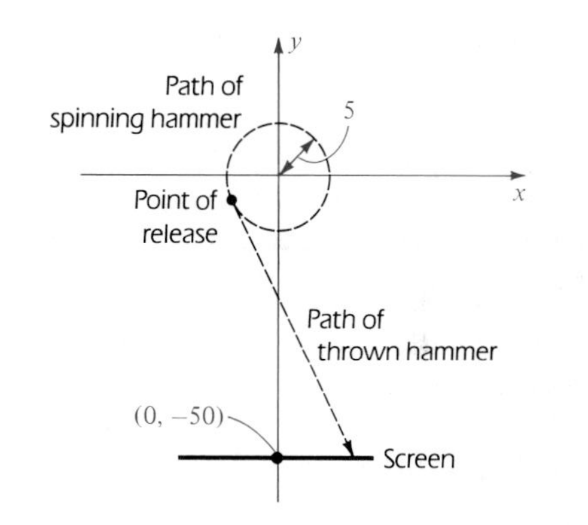

1.6 Review

Define or discuss each of the following.

1 Integers

2 Rational and irrational numbers

3 Coordinate line

4 Set

5 Element of a set

6 Variable

7 Domain of a variable

8 Constant

9 A number a is greater than a number b
10 A number a is less than a number b
11 Intervals
12 Absolute value of a real number
13 The distance between points on a coordinate line
14 Solution of an equation
15 Identity
16 Equivalent equations
17 Quadratic formula
18 Exponent
19 Laws of Exponents
20 Principal nth root
21 Radical notation
22 Laws of Radicals
23 Rationalizing a denominator
24 Rational exponents
25 Algebraic expression
26 Polynomial
27 Degree of a polynomial
28 Prime polynomial
29 Irreducible polynomial
30 Rational expression
31 Properties of inequalities
32 Solution of an inequality
33 Equivalent inequalities
34 Ordered pair
35 Rectangular coordinate system in a plane
36 Coordinate axes
37 Quadrants
38 Coordinates of a point
39 Distance formula
40 Midpoint formula
41 Graph of an equation in x and y
42 Tests for symmetry
43 Equation of a circle
44 Unit circle
45 Slope of a line
46 Point-Slope Form
47 Slope-Intercept Form
48 Linear equation in x and y

Exercises 1.6

1 Replace the symbol $\square$ with either $<$, $>$, or $=$.

(a) $-0.1 \square -0.01$ (b) $\sqrt{9} \square -3$

(c) $\frac{1}{6} \square 0.166$

2 Express in terms of inequalities:

(a) x is negative.

(b) a is between $\frac{1}{2}$ and $\frac{1}{3}$.

(c) The absolute value of x is not greater than 4.

3 Rewrite without using the absolute value symbol:

(a) $|-7|$

(b) $|-5|/(-5)$

(c) $|3^{-1} - 2^{-1}|$

4 If points A, B, and C on a coordinate line have coordinates -8, 4, and -3, respectively, find the following:

(a) $d(A, C)$ (b) $d(C, A)$

(c) $d(B, C)$

Use the Quadratic Formula to find the solutions of the equations in Exercises 5–6.

5 $x^2 + 5x - 12 = 0$

6 $3x^2 + 4x - 5 = 0$

7 A closed cylindrical oil drum of height 4 feet is to be constructed so that the total surface area is 10π ft^2. Find the diameter of the drum.

8 A boy throws a baseball straight upward with an initial speed of 64 feet per second. The number of feet s above the ground after t seconds is given by $s = -16t^2 + 64t$.

(a) When will the baseball be 48 feet above the ground?

(b) When will it hit the ground?

(c) What is its maximum height?

Simplify the expressions in Exercises 9–16.

9 $(3a^2b)^2(2ab^3)$

10 $\dfrac{6r^3y^2}{2r^5y}$

11 $\dfrac{(3x^2y^{-3})^{-2}}{x^{-5}y}$

12 $\left(\dfrac{a^{2/3}b^{3/2}}{a^2b}\right)^6$

13 $\sqrt[3]{(x^4y^{-1})^6}$

14 $\sqrt[3]{8x^5y^3z^4}$

15 $\dfrac{1-\sqrt{x}}{1+\sqrt{x}}$

16 $\sqrt{\dfrac{a^2b^3}{c}}$

Factor the expressions in Exercises 17–22.

17 $28x^2 + 4x - 9$

18 $16a^4 + 24a^2b^2 + 9b^4$

19 $2c^3 - 12c^2 + 3c - 18$

20 $u^3v^4 - u^6v$

21 $8x^3 + 64y^3$

22 $x^4 - 8x^3 + 16x^2$

Simplify the expressions in Exercises 23–32.

23 $(3x^3 - 4x^2 + x - 7) + (x^4 - 2x^3 + 3x^2 + 5)$

24 $(x + 4)(x + 3) - (2x - 1)(x - 5)$

25 $(3y^3 - 2y^2 + y + 4)(y^2 - 3)$

26 $(4x - 5)(2x^2 + 3x - 7)$

27 $\dfrac{6x^2 - 7x - 5}{4x^2 + 4x + 1}$

28 $\dfrac{r^3 - t^3}{r^2 - t^2}$

29 $\dfrac{6x^2 - 5x - 6}{x^2 - 4} \div \dfrac{2x^2 - 3x}{x + 2}$

30 $\dfrac{2}{4x - 5} - \dfrac{5}{10x + 1}$

31 $\dfrac{7}{x + 2} + \dfrac{3x}{(x + 2)^2} - \dfrac{5}{x}$

32 $\dfrac{(4 - x^2)(\frac{1}{3})(6x + 1)^{-2/3}(6) - (6x + 1)^{1/3}(-2x)}{(4 - x^2)^2}$

Solve the inequalities in Exercises 33–38 and express the solutions in terms of intervals.

33 $-\dfrac{1}{2} < \dfrac{2x + 3}{5} < \dfrac{3}{2}$

34 $10 - 7x < 4 + 2x$

35 $|16 - 3x| \geq 5$

36 $|4x + 7| < 21$

37 $10x^2 + 11x > 6$

38 $\dfrac{3}{2x + 3} < \dfrac{1}{x - 2}$

39 Boyle's Law for a certain gas states that $pv = 200$ where p denotes the pressure (lb/in^2) and v denotes the volume (in^3). If $25 \leq v \leq 50$, what is the corresponding range for p?

40 The *Lorentz contraction formula* in relativity theory relates the length L of an object moving at a velocity of V miles per second with respect to an observer to its length L_0 at rest. If c is the speed of light, then

$$L^2 = L_0^2\left(1 - \frac{V^2}{c^2}\right).$$

For what velocities will L be less than $\frac{1}{2}L_0$? State your answer in terms of c.

41 Plot the points $A(3, 1)$, $B(-5, -3)$, and $C(4, -1)$, and prove that they are vertices of a right triangle. What is the area of the triangle?

42 Given points $P(-5, 9)$ and $Q(-8, -7)$, find (a) the midpoint of the segment PQ, and (b) a point T such that Q is the midpoint of PT.

43 Describe the set of all points (x, y) in a coordinate plane such that $y/x < 0$.

44 Find the slope of the line through $C(11, -5)$ and $D(-8, 6)$.

45 Prove that the points $A(-3, 1)$, $B(1, -1)$, $C(4, 1)$, and $D(3, 5)$ are vertices of a trapezoid.

46 Find an equation of the circle that has center $C(7, -4)$ and passes through the point $Q(-3, 3)$.

47 Find an equation of the circle that has center $C(-5, -1)$ and is tangent to the line $x = 4$.

48 Express the equation $8x + 3y - 24 = 0$ in Slope-Intercept Form.

49 Find an equation of the line through $A(\frac{1}{2}, -\frac{1}{3})$ that is (a) parallel to the line $6x + 2y + 5 = 0$; (b) perpendicular to the line $6x + 2y + 5 = 0$.

50 Find an equation of the line that has x-intercept -3 and passes through the center of the circle that has the equation $x^2 + y^2 - 4x + 10y + 26 = 0$.

Sketch the graphs of the equations in Exercises 51–60.

51 $2y + 5x - 8 = 0$

52 $x = 3y + 4$

53 $x + 5 = 0$

54 $2y - 7 = 0$

55 $y = \sqrt{1 - x}$

56 $3x - 7y^2 = 0$

57 $9y^2 + 2x = 0$

58 $y^2 = 16 - x^2$

59 $x^2 + y^2 + 4x - 16y + 64 = 0$

60 $x^2 + y^2 - 8x = 0$

CHAPTER 2

Functions

One of the most important concepts in mathematics is that of *function*. ■ Indeed, without the notion of function, little progress could be made in mathematics or in any area of science. ■ In this chapter we discuss many important characteristics of functions and their graphs.

2.1 Definition of Function

The notion of **correspondence** occurs frequently in everyday life. For example, to each book in a library there corresponds the number of pages in the book. As another example, to each human being there corresponds a birth date. To cite a third example, if the temperature of the air is recorded throughout a day, then at each instant of time there is a corresponding temperature. These examples of correspondence involve two sets D and E. In our first example D denotes the set of books in a library and E the set of positive integers. For each book x in D there corresponds a positive integer y in E, namely the number of pages in the book.

FIGURE 2.1

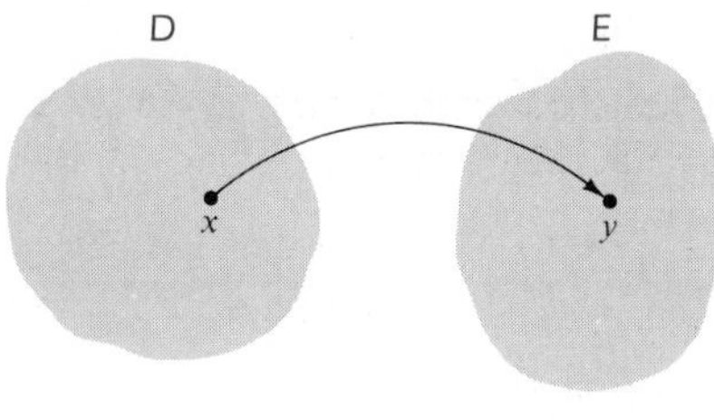

We sometimes depict correspondences by diagrams of the type shown in Figure 2.1, where the sets D and E are represented by points within regions in a plane. The curved arrow indicates that the element y of E corresponds to the element x of D. The two sets may have elements in common. As a matter of fact, we often have $D = E$.

Our examples indicate that *to each x in D there corresponds one and only one y in E*; that is, *y is unique* for a given x. However, the same element of E may correspond to different elements of D. For example, two different books may have the same number of pages, two different people may have the same birthday, and so on.

In most of our work D and E will be sets of numbers. To illustrate, let D and E both denote the set $\mathbb{R}$ of real numbers, and to each real number x let us assign its square x^2. Thus, to 3 we assign 9, to -5 we assign 25, and to $\sqrt{2}$, the number 2. This gives us a correspondence from $\mathbb{R}$ to $\mathbb{R}$.

Each of the preceding examples of a correspondence is a *function.*

DEFINITION

> A **function** f from a set D to a set E is a correspondence that assigns to each element x of D a unique element y of E.

FIGURE 2.2

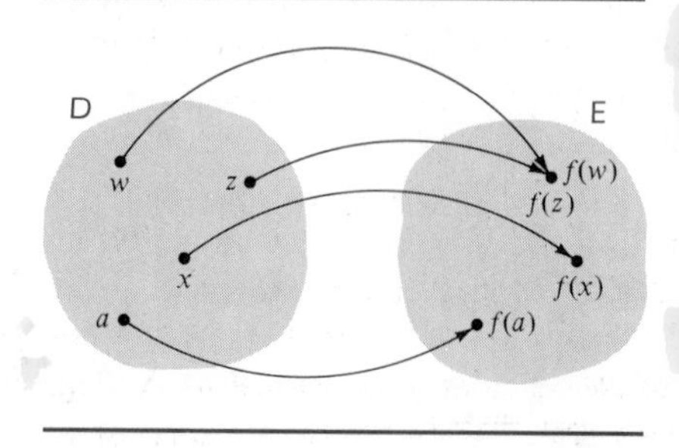

The element y of E is called the **value** of f at x and is denoted by $f(x)$ (read "f of x"). The set D is called the **domain** of the function. The **range** of f consists of all possible values $f(x)$, where x is in D.

We may now sketch the diagram in Figure 2.2. The curved arrows indicate that the elements $f(x)$, $f(w)$, $f(z)$, and $f(a)$ of E correspond to the elements x, w, z, and a of D. It is important to remember that *to each x in D there is assigned precisely one value $f(x)$ in E;* however, different elements of D, such as w and z in Figure 2.2, may have the same value in E.

The symbols

$$D \xrightarrow{f} E, \quad \text{or}$$

signify that f is a function from D to E. Beginning students are sometimes confused by the notations f and $f(x)$. Remember that f is used to represent the function. It is neither in D nor in E. However, $f(x)$ is an element of E, namely the element that f assigns to x.

Equality of two functions f and g from D to E is defined as follows:

$$f = g \quad \text{if and only if} \quad f(x) = g(x) \quad \text{for every } x \text{ in } D.$$

For example, if $g(x) = \frac{1}{2}(2x^2 - 6) + 3$ and $f(x) = x^2$ for all x in $\mathbb{R}$, then $g = f$, since $\frac{1}{2}(2x^2 - 6) + 3 = x^2$ for every x.

EXAMPLE 1 Let f be the function with domain $\mathbb{R}$ such that $f(x) = x^2$ for every x in $\mathbb{R}$. Find $f(-6)$, $f(\sqrt{3})$, and $f(a)$ for any real number a. What is the range of f?

SOLUTION We may find values of f by substituting for x in the equation $f(x) = x^2$. Thus,

$$f(-6) = (-6)^2 = 36, \quad f(\sqrt{3}) = (\sqrt{3})^2 = 3, \quad \text{and} \quad f(a) = a^2.$$

By definition, the range of f consists of all numbers of the form $f(a) = a^2$, where a is in $\mathbb{R}$. Since the square of every real number is nonnegative, the range is contained in the set of all nonnegative real numbers. Moreover, every nonnegative real number c is a value of f since $f(\sqrt{c}) = (\sqrt{c})^2 = c$. Hence, the range of f is the set of all nonnegative real numbers. ■

If a function is defined as in Example 1, the symbols used for the function and variable are immaterial; that is, expressions such as $f(x) = x^2$, $f(s) = s^2$, $g(t) = t^2$, and $k(r) = r^2$ all define the same function. This is true because if a is any number in the domain, then the same value a^2 is obtained no matter which expression is employed.

In the remainder of our work the phrase "f is a function" will mean that the domain and range are sets of real numbers. If a function is defined by means of an expression, as in Example 1, and the domain D is not stated explicitly, then D is considered to be the totality of real numbers x such that $f(x)$ is real. To illustrate, if $f(x) = \sqrt{x - 2}$, then the domain is assumed to be the set of real numbers x such that $\sqrt{x - 2}$ is real; that is, $x - 2 \geq 0$, or $x \geq 2$. Thus, the domain is the infinite interval $[2, \infty)$. If x is in the domain, we say that f **is defined at** x, or that $f(x)$ **exists.** If a set S is contained in the domain, we often say that f **is defined on** S. The terminology f **is undefined at** x means that x is not in the domain of f.

EXAMPLE 2

If $g(x) = \dfrac{\sqrt{4 + x}}{1 - x}$, find

(a) the domain of g. (b) $g(5)$, $g(-2)$, $g(-a)$, $-g(a)$.

SOLUTION

(a) The fractional expression is a real number if and only if the radicand is nonnegative and the denominator is different from 0. Thus, $g(x)$ exists if and only if

$$4 + x \geq 0 \quad \text{and} \quad 1 - x \neq 0,$$

or equivalently,

$$x \geq -4 \quad \text{and} \quad x \neq 1.$$

The domain may be expressed in terms of intervals as $[-4, 1) \cup (1, \infty)$.

(b) To find values of g, we substitute for x as follows:

$$g(5) = \frac{\sqrt{4 + 5}}{1 - 5} = \frac{\sqrt{9}}{-4} = -\frac{3}{4}$$

$$g(-2) = \frac{\sqrt{4 + (-2)}}{1 - (-2)} = \frac{\sqrt{2}}{3}$$

$$g(-a) = \frac{\sqrt{4 + (-a)}}{1 - (-a)} = \frac{\sqrt{4 - a}}{1 + a}$$

$$-g(a) = -\frac{\sqrt{4 + a}}{1 - a} = \frac{\sqrt{4 + a}}{a - 1}$$

■

Manipulations of the type given in the next example occur in calculus.

EXAMPLE 3 Suppose $f(x) = x^2 + 3x - 2$. If a and h are real numbers, and $h \neq 0$, find and simplify

$$\frac{f(a + h) - f(a)}{h}.$$

SOLUTION Since

$$\begin{aligned} f(a + h) &= (a + h)^2 + 3(a + h) - 2 \\ &= (a^2 + 2ah + h^2) + (3a + 3h) - 2 \end{aligned}$$

and $$f(a) = a^2 + 3a - 2,$$

we see that

$$\frac{f(a+h) - f(a)}{h} = \frac{[(a^2 + 2ah + h^2) + (3a + 3h) - 2] - (a^2 + 3a - 2)}{h}$$

$$= \frac{2ah + h^2 + 3h}{h}$$

$$= 2a + h + 3. \quad ■$$

FIGURE 2.3

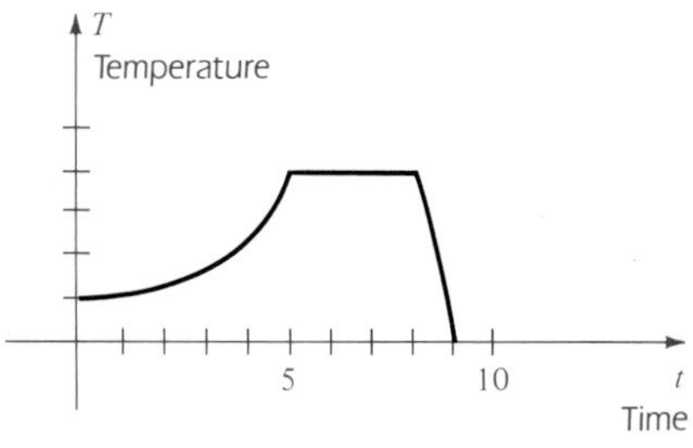

Graphs are often used to describe the variation of physical quantities. For example, a scientist may use the graph in Figure 2.3 to indicate the temperature T of a certain solution at various times t during an experiment. The sketch shows that the temperature increased gradually from time $t = 0$ to time $t = 5$, did not change between $t = 5$ and $t = 8$, and then decreased rapidly from $t = 8$ to $t = 9$. This visual aid reveals the variation of T more clearly than a long table of numerical values would.

Similarly, if f is a function, we may use a graph to indicate the change in $f(x)$ as x varies through the domain of f. Specifically, we have the following.

DEFINITION

The **graph of a function** f is the set of all points $(x, f(x))$ in a coordinate plane such that x is in the domain of f.

FIGURE 2.4

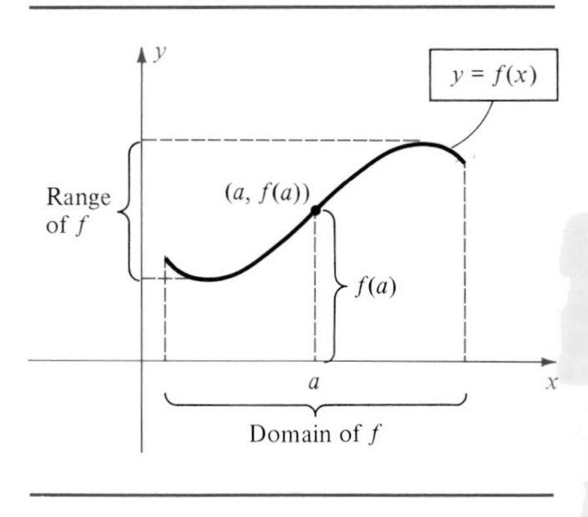

Note that the graph of f is the same as the graph of the equation $y = f(x)$ for x in the domain of f. As shown in Figure 2.4, we often attach the label $y = f(x)$ to a sketch of the graph. If $P(a, b)$ is a point on the graph, then the y-coordinate b is the function value $f(a)$. The figure exhibits the domain of f (the set of possible values of x) and the range of f (the corresponding values of y). Although we have pictured the domain and range as closed intervals, they may be infinite intervals or other sets of real numbers.

It is important to note that since there is a unique value $f(a)$ for each a in the domain, only *one* point on the graph has x-coordinate a. Thus, *every vertical line intersects the graph of a function in at most one point.* Consequently, the graph of a function cannot be a figure such as a circle, in which a vertical line may intersect the graph in several points.

The x-intercepts of the graph of a function f are the solutions of the equation $f(x) = 0$. These numbers are called the **zeros** of the function. The y-intercept of the graph is $f(0)$, if it exists.

FIGURE 2.5

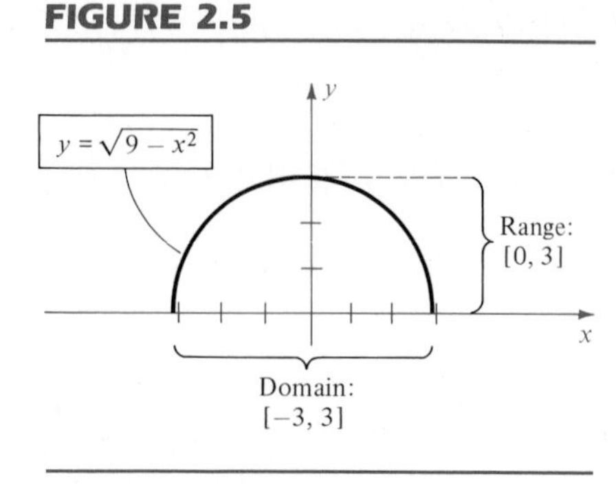

EXAMPLE 4 If $f(x) = \sqrt{9 - x^2}$, sketch the graph of f and find the domain and range of f.

SOLUTION By definition, the graph of f is the graph of the equation $y = \sqrt{9 - x^2}$. We know from our work with circles in Section 1.4 that the graph of $x^2 + y^2 = 9$ is a circle of radius 3 with center at the origin. Solving the equation $x^2 + y^2 = 9$ for y gives us $y = \pm\sqrt{9 - x^2}$. It follows that the graph of f is the *upper half* of the circle, as illustrated in Figure 2.5. Referring to the figure, we see that the domain of f is the closed interval $[-3, 3]$, and the range of f is the interval $[0, 3]$. ■

FIGURE 2.6

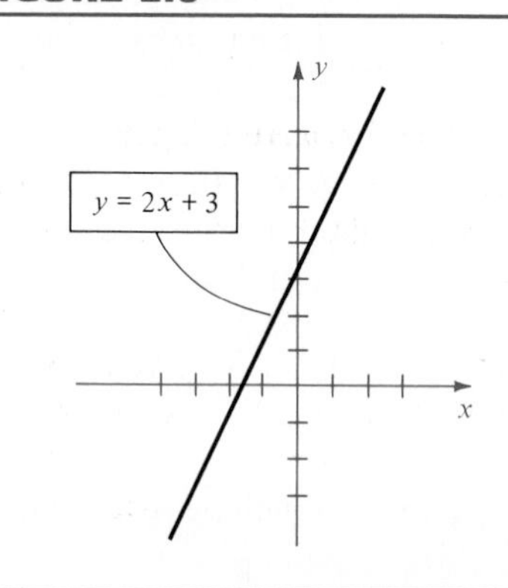

EXAMPLE 5 If $f(x) = 2x + 3$, sketch the graph of f and find the domain and range of f.

SOLUTION The graph of f is the graph of the equation $y = 2x + 3$ and hence, from Section 1.5, is a line having slope 2 and y-intercept 3, as illustrated in Figure 2.6. Since the values of x and y may be any real numbers, both the domain and range of f are $\mathbb{R}$. ■

If, as illustrated in Example 5, $f(x) = ax + b$ for some constants a and b, then the graph of f is a line and f is called a **linear function.**

If $f(x) = x$ for every x in the domain D of f, then f is called the **identity function** on D. A function f is a **constant function** if there is some number c in the range such that $f(x) = c$ for every x in the domain.

The types of functions described in the next definition occur frequently.

DEFINITION

A function f with domain D is
(i) **even** if $f(-x) = f(x)$ for every x in D,
(ii) **odd** if $f(-x) = -f(x)$ for every x in D.

EXAMPLE 6

(a) If $f(x) = 3x^4 - 2x^2 + 5$, show that f is an even function.

(b) If $g(x) = 2x^5 - 7x^3 + 4x$, show that f is an odd function.

SOLUTION If x is any real number, then

(a)
$$\begin{aligned} f(-x) &= 3(-x)^4 - 2(-x)^2 + 5 \\ &= 3x^4 - 2x^2 + 5 = f(x) \end{aligned}$$

and hence f is even.

(b)
$$g(-x) = 2(-x)^5 - 7(-x)^3 + 4(-x)$$
$$= -2x^5 + 7x^3 - 4x$$
$$= -(2x^5 - 7x^3 + 4x) = -g(x)$$

Thus, g is odd. ■

The next theorem gives useful information about the graphs of even or odd functions.

THEOREM ON SYMMETRY

(i) The graph of an even function is symmetric with respect to the y-axis.

(ii) The graph of an odd function is symmetric with respect to the origin.

PROOF If f is even, then $f(-x) = f(x)$, and hence the equation $y = f(x)$ is not changed if $-x$ is substituted for x. Statement (i) now follows from Symmetry Test (i) of Section 1.4. The proof of (ii) is left to the reader. □

Note that if $f(x) = \sqrt{9 - x^2}$, then f is even and hence its graph is symmetric with respect to the y-axis (see Figure 2.5). If $f(x) = \frac{1}{4}x^3$, then f is odd and the graph is symmetric with respect to the origin (see Figure 1.34).

Several other important properties of graphs of functions will be discussed in the next section.

Many formulas that occur in mathematics and the sciences determine functions. As an illustration, the formula $A = \pi r^2$ for the area A of a circle of radius r assigns to each positive real number r a unique value of A. This determines a function f such that $f(r) = \pi r^2$, and we may write $A = f(r)$. The letter r, which represents an arbitrary number from the domain of f, is often called an **independent variable.** The letter A, which represents a number from the range of f, is called a **dependent variable,** since its value depends on the number assigned to r. If two variables r and A are related in this manner, it is customary to use the phrase "A is a function of r." To cite another example, if an automobile travels at a uniform rate of 50 miles per hour, then the distance d (miles) traveled in time t (hours) is given by $d = 50t$, and hence, the distance d is a function of time t.

EXAMPLE 7 Express the radius of a circle as a function of its area.

SOLUTION If A and r denote the area and radius, then

$$A = \pi r^2 \quad \text{and} \quad r^2 = \frac{A}{\pi}.$$

Since both A and r are positive,

$$r = \sqrt{\frac{A}{\pi}}.$$

This shows that r is a function of A, since to each value of A there corresponds a unique value $\sqrt{A/\pi}$ of r. ■

EXAMPLE 8 A steel storage tank for propane gas is to be constructed in the shape of a right circular cylinder of altitude 10 feet with a hemisphere attached to each end. The radius r is yet to be determined. Express the volume V of the tank as a function of r.

FIGURE 2.7

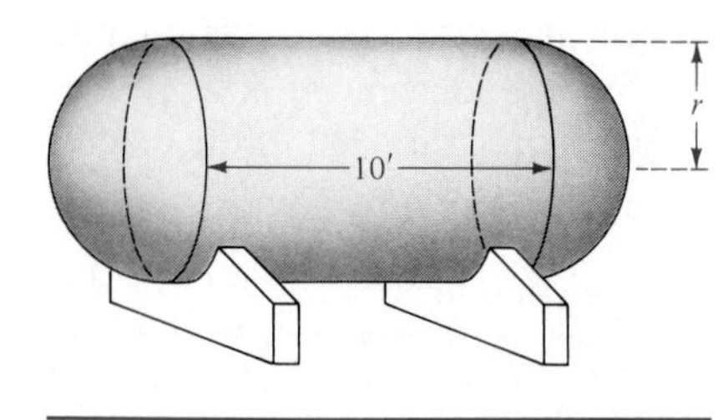

SOLUTION The tank is sketched in Figure 2.7. The volume of the cylindrical part of the tank may be found by multiplying the altitude 10 by the area πr^2 of the base of the cylinder. This gives us

$$\text{volume of cylinder} = 10(\pi r^2) = 10\pi r^2.$$

The two hemispherical ends, taken together, form a sphere of radius r. Using the formula for the volume of a sphere, we obtain

$$\text{volume of the two ends} = \tfrac{4}{3}\pi r^3.$$

Thus, the volume V of the tank is

$$V = \tfrac{4}{3}\pi r^3 + 10\pi r^2.$$

FIGURE 2.8

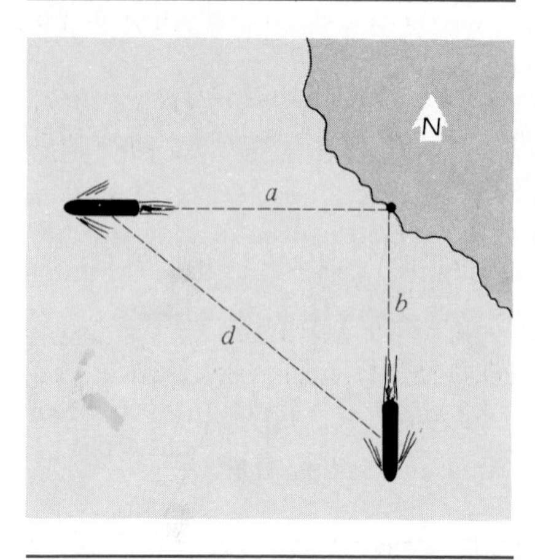

We may write this in the factored form

$$V = \tfrac{1}{3}\pi r^2(4r + 30) = \tfrac{2}{3}\pi r^2(2r + 15).$$ ■

EXAMPLE 9 Two ships leave port at the same time, one sailing west at a rate of 17 miles per hour and the other sailing south at 12 miles per hour. If t is the time (in hours) after their departure, express the distance d between the ships as a function of t.

SOLUTION To help visualize the problem we begin by drawing a picture and labeling it, as in Figure 2.8. Using the Pythagorean Theorem, we obtain

$$d^2 = a^2 + b^2$$

or

$$d = \sqrt{a^2 + b^2}.$$

Since distance = (rate)(time), and since the rates are 17 and 12, respectively,

$$a = 17t \quad \text{and} \quad b = 12t.$$

Substitution in $d = \sqrt{a^2 + b^2}$ gives us

$$d = \sqrt{(17t)^2 + (12t)^2} = \sqrt{289t^2 + 144t^2} = \sqrt{433t^2}.$$

Thus the functional relationship is

$$d = \sqrt{433}t.$$

An *approximate* formula for expressing d as a function of t is $d \approx (20.8)t$.

■

The concept of ordered pair can be used to obtain an alternate approach to functions. We first observe that a function f from D to E determines the following set W of ordered pairs:

$$W = \{(x, f(x)) : x \text{ is in } D\}.$$

Thus, W is the totality of ordered pairs for which the first number x is in D and the second number is the function value $f(x)$. In Example 1, where $f(x) = x^2$, W consists of all ordered pairs of the form (x, x^2). It is important to note that, for each x, there is exactly one ordered pair (x, y) in W having x in the first position.

Conversely, if we begin with a set W of ordered pairs such that each x in D appears exactly once in the first position of an ordered pair, then W determines a function. Specifically, for each x in D there is a unique pair (x, y) in W, and by letting y correspond to x, we obtain a function with domain D. The range consists of all real numbers y that appear in the second position of the ordered pairs.

It follows from the preceding discussion that the next statement could also be used as a definition of function. We prefer, however, to think of it as an alternate approach to this concept.

ALTERNATE DEFINITION OF A FUNCTION

A function with domain D is a set W of ordered pairs such that, for each x in D, there is exactly one ordered pair (x, y) in W having x in the first position.

In terms of the preceding definition, the ordered pairs $(x, x^2 + 3x - 2)$ determine the function of Example 3, where we had $f(x) = x^2 + 3x - 2$.

Exercises 2.1

1 If $f(x) = 2x^2 - 3x + 4$, find $f(1)$, $f(-1)$, $f(0)$, and $f(2)$.

2 If $f(x) = x^3 + 5x^2 - 1$, find $f(2)$, $f(-2)$, $f(0)$, and $f(-1)$.

3 If $f(x) = \sqrt{x-1} + 2x$, find $f(1)$, $f(3)$, $f(5)$, and $f(10)$.

4 If $f(x) = \dfrac{x}{x-2}$, find $f(1)$, $f(3)$, $f(-2)$, and $f(0)$.

In Exercises 5–8 find each of the following, where a, b, and h are real numbers:

(a) $f(a)$ (b) $f(-a)$

(c) $-f(a)$ (d) $f(a+h)$

(e) $f(a) + f(h)$

(f) $\dfrac{f(a+h) - f(a)}{h}$, provided $h \neq 0$

5 $f(x) = 5x - 2$ **6** $f(x) = 3 - 4x$

7 $f(x) = 2x^2 - x + 3$ **8** $f(x) = x^3 - 2x$

In Exercises 9–12 find the following:

(a) $g(1/a)$ (b) $1/g(a)$ (c) $g(a^2)$

(d) $(g(a))^2$ (e) $g(\sqrt{a})$ (f) $\sqrt{g(a)}$

9 $g(x) = 3x^2$ **10** $g(x) = 3x - 8$

11 $g(x) = \dfrac{2x}{x^2+1}$ **12** $g(x) = \dfrac{x^2}{x+1}$

In Exercises 13–20 find the domain of the function f.

13 $f(x) = \sqrt{3x-5}$ **14** $f(x) = \sqrt{7-2x}$

15 $f(x) = \sqrt{4-x^2}$ **16** $f(x) = \sqrt{x^2-9}$

17 $f(x) = \dfrac{x+1}{x^3-9x}$ **18** $f(x) = \dfrac{4x+7}{6x^2+13x-5}$

19 $f(x) = \dfrac{\sqrt{x}}{2x^2-11x+12}$ **20** $f(x) = \dfrac{x^3-1}{x^2-1}$

In Exercises 21–26 sketch the graph and find the domain and range of f.

21 $f(x) = 5 - 7x$ **22** $f(x) = \sqrt{x} + 2$

23 $f(x) = \sqrt{x-3}$ **24** $f(x) = 9 - x^2$

25 $f(x) = 1/x$ **26** $f(x) = 1/x^2$

In Exercises 27–36 determine whether f is even, odd, or neither even nor odd.

27 $f(x) = 3x^3 - 4x$ **28** $f(x) = 7x^6 - x^4 + 7$

29 $f(x) = 9 - 5x^2$ **30** $f(x) = 2x^5 - 4x^3$

31 $f(x) = 2$ **32** $f(x) = 2x^3 + x^2$

33 $f(x) = 2x^2 - 3x + 4$ **34** $f(x) = \sqrt{x^2+1}$

35 $f(x) = \sqrt[3]{x^3-4}$ **36** $f(x) = |x| + 5$

37 An open box is to be made from a rectangular piece of cardboard having dimensions 20 inches × 30 inches by cutting out identical squares of area x^2 from each corner and turning up the sides (see figure). Express the volume V of the box as a function of x.

FIGURE FOR EXERCISE 37

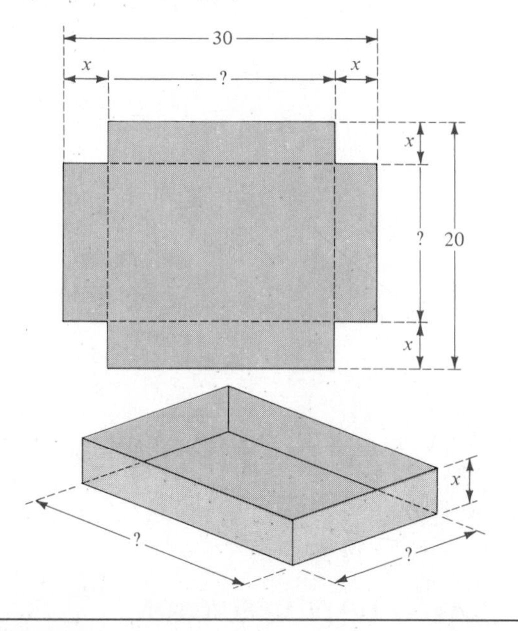

38 An aquarium of height 1.5 feet is to have a volume of 6 ft^3. Let x denote the length of the base, and let y denote the width (see figure).

(a) Express y as a function of x.

(b) Express the total number of square feet of glass needed as a function of x.

FIGURE FOR EXERCISE 38

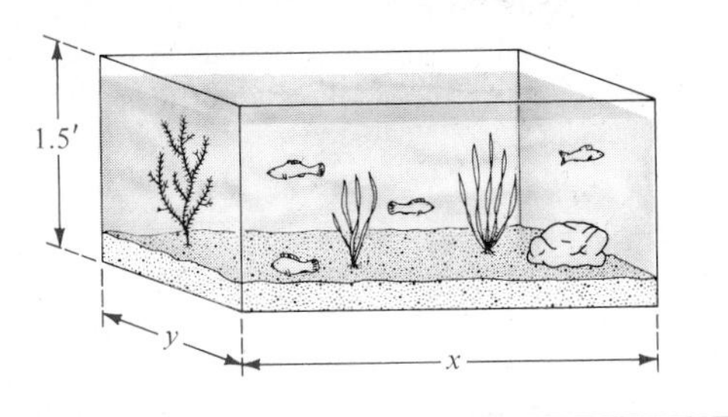

39 A small office building is to contain 500 ft^2 of floor space. The simple floor plans are shown in the figure.

(a) Express the length y of the building as a function of the width x.

(b) If the walls cost \$100 per running foot, express the cost C of the walls as a function of the width x. (Disregard the wall space above the door.)

FIGURE FOR EXERCISE 39

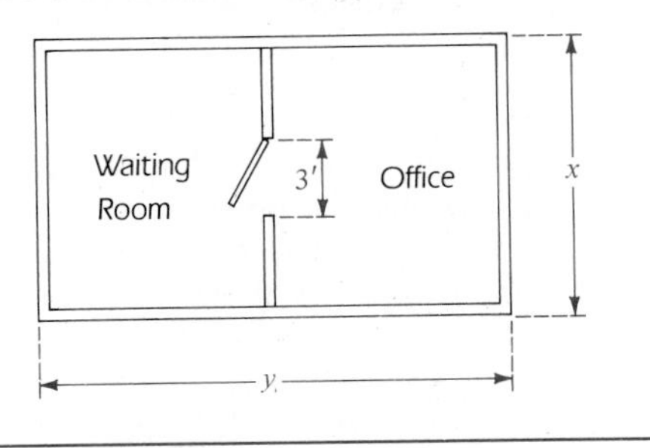

40 A meteorologist is inflating a spherical weather balloon with helium gas.

(a) Express the surface area of the balloon as a function of the volume of gas it contains.

(b) What is the surface area if the volume is 12 ft^3?

41 A hot-air balloon is released at 1:00 P.M. and rises vertically at a rate of 2 meters per second. An observation point is situated 100 meters from a point on the ground directly below the balloon (see figure). If t denotes the time (in seconds) after 1:00 P.M., express the distance d between the balloon and the observation point as a function of t.

FIGURE FOR EXERCISE 41

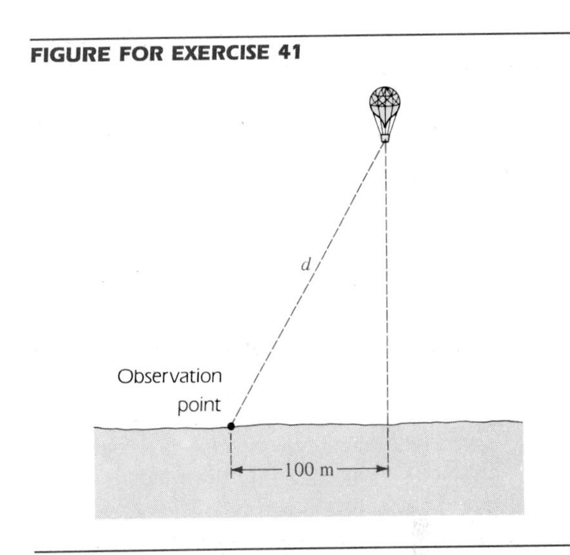

42 Triangle ABC is inscribed in a semicircle of diameter 15 (see figure).

(a) If x denotes the length of side AC, express the length y of side BC as a function of x.

(b) Express the area of triangle ABC as a function of x, and find the domain of this function.

FIGURE FOR EXERCISE 42

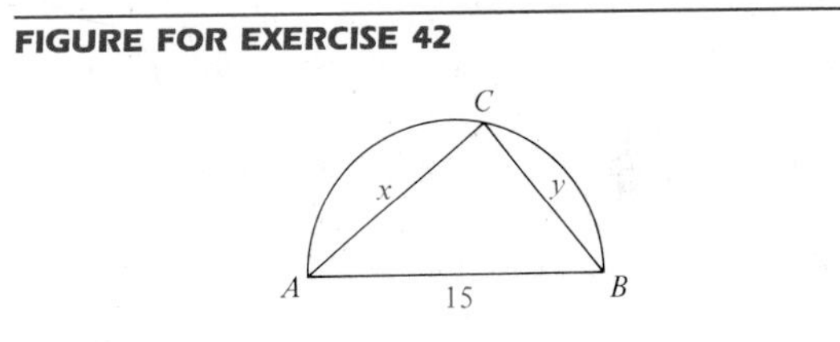

43 From an exterior point P that is h units from a circle of radius r, a tangent line is drawn to the circle (see figure).

FIGURE FOR EXERCISE 43

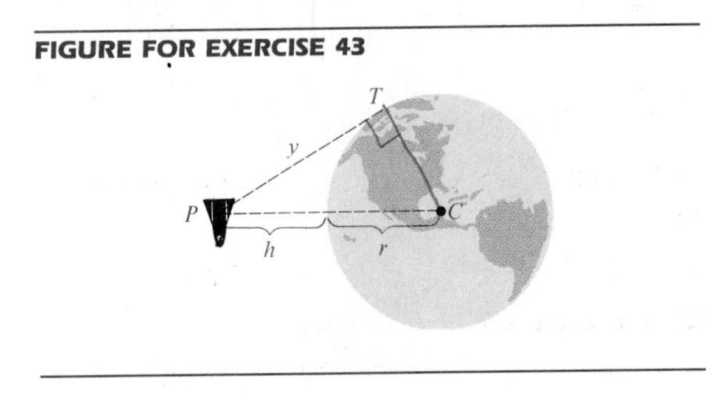

Let y denote the distance from the point P to the point of tangency T.

(a) Express y as a function of h. (*Hint:* If C is the center of the circle, then PT is perpendicular to CT.)

(b) If r is the radius of the earth, and h is the altitude of a space shuttle, then we can derive a formula for the maximum distance (to the earth) that an astronaut can see from the shuttle. In particular, if $h = 200$ miles and $r \approx 4000$ miles, approximate y.

44 The accompanying figure illustrates the apparatus for a tightrope walker. Two poles are set 50 feet apart, but the point of attachment P for the rope is yet to be determined.

(a) Express the length L of the rope as a function of the distance x from point P to the ground.

(b) If the total walk is be 75 feet, determine the height of the point of attachment P.

FIGURE FOR EXERCISE 44

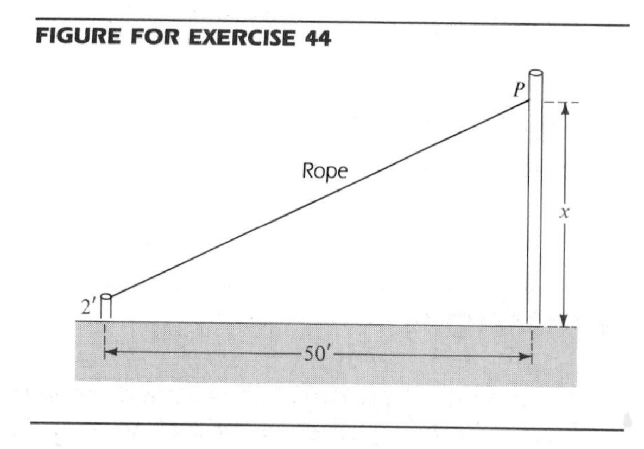

45 A steel storage tank for propane gas is to be constructed in the shape of a right circular cylinder of altitude 10 feet with a hemisphere attached to each end. The radius r is yet to be determined. Express the surface area S of the tank as a function of r. (Compare Example 8.)

46 A company sells running shoes to dealers at a rate of \$20 per pair if less than 50 pairs are ordered. If a dealer orders 50 or more pairs (up to 600), the price per pair is reduced at a rate of 2 cents times the number ordered. Let A denote the amount of money received when x pairs are ordered. Express A as a function of x.

47 The relative positions of an aircraft runway and a 20-foot tall control tower are shown in the figure. The beginning of the runway is at a perpendicular distance of 300 feet from the base of the tower. If x denotes the distance an airplane has moved down the runway, express the distance d between the airplane and the control booth as a function of x.

FIGURE FOR EXERCISE 47

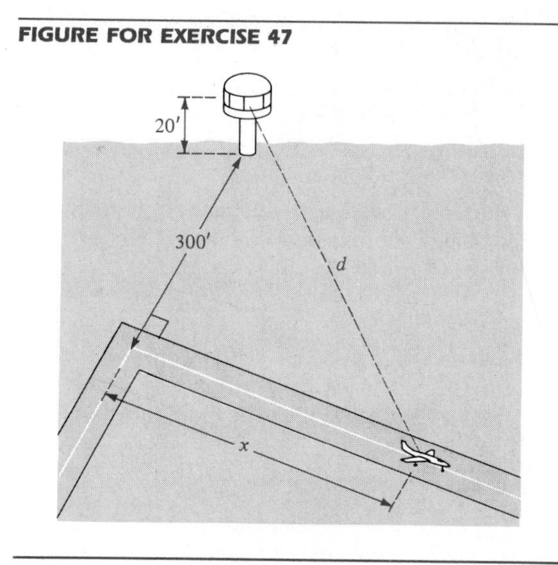

48 A man in a rowboat that is 2 miles from the nearest point A on a straight shoreline wishes to reach a house located at a point B that is 6 miles further downshore (see figure). He plans to row to a point P that is between A and B and is x miles from the house, and then walk the remainder of the distance. Suppose he can row at a rate of 3 mph and can walk at a rate of 5 mph. If T is the total time required to reach the house, express T as a function of x.

FIGURE FOR EXERCISE 48

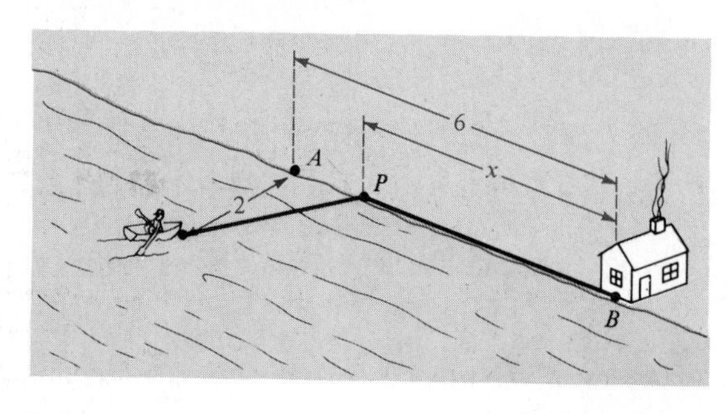

49 A right circular cylinder of radius r and height h is inscribed in a cone of altitude 12 and base radius 4, as illustrated in the figure.

(a) Express h as a function of r. (*Hint:* Use similar triangles.)

(b) Express the volume V of the cylinder as a function of r.

FIGURE FOR EXERCISE 49

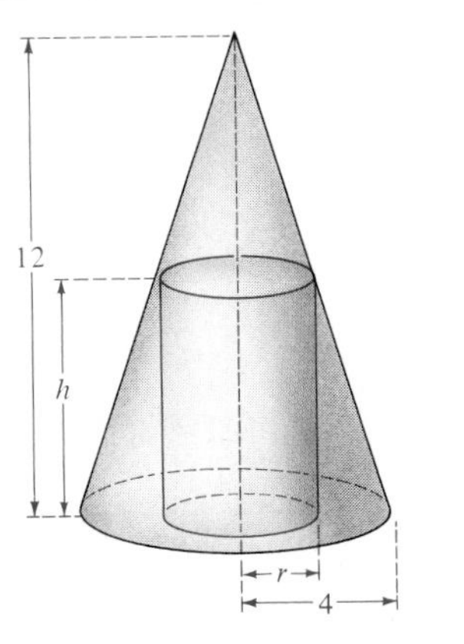

50 Suppose a point $P(x, y)$ is moving along the parabola $y = x^2$. Let d denote the distance from the point $A(3, 1)$ to P. Express d as a function of x.

In Exercises 51–60 determine whether the set W of ordered pairs is a function in the sense of the alternate definition of function on page 69.

51 $W = \{(x, y): 2y = x^2 + 5\}$

52 $W = \{(x, y): x = 3y + 2\}$

53 $W = \{(x, y): x^2 + y^2 = 4\}$

54 $W = \{(x, y): y^2 - x^2 = 1\}$

55 $W = \{(x, y): y = 3\}$

56 $W = \{(x, y): x = y\}$

57 $W = \{(x, y): xy = 0\}$

58 $W = \{(x, y): x + y = 0\}$

59 $W = \{(x, y): |y| = |x|\}$

60 $W = \{(x, y): y < x\}$

2.2 Graphs of Functions

If $f(x) = 2x + 3$, then the graph of f is a line of slope 2, and the graph rises as x increases. A function of this type is said to be *increasing*. For certain functions $f(x_1) > f(x_2)$ whenever $x_1 < x_2$. In this case the graph of f falls as x increases, and the function is called a *decreasing* function. In general, we shall speak of functions that increase or decrease on certain intervals, as in the following definition.

DEFINITION

Let a function f be defined on an interval I and let x_1, x_2 be numbers in I.

(i) f is **increasing** on I if $f(x_1) < f(x_2)$ whenever $x_1 < x_2$.

(ii) f is **decreasing** on I if $f(x_1) > f(x_2)$ whenever $x_1 < x_2$.

(iii) f is **constant** on I if $f(x_1) = f(x_2)$ for every x_1 and x_2.

Geometric interpretations of this definition are given in Figure 2.9, where the interval I is not indicated. Note that if f is constant, then the graph is part of a horizontal line.

FIGURE 2.9

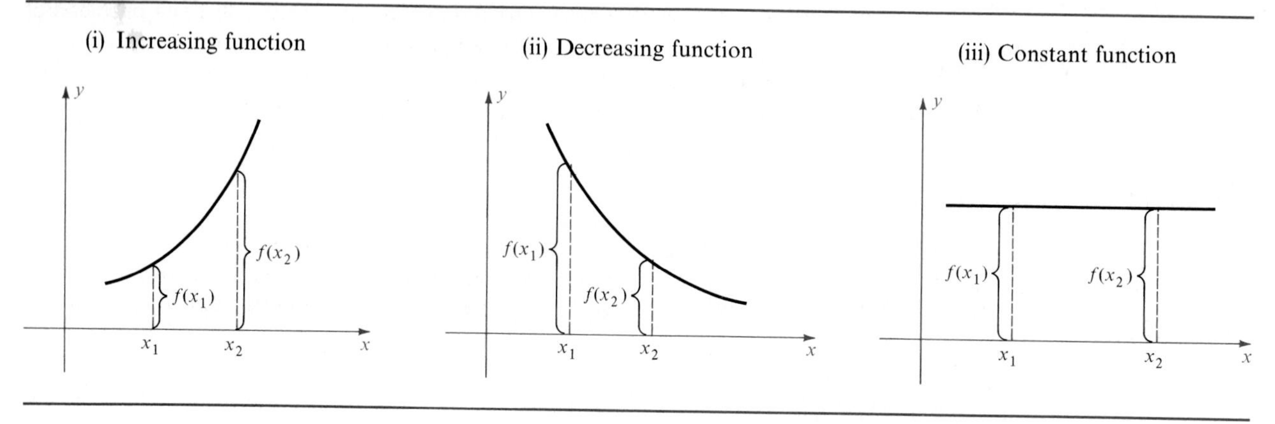

We shall use the phrases "f is increasing" and "$f(x)$ is increasing" interchangeably. This will also be done for the term "decreasing."

FIGURE 2.10

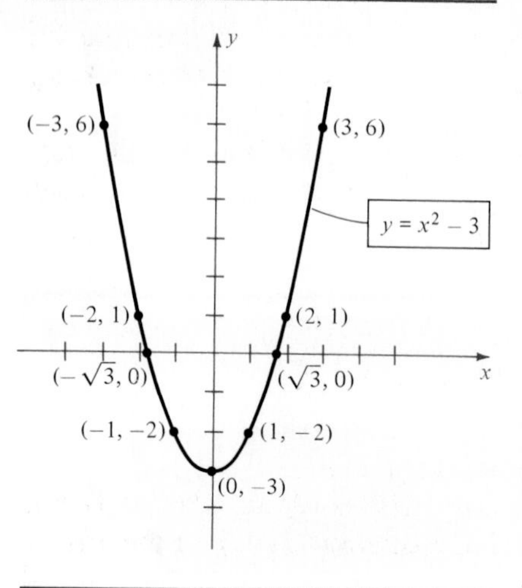

EXAMPLE 1 If $f(x) = x^2 - 3$, sketch the graph of f and find the domain and range of f. Determine the intervals on which f is increasing or decreasing.

SOLUTION Coordinates (x, y) of some points on the graph of f are listed in the following table, where $y = x^2 - 3$.

x	-3	-2	-1	0	1	2	3
y	6	1	-2	-3	-2	1	6

The x-intercepts are the solutions of the equation $f(x) = 0$, that is, of $x^2 - 3 = 0$. Thus the x-intercepts are $\pm\sqrt{3}$. The y-intercept is $f(0) = -3$. Plotting points and using the x-intercepts leads to the sketch in Figure 2.10.

Since x can have any value, the domain of f is $\mathbb{R}$. The range of f consists of all real numbers y such that $y \geq -3$, and hence is the interval $[-3, \infty)$.

We see from the graph that f is decreasing on $(-\infty, 0]$ and that f is increasing on $[0, \infty)$. ■

Referring to Figure 2.10, we see that $f(x) = x^2 - 3$ takes on its least value at $x = 0$. This smallest value, -3, is called the **minimum value** of f. The corresponding point $(0, -3)$ is the lowest point on the graph. Clearly, $f(x)$ does not attain a **maximum value,** that is, a *largest* value.

The solution to Example 1 could have been shortened by observing that since $(-x)^2 - 3 = x^2 - 3$, f is an even function, and hence the graph is symmetric with respect to the y-axis (see page 67).

FIGURE 2.11

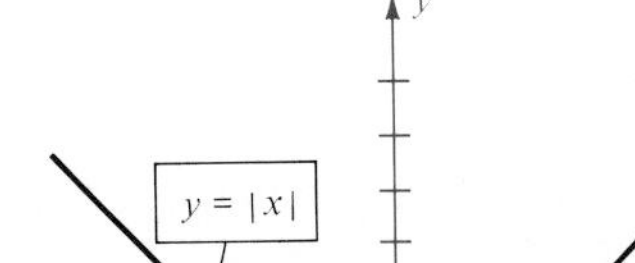

EXAMPLE 2 Sketch the graph of f for $f(x) = |x|$, and find the domain and range of f. Determine the intervals on which f is increasing or decreasing.

SOLUTION If $x \geq 0$, then $f(x) = x$ and hence the points (x, x) in the first quadrant are on the graph of f. Some special cases are (0, 0), (1, 1), (2, 2), (3, 3), and (4, 4). Since $|-x| = |x|$ we see that f is an even function and hence the graph is symmetric with respect to the y-axis. Plotting points and using symmetry leads to the sketch in Figure 2.11.

The domain of f is $\mathbb{R}$, and the range is $[0, \infty)$. As in Example 1, this function is decreasing on $(-\infty, 0]$ and increasing on $[0, \infty)$. ■

EXAMPLE 3 Sketch the graph of f if $f(x) = \sqrt{x - 1}$ and find the domain and range of f. Determine where f is increasing or decreasing.

FIGURE 2.12

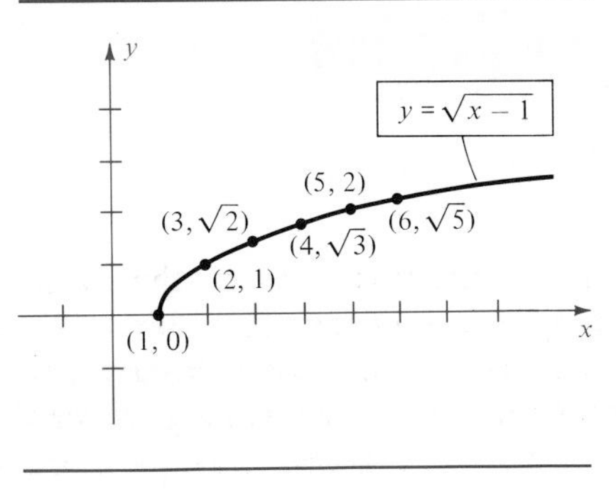

SOLUTION The domain of f consists of all real numbers x such that $x \geq 1$ and hence is the interval $[1, \infty)$. The following table lists some points (x, y) on the graph, where $y = \sqrt{x - 1}$.

x	1	2	3	4	5	6
y	0	1	$\sqrt{2}$	$\sqrt{3}$	2	$\sqrt{5}$

Plotting points leads to the sketch in Figure 2.12.

We see from the graph that the range of f is $[0, \infty)$. The function is increasing throughout its domain. The x-intercept is 1, and there is no y-intercept. ■

FIGURE 2.13

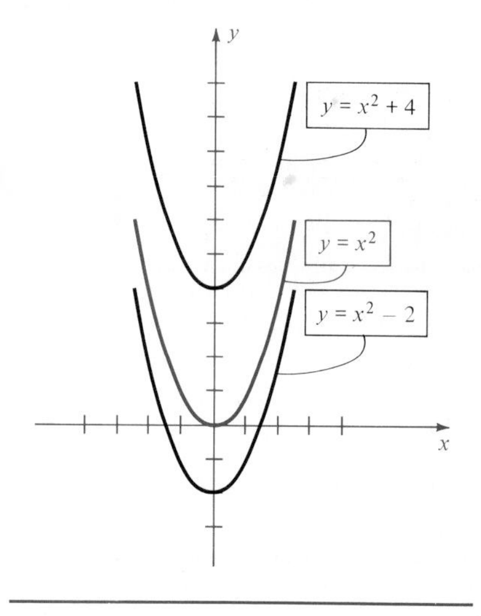

EXAMPLE 4 Given $f(x) = x^2 + c$, sketch the graph of f if $c = 4$; if $c = -2$.

SOLUTION We shall sketch both graphs on the same coordinate axes. The graph of $y = x^2$ was sketched in Figure 1.31, and, for reference, is represented in color in Figure 2.13. To find the graph of $y = x^2 + 4$ we may simply add 4 to the y-coordinate of each point on the graph of $y = x^2$. This amounts to *shifting* the graph of $y = x^2$ *upward* 4 units as shown in the figure. For $c = -2$ we decrease y-coordinates by 2 and, hence, the graph of $y = x^2 - 2$ may be obtained by shifting the graph of $y = x^2$ *downward* 2 units. Each graph is a parabola symmetric with respect to the y-axis. To determine the correct position of each graph, it is advisable to plot several points. ■

The graphs in the preceding example illustrate **vertical shifts** of the graph of $y = x^2$ and are special cases of the following general rules.

VERTICAL SHIFTS OF GRAPHS ($c > 0$)

To obtain the graph of:	shift the graph of $y = f(x)$:
$y = f(x) - c$	c units downward
$y = f(x) + c$	c units upward

FIGURE 2.14

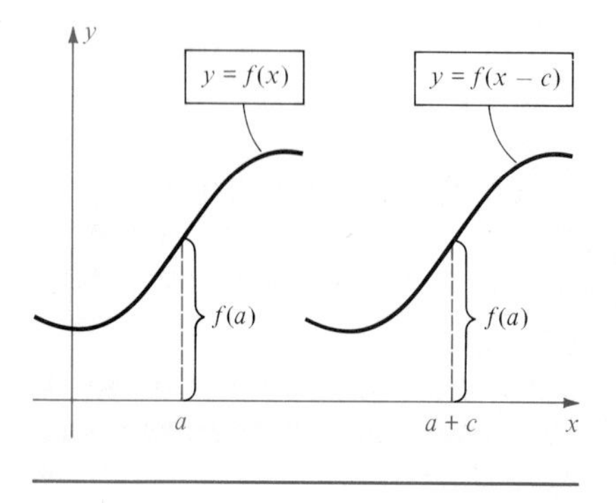

Similar rules can be stated for **horizontal shifts.** Specifically, if $c > 0$, consider the graphs of $y = f(x)$ and $y = f(x - c)$ sketched on the same coordinate axes, as illustrated in Figure 2.14.

Since $f(a) = f(a + c - c)$, we see that the point with x-coordinate a on the graph of $y = f(x)$ has the same y-coordinate as the point with x-coordinate $a + c$ on the graph of $y = f(x - c)$. This implies that the graph of $y = f(x - c)$ can be obtained by shifting the graph of $y = f(x)$ to the right c units. Similarly, the graph of $y = f(x + c)$ can be obtained by shifting the graph of f to the left c units. These rules are listed for reference in the next box.

FIGURE 2.15

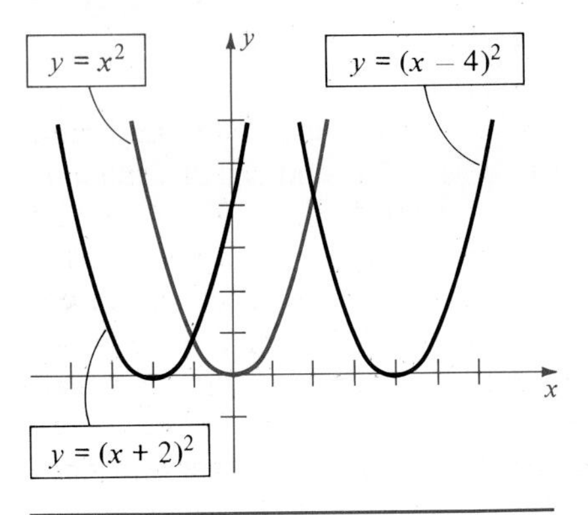

HORIZONTAL SHIFTS OF GRAPHS ($c > 0$)

To obtain the graph of:	shift the graph of $y = f(x)$:
$y = f(x - c)$	c units to the right
$y = f(x + c)$	c units to the left

EXAMPLE 5 Sketch the graph of f if

(a) $f(x) = (x - 4)^2$ (b) $f(x) = (x + 2)^2$

SOLUTION The graph of $y = x^2$ is sketched in color in Figure 2.15. According to the rules for horizontal shifts, shifting this graph to the right 4 units gives us the graph of $y = (x - 4)^2$, whereas shifting to the left 2 units leads to the graph of $y = (x + 2)^2$. Students who are not convinced of the validity of this technique are urged to plot several points on each graph. ■

FIGURE 2.16

(i)

$y = 4x^2$ $y = x^2$

To obtain the graph of $y = cf(x)$ for some real number c, we may *multiply* the y-coordinates of points on the graph of $y = f(x)$ by c. For example, if $y = 2f(x)$, we double y-coordinates, or if $y = \frac{1}{2}f(x)$, we multiply each y-coordinate by $\frac{1}{2}$. If $c > 0$ (and $c \neq 1$) we shall refer to this procedure as **stretching** the graph of $y = f(x)$.

(ii)

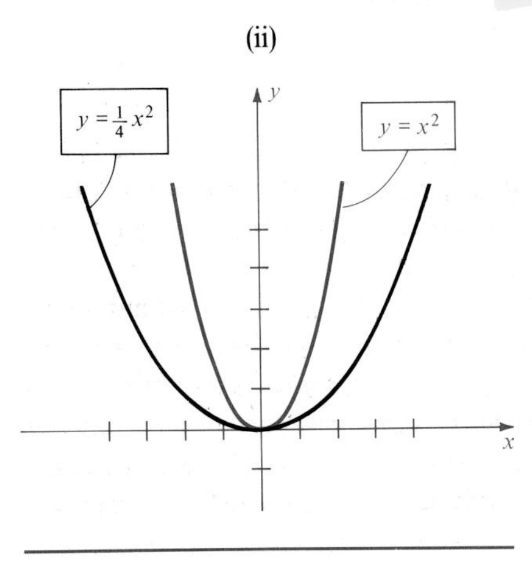

EXAMPLE 6 Sketch the graph of (a) $y = 4x^2$; (b) $y = \frac{1}{4}x^2$.

SOLUTION

(a) To sketch the graph of $y = 4x^2$ we may refer to the graph of $y = x^2$ (shown in color in Figure 2.16(i)) and multiply the y-coordinate of each point by 4. This gives us a narrower parabola that is sharper at the vertex, as illustrated in (i) of the figure. To obtain the correct shape, several points such as $(0, 0)$, $(\frac{1}{2}, 1)$, and $(1, 4)$ should be plotted.

(b) The graph of $y = \frac{1}{4}x^2$ may be sketched by multiplying y-coordinates of points on the graph of $y = x^2$ by $\frac{1}{4}$. The graph is a wider parabola that is flatter at the vertex, as shown in Figure 2.16(ii). ■

The graph of $y = -f(x)$ may be obtained by multiplying the y-coordinate of each point on the graph of $y = f(x)$ by -1. Thus, every point (a, b) on the graph of $y = f(x)$ that lies above the x-axis determines a point $(a, -b)$ on the graph of $y = -f(x)$ that lies below the x-axis. Similarly, if (c, d) lies below the x-axis (that is, $d < 0$), then $(c, -d)$ lies above the x-axis. The graph of $y = -f(x)$ is called a **reflection** of the graph of $y = f(x)$ through the x-axis.

FIGURE 2.17

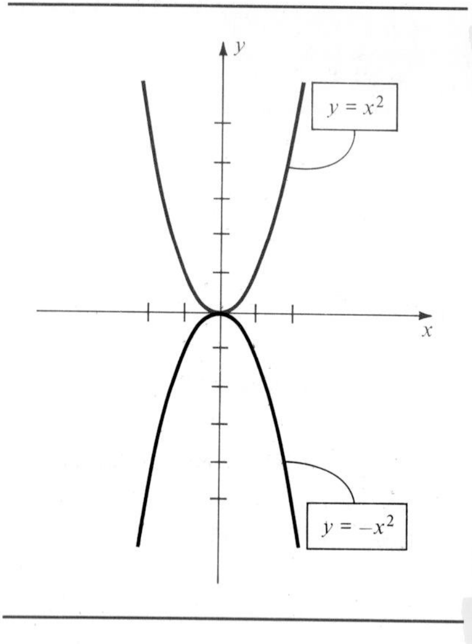

EXAMPLE 7 Sketch the graph of $y = -x^2$.

SOLUTION The graph may be found by plotting points; however, since the graph of $y = x^2$ is well known, we sketch it in color, as in Figure 2.17, and then multiply y-coordinates of points by -1. This gives us the reflection through the x-axis indicated in the figure. ■

Sometimes functions are described in terms of more than one expression, as in the next examples. We shall call such functions **piecewise-defined functions.**

EXAMPLE 8 Sketch the graph of the piecewise-defined function f given by

FIGURE 2.18

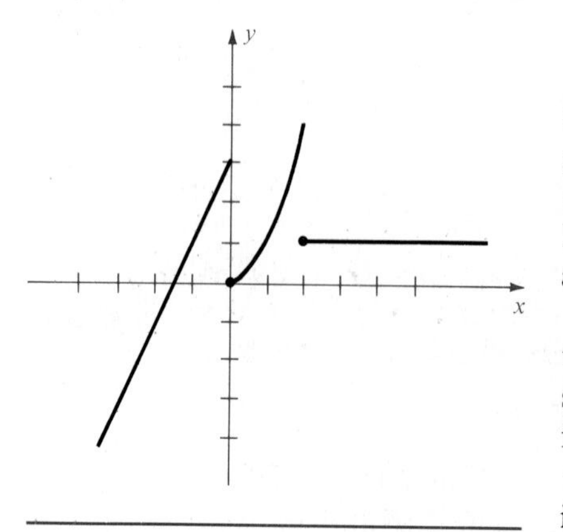

$$f(x) = \begin{cases} 2x + 3 & \text{if } x < 0 \\ x^2 & \text{if } 0 \le x < 2 \\ 1 & \text{if } x \ge 2 \end{cases}$$

SOLUTION If $x < 0$, then $f(x) = 2x + 3$. This means that if x is negative, the expression $2x + 3$ should be used to find function values. Consequently, if $x < 0$, then the graph of f coincides with the graph in Figure 2.6 and we sketch that portion of the graph to the left of the y-axis as indicated in Figure 2.18.

If $0 \le x < 2$, we use x^2 to find values of f, and therefore this part of the graph of f coincides with the graph of the equation $y = x^2$. We then sketch the part of the graph of f between $x = 0$ and $x = 2$, as indicated in Figure 2.18.

Finally, if $x \ge 2$, the values of f are always 1. That part of the graph is the horizontal half-line illustrated in Figure 2.18. ■

FIGURE 2.19

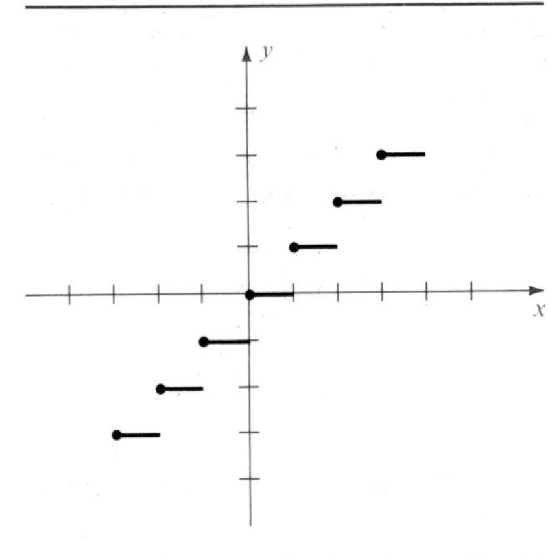

EXAMPLE 9 If x is any real number, then there exist consecutive integers n and $n+1$ such that $n \leq x < n+1$. Let f be the function defined as follows: If $n \leq x < n+1$, then $f(x) = n$. Sketch the graph of f.

SOLUTION The x- and y-coordinates of some points on the graph may be listed as follows:

Values of x	$f(x)$
...	...
$-2 \leq x < -1$	-2
$-1 \leq x < 0$	-1
$0 \leq x < 1$	0
$1 \leq x < 2$	1
$2 \leq x < 3$	2
...	...

Since $f(x)$ does not change when x is between successive integers, the corresponding part of the graph is a segment of a horizontal line. Part of the graph is sketched in Figure 2.19. The graph continues indefinitely to the right and to the left. ■

The symbol $[\![x]\!]$ is often used to denote the largest integer n such that $n \leq x$. For example $[\![1.6]\!] = 1$, $[\![\sqrt{5}]\!] = 2$, $[\![\pi]\!] = 3$, and $[\![-3.5]\!] = -4$. Using this notation, the function f of Example 9 may be defined by $f(x) = [\![x]\!]$. It is customary to refer to f as the **greatest integer function.**

Exercises 2.2

In Exercises 1–20 sketch the graph of f, find the domain and range of f, and describe the intervals in which f is increasing, decreasing, or constant.

1 $f(x) = 5x$

2 $f(x) = -3x$

3 $f(x) = -4x + 2$

4 $f(x) = 4x - 2$

5 $f(x) = 3 - x^2$

6 $f(x) = 4x^2 + 1$

7 $f(x) = 2x^2 - 4$

8 $f(x) = \frac{1}{100}x^2$

9 $f(x) = \sqrt{x+4}$

10 $f(x) = \sqrt{4-x}$

11 $f(x) = \sqrt{x} + 2$

12 $f(x) = 2 - \sqrt{x}$

13 $f(x) = 4/x$

14 $f(x) = 1/x^2$

15 $f(x) = |x - 2|$

16 $f(x) = |x + 2|$

17 $f(x) = |x| - 2$

18 $f(x) = |x| + 2$

19 $f(x) = \dfrac{x}{|x|}$

20 $f(x) = x + |x|$

In Exercises 21–30 sketch, on the same coordinate axes, the graphs of f for the stated values of c. (Make use of vertical shifts, horizontal shifts, stretching, or reflecting).

21 $f(x) = 3x + c;\quad c = 0,\ c = 2,\ c = -1$

22 $f(x) = -2x + c;\quad c = 0, c = 1, c = -3$

23 $f(x) = x^3 + c;\quad c = 0, c = 1, c = -2$

24 $f(x) = -x^3 + c$; $c = 0,\ c = 2,\ c = -1$

25 $f(x) = \sqrt{4 - x^2} + c$; $c = 0,\ c = 4,\ c = -3$

26 $f(x) = c - |x|$; $c = 0,\ c = 5,\ c = -2$

27 $f(x) = 3(x - c)$; $c = 0,\ c = 2,\ c = 3$

28 $f(x) = -2(x - c)^2$; $c = 0,\ c = 3,\ c = 1$

29 $f(x) = (x + c)^3$; $c = 0,\ c = 2,\ c = -2$

30 $f(x) = c\sqrt{9 - x^2}$; $c = 0,\ c = 2,\ c = 3$

31 The graph of a function f with domain $0 \le x \le 4$ is shown in the accompanying figure. Sketch the graph of each of the following:

(a) $y = f(x + 2)$ (b) $y = f(x - 2)$

(c) $y = f(x) + 2$ (d) $y = f(x) - 2$

(e) $y = 2f(x)$ (f) $y = \frac{1}{2}f(x)$

(g) $y = -2f(x)$ (h) $y = f(x - 3) + 1$

FIGURE FOR EXERCISE 31

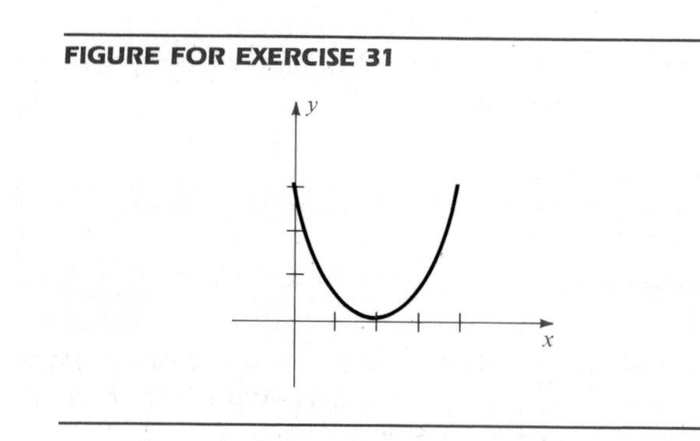

32 The graph of a function f with domain $0 \le x \le 4$ is shown in the accompanying figure. Sketch the graph of each of the following:

(a) $y = f(x - 1)$ (b) $y = f(x + 3)$

(c) $y = f(x) - 2$ (d) $y = f(x) + 1$

(e) $y = 3f(x)$ (f) $y = \frac{1}{2}f(x)$

(g) $y = f(x + 2) - 2$ (h) $y = f(2x)$

FIGURE FOR EXERCISE 32

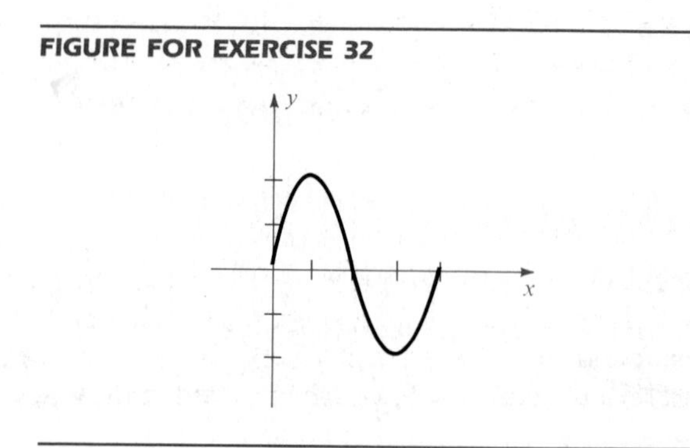

In Exercises 33–40 sketch the graph of the piecewise-defined function f.

33 $f(x) = \begin{cases} 2 & \text{if } x < 0 \\ -1 & \text{if } x \ge 0 \end{cases}$

34 $f(x) = \begin{cases} -1 & \text{if } x \text{ is an integer} \\ 1 & \text{if } x \text{ is not an integer} \end{cases}$

35 $f(x) = \begin{cases} 3 & \text{if } x < -3 \\ -x & \text{if } -3 \le x \le 3 \\ -3 & \text{if } x > 3 \end{cases}$

36 $f(x) = \begin{cases} x & \text{if } x < 0 \\ -2 & \text{if } 0 \le x < 1 \\ x^2 & \text{if } x \ge 1 \end{cases}$

37 $f(x) = \begin{cases} x^2 & \text{if } x \le -1 \\ x^3 & \text{if } |x| < 1 \\ 2x & \text{if } x \ge 1 \end{cases}$

38 $f(x) = \begin{cases} x & \text{if } x \le 1 \\ -x^2 & \text{if } 1 < x < 2 \\ x & \text{if } x \ge 2 \end{cases}$

39 $f(x) = \begin{cases} \dfrac{x^2 - 4}{x - 2} & \text{if } x \ne 2 \\ 3 & \text{if } x = 2 \end{cases}$

40 $f(x) = \begin{cases} \dfrac{x^2 - 1}{1 - x} & \text{if } x \ne 1 \\ 2 & \text{if } x = 1 \end{cases}$

41 If $[\![x]\!]$ denotes values of the greatest integer function, sketch the graph of f in each of the following:

(a) $f(x) = [\![x - 3]\!]$ (b) $f(x) = 2[\![x]\!]$

(c) $f(x) = -[\![x]\!]$

42 Explain why the graph of the equation $y^2 = x$ is not the graph of a function.

2.3 Quadratic Functions

Among the most important functions in mathematics are those defined as follows:

DEFINITION

A function f is a **polynomial function** if

$$f(x) = a_n x^n + a_{n-1}x^{n-1} + \cdots + a_1 x + a_0$$

where the coefficients $a_0, a_1, \ldots, a_n$ are real numbers and the exponents are nonnegative integers.

If $a_n \neq 0$ in the preceding definition, we say that f has **degree *n*.** Note that a polynomial function of degree 1 is a linear function. If the degree is 2, then, as in the next definition, f is called a *quadratic function.*

DEFINITION

A function f is a **quadratic function** if

$$f(x) = ax^2 + bx + c$$

where a, b, and c are real numbers and $a \neq 0$.

If $b = c = 0$ in the preceding definition, then $f(x) = ax^2$, and the graph is a parabola with vertex at the origin, opening upward if $a > 0$ or downward if $a < 0$ (see Figures 2.16 and 2.17). If $b = 0$ and $c \neq 0$, then

$$f(x) = ax^2 + c,$$

and from the discussion of vertical shifts in Section 2.2, the graph is a parabola with vertex at the point $(0, c)$ on the y-axis. Some typical graphs are illustrated in Figure 2.13. Another is given in the next example.

EXAMPLE 1 Sketch the graph of f if $f(x) = -\frac{1}{2}x^2 + 4$.

SOLUTION The graph of $y = -\frac{1}{2}x^2$ is similar in shape to the graph of $y = -x^2$, sketched in Figure 2.17, but is somewhat wider. The graph of $y = -\frac{1}{2}x^2 + 4$ may be found by shifting the graph of $y = -\frac{1}{2}x^2$ upward 4 units. Coordinates of several points on the graph are listed in the following table.

FIGURE 2.20

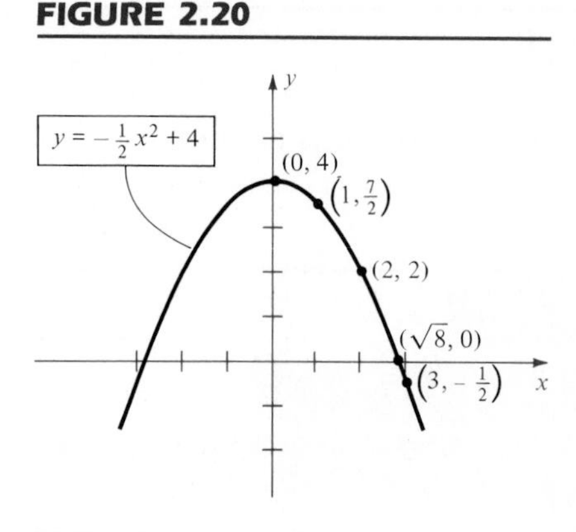

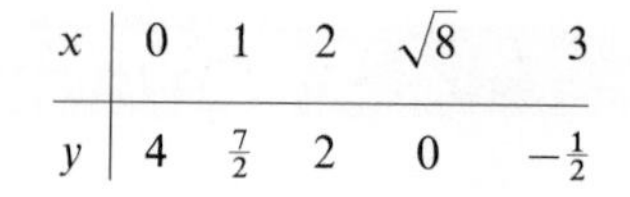

x	0	1	2	$\sqrt{8}$	3
y	4	$\frac{7}{2}$	2	0	$-\frac{1}{2}$

Plotting and using symmetry with respect to the y-axis gives us the sketch in Figure 2.20. ■

If $f(x) = ax^2 + bx + c$ and $b \neq 0$, then by completing the square we can change the form of $f(x)$ to

$$f(x) = a(x - h)^2 + k$$

for some real numbers h and k. This technique is illustrated in the next example. As we shall see, the graph of f can be readily obtained from this new form.

EXAMPLE 2 If $f(x) = 3x^2 + 24x + 50$, express $f(x)$ in the form $a(x - h)^2 + k$.

SOLUTION Before completing the square *it is essential that we factor out the coefficient of* x^2 *from the first two terms* of $f(x)$ as follows:

$$\begin{aligned} f(x) &= 3x^2 + 24x + 50 \\ &= 3(x^2 + \ 8x \qquad) + 50. \end{aligned}$$

We may now complete the square for the expression $x^2 + 8x$ by adding the square of one-half the coefficient of x, that is, $(\frac{8}{2})^2$, or 16. However, if we add 16 to the expression within parentheses, then, because of the factor 3, we are actually adding 48 to $f(x)$. Hence, we must compensate by subtracting 48:

$$\begin{aligned} f(x) &= 3(x^2 + 8x \qquad) + 50 \\ &= 3(x^2 + 8x + 16) + 50 - 48 \\ &= 3(x + 4)^2 + 2, \end{aligned}$$

which has the desired form, with $a = 3$, $h = -4$, and $k = 2$. ■

If $f(x) = ax^2 + bx + c$, then by completing the square as in Example 2, we see that the graph of f is the same as the graph of an equation of the form

$$y = a(x - h)^2 + k.$$

From the discussion of horizontal shifts in Section 2.2, we can find the graph of $y = a(x - h)^2$ by shifting the graph of $y = ax^2$ either to the left or right, depending on the sign of h. Thus, $y = a(x - h)^2$ is an equation of a parabola that has vertex $(h, 0)$ and a vertical axis. A typical graph is sketched in Figure 2.21(i) for the case in which both a and h are positive.

FIGURE 2.21

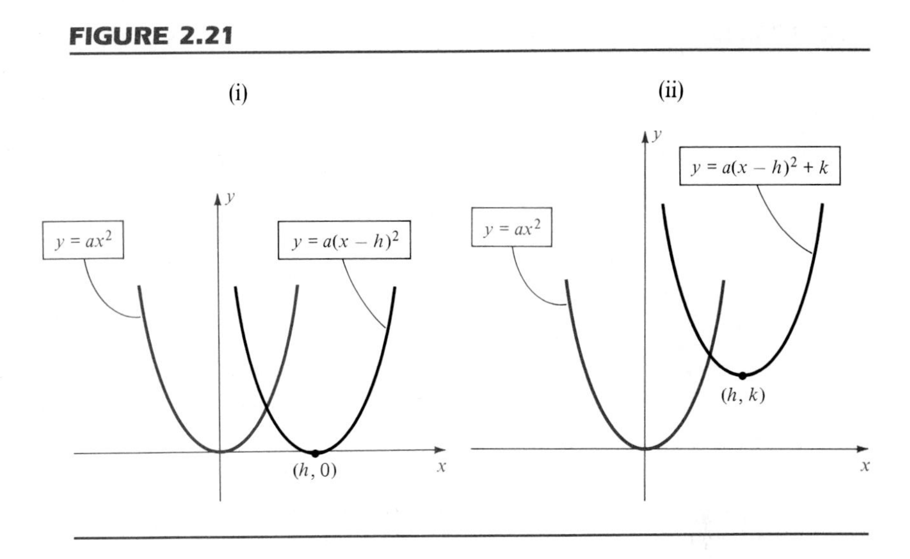

Since the graph of $y = a(x - h)^2 + k$ can be obtained from that of $y = a(x - h)^2$ by a *vertical* shift of $|k|$ units, it follows that *the graph of a quadratic function f is a parabola* that has vertex (h, k) and a vertical axis. The sketch in Figure 2.21(ii) illustrates one possible graph. Observe that since (h, k) is the lowest (or highest) point on the parabola, $f(x)$ has its minimum (or maximum) value at $x = h$. This value is $f(h) = k$.

We have derived the following equation of a parabola that has vertex (h, k) and a vertical axis.

STANDARD EQUATION OF A PARABOLA (VERTICAL AXIS)

$$y - k = a(x - h)^2$$

This parabola opens upward if $a > 0$ or downward if $a < 0$.

EXAMPLE 3 Sketch the graph of f if $f(x) = 2x^2 - 6x + 4$ and find the minimum value of $f(x)$.

SOLUTION The graph of f is a parabola and is the same as the graph of $y = 2x^2 - 6x + 4$. We begin by completing the square:

$$\begin{aligned} y &= 2x^2 - 6x + 4 \\ &= 2(x^2 - 3x \qquad) + 4 \\ &= 2(x^2 - 3x + \tfrac{9}{4}) + (4 - \tfrac{9}{2}) \\ &= 2(x - \tfrac{3}{2})^2 - \tfrac{1}{2} \end{aligned}$$

FIGURE 2.22

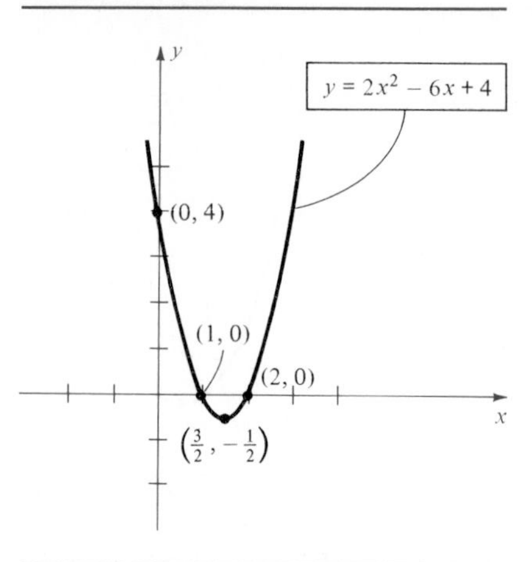

If we write the last equation as

$$y + \tfrac{1}{2} = 2(x - \tfrac{3}{2})^2$$

and compare it with the standard equation of a parabola, we see that $h = \frac{3}{2}$ and $k = -\frac{1}{2}$. Consequently, the vertex (h, k) of the parabola is $(\frac{3}{2}, -\frac{1}{2})$. Since $a = 2 > 0$, the parabola opens upward. The y-intercept is $f(0) = 4$. To find the x-intercepts we solve $2x^2 - 6x + 4 = 0$, or the equivalent equation $(2x - 2)(x - 2) = 0$, obtaining $x = 1$ and $x = 2$. Plotting the vertex together with the x- and y-intercepts provides enough points for a reasonably accurate sketch (see Figure 2.22). The minimum value of f occurs at the vertex, and hence is $f(\frac{3}{2}) = -\frac{1}{2}$. ■

EXAMPLE 4 Sketch the graph of f if $f(x) = 8 - 2x - x^2$ and find the maximum value of $f(x)$.

SOLUTION We know that the graph is a parabola with a vertical axis. Let us set $y = f(x)$ and find the vertex by completing the square:

$$\begin{aligned} y &= -x^2 - 2x + 8 \\ &= -(x^2 + 2x) + 8 \\ &= -(x^2 + 2x + 1) + 8 + 1 \\ &= -(x + 1)^2 + 9. \end{aligned}$$

If we now write $\quad y - 9 = -(x + 1)^2$

and compare this equation with the standard equation of a parabola, we see that $h = -1$, $k = 9$, and hence the vertex is $(-1, 9)$. Since $a = -1 < 0$, the parabola opens downward. To find the x-intercepts we solve the equation $8 - 2x - x^2 = 0$, or, equivalently, $x^2 + 2x - 8 = 0$. Factoring gives us $(x + 4)(x - 2) = 0$, and hence the intercepts are $x = -4$ and $x = 2$. The y-intercept is 8. Using this information gives us the sketch in Figure 2.23. The maximum value of f occurs at the vertex and hence is $f(-1) = 9$. ■

FIGURE 2.23

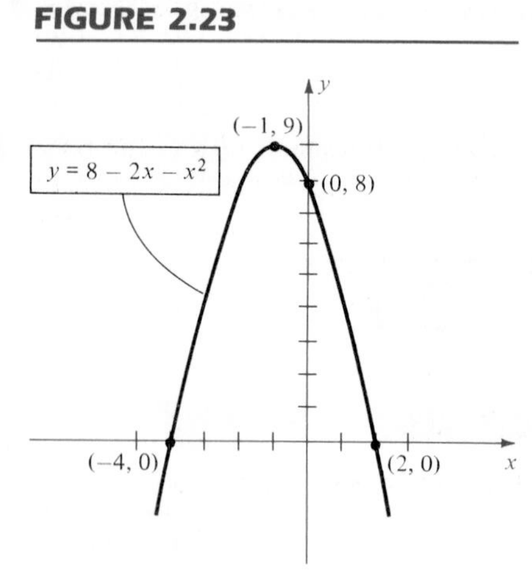

The x-intercepts of the graph of $y = ax^2 + bx + c$ are the solutions of the quadratic equation $ax^2 + bx + c = 0$ and hence are $(-b \pm \sqrt{b^2 - 4ac})/2a$. If $b^2 - 4ac > 0$, the equation has two real and unequal solutions and the graph has two x-intercepts. If $b^2 - 4ac = 0$, the equation has one (double) solution, and the graph is tangent to the x-axis. If $b^2 - 4ac < 0$, the equation has no real solutions, and the graph has no x-intercepts. We have illustrated the three cases in Figure 2.24 for $a > 0$. A similar situation occurs if $a < 0$, but in this case the parabolas open downward.

FIGURE 2.24

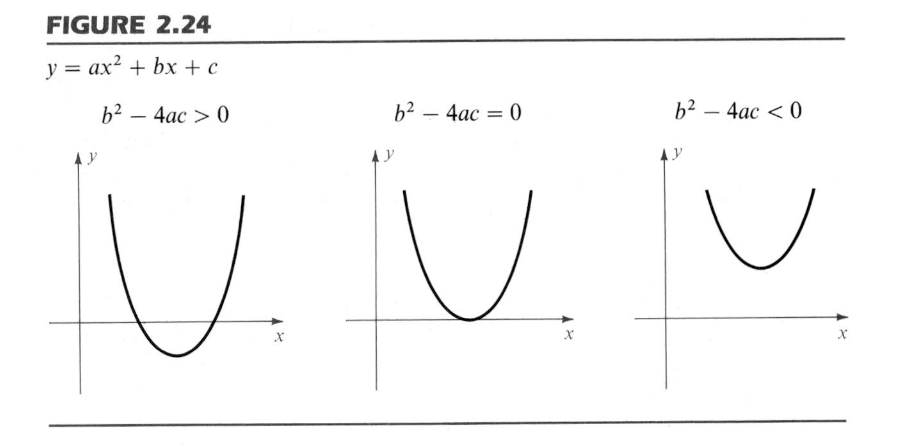

We can solve certain applied problems by finding maximum or minimum values of quadratic functions. The next example is one illustration.

EXAMPLE 5 A long rectangular sheet of metal, 12 inches wide, is to be made into a rain gutter by turning up two sides so that they are perpendicular to the sheet. How many inches should be turned up to give the gutter its greatest capacity?

FIGURE 2.25

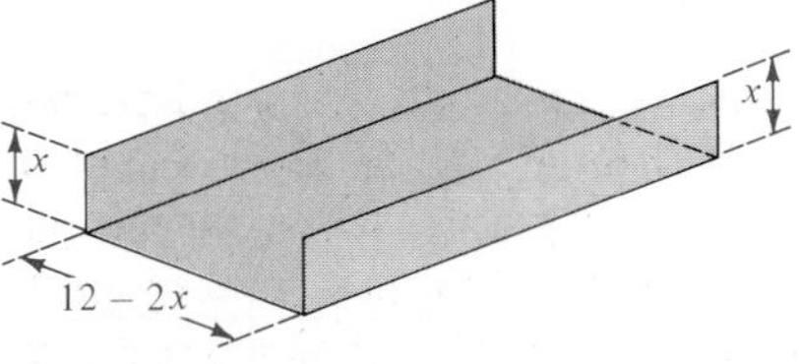

SOLUTION The gutter is illustrated in Figure 2.25, where since x denotes the number of inches turned up on each side, the width of the base of the gutter is expressed as $12 - 2x$ inches. The capacity will be greatest when the area of the rectangle with sides of lengths x and $12 - 2x$ has its greatest value. Letting $f(x)$ denote this area, we have

$$\begin{aligned} f(x) &= x(12 - 2x) \\ &= 12x - 2x^2 \\ &= -2x^2 + 12x. \end{aligned}$$

We may find the maximum value of $f(x)$ by completing a square:

$$\begin{aligned} f(x) &= -2(x^2 - 6x) \\ &= -2(x^2 - 6x + 9) + 18 \\ &= -2(x - 3)^2 + 18. \end{aligned}$$

The graph of f is the graph of the equation

$$y = -2(x - 3)^2 + 18 \quad \text{or} \quad y - 18 = -2(x - 3)^2$$

and hence is a parabola that opens downward. The largest value of $y = f(x)$ occurs at the vertex (3, 18); that is at $x = 3$. Thus, 3 inches should be turned up to achieve maximum capacity. ■

Parabolas (and hence quadratic functions) occur in applications of mathematics to the physical world. For example, it can be shown that if a projectile is fired, and we assume that it is acted upon only by the force of gravity (that is, air resistance and other outside factors are ignored), then the path of the projectile is parabolic. Properties of parabolas are used in the design of mirrors for telescopes and searchlights and in the construction of radar antenna.

Exercises 2.3

1 If $f(x) = ax^2 + 2$, sketch the graph of f for each value of a.

(a) $a = 2$ (b) $a = 5$

(c) $a = \frac{1}{2}$ (d) $a = -3$

2 If $f(x) = 4x^2 + c$, sketch the graph of f for each value of c.

(a) $c = 2$ (b) $c = 5$

(c) $c = \frac{1}{2}$ (d) $c = -3$

In Exercises 3–6 sketch the graph of f and find the vertex.

3 $f(x) = 4x^2 - 9$

4 $f(x) = 2x^2 + 3$

5 $f(x) = 9 - 4x^2$

6 $f(x) = 16 - 9x^2$

In Exercises 7–10 use the Quadratic Formula to find the zeros of f.

7 $f(x) = 4x^2 - 11x - 3$

8 $f(x) = 25x^2 + 10x + 1$

9 $f(x) = 9x^2 - 12x + 4$

10 $f(x) = 10x^2 + x - 21$

In Exercises 11–14 use the technique of completing the square to express $f(x)$ in the form $a(x - h)^2 + k$, and then find the maximum or minimum value of $f(x)$.

11 $f(x) = 2x^2 - 16x + 23$

12 $f(x) = 3x^2 - 12x + 7$

13 $f(x) = -5x^2 - 10x + 3$

14 $f(x) = -2x^2 - 12x - 12$

In Exercises 15–24 sketch the graph of f and then find the maximum or minimum value of $f(x)$.

15 $f(x) = x^2 + 5x + 4$

16 $f(x) = x^2 - 6x$

17 $f(x) = 8x - 12 - x^2$

18 $f(x) = 10 + 3x - x^2$

19 $f(x) = x^2 + x + 3$

20 $f(x) = x^2 + 2x + 5$

21 $f(x) = 3x^2 - 12x + 16$

22 $f(x) = 2x^2 + 12x + 13$

23 $f(x) = -5x^2 - 10x - 4$

24 $f(x) = -3x^2 + 24x - 42$

25 Flights of leaping animals typically form parabolas. The figure illustrates a frog jump superimposed on a rectangular coordinate system. The length of the leap is 9 feet and the maximum height off the ground is 3 feet. Find a quadratic function f that specifies the path of the frog.

FIGURE FOR EXERCISE 25

26 In the 1940s, the human cannonball stunt was performed regularly by Emmanuel Zacchini for The Ringling Brothers and Barnum & Bailey Circus. The tip of the cannon rose 15 feet off the ground, and the total horizontal distance traveled was 175 feet. When the cannon is aimed at a 45-degree angle, the equation of the parabolic flight has the form $y = ax^2 + x + c$ (see figure).

(a) Find values for a and c that correspond to the given information.

(b) Find the maximum height attained by the human cannonball.

FIGURE FOR EXERCISE 26

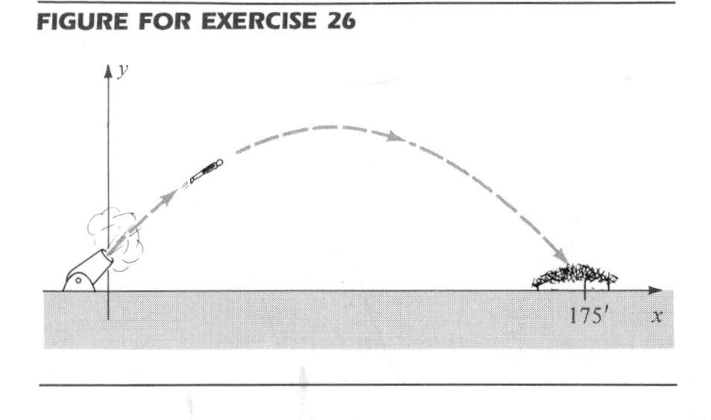

27 One section of a suspension bridge has its weight uniformly distributed between twin towers that are 400 feet apart and rise 90 feet above the horizontal roadway (see figure). A cable strung between the tops of the towers has the shape of a parabola, and its center point is 10 feet above the roadway. Suppose coordinate axes are introduced as shown in the figure.

(a) Find an equation for the parabola.

(b) Nine equally spaced vertical cables are used to support the bridge (see figure). Find the total length of these supports.

FIGURE FOR EXERCISE 27

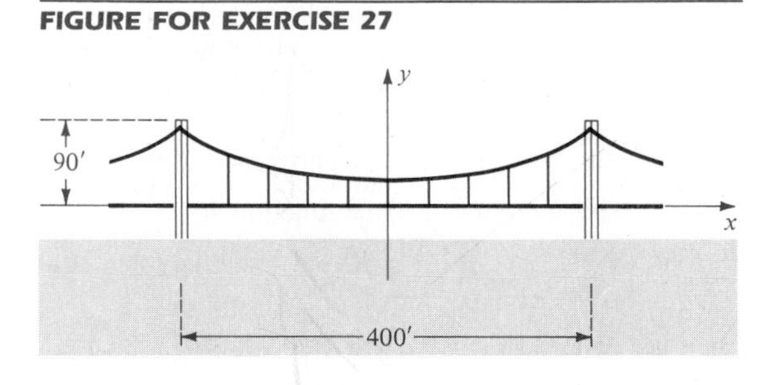

28 Traffic engineers are designing a stretch of highway that will connect a horizontal highway with one having a 20% grade (i.e., slope $\frac{1}{5}$) as illustrated in the figure. The smooth transition is to take place over a horizontal distance of 800 feet using a parabolic piece of highway to connect points A and B. If the equation of the parabolic segment is $y = ax^2 + bx + c$, it can be shown that the slope of the tangent line at the point $P(x, y)$ on the parabola is given by $m = 2ax + b$.

(a) Find an equation of the parabola that has a tangent line of slope 0 at A and $\frac{1}{5}$ at B.

(b) Find the coordinates of B.

FIGURE FOR EXERCISE 28

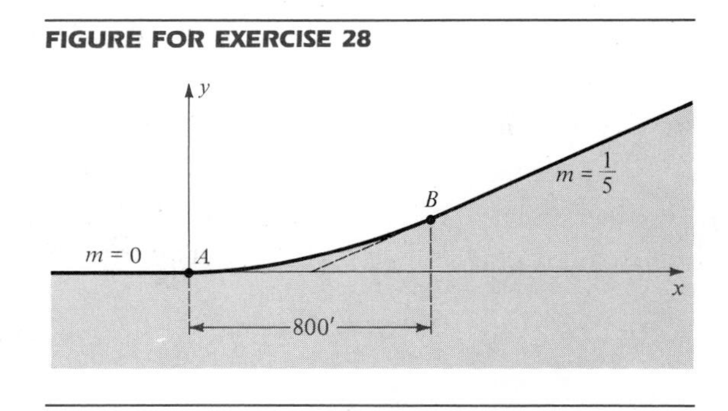

29 Find an equation of a parabola that has a vertical axis and passes through the points $A(2, 3)$, $B(-1, 6)$, and $C(1, 0)$.

30 Prove that there is exactly one line of a given slope m that intersects the parabola $x^2 = 4py$ in exactly one point, and that its equation is $y = mx - pm^2$.

31 A person standing on the top of a building projects an object directly upward with a velocity of 144 feet per second. Its height $s(t)$ in feet above the ground after t seconds is given by $s(t) = -16t^2 + 144t + 100$. What is its maximum height? What is the height of the building?

32 A toy rocket is shot straight up into the air with an initial velocity of v_0 feet per second, and its height $s(t)$ in feet above the ground after t seconds is given by $s(t) = -16t^2 + v_0t$.

(a) The rocket hits the ground after 12 seconds. What is the initial velocity v_0?

(b) What is the maximum height attained by the rocket?

33 The growth rate y (in pounds per month) of infants is related to their present weight x (in pounds) by the formula $y = cx(21 - x)$ for some constant $c > 0$. At what weight is the rate of growth a maximum?

34 The gasoline mileage y (in miles per gallon) of a small economy car is related to the velocity v (in mph) by the formula

$$y = -\tfrac{1}{30}v^2 + 2.5v \quad \text{for } 0 < v < 70.$$

What is the most economical speed for a long trip?

35 One thousand feet of chain-link fence is to be used to construct six cages for a zoo exhibit. The design is shown in the figure.

(a) Express the width y as a function of the length x.

(b) Express the total enclosed area A of the exhibit as a function of x.

(c) Find the dimensions that maximize the enclosed area.

FIGURE FOR EXERCISE 35

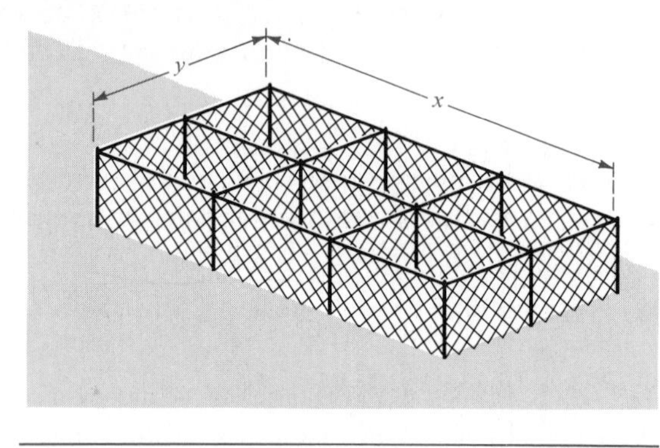

36 A man wishes to put a fence around a rectangular field and then subdivide the field into three smaller rectangular plots by placing two fences parallel to one of the sides. If he can afford only 1000 yards of fencing, what dimensions will give the maximum rectangular area?

37 A piece of wire 24 inches long is bent into the shape of a rectangle having width x and length y.

(a) Express y as a function of x.

(b) Express the area A of the rectangle as a function of x.

(c) Prove that the area A is greatest if the rectangle is a square.

38 A company sells running shoes to dealers at a rate of \$20 per pair if less than 50 pairs are ordered. If a dealer orders 50 or more pairs (up to 600), the price per pair is reduced at a rate of 2 cents times the number ordered. What size order will produce the maximum amount of money for the company?

39 A boy tosses a baseball from the edge of a plateau down a hill as illustrated in the figure. The ball, thrown at an angle of 45 degrees, lands 50 feet down the hill, which is defined by the line $4y + 3x = 0$. Using calculus, it can be shown that the path of the baseball is given by the equation $y = ax^2 + x + c$ for some constants a and c.

(a) Ignoring the height of the boy, find an equation for the path.

(b) What is the maximum height of the ball *off the ground*?

FIGURE FOR EXERCISE 39

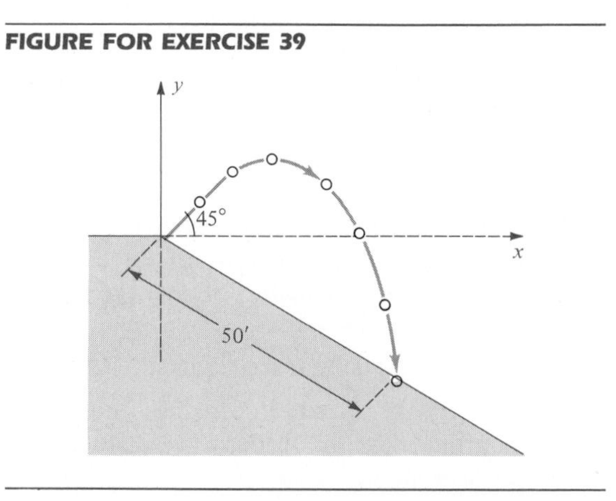

40 A cable television firm presently serves 5000 households and charges \$20 per month. A marketing survey indicates that each decrease of \$1 in the monthly charge will result in 500 new customers. Let $R(x)$ denote the total monthly revenue when the monthly charge is x dollars.

(a) Determine the revenue function R.

(b) Sketch the graph of R and find the value of x that results in maximum monthly revenue.

2.4 Operations on Functions

Functions are often defined in terms of sums, differences, products, and quotients of various expressions. For example, if

$$h(x) = x^2 + \sqrt{5x + 1},$$

we may regard $h(x)$ as a sum of values of the simpler functions f and g defined by

$$f(x) = x^2 \quad \text{and} \quad g(x) = \sqrt{5x + 1}.$$

It is natural to refer to the function h as the *sum* of f and g.

In general, suppose f and g are *any* functions. Let I be *the intersection of their domains;* that is, the numbers *common* to both domains. The **sum** of f and g is the function h defined by

$$h(x) = f(x) + g(x)$$

where x is in I.

It is convenient to denote h by the symbol $f + g$. Since f and g are functions, not numbers, the $+$ used between f and g is not to be considered as addition of real numbers. It is used to indicate that the value of x for $f + g$ is $f(x) + g(x)$; that is,

$$(f + g)(x) = f(x) + g(x).$$

Similarly, the **difference** $f - g$ and the **product** fg of f and g are defined by

$$(f - g)(x) = f(x) - g(x) \quad \text{and} \quad (fg)(x) = f(x)g(x)$$

where x is in I. Finally, the **quotient** f/g of f by g is given by

$$\left(\frac{f}{g}\right)(x) = \frac{f(x)}{g(x)}$$

where x is in I and $g(x) \neq 0$.

EXAMPLE 1 If $f(x) = \sqrt{4 - x^2}$ and $g(x) = 3x + 1$, find the sum, difference, and product of f and g, and the quotient of f by g.

SOLUTION The domain of f is the closed interval $[-2, 2]$ and the domain of g is $\mathbb{R}$. Consequently, the intersection of their domains is

$[-2, 2]$, and the required functions are given by

$$(f+g)(x) = \sqrt{4-x^2} + (3x+1), \qquad -2 \le x \le 2$$
$$(f-g)(x) = \sqrt{4-x^2} - (3x+1), \qquad -2 \le x \le 2$$
$$(fg)(x) = \sqrt{4-x^2}\,(3x+1), \qquad -2 \le x \le 2$$
$$\left(\frac{f}{g}\right)(x) = \frac{\sqrt{4-x^2}}{3x+1}, \qquad -2 \le x \le 2, \quad x \ne -\tfrac{1}{3}.$$ ■

A polynomial function (see page 81) may be thought of as a sum of functions whose values are of the form cx^k, where c is a real number and k is a nonnegative integer. A function f is called **algebraic** if it can be expressed in terms of sums, differences, products, quotients, or roots of polynomial functions. For example, if

$$f(x) = 5x^4 - 2\sqrt[3]{x} + \frac{x(x^2+5)}{\sqrt{x^3+\sqrt{x}}},$$

then f is an algebraic function. Functions that are not algebraic are termed **transcendental.** The exponential, logarithmic, and trigonometric functions considered in Chapters 4 and 5 are examples of transcendental functions.

We shall conclude this section by describing an important method of using two functions f and g to obtain a third function. Suppose D, E, and K are sets of real numbers, and let f be a function from D to E and g a function from E to K. Using arrow notation we have

$$D \xrightarrow{f} E \xrightarrow{g} K$$

We shall use f and g to define a function from D to K.

For every x in D, the number $f(x)$ is in E. Since the domain of g is E, we may then find the number $g(f(x))$ in K. By associating $g(f(x))$ with x, we obtain a function from D to K called the *composite function* of g by f. This is illustrated pictorially in Figure 2.26, where the dashes indicate the correspondence we have defined from D to K.

FIGURE 2.26

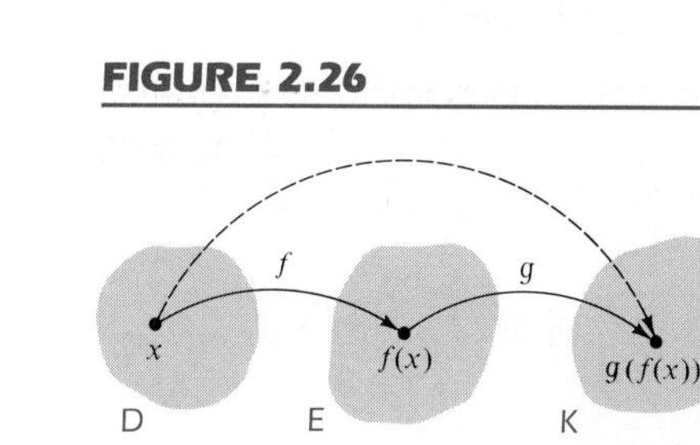

We sometimes use an operational symbol $\circ$ and denote a composite function by $g \circ f$. The following definition summarizes our discussion.

DEFINITION

Let f be a function from D to E and let g be a function from E to K. The **composite function $g \circ f$** is the function from D to K defined by

$$(g \circ f)(x) = g(f(x))$$

for every x in D.

If the domain of g is a *subset* E' of E, then the domain of $g \circ f$ consists of all x in D such that $f(x)$ is in E'. (See Example 3.)

EXAMPLE 2 If $f(x) = x^3$ and $g(x) = 5x^2 + 2x + 1$, find $(g \circ f)(x)$.

SOLUTION By definition,

$$(g \circ f)(x) = g(f(x)) = g(x^3).$$

Since $g(x^3)$ means that x^3 should be substituted for x in the expression for $g(x)$, we have

$$g(x^3) = 5(x^3)^2 + 2(x^3) + 1.$$

Consequently,
$$(g \circ f)(x) = 5x^6 + 2x^3 + 1.$$
■

EXAMPLE 3 If the functions f and g are defined by $f(x) = x - 2$ and $g(x) = 5x + \sqrt{x}$, find $(g \circ f)(x)$ and the domain of $g \circ f$.

SOLUTION Formal substitutions give us the following:

$$\begin{aligned}
(g \circ f)(x) &= g(f(x)) && \text{(definition of } g \circ f\text{)} \\
&= g(x - 2) && \text{(definition of } f\text{)} \\
&= 5(x - 2) + \sqrt{x - 2} && \text{(definition of } g\text{)} \\
&= 5x - 10 + \sqrt{x - 2} && \text{(simplifying)}
\end{aligned}$$

The domain of f is the set of all real numbers; however, the last equality implies that $(g \circ f)(x)$ is a real number only if $x \geq 2$. Thus, the domain of the composite function $g \circ f$ is the interval $[2, \infty)$. ■

Given f and g, it may also be possible to find $(f \circ g)(x) = f(g(x))$ as illustrated in the next example.

EXAMPLE 4 If $f(x) = x^2 - 1$ and $g(x) = 3x + 5$, find $(f \circ g)(x)$ and $(g \circ f)(x)$.

SOLUTION

$$\begin{aligned}
(f \circ g)(x) &= f(g(x)) && \text{(definition of } f \circ g\text{)} \\
&= f(3x + 5) && \text{(definition of } g\text{)} \\
&= (3x + 5)^2 - 1 && \text{(definition of } f\text{)} \\
&= 9x^2 + 30x + 24 && \text{(simplifying)}
\end{aligned}$$

Similarly,

$$\begin{aligned} (g \circ f)(x) &= g(f(x)) && \text{(definition of } g \circ f) \\ &= g(x^2 - 1) && \text{(definition of } f) \\ &= 3(x^2 - 1) + 5 && \text{(definition of } g) \\ &= 3x^2 + 2 && \text{(simplifying)} \end{aligned}$$

■

We see from Example 4 that $f(g(x))$ and $g(f(x))$ are not always the same, that is, $f \circ g \neq g \circ f$. In certain cases it may happen that equality *does* occur. Of major importance is the case where $f(g(x))$ and $g(f(x))$ are not only identical, but both are equal to x. Of course, f and g must be very special functions for this to happen. In the next section we shall indicate the manner in which these functions will be restricted.

In certain applications, it is necessary to express a quantity y as a function of time t. The following example illustrates that it is often easier to introduce a third variable x, then express x as a function of t—that is, $x = g(t)$. Next express y as a function of x,—that is, $y = f(x)$—and finally form the composite function given by $y = f(x) = f(g(t))$.

EXAMPLE 5 A spherical toy balloon is being inflated with helium gas. If the radius of the balloon is changing at a rate of 1.5 cm/sec, express the volume V of the balloon as a function of time t (seconds).

SOLUTION Let x denote the radius of the balloon. If we assume that the radius is 0 initially, then after t seconds

$$x = 1.5t \qquad \text{(radius of balloon after } t \text{ seconds)}.$$

To illustrate, after 1 second the radius is 1.5 cm; after 2 seconds it is 3.0 cm; after 3 seconds it is 4.5 cm; and so on.

Next we write

$$V = \tfrac{4}{3}\pi x^3 \qquad \text{(volume of a sphere of radius } x).$$

This gives us a composite function relationship in which V is a function of x, and x is a function of t. By substitution, we obtain

$$V = \tfrac{4}{3}\pi x^3 = \tfrac{4}{3}\pi(1.5t)^3 = \tfrac{4}{3}\pi(\tfrac{3}{2}t)^3 = \tfrac{4}{3}\pi(\tfrac{27}{8}t^3).$$

Simplifying, we obtain the following formula for V as a function of t:

$$V = \tfrac{9}{2}\pi t^3.$$

■

Exercises 2.4

In Exercises 1–6, find the sum, difference, and product of f and g, and the quotient of f by g.

1 $f(x) = 3x^2,\ g(x) = 1/(2x - 3)$

2 $f(x) = \sqrt{x+3},\ g(x) = \sqrt{x+3}$

3 $f(x) = x + (1/x),\ g(x) = x - (1/x)$

4 $f(x) = x^3 + 3x,\ g(x) = 3x^2 + 1$

5 $f(x) = 2x^3 - x + 5,\ g(x) = x^2 + x + 2$

6 $f(x) = 7x^4 + x^2 - 1,\ g(x) = 7x^4 - x^3 + 4x$

In Exercises 7–24 find $(f \circ g)(x)$ and $(g \circ f)(x)$.

7 $f(x) = 3x + 2,\ g(x) = 2x - 1$

8 $f(x) = x + 4,\ g(x) = 5x - 3$

9 $f(x) = 4x^2 - 5,\ g(x) = 3x$

10 $f(x) = 7x + 1,\ g(x) = 2x^2$

11 $f(x) = 3x^2 + 2x,\ g(x) = 2x - 1$

12 $f(x) = 5x + 7,\ g(x) = 4 - x^2$

13 $f(x) = x - 1,\ g(x) = x^3$

14 $f(x) = x^3 + 5x,\ g(x) = 4x$

15 $f(x) = x^2 + 9x,\ g(x) = \sqrt{x+9}$

16 $f(x) = \sqrt[3]{x^2+1},\ g(x) = x^3 + 1$

17 $f(x) = \dfrac{1}{2x-5},\ g(x) = \dfrac{1}{x^2}$

18 $f(x) = \dfrac{x}{x+1},\ g(x) = x - 1$

19 $f(x) = |x|,\ g(x) = -5$

20 $f(x) = 7,\ g(x) = 10$

21 $f(x) = x^2,\ g(x) = 1/x^2$

22 $f(x) = \dfrac{1}{x+1},\ g(x) = x + 1$

23 $f(x) = 2x - 3,\ g(x) = \dfrac{x+3}{2}$

24 $f(x) = x^3 - 1,\ g(x) = \sqrt[3]{x+1}$

Use the method of Example 5 to solve Exercises 25–30.

25 A fire has started in a dry open field and spreads in the form of a circle. If the radius of this circle increases at the rate of 6 feet per minute, express the total fire area as a function of time t (in minutes).

26 A 100-foot-long cable of diameter 4 inches is submerged in seawater. Due to corrosion, the surface area of the cable decreases at the rate of 750 in^2 per year. Express the diameter of the cable as a function of time. (Ignore corrosion at the ends of the cable.)

27 A hot-air balloon rises vertically as a rope attached to the base of the balloon is released at the rate of 5 feet per second. The pulley that releases the rope is 20 feet from a platform where passengers board the balloon. Express the height of the balloon as a function of time.

FIGURE FOR EXERCISE 27

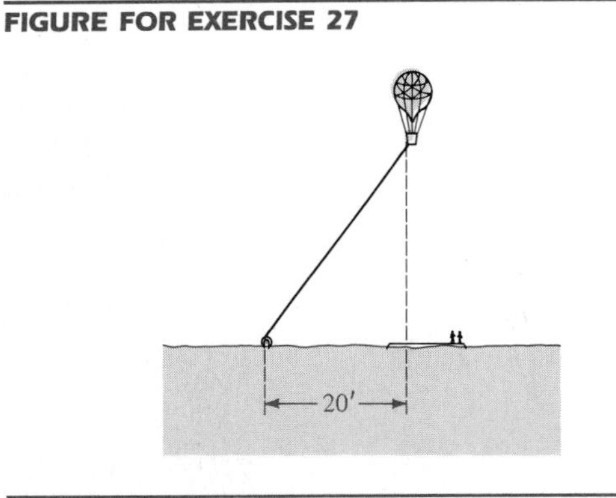

28 The diameter d of a cube is the distance between two opposite vertices. Express d as a function of the edge x of the cube. (*Hint:* First express the diagonal y of a face as a function of x.)

29 Refer to Exercise 44 of Section 2.1. The tightrope walker moves up the rope at a steady rate of 1 foot per second. If the rope is attached 30 feet up the pole, express the height h of the walker above the ground as a function of time t. (*Hint:* Let d denote the total distance traveled along the wire. First express d as a function of t, and then h as a function of d.)

30 Refer to Exercise 47 of Section 2.1. When the airplane is 500 feet down the runway, it has reached and will maintain a speed of 68 feet per second (or about 100 mph) until takeoff. Express the distance of the plane from the control tower as a function of time t (in seconds). (*Hint:* In the figure, first write x as a function of t.)

2.5 Inverse Functions

A function f may have the same value for different numbers in its domain. For example, if $f(x) = x^2$, then $f(2) = 4$ and $f(-2) = 4$, but $2 \neq -2$. To define *the inverse of a function*, it is essential that different numbers in the domain *always* give different values of f. Such functions are called *one-to-one*.

DEFINITION

A function f with domain D and range E is a **one-to-one function** if whenever $a \neq b$ in D, then $f(a) \neq f(b)$ in E.

EXAMPLE 1

(a) If $f(x) = 3x + 2$, prove that f is one-to-one.

(b) If $g(x) = x^4 + 2x^2$ prove that g is not one-to-one.

SOLUTION

(a) If $a \neq b$, then $3a \neq 3b$ and hence $3a + 2 \neq 3b + 2$, or $f(a) \neq f(b)$. Thus, f is one-to-one.

(b) The function g is not one-to-one, since different numbers in the domain may have the same value. For example, although $-1 \neq 1$, both $g(-1)$ and $g(1)$ are equal to 3. ■

THEOREM

(i) If f is an increasing function throughout its domain, then f is one-to-one.

(ii) If f is a decreasing function throughout its domain, then f is one-to-one.

PROOF Suppose f is an increasing function. If a and b are in the domain of f and $a \neq b$, then either $a < b$ or $b < a$, and hence either

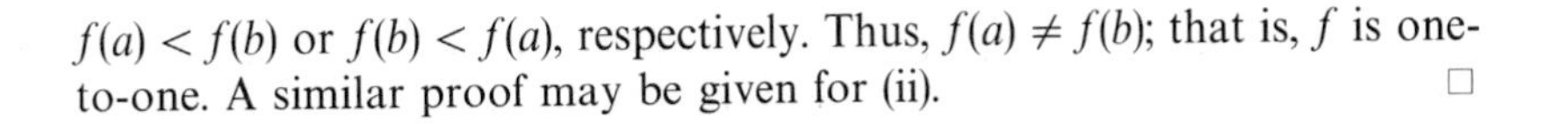

$f(a) < f(b)$ or $f(b) < f(a)$, respectively. Thus, $f(a) \neq f(b)$; that is, f is one-to-one. A similar proof may be given for (ii). □

FIGURE 2.27

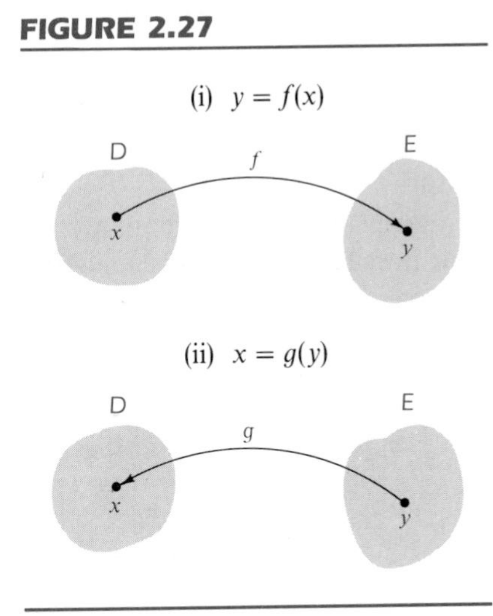

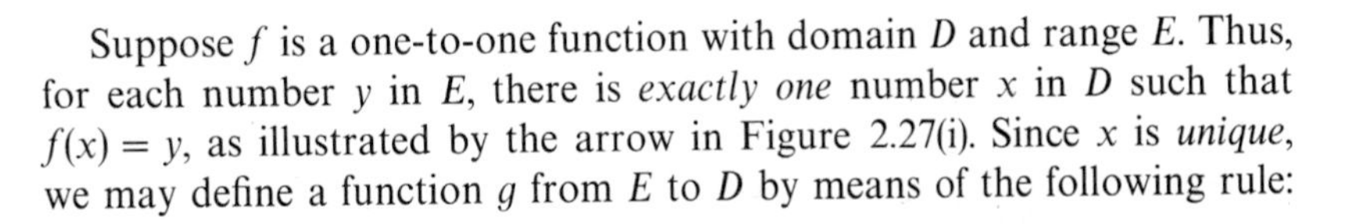

Suppose f is a one-to-one function with domain D and range E. Thus, for each number y in E, there is *exactly one* number x in D such that $f(x) = y$, as illustrated by the arrow in Figure 2.27(i). Since x is *unique*, we may define a function g from E to D by means of the following rule:

$$g(y) = x.$$

(See Figure 2.27(ii)). In words, *g reverses the correspondence* given by f. Since

$$g(y) = x \quad \text{and} \quad f(x) = y$$

for x in D and y in E, we see, by substitution, that

$$g(f(x)) = x \quad \text{and} \quad f(g(y)) = y.$$

The last two equations may be written

$$(g \circ f)(x) = x \quad \text{and} \quad (f \circ g)(y) = y,$$

which means that $g \circ f$ and $f \circ g$ are the identity functions on D and E, respectively (see page 66); because of this, we call g the *inverse function* of f and denote it by f^{-1}, as indicated in the next definition.

DEFINITION

> Let f be a one-to-one function with domain D and range E. A function f^{-1} with domain E and range D is called the **inverse function of f** if
>
> $$f^{-1}(f(x)) = x \quad \text{for every } x \text{ in } D$$
>
> and $$f(f^{-1}(y)) = y \quad \text{for every } y \text{ in } E.$$

The -1 used in the notation f^{-1} should not be mistaken for an exponent; that is, $f^{-1}(y)$ *does not mean* $1/[f(y)]$. The reciprocal $1/[f(y)]$ may be denoted by $[f(y)]^{-1}$.

It is important to note that *a function f that either increases or decreases throughout its domain has an inverse function,* since by the previous theorem such functions are one-to-one.

When we considered functions in previous sections we usually let x denote an arbitrary number in the domain. Similarly, for the inverse function f^{-1}, we may wish to consider $f^{-1}(x)$, *where x is in the domain E of* f^{-1}. In this event, the two formulas in our definition are written

$$f^{-1}(f(x)) = x \quad \text{for every } x \text{ in } D$$

and

$$f(f^{-1}(x)) = x \quad \text{for every } x \text{ in } E.$$

The diagrams in Figure 2.27 contain a hint for finding the inverse of a one-to-one function in certain cases. If possible, we *solve the equation* $y = f(x)$ *for x in terms of y*, obtaining an equation of the form $x = g(y)$. If the two conditions $g(f(x)) = x$ and $f(g(x)) = x$ are true for all x in the domains of f and g, respectively, then g is the required inverse function f^{-1}. The following three guidelines summarize this procedure, where in guideline 2, in anticipation of finding f^{-1}, we write $x = f^{-1}(y)$ instead of $x = g(y)$.

GUIDELINES: Finding f^{-1} in Simple Cases

1. Verify that f is a one-to-one function (or that f is increasing or decreasing) throughout its domain.
2. Solve the equation $y = f(x)$ for x in terms of y, obtaining an equation of the form $x = f^{-1}(y)$.
3. Verify the two conditions

$$f^{-1}(f(x)) = x \quad \text{and} \quad f(f^{-1}(x)) = x$$

for all x in the domains of f and f^{-1}, respectively. ■■■

The success of this method depends on the nature of the equation $y = f(x)$, since we must be able to solve for x in terms of y. That is what we mean by "simple cases" in the heading of the guidelines.

EXAMPLE 2 If $f(x) = 3x - 5$, find the inverse function of f.

SOLUTION We shall follow the three guidelines. First, we note that the graph of the linear function f is a line of slope 3, and hence f is increasing throughout $\mathbb{R}$. Thus the inverse function f^{-1} exists. Moreover, since the domain and range of f is $\mathbb{R}$, the same is true for f^{-1}.

As in guideline 2, we consider

$$y = 3x - 5$$

and then solve for x in terms of y, obtaining

$$x = \frac{y + 5}{3}.$$

We now formally let

$$f^{-1}(y) = \frac{y + 5}{3}.$$

Since the symbol used for the variable is immaterial, we may also write

$$f^{-1}(x) = \frac{x + 5}{3}.$$

Finally, we verify that the two conditions

$$f^{-1}(f(x)) = x \quad \text{and} \quad f(f^{-1}(x)) = x$$

are fulfilled. Thus,

$$\begin{aligned} f^{-1}(f(x)) &= f^{-1}(3x - 5) && \text{(definition of } f\text{)} \\ &= \frac{(3x - 5) + 5}{3} && \text{(definition of } f^{-1}\text{)} \\ &= x && \text{(simplifying)} \end{aligned}$$

Also,

$$\begin{aligned} f(f^{-1}(x)) &= f\left(\frac{x + 5}{3}\right) && \text{(definition of } f^{-1}\text{)} \\ &= 3\left(\frac{x + 5}{3}\right) - 5 && \text{(definition of } f\text{)} \\ &= x && \text{(simplifying)} \end{aligned}$$

This proves that the inverse function of f is given by

$$f^{-1}(x) = \frac{x + 5}{3}.$$

■

FIGURE 2.28

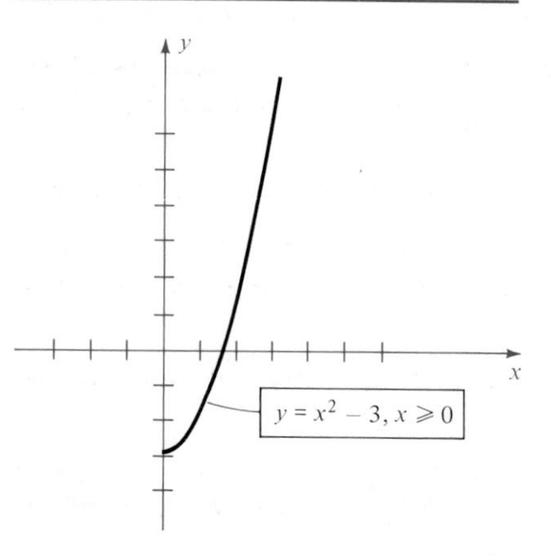

EXAMPLE 3 Find the inverse function of f if $f(x) = x^2 - 3$ and $x \geq 0$.

SOLUTION The graph of f is sketched in Figure 2.28. The domain D is $[0, \infty)$, and the range E is $[-3, \infty)$. Since f is increasing on D, it has an inverse function f^{-1} that has domain E and range D.

As in guideline 2, we consider the equation

$$y = x^2 - 3.$$

and solve for x, obtaining

$$x = \pm\sqrt{y + 3}.$$

Since x is nonnegative, we reject $x = -\sqrt{y + 3}$ and let

$$f^{-1}(y) = \sqrt{y + 3} \quad \text{or, equivalently,} \quad f^{-1}(x) = \sqrt{x + 3}.$$

Finally, we verify that $f^{-1}(f(x)) = x$ for x in $D = [0, \infty)$, and that $f(f^{-1}(x)) = x$ for x in $E = [-3, \infty)$. Thus,

$$f^{-1}(f(x)) = f^{-1}(x^2 - 3) = \sqrt{(x^2 - 3) + 3} = \sqrt{x^2} = x \quad \text{if } x \geq 0,$$

and

$$f(f^{-1}(x)) = f(\sqrt{x + 3}) = (\sqrt{x + 3})^2 - 3 = (x + 3) - 3 = x \quad \text{if } x \geq -3.$$

This proves that the inverse function is given by

$$f^{-1}(x) = \sqrt{x + 3} \quad \text{for } x \geq -3.$$

■

FIGURE 2.29

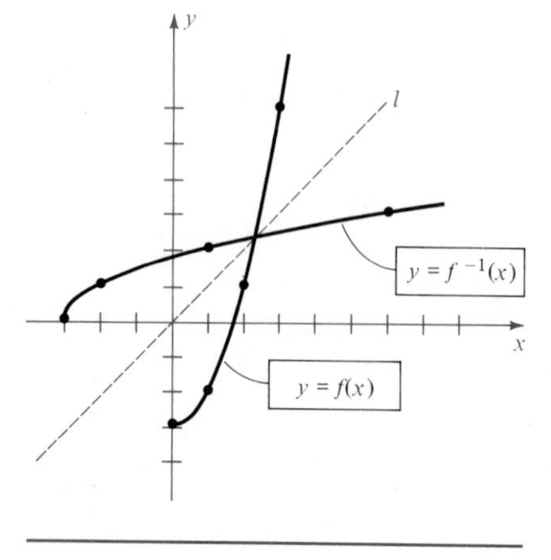

An interesting relationship exists between the graphs of a function f and its inverse function f^{-1}. We first note that $b = f(a)$ means the same thing as $a = f^{-1}(b)$. These equations imply that the point (a, b) is on the graph of f if and only if the point (b, a) is on the graph of f^{-1}. As an illustration, in Example 3 we found that the functions f and f^{-1} given by

$$f(x) = x^2 - 3 \quad \text{and} \quad f^{-1}(x) = \sqrt{x + 3}$$

are inverse functions of one another, provided that x is suitably restricted. Some points on the graph of f are $(0, -3)$, $(1, -2)$, $(2, 1)$, and $(3, 6)$. Corresponding points on the graph of f^{-1} are $(-3, 0)$, $(-2, 1)$, $(1, 2)$, and $(6, 3)$. The graphs of f and f^{-1} are sketched on the same coordinate axes in Figure 2.29. If the page is folded along the line l that bisects quadrants I and III (as indicated by the dashes in the figure), then the graphs of f and f^{-1} coincide. Note that an equation for l is $y = x$. The two graphs are said to be *reflections* of one another through the line l (or *symmetric* with respect to l). This is typical of the graph of every function f that has an inverse function f^{-1}. (See Exercise 34.)

Exercises 2.5

In Exercises 1–12 determine whether the function f is one-to-one.

1 $f(x) = 2x + 9$

2 $f(x) = 1/(7x + 9)$

3 $f(x) = 5 - 3x^2$

4 $f(x) = 2x^2 - x - 3$

5 $f(x) = \sqrt{x}$

6 $f(x) = x^3$

7 $f(x) = |x|$

8 $f(x) = 4$

9 $f(x) = x^2 - 3x + 2$

10 $f(x) = \sqrt[3]{x}$

11 $f(x) = 1/x$

12 $f(x) = \sqrt{9 - x^2}$

In Exercises 13–16 prove that f and g are inverse functions of one another and sketch the graphs of f and g on the same coordinate plane.

13 $f(x) = 7x + 5;\quad g(x) = (x - 5)/7$

14 $f(x) = x^2 - 1,\ x \geq 0;\quad g(x) = \sqrt{x + 1},\ x \geq -1$

15 $f(x) = \sqrt{2x - 4},\ x \geq 2;\quad g(x) = \frac{1}{2}(x^2 + 4),\ x \geq 0$

16 $f(x) = x^3 + 1;\quad g(x) = \sqrt[3]{x - 1}$

In Exercises 17–30 find the inverse function of f.

17 $f(x) = 4x - 3$

18 $f(x) = 9 - 7x$

19 $f(x) = \dfrac{1}{2x + 5},\ x > -\frac{5}{2}$

20 $f(x) = \dfrac{1}{3x - 1},\ x > \frac{1}{3}$

21 $f(x) = 9 - x^2,\ x \geq 0$

22 $f(x) = 4x^2 + 1,\ x \geq 0$

23 $f(x) = 5x^3 - 2$

24 $f(x) = 7 - 2x^3$

25 $f(x) = \sqrt{3x - 5},\ x \geq \frac{5}{3}$

26 $f(x) = \sqrt{4 - x^2},\ 0 \leq x \leq 2$

27 $f(x) = \sqrt[3]{x} + 8$

28 $f(x) = (x^3 + 1)^5$

29 $f(x) = x$

30 $f(x) = -x$

31 (a) Prove that the linear function defined by $f(x) = ax + b$ for $a \neq 0$ has an inverse function, and find $f^{-1}(x)$.

(b) Does a constant function have an inverse? Explain.

32 If f is a one-to-one function with domain D and range E, prove that f^{-1} is a one-to-one function with domain E and range D.

33 Prove that a one-to-one function has only one inverse function.

34 Establish the fact that the graph of f^{-1} is the reflection of the graph of f through the line $y = x$ by verifying each of the following:

(a) If $P(a, b)$ is on the graph of f, then $Q(b, a)$ is on the graph of f^{-1}.

(b) The midpoint of line segment PQ is on the line $y = x$.

(c) The line PQ is perpendicular to the line $y = x$.

The graphs of one-to-one functions are shown in Exercises 35–38. Use the reflection property to sketch the graph of f^{-1}. Determine the domain and range of (a) f; (b) f^{-1}.

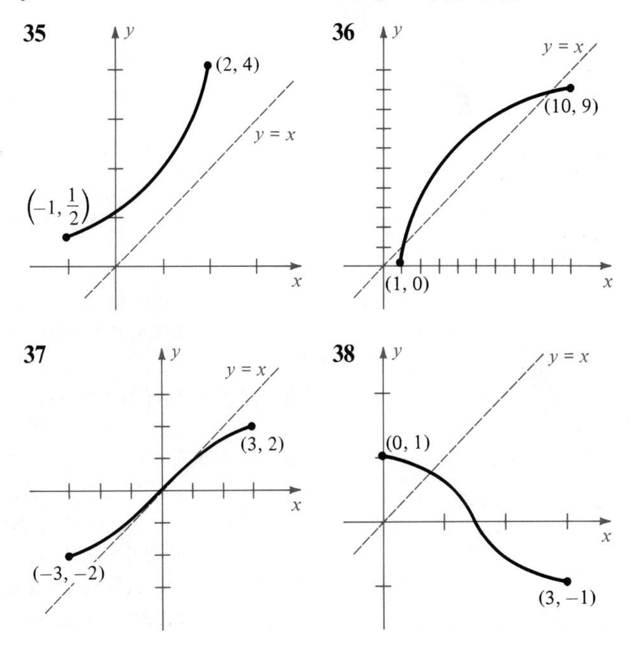

2.6 Review

Define or discuss each of the following.

1 Function
2 Domain and range of a function
3 Graph of a function
4 Linear function
5 Identity function
6 Constant function
7 Even function
8 Odd function
9 Increasing function
10 Decreasing function
11 Vertical shifts of graphs
12 Horizontal shifts of graphs
13 Stretching of graphs
14 Reflections of graphs
15 Piecewise-defined function
16 Greatest integer function
17 Polynomial function
18 Quadratic function
19 Graph of a quadratic function
20 Maximum or minimum values of quadratic functions
21 Operations on functions
22 Algebraic function
23 Transcendental function
24 Composite function of two functions
25 One-to-one function
26 Inverse function

Exercises 2.6

1 If $f(x) = x/\sqrt{x+3}$ find (a)–(g):

(a) $f(1)$ (b) $f(-1)$ (c) $f(0)$
(d) $f(-x)$ (e) $-f(x)$ (f) $f(x^2)$
(g) $(f(x))^2$

2 Find the domain and range of f if:

(a) $f(x) = \sqrt{3x-4}$ (b) $f(x) = 1/(x+3)^2$

In Exercises 3–10 sketch the graph of f. Find the domain, the range, and the intervals in which f is increasing or decreasing.

3 $f(x) = \dfrac{1-3x}{2}$

4 $f(x) = |x+3|$

5 $f(x) = 1 - \sqrt{x+1}$

6 $f(x) = \sqrt{2-x}$

7 $f(x) = 1000$

8 $f(x) = 9 - x^2$

9 $f(x) = x^2 + 6x + 16$

10 $f(x) = \begin{cases} x^2 & \text{if } x < 0 \\ 3x & \text{if } 0 \le x < 2 \\ 6 & \text{if } x \ge 2 \end{cases}$

11 Sketch the graphs of the following equations, making use of shifting, stretching, or reflecting.

(a) $y = \sqrt{x}$ (b) $y = \sqrt{x+4}$
(c) $y = \sqrt{x} + 4$ (d) $y = 4\sqrt{x}$
(e) $y = \frac{1}{4}\sqrt{x}$ (f) $y = -\sqrt{x}$

12 Determine whether f is even, odd, or neither even nor odd:

(a) $f(x) = \sqrt[3]{x^3 + 4x}$ (b) $f(x) = \sqrt[3]{3x^2 - x^3}$
(c) $f(x) = \sqrt[3]{x^4 + 3x^2 + 5}$

In Exercises 13 and 14 find the maximum or minimum value of $f(x)$ by completing the square. Sketch the graph of f.

13 $f(x) = 5x^2 + 30x + 49$

14 $f(x) = -3x^2 + 30x - 82$

In Exercises 15 and 16 find $(f \circ g)(x)$ and $(g \circ f)(x)$.

15 $f(x) = 2x^2 - 5x + 1, \quad g(x) = 3x + 2$

16 $f(x) = \sqrt{3x + 2}$, $g(x) = 1/x^2$

In Exercises 17 and 18 find $f^{-1}(x)$ and sketch the graphs of f and f^{-1} on the same coordinate plane.

17 $f(x) = 10 - 15x$

18 $f(x) = 9 - 2x^2$, $x \leq 0$

19 If the altitude and radius of a right circular cylinder are equal, express the volume V as a function of the circumference C of the base.

20 A company plans to manufacture a container having the shape of a right circular cylinder, open at the top, and having a capacity of 24π in^3. If the cost of the material for the bottom is 30 cents per in^2 and that for the curved sides is 10 cents per in^2, express the total cost C of the material as a function of the radius r of the base of the container.

21 An open rectangular storage shelter consisting of two vertical sides, 4 feet wide and a flat roof is to be attached to an existing structure as illustrated in the figure. The flat roof is made of tin and costs \$5 per square foot while the other two sides are made of plywood costing \$2 per square foot.

(a) If \$400 is available for construction, express the length y as a function of the height x.

(b) Express the volume V inside the shelter as a function of x.

(c) Find the design that maximizes the space inside the shelter.

FIGURE FOR EXERCISE 21

22 Water in a paper conical filter drips into a cup as shown in the figure. Suppose 5 in^3 of water is poured into the filter. Let x denote the height of the water in the filter and let y denote the height of the water in the cup.

(a) Express the radius r (see figure) as a function of x. (*Hint:* Use similar triangles.)

(b) Express the height y of the water in the cup as a function of x. (*Hint:* What is the sum of the two volumes shown in the figure?)

FIGURE FOR EXERCISE 22

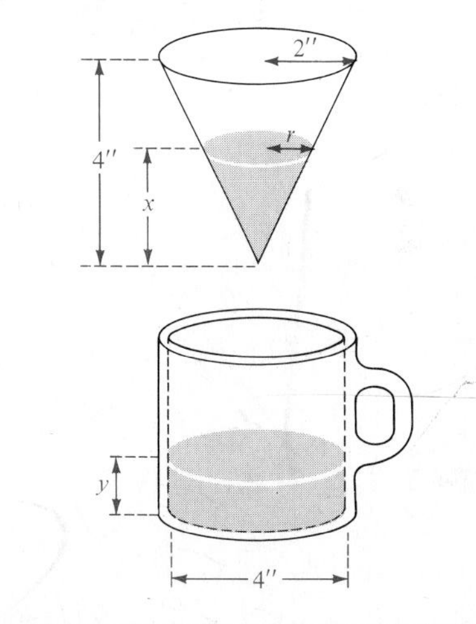

23 A cross section of a rectangular pool of dimensions 80 feet by 40 feet is shown in the figure. The pool is being filled with water at the rate of 10 ft^3 per minute.

(a) Express the volume V as a function of depth h at the deep end for $0 \leq h \leq 6$ and then for $6 \leq h \leq 9$.

(b) Express V as a function of time t. Then express h as a function of t.

FIGURE FOR EXERCISE 23

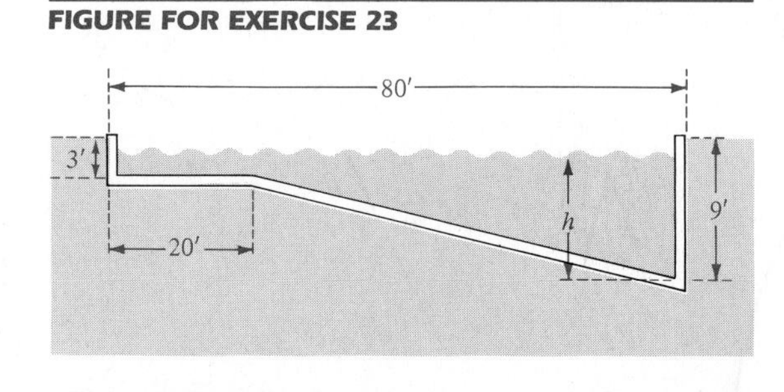

24 An automobile presently gets 20 miles per gallon but is in need of a tune-up that will cost \$50. The tune-up will improve gasoline mileage by 10%.

(a) If gasoline costs \$1.25 per gallon, find a linear function that gives the cost C of driving x miles without the tune-up.

(b) Find a linear function that gives the cost C of driving x miles with the tune-up.

(c) How many miles must the automobile be driven before the tune-up saves money?

25 The interior of a half-mile race track consists of a rectangle with semicircles at two opposite ends. Find the dimensions that will maximize the area of the rectangle.

26 At 1:00 P.M. ship A is 30 miles due south of ship B and is sailing north at a rate of 15 mph. If ship B is sailing west at a rate of 10 mph, what is the time at which the distance between the ships is minimal (see figure)?

FIGURE FOR EXERCISE 26

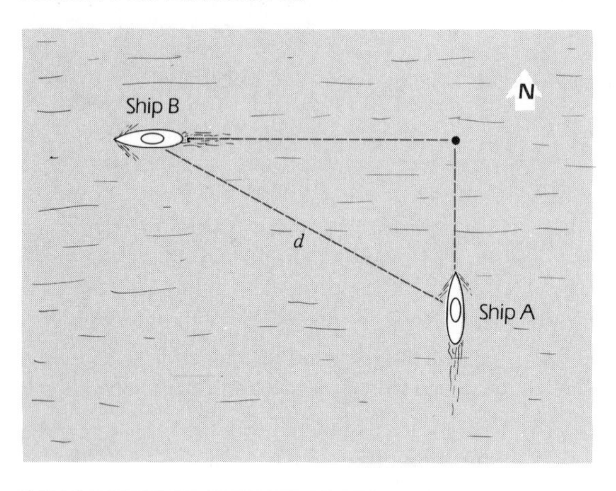

27 A rocket is fired up a hillside, following a path given by $y = -0.016x^2 + 1.6x$. The hillside has slope $\frac{1}{5}$ as illustrated in the figure.

(a) Where does the rocket land?

(b) Find the maximum height of the rocket *above the ground*.

FIGURE FOR EXERCISE 27

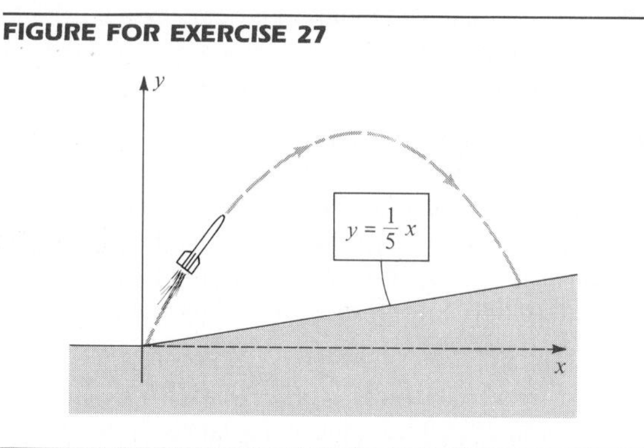

28 When a particular basketball player leaps straight up for a dunk, his distance $f(t)$ (in feet) off the ground after t seconds is given by

$$f(t) = -\tfrac{1}{2}gt^2 + 16t.$$

(a) If $g = 32$, what is the player's *hang time;* that is, what is the total number of seconds that the player is in the air?

(b) Find the player's *vertical leap*; that is, find the maximum distance of his feet from the floor.

(c) On the moon $g = \frac{32}{6}$. Rework parts (a) and (b) for a player on the moon.

CHAPTER 3

Polynomial and Rational Functions

Polynomial functions are the most basic functions in algebra. ■ Techniques for sketching their graphs are discussed in the first section. ■ Then we turn our attention to division and study methods for finding zeros of polynomial functions. ■ Finally, we consider quotients of polynomial functions, that is, *rational functions.*

3.1 Graphs of Polynomial Functions of Degree Greater Than 2

Let f be a polynomial function of degree n; that is,

$$f(x) = a_n x^n + a_{n-1}x^{n-1} + \cdots + a_1 x + a_0$$

for some $a_n \neq 0$. The domain of f is $\mathbb{R}$. If the degree is odd, then the range of f is also $\mathbb{R}$; however, if the degree is even, then the range is an infinite interval of the form $(-\infty, a]$ or $[a, \infty)$. These facts are illustrated by the graphs in this section.

Recall that if $f(c) = 0$, then c is a **zero** of f, or of $f(x)$. We also call c a **solution,** or **root,** of the equation $f(x) = 0$. The zeros of f are the x-intercepts of the graph of f.

If a polynomial function has degree 0, then $f(x) = a$ for some nonzero real number a, and the graph is a horizontal line. Graphs of polynomial functions of degree 1 (linear functions) are lines. Polynomial functions of degree 2 (quadratic functions) have parabolas for their graphs. In this section we shall study graphs of polynomial functions of degree greater than 2.

If f has degree n, and all the coefficients except a_n are zero, then

$$f(x) = ax^n \quad \text{for some } a = a_n \neq 0.$$

In this case, if $n = 1$, the graph of f is a line that passes through the origin, whereas if $n = 2$, the graph is a parabola with vertex at the origin. Several illustrations with $n = 3$ are given in the following example.

EXAMPLE 1 Sketch the graph of f if (a) $f(x) = \frac{1}{2}x^3$; (b) $f(x) = -\frac{1}{2}x^3$.

SOLUTION

(a) The following table lists several points on the graph of $y = \frac{1}{2}x^3$.

x	0	$\frac{1}{2}$	1	$\frac{3}{2}$	2	$\frac{5}{2}$
y	0	$\frac{1}{16} \approx 0.06$	$\frac{1}{2}$	$\frac{27}{16} \approx 1.7$	4	$\frac{125}{16} \approx 7.8$

Since f is an odd function, the graph of f is symmetric with respect to the origin (see Section 2.1), and hence the points $(-\frac{1}{2}, -\frac{1}{16})$, $(-1, -\frac{1}{2})$, and so on are also on the graph. The graph is sketched in Figure 3.1(i).

FIGURE 3.1

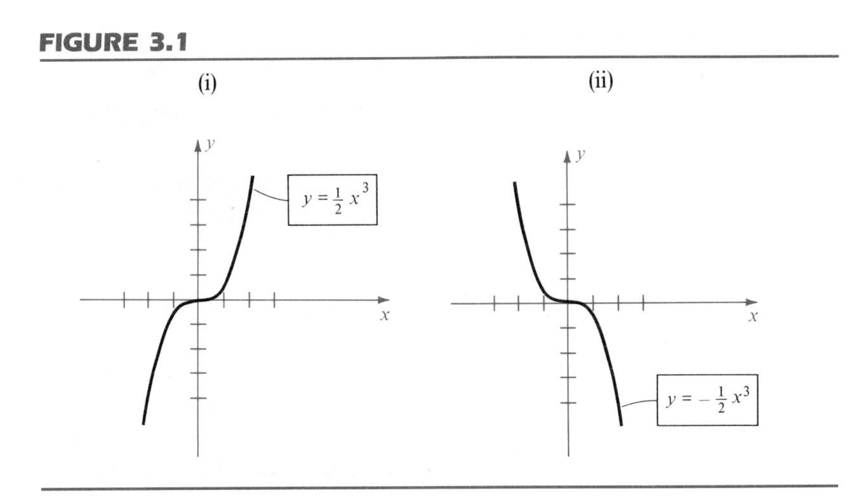

(b) If $y = -\frac{1}{2}x^3$, the graph can be obtained from that in part (a) by multiplying all y-coordinates by -1. Reflecting the graph in part (a) through the x-axis, we obtain the sketch shown in Figure 3.1(ii). ■

In general, if $f(x) = ax^3$, then increasing the absolute value of the coefficient a results in a graph that rises or falls more sharply. For example, if $f(x) = 10x^3$, then $f(1) = 10$, $f(2) = 80$, and $f(-2) = -80$. The effect of using a larger exponent and holding a fixed as, for example, in $f(x) = \frac{1}{2}x^5$, also leads to a graph that rises or falls more rapidly if $|x| > 1$.

If $f(x) = ax^n$ and n is an *even* integer, then the graph of f is symmetric with respect to the y-axis as illustrated in Figure 3.2 for the case $|a| = 1$. Note that as the exponent increases, the graph becomes flatter at the origin. It also rises (or falls) more rapidly if we let $|x|$ increase through values greater than 1.

FIGURE 3.2

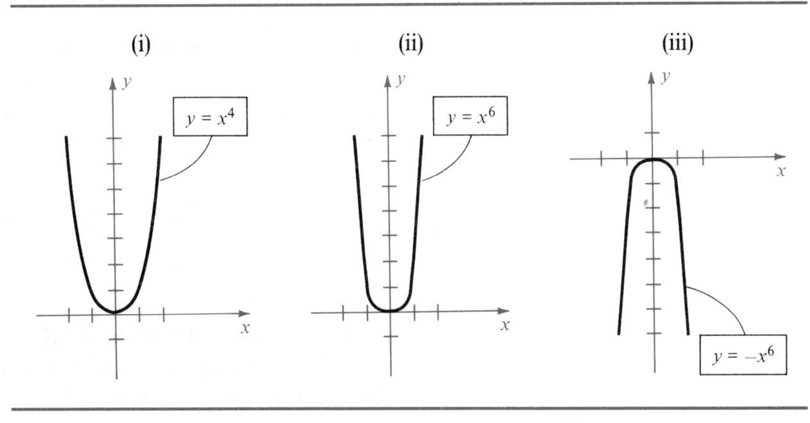

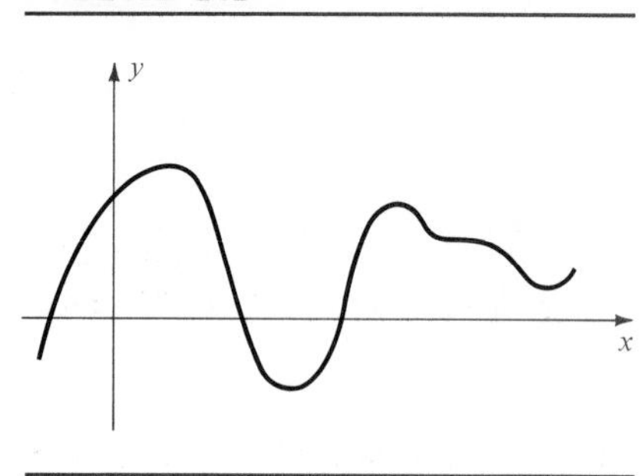

A complete analysis of graphs of polynomial functions of degree greater than 2 requires methods that are used in calculus. As the degree increases, the graphs usually become more complicated. However, they always have a smooth appearance with a number of *peaks* (high points) and *valleys* (low points) as illustrated in Figure 3.3. The points at which peaks and valleys occur are sometimes called **turning points** for the graph of f. At a turning point, f changes from an increasing function to a decreasing function, or vice versa.

A crude method for obtaining a rough sketch of the graph of a polynomial function is to plot many points and then fit a curve to the resulting configuration; however, this is usually an extremely tedious procedure. This method is based on a property of polynomial functions called **continuity.** Continuity, which is studied extensively in calculus, implies that a small change in x produces a small change in $f(x)$. The next theorem specifies another important property of polynomial functions. The proof requires advanced mathematical methods.

INTERMEDIATE VALUE THEOREM FOR POLYNOMIAL FUNCTIONS

> If f is a polynomial function and $f(a) \neq f(b)$ for $a < b$, then f takes on every value between $f(a)$ and $f(b)$ in the interval $[a, b]$.

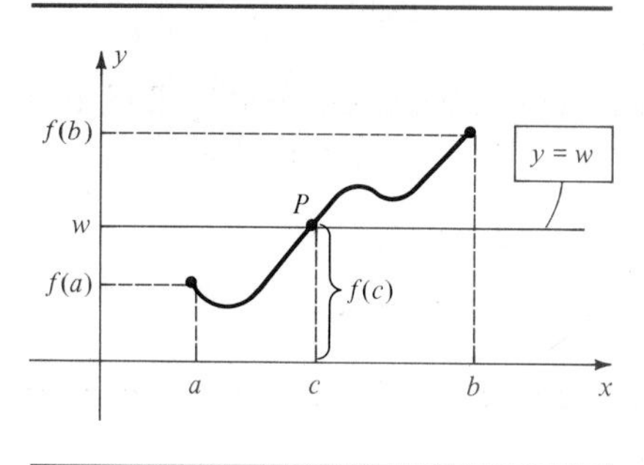

The Intermediate Value Theorem states that if w is any number between $f(a)$ and $f(b)$, then there is a number c between a and b such that $f(c) = w$. If the graph of the polynomial function f is regarded as extending continuously from the point $(a, f(a))$ to the point $(b, f(b))$, as illustrated in Figure 3.4, then for any number w between $f(a)$ and $f(b)$ it appears that a horizontal line with y-intercept w should intersect the graph in at least one point P. The x-coordinate c of P is a number such that $f(c) = w$.

A corollary to the Intermediate Value Theorem is that if $f(a)$ and $f(b)$ have opposite signs, then there is at least one number c between a and b such that $f(c) = 0$; that is, f has a zero at c. This implies that if the point $(a, f(a))$ on the graph of a polynomial function lies below the x-axis, and the point $(b, f(b))$ lies above the x-axis, or vice versa, then the graph crosses the x-axis at least once between the points $(a, 0)$ and $(b, 0)$. A by-product of this fact is that if c and d are *successive* zeros of $f(x)$, that is, there are no other zeros between c and d, then $f(x)$ *does not change sign on the interval* (c, d). Thus, if we choose any number k such that $c < k < d$, and if $f(k)$ is positive, then $f(x)$ is positive throughout (c, d). Similarly, if $f(k)$ is negative, then $f(x)$ is negative throughout (c, d). We shall call $f(k)$ a **test value** for $f(x)$ on the interval (c, d). Test values may also be used on infinite intervals of the form $(-\infty, a)$ or (a, ∞), provided that $f(x)$ has no zeros on these intervals. The use of test values in graphing is similar to that used for inequalities in Section 1.3.

EXAMPLE 2 If $f(x) = x^3 + x^2 - 4x - 4$, determine all values of x such that $f(x) > 0$, and all values of x such that $f(x) < 0$. Use this information to help sketch the graph of f.

SOLUTION The graph of f lies above the x-axis for values of x such that $f(x) > 0$, whereas the graph lies below the x-axis if $f(x) < 0$. We may factor $f(x)$ by grouping terms as follows:

$$\begin{aligned} f(x) &= (x^3 + x^2) - (4x + 4) \\ &= x^2(x + 1) - 4(x + 1) \\ &= (x^2 - 4)(x + 1) \\ &= (x + 2)(x - 2)(x + 1) \end{aligned}$$

FIGURE 3.5

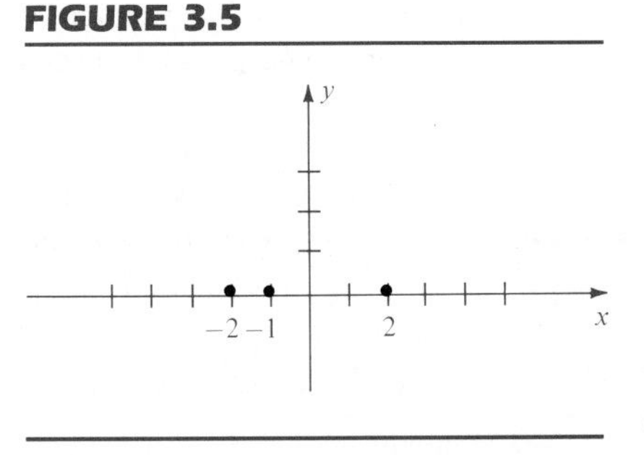

We see from the last equation that the zeros of $f(x)$ (the x-intercepts of the graph) are -2, -1, and 2. The corresponding points on the graph (see Figure 3.5) divide the x-axis into four parts, and we consider the open intervals

$$(-\infty, -2), \quad (-2, -1), \quad (-1, 2), \quad \text{and} \quad (2, \infty).$$

The sign of $f(x)$ in each of these intervals can be determined by finding a suitable test value. Thus, if we choose -3 in $(-\infty, -2)$, then

$$\begin{aligned} f(-3) &= (-3)^3 + (-3)^2 - 4(-3) - 4 \\ &= -27 + 9 + 12 - 4 = -10. \end{aligned}$$

Since the test value $f(-3) = -10$ is negative, $f(x)$ is negative throughout the interval $(-\infty, -2)$.

If we choose $-\frac{3}{2}$ in the interval $(-2, -1)$, then the corresponding test value is

$$\begin{aligned} f(-\tfrac{3}{2}) &= (-\tfrac{3}{2})^3 + (-\tfrac{3}{2})^2 - 4(-\tfrac{3}{2}) - 4 \\ &= -\tfrac{27}{8} + \tfrac{9}{4} + 6 - 4 = \tfrac{7}{8}. \end{aligned}$$

Since $f(-\frac{3}{2}) = \frac{7}{8}$ is positive, $f(x)$ is positive throughout $(-2, -1)$. The following table summarizes these facts and lists suitable test values for the remaining two intervals.

FIGURE 3.6

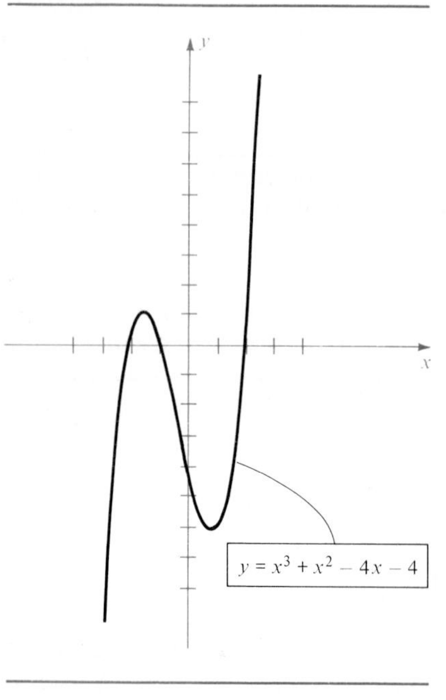

Interval	$(-\infty, -2)$	$(-2, -1)$	$(-1, 2)$	$(2, \infty)$
Test value	$f(-3) = -10$	$f(-\frac{3}{2}) = \frac{7}{8}$	$f(0) = -4$	$f(3) = 20$
Sign of $f(x)$	$-$	$+$	$-$	$+$
Position of graph	Below x-axis	Above x-axis	Below x-axis	Above x-axis

Using the information from the table and plotting several points gives us the graph in Figure 3.6. To find the turning points of the graph it is necessary to use methods developed in calculus. ■

The graph of every polynomial function of degree 3 has an S-shaped appearance similar to that shown in Figure 3.6, or it has an inverted version of that graph if the coefficient of x^3 is negative. However, sometimes the graph may have only one x-intercept or the S shape may be elongated, as in Figure 3.1.

EXAMPLE 3 If $f(x) = x^4 - 4x^3 + 3x^2$, determine all values of x such that $f(x) > 0$, and all x such that $f(x) < 0$. Sketch the graph of f.

SOLUTION We shall follow the same steps used in the solution of Example 2. Thus, we begin by factoring $f(x)$:

$$\begin{aligned} f(x) &= x^2(x^2 - 4x + 3) \\ &= x^2(x - 1)(x - 3). \end{aligned}$$

The zeros of $f(x)$, that is, the x-intercepts of the graph are, in *increasing* order,

$$0, \quad 1, \quad \text{and} \quad 3.$$

The corresponding points on the graph divide the x-axis into four parts, and we consider the open intervals

$$(-\infty, 0), \quad (0, 1), \quad (1, 3), \quad (3, \infty).$$

We next determine the sign of $f(x)$ in each interval by using suitable test values. The following table summarizes results. You should check each entry.

FIGURE 3.7

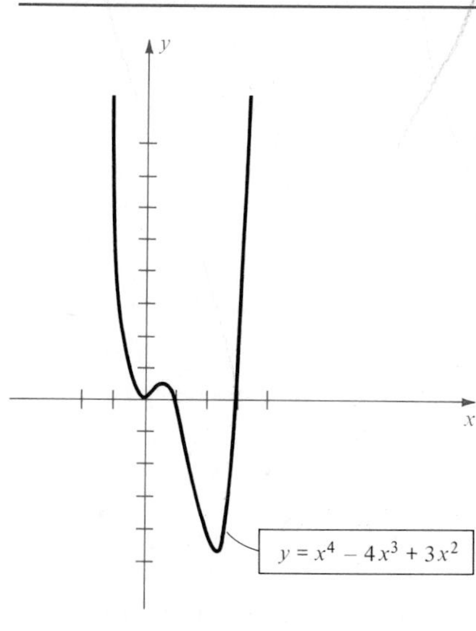

Interval	$(-\infty, 0)$	$(0, 1)$	$(1, 3)$	$(3, \infty)$
Test value	$f(-1) = 8$	$f(\frac{1}{2}) = \frac{5}{16}$	$f(2) = -4$	$f(4) = 48$
Sign of $f(x)$	+	+	−	+
Position of graph	Above x-axis	Above x-axis	Below x-axis	Above x-axis

Making use of the information in the table and plotting several points gives us the sketch in Figure 3.7. ■

EXAMPLE 4 Show that $f(x) = x^5 + 2x^4 - 6x^3 + 2x - 3$ has a zero between 1 and 2.

SOLUTION Substitution for x gives us

$$f(1) = 1 + 2 - 6 + 2 - 3 = -4$$
$$f(2) = 32 + 32 - 48 + 4 - 3 = 17.$$

Since $f(1)$ and $f(2)$ have opposite signs, it follows from the Intermediate Value Theorem that $f(c) = 0$ for some real number c between 1 and 2. ■

The preceding example illustrates a scheme for locating zeros of polynomials. By using a method of *successive approximation*, each zero can be approximated to any degree of accuracy (see, for example, Calculator Exercises 1–10).

Exercises 3.1

1 If $f(x) = ax^3 + 2$, sketch the graph of f for each value of a.

(a) $a = 2$ (b) $a = 4$
(c) $a = \frac{1}{4}$ (d) $a = -2$

2 If $f(x) = 2x^3 + c$, sketch the graph of f for each value of c.

(a) $c = 2$ (b) $c = 4$
(c) $c = \frac{1}{4}$ (d) $c = -2$

In Exercises 3–18 determine all x such that $f(x) > 0$, and all x such that $f(x) < 0$. Sketch the graph of f.

3 $f(x) = \frac{1}{2}x^3 - 4$ **4** $f(x) = -\frac{1}{4}x^3 - 16$

5 $f(x) = \frac{1}{8}x^4 + 2$ **6** $f(x) = 1 - x^5$

7 $f(x) = x^3 - 9x$ **8** $f(x) = 16x - x^3$

9 $f(x) = -x^3 - x^2 + 2x$

10 $f(x) = x^3 + x^2 - 12x$

11 $f(x) = (x + 4)(x - 1)(x - 5)$

12 $f(x) = (x + 2)(x - 3)(x - 4)$

13 $f(x) = x^4 - 16$ **14** $f(x) = 16 - x^4$

15 $f(x) = -x^4 - 3x^2 + 4$

16 $f(x) = x^4 - 7x^2 - 18$

17 $f(x) = x(x-2)(x+1)(x+3)$

18 $f(x) = x(x+1)^2(x-3)(x-5)$

19 If $f(x)$ is a polynomial, and if the coefficients of all odd powers of x are 0, show that f is an even function.

20 If $f(x)$ is a polynomial, and if the coefficients of all even powers of x are 0, show that f is an odd function.

21 If $f(x) = 3x^3 - kx^2 + x - 5k$, find a number k such that the graph of f contains the point $(-1, 4)$.

22 If one zero of $f(x) = x^3 - 2x^2 - 16x + 16k$ is 2, find two other zeros.

In Exercises 23–28 show that f has a zero between a and b.

23 $f(x) = x^3 - 4x^2 + 3x - 2;\quad a = 3,\ b = 4$

24 $f(x) = 2x^3 + 5x^2 - 3;\quad a = -3,\ b = -2$

25 $f(x) = -x^4 + 3x^3 - 2x + 1;\quad a = 2,\ b = 3$

26 $f(x) = 2x^4 + 3x - 2;\quad a = \frac{1}{2},\ b = \frac{3}{4}$

27 $f(x) = x^5 + x^3 + x^2 + x + 1;\quad a = -\frac{1}{2},\ b = -1$

28 $f(x) = x^5 - 3x^4 - 2x^3 + 3x^2 - 9x - 6;\quad a = 3,\ b = 4$

29 The third-degree Legendre polynomial

$$P(x) = \tfrac{1}{2}(5x^3 - 3x)$$

occurs in the solution of heat transfer problems in physics and engineering. Sketch the graph of P after first determining where $P(x) > 0$ and $P(x) < 0$.

30 The fourth-degree Chebyshev polynomial

$$f(x) = 8x^4 - 8x^2 + 1$$

occurs in statistical studies. Determine where $f(x) > 0$. (*Hint:* Let $z = x^2$ and use the Quadratic Formula.)

31 An open box is to be made from a rectangular piece of cardboard having dimensions 20 inches × 30 inches by cutting out identical squares of area x^2 from each corner and turning up the sides (see Exercise 37 of Section 2.1). Show that the volume of the box is given by the third-degree polynomial

$$V(x) = x(20 - 2x)(30 - 2x).$$

For what positive values of x is $V(x) > 0$? Sketch the graph of V for $x > 0$.

32 The frame for a shipping crate is to be constructed from 24 feet of 2 × 2 lumber.

(a) If the crate is to have square ends of side x feet, express the volume V of the crate as a function of x. (See figure.)

(b) Sketch the graph of V for $x > 0$.

FIGURE FOR EXERCISE 32

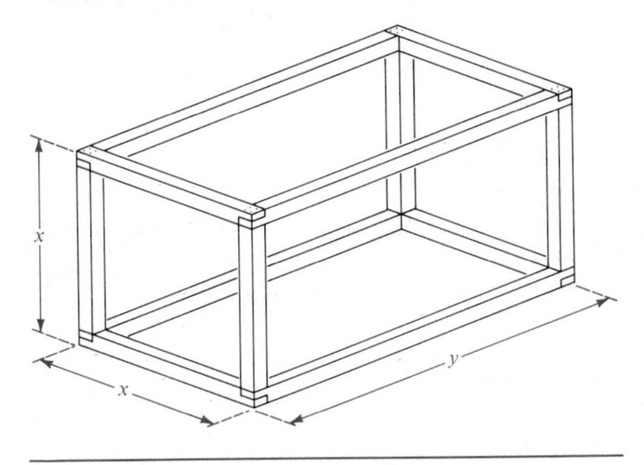

33 A meteorologist determines that the temperature T (in °F) on a certain cold winter day was given by

$$T = 0.05t(t - 12)(t - 24)$$

where t is the time (in hours) and $t = 0$ corresponds to 6 A.M.

(a) When was $T > 0$ and when was $T < 0$? Sketch a graph that depicts the temperature T for $0 \le t \le 24$.

(b) Show that the temperature was 32 °F sometime between 12 noon and 1 P.M. (*Hint:* Use the Intermediate Value Theorem.)

34 A diver stands on the very end of a diving board before beginning a dive. It is known that the deflection d of the board at a position s feet from the stationary end is given by the third-degree polynomial

$$d = cs^2(3L - s) \quad \text{for} \quad 0 \le s \le L,$$

where L is the length of the board and c is a positive constant that depends on the weight of the diver and on the physical properties of the board (see figure). Suppose the board is 10 feet long.

(a) If the deflection at the end of the board is 1 foot, find c.

(b) Show that the deflection is $\frac{1}{2}$ foot somewhere between $s = 6.5$ and $s = 6.6$.

FIGURE FOR EXERCISE 34

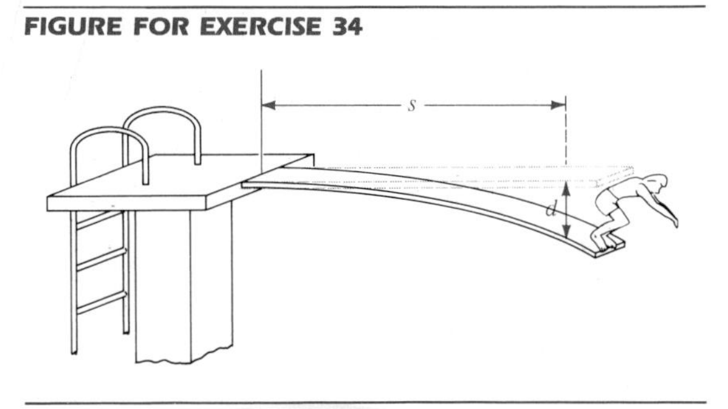

35 A herd of 100 deer is introduced to a small island. The herd at first increases rapidly, but eventually the food resources of the island dwindle and the population declines. Suppose that the number $N(t)$ of deer after t years is given by

$$N(t) = -t^4 + 21t^2 + 100.$$

(a) Determine the positive values of t for which $N(t) > 0$. Does the population become extinct? If so, when?

(b) Sketch the graph of N for $t > 0$.

36 It can be shown by means of calculus that the rate R at which the deer population in Exercise 35 grows (or declines) at time t is given by

$$R = -4t^3 + 42t \text{ (deer per year)}$$

(a) When does the population cease to grow?

(b) Determine the positive values of t for which $R > 0$.

Calculator Exercises 3.1

If a zero of a polynomial lies between two integers, then a method of *successive approximation* may be used to approximate the zero to any degree of accuracy. One such method is illustrated in Exercise 1.

1 If $f(x) = x^3 - 3x + 1$:

(a) Show that f has a zero between 1 and 2.

(b) By increasing values of x by tenths, show that f has a zero between 1.5 and 1.6.

(c) By increasing values of x by hundredths, show that f has a zero between 1.53 and 1.54.

In (c) of Exercise 1, a zero of $x^3 - 3x + 1$ is said to be *isolated between successive hundredths* in the interval $[1, 2]$. In Exercises 2–6 isolate a zero of $f(x)$ between successive hundredths in the interval $[a, b]$.

2 $f(x) = x^3 + 5x - 3$; $a = 0$, $b = 1$

3 $f(x) = 2x^3 - 4x^2 - 3x + 1$; $a = 2$, $b = 3$

4 $f(x) = x^4 - 4x^3 + 3x^2 - 8x + 2$; $a = 3$, $b = 4$

5 $f(x) = x^5 - 2x^2 + 4$; $a = -2$, $b = -1$

6 $f(x) = x^4 - 2x^3 + 10x - 25$; $a = 2$, $b = 3$

In Exercises 7–10 isolate a zero of $f(x)$ between successive thousandths in the interval $[a, b]$.

7 $f(x) = x^4 + 2x^3 - 5x^2 + 1$; $a = 1$, $b = 2$

8 $f(x) = x^4 - 5x^2 + 2x - 5$; $a = 2$, $b = 3$

9 $f(x) = x^5 + x^2 - 9x - 3$; $a = -2$, $b = -1$

10 $f(x) = x^5 + 3x^4 - x^3 + 2x^2 + 6x - 2$; $a = -4$, $b = -3$

3.2 Division of Polynomials

In the following discussion, symbols such as $f(x)$ and $g(x)$ will be used to denote polynomials in x. If a polynomial $g(x)$ is a factor of a polynomial $f(x)$, then $f(x)$ is said to be **divisible** by $g(x)$. For example, the polynomial

$x^4 - 16$ is divisible by $x^2 - 4$, by $x^2 + 4$, by $x + 2$, and by $x - 2$; but $x^2 + 3x + 1$ is not a factor of $x^4 - 16$. However, by *long division*, we can write

$$
\begin{array}{r|lllll}
\multicolumn{1}{r}{} & & & x^2 & -\,3x & +\,8 \\ \cline{2-6}
x^2+3x+1 & x^4 & & & & -16 \\
 & x^4 & +\,3x^3 & +\,x^2 & & \\ \cline{2-4}
 & & -\,3x^3 & -\,x^2 & & \\
 & & -\,3x^3 & -\,9x^2 & -\,3x & \\ \cline{3-5}
 & & & 8x^2 & +\,3x & -\,16 \\
 & & & 8x^2 & +\,24x & +\,8 \\ \cline{4-6}
 & & & & -\,21x & -\,24
\end{array}
$$

The polynomial $x^2 - 3x + 8$ is called the **quotient** and $-21x - 24$ is the **remainder.**

Note that the long division process ends when we arrive at a polynomial (the remainder) that either is 0 or has smaller degree than the divisor. The result of this division is often written

$$\frac{x^4 - 16}{x^2 + 3x + 1} = (x^2 - 3x + 8) + \left(\frac{-21x - 24}{x^2 + 3x + 1}\right).$$

Multiplying both sides of the equation by $x^2 + 3x + 1$, we obtain

$$x^4 - 16 = (x^2 + 3x + 1)(x^2 - 3x + 8) + (-21x - 24).$$

This example illustrates the following theorem, which we state without proof.

DIVISION ALGORITHM FOR POLYNOMIALS

> If $f(x)$ and $g(x)$ are polynomials and if $g(x) \neq 0$, then there exist unique polynomials $q(x)$ and $r(x)$ such that
>
> $$f(x) = g(x)q(x) + r(x)$$
>
> where either $r(x) = 0$, or the degree of $r(x)$ is less than the degree of $g(x)$. The polynomial $q(x)$ is called the **quotient,** and $r(x)$ is the **remainder** in the division of $f(x)$ by $g(x)$.

An interesting special case occurs if $f(x)$ is divided by a polynomial of the form $x - c$ where c is a real number. If $x - c$ is a factor of $f(x)$, then

$$f(x) = (x - c)q(x)$$

for some polynomial $q(x)$; that is, the remainder $r(x)$ is 0. If $x - c$ is not a factor of $f(x)$, then the degree of the remainder $r(x)$ is less than the degree of $x - c$, and hence $r(x)$ must have degree 0. This, in turn, means that the remainder is a nonzero number. Consequently, in all cases we have

$$f(x) = (x - c)q(x) + d$$

where the remainder d is some real number (possibly $d = 0$). If c is substituted for x in the equation $f(x) = (x - c)q(x) + d$, we obtain

$$f(c) = (c - c)q(c) + d,$$

which reduces to $f(c) = d$. This proves the following theorem.

REMAINDER THEOREM

If a polynomial $f(x)$ is divided by $x - c$, then the remainder is $f(c)$.

EXAMPLE 1 If $f(x) = x^3 - 3x^2 + x + 5$, use the Remainder Theorem to find $f(2)$.

SOLUTION According to the Remainder Theorem, $f(2)$ is the remainder when $f(x)$ is divided by $x - 2$. By long division,

$$\begin{array}{r|l} & x^2 - x - 1 \\ \hline x - 2 & x^3 - 3x^2 + x + 5 \\ & x^3 - 2x^2 \\ \hline & \quad -x^2 + x \\ & \quad -x^2 + 2x \\ \hline & \qquad -x + 5 \\ & \qquad -x + 2 \\ \hline & \qquad\quad 3 \end{array}$$

Hence, $f(2) = 3$. We may check this fact by direct substitution. Thus, $f(2) = 2^3 - 3(2)^2 + 2 + 5 = 3$. ■

FACTOR THEOREM

A polynomial $f(x)$ has a factor $x - c$ if and only if $f(c) = 0$.

PROOF By the Remainder Theorem, $f(x) = (x - c)q(x) + f(c)$ for some quotient $q(x)$. If $f(c) = 0$, then $f(x) = (x - c)q(x)$; that is, $x - c$ is a factor of $f(x)$. Conversely, if $x - c$ is a factor, then the remainder upon division of $f(x)$ by $x - c$ must be 0, and hence, by the Remainder Theorem, $f(c) = 0$. □

The Factor Theorem is useful for finding factors of polynomials, as illustrated in the next example.

EXAMPLE 2 Show that $x - 2$ is a factor of the polynomial

$$f(x) = x^3 - 4x^2 + 3x + 2.$$

SOLUTION Since $f(2) = 8 - 16 + 6 + 2 = 0$, it follows from the Factor Theorem that $x - 2$ is a factor of $f(x)$. Of course, another method of solution would be to divide $f(x)$ by $x - 2$ and show that the remainder is 0. The quotient in the division would be another factor of $f(x)$. ■

EXAMPLE 3 Find a polynomial $f(x)$ of degree 3 that has zeros 2, -1, and 3.

SOLUTION By the Factor Theorem, $f(x)$ has factors $x - 2$, $x + 1$, and $x - 3$. We may then write

$$f(x) = a(x - 2)(x + 1)(x - 3)$$

where any nonzero value may be assigned to a. If we let $a = 1$ and multiply, we obtain

$$f(x) = x^3 - 4x^2 + x + 6.$$ ■

To apply the Remainder Theorem it is necessary to divide a given polynomial by $x - c$. A method called **synthetic division** may be used to simplify this work. The following rules state how to proceed. The method can be justified by a careful (and lengthy) comparison with the method of long division.

RULES: Synthetic Division of $a_nx^n + a_{n-1}x^{n-1} + \cdots + a_1x + a_0$ by $x - c$

1 Begin with the following display, supplying zeros for any missing coefficients in the given polynomial.

$$\begin{array}{c|cccccc} c & a_n & a_{n-1} & a_{n-2} & \cdots & a_1 & a_0 \\ & & & & & & \\ \hline & a_n & & & & & \end{array}$$

2 Multiply a_n by c and place the product ca_n underneath a_{n-1} as indicated by the arrow in the following display. (This arrow, and others, is used only to help clarify these rules, and will not appear in *specific* synthetic divisions.) Next find the sum $b_1 = a_{n-1} + ca_n$ and place it below the line as shown.

$$\begin{array}{c|cccccccc} c & a_n & a_{n-1} & a_{n-2} & \cdots & & a_1 & a_0 \\ & & ca_n & cb_1 & cb_2 & \cdots & cb_{n-2} & cb_{n-1} \\ \hline & a_n & b_1 & b_2 & \cdots & b_{n-2} & b_{n-1} & r \end{array}$$

3 Multiply b_1 by c and place the product cb_1 underneath a_{n-2} as indicated by another arrow. Next find the sum $b_2 = a_{n-2} + cb_1$ and place it below the line as shown.

4 Continue this process, as indicated by the arrows, until the final sum $r = a_0 + cb_{n-1}$ is obtained. The numbers

$$a_n, \quad b_1, \quad b_2, \quad \ldots, \quad b_{n-2}, \quad b_{n-1}$$

are the coefficients of the quotient $q(x)$; that is,

$$q(x) = a_n x^{n-1} + b_1 x^{n-2} + \cdots + b_{n-2}x + b_{n-1},$$

and r is the remainder. ■ ■ ■

The following examples illustrate synthetic division for some special cases.

EXAMPLE 4 Use synthetic division to find the quotient and remainder if $2x^4 + 5x^3 - 2x - 8$ is divided by $x + 3$.

SOLUTION Since the divisor is $x + 3$, the c in the expression $x - c$ is -3. Hence, the synthetic division takes this form:

$$\begin{array}{r|rrrrr} -3 & 2 & 5 & 0 & -2 & -8 \\ & & -6 & 3 & -9 & 33 \\ \hline & 2 & -1 & 3 & -11 & 25 \end{array}$$

The first four numbers in the third row are the coefficients of the quotient $q(x)$ and the last number is the remainder r. Thus,

$$q(x) = 2x^3 - x^2 + 3x - 11 \quad \text{and} \quad r = 25.$$ ■

Synthetic division can be used to find values of polynomial functions, as illustrated in the next example.

EXAMPLE 5 If $f(x) = 3x^5 - 38x^3 + 5x^2 - 1$, use synthetic division to find $f(4)$.

SOLUTION By the Remainder Theorem, $f(4)$ is the remainder when $f(x)$ is divided by $x - 4$. Dividing synthetically, we obtain

4	3	0	−38	5	0	−1
		12	48	40	180	720
	3	12	10	45	180	719

Consequently, $f(4) = 719$. ■

Synthetic division may be employed to help find zeros of polynomials. By the method illustrated in the preceding example, $f(c) = 0$ if and only if the remainder in the synthetic division by $x - c$ is 0.

EXAMPLE 6 Show that -11 is a zero of the polynomial

$$f(x) = x^3 + 8x^2 - 29x + 44.$$

SOLUTION Dividing synthetically by $x - (-11) = x + 11$ gives us

−11	1	8	−29	44
		−11	33	−44
	1	−3	4	0

Thus, $f(-11) = 0$. ■

Example 6 shows that the number -11 is a solution of the equation $x^3 + 8x^2 - 29x + 44 = 0$. In Section 3.5 we shall use synthetic division to find rational solutions of equations.

Exercises 3.2

In Exercises 1–6 find the quotient $q(x)$ and the remainder $r(x)$ if $f(x)$ is divided by $g(x)$.

1 $f(x) = x^4 + 3x^3 - 2x + 5,\ g(x) = x^2 + 2x - 4$

2 $f(x) = 4x^3 - x^2 + x - 3,\ g(x) = x^2 - 5x$

3 $f(x) = 5x^3 - 2x,\ g(x) = 2x^2 + 1$

4 $f(x) = 3x^4 - x^3 - x^2 + 3x + 4,\ g(x) = 2x^3 - x + 4$

5 $f(x) = 7x^3 - 5x + 2,\ g(x) = 2x^4 - 3x^2 + 9$

6 $f(x) = 10x - 4,\ g(x) = 8x^2 - 5x + 17$

In Exercises 7–16 use synthetic division to find the quotient and remainder assuming the first polynomial is divided by the second.

7 $2x^3 - 3x^2 + 4x - 5,\ x - 2$

8 $3x^3 - 4x^2 - x + 8,\ x + 4$

9 $x^3 - 8x - 5,\ x + 3$

10 $5x^3 - 6x^2 + 15,\ x - 4$

11 $3x^5 + 6x^2 + 7,\ x + 2$

12 $-2x^4 + 10x - 3,\ x - 3$

13 $4x^4 - 5x^2 + 1,\ x - \frac{1}{2}$

14 $9x^3 - 6x^2 + 3x - 4,\ x - \frac{1}{3}$

15 $x^n - 1,\ x - 1$ where n is any positive integer

16 $x^n + 1,\ x + 1$ where n is any positive integer

In Exercises 17–28 use the Remainder Theorem to find $f(c)$.

17 $f(x) = 2x^3 - x^2 - 5x + 3,\ c = 4$

18 $f(x) = 4x^3 - 3x^2 + 7x + 10,\ c = 3$

19 $f(x) = x^4 + 5x^3 - x^2 + 5,\ c = -2$

20 $f(x) = x^4 - 7x^2 + 2x - 8,\ c = -3$

21 $f(x) = x^6 - 3x^4 + 4,\ c = \sqrt{2}$

22 $f(x) = x^5 - x^4 + x^3 - x^2 + x - 1,\ c = -1$

23 $f(x) = x^4 - 4x^3 + x^2 - 3x - 5,\ c = 2$

24 $f(x) = 0.3x^3 + 0.04x - 0.034,\ c = -0.2$

25 $f(x) = x^6 - x^5 + x^4 - x^3 + x^2 - x + 1,\ c = 4$

26 $f(x) = 8x^5 - 3x^2 + 7,\ c = \frac{1}{2}$

27 $f(x) = x^2 + 3x - 5,\ c = 2 + \sqrt{3}$

28 $f(x) = x^3 - 3x^2 - 8,\ c = 1 + \sqrt{2}$

In Exercises 29–32 use synthetic division to show that c is a zero of $f(x)$.

29 $f(x) = 3x^4 + 8x^3 - 2x^2 - 10x + 4,\ c = -2$

30 $f(x) = 4x^3 - 9x^2 - 8x - 3,\ c = 3$

31 $f(x) = 4x^3 - 6x^2 + 8x - 3,\ c = \frac{1}{2}$

32 $f(x) = 27x^4 - 9x^3 + 3x^2 + 6x + 1,\ c = -\frac{1}{3}$

33 Determine k so that $f(x) = x^3 + kx^2 - kx + 10$ is divisible by $x + 3$.

34 Determine all values of k such that $f(x) = k^2x^3 - 4kx - 3$ is divisible by $x - 1$.

35 Use the Factor Theorem to show that $x - 2$ is a factor of $f(x) = x^4 - 3x^3 - 2x^2 + 5x + 6$.

36 Show that $x + 2$ is a factor of $f(x) = x^{12} - 4096$.

37 Prove that $f(x) = 3x^4 + x^2 + 5$ has no factor of the form $x - c$ where c is a real number.

38 Find the remainder if the polynomial $3x^{100} + 5x^{85} - 4x^{38} + 2x^{17} - 6$ is divided by $x + 1$.

39 Use the Factor Theorem to prove that $x - y$ is a factor of $x^n - y^n$ for all positive integers n. Assuming n is even, show that $x + y$ is also a factor of $x^n - y^n$.

40 Assuming n is an odd positive integer, prove that $x + y$ is a factor of $x^n + y^n$.

3.3 Complex Numbers

One of the reasons that real numbers are important is that we need them to find solutions of equations. To illustrate, the nonnegative integers 0, 1, 2, 3, . . . do not contain solutions of equations such as $x + 5 = 0$. To remedy this defect, we need the system of integers, which also contains negatives $-1, -2, -3, \ldots$. In this expanded number system we find the solution -5 for the equation $x + 5 = 0$.

Similarly, to solve the equation $3x + 5 = 0$, we must enlarge the set of integers to the rational numbers, thereby obtaining the solution $x = -\frac{5}{3}$.

The set of rational numbers is still not large enough to solve every equation, for example, $x^2 = 5$ has no rational solutions. Thus, we must again expand our number system to include irrational numbers, such as $\sqrt{5}$ and $-\sqrt{5}$. This leads us to the real number system $\mathbb{R}$, which contains all rational and irrational numbers.

There is one more defect to overcome. Since squares of real numbers are never negative, $\mathbb{R}$ does not contain solutions of equations of the form $x^2 = -5$. To remedy this situation, we shall invent a larger number system $\mathbb{C}$, called the **complex number system,** which contains $\mathbb{R}$ and also contains numbers whose squares are negative.

We begin by introducing the **imaginary unit,** denoted by the letter i, which has the following property:

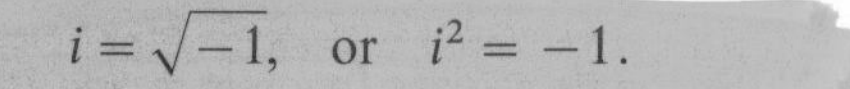

$$i = \sqrt{-1}, \quad \text{or} \quad i^2 = -1.$$

Obviously, i is not a real number. It is a new mathematical entity that will enable us to obtain the number system $\mathbb{C}$, which contains solutions of every algebraic equation.

Since i, together with $\mathbb{R}$, is to be contained in $\mathbb{C}$, it is necessary to consider products of the form bi, where b is a real number, and also expressions of the form $a + bi$, where both a and b are real numbers. This is the motivation for the next definition.

DEFINITION

> A **complex number** is an expression of the form $a + bi$, where a and b are real numbers and $i^2 = -1$.

Equality, addition, and multiplication of two complex numbers $a + bi$ and $c + di$ are defined as follows:

$$a + bi = c + di \quad \text{if and only if} \quad a = c \quad \text{and} \quad b = d.$$

$$(a + bi) + (c + di) = (a + c) + (b + d)i$$

$$(a + bi)(c + di) = (ac - bd) + (ad + bc)i$$

It is unnecessary to memorize the preceding definitions of addition and multiplication. Indeed, when working with complex numbers, *we may treat all symbols as though they represented real numbers, with exactly one exception: wherever the symbol i^2 appears, it may be replaced by* -1. Note that if we use this technique for multiplication, we obtain

$$\begin{aligned}(a+bi)(c+di) &= (a+bi)c + (a+bi)(di)\\ &= ac + (bi)c + a(di) + (bi)(di)\\ &= ac + (bc)i + (ad)i + (bd)(i^2)\\ &= ac + (bc)i + (ad)i + (bd)(-1)\\ &= ac + (bd)(-1) + (ad)i + (bc)i\\ &= (ac - bd) + (ad + bc)i\end{aligned}$$

This agrees with our definition of multiplication.

EXAMPLE 1 Write each expression in the form $a + bi$, where a and b are real numbers.

(a) $(3 + 4i) + (2 + 5i)$ (b) $(3 + 4i)(2 + 5i)$

SOLUTION

(a) $$(3+4i)+(2+5i) = (3+2)+(4+5)i = 5+9i$$

(b) $$\begin{aligned}(3+4i)(2+5i) &= (3+4i)2 + (3+4i)(5i)\\ &= 6 + 8i + 15i + 20i^2\\ &= 6 + 20(-1) + 23i\\ &= -14 + 23i\end{aligned}$$ ■

The set $\mathbb{R}$ of real numbers may be identified with complex numbers of the form $a + 0i$. It is also convenient to denote the complex number $0 + bi$ by bi. Using these conventions we see that

$$(a + 0i) + (0 + bi) = (a + 0) + (0 + b)i = a + bi.$$

Thus, $a + bi$ may be regarded as the sum of two complex numbers a and bi (that is, $a + 0i$ and $0 + bi$).

The **identity element** relative to addition is 0 (or, equivalently, $0 + 0i$); that is,

$$(a + bi) + 0 = a + bi$$

for every complex number $a + bi$.

If $(-a) + (-b)i$ is added to $a + bi$, we obtain 0. This implies that $(-a) + (-b)i$ is the **additive inverse** of $a + bi$; that is,

$$-(a + bi) = (-a) + (-b)i.$$

Subtraction of complex numbers is defined using additive inverses:

$$(a + bi) - (c + di) = (a + bi) + [-(c + di)].$$

Since $-(c + di) = (-c) + (-d)i$, it follows that

$$(a + bi) - (c + di) = (a - c) + (b - d)i.$$

If c, d, and k are real numbers, then,

$$k(c + di) = (k + 0i)(c + di) = (kc - 0d) + (kd + 0c)i;$$

that is,

$$k(c + di) = kc + (kd)i.$$

One illustration of this formula is

$$3(5 + 2i) = 15 + 6i.$$

If, as in the next example, we are asked to write an expression in the form $a + bi$, we shall also accept the form $a - di$, since $a - di = a + (-d)i$.

EXAMPLE 2 Write each expression in the form $a + bi$.

(a) $4(2 + 5i) - (3 - 4i)$ (b) $(4 - 3i)(2 + i)$

(c) $i(3 - 2i)^2$ (d) i^{51}

SOLUTION

(a) $$4(2 + 5i) - (3 - 4i) = 8 + 20i - 3 + 4i = 5 + 24i$$

(b) $$(4 - 3i)(2 + i) = 8 - 6i + 4i - 3i^2 = 11 - 2i$$

(c) $$i(3 - 2i)^2 = i(9 - 12i + 4i^2) = i(5 - 12i) = 5i - 12i^2 = 12 + 5i$$

(d) Taking successive powers of i, we obtain $i^1 = i$, $i^2 = -1$, $i^3 = -i$, $i^4 = 1$, and then the cycle starts over: $i^5 = i$, $i^6 = i^2 = -1$, etc. In particular, $i^{51} = i^{48}i^3 = (i^4)^{12}i^3 = (1)^{12}i^3 = i^3 = -i$. ■

The complex number $a - bi$ is called the **conjugate** of the complex number $a + bi$. Since

$$(a + bi) + (a - bi) = 2a$$

and

$$(a + bi)(a - bi) = a^2 + b^2,$$

we see that *the sum and product of a complex number and its conjugate are real numbers.* Conjugates are useful for finding the multiplicative inverse $1/(a + bi)$ of $a + bi$, or for simplifying the quotient $(a + bi)/(c + di)$ of two complex numbers, as illustrated in the next example.

EXAMPLE 3 Express each fraction in the form $a + bi$.

(a) $\dfrac{1}{9 + 2i}$ (b) $\dfrac{7 - i}{3 - 5i}$

SOLUTION We multiply numerator and denominator by the conjugate of the denominator, as follows:

(a)
$$\frac{1}{9 + 2i} = \frac{1}{9 + 2i} \cdot \frac{9 - 2i}{9 - 2i} = \frac{9 - 2i}{81 + 4} = \frac{9}{85} - \frac{2}{85}i$$

(b)
$$\frac{7 - i}{3 - 5i} = \frac{7 - i}{3 - 5i} \cdot \frac{3 + 5i}{3 + 5i} = \frac{21 - 3i + 35i - 5i^2}{9 - 25i^2}$$
$$= \frac{26 + 32i}{34} = \frac{13}{17} + \frac{16}{17}i$$

■

It is easy to see that if p is any positive real number, then the equation $x^2 = -p$ has solutions in $\mathbb{C}$. As a matter of fact, one solution is $i\sqrt{p}$, since

$$(i\sqrt{p})^2 = i^2(\sqrt{p})^2 = (-1)p = -p.$$

Similarly, $-i\sqrt{p}$ is also a solution.

The next definition is motivated by the fact that $(i\sqrt{p})^2 = -p$.

DEFINITION

> If p is a positive real number, then the **principal square root** of $-p$ is denoted by $\sqrt{-p}$ and is defined by
>
> $$\sqrt{-p} = i\sqrt{p}.$$

We also express this as $\sqrt{-p} = \sqrt{p}\,i$; however, care must be taken so as *not* to write $\sqrt{pi}$ when $\sqrt{p}\,i$ is intended. Some examples of principal square roots are

$$\sqrt{-9} = i\sqrt{9} = i(3) = 3i, \qquad \sqrt{-5} = i\sqrt{5}, \qquad \sqrt{-1} = i\sqrt{1} = i.$$

The radical sign must be used with caution when the radicand is negative. For example, the formula $\sqrt{a}\sqrt{b} = \sqrt{ab}$, which holds for positive real numbers, is not true when a and b are both negative. To illustrate,

$$\sqrt{-3}\sqrt{-3} = (i\sqrt{3})(i\sqrt{3}) = i^2(\sqrt{3})^2 = (-1)3 = -3,$$

whereas
$$\sqrt{(-3)(-3)} = \sqrt{9} = 3.$$

Hence,
$$\sqrt{-3}\sqrt{-3} \neq \sqrt{(-3)(-3)}.$$

However, if only *one* of a or b is negative, then $\sqrt{a}\sqrt{b} = \sqrt{ab}$. In general, we shall not apply laws of radicals if radicands are negative. Instead, we shall change the form of radicals before performing any operations, as illustrated in the next example.

EXAMPLE 4 Express $(5 - \sqrt{-9})(-1 + \sqrt{-4})$ in the form a + *bi*.

SOLUTION

$$\begin{aligned}(5 - \sqrt{-9})(-1 + \sqrt{-4}) &= (5 - i\sqrt{9})(-1 + i\sqrt{4})\\ &= (5 - 3i)(-1 + 2i)\\ &= -5 + 3i + 10i - 6i^2\\ &= -5 + 13i + 6 = 1 + 13i\end{aligned}$$

■

If a, b, and c are real numbers such that $b^2 - 4ac \geq 0$, and if $a \neq 0$, then the solutions of the quadratic equation $ax^2 + bx + c = 0$ are

$$\frac{-b + \sqrt{b^2 - 4ac}}{2a} \quad \text{and} \quad \frac{-b - \sqrt{b^2 - 4ac}}{2a}.$$

If $b^2 - 4ac \geq 0$, the solutions are real. If $b^2 - 4ac < 0$, then the solutions are two complex numbers that are conjugates of one another.

EXAMPLE 5 Find the solutions of the equation $5x^2 + 2x + 1 = 0$.

SOLUTION By the Quadratic Formula,

$$x = \frac{-2 \pm \sqrt{4 - 20}}{10} = \frac{-2 \pm \sqrt{-16}}{10} = \frac{-2 \pm 4i}{10}.$$

Dividing numerator and denominator by 2, we see that the solutions of the equation are $-\frac{1}{5} + \frac{2}{5}i$ and $-\frac{1}{5} - \frac{2}{5}i$. ■

EXAMPLE 6 Find the solutions of the equation $x^3 - 1 = 0$.

SOLUTION The equation $x^3 - 1 = 0$ may be written

$$(x - 1)(x^2 + x + 1) = 0.$$

Setting each factor equal to zero and solving the resulting equations, we obtain the solutions

$$1, \quad \frac{-1 \pm \sqrt{1-4}}{2},$$

or, equivalently,

$$1, \quad -\frac{1}{2} + \frac{\sqrt{3}}{2}i, \quad -\frac{1}{2} - \frac{\sqrt{3}}{2}i.$$ ■

The three solutions of $x^3 - 1 = 0$ are called the **cube roots of unity.** It can be shown that if n is any positive integer, the equation $x^n - 1 = 0$ has n distinct complex solutions, called the ***n*th roots of unity.**

Exercises 3.3

In Exercises 1–32 write the expression in the form $a + bi$.

1 $(3 + 2i) + (-5 + 4i)$

2 $(8 - 5i) + (2 - 3i)$

3 $(-4 + 5i) + (2 - i)$

4 $(5 + 7i) + (-8 - 4i)$

5 $(16 + 10i) - (9 + 15i)$

6 $(2 - 6i) - (7 + 2i)$

7 $7 - (3 - 7i)$

8 $-9 + (5 + 9i)$

9 $5i - (6 + 2i)$

10 $(10 + 7i) - 12i$

11 $(4 + 3i)(-1 + 2i)$

12 $(3 - 6i)(2 + i)$

13 $(-7 + i)(-3 + i)$

14 $(5 + 2i)^2$

15 $(2 + 3i)^2$

16 $7i(13 + 8i)$

17 $-9i(4 - 8i)$

18 $(1 + i)^3$

19 $(3 + i)^2(3 - i)^2$

20 $-3(-6 + 12i)$

21 i^{30}

22 i^{25}

23 $\dfrac{1}{3 + 2i}$

24 $\dfrac{1}{5 + 8i}$

25 $\dfrac{7}{5 - 6i}$

26 $\dfrac{-3}{2 - 5i}$

27 $\dfrac{4 - 3i}{2 + 4i}$

28 $\dfrac{4 + 3i}{-1 + 2i}$

29 $\dfrac{6 + 4i}{1 - 5i}$

30 $\dfrac{7 - 6i}{-5 - i}$

31 $\dfrac{21 - 7i}{i}$

32 $\dfrac{10 + 9i}{-3i}$

In Exercises 33 and 34 find x and y.

33 $(x - y) + 3i = 7 + yi$

34 $8 + (3x + y)i = 2x - 4i$

Find the solutions of the equations in Exercises 35–48.

35 $x^2 - 3x + 10 = 0$

36 $x^2 - 5x + 20 = 0$

37 $x^2 + 2x + 5 = 0$

38 $x^2 + 3x + 6 = 0$

39 $4x^2 + x + 3 = 0$

40 $-3x^2 + x - 5 = 0$

41 $x^3 - 125 = 0$

42 $x^3 + 27 = 0$

43 $x^6 - 64 = 0$

44 $x^4 = 81$

45 $4x^4 + 25x^2 + 36 = 0$

46 $27x^4 + 21x^2 + 4 = 0$

47 $x^3 + 3x^2 + 4x = 0$

48 $8x^3 - 12x^2 + 2x - 3 = 0$

49 If $z = a + bi$ is any complex number, its conjugate is often denoted by $\bar{z}$, that is, $\bar{z} = a - bi$. Prove each of the following.

(a) $\overline{z + w} = \bar{z} + \bar{w}$

(b) $\overline{z \cdot w} = \bar{z} \cdot \bar{w}$

(c) $\overline{z^2} = (\bar{z})^2$, $\overline{z^3} = (\bar{z})^3$, and $\overline{z^4} = (\bar{z})^4$

(d) $\bar{z} = z$ if and only if z is real.

50 Refer to Exercise 49. Prove that $\overline{z - w} = \bar{z} - \bar{w}$ and $\bar{\bar{z}} = z$.

3.4 The Zeros of a Polynomial

The Factor and Remainder Theorems can be extended to the system of complex numbers. Thus, a complex number $c = a + bi$ is a zero of a polynomial $f(x)$; that is, $f(c) = 0$ if and only if $x - c$ is a factor of $f(x)$. Except in special cases, zeros of polynomials are very difficult to find. For example, $f(x) = x^5 - 3x^4 + 4x^3 + 4x - 10$ has no obvious zeros. Moreover, there is no formula that can be used to find the zeros. In spite of the practical difficulty of determining zeros of polynomials, it is possible to make some headway concerning the *theory* of such zeros. The results in this section form the basis for work in what is known as *The Theory of Equations.*

FUNDAMENTAL THEOREM OF ALGEBRA

If a polynomial $f(x)$ has positive degree and complex coefficients, then $f(x)$ has at least one complex zero.

The usual proof of this theorem requires results from the field of mathematics called *functions of a complex variable.* In turn, a prerequisite for studying this field is a strong background in calculus. The first proof of the Fundamental Theorem of Algebra was given by the German mathematician Carl Friedrich Gauss (1777–1855), who is considered by many to be the greatest mathematician of all time.

As a special case of the Fundamental Theorem, if all the coefficients of $f(x)$ are real, then $f(x)$ has at least one complex zero. If $a + bi$ is a complex zero, it may happen that $b = 0$, in which case we refer to the number as a **real zero.** If the Fundamental Theorem is combined with the Factor Theorem, the following useful corollary is obtained.

COROLLARY

Every polynomial of positive degree has a factor of the form $x - c$ where c is a complex number.

The corollary enables us, at least in theory, to express every polynomial $f(x)$ of positive degree as a product of polynomials of degree 1. If $f(x)$ has degree $n > 0$, then applying the corollary gives us

$$f(x) = (x - c_1)f_1(x)$$

where c_1 is a complex number and $f_1(x)$ is a polynomial of degree $n - 1$. If $n - 1 > 0$, we may apply the corollary again, obtaining

$$f_1(x) = (x - c_2)f_2(x)$$

where c_2 is a complex number and $f_2(x)$ is a polynomial of degree $n - 2$. Hence,

$$f(x) = (x - c_1)(x - c_2)f_2(x).$$

Continuing this process, after n steps we arrive at a polynomial $f_n(x)$ of degree 0. Thus, $f_n(x) = a$ for some nonzero number a, and we may write

$$f(x) = a(x - c_1)(x - c_2) \cdots (x - c_n)$$

where each complex number c_j is a zero of $f(x)$. Evidently, the leading coefficient of the polynomial on the right in the last equation is a. It follows that a is the leading coefficient of $f(x)$. We have proved the following theorem.

THEOREM

If $f(x)$ is a polynomial of degree $n > 0$, then there exist n complex numbers $c_1, c_2, \ldots, c_n$ such that

$$f(x) = a(x - c_1)(x - c_2) \cdots (x - c_n)$$

where a is the leading coefficient of $f(x)$. Each number c_j is a zero of $f(x)$.

COROLLARY

> A polynomial of degree $n > 0$ has at most n different complex zeros.

PROOF We shall give an indirect proof. Suppose $f(x)$ has *more* than n different complex zeros. Let us choose $n + 1$ of these zeros and label them $c_1, c_2, \ldots, c_n$, and c. We may use the c_j to obtain the factorization indicated in the statement of the theorem. Substituting c for x and using the fact that $f(c) = 0$, we obtain

$$0 = a(c - c_1)(c - c_2) \cdots (c - c_n).$$

However, each factor on the right side is different from zero because $c \neq c_j$ for every j. Since the product of nonzero numbers cannot equal zero, we have a contradiction. □

EXAMPLE 1 Find a polynomial $f(x)$ of degree 3 with zeros $2, -1$, and 3 that has the value 5 at $x = 1$.

SOLUTION By the Factor Theorem $f(x)$ has factors $x - 2$, $x + 1$, and $x - 3$. No other factors of degree 1 exist, since, by the Factor Theorem, another linear factor $x - c$ would produce a fourth zero of $f(x)$ in violation of the last corollary. Hence, $f(x)$ has the form

$$f(x) = a(x - 2)(x + 1)(x - 3)$$

for some number a. If $f(x)$ has the value 5 at $x = 1$, then $f(1) = 5$; that is,

$$a(1 - 2)(1 + 1)(1 - 3) = 5 \quad \text{or} \quad 4a = 5.$$

Consequently, $a = \frac{5}{4}$ and

$$f(x) = \tfrac{5}{4}(x - 2)(x + 1)(x - 3).$$

Multiplying the four factors we obtain

$$f(x) = \tfrac{5}{4}x^3 - 5x^2 + \tfrac{5}{4}x + \tfrac{15}{2}.$$ ■

The numbers $c_1, c_2, \ldots, c_n$ in the preceding theorem are not necessarily all different. To illustrate, the polynomial $f(x) = x^3 + x^2 - 5x + 3$ has the factorization

$$f(x) = (x + 3)(x - 1)(x - 1).$$

If a factor $x - c$ occurs m times in the factorization, then c is called a **zero of multiplicity *m*** of $f(x)$, or a **root of multiplicity *m*** of the equation

$f(x) = 0$. In the preceding illustration, 1 is a zero of multiplicity 2 and -3 is a zero of multiplicity 1.

As another illustration, if

$$f(x) = (x - 2)(x - 4)^3(x + 1)^2,$$

then $f(x)$ has degree 6 and possesses three distinct zeros 2, 4, and -1, where 2 has multiplicity 1, 4 has multiplicity 3, and -1 has multiplicity 2.

If $f(x)$ is a polynomial of degree n and $f(x) = a(x - c_1)(x - c_2) \cdots (x - c_n)$, then the n complex numbers $c_1, c_2, \ldots, c_n$ are zeros of $f(x)$. If a zero of multiplicity m is counted as m zeros, this tells us that $f(x)$ has at least n zeros (not necessarily all different). Combining this with the fact that $f(x)$ has at most n zeros gives us the next result.

THEOREM

> If $f(x)$ is a polynomial of degree $n > 0$ and if a zero of multiplicity m is counted m times, then $f(x)$ has precisely n zeros.

EXAMPLE 2 Express $f(x) = x^5 - 4x^4 + 13x^3$ as a product of linear factors, and list the five zeros of $f(x)$.

SOLUTION We begin by writing

$$f(x) = x^3(x^2 - 4x + 13).$$

By the Quadratic Formula, the zeros of the polynomial $x^2 - 4x + 13$ are

$$\frac{4 \pm \sqrt{16 - 52}}{2} = \frac{4 \pm \sqrt{-36}}{2} = \frac{4 \pm 6i}{2} = 2 \pm 3i.$$

Hence, by the Factor Theorem, $x^2 - 4x + 13$ has factors $x - (2 + 3i)$ and $x - (2 - 3i)$, and we obtain the desired factorization

$$f(x) = x \cdot x \cdot x \cdot (x - 2 - 3i)(x - 2 + 3i).$$

Since $x - 0$ occurs as a factor three times, the number 0 is zero of multiplicity three, and the five zeros of $f(x)$ are 0, 0, 0, $2 + 3i$, and $2 - 3i$. ■

The next theorem may be used to obtain information about the zeros of a polynomial $f(x)$ with real coefficients. In the statement of the theorem it is assumed that the terms of $f(x)$ are arranged in order of decreasing powers of x, and that terms with zero coefficients are deleted. It is also assumed that the **constant term,** that is, the term that does not contain x, is different from 0. We say there is a **variation of sign** in $f(x)$ if two consecutive coefficients have opposite signs. To illustrate, the polynomial

$$f(x) = 2x^5 - 7x^4 + 3x^2 + 6x - 5$$

has three variations in sign, since there is one variation from $2x^5$ to $-7x^4$, a second from $-7x^4$ to $3x^2$, and a third from $6x$ to -5.

The theorem also refers to the variations of sign in $f(-x)$. Using the previous illustration, note that

$$\begin{aligned} f(-x) &= 2(-x)^5 - 7(-x)^4 + 3(-x)^2 + 6(-x) - 5 \\ &= -2x^5 - 7x^4 + 3x^2 - 6x - 5. \end{aligned}$$

Hence, there are two variations of sign in $f(-x)$: one from $-7x^4$ to $3x^2$, and a second from $3x^2$ to $-6x$.

DESCARTES' RULE OF SIGNS

Let $f(x)$ be a polynomial with real coefficients and nonzero constant term.

(i) The number of positive real solutions of the equation $f(x) = 0$ either is equal to the number of variations of sign in $f(x)$ or is less than that number by an even integer.

(ii) The number of negative real solutions of the equation $f(x) = 0$ either is equal to the number of variations of sign in $f(-x)$ or is less than that number by an even integer.

The proof of Descartes' Rule is rather technical and will not be given in this text.

EXAMPLE 3 Discuss the number of possible positive and negative real solutions, and nonreal complex solutions of the equation

$$2x^5 - 7x^4 + 3x^2 + 6x - 5 = 0.$$

SOLUTION The polynomial $f(x)$ on the left side of the equation is the same as the one given in the illustration preceding the statement of Descartes' Rule. Since there are three variations of sign in $f(x)$, the equation has either three positive real solutions or one positive real solution.

Since $f(-x) = -2x^5 - 7x^4 + 3x^2 - 6x - 5$ has two variations of sign, the given equation has either two negative solutions or no negative solutions. The solutions that are not real numbers are complex numbers of the form $a + bi$ where a and b are real with $b \neq 0$. The following table summarizes the various possibilities that can occur for solutions of the equation.

Number of positive real solutions	1	1	3	3
Number of negative real solutions	0	2	0	2
Number of nonreal, complex solutions	4	2	2	0
Total number of solutions	5	5	5	5

■

Descartes' Rule stipulates that the constant term of the polynomial is different from 0. If the constant term is 0, as in the equation

$$x^4 - 3x^3 + 2x^2 - 5x = 0,$$

then we may write

$$x(x^3 - 3x^2 + 2x - 5) = 0.$$

In this case $x = 0$ is a solution, and Descartes' Rule may be applied to $x^3 - 3x^2 + 2x - 5 = 0$.

EXAMPLE 4 Discuss the nature of the roots of the equation

$$3x^5 + 4x^3 + 2x - 5 = 0.$$

SOLUTION The polynomial $f(x)$ on the left side of the equation has one variation of sign and hence, by (i) of Descartes' Rule, the equation has precisely one positive real root. Since

$$\begin{aligned} f(-x) &= 3(-x)^5 + 4(-x)^3 + 2(-x) - 5 \\ &= -3x^5 - 4x^3 - 2x - 5, \end{aligned}$$

$f(-x)$ has no variations of sign and hence, by (ii) of Descartes' Rule, there are no negative real roots. Thus, the equation has one real root and four nonreal, complex roots. ■

When applying Descartes' Rule, we count roots of multiplicity k as k roots. For example, given $x^2 - 2x + 1 = 0$, the polynomial $x^2 - 2x + 1$ has two variations of sign and hence the equation has either two positive real roots or none. The factored form of the equation is $(x - 1)^2 = 0$, and hence 1 is a root of multiplicity 2.

We shall conclude this section with a discussion of *bounds* for the real solutions of an equation $f(x) = 0$ where $f(x)$ is a polynomial with real coefficients. By definition, a real number b is an **upper bound** for the solutions if no solution is greater than b. A real number a is a **lower bound** for the solutions if no solution is less than a. Thus, if r is a real solution of $f(x) = 0$, then $a \le r \le b$. Note that upper and lower bounds are not unique, since any number greater than b is also an upper bound, and any number less than a is a lower bound.

We may use synthetic division to find upper and lower bounds for the solutions of $f(x) = 0$. Recall that if $f(x)$ is divided synthetically by $x - c$, then the third row that appears in the division process consists of the coefficients of the quotient $q(x)$ together with the remainder $f(c)$. The following theorem indicates how this third row may be used to find upper and lower bounds for the real solutions.

BOUNDS FOR REAL ZEROS OF POLYNOMIALS

Suppose that $f(x)$ is a polynomial with real coefficients and positive leading coefficient and that $f(x)$ is divided synthetically by $x - c$.

(i) If $c > 0$ and if all numbers in the third row of the division process are either positive or zero, then c is an upper bound for the real solutions of the equation $f(x) = 0$.

(ii) If $c < 0$ and if the numbers in the third row of the division process are alternately positive and negative (where a 0 in the third row is considered to be either positive or negative), then c is a lower bound for the real solutions of the equation $f(x) = 0$.

A general proof of this theorem can be patterned after the solution given in the next example.

EXAMPLE 5 Find upper and lower bounds for the real solutions of the equation

$$2x^3 + 5x^2 - 8x - 7 = 0.$$

SOLUTION If we divide synthetically by $x - 1$ and $x - 2$, we obtain

$$\begin{array}{r|rrrr} 1 & 2 & 5 & -8 & -7 \\ & & 2 & 7 & -1 \\ \hline & 2 & 7 & -1 & -8 \end{array} \qquad \begin{array}{r|rrrr} 2 & 2 & 5 & -8 & -7 \\ & & 4 & 18 & 20 \\ \hline & 2 & 9 & 10 & 13 \end{array}$$

Since all numbers in the third row of the synthetic division by $x - 2$ are positive, it follows from (i) of the preceding theorem that 2 is an upper bound for the real solutions of the equation. This fact is also evident if we express the division by $x - 2$ in the Division Algorithm form

$$2x^3 + 5x^2 - 8x - 7 = (x - 2)(2x^2 + 9x + 10) + 13.$$

If $x > 2$, then it is obvious that the right side of the equation is positive, and hence is not zero. Consequently, $2x^3 + 5x^2 - 8x - 7$ is not zero if $x > 2$.

After some trial-and-error attempts using $x - (-1)$ and $x - (-2)$, we find that synthetic division by $x - (-3)$ and $x - (-4)$ gives us

$$\begin{array}{r|rrrr} -3 & 2 & 5 & -8 & -7 \\ & & -6 & 3 & 15 \\ \hline & 2 & -1 & -5 & 8 \end{array} \qquad \begin{array}{r|rrrr} -4 & 2 & 5 & -8 & -7 \\ & & -8 & 12 & -16 \\ \hline & 2 & -3 & 4 & -23 \end{array}$$

Since the numbers in the third row of the synthetic division by $x - (-4)$ are alternately positive and negative, it follows from (ii) of the preceding theorem that -4 is a lower bound for the real solutions. This can also be proved by expressing the division by $x + 4$ in the form

$$2x^3 + 5x^2 - 8x - 7 = (x + 4)(2x^2 - 3x + 4) - 23.$$

If $x < -4$, then the right side of this equation is negative (Why?), and therefore is not zero. Hence, $2x^3 + 5x^2 - 8x - 7$ is not zero if $x < -4$.

It follows that all real solutions of the given equation lie in the interval $[-4, 2]$. ■

Exercises 3.4

In Exercises 1–6 find a polynomial $f(x)$ of degree 3 with the indicated zeros and satisfying the given conditions.

1 $5, -2, -3;\ f(2) = 4$

2 $4, 1, -6;\ f(5) = 2$

3 $2, 3, 1;\ f(0) = 12$

4 $\sqrt{2}, \pi, 0;\ f(0) = 0$

5 $2 + i, 2 - i, -4;\ f(1) = 3$

6 $1 + 2i, 1 - 2i, 5;\ f(-2) = 1$

7 Find a polynomial of degree 4 such that both -2 and 3 are zeros of multiplicity 2.

8 Find a polynomial of degree 5 such that -2 is a zero of multiplicity 3 and 4 is a zero of multiplicity 2.

9 Find a polynomial $f(x)$ of degree 8 such that 2 is a zero of multiplicity 3, 0 is a zero of multiplicity 5, and $f(3) = 54$.

10 Find a polynomial $f(x)$ of degree 7 such that 1 is a zero of multiplicity 2, -1 is a zero of multiplicity 2, 0 is a zero of multiplicity 3, and $f(2) = 36$.

In Exercises 11–18 find the zeros of the polynomials and state the multiplicity of each zero.

11 $f(x) = (x + 4)^3(3x - 4)$

12 $f(x) = (x - 5)^2(4x + 7)^3$

13 $f(x) = 2x^5 - 8x^4 - 10x^3$

14 $f(x) = (4x^2 - 5)^2$

15 $f(x) = (9x^2 - 25)^4(x^2 + 16)$

16 $f(x) = (2x^2 + 13x - 7)^3$

17 $f(x) = (x^2 + x - 2)^2(x^2 - 4)$

18 $f(x) = 4x^6 + x^4$

19 Show that -3 is a zero of multiplicity 2 of the polynomial $f(x) = x^4 + 7x^3 + 13x^2 - 3x - 18$, and express $f(x)$ as a product of linear factors.

20 Show that 4 is a zero of multiplicity 2 of the polynomial $f(x) = x^4 - 9x^3 + 22x^2 - 32$, and express $f(x)$ as a product of linear factors.

21 Show that 1 is a zero of multiplicity 5 of the polynomial $f(x) = x^6 - 4x^5 + 5x^4 - 5x^2 + 4x - 1$, and express $f(x)$ as a product of linear factors.

22 Show that -1 is a zero of multiplicity 4 of the polynomial $f(x) = x^5 + x^4 - 6x^3 - 14x^2 - 11x - 3$, and express $f(x)$ as a product of linear factors.

In Exercises 23–30 use Descartes' Rule of Signs to determine the number of possible positive, negative, and nonreal complex solutions of the equation.

23 $4x^3 - 6x^2 + x - 3 = 0$

24 $5x^3 - 6x - 4 = 0$

25 $4x^3 + 2x^2 + 1 = 0$

26 $3x^3 - 4x^2 + 3x + 7 = 0$

27 $3x^4 + 2x^3 - 4x + 2 = 0$

28 $2x^4 - x^3 + x^2 - 3x + 4 = 0$

29 $x^5 + 4x^4 + 3x^3 - 4x + 2 = 0$

30 $2x^6 + 5x^5 + 2x^2 - 3x + 4 = 0$

In Exercises 31–36 find the smallest and largest integers that are upper and lower bounds, respectively, for the real solutions of the equation.

31 $x^3 - 4x^2 - 5x + 7 = 0$

32 $2x^3 - 5x^2 + 4x - 8 = 0$

33 $x^4 - x^3 - 2x^2 + 3x + 6 = 0$

34 $2x^4 - 9x^3 - 8x - 10 = 0$

35 $2x^5 - 13x^3 + 2x - 5 = 0$

36 $3x^5 + 2x^4 - x^3 - 8x^2 - 7 = 0$

37 Let $f(x)$ and $g(x)$ be polynomials of degree not greater than n, where n is a positive integer. Show that if $f(x)$ and $g(x)$ are equal in value for more than n distinct values of x, then $f(x)$ and $g(x)$ are identical; that is, coefficients of like powers are the same. (*Hint:* Write

$$f(x) = a_nx^n + a_{n-1}x^{n-1} + \cdots + a_1x + a_0$$
$$g(x) = b_nx^n + b_{n-1}x^{n-1} + \cdots + b_1x + b_0$$

and consider

$$h(x) = f(x) - g(x) = (a_n - b_n)x^n + \cdots + (a_0 - b_0).$$

Then show that $h(x)$ has more than n distinct zeros and conclude that $a_j = b_j$ for all j.)

3.5 Complex and Rational Zeros of Polynomials

Example 2 of the preceding section illustrates an interesting fact about polynomials with real coefficients: The two complex zeros $2 + 3i$ and $2 - 3i$ of $x^5 - 4x^4 + 13x^3$ are conjugates of one another. The relationship is not accidental, since the following general result is true.

THEOREM

If a polynomial $f(x)$ of degree $n > 0$ has real coefficients, and if z is a complex zero of $f(x)$, then the conjugate $\bar{z}$ of z is also a zero of $f(x)$.

PROOF We may write

$$f(x) = a_n x^n + a_{n-1}x^{n-1} + \cdots + a_1 x + a_0$$

where each coefficient a_k is a real number and $a_n \neq 0$. If $f(z) = 0$, then

$$a_n z^n + a_{n-1}z^{n-1} + \cdots + a_1 z + a_0 = 0.$$

If two complex numbers are equal, then so are their conjugates. Hence, the conjugate of the left side of the last equation equals the conjugate of the right side; that is,

$$\overline{a_n z^n + a_{n-1}z^{n-1} + \cdots + a_1 z + a_0} = \overline{0} = 0$$

where the fact that $\overline{0} = 0$ follows from $\overline{0} = \overline{0 + 0i} = 0 - 0i = 0$.

It is not difficult to prove that if z and w are complex numbers, then $\overline{z + w} = \overline{z} + \overline{w}$. More generally, the conjugate of *any* sum of complex numbers is the sum of the conjugates. Consequently,

$$\overline{a_n z^n} + \overline{a_{n-1}z^{n-1}} + \cdots + \overline{a_1 z} + \overline{a_0} = 0.$$

It can also be shown that if z and w are complex numbers, then $\overline{z \cdot w} = \overline{z} \cdot \overline{w}$, $\overline{z^n} = \overline{z}^n$ for every positive integer n, and $\overline{z} = z$ if and only if z is real. (See Exercise 49 of Section 3.3.) Thus, for every j and k,

$$\overline{a_j z^k} = \overline{a_j} \cdot \overline{z^k} = \overline{a_j} \cdot \overline{z}^k = a_j \overline{z}^k,$$

and therefore,

$$a_n \overline{z}^n + a_{n-1}\overline{z}^{n-1} + \cdots + a_1 \overline{z} + a_0 = 0.$$

The last equation states that $f(\overline{z}) = 0$, which completes the proof. □

EXAMPLE 1 Find a polynomial $f(x)$ of degree 4 that has real coefficients and zeros $2 + i$ and $-3i$.

SOLUTION By the preceding theorem, $f(x)$ must also have zeros $2 - i$ and $3i$. Applying the Factor Theorem, $f(x)$ has factors $x-(2+i)$, $x-(2-i)$, $x - (-3i)$, and $x - (3i)$. Multiplying those factors gives us a polynomial of the required type:

$$\begin{aligned} f(x) &= [x - (2 + i)][x - (2 - i)](x - 3i)(x + 3i) \\ &= (x^2 - 4x + 5)(x^2 + 9) \\ &= x^4 - 4x^3 + 14x^2 - 36x + 45. \end{aligned}$$

■

If a polynomial with real coefficients is factored as on page 125, some of the factors $x - c_k$ may have a complex coefficient c_k. However, it is always possible to obtain a factorization into polynomials with real coefficients, as stated in the next theorem.

THEOREM

> Every polynomial with real coefficients and positive degree n can be expressed as a product of linear and quadratic polynomials with real coefficients, where the quadratic factors have no real zeros.

PROOF Since $f(x)$ has precisely n complex zeros $c_1, c_2, \ldots, c_n$, we may write

$$f(x) = a(x - c_1)(x - c_2) \cdots (x - c_n)$$

where a is the leading coefficient of $f(x)$. Of course, some of the zeros may be real. In such cases we obtain the linear factors referred to in the statement of the theorem. If a zero c_k is not real, then by the preceding theorem the conjugate $\overline{c_k}$ is also a zero of $f(x)$ and hence must be one of the numbers $c_1, c_2, \ldots, c_n$. This implies that both $x - c_k$ and $x - \overline{c_k}$ appear in the factorization of $f(x)$. If those factors are multiplied, we obtain

$$(x - c_k)(x - \overline{c_k}) = x^2 - (c_k + \overline{c_k})x + c_k\overline{c_k},$$

which has *real* coefficients, since $c_k + \overline{c_k}$ and $c_k\overline{c_k}$ are real numbers (see page 120). Thus, the complex zeros of $f(x)$ and their conjugates give rise to quadratic polynomials that are irreducible over $\mathbb{R}$. This completes the proof. □

EXAMPLE 2 Express $x^4 - 2x^2 - 3$ as a product (a) of linear polynomials, and (b) of linear and quadratic polynomials with real coefficients that are irreducible over $\mathbb{R}$.

SOLUTION

(a) We can find the zeros of the given polynomial by solving the equation $x^4 - 2x^2 - 3 = 0$, which may be regarded as quadratic in x^2. Solving for x^2 by means of the Quadratic Formula, we obtain

$$x^2 = \frac{2 \pm \sqrt{4 + 12}}{2} = \frac{2 \pm 4}{2}.$$

Thus, $x^2 = 3$ and $x^2 = -1$. Hence, the zeros are $\sqrt{3}$, $-\sqrt{3}$, i, and $-i$, and we obtain the factorization

$$x^4 - 2x^2 - 3 = (x - \sqrt{3})(x + \sqrt{3})(x - i)(x + i).$$

(b) Multiplying the last two factors in the preceding factorization gives us

$$x^4 - 2x^2 - 3 = (x - \sqrt{3})(x + \sqrt{3})(x^2 + 1),$$

which is of the form stated in the preceding theorem.

The solution of this example could also have been obtained by factoring the original expression without first finding the zeros. Thus,

$$\begin{aligned} x^4 - 2x^2 - 3 &= (x^2 - 3)(x^2 + 1) \\ &= (x + \sqrt{3})(x - \sqrt{3})(x + i)(x - i). \end{aligned}$$

■

We have already pointed out that it is generally very difficult to find the zeros of a polynomial of high degree. However, if all the coefficients are integers or rational numbers, there is a method for finding the *rational* zeros, if they exist. The method is a consequence of the following theorem.

THEOREM ON RATIONAL ZEROS

> Suppose that $f(x) = a_nx^n + a_{n-1}x^{n-1} + \cdots + a_1x + a_0$ is a polynomial with integral coefficients. If c/d is a rational zero of $f(x)$, where c and d have no common prime factors and $c > 0$, then c is a factor of a_0 and d is a factor of a_n.

PROOF Let us show that c is a factor of a_0. If $c = 1$, the theorem follows at once, since 1 is a factor of *any* number. Now suppose that $c \neq 1$. In this case $c/d \neq 1$, for if $c/d = 1$, we obtain $c = d$, and since c and d have no prime factor in common, this implies that $c = d = 1$, a contradiction. Hence, in the following discussion we have $c \neq 1$ and $c \neq d$.

Since $f(c/d) = 0$,

$$a_n(c^n/d^n) + a_{n-1}(c^{n-1}/d^{n-1}) + \cdots + a_1(c/d) + a_0 = 0.$$

Multiplying by d^n and then adding $-a_0d^n$ to both sides, we obtain

$$a_nc^n + a_{n-1}c^{n-1}d + \cdots + a_1cd^{n-1} = -a_0d^n$$

or

$$c(a_nc^{n-1} + a_{n-1}c^{n-2}d + \cdots + a_1d^{n-1}) = -a_0d^n.$$

This shows that c is a factor of the integer a_0d^n. If c is factored into primes, such as $c = p_1p_2 \cdots p_k$, then each prime p_j is also a factor of a_0d^n. However, by hypothesis, none of the p_j is a factor of d. This implies that each p_j is a factor of a_0, that is, c is a factor of a_0. A similar argument may be used to prove that d is a factor of a_n. □

The technique of using the preceding theorem for finding rational solutions of equations with integral coefficients is illustrated in the following example.

EXAMPLE 3 Find all rational solutions of the equation

$$3x^4 + 14x^3 + 14x^2 - 8x - 8 = 0.$$

SOLUTION The problem is equivalent to finding the rational zeros of the polynomial on the left side of the equation. According to the preceding theorem, if c/d is a rational zero and $c > 0$, then c is a divisor of -8 and d is a divisor of 3. Hence, the possible choices for c are 1, 2, 4, and 8, and the choices for d are ± 1 and ± 3. Consequently, any rational roots are included among the numbers ± 1, ± 2, ± 4, ± 8, $\pm \frac{1}{3}$, $\pm \frac{2}{3}$, $\pm \frac{4}{3}$, and $\pm \frac{8}{3}$. We can reduce the number of possibilities by finding upper and lower bounds for the real solutions; however, we shall not do so here. It is necessary to check to see which of the numbers, if any, are zeros. Synthetic division is the appropriate method for this task. After perhaps many trial-and-error attempts, we obtain:

$$\begin{array}{r|rrrrr} -2 & 3 & 14 & 14 & -8 & -8 \\ & & -6 & -16 & 4 & 8 \\ \hline & 3 & 8 & -2 & -4 & 0 \end{array}$$

This result shows that -2 is a zero. Moreover, the synthetic division provides the coefficients of the quotient in the division of the polynomial by $x + 2$. Hence, we have the following factorization of the given polynomial:

$$(x + 2)(3x^3 + 8x^2 - 2x - 4).$$

The remaining solutions of the equation must be zeros of the second factor, and therefore we may use that polynomial to check for solutions. Again proceeding by trial and error, we ultimately find that synthetic division by $x + \frac{2}{3}$ gives us the following:

$$\begin{array}{r|rrrr} -\frac{2}{3} & 3 & 8 & -2 & -4 \\ & & -2 & -4 & 4 \\ \hline & 3 & 6 & -6 & 0 \end{array}$$

Therefore, $-\frac{2}{3}$ is a zero.

The remaining zeros are solutions of the equation $3x^2 + 6x - 6 = 0$, or equivalently, $x^2 + 2x - 2 = 0$. By the Quadratic Formula, this equation has solutions

$$\frac{-2 \pm \sqrt{4 - 4(-2)}}{2} = \frac{-2 \pm \sqrt{12}}{2} = \frac{-2 \pm 2\sqrt{3}}{2}.$$

Hence, the given polynomial has two rational roots, -2 and $-\frac{2}{3}$, and two irrational roots, $-1 + \sqrt{3}$ and $-1 - \sqrt{3}$. ■

The Theorem on Rational Zeros may be applied to equations with rational coefficients. We merely multiply both sides of the equation by the least common denominator of all the coefficients to obtain an equation with integral coefficients, and then proceed as in Example 3.

EXAMPLE 4 Find all rational solutions of the equation

$$\tfrac{2}{3}x^4 + \tfrac{1}{2}x^3 - \tfrac{5}{4}x^2 - x - \tfrac{1}{6} = 0.$$

SOLUTION Multiplying both sides of the equation by 12, we obtain the equivalent equation

$$8x^4 + 6x^3 - 15x^2 - 12x - 2 = 0.$$

If c/d is a rational solution, then the choices for c are 1 and 2, and the choices for d are ± 1, ± 2, ± 4, and ± 8. Hence, the only possible rational roots are ± 1, ± 2, $\pm \frac{1}{2}$, $\pm \frac{1}{4}$, and $\pm \frac{1}{8}$. Trying various possibilities we obtain, by synthetic division,

$$\begin{array}{r|rrrrr} -\frac{1}{2} & 8 & 6 & -15 & -12 & -2 \\ & & -4 & -1 & 8 & 2 \\ \hline & 8 & 2 & -16 & -4 & 0 \end{array}$$

Hence, $-\frac{1}{2}$ is a solution. Using synthetic division on the coefficients of the quotient gives us

$$\begin{array}{r|rrrr} -\frac{1}{4} & 8 & 2 & -16 & -4 \\ & & -2 & 0 & 4 \\ \hline & 8 & 0 & -16 & 0 \end{array}$$

Consequently, $-\frac{1}{4}$ is a solution. The last synthetic division gave us the quotient $8x^2 - 16$. Setting this equal to zero and solving, we obtain $x^2 = 2$, or $x = \pm\sqrt{2}$. Thus, the given equation has rational solutions $-\frac{1}{2}$, $-\frac{1}{4}$ and irrational solutions $\sqrt{2}$, $-\sqrt{2}$. ■

EXAMPLE 5 A grain silo has the shape of a right circular cylinder with a hemisphere attached to the top. If the total height of the structure is 30 feet, what is the radius of the cylinder that will result in a total volume of 1008π cubic feet?

FIGURE 3.8

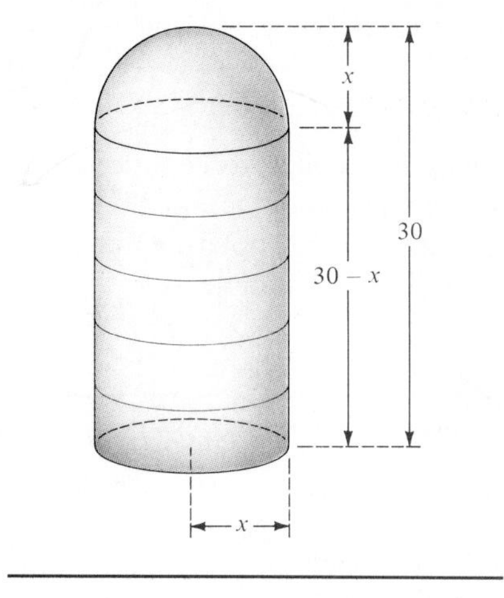

SOLUTION Let x denote the radius of the cylinder. A sketch of the silo, appropriately labeled, is shown in Figure 3.8. Since the volume of the cylinder is $\pi x^2(30 - x)$, and the volume of the hemisphere is $\frac{2}{3}\pi x^3$, we must solve the equation

$$\pi x^2(30 - x) + \tfrac{2}{3}\pi x^3 = 1008\pi.$$

Each of the following is equivalent to the preceding equation:

$$\begin{aligned} 3x^2(30 - x) + 2x^3 &= 3024 \\ 90x^2 - 3x^3 + 2x^3 &= 3024 \\ 90x^2 - x^3 &= 3024 \\ x^3 - 90x^2 + 3024 &= 0 \end{aligned}$$

To look for rational roots, we first factor 3024 into primes, obtaining $3024 = 2^4 \cdot 3^3 \cdot 7$. It follows that some of the positive factors of 3024 are

$$1,\ 2,\ 3,\ 4,\ 6,\ 8,\ 9,\ 12, \ldots$$

Dividing synthetically, we eventually arrive at

$$\begin{array}{r|rrrr} 6 & 1 & -90 & 0 & 3024 \\ & & 6 & -504 & -3024 \\ \hline & 1 & -84 & -504 & 0 \end{array}$$

Thus, 6 is a solution of the equation $x^3 - 90x^2 + 3024 = 0$, and the desired radius is 6 feet.

The remaining two solutions of the equation can be found by solving $x^2 - 84x - 504 = 0$. It is not difficult to show that neither of the solutions of this quadratic equation satisfy the conditions of the problem. ■

The discussion in this section gives no practical information about finding the irrational zeros of polynomials. The examples we have worked are not typical of problems encountered in applications. Indeed, polynomials with rational coefficients often have *no* rational zeros. A standard way to approximate irrational zeros is to use a calculus technique called Newton's Method. In practice, computers have taken over the task of approximating irrational solutions of equations.

Exercise 3.5

In Exercises 1–8 find a polynomial with real coefficients that has the given zero (or zeros) and degree.

1 $4 + i$; degree 2

2 $3 - 3i$; degree 2

3 $4, 3 - 2i$; degree 3

4 $-2, 1 + 5i$; degree 3

5 $1 + 4i, 2 - i$; degree 4

6 $3, 0, 8 - 7i$; degree 4

7 $5i, 1 + i, 0$; degree 5

8 $i, 2i, 3i$; degree 6

9 Does there exist a polynomial of degree 3 with real coefficients that has zeros $1, -1$, and i? Justify your answer.

10 The complex number i is a zero of the polynomial $f(x) = x^3 - ix^2 + 2ix + 2$; however, the conjugate $-i$ of i is not a zero. Why doesn't this contradict the first theorem of this section?

In Exercises 11–28 find all solutions of the given equations.

11 $x^3 - x^2 - 10x - 8 = 0$

12 $x^3 + x^2 - 14x - 24 = 0$

13 $2x^3 - 3x^2 - 17x + 30 = 0$

14 $12x^3 + 8x^2 - 3x - 2 = 0$

15 $x^4 + 3x^3 - 30x^2 - 6x + 56 = 0$

16 $3x^5 - 10x^4 - 6x^3 + 24x^2 + 11x - 6 = 0$

17 $2x^3 - 7x^2 - 10x + 24 = 0$

18 $2x^3 - 3x^2 - 8x - 3 = 0$

19 $6x^3 + 19x^2 + x - 6 = 0$

20 $6x^3 + 5x^2 - 17x - 6 = 0$

21 $8x^3 + 18x^2 + 45x + 27 = 0$

22 $3x^3 - x^2 + 11x - 20 = 0$

23 $x^4 - x^3 - 9x^2 - 3x - 36 = 0$

24 $3x^4 + 16x^3 + 28x^2 + 31x + 30 = 0$

25 $9x^4 + 15x^3 - 20x^2 - 20x + 16 = 0$

26 $15x^4 + 4x^3 + 11x^2 + 4x - 4 = 0$

27 $4x^5 + 12x^4 - 41x^3 - 99x^2 + 10x + 24 = 0$

28 $4x^5 + 24x^4 - 13x^3 - 174x^2 + 9x + 270 = 0$

Show that the equations in Exercises 29–36 have no rational roots.

29 $x^3 + 3x^2 - 4x + 6 = 0$

30 $3x^3 - 4x^2 + 7x + 5 = 0$

31 $x^5 - 3x^3 + 4x^2 + x - 2 = 0$

32 $2x^5 + 3x^3 + 7 = 0$

33 $3x^4 + 7x^3 + 3x^2 - 8x - 10 = 0$

34 $2x^4 - 3x^3 + 6x^2 - 24x + 5 = 0$

35 $8x^4 + 16x^3 - 26x^2 - 12x + 15 = 0$

36 $5x^5 + 2x^4 + x^3 - 10x^2 - 4x - 2 = 0$

37 If n is an odd positive integer, prove that a polynomial of degree n with real coefficients has at least one real zero.

38 Show that the theorem on page 132 is not necessarily true if $f(x)$ has complex coefficients.

39 Complete the proof of the Theorem on Rational Zeros by showing that d is a factor of a_n.

40 If a polynomial of the form

$$x^n + a_{n-1}x^{n-1} + \cdots + a_1x + a_0,$$

where each a_k is an integer, has a rational root r, show that r is an integer and is a factor of a_0.

41 An open box is to be made from a rectangular piece of cardboard having dimensions 20 inches × 30 inches by removing squares of area x^2 from each corner and turning up the sides. Show that there are two boxes that have a volume of 1000 in^3. Which box has the smaller surface area? (Compare Exercise 31 of Section 3.1.)

42 The frame for a shipping crate is to be constructed from 24 feet of 2 × 2 lumber. Assuming the crate is to have square ends of length x feet, determine the value(s) of x that result in a volume of 4 ft^3. (Compare Exercise 32 of Section 3.1.)

43 A meteorologist determines that the temperature T (in °F) on a certain cold winter day was given by $T = 0.05t(t - 12)(t - 24)$ for $0 \le t \le 24$, where t is time in hours and $t = 0$ corresponds to 6 A.M. At what time(s) of the day was the temperature 32 °F? (Compare Exercise 33 of Section 3.1.)

44 A herd of 100 deer is introduced to a small island. Assuming the number $N(t)$ of deer after t years is given by $N(t) = -t^4 + 21t^2 + 100$ (for $t > 0$), determine when the herd size exceeds 180. (Compare Exercise 35 of Section 3.1.)

45 A right triangle has area 30 ft^2 and a hypotenuse that is 1 foot longer than one of the legs.

(a) If x denotes the length of this leg, show that $2x^3 + x^2 - 3600 = 0$.

(b) Show that there is a single positive root of the equation in part (a) and that this root is less than 13.

(c) Find the dimensions of the triangle.

46 A storage tank for propane gas is to be constructed in the shape of a right circular cylinder of altitude 10 feet with a hemisphere attached to each end. Determine the radius x, so that the resulting volume is 27π ft^3. (Compare Example 8 of Section 2.1.)

47 A storage shelter is to be constructed in the shape of a cube with a triangular prism forming the roof (see figure). The length x of a side of the cube is yet to be determined.

(a) If the total height of the structure is 6 feet, show that its volume V is given by $V = x^3 + \frac{1}{2}x^2(6 - x)$.

(b) Determine x so that the volume is 80 ft^3.

FIGURE FOR EXERCISE 47

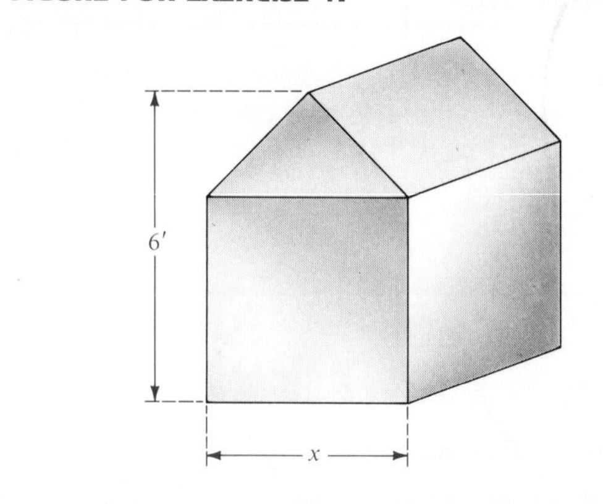

48 A canvas camping tent has the shape of a pyramid with a square base. An 8-foot pole will form the center support as illustrated in the figure. Find the length x of a side of the base so that the total canvas needed for the sides and bottom is 384 ft^2.

FIGURE FOR EXERCISE 48

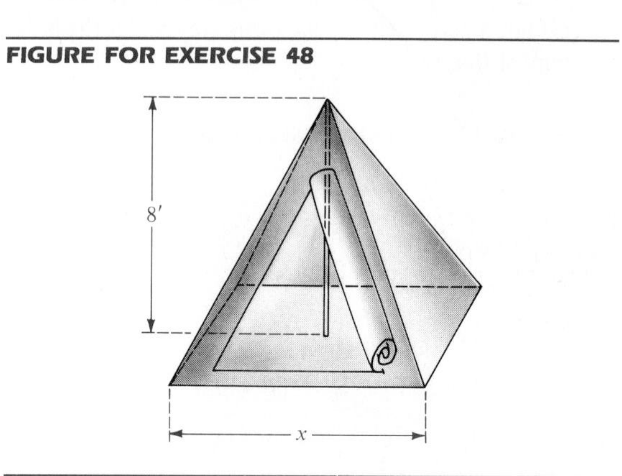

3.6 Rational Functions

A function f is a **rational function** if, for all x in its domain,

$$f(x) = \frac{g(x)}{h(x)}$$

where $g(x)$ and $h(x)$ are polynomials. The zeros of the numerator and denominator are of major importance. *Throughout this section we shall assume that $g(x)$ and $h(x)$ have no common factors* and hence no common zeros. If $g(c) = 0$, then $f(c) = 0$. However, if $h(c) = 0$, then $f(c)$ is undefined. As indicated in the next example, the behavior of $f(x)$ requires special attention when x is near a zero of the denominator $h(x)$.

EXAMPLE 1 Sketch the graph of f if

$$f(x) = \frac{1}{x-2}.$$

SOLUTION The numerator 1 is never zero and hence $f(x)$ has no zeros. This means that the graph has no x-intercepts.

The denominator $x - 2$ is zero at $x = 2$. If x is close to 2, and $x > 2$, then $f(x)$ is large. For example,

$$f(2.1) = \frac{1}{2.1-2} = \frac{1}{0.1} = 10$$

$$f(2.01) = \frac{1}{2.01-2} = \frac{1}{0.01} = 100$$

$$f(2.001) = \frac{1}{2.001-2} = \frac{1}{0.001} = 1000.$$

If x is close to 2, and $x < 2$, then $|f(x)|$ is large, but $f(x)$ is negative. Thus,

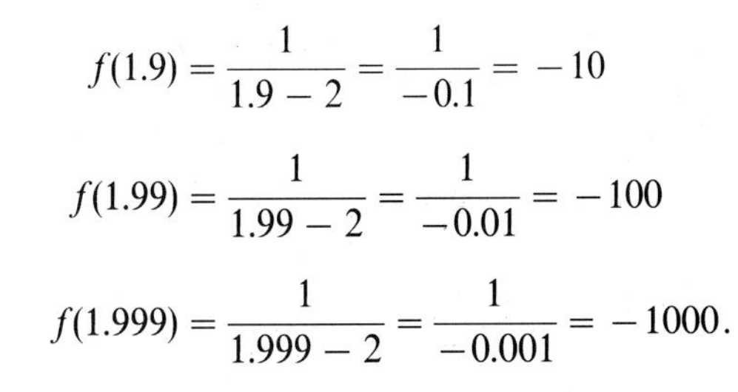

$$f(1.9) = \frac{1}{1.9-2} = \frac{1}{-0.1} = -10$$

$$f(1.99) = \frac{1}{1.99-2} = \frac{1}{-0.01} = -100$$

$$f(1.999) = \frac{1}{1.999-2} = \frac{1}{-0.001} = -1000.$$

Several other values of $f(x)$ are displayed in the following table:

x	-8	-1	0	1	3	4	12
$f(x)$	$-\frac{1}{10}$	$-\frac{1}{3}$	$-\frac{1}{2}$	-1	1	$\frac{1}{2}$	$\frac{1}{10}$

Observe that as $|x|$ increases, $f(x)$ approaches zero. Plotting points and paying attention to what happens near $x = 2$ gives us the sketch in Figure 3.9. ■

FIGURE 3.9

$$y = \frac{1}{x-2}$$

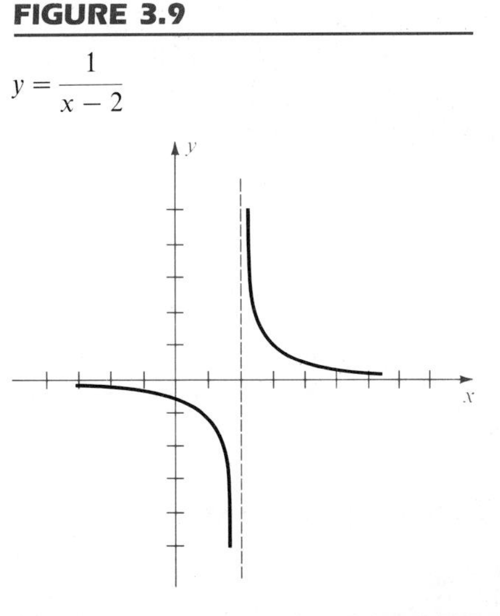

In Example 1, $f(x)$ can be made as large as desired by choosing x close to 2 (and $x > 2$). We denote this fact by writing

$$f(x) \to \infty \quad \text{as} \quad x \to 2^+.$$

We say that $f(x)$ *increases without bound* (or $f(x)$ *becomes positively infinite*) *as x approaches 2 from the right.* It is important to remember that the symbol ∞ (read "infinity") does not represent a real number, but is used merely as an abbreviation for certain types of functional behavior.

For the case $x < 2$ we write

$$f(x) \to -\infty \quad \text{as} \quad x \to 2^-$$

and say that $f(x)$ *decreases without bound* (or $f(x)$ *becomes negatively infinite*) *as x approaches 2 from the left.*

In general, the notation $x \to a^+$ will signify that x approaches a from the *right*, that is, through values *greater* than a. The symbol $x \to a^-$ will mean that x approaches a from the *left*, that is, through values *less* than a. Some illustrations of the manner in which a function f may increase or decrease without bound, together with the notation used, are shown in Figure 3.10. In the figure, a is pictured as positive, but we can also have $a \leq 0$.

FIGURE 3.10

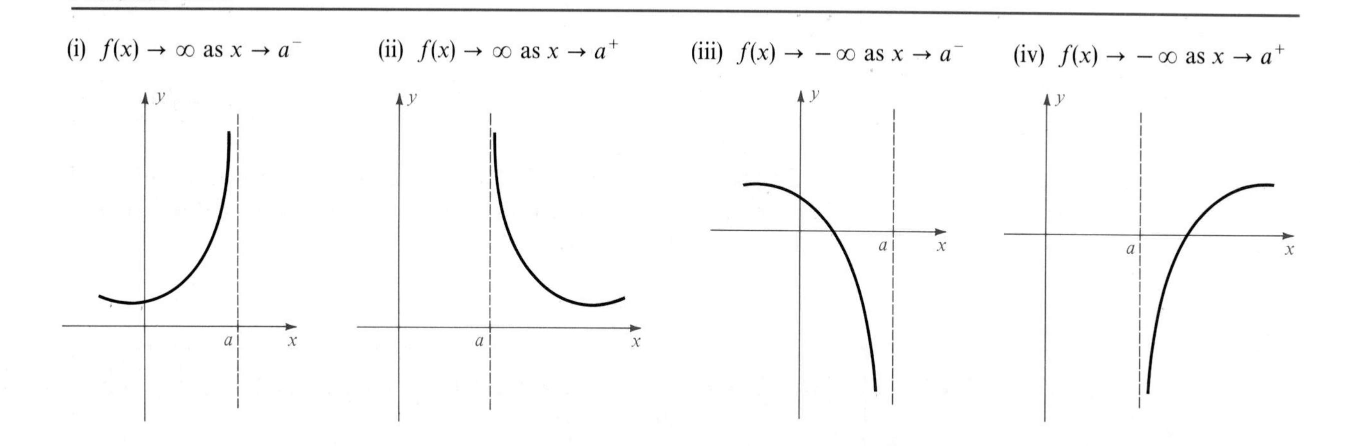

DEFINITION

The line $x = a$ is a **vertical asymptote** for the graph of a function f if

$$f(x) \to \infty \quad \text{or} \quad f(x) \to -\infty$$

as x approaches a from either the left or right.

In Figure 3.10 the dashed lines represent vertical asymptotes. Note that in Figure 3.9 the line $x = 2$ is a vertical asymptote for the graph of $y = 1/(x - 2)$.

Vertical asymptotes are common for graphs of rational functions. Indeed, *if the number a is a zero of the denominator* $h(x)$, *then the graph of* $f(x) = g(x)/h(x)$ *has the vertical asymptote* $x = a$.

We are also interested in values of $f(x)$ when $|x|$ is large. As an illustration, consider Example 1, where $f(x) = 1/(x - 2)$. If we assign very large values to x, then $f(x)$ is close to 0. Thus,

$$f(1002) = \frac{1}{1000} = 0.001 \quad \text{and} \quad f(1{,}000{,}002) = \frac{1}{1{,}000{,}000} = 0.000001.$$

Moreover, we can make $f(x)$ as close to 0 as we desire by choosing x sufficiently large. This is expressed symbolically by

$$f(x) \to 0 \quad \text{as} \quad x \to \infty,$$

which is read $f(x)$ *approaches* 0 *as x increases without bound* (or *as x becomes positively infinite*).

Similarly, in Example 1, we write

$$f(x) \to 0 \quad \text{as} \quad x \to -\infty,$$

which is read $f(x)$ *approaches* 0 *as x decreases without bound* (or *as x becomes negatively infinite*).

In Example 1 the line $y = 0$, that is, the x-axis, is called a *horizontal asymptote* for the graph. In general, we have the following definition. The notation should be self-evident.

DEFINITION

> The line $y = c$ is a **horizontal asymptote** for the graph of a function f if
>
> $$f(x) \to c \quad \text{as} \quad x \to \infty \quad \text{or as} \quad x \to -\infty.$$

Some typical horizontal asymptotes (for $x \to \infty$) are illustrated in Figure 3.11. The manner in which the graph "approaches" the line $y = c$ may vary, depending on the nature of the function. Similar sketches may be made for the case $x \to -\infty$. Note that, as in the third sketch, the graph of f may cross a horizontal asymptote.

FIGURE 3.11

$f(x) \to c$ as $x \to \infty$

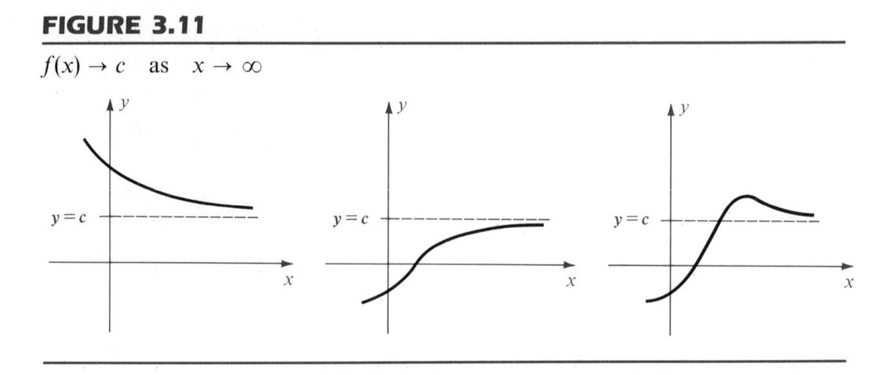

The following theorem is useful for locating horizontal asymptotes for the graph of a rational function.

THEOREM ON HORIZONTAL ASYMPTOTES

Let $f(x) = \dfrac{a_n x^n + a_{n-1}x^{n-1} + \cdots + a_1 x + a_0}{b_k x^k + b_{k-1}x^{k-1} + \cdots + b_1 x + b_0}$.

(i) If $n < k$, then the x-axis is a horizontal asymptote for the graph of f.

(ii) If $n = k$, then the line $y = a_n/b_k$ is a horizontal asymptote.

(iii) If $n > k$, the graph of f has no horizontal asymptote.

Proofs of (i) and (ii) of this theorem may be patterned after the solution to the following example. A similar argument can be given in case (iii).

EXAMPLE 2 Find the horizontal asymptotes for the graph of f if:

(a) $f(x) = \dfrac{3x - 1}{x^2 - x - 6}$ (b) $f(x) = \dfrac{5x^2 + 1}{3x^2 - 4}$

SOLUTION

(a) The degree of the numerator $3x - 1$ is less than the degree of the denominator $x^2 - x - 6$, and hence by (i) of the theorem, the x-axis is a horizontal asymptote. To verify this directly, we divide numerator and denominator of the quotient by x^2, obtaining

$$f(x) = \frac{\left(\dfrac{3x - 1}{x^2}\right)}{\left(\dfrac{x^2 - x - 6}{x^2}\right)} = \frac{\dfrac{3}{x} - \dfrac{1}{x^2}}{1 - \dfrac{1}{x} - \dfrac{6}{x^2}}, \qquad x \neq 0.$$

If x is very large, then both $1/x$ and $1/x^2$ are close to 0, and hence,

$$f(x) \approx \frac{0 - 0}{1 - 0 - 0} = \frac{0}{1} = 0.$$

Thus, $$f(x) \to 0 \quad \text{as} \quad x \to \infty.$$

Since $f(x)$ is the y-coordinate of a point on the graph, this means that the x-axis is a horizontal asymptote.

(b) If $f(x) = (5x^2 + 1)/(3x^2 - 4)$, then the numerator and denominator have the same degree, and hence by (ii) of the theorem, the line $y = \frac{5}{3}$ is a horizontal asymptote. This may be proved directly by dividing numerator and denominator of $f(x)$ by x^2, obtaining

$$f(x) = \frac{5 + \dfrac{1}{x^2}}{3 - \dfrac{4}{x^2}}.$$

Since $1/x^2 \to 0$ as $x \to \infty$, we see that

$$f(x) \to \frac{5 + 0}{3 - 0} = \frac{5}{3} \quad \text{as} \quad x \to \infty. \qquad \blacksquare$$

We shall next list some guidelines for sketching the graph of a rational function. Their use will be illustrated in Examples 3, 4, and 5.

GUIDELINES: Sketching the Graph of $f(x) = \dfrac{g(x)}{h(x)}$ where $g(x)$ and $h(x)$ are Polynomials That Have No Common Factor

STEP 1 Find the real zeros of the numerator $g(x)$ and use them to plot the points corresponding to the x-intercepts.

STEP 2 Find the real zeros of the denominator $h(x)$. For each zero a, the line $x = a$ is a vertical asymptote. Represent $x = a$ with dashes.

STEP 3 Find the sign of $f(x)$ in each of the intervals determined by the zeros of $g(x)$ and $h(x)$. Use these signs to determine whether the graph lies above or below the x-axis in each interval.

STEP 4 If $x = a$ is a vertical asymptote, use the information in Step 3 to determine whether $f(x) \to \infty$ or $f(x) \to -\infty$ for each case:

$$\text{(i)} \quad x \to a^-; \qquad \text{(ii)} \quad x \to a^+$$

Make note of this by sketching a portion of the graph on each side of $x = a$.

STEP 5 Use the information in Step 3 to determine the manner in which the graph intersects the x-axis.

STEP 6 Apply the Theorem on Horizontal Asymptotes. If there is a horizontal asymptote, represent it with dashes.

STEP 7 Sketch the graph, using the information found in the preceding steps and plotting points wherever necessary. ■■■

EXAMPLE 3 Sketch the graph of f if

$$f(x) = \frac{x - 1}{x^2 - x - 6}.$$

SOLUTION We begin by factoring the denominator as follows:

$$f(x) = \frac{x - 1}{(x + 2)(x - 3)}.$$

We shall obtain the graph by following the steps listed in the guidelines.

FIGURE 3.12

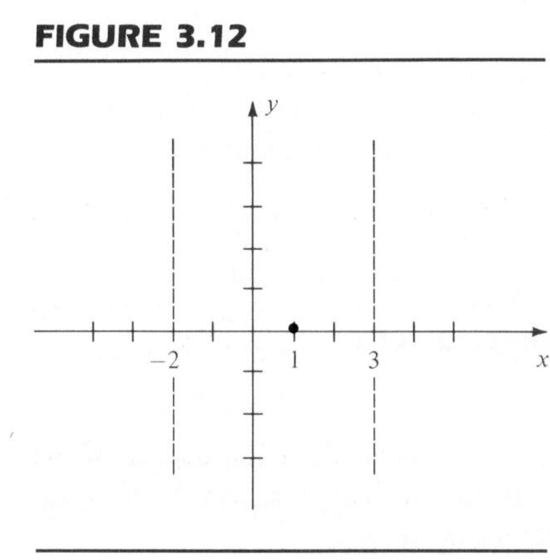

Step 1: The numerator $x - 1$ has the zero 1, and we plot the point $(1, 0)$ on the graph, as shown in Figure 3.12.

Step 2: The denominator has zeros -2 and 3. Hence, the lines $x = -2$ and $x = 3$ are vertical asymptotes, and we represent them with dashes, as in Figure 3.12.

Step 3: The zeros -2, 1, and 3 of the numerator and denominator of $f(x)$ determine the following intervals:

$$(-\infty, -2), \quad (-2, 1), \quad (1, 3), \quad \text{and} \quad (3, \infty).$$

Since $f(x)$ is a quotient of two polynomials, it follows from our work in Section 3.1 that $f(x)$ is always positive or always negative throughout each interval. Using test values to determine the sign of $f(x)$, we arrive at the following table.

Interval	$(-\infty, -2)$	$(-2, 1)$	$(1, 3)$	$(3, \infty)$
Test value	$f(-3) = -\frac{2}{3}$	$f(0) = \frac{1}{6}$	$f(2) = -\frac{1}{4}$	$f(4) = \frac{1}{2}$
Sign of $f(x)$	$-$	$+$	$-$	$+$
Position of graph	Below x-axis	Above x-axis	Below x-axis	Above x-axis

Step 4: We shall use the fourth row of the table in Step 3 to investigate the behavior of $f(x)$ near each vertical asymptote.

(a) Consider the vertical asymptote $x = -2$. Since the graph lies *below* the x-axis throughout the interval $(-\infty, -2)$, it follows that

$$f(x) \to -\infty \quad \text{as} \quad x \to -2^-.$$

Since the graph lies *above* the x-axis throughout the interval $(-2, 1)$, it follows that

$$f(x) \to \infty \quad \text{as} \quad x \to -2^+.$$

FIGURE 3.13

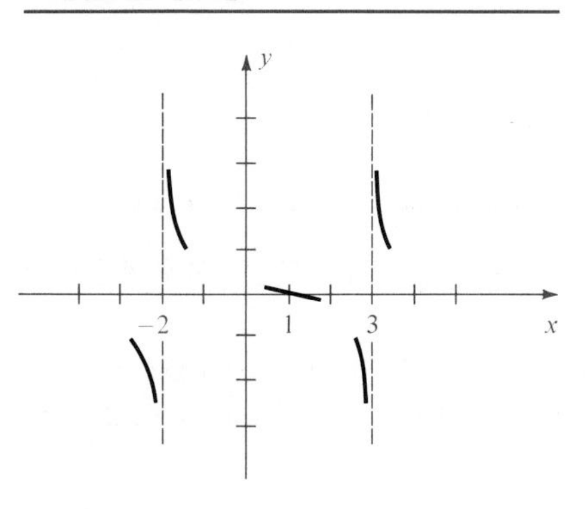

We note these facts in Figure 3.13 by sketching portions of the graph on each side of the line $x = -2$.

(b) Consider the vertical asymptote $x = 3$. The graph lies *below* the x-axis throughout the interval $(1, 3)$, and hence

$$f(x) \to -\infty \quad \text{as} \quad x \to 3^-.$$

The graph lies *above* the x-axis throughout $(3, \infty)$, and hence

$$f(x) \to \infty \quad \text{as} \quad x \to 3^+.$$

We note these facts in Figure 3.13 by sketching portions of the graph on each side of $x = 3$.

Step 5: Referring to the fourth row of the table in Step 3, we see that the graph crosses the x-axis at $(1, 0)$ in a manner similar to that illustrated in Figure 3.13.

Step 6: The degree of the numerator $x - 1$ is less than the degree of the denominator $x^2 - x - 6$. Hence, by (i) of the Theorem on Horizontal Asymptotes, the x-axis is a horizontal asymptote.

FIGURE 3.14

$y = \frac{x-1}{x^2 - x - 6}$

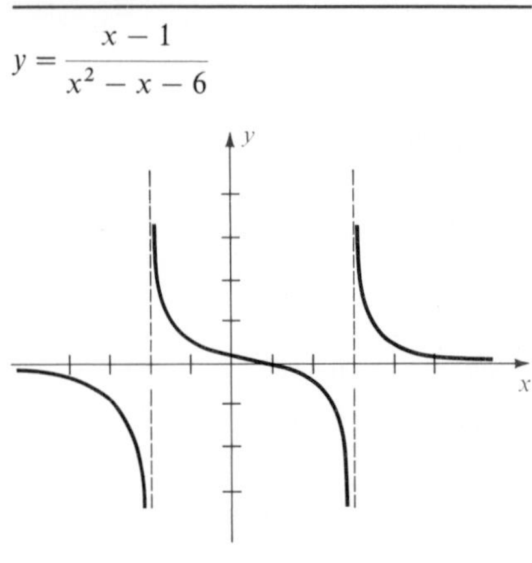

Step 7: Using the information found in Steps 4, 5, and 6, and plotting several points, we obtain the sketch in Figure 3.14. ■

EXAMPLE 4 Sketch the graph of f if

$$f(x) = \frac{x^2}{x^2 - x - 2}.$$

SOLUTION Factoring the denominator gives us

$$f(x) = \frac{x^2}{(x+1)(x-2)}.$$

We shall again follow the guidelines listed earlier.

Step 1: The numerator x^2 has 0 as a zero, and hence the graph intersects the x-axis at (0, 0), as shown in Figure 3.15.

FIGURE 3.15

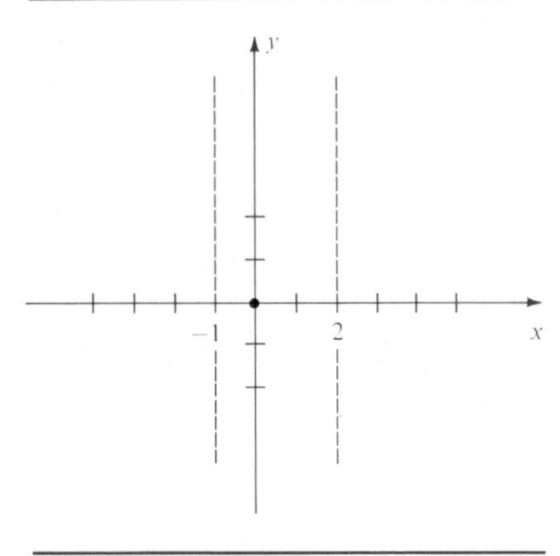

Step 2: Since the denominator has zeros -1 and 2, the lines $x = -1$ and $x = 2$ are vertical asymptotes, and we represent them with dashes, as in Figure 3.15.

Step 3: The intervals determined by the zeros in Steps 1 and 2 are

$$(-\infty, -1), \quad (-1, 0), \quad (0, 2), \quad \text{and} \quad (2, \infty).$$

Following the procedure used in Step 3 of Example 3, we arrive at the following table.

Interval	$(-\infty, -1)$	$(-1, 0)$	$(0, 2)$	$(2, \infty)$
Test value	$f(-2) = 1$	$f(-\frac{1}{2}) = -\frac{1}{5}$	$f(1) = -\frac{1}{2}$	$f(3) = \frac{9}{4}$
Sign of $f(x)$	+	−	−	+
Position of graph	Above x-axis	Below x-axis	Below x-axis	Above x-axis

Step 4: We refer to the fourth row of the table in Step 3 and proceed as follows:

(a) Consider the vertical asymptote $x = -1$. Since the graph is above the x-axis in $(-\infty, -1)$, it follows that

$$f(x) \to \infty \quad \text{as} \quad x \to -1^-.$$

FIGURE 3.16

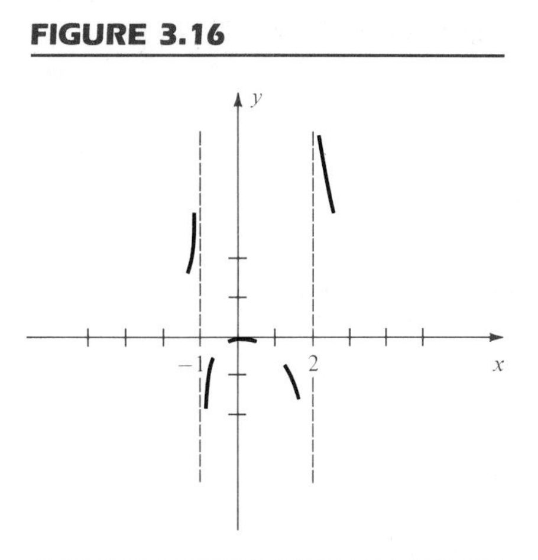

Since the graph is below the x-axis in $(-1, 0)$, we have

$$f(x) \to -\infty \quad \text{as} \quad x \to -1^{+}.$$

We note these facts in Figure 3.16 by sketching portions of the graph on each side of $x = -1$.

(b) Consider the vertical asymptote $x = 2$. The graph is below the x-axis in the interval $(0, 2)$, and hence

$$f(x) \to -\infty \quad \text{as} \quad x \to 2^{-}.$$

The graph is above the x-axis in $(2, \infty)$, and hence

$$f(x) \to \infty \quad \text{as} \quad x \to 2^{+}.$$

These facts are noted in Figure 3.16.

Step 5: Referring to the table in Step 3, we see that the graph lies below the x-axis in both of the intervals $(-1, 0)$ and $(0, 2)$. Consequently, the graph intersects, but does not cross, the x-axis at $(0, 0)$.

FIGURE 3.17

$y = \dfrac{x^2}{x^2 - x - 2}$

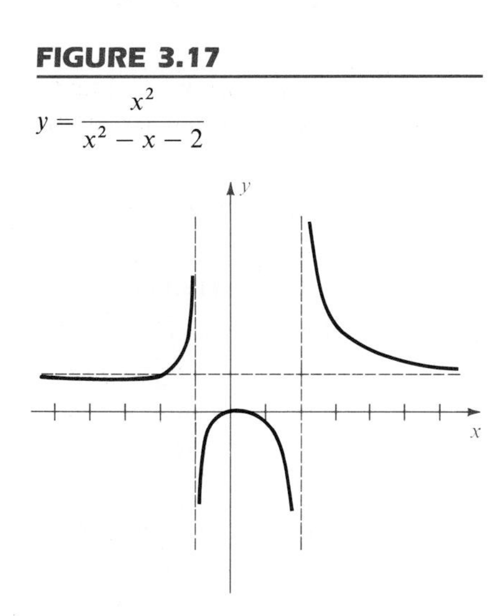

Step 6: The numerator and denominator of $f(x)$ have the same degree, and both leading coefficients are 1. Hence, by (ii) of the Theorem on Horizontal Asymptotes, the line $y = \frac{1}{1} = 1$ is a horizontal asymptote. We sketch this line with dashes, as in Figure 3.17.

Step 7: Using the information found in Steps 4, 5, and 6, and plotting several points, we obtain the graph sketched in Figure 3.17. The graph intersects the horizontal asymptote at $x = -2$; this may be verified by solving the equation $x^2/(x^2 - x - 2) = 1$. The fact that the graph lies below the horizontal asymptote if $x < -2$ and above it if $-2 < x < -1$ may be verified by plotting points. ■

EXAMPLE 5 Sketch the graph of f if

$$f(x) = \frac{2x^4}{x^4 + 1}.$$

SOLUTION In this solution we shall not formally write down each step in the guidelines. Note that since $f(-x) = f(x)$, the function is even, and hence the graph is symmetric with respect to the y-axis.

The graph intersects the x-axis at $(0, 0)$. Since the denominator of $f(x)$ has no real zeros, the graph has no vertical asymptotes.

The numerator and denominator of $f(x)$ have the same degree. Since the leading coefficients are 2 and 1, respectively, it follows from (ii) of the

FIGURE 3.18

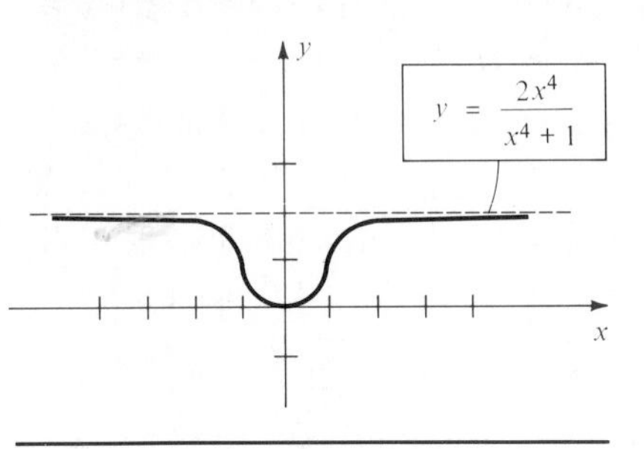

Theorem on Horizontal Asymptotes, that the line $y = \frac{2}{1} = 2$ is a horizontal asymptote. We represent this line with dashes in Figure 3.18.

Plotting several points and making use of the symmetry with respect to the y-axis leads to the sketch in Figure 3.18. ■

If $f(x) = g(x)/h(x)$ for polynomials $g(x)$ and $h(x)$, and *if the degree of* $g(x)$ *is one greater than the degree of* $h(x)$, then the graph of f has an **oblique asymptote** $y = ax + b$; that is, the graph approaches this line as $x \to \infty$ or as $x \to -\infty$. To find the oblique asymptote we may use division to express $f(x)$ in the form

$$f(x) = \frac{g(x)}{h(x)} = (ax + b) + \frac{r(x)}{h(x)}$$

where either $r(x) = 0$ or the degree of $r(x)$ is less than the degree of $h(x)$. It follows from (i) of the Theorem on Horizontal Asymptotes that

$$\frac{r(x)}{h(x)} \to 0 \quad \text{as} \quad x \to \infty \quad \text{or as} \quad x \to -\infty.$$

Consequently, $f(x)$ gets closer and closer to $ax + b$ as $|x|$ increases without bound. The next example illustrates a special case of this procedure.

EXAMPLE 6 Find all the asymptotes and sketch the graph of f if

$$f(x) = \frac{x^2 - 9}{2x - 4}.$$

SOLUTION A vertical asymptote occurs if $2x - 4 = 0$, that is, if $x = 2$.

The degree of the numerator of $f(x)$ is greater than the degree of the denominator. Hence by (iii) of the Theorem on Horizontal Asymptotes, there is no horizontal asymptote. However, since the degree of the numerator $x^2 - 9$ is *one* greater than the degree of the denominator $2x - 4$, the graph has an oblique asymptote. Dividing, we obtain

$$\begin{array}{r|l} & \tfrac{1}{2}x + 1 \\ \hline 2x - 4 & x^2 \qquad\quad - 9 \\ & x^2 - 2x \\ \cline{2-2} & \qquad 2x - 9 \\ & \qquad 2x - 4 \\ \cline{2-2} & \qquad\quad -5 \end{array}$$

Therefore, $$\frac{x^2 - 9}{2x - 4} = \left(\frac{1}{2}x + 1\right) - \frac{5}{2x - 4}.$$

As we indicated in the discussion preceding this example, the line $y = \frac{1}{2}x + 1$ is an oblique asymptote. This line and the vertical asymptote $x = 2$ are sketched (with dashes) in (i) of Figure 3.19.

FIGURE 3.19

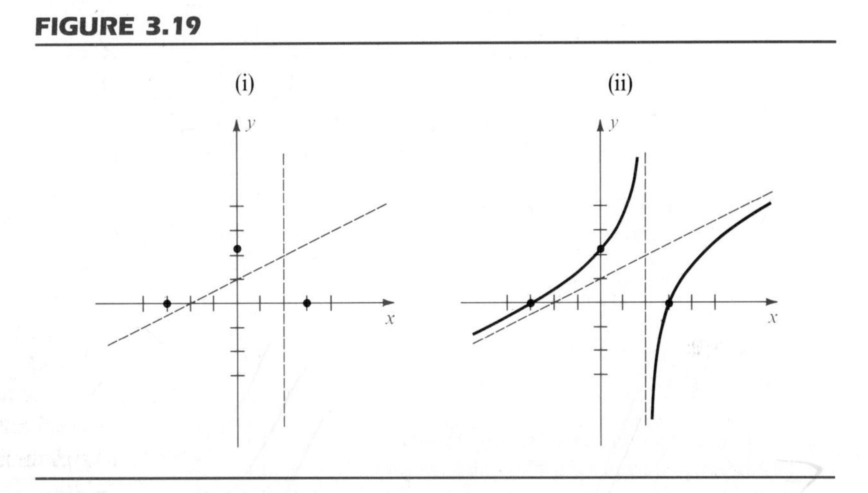

The x-intercepts of the graph are the solutions of the equation $x^2 - 9 = 0$, and hence are 3 and -3. The y-intercept is $f(0) = \frac{9}{4}$. The corresponding points are plotted in (i) of Figure 3.19. It is now easy to show that the graph has the shape indicated in (ii) of Figure 3.19. ■

Graphs of rational functions may become increasingly complicated as the degrees of the polynomials in the numerator and denominator increase. Techniques developed in calculus must be employed for a thorough treatment of such graphs.

Exercise 3.6

Sketch the graph of f in Exercises 1–20.

1 $f(x) = \dfrac{1}{x + 2}$

2 $f(x) = \dfrac{1}{x + 3}$

3 $f(x) = \dfrac{-2}{x + 4}$

4 $f(x) = \dfrac{-3}{x - 1}$

5 $f(x) = \dfrac{x}{x - 5}$

6 $f(x) = \dfrac{x}{3x + 2}$

7 $f(x) = \dfrac{4}{(x - 1)^2}$

8 $f(x) = \dfrac{-1}{(x + 2)^2}$

9 $f(x) = \dfrac{1}{x^2 - 4}$

10 $f(x) = \dfrac{2}{x^2 + x - 2}$

11 $f(x) = \dfrac{5x}{4 - x^2}$

12 $f(x) = \dfrac{x^2}{x^2 - 4}$

13 $f(x) = \dfrac{x^2}{x^2 - 7x + 10}$

14 $f(x) = \dfrac{x}{x^2 - x - 6}$

15 $f(x) = \dfrac{3x + 2}{x}$

16 $f(x) = \dfrac{x^2 - 4}{x^2}$

17 $f(x) = \dfrac{4}{x^2 + 4}$

18 $f(x) = \dfrac{3x}{x^2 + 1}$

19 $f(x) = \dfrac{1}{x^3 + x^2 - 6x}$

20 $f(x) = \dfrac{x^2 - x}{16 - x^2}$

In Exercises 21–24 find the vertical and oblique asymptotes, and sketch the graph of f.

21 $f(x) = \dfrac{x^2 - x - 6}{x + 1}$

22 $f(x) = \dfrac{2x^2 - x - 3}{x - 2}$

23 $f(x) = \dfrac{8 - x^3}{2x^2}$

24 $f(x) = \dfrac{x^3 + 1}{x^2 - 9}$

25 A cylindrical container for storing radioactive waste is to be constructed from lead. This container must be 6 inches thick (see figure). The volume of the outside cylinder shown in the figure is to be 16π ft^3.

(a) Express the height h of the inside cylinder as a function of the inside radius r.

(b) Show that the inside volume is given by a rational function V such that

$$V(r) = \pi r^2 \left[\frac{16}{(r + 0.5)^2} - 1 \right].$$

FIGURE FOR EXERCISE 25

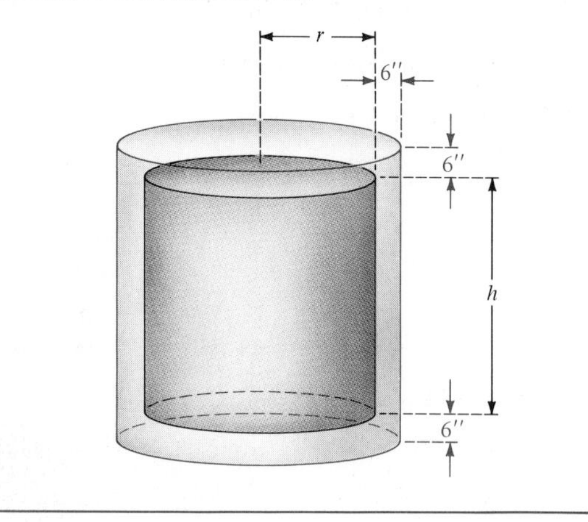

(c) What values of r must be excluded in the formula for $V(r)$?

26 Young's Rule is a formula that is used for modifying adult drug dosage levels for young children. If a denotes the adult dose (in mg), and if t is the age of the child (in years), then the child's dose y is given by $y = ta/(t + 12)$. Sketch the graph of this equation for $t > 0$.

27 Salt water of concentration 0.1 pounds of salt per gallon flows into a large tank that initially contains 50 gallons of pure water.

(a) If the flow rate of salt water into the tank is 5 gallons per minute, what is the volume $V(t)$ of water and the amount $A(t)$ of salt in the tank at time t?

(b) Show that the salt concentration at time t is given by $c(t) = t/(10t + 100)$.

(c) Discuss the behavior of $c(t)$ as $t \to \infty$.

28 An important problem in fishery science is predicting the next years' adult breeding population R (the recruits) from the number S that are presently spawning. For some species (such as North Sea herring), the relationship between R and S takes the form $R = aS/(S + b)$. What is the interpretation of the constant a? Conclude that for large values of S, recruitment is more or less constant.

29 Coulomb's Law in electricity asserts that the force of attraction F between two charged particles is inversely proportional to the square of the distance between the particles and directly proportional to the product of the charges. Suppose a particle of charge $+1$ is placed on a coordinate line between two particles of charge -1 as shown in the figure.

(a) Show that the net force acting on the particle of charge $+1$ is given by

$$F(x) = -\frac{k}{x^2} + \frac{k}{(x - 2)^2}$$

for some $k > 0$.

(b) Let $k = 1$ and sketch the graph of F for $0 < x < 2$.

FIGURE FOR EXERCISE 29

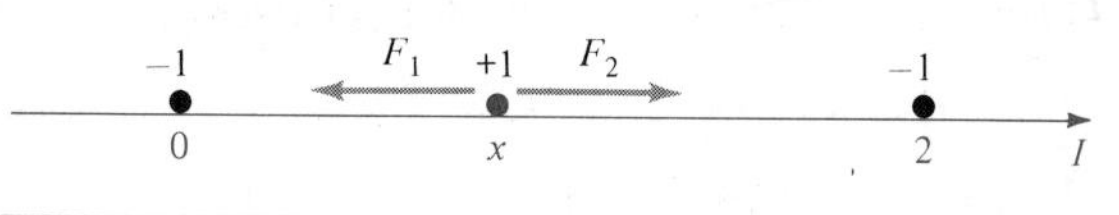

30 Biomathematicians have proposed many different functions for describing the effect of light on the rate at which photosynthesis can take place. If the function is to be realistic, then it must exhibit the *photoinhibition effect;* that is, the rate of production P of photosynthesis must decrease to 0 as the light intensity I reaches high levels (see figure). Which of the following functions might be used and which may not be used? Why?

(a) $P = \dfrac{aI}{b + I}$ (b) $P = \dfrac{aI}{b + I^2}$

FIGURE FOR EXERCISE 30

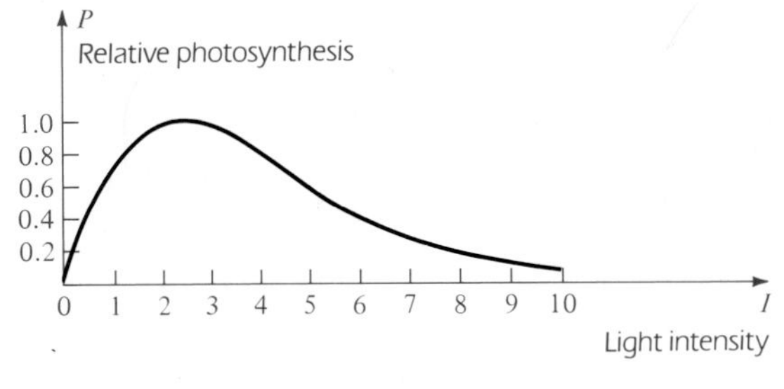

3.7 Review

Define or discuss each of the following

1. Intermediate Value Theorem
2. Graphs of polynomial functions of degree greater than 2
3. Division Algorithm for Polynomials
4. Remainder Theorem
5. Factor Theorem
6. Synthetic division
7. Complex numbers
8. Conjugate of a complex number
9. Fundamental Theorem of Algebra
10. Multiplicity of a zero of a polynomial
11. The number of zeros of a polynomial
12. Descartes' Rule of Signs
13. Upper and lower bounds for solutions of an equation
14. Rational zeros of a polynomial function
15. Rational function
16. Graph of a rational function
17. Vertical and horizontal asymptotes
18. Oblique asymptotes

Exercises 3.7

Sketch the graphs of the equations in Exercises 1 and 2.

1 $4y = (x + 2)(x - 1)^2(3 - x)$

2 $y = \frac{1}{15}(x^5 - 20x^3 + 64x)$

In Exercises 3–12 sketch the graph of f.

3 $f(x) = (x + 2)^3$

4 $f(x) = 2x^2 + x^3 - x^4$

5 $f(x) = x^3 + 2x^2 - 8x$

6 $f(x) = x^6 - 32$

7 $f(x) = \dfrac{-2}{(x+1)^2}$ **8** $f(x) = \dfrac{1}{(x-1)^3}$

9 $f(x) = \dfrac{3x^2}{16 - x^2}$

10 $f(x) = \dfrac{x}{(x+5)(x^2 - 5x + 4)}$

11 $f(x) = \dfrac{x^2 + 2x - 8}{x + 3}$ **12** $f(x) = \dfrac{x^4 - 16}{x^3}$

In Exercises 13–16 find the quotient and remainder if $f(x)$ is divided by $g(x)$.

13 $f(x) = 3x^5 - 4x^3 + x + 5$, $g(x) = x^3 - 2x + 7$

14 $f(x) = 7x^2 + 3x - 10$, $g(x) = x^3 - x^2 + 10$

15 $f(x) = 9x + 4$, $g(x) = 2x - 5$

16 $f(x) = 4x^3 - x^2 + 2x - 1$, $g(x) = x^2$

17 If $f(x) = -4x^4 + 3x^3 - 5x^2 + 7x - 10$, use the Remainder Theorem to find $f(-2)$.

18 Use the Remainder Theorem to prove that $x - 3$ is a factor of $2x^4 - 5x^3 - 4x^2 + 9$.

In Exercises 19 and 20 use synthetic division to find the quotient and remainder if $f(x)$ is divided by $g(x)$.

19 $f(x) = 6x^5 - 4x^2 + 8$, $g(x) = x + 2$

20 $f(x) = 2x^3 + 5x^2 - 2x + 1$, $g(x) = x - \sqrt{2}$

In Exercises 21 and 22 find polynomials with real coefficients that have the indicated zeros and degrees, and that satisfy the given conditions.

21 $-3 + 5i$, -1; degree 3; $f(1) = 4$

22 $1 - i$, 3, 0; degree 4; $f(2) = -1$

23 Find a polynomial of degree 7 such that -3 is a zero of multiplicity 2 and 0 is a zero of multiplicity 5.

24 Show that 2 is a zero of multiplicity 3 of the polynomial $x^5 - 4x^4 - 3x^3 + 34x^2 - 52x + 24$ and express this polynomial as a product of linear factors.

In Exercises 25 and 26 find the zeros of the polynomials and state the multiplicity of each zero.

25 $(x^2 - 2x + 1)^2(x^2 + 2x - 3)$

26 $x^6 + 2x^4 + x^2$

In Exercises 27 and 28 (a) use Descartes' Rule of Signs to determine the number of positive, negative, and nonreal complex solutions; and (b) find the smallest and largest integers that are upper and lower bounds, respectively, of the real solutions.

27 $2x^4 - 4x^3 + 2x^2 - 5x - 7 = 0$

28 $x^5 - 4x^3 + 6x^2 + x + 4 = 0$

29 Prove that $7x^6 + 2x^4 + 3x^2 + 10$ has no real zeros.

Find all solutions of the equations in Exercises 30–32.

30 $x^4 + 9x^3 + 31x^2 + 49x + 30 = 0$

31 $16x^3 - 20x^2 - 8x + 3 = 0$

32 $x^4 - 7x^2 + 6 = 0$

33 The cost $C(x)$ of cleaning up x percent of an oil spill that has washed ashore increases greatly as x approaches 100. Suppose that

$$C(x) = \frac{20x}{101 - x} \quad \text{(thousand dollars)}.$$

(a) Compare $C(100)$ to $C(90)$.

(b) Sketch the graph of C for $0 < x < 100$.

CHAPTER 4

Exponential and Logarithmic Functions

Exponential and logarithmic functions have applications in almost every field of human endeavor. ■ They are especially useful in the study of chemistry, biology, physics, and engineering to describe the manner in which quantities vary. ■ In this chapter we shall examine properties of these functions and consider many of their applications in everyday life.

4.1 Exponential Functions

Throughout this section the letter a will denote a positive real number. In Chapter 1 we defined a^r for every rational number r as follows: if m and n are integers with $n > 0$, then $a^{m/n} = \sqrt[n]{a^m}$. Using methods developed in calculus, we can define a^x for every *real* number x. To illustrate, for a^π we could use the nonterminating decimal representation 3.1415926 . . . for π and consider the following *rational* powers of a:

$$a^3, \quad a^{3.1}, \quad a^{3.14}, \quad a^{3.141}, \quad a^{3.1415}, \quad a^{3.14159}, \quad \ldots$$

If a^x is properly defined, then each successive power gets closer to a^π. In this chapter we shall assume that a^x can be obtained in similar fashion for every real number x and that the Laws of Exponents are valid in this more general setting.

It can be shown that the following result about rational exponents is also true for *real* exponents.

THEOREM

If a is a real number such that $a > 1$, then:
(i) $a^r > 1$ for every positive rational number r,
(ii) if r and s are rational numbers such that $r < s$, then $a^r < a^s$.

PROOF

(i) Multiplying both sides of the inequality $a > 1$ by a, we obtain $a^2 > a$ and hence $a^2 > a > 1$. Thus, $a^2 > 1$. Multiplying both sides of $a^2 > 1$ by a gives us $a^3 > a$ and hence $a^3 > 1$. Continuing this process, we see that $a^p > 1$ for every positive integer p. (A rigorous proof of this fact requires the method of mathematical induction.) Similarly, if $0 < a \leq 1$ it follows that $a^q \leq 1$ for every positive integer q.

Now consider $r = p/q$ for positive integers p and q. If it were true that $a^{p/q} \leq 1$, then from the previous discussion, $(a^{p/q})^q \leq 1$ or $a^p \leq 1$, which contradicts the fact that $a^p > 1$ for every positive integer p. Consequently, $a^r > 1$ for every positive rational number r.

(ii) If r and s are rational numbers such that $r < s$, then $s - r$ is a positive rational number, and hence from part (i), $1 < a^{s-r}$. Multiplying both sides of the last inequality by a^r,

$$a^r < (a^{s-r})a^r \quad \text{or} \quad a^r < a^s \qquad \square$$

Since to each real number x there corresponds a unique real number a^x, we can define a function as follows:

DEFINITION

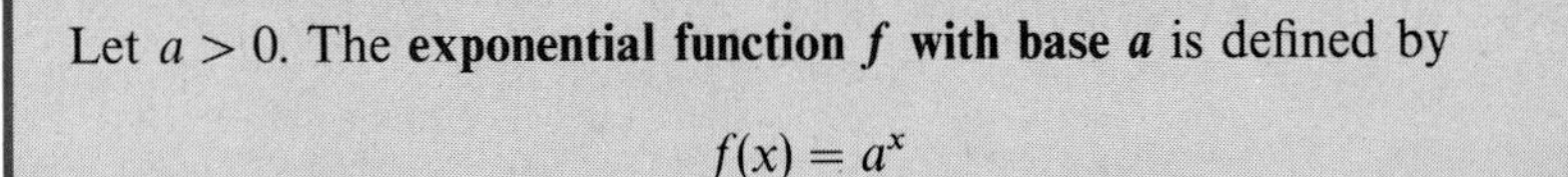

Let $a > 0$. The **exponential function f with base a** is defined by

$$f(x) = a^x$$

for every real number x.

FIGURE 4.1

If $a > 1$, and if x_1 and x_2 are real numbers such that $x_1 < x_2$, then $a^{x_1} < a^{x_2}$; that is, $f(x_1) < f(x_2)$. This means that if $a > 1$, then the exponential function f with base a is increasing throughout $\mathbb{R}$. It can also be shown that if $0 < a < 1$, then f is decreasing throughout $\mathbb{R}$.

EXAMPLE 1 Sketch the graph of f if $f(x) = 2^x$.

SOLUTION Coordinates of some points on the graph of $y = 2^x$ are listed in the following table.

x	-3	-2	-1	0	1	2	3	4
y	$\frac{1}{8}$	$\frac{1}{4}$	$\frac{1}{2}$	1	2	4	8	16

Plotting points and using the fact that f is increasing gives us the sketch in Figure 4.1. ■

For any $a > 1$, the graph of the exponential function with base a has the general appearance of the graph in Figure 4.2(i); however, the exact

FIGURE 4.2

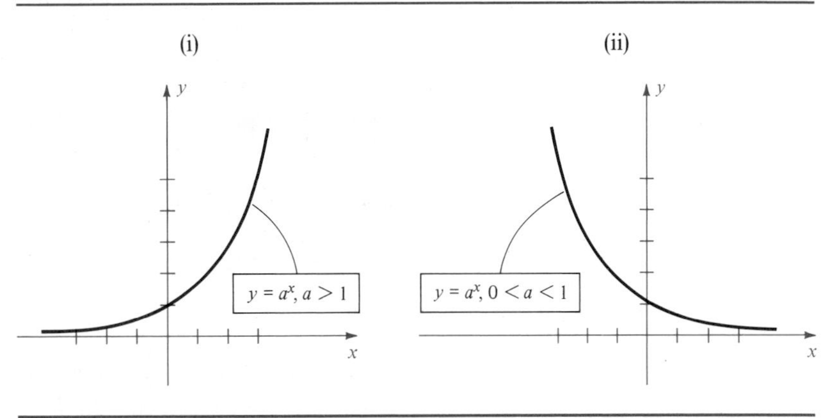

shape depends on the value of a. If $0 < a < 1$, the graph has the appearance illustrated in (ii) of the figure. In both cases the domain of f is $\mathbb{R}$ and the range is the set of positive real numbers.

Since $a^0 = 1$, the y-intercept is always 1. If $a > 1$, then as x decreases through negative values, the graph approaches the x-axis but never intersects it, since $a^x > 0$ for all x. This means that the x-axis is a *horizontal asymptote* for the graph. As x increases through positive values, the graph rises very rapidly. Indeed, given $f(x) = 2^x$, if we begin with $x = 0$ and consider successive unit changes in x, then the corresponding changes in y are 1, 2, 4, 8, 16, 32, 64, and so on. This type of variation is characteristic of the **exponential law of growth.** In this case f is called a **growth function.** Figure 4.2(ii) illustrates **exponential decay.**

FIGURE 4.3

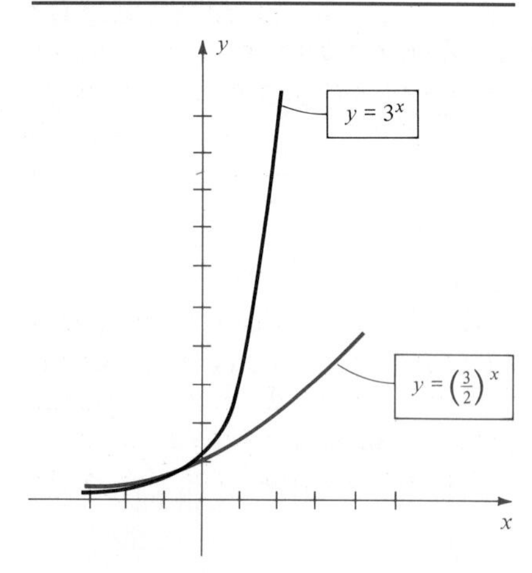

EXAMPLE 2 If $f(x) = (\frac{3}{2})^x$ and $g(x) = 3^x$, sketch the graphs of f and g on the same coordinate plane.

SOLUTION The following table displays coordinates of several points on the graphs.

x	-2	-1	0	1	2	3	4
$(\frac{3}{2})^x$	$\frac{4}{9} \approx .4$	$\frac{2}{3} \approx .7$	1	$\frac{3}{2}$	$\frac{9}{4} \approx 2.3$	$\frac{27}{8} \approx 3.4$	$\frac{81}{16} \approx 5.1$
3^x	$\frac{1}{9} \approx .1$	$\frac{1}{3} \approx .3$	1	3	9	27	81

Plotting points we obtain Figure 4.3, in which color has been used for the graph of f to distinguish it from the graph of g. ■

Example 2 illustrates the fact that if $1 < a < b$, then $a^x < b^x$ for positive values of x and $b^x < a^x$ for negative values of x. In particular, since $\frac{3}{2} < 2 < 3$, the graph of $y = 2^x$ in Example 1 lies between the graphs of f and g in Example 2.

FIGURE 4.4

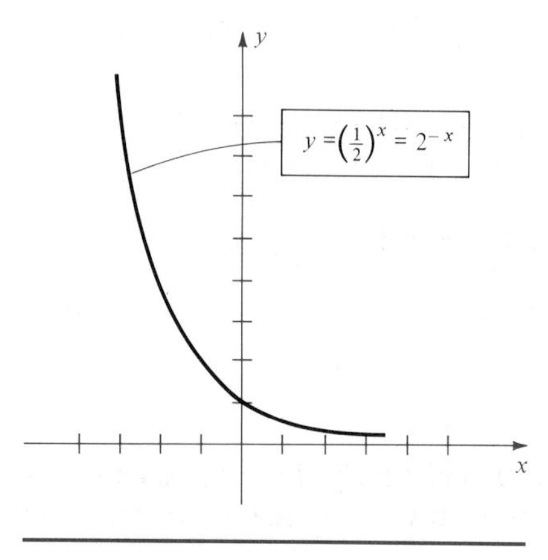

EXAMPLE 3 Sketch the graph of the equation $y = (\frac{1}{2})^x$.

SOLUTION Some points on the graph may be obtained from the following table:

x	-3	-2	-1	0	1	2	3
$(\frac{1}{2})^x$	8	4	2	1	$\frac{1}{2}$	$\frac{1}{4}$	$\frac{1}{8}$

The graph is sketched in Figure 4.4. Since $(\frac{1}{2})^x = 2^{-x}$, the graph is the same as the graph of the equation $y = 2^{-x}$. ■

In advanced mathematics and applications it is often necessary to consider functions such that $f(x) = a^p$, where p is some expression in x. The next example illustrates the case $p = -x^2$.

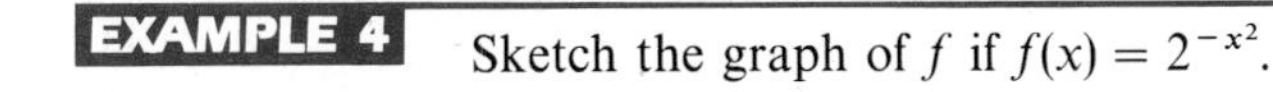

EXAMPLE 4 Sketch the graph of f if $f(x) = 2^{-x^2}$.

SOLUTION If we rewrite $f(x)$ as

$$f(x) = \frac{1}{2^{(x^2)}}$$

then it is evident that as $|x|$ increases, the point $(x, f(x))$ approaches the x-axis. Thus, the x-axis is a horizontal asymptote for the graph. The maximum value of $f(x)$ occurs at $x = 0$. Since f is an even function, the graph is symmetric with respect to the y-axis. Several points on the graph are $(0, 1)$, $(1, \frac{1}{2})$, and $(2, \frac{1}{16})$. Plotting and using symmetry gives us the sketch in Figure 4.5. Functions similar to f arise in the branch of mathematics called *probability*. ■

FIGURE 4.5

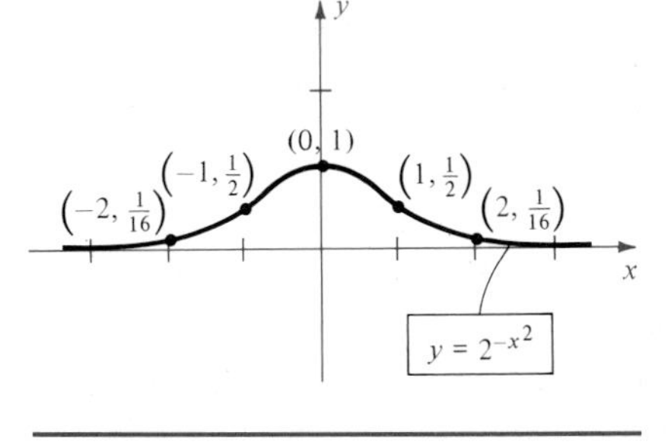

APPLICATION: Bacterial Growth

Exponential functions occur in the study of the growth of certain populations. As an illustration, it might be observed experimentally that the number of bacteria in a culture doubles every hour. If 1000 bacteria are present at the start of the experiment, then the experimenter would obtain the readings listed below, where t is the time in hours and $f(t)$ is the bacteria count at time t.

t (time)	0	1	2	3	4
$f(t)$ (bacteria count)	1000	2000	4000	8000	16,000

It appears that $f(t) = (1000)2^t$. With this formula we can predict the number of bacteria present at any time t. For example, at $t = 1.5 = \frac{3}{2}$,

$$f(t) = (1000)2^{3/2} \approx 2828.$$

The graph of f is sketched in Figure 4.6. ■ ■ ■

FIGURE 4.6

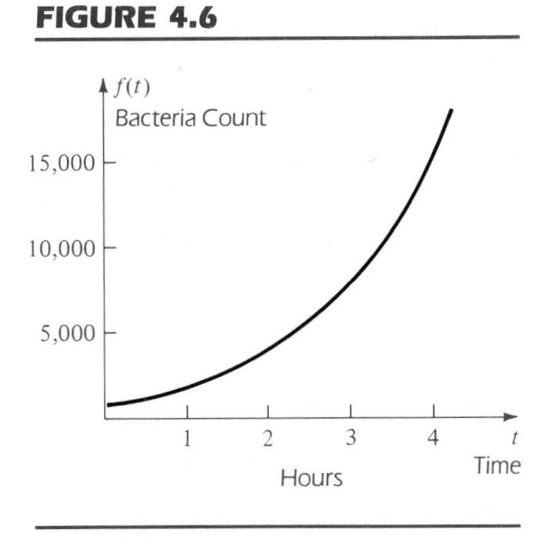

APPLICATION: Radioactive Decay

Certain physical quantities *decrease* exponentially. In such cases if a is the base of the exponential function, then $0 < a < 1$. One of the most

FIGURE 4.7

Decay of polonium

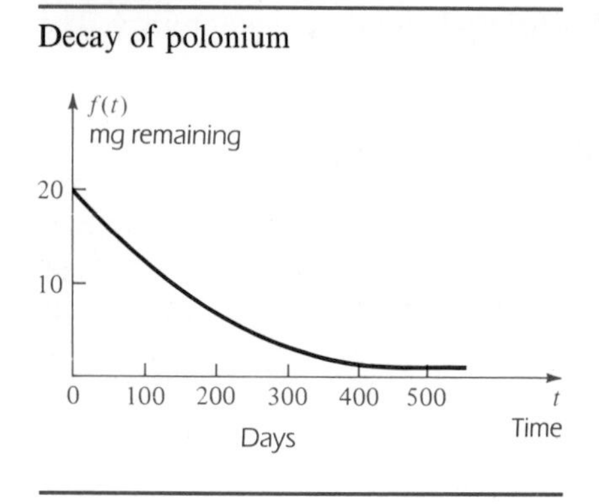

common examples is the decay of a radioactive substance. As an illustration, the polonium isotope ^{210}Po has a half-life of approximately 140 days; that is, given any amount, one-half of it will disintegrate in 140 days. If 20 mg of ^{210}Po is present initially, then the following table indicates the amount remaining after various intervals of time.

t (days)	0	140	280	420	560
Amount remaining (mg)	20	10	5	2.5	1.25

The sketch in Figure 4.7 illustrates the exponential nature of the disintegration. ■■■

APPLICATION: Compound Interest

Compound interest provides a good illustration of exponential growth. If a sum of money P, called the **principal,** is invested at a *simple* interest rate r, then the interest at the end of one interest period is the product Pr when r is expressed as a decimal. For example, if $P = \$1000$ and the interest rate is 9% per year, then $r = 0.09$, and the interest at the end of one year is \$1000(0.09), or \$90.

If the interest is reinvested at the end of this period, then the new principal is

$$P + Pr \quad \text{or} \quad P(1 + r).$$

Note that to find the new principal we multiply the original principal by $(1 + r)$. In the preceding illustration the new principal is \$1000(1.09), or \$1090.

After another time period has elapsed, the new principal may be found by multiplying $P(1 + r)$ by $(1 + r)$. Thus, the principal after two time periods is $P(1 + r)^2$. If we continue to reinvest, the principal after three periods is $P(1 + r)^3$; after four it is $P(1 + r)^4$; and in general, the amount A invested after k time periods is

$$A = P(1 + r)^k.$$

Interest accumulated by means of this formula is called **compound interest.** Note that A is expressed in terms of an exponential function with base $1 + r$. The time period may vary and may be measured in years, months, weeks, days, or any other suitable unit of time. When applying the formula for A, remember that r is the interest rate per time period expressed as a decimal. For example, if the rate is stated as 6% *per year compounded* *monthly*, then the rate per month is $\frac{6}{12}\%$, or equivalently, 0.5%. Thus, $r = 0.005$ and k is the number of months. If \$100 is invested at this rate, then the formula for A is

$$A = 100(1 + 0.005)^k = 100(1.005)^k$$

Generally, suppose that r is the yearly interest rate (expressed as a decimal) and that interest is compounded n times per year. The interest rate per time period is r/n. If the principal P is invested for t years, then the number of interest periods is nt, and the amount A after t years is given by the following formula.

COMPOUND INTEREST FORMULA

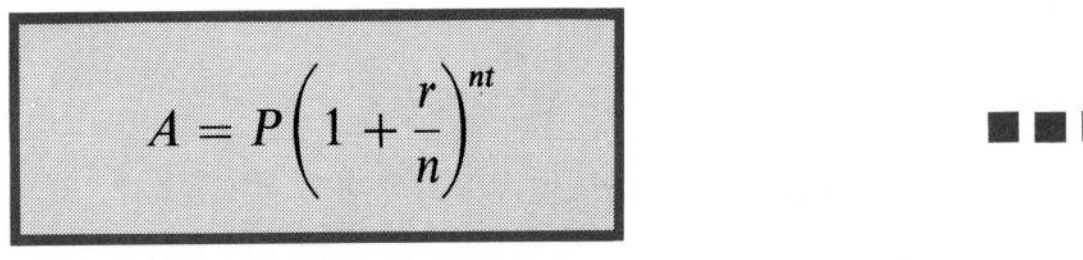

$$A = P\left(1 + \frac{r}{n}\right)^{nt}$$

EXAMPLE 5 Suppose that \$1000 is invested at an interest rate of 9% compounded monthly. Find the new amount of principal after 5 years; after 10 years; after 15 years. Illustrate graphically the growth of the investment.

SOLUTION Applying the Compound Interest Formula with $r = 0.09$, $n = 12$, and $P = \$1000$, the amount after t years is

$$A = 1000\left(1 + \frac{0.09}{12}\right)^{12t} = 1000(1.0075)^{12t}.$$

Substituting $t = 5$, 10, and 15, and using a calculator, we obtain the following amounts:

After 5 years: $A = 1000(1.0075)^{60} = \$1565.68$

After 10 years: $A = 1000(1.0075)^{120} = \$2451.36$

After 15 years: $A = 1000(1.0075)^{180} = \$3838.04$

The exponential nature of the increase is indicated by the fact that during the first five years, the growth in the investment is \$565.68; during the second five-year period, the growth is \$885.68; and during the last five-year period, it is \$1368.68.

The sketch in Figure 4.8 illustrates the growth of \$1000 invested over a period of 15 years. ■

FIGURE 4.8

Compound interest: $A = 1000(1.0075)^{12t}$

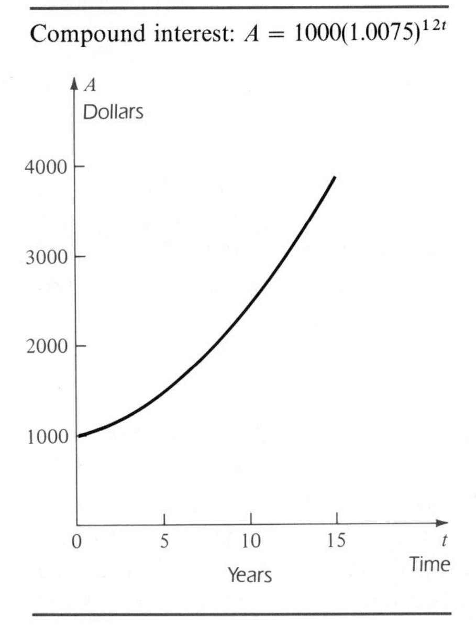

Exercises 4.1

In Exercises 1–24 sketch the graph of the function f.

1 $f(x) = 4^x$ **2** $f(x) = 5^x$

3 $f(x) = 10^x$ **4** $f(x) = 8^x$

5 $f(x) = 3^{-x}$ **6** $f(x) = 4^{-x}$

7 $f(x) = -2^x$ **8** $f(x) = -3^x$

9 $f(x) = 4 - 2^{-x}$ **10** $f(x) = 2 + 3^{-x}$

11 $f(x) = (\frac{2}{3})^x$ **12** $f(x) = (\frac{3}{4})^{-x}$

13 $f(x) = (\frac{5}{2})^{-x}$ **14** $f(x) = (\frac{4}{3})^x$

15 $f(x) = 2^{|x|}$ **16** $f(x) = 2^{-|x|}$

17 $f(x) = 2^{x+3}$ **18** $f(x) = 3^{x+2}$

19 $f(x) = 2^{3-x}$ **20** $f(x) = 3^{-2-x}$

21 $f(x) = 3^{1-x^2}$ **22** $f(x) = 2^{-(x+1)^2}$

23 $f(x) = 3^x + 3^{-x}$ **24** $f(x) = 3^x - 3^{-x}$

25 One hundred elk, each one year old, are introduced into a game preserve. The number $N(t)$ still alive after t years is predicted to be $N(t) = 100(0.9)^t$. Estimate the number of elk still alive after (a) 1 year; (b) 5 years; (c) 10 years.

26 A drug is eliminated from the body through urine. The initial dose is 10 mg and the amount $A(t)$ in the body t hours later is given by $A(t) = 10(0.8)^t$.

(a) Estimate the amount of the drug in the body 8 hours after the initial dose.

(b) What percentage of the drug still in the body is eliminated each hour?

27 The number of bacteria in a certain culture increased from 600 to 1800 between 7:00 A.M. and 9:00 A.M. Assuming exponential growth and using methods of calculus, it can be shown that the number $f(t)$ of bacteria t hours after 7:00 A.M. was given by $f(t) = 600(3)^{t/2}$.

(a) Estimate the number of bacteria in the culture at 8:00 A.M.; 10:00 A.M.; 11:00 A.M.

(b) Sketch the graph of f from $t = 0$ to $t = 4$.

28 According to Newton's Law of Cooling, the rate at which an object cools is directly proportional to the difference in temperature between the object and the surrounding medium. If a certain object cools from 125° to 100° in 30 minutes when surrounded by air that has a temperature of 75°, then its temperature $f(t)$ after t hours of cooling is given by $f(t) = 50(2)^{-2t} + 75$.

(a) Assuming $t = 0$ corresponds to 1:00 P.M., approximate to the nearest tenth of a degree the temperature at 2:00 P.M., 3:30 P.M., and 4:00 P.M.

(b) Sketch the graph of f from $t = 0$ to $t = 4$.

29 The radioactive isotope ^{210}Bi has a half-life of 5 days; that is, the number of radioactive particles will decrease to one-half the number in 5 days. If there are 100 mg of ^{210}Bi present at $t = 0$, then the amount $f(t)$ remaining after t days is given by $f(t) = 100(2)^{-t/5}$.

(a) How much ^{210}Bi remains after 5 days? 10 days? 12.5 days?

(b) Sketch the graph of f from $t = 0$ to $t = 30$.

30 An important problem in oceanography is to determine the amount of light that can penetrate to various ocean depths. The Beer-Lambert Law asserts that an exponential function I such that $I(x) = I_0a^x$ should be used to model this phenomenon. Assuming $I(x) = 10(0.4)^x$ is the amount of light (in calories/cm²/second) reaching a depth of x meters,

(a) What is the amount of light at a depth of 2 meters?

(b) Sketch the graph of I from $x = 0$ to $x = 5$.

31 The half-life of radium is 1600 years; that is, given any quantity, one-half of it will disintegrate in 1600 years. If the initial amount is q_0 milligrams, then the quantity $q(t)$ remaining after t years is given by $q(t) = q_0 2^{kt}$. Find k.

32 If 10 grams of salt are added to a quantity of water, then the amount $q(t)$ that is undissolved after t minutes is given by $q(t) = 10(\frac{4}{5})^t$. Sketch a graph that shows the value $q(t)$ at any time from $t = 0$ to $t = 10$.

33 If $1000 is invested at a rate of 12% per year compounded monthly, what is the principal after (a) 1 month? (b) 2 months? (c) 6 months? (d) 1 year?

34 If a savings fund pays interest at a rate of 10% compounded semiannually, how much money invested now will amount to $5000 after one year?

35 If a certain make of automobile is purchased for C dollars, then its trade-in value $v(t)$ at the end of t years is given by $v(t) = 0.78C(0.85)^{t-1}$. If the original cost is \$10,000, calculate to the nearest dollar the value after (a) 1 year; (b) 4 years; (c) 7 years.

36 If the value of real estate increases at a rate of 10% per year, then after t years the value V of a house purchased for P dollars is given by $V = P(1.1)^t$. If a house is purchased for \$80,000 in 1986, what will it be worth in 1990?

37 Why was $a < 0$ ruled out in the discussion of a^x?

38 Prove that if $0 < a < 1$ and r and s are rational numbers such that $r < s$, then $a^r > a^s$.

39 How does the graph of $y = a^x$ compare with the graph of $y = -a^x$?

40 If $a > 1$, how does the graph of $y = a^x$ compare with the graph of $y = a^{-x}$?

Calculator Exercises 4.1

In Exercises 1 and 2 sketch the graph for $-3 \leq x \leq 3$ by choosing values of x at intervals of length 0.5; that is, $x = -3, -2.5, -2, -1.5$, and so on.

1 $y = (1.8)^x$ **2** $y = (2.3)^{-x^2}$

Solve Exercises 3–8 by using the Compound Interest Formula.

3 Assuming \$1000 is invested at an interest rate of 6% per year compounded quarterly, find the principal at the end of (a) one year; (b) two years; (c) five years; (d) ten years.

4 Rework Exercise 3 for an interest rate of 6% per year compounded monthly.

5 If \$10,000 is invested at a rate of 9% per year compounded semiannually, how long will it take for the principal to exceed (a) \$15,000? (b) \$20,000? (c) \$30,000?

6 A certain department store requires its credit card customers to pay interest at the rate of 18% per year, compounded monthly, on any unpaid bills. If a man buys a television set for \$500 on credit and then makes no payments for one year, how much does he owe at the end of the year?

7 A boy deposits \$500 in a savings account that pays interest at a rate of 6% per year compounded weekly. How much is in the account after one year?

8 A savings fund pays interest at the rate of 9% compounded daily. How much should be invested in order to have \$2000 at the end of 10 weeks?

4.2 The Natural Exponential Function

At the end of Section 4.1 we discussed the *Compound Interest Formula*

$$A = P\left(1 + \frac{r}{n}\right)^{nt}$$

where P is the principal invested, r is the interest rate (expressed as a decimal), n is the number of interest periods per year, and t is the number of years that the principal is invested. The next example illustrates what happens if the rate and total time invested are fixed, but the *time period* for compounding interest is varied.

EXAMPLE 1 Suppose \$1000 is invested at a compound interest rate of 9%. Find the new amount of principal after one year if the interest is compounded monthly; weekly; daily; hourly; each minute.

SOLUTION If we let $P = \$1000$, $t = 1$, and $r = 0.09$ in the Compound Interest Formula, then

$$A = 1000\left(1 + \frac{0.09}{n}\right)^n$$

for n interest periods per year. To find the desired amounts, we let n have the following values:

$$12, \quad 52, \quad 365, \quad 8760, \quad 525{,}600.$$

We have assumed there are 365 days in a year and, hence, $(365)(24) = 8760$ hours, and $(8760)(60) = 525{,}600$ minutes. (Actually, in business transactions an investment year is considered to be 360 days). Using the Compound Interest Formula (and a calculator), we obtain the following table:

Time period for compounding interest	Amount of principal after one year
Month	$1000\left(1 + \frac{0.09}{12}\right)^{12} = \$1093.81$
Week	$1000\left(1 + \frac{0.09}{52}\right)^{52} = \$1094.09$
Day	$1000\left(1 + \frac{0.09}{365}\right)^{365} = \$1094.16$
Hour	$1000\left(1 + \frac{0.09}{8760}\right)^{8760} = \$1094.17$
Minute	$1000\left(1 + \frac{0.09}{525{,}600}\right)^{525{,}600} = \$1094.17$

■

Note that, in the preceding example, after a certain time period is reached, the number of interest periods per year has little effect on the final amount. If interest had been compounded each *second*, the result would still be \$1094.17. Thus, the amount approaches a fixed value as n increases. Interest is said to be **compounded continuously** if the number n of time periods per year increases without bound. Evidently, if we allow this to happen in Example 1, the amount of principal after one year is the same as that obtained for a time period of one hour, or of one minute.

If we let $P = 1$, $r = 1$, and $t = 1$ in the Compound Interest Formula, we obtain

$$A = \left(1 + \frac{1}{n}\right)^n.$$

The expression on the right of the equation occurs in the study of calculus. In Example 1, in a similar situation, as n increased, A approached a limiting value. The same phenomenon occurs here, as illustrated by the following table, which was obtained using a calculator.

n	Approximation to $\left(1 + \frac{1}{n}\right)^n$
1	2.0000000
10	2.5937425
100	2.7048138
1000	2.7169238
10,000	2.7181459
100,000	2.7182546
1,000,000	2.7182818
10,000,000	2.7182818

It can be proved that as n increases, $[1 + (1/n)]^n$ gets closer and closer to a certain irrational number, denoted by e. As indicated by the values in the table, e can be assigned the following decimal approximation:

FIGURE 4.9

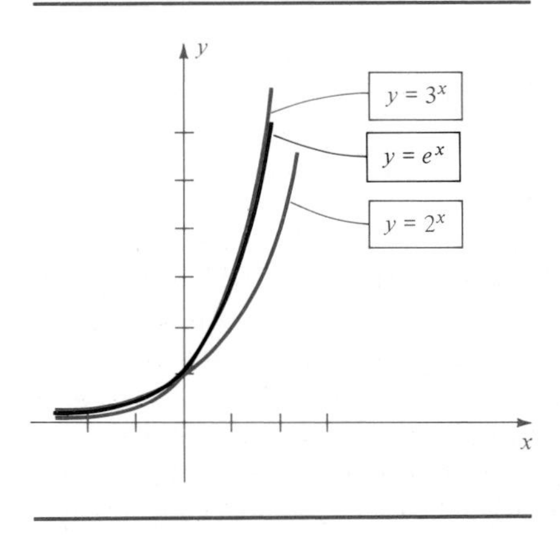

$$e \approx 2.71828$$

The number e arises naturally in the investigation of many physical phenomena. For this reason, the function f defined by $f(x) = e^x$ is called the **natural exponential function.** It is one of the most important functions that occurs in advanced mathematics and applications. Since $2 < e < 3$, the graph of $y = e^x$ lies "between" the graphs of $y = 2^x$ and $y = 3^x$, as shown in Figure 4.9.

A brief table of values of e^x and e^{-x} is given in Table 2 of Appendix II. Some calculators have an $\boxed{e^x}$ key for approximating values of the natural exponential function. There may also be a $\boxed{y^x}$ key that can be used for any positive base y. Approximations to e^x can then be found by calculating $(2.71828)^x$.

EXAMPLE 2 Verify the graph of $y = e^x$ sketched in Figure 4.9 by plotting a sufficient number of points.

SOLUTION The following table may be obtained by using a calculator or Table 2 of Appendix II, and rounding off values of e^x to two decimal places.

x	-2.0	-1.5	-1.0	-0.5	0	0.5	1.0	1.5	2.0
e^x (approx.)	0.14	0.22	0.37	0.61	1.00	1.65	2.72	4.48	7.39

Plotting points and using the fact that the exponential function with base e is increasing leads to the graph of $y = e^x$ sketched in Figure 4.9. ■

EXAMPLE 3 Sketch the graph of f if

$$f(x) = \frac{e^x + e^{-x}}{2}.$$

SOLUTION Note that f is an even function, because

$$f(-x) = \frac{e^{-x} + e^{-(-x)}}{2} = \frac{e^{-x} + e^x}{2} = f(x).$$

Thus, the graph is symmetric with respect to the y-axis (see page 67). Using a calculator or Table 2 of Appendix II, we obtain the following approximations to $f(x)$:

x	0	0.5	1.0	1.5	2.0
$f(x)$ (approx.)	1	1.13	1.54	2.35	3.76

Plotting points and using symmetry with respect to the y-axis gives us the sketch in Figure 4.10. ■

FIGURE 4.10

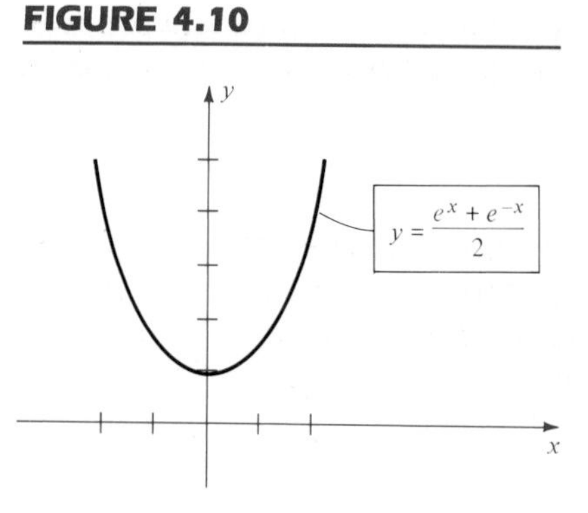

APPLICATION: Flexible Cables

The function f of Example 3 is important in applied mathematics and engineering, where it is called the **hyperbolic cosine function.** This function

FIGURE 4.11

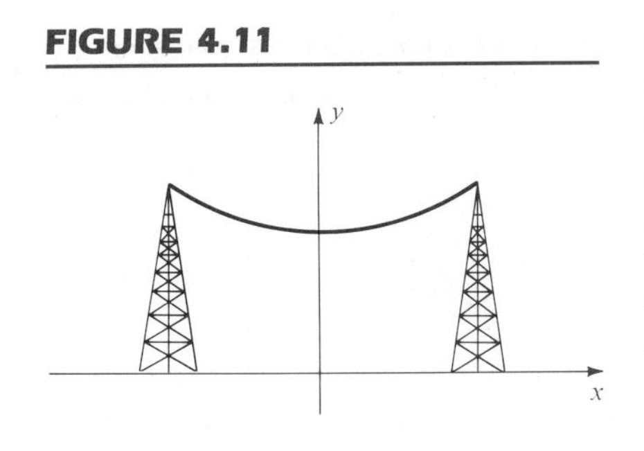

can be used to describe the shape of a uniform flexible cable, or chain, whose ends are supported from the same height. This is often the case for telephone or power lines, as illustrated in Figure 4.11. If we introduce a coordinate system as indicated in the figure, then it can be shown that an equation that corresponds to the shape of the cable is $y = (a/2)(e^{x/a} + e^{-x/a})$ where a is a real number. The graph is called a **catenary,** after the Latin word for *chain.* Note that the function in Example 3 is the special case in which $a = 1$. ■■■

APPLICATION: Radiotherapy

One of the many fields in which exponential functions with base e play an important role is *radiotherapy*, the treatment of tumors by radiation. Of major interest is the fraction of a tumor population that survives a treatment. This *surviving fraction* depends not only on the energy and nature of the radiation, but also on the depth, size, and characteristics of the tumor itself. The exposure to radiation may be thought of as a number of potentially damaging events, where only one "hit" is required to kill a tumor cell. Suppose that each cell has exactly one "target" that must be hit. If k denotes the average target size of a tumor cell, and if x is the number of damaging events (the *dose*), then the surviving fraction $f(x)$ is

$$f(x) = e^{-kx}.$$

This is called the *one-target–one-hit surviving fraction.*

Next, suppose that each cell has n targets and that hitting any one of the targets results in the death of a cell. In this case, the *n-target–one-hit surviving fraction* is

$$f(x) = 1 - (1 - e^{-kx})^n.$$

FIGURE 4.12

Surviving fraction of tumor cells after a radiation treatment

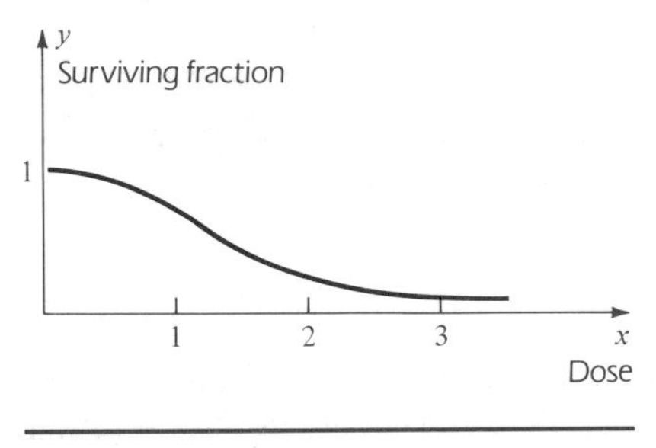

The graph of f may be analyzed to determine what effect increasing the dosage x will have on decreasing the surviving fraction of tumor cells. Note that $f(0) = 1$; that is, if there is no dose, then all cells survive. As a special case, if $k = 1$ and $n = 2$, then

$$\begin{aligned} f(x) &= 1 - (1 - e^{-x})^2 = 1 - (1 - 2e^{-x} + e^{-2x}) \\ &= 2e^{-x} - e^{-2x}. \end{aligned}$$

A complete analysis of the graph of f requires methods of calculus. It can be shown that the graph has the shape indicated in Figure 4.12. The "shoulder" on the curve near the point (0, 1) represents the threshold nature of the treatment; that is, a small dose results in very little tumor elimination. Note that for a large x, an increase in dosage has little effect on the surviving fraction. To determine the ideal dose that should be administered to a patient, specialists in radiation therapy must also take into account the number of healthy cells that are killed during a treatment.

Problems of the type illustrated in the next example occur in the study of calculus. (See also Exercises 9–12.)

EXAMPLE 4 Find the zeros of f if $f(x) = x^2(-2e^{-2x}) + 2xe^{-2x}$.

SOLUTION We may factor $f(x)$ as follows:

$$\begin{aligned} f(x) &= 2xe^{-2x} - 2x^2e^{-2x} \\ &= 2xe^{-2x}(1 - x). \end{aligned}$$

To find the zeros of f, we must solve the equation $f(x) = 0$. Since $e^{-2x} > 0$ for all x, it follows that $f(x) = 0$ if and only if $x = 0$ or $1 - x = 0$. Thus, the zeros of f are 0 and 1.

Exercises 4.2

In Exercises 1–7 use your knowledge of the graph of $y = e^x$ to help you sketch the graph of f.

1 (a) $f(x) = e^{-x}$ (b) $f(x) = -e^x$

2 (a) $f(x) = e^{2x}$ (b) $f(x) = 2e^x$

3 (a) $f(x) = e^{x+4}$ (b) $f(x) = e^x + 4$

4 (a) $f(x) = e^{-2x}$ (b) $f(x) = -2e^x$

5 $f(x) = \dfrac{e^{2x} + e^{-2x}}{2}$

6 $f(x) = \dfrac{e^x - e^{-x}}{2}$

7 $f(x) = \dfrac{2}{e^x + e^{-x}}$
(*Hint:* Take reciprocals of y-coordinates in Example 3.)

8 In statistics the **normal distribution function** is defined by

$$f(x) = \frac{1}{\sigma\sqrt{2\pi}} e^{(-1/2)[(x-\mu)/\sigma]^2}$$

for real numbers μ and $\sigma > 0$. (μ is called the *mean* and σ is called the *variance* of the distribution.) Sketch the graph of f for the case $\sigma = 1$ and $\mu = 0$.

In Exercises 9–12 find the zeros of f.

9 $f(x) = xe^x + e^x$

10 $f(x) = -x^2e^{-x} + 2xe^{-x}$

11 $f(x) = x^3(4e^{4x}) + 3x^2e^{4x}$

12 $f(x) = x^2(2e^{2x}) + 2xe^{2x} + e^{2x} + 2xe^{2x}$

Simplify the expressions in Exercises 13 and 14.

13 $\dfrac{(e^x + e^{-x})(e^x + e^{-x}) - (e^x - e^{-x})(e^x - e^{-x})}{(e^x + e^{-x})^2}$

14 $\dfrac{(e^x - e^{-x})^2 - (e^x + e^{-x})^2}{(e^x + e^{-x})^2}$

Calculator Exercises 4.2

1 An exponential function W such that $W(t) = W_0e^{kt}$ (for $k > 0$) describes the first month of growth for crops such as maize, cotton, and soybeans. Here $W(t)$ is the total weight in mg, W_0 is the weight on the day of emergence, and t is the time in days. If, for a species of soybean, $k = 0.2$ and $W_0 = 68$ mg, predict the weight at the end of the month ($t = 30$).

2 Refer to Exercise 1. It is often difficult to measure the weight W_0 of the plant when it first emerges from the soil. If, for a species of cotton, $k = 0.21$ and $W(10) = 575$ mg, estimate W_0.

3 The 1980 population of the United States was approximately 227 million, and the population has been growing at a rate of 0.7% per year. It is possible to show, using calculus, that the population $N(t)$, t years later, may be approximated by $N(t) = 227e^{0.007t}$. If this growth trend continues, predict the population in the year 2000.

4 The 1980 population estimate for India was 651 million, and the population has been growing at a rate of about 2% per year. The population $N(t)$, t years later, may be approximated by $N(t) = 651e^{0.02t}$. Assuming that this rapid growth rate continues, estimate the population of India in the year 2000.

5 In fishery science, the collection of fish that results from one annual reproduction is referred to as a *cohort*. It is usually assumed that the number $N(t)$ still alive after t years is given by an exponential function. For Pacific halibut, $N(t) = N_0e^{-0.2t}$ where N_0 is the initial size of the cohort. What percentage of the original number is still alive after 10 years?

6 The radioactive tracer ^{51}Cr can be used to locate the position of the placenta in a pregnant woman. Often the tracer must be ordered from a medical lab. If A_0 units (microcuries) are shipped, then because of radioactive decay, the number of units $A(t)$ present after t days is given by $A(t) = A_0e^{-0.0249t}$. If 35 units are shipped and it takes 2 days for the tracer to arrive, how many units are then available for the test? If 35 units are needed for the test, how many units should be shipped?

7 In 1966, the International Whaling Commission protected the world population of blue whales from hunting. In 1978, the population in the southern hemisphere was thought to number 5000. Now without predators and with an abundant food supply, the population $N(t)$ is expected to grow exponentially according to the formula $N(t) = 5000e^{0.047t}$ where t is in years. Predict the population in (a) 1990; (b) 2000.

8 The length (in cm) of many common commercial fish t years old is closely approximated by a von-Bertalanffy growth function $f(t) = a(1 - be^{-kt})$ where a, b, and k are constants.

(a) For Pacific halibut, $a = 200$, $b = 0.956$, and $k = 0.18$. Estimate the length of a typical 10-year-old halibut.

(b) What is the interpretation of the constant a in the formula?

9 Under certain conditions the atmospheric pressure p (in inches) at altitude h feet is given by $p = 29e^{-0.000034h}$. What is the pressure at an altitude of 40,000 feet?

10 Starting with c milligrams of the polonium isotope ^{210}Po, the amount remaining after t days may be approximated by $A = ce^{-0.00495t}$. If the initial amount is 50 milligrams, find, to the nearest hundredth, the amount remaining after (a) 30 days; (b) 180 days; (c) 365 days.

4.3 Logarithmic Functions

If $f(x) = a^x$ and $a > 1$, then f is increasing throughout $\mathbb{R}$, whereas if $0 < a < 1$, then f is decreasing (see Figure 4.2). Thus, if $a > 0$ and $a \neq 1$, then f is a one-to-one function and hence has an inverse function f^{-1}

(see Section 2.5). The inverse of the exponential function with base a is called the **logarithmic function with base a** and is denoted by $\log_a$. Its values are denoted by $\log_a(x)$ or $\log_a x$, read "**the logarithm of x with base a.**" Since

$$y = f^{-1}(x) \quad \text{if and only if} \quad x = f(y),$$

the definition of $\log_a$ may be expressed as follows:

DEFINITION OF $\log_a x$

$$y = \log_a x \quad \text{if and only if} \quad x = a^y.$$

Since the domain and range of the exponential function with base a are $\mathbb{R}$ and the positive real numbers, respectively, the domain of its inverse $\log_a$ is the positive real numbers and the range is $\mathbb{R}$. Thus, in the definition, $x > 0$ and y is in $\mathbb{R}$.

Note that

$$\text{if} \quad y = \log_a x, \quad \text{then} \quad x = a^y = a^{\log_a x}.$$

In words, *$\log_a x$ is the exponent to which a must be raised in order to obtain x.* As illustrations,

$$\begin{aligned} \log_2 8 &= 3 && \text{since} \quad 2^3 = 8 \\ \log_5 \tfrac{1}{25} &= -2 && \text{since} \quad 5^{-2} = \tfrac{1}{25} \\ \log_{10} 10{,}000 &= 4 && \text{since} \quad 10^4 = 10{,}000. \end{aligned}$$

The next theorem is an immediate consequence of the definition of logarithm.

THEOREM

(i) $a^{\log_a x} = x$ for every $x > 0$
(ii) $\log_a a = 1$
(iii) $\log_a 1 = 0$

We have already proved (i). To prove (ii) and (iii) it is sufficient to note that $a^1 = a$ and $a^0 = 1$, respectively.

EXAMPLE 1 Find s if:

(a) $\log_4 2 = s$ (b) $\log_5 s = 2$ (c) $\log_s 8 = 3$

SOLUTION

(a) If $\log_4 2 = s$, then $4^s = 2$ and hence $s = \frac{1}{2}$.

(b) If $\log_5 s = 2$, then $5^2 = s$ and hence $s = 25$.

(c) If $\log_s 8 = 3$, then $s^3 = 8$ and hence $s = \sqrt[3]{8} = 2$. ■

EXAMPLE 2 Solve the equation $\log_4 (5 + x) = 3$.

SOLUTION If $\log_4 (5 + x) = 3$, then by the definition of logarithm,

$$5 + x = 4^3 \quad \text{or} \quad 5 + x = 64.$$

Hence, the solution is $x = 59$. ■

The following laws are fundamental for all work with logarithms of positive real numbers u and w.

LAWS OF LOGARITHMS

(i) $\log_a (uw) = \log_a u + \log_a w$

(ii) $\log_a (u/w) = \log_a u - \log_a w$

(iii) $\log_a (u^c) = c \log_a u$ for every real number c

PROOF To prove (i) we begin by letting

$$r = \log_a u \quad \text{and} \quad s = \log_a w.$$

Applying the definition of logarithm, $a^r = u$ and $a^s = w$. Consequently,

$$a^r a^s = uw$$

and hence,

$$a^{r+s} = uw.$$

By the definition of logarithm, the last equation is equivalent to

$$r + s = \log_a (uw).$$

Since $r = \log_a u$ and $s = \log_a w$, we obtain

$$\log_a u + \log_a w = \log_a uw.$$

This completes the proof of Law (i).

To prove (ii) we begin as in the proof of (i), but divide a^r by a^s, obtaining

$$\frac{a^r}{a^s} = \frac{u}{w} \quad \text{or} \quad a^{r-s} = \frac{u}{w}.$$

Using the definition of logarithm, we may write the last equation as

$$r - s = \log_a (u/w).$$

Substituting for r and s gives us

$$\log_a u - \log_a w = \log_a (u/w)$$

This proves Law (ii).

Finally, if c is any real number, then

$$(a^r)^c = u^c \quad \text{or} \quad a^{cr} = u^c.$$

By the definition of logarithm, the last equality implies that

$$cr = \log_a u^c.$$

Substituting for r, we obtain

$$c \log_a u = \log_a u^c.$$

This proves Law (iii). □

The following examples illustrate uses of the Laws of Logarithms.

EXAMPLE 3 If $\log_a 3 = 0.4771$ and $\log_a 2 = 0.3010$, find:

(a) $\log_a 6$ (b) $\log_a \frac{3}{2}$ (c) $\log_a \sqrt{2}$ (d) $\dfrac{\log_a 3}{\log_a 2}$

SOLUTION

(a) Since $6 = 2 \cdot 3$, we may use Law (i) to obtain

$$\begin{aligned} \log_a 6 = \log_a (2 \cdot 3) &= \log_a 2 + \log_a 3 \\ &= 0.4771 + 0.3010 = 0.7781. \end{aligned}$$

(b) By Law (ii),

$$\begin{aligned} \log_a \tfrac{3}{2} &= \log_a 3 - \log_a 2 \\ &= 0.4771 - 0.3010 = 0.1761. \end{aligned}$$

(c) Using Law (iii),

$$\log_a \sqrt{2} = \log_a 2^{1/2} = \tfrac{1}{2}\log_a 2$$
$$= \tfrac{1}{2}(0.3010) = 0.1505.$$

(d) There is no law of logarithms that allows us to simplify $(\log_a 3)/(\log_a 2)$. Consequently, we *divide* 0.4771 by 0.3010, obtaining the approximation 1.585. It is important to notice the difference between this problem and part (b). ■

EXAMPLE 4 Solve the following equations:

(a) $\log_2 (2x + 3) = \log_2 11 + \log_2 3$

(b) $\log_4 (x + 6) - \log_4 10 = \log_4 (x - 1) - \log_4 2$

SOLUTION

(a) Using Law (i), we may write the equation as

$$\log_2 (2x + 3) = \log_2 (11 \cdot 3)$$

or

$$\log_2 (2x + 3) = \log_2 33.$$

Since the bases are equal, we must have

$$2x + 3 = 33 \quad \text{or} \quad 2x = 30.$$

Hence, the solution is $x = 15$.

(b) The given equation is equivalent to

$$\log_4 (x + 6) - \log_4 (x - 1) = \log_4 10 - \log_4 2.$$

Applying Law (ii),

$$\log_4 \left(\frac{x + 6}{x - 1}\right) = \log_4 \frac{10}{2} = \log_4 5$$

and hence,

$$\frac{x + 6}{x - 1} = 5.$$

The last equation implies that

$$x + 6 = 5x - 5 \quad \text{or} \quad 4x = 11.$$

Thus, the solution is $x = \frac{11}{4}$. ■

Extraneous solutions sometimes occur in the process of solving equations that involve logarithms, as illustrated in the next example.

EXAMPLE 5 Solve the equation $2 \log_7 x = \log_7 36$.

SOLUTION Applying Law (iii), we obtain $2 \log_7 x = \log_7 x^2$, and substitution in the given equation leads to

$$\log_7 x^2 = \log_7 36.$$

Consequently, $x^2 = 36$ and, hence, either $x = 6$ or $x = -6$. However, $x = -6$ is not a solution of the original equation since x must be positive in order for $\log_7 x$ to exist. Thus, the only solution is $x = 6$.

The preceding difficulty could have been avoided by writing the given equation as

$$\log_7 x = \tfrac{1}{2} \log_7 36 = \log_7 36^{1/2} = \log_7 6$$

and, therefore, $x = 6$. ■

The Laws of Logarithms are often used as in the following two examples.

EXAMPLE 6

Express $\log_a \dfrac{x^3 \sqrt{y}}{z^2}$ in terms of the logarithms of x, y, and z.

SOLUTION Writing $\sqrt{y}$ as $y^{1/2}$ and using the three Laws of Logarithms, we obtain

$$\begin{aligned} \log_a \frac{x^3 y^{1/2}}{z^2} &= \log_a (x^3 y^{1/2}) - \log_a z^2 \\ &= \log_a x^3 + \log_a y^{1/2} - \log_a z^2 \\ &= 3 \log_a x + \tfrac{1}{2} \log_a y - 2 \log_a z. \end{aligned}$$

■

EXAMPLE 7 Express in terms of one logarithm:

$$\tfrac{1}{3} \log_a (x^2 - 1) - \log_a y - 4 \log_a z.$$

SOLUTION Using the Laws of Logarithms we have

$$\begin{aligned} \tfrac{1}{3} \log_a (x^2 - 1) - \log_a y - 4 \log_a z &= \log_a (x^2 - 1)^{1/3} - \log_a y - \log_a z^4 \\ &= \log_a \sqrt[3]{x^2 - 1} - (\log_a y + \log_a z^4) \\ &= \log_a \sqrt[3]{x^2 - 1} - \log_a yz^4 \\ &= \log_a \frac{\sqrt[3]{x^2 - 1}}{yz^4}. \end{aligned}$$

■

It is important to note that there are no laws for expressing $\log_a (u + w)$ or $\log_a (u - w)$ in terms of simpler logarithms. It is evident that

$$\log_a (u + w) \neq \log_a u + \log_a w,$$

since the latter sum equals $\log_a (uw)$. Similarly,

$$\log_a (u - w) \neq \log_a u - \log_a w.$$

Logarithmic functions occur frequently in applications. Indeed, if two variables u and v are related such that u is an exponential function of v, then v is a logarithmic function of u.

EXAMPLE 8 The number N of bacteria in a certain culture after t hours is given by $N = (1000)2^t$. Express t as a logarithmic function of N with base 2.

SOLUTION If $N = (1000)2^t$, then $2^t = \dfrac{N}{1000}$. Changing to logarithmic form,

$$t = \log_2 \frac{N}{1000}.$$ ■

Exercises 4.3

Change the equations in Exercises 1–8 to logarithmic form.

1 $4^3 = 64$ **2** $3^5 = 243$

3 $2^7 = 128$ **4** $5^3 = 125$

5 $10^{-3} = 0.001$ **6** $10^{-2} = 0.01$

7 $t^r = s$ **8** $v^w = u$

Change the equations in Exercises 9–16 to exponential form.

9 $\log_{10} 1000 = 3$ **10** $\log_3 81 = 4$

11 $\log_3 \frac{1}{243} = -5$ **12** $\log_4 \frac{1}{64} = -3$

13 $\log_7 1 = 0$ **14** $\log_9 1 = 0$

15 $\log_t r = p$ **16** $\log_v w = q$

Find the numbers in Exercises 17–22.

17 $\log_4 \frac{1}{16}$ **18** $\log_2 32$

19 $\log_{10} 100$ **20** $\log_8 64$

21 $10^{\log_{10} 5}$ **22** $\log_{10} 0.0001$

Find the solutions of the equations in Exercises 23–34.

23 $\log_3 (x - 4) = 2$ **24** $\log_2 (x - 5) = 4$

25 $\log_9 x = \frac{3}{2}$ **26** $\log_4 x = -\frac{3}{2}$

27 $\log_5 x^2 = -2$ **28** $\log_{10} x^2 = -4$

29 $\log_6 (2x - 3) = \log_6 12 - \log_6 3$

30 $2 \log_3 x = 3 \log_3 5$

31 $\log_2 x - \log_2 (x + 1) = 3 \log_2 4$

32 $\log_5 x + \log_5 (x + 6) = \frac{1}{2} \log_5 9$

33 $\log_{10} x^2 = \log_{10} x$

34 $\frac{1}{2} \log_5 (x - 2) = 3 \log_5 2 - \frac{3}{2} \log_5 (x - 2)$

In Exercises 35–42 express the logarithm in terms of logarithms of x, y, and z.

35 $\log_a \dfrac{x^2y}{z^3}$

36 $\log_a \dfrac{x^3y^2}{z^5}$

37 $\log_a \dfrac{\sqrt{x}z^2}{y^4}$

38 $\log_a x \sqrt[3]{\dfrac{y^2}{z^4}}$

39 $\log_a \sqrt[3]{\dfrac{x^2}{yz^5}}$

40 $\log_a \dfrac{\sqrt{x}y^6}{\sqrt[3]{z^2}}$

41 $\log_a \sqrt{x\sqrt{yz^3}}$

42 $\log_a \sqrt[3]{x^2y\sqrt{z}}$

In Exercises 43–46 write the expression as one logarithm.

43 $2 \log_a x + \frac{1}{3} \log_a (x - 2) - 5 \log_a (2x + 3)$

44 $5 \log_a x - \frac{1}{2} \log_a (3x - 4) + 3 \log_a (5x + 1)$

45 $\log_a (y^2x^3) - 2 \log_a x\sqrt[3]{y} + 3 \log_a \left(\dfrac{x}{y}\right)$

46 $2 \log_a \dfrac{y^3}{x} - 3 \log_a y + \frac{1}{2} \log_a x^4y^2$

47 Starting with q_0 milligrams of pure radium, the amount q remaining after t years is $q = q_0(2)^{-t/1600}$. Use logarithms with base 2 to solve for t in terms of q and q_0.

48 The radioactive isotope ^{210}Bi disintegrates according to the law $Q = k(2)^{-t/5}$ where t is in days. Use logarithms with base 2 to solve for t in terms of Q and k.

49 The number N of bacteria in a certain culture at time t is given by $N = 10^4(3)^t$. Use logarithms with base 3 to solve for t in terms of N.

50 When a linear dimension x (such as height) of an organism is related to volume or weight y, the relationship between $\log_a y$ and $\log_a x$ is often linear, in the sense that $\log_a y = k \log_a x + \log_a b$, for some constants k and $b > 0$. Rewrite this equation using laws of logarithms and show that y and x are related by an equation of the form $y = cx^d$ for some constants c and d.

4.4 Graphs of Logarithmic Functions

Since the logarithmic function $\log_a$ is the inverse of the exponential function with base a, the graph of $y = \log_a x$ can be obtained by reflecting the graph of $y = a^x$ through the line $y = x$ that bisects quadrants I and III (see page 98). This is illustrated in Figure 4.13 for the case $a > 1$.

FIGURE 4.13

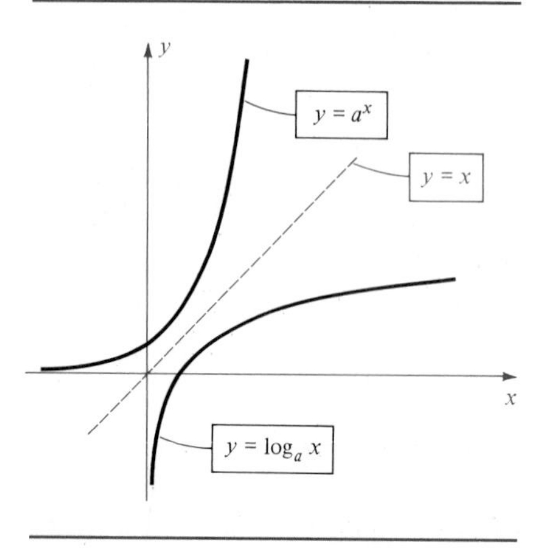

We can also sketch the graph by plotting points. Since

$$y = \log_a x \quad \text{if and only if} \quad x = a^y,$$

coordinates of points on the graph of $y = \log_a x$ may be found by using the equation $x = a^y$. This leads to the following table:

y	-3	-2	-1	0	1	2	3
x	$\frac{1}{a^3}$	$\frac{1}{a^2}$	$\frac{1}{a}$	1	a	a^2	a^3

If $a > 1$, we obtain the sketch in Figure 4.14(i). In this case f is an increasing function throughout its domain. If $0 < a < 1$, then the graph has the general shape shown in Figure 4.14(ii) and hence, f is a decreasing function.

FIGURE 4.14

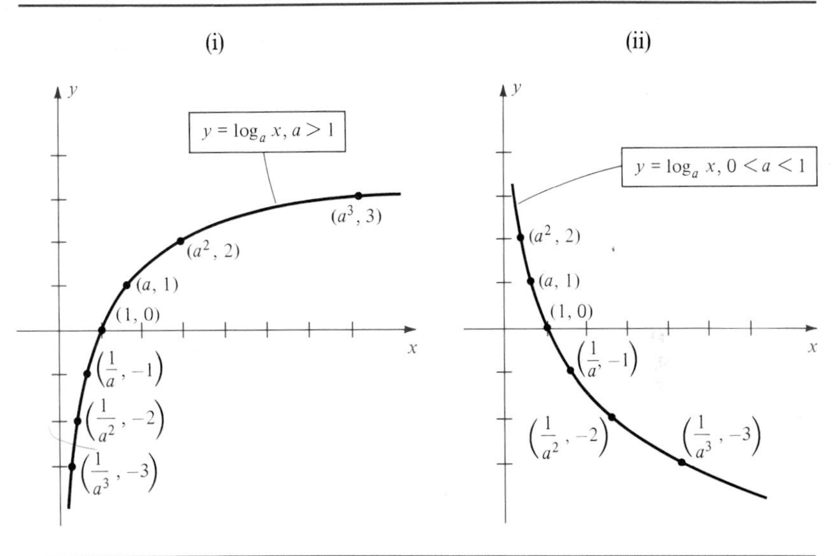

Functions defined in terms of $\log_a p$, for some expression p that involves x, often occur in mathematics and applications. Functions of this type are classified as members of the logarithmic family; however, the graphs may differ from those sketched in Figures 4.13 and 4.14, as illustrated in the following examples.

FIGURE 4.15

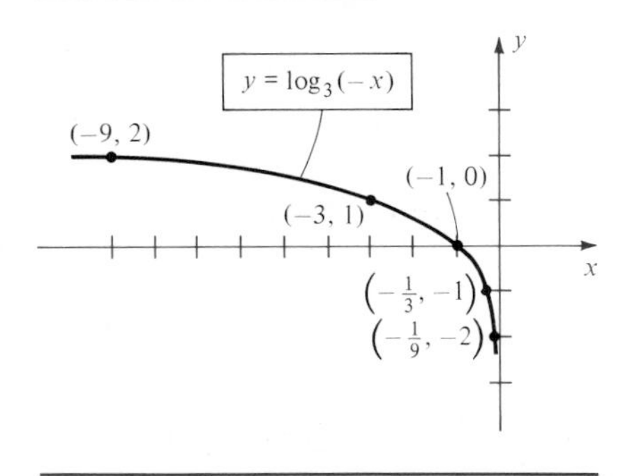

EXAMPLE 1 Sketch the graph of f if $f(x) = \log_3(-x)$ for $x < 0$.

SOLUTION If $x < 0$, then $-x > 0$, and hence, $\log_3(-x)$ exists. We wish to sketch the graph of the equation $y = \log_3(-x)$, which, by the definition of logarithm, is equivalent to $3^y = -x$. Thus, to find points on the graph of f, we may substitute for y in the equation $x = -(3^y)$. The following table displays the coordinates of several such points:

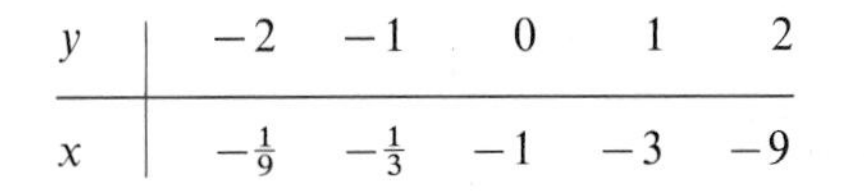

y	-2	-1	0	1	2
x	$-\frac{1}{9}$	$-\frac{1}{3}$	-1	-3	-9

Plotting these points leads to the sketch in Figure 4.15. ■

FIGURE 4.16

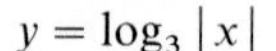

$y = \log_3 |x|$

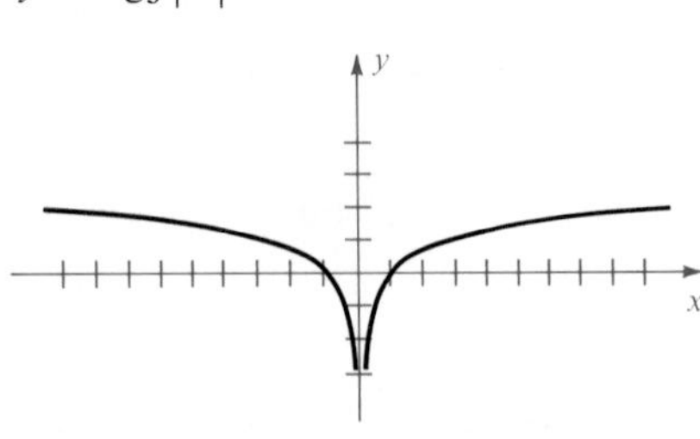

EXAMPLE 2 Sketch the graph of the equation $y = \log_3 |x|$ if $x \neq 0$.

SOLUTION Since $|x| > 0$ for all $x \neq 0$, the graph includes points corresponding to negative values of x as well as to positive values. If $x > 0$, then $|x| = x$ and hence, to the right of the y-axis the graph coincides with the graph of $y = \log_3 x$, or equivalently, $x = 3^y$. If $x < 0$, then $|x| = -x$ and the graph is the same as that of $y = \log_3 (-x)$ (see Example 1). The graph is sketched in Figure 4.16. Note that the graph is symmetric with respect to the y-axis. ■

EXAMPLE 3 Sketch the graph of

(a) $y = \log_3 (x - 2)$ (b) $y = \log_3 x - 2$.

SOLUTION

(a) We can obtain the graph of $y = \log_3 (x - 2)$ by shifting the graph of $y = \log_3 x$ two units to the right. (See the discussion of *horizontal shifts* in Section 2.2). Since the graph of $y = \log_3 x$ is the part of the graph in Figure 4.16 that lies to the right of the y-axis, this leads to the sketch in Figure 4.17(i).

(b) The graph of $y = \log_3 x - 2$ can be obtained by shifting the graph of $y = \log_3 x$ two units downward. (See *vertical shifts*, Section 2.2). This leads to the sketch in Figure 4.17(ii). Note that the x-intercept is given by $\log_3 x = 2$, or $x = 3^2 = 9$.

FIGURE 4.17

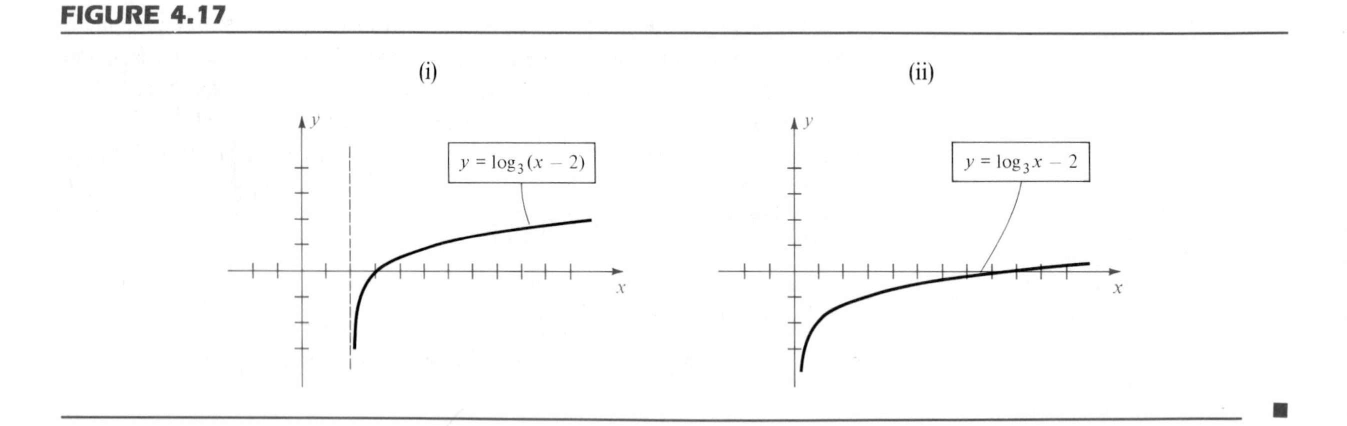

■

Exercises 4.4

Sketch the graph of f in Exercises 1–20.

1 $f(x) = \log_2 x$

2 $f(x) = \log_5 x$

3 $f(x) = \log_4 x$

4 $f(x) = \log_{10} x$

5 $f(x) = \log_2 (x + 3)$

6 $f(x) = \log_2 (x - 3)$

7 $f(x) = \log_2 x + 3$

8 $f(x) = \log_2 x - 3$

9 $f(x) = \log_3 (3x)$

10 $f(x) = \log_2 (x^2)$

11 $f(x) = 3 \log_3 x$

12 $f(x) = \log_2 (x^3)$

13 $f(x) = \log_2 \sqrt{x}$

14 $f(x) = \log_2 |x|$

15 $f(x) = \log_3 (1/x)$

16 $f(x) = \log_3 (2 - x)$

17 $f(x) = 1/(\log_3 x)$

18 $f(x) = |\log_2 x|$

19 $f(x) = \log_2 |x - 5|$

20 $f(x) = \log_3 |x + 1|$

4.5 Common and Natural Logarithms

Before electronic calculators were invented, logarithms with base 10 were used for complicated numerical computations involving products, quotients, and powers of real numbers. Base 10 was employed because it is well suited for numbers that are expressed in decimal form. Logarithms with base 10 are called **common logarithms.** The symbol $\log x$ is used as an abbreviation for $\log_{10} x$. Thus, we have the following definition.

DEFINITION OF COMMON LOGARITHMS

$$\log x = \log_{10} x \quad \text{for every } x > 0.$$

Since inexpensive calculators are now available, there is little need for common logarithms as a tool for computational work. However, base 10 does occur in applications, and hence many calculators have a [log] key that can be used to approximate common logarithms. Appendix II contains a table of common logarithms that may be used if a calculator either is not available or is inoperative. The use of the table of common logarithms is explained in Appendix I.

EXAMPLE 1 Using the *Richter scale*, the magnitude R of an earthquake of intensity I may be found by means of the formula

$$R = \log \frac{I}{I_0}$$

where I_0 is a certain minimum intensity.

(a) Find R assuming the intensity of an earthquake is $(1000)I_0$.

(b) Express I in terms of R and I_0.

SOLUTION

(a) If $I = (1000)I_0$, then

$$R = \log \frac{I}{I_0} = \log \frac{(1000)I_0}{I_0} = \log (1000).$$

We could use a calculator to find log (1000) or write

$$\log (1000) = \log 10^3 = 3 \log 10 = 3(1) = 3$$

where we have used the fact that $\log_a a = 1$. Hence $R = 3$.

(b) By the definition of logarithm with base 10,

$$\text{if} \quad R = \log \frac{I}{I_0}, \quad \text{then} \quad \frac{I}{I_0} = 10^R.$$

Thus, $$I = I_0(10^R).$$ ■

In Section 4.2 we defined the natural exponential function f by means of the equation $f(x) = e^x$. The logarithmic function with base e is called the **natural logarithmic function.** We use **ln *x*** (read "*ell-en of x*") as an abbreviation for $\log_e x$ and refer to it as the **natural logarithm of *x*.** Thus *the natural logarithmic and natural exponential functions are inverse functions of one another.* Let us state the definition for reference.

DEFINITION OF NATURAL LOGARITHMS

$$\ln x = \log_e x \quad \text{for every} \quad x > 0.$$

Since $e \approx 3$, the graph of $y = \ln x$ is similar in appearance to the graph of $y = \log_3 x$. The Laws of Logarithms for natural logarithms are as follows.

LAWS OF NATURAL LOGARITHMS

(i) $\ln (uv) = \ln u + \ln v$

(ii) $\ln \frac{u}{v} = \ln u - \ln v$

(iii) $\ln u^c = c \ln u$

If we substitute e for a in the theorem on page 170, we obtain the following for $x > 0$:

$$e^{\ln x} = x, \qquad \ln e = 1, \qquad \ln 1 = 0.$$

Many calculators have a key labeled $\boxed{\ln x}$ that can be used to approximate natural logarithms. A short table of values of $\ln x$ is given in Table 3 of Appendix II.

EXAMPLE 2 Use a calculator to approximate $\log 436$; $\ln 436$; $\log 0.0436$; and $\ln 0.0436$.

SOLUTION Entering the indicated numbers and pressing the appropriate keys, we obtain the following approximations:

$$\log 436 \approx 2.6394865$$

$$\ln 436 \approx 6.0776422$$

$$\log 0.0436 \approx -1.3605135$$

$$\ln 0.0436 \approx -3.1326981$$ ■

EXAMPLE 3 Newton's Law of Cooling states that the rate at which an object cools is directly proportional to the difference in temperature between the object and its surrounding medium. Newton's Law can be used to show that under certain conditions the temperature T of an object at time t is given by $T = 75e^{-2t}$. Express t as a function of T.

SOLUTION The equation $T = 75e^{-2t}$ may be written

$$e^{-2t} = \frac{T}{75}.$$

Using natural logarithms gives us

$$-2t = \log_e \frac{T}{75} = \ln \frac{T}{75}.$$

Consequently,

$$t = -\tfrac{1}{2} \ln \frac{T}{75} \quad \text{or} \quad t = -\tfrac{1}{2}[\ln T - \ln 75].$$ ■

EXAMPLE 4 If a beam of light that has intensity k is projected vertically downward into water, then its intensity $I(x)$ at a depth of x meters is $I(x) = ke^{-1.4x}$. At what depth is the intensity one-half its value at the surface?

SOLUTION At the surface $x = 0$, and the intensity is

$$I(0) = ke^0 = k.$$

We wish to find the value of x such that $I(x) = \frac{1}{2}k$; that is,

$$ke^{-1.4x} = \tfrac{1}{2}k \quad \text{or} \quad e^{-1.4x} = \tfrac{1}{2}.$$

Using natural logarithms,

$$-1.4x = \ln \tfrac{1}{2}$$

or

$$-1.4x = \ln 1 - \ln 2 = 0 - \ln 2 = -\ln 2.$$

Hence,

$$x = \frac{-\ln 2}{-1.4} \approx \frac{-0.693}{-1.4} \approx 0.5 \text{ meter.}$$

■

To solve certain problems, it is necessary to find x when given either $\log x$ or $\ln x$. One way to accomplish this is by using the inverse function key [INV]. If we first press [INV] and then press [log], we obtain the *inverse logarithmic function* $\log^{-1}$. Recall that $\log^{-1}$ is the exponential function with base 10. Since $\log^{-1}(\log x) = x$, we can obtain x by entering $\log x$ and then pressing, successively, [INV] and [log]. Similarly, given $\ln x$, we can find x by entering $\ln x$ and pressing [INV] [ln x]. This procedure is illustrated in the next example.

EXAMPLE 5 Approximate x to three decimal places if:

(a) $\log x = 1.7959$ (b) $\ln x = 4.7$

SOLUTION

(a) Given $\log x = 1.7959$,

Enter: 1.7959

Press [INV] [log]: 62.502876

Thus, $x \approx 62.503$.

(b) Given $\ln x = 4.7$,

Enter: 4.7

Press [INV] [ln x]: 109.94717

Hence, $x \approx 109.947$. ■

Alternatively, Example 5(a) could be solved by noting that since common logarithms have base 10,

$$\text{if} \quad \log x = 1.7959, \quad \text{then} \quad x = 10^{1.7959}.$$

The number x can then be approximated using a [y^x] key.

Similarly, part (b) could be solved by using the fact that natural logarithms have base e. Specifically,

$$\text{if} \quad \ln x = 4.7, \quad \text{then} \quad x = e^{4.7}.$$

Table 2 in Appendix II or a calculator could then be used to approximate x.

Finally, it is sometimes necessary to *change the base* of a logarithm by expressing $\log_b u$ in terms of $\log_a u$, for some positive real number $b \neq 1$. We begin with the equivalent equations

$$v = \log_b u \quad \text{and} \quad b^v = u.$$

Taking the logarithm, base a, of both sides of the second equation gives us

$$\log_a b^v = \log_a u,$$

or equivalently,

$$v \log_a b = \log_a u.$$

Solving for v (that is, $\log_b u$), we obtain formula (i) in the next box.

CHANGE OF BASE FORMULAS

$$\text{(i)} \quad \log_b u = \frac{\log_a u}{\log_a b} \qquad \text{(ii)} \quad \log_b a = \frac{1}{\log_a b}$$

To obtain formula (ii) we let $u = a$ in (i) and use the fact that $\log_a a = 1$. If we let $a = e$ and $b = 10$ in the change of base formulas we obtain the following special cases:

$$\log u = \frac{\ln u}{\ln 10} \quad \text{and} \quad \log e = \frac{1}{\ln 10}.$$

Exercises 4.5

In Exercises 1–12 approximate x to three significant figures.

1 $\log x = 3.6274$

2 $\log x = 1.8965$

3 $\log x = 0.9469$

4 $\log x = 4.9680$

5 $\log x = -1.6253$

6 $\log x = -2.2118$

7 $\ln x = 2.3$

8 $\ln x = 3.7$

9 $\ln x = 0.05$

10 $\ln x = 0.95$

11 $\ln x = -1.6$

12 $\ln x = -5$

13 Using the Richter scale formula $R = \log (I/I_0)$, find the magnitude of an earthquake that has intensity

(a) 100 times that of I_0;
(b) 10,000 times that of I_0;
(c) 100,000 times that of I_0.

14 Refer to Exercise 13. The largest recorded magnitudes of earthquakes have been between 8 and 9 on the Richter scale. Find the corresponding intensities in terms of I_0.

15 The following formula, valid for earthquakes in the eastern U.S., relates the magnitude R of the earthquake to the surrounding area A (in square miles) that is affected by the quake:

$$R = 2.3 \log (A + 34{,}000) - 7.5.$$

Solve for A in terms of R.

16 For the western U.S., the area-magnitude formula (see Exercise 15) takes the form

$$R = 2.3 \log (A + 3000) - 5.1.$$

If A_1 is the area affected by an earthquake of magnitude R in the west, while A_2 is the area affected by a similar quake in the east, find a formula for A_1/A_2 in terms of R.

17 The loudness of a sound, as experienced by the human ear, is based upon intensity levels. A formula used for finding the intensity level α, in decibels, that corresponds to a sound intensity I is $\alpha = 10 \log (I/I_0)$ where I_0 is a special value of I agreed to be the weakest sound that can be detected by the ear under certain conditions. Find α if:

(a) I is 10 times as great as I_0;

(b) I is 1000 times as great as I_0;

(c) I is 10,000 times as great as I_0. (This is the intensity level of the average voice.)

18 A sound intensity level of 140 decibels produces pain in the average human ear. Approximately how many times greater than I_0 must I be in order for α to reach this level? (Refer to Exercise 17.)

19 Chemists use a number denoted by pH to describe quantitatively the acidity or basicity of solutions. By definition,

$$\text{pH} = -\log [\text{H}^+]$$

where $[\text{H}^+]$ is the hydrogen ion concentration in moles per liter. Approximate the pH of each substance:

(a) vinegar: $[\text{H}^+] \approx 6.3 \times 10^{-3}$;

(b) carrots: $[\text{H}^+] \approx 1.0 \times 10^{-5}$;

(c) sea water: $[\text{H}^+] \approx 5.0 \times 10^{-9}$.

20 Approximate the hydrogen ion concentration $[\text{H}^+]$ in each of the following substances. (Refer to Exercise 19.)

(a) apples: pH ≈ 3.0 (b) beer: pH ≈ 4.2

(c) milk: pH ≈ 6.6

21 A solution is considered acidic if $[\text{H}^+] > 10^{-7}$ or basic if $[\text{H}^+] < 10^{-7}$. What are the corresponding inequalities involving pH?

22 Many solutions have a pH between 1 and 14. What is the corresponding range of the hydrogen ion content $[\text{H}^+]$?

23 The current I at time t in a certain electrical circuit is given by $I = 20e^{-Rt/L}$, where R and L denote the resistance and inductance, respectively. Use natural logarithms to solve for t in terms of the remaining variables.

24 An electrical condenser with initial charge Q_0 is allowed to discharge. After t seconds the charge Q is $Q = Q_0e^{kt}$ where k is a constant. Use natural logarithms to solve for t in terms of Q_0, Q, and k.

25 Under certain conditions the atmospheric pressure p at altitude h is given by $p = 29e^{-0.000034h}$. Use natural logarithms to solve for h as a function of p.

26 If p denotes the selling price (in dollars) of a commodity, and x is the corresponding demand (in number sold per day), then frequently the relationship between p and x is $p = p_0e^{-ax}$ where p_0 and a are positive constants. Express x as a function of p.

27 The *Ehrenberg relation* $\ln W = \ln 2.4 + (1.84)h$ is an empirically based formula relating the height h (in meters) to the weight W (in kg) for children aged 5 through 13 years old. The formula has been verified in many different countries. Express W as a function of h.

28 A rocket of mass m_1 is filled with fuel of initial mass m_2. Assuming frictional forces are neglected, the total mass m of the rocket at time t is related to its upward velocity by the formula $v = -a \ln m + b$ for some constants a and b. At time $t = 0$ we have $v = 0$ and $m = m_1 + m_2$. At burnout $m = m_1$. Using this information, find a formula for the velocity of the rocket at burnout. (Write the formula in terms of one logarithm.)

29 If n is the average number of earthquakes (worldwide) in a given year with magnitude (on the Richter scale) between R and $R + 1$, then $\log n = 7.7 - (0.9)R$.

(a) Solve for n in terms of R.

(b) Find n if $R = 4$, 5, and 6.

30 The energy E (in ergs) released during an earthquake of magnitude R is given by the formula $\log E = 1.4 + (1.5)R$.

(a) Solve for E in terms of R.

(b) Find the energy released during the famous Alaskan quake of 1964, which measured 8.4 on the Richter scale.

31 A certain radioactive substance decays according to the formula $q(t) = q_0 e^{-0.0063t}$ where q_0 is the initial amount of the substance and t is the time in days. Approximate its half-life, that is, the number of days it takes for half of the substance to decay.

32 The air pressure $p(h)$, in pounds per in^2, at an altitude of h feet above sea level may be approximated by the formula $p(h) = 14.7e^{-0.0000385h}$. At approximately what altitude h is the air pressure (a) 10 pounds per in^2? (b) one-half its value at sea level?

33 Use the Compound Interest Formula to determine how long it will take a sum of money to double if it is invested at a rate of 6% per year compounded monthly.

34 If interest is compounded continuously at the rate of 10% per year, the compound interest formula takes the form $A = Pe^{0.1t}$. A man has determined that he will need to deposit about \$5600 to generate \$25,000 for his son's college education. To provide a cushion for unexpected expenses, he decides to deposit \$6000. After how many years will this initial deposit have grown to \$25,000?

35 The population $N(t)$ (in millions) of the United States t years after 1980 may be approximated by the formula $N(t) = 227e^{0.007t}$. When will the population be twice what it was in 1980?

36 The population $N(t)$ of India (in millions) t years after 1980 may be approximated by $N(t) = 651e^{0.02t}$. When will the population grow to one billion?

4.6 Exponential and Logarithmic Equations

If variables in equations appear as exponents or logarithms, the equations are often called **exponential** or **logarithmic equations,** respectively. The following examples illustrate techniques for solving such equations.

EXAMPLE 1 Solve the equation $3^x = 21$.

SOLUTION Taking the common logarithm of both sides and using (iii) of the Laws of Logarithms, we obtain

$$\log(3^x) = \log 21$$

$$x \log 3 = \log 21$$

$$x = \frac{\log 21}{\log 3}.$$

If an approximation is desired, we may use a calculator or Table 1 in Appendix II to obtain

$$x \approx \frac{1.3222}{0.4771} \approx 2.77$$

A partial check on the solution is to note that since $3^2 = 9$ and $3^3 = 27$, the number x such that $3^x = 21$ should lie between 2 and 3, somewhat closer to 3 than to 2.

We could also have solved for x by using natural logarithms. In this case, a calculator yields

$$\begin{aligned} x = \frac{\ln 21}{\ln 3} &\approx \frac{3.0445224}{1.0986123} \\ &\approx 2.7712437 \approx 2.77. \end{aligned}$$

■

EXAMPLE 2 Solve the equation $5^{2x+1} = 6^{x-2}$.

SOLUTION If we take the common logarithm of both sides and use (iii) of the Laws of Logarithms, we obtain

$$(2x + 1) \log 5 = (x - 2) \log 6.$$

We may now solve for x:

$$\begin{aligned} 2x \log 5 + \log 5 &= x \log 6 - 2 \log 6 \\ 2x \log 5 - x \log 6 &= -\log 5 - 2 \log 6 \\ x(2 \log 5 - \log 6) &= -(\log 5 + \log 6^2) \\ x &= \frac{-(\log 5 + \log 36)}{2 \log 5 - \log 6} \\ &= \frac{-\log (5 \cdot 36)}{\log 5^2 - \log 6}. \end{aligned}$$

Thus,
$$x = -\frac{\log 180}{\log \frac{25}{6}}.$$

If an approximation to the solution is desired, we could proceed as in Example 1:

$$x \approx -\frac{2.2553}{0.6198} \approx -3.64.$$

Natural logarithms could also have been used:

$$x = -\frac{\ln 180}{\ln \frac{25}{6}} \approx -\frac{5.1929569}{1.4271164} \approx -3.64.$$

■

EXAMPLE 3 Solve the equation $\log(5x - 1) - \log(x - 3) = 2$.

SOLUTION The equation may be written

$$\log \frac{5x - 1}{x - 3} = 2.$$

Using the definition of logarithm with $a = 10$ gives us

$$\frac{5x - 1}{x - 3} = 10^2.$$

Consequently,

$$5x - 1 = 10^2(x - 3) = 100x - 300 \quad \text{or} \quad 299 = 95x.$$

Hence,

$$x = \tfrac{299}{95}.$$

We leave it to the reader to check that this is the solution of the equation.

■

EXAMPLE 4 Solve the equation

$$\frac{5^x - 5^{-x}}{2} = 3.$$

SOLUTION Multiplying both sides of the equation by 2 gives us

$$5^x - 5^{-x} = 6.$$

If we now multiply both sides by 5^x, we obtain

$$5^{2x} - 1 = 6(5^x),$$

which may be written

$$(5^x)^2 - 6(5^x) - 1 = 0.$$

Letting $u = 5^x$ leads to the quadratic equation

$$u^2 - 6u - 1 = 0$$

in the variable u. Applying the Quadratic Formula,

$$u = \frac{6 \pm \sqrt{36 + 4}}{2} = 3 \pm \sqrt{10},$$

that is, $5^x = 3 \pm \sqrt{10}$. Since 5^x is never negative, the number $3 - \sqrt{10}$ must be discarded; therefore,

$$5^x = 3 + \sqrt{10}.$$

Taking the common logarithm of both sides and using (iii) of the Laws of Logarithms,

$$x \log 5 = \log (3 + \sqrt{10}) \quad \text{or} \quad x = \frac{\log (3 + \sqrt{10})}{\log 5}.$$

To obtain an approximate solution, we may write $3 + \sqrt{10} \approx 6.16$ and use Table 1 or a calculator. This gives us

$$x \approx \frac{\log 6.16}{\log 5} \approx \frac{0.7896}{0.6990} \approx 1.13.$$

Natural logarithms could also have been used to obtain

$$x = \frac{\ln (3 + \sqrt{10})}{\ln 5}.$$

■

EXAMPLE 5 The Beer-Lambert Law asserts that the amount of light I that penetrates to a depth of x meters in an ocean is given by $I = I_0 a^x$, where $0 < a < 1$ and I_0 is the amount of light at the surface.

(a) Solve for x in terms of I, I_0, and a by using (i) common logarithms; (ii) natural logarithms.

(b) If $a = \frac{1}{4}$, what is the depth at which $I = 0.01 I_0$? (This depth determines the zone where photosynthesis can take place.)

SOLUTION

(a) Taking the common logarithm of both sides of the equation $I = I_0 a^x$ and using laws of logarithms, we obtain

$$\log I = \log (I_0 a^x) = \log I_0 + \log a^x = \log I_0 + x \log a.$$

Hence,
$$x \log a = \log I - \log I_0 = \log (I/I_0)$$

and $$x = \frac{\log (I/I_0)}{\log a}.$$

If natural logarithms are used, then

$$x = \frac{\ln (I/I_0)}{\ln a}$$

(b) Letting $I = 0.01I_0$ and $a = \frac{1}{4}$ in the formula for x obtained in part (a),

$$x = \frac{\log (0.01I_0/I_0)}{\log \frac{1}{4}} = \frac{\log (0.01)}{\log 1 - \log 4} = \frac{\log 10^{-2}}{-\log 4}$$

Using a calculator or Table 1,

$$x \approx \frac{-2}{-0.6021} \approx 3.32 \text{ meters}.$$ ■

Exercises 4.6

In Exercises 1–14 (a) find the solutions of the equations without using a calculator or a table; (b) find two-decimal-place approximations to the solutions.

1 $10^x = 7$

2 $5^x = 8$

3 $4^x = 3$

4 $10^x = 6$

5 $3^{4-x} = 5$

6 $(\frac{1}{3})^x = 100$

7 $3^{x+4} = 2^{1-3x}$

8 $4^{2x+3} = 5^{x-2}$

9 $2^{-x} = 8$

10 $2^{-x^2} = 5$

11 $\log x = 1 - \log (x - 3)$

12 $\log (5x + 1) = 2 + \log (2x - 3)$

13 $\log (x^2 + 4) - \log (x + 2) = 3 + \log (x - 2)$

14 $\log (x - 4) - \log (3x - 10) = \log (1/x)$

Solve the equations in Exercises 15–20 without using a calculator or a table.

15 $\log (x^2) = (\log x)^2$

16 $\log \sqrt{x} = \sqrt{\log x}$

17 $\log (\log x) = 2$

18 $\log \sqrt{x^3 - 9} = 2$

19 $x^{\sqrt{\log x}} = 10^8$

20 $\log (x^3) = (\log x)^3$

In Exercises 21–24 use common logarithms to solve for x in terms of y.

21 $y = \dfrac{10^x + 10^{-x}}{2}$

22 $y = \dfrac{10^x - 10^{-x}}{2}$

23 $y = \dfrac{10^x - 10^{-x}}{10^x + 10^{-x}}$

24 $y = \dfrac{1}{10^x - 10^{-x}}$

In Exercises 25–28 use natural logarithms to solve for x in terms of y.

25 $y = \dfrac{e^x - e^{-x}}{2}$

26 $y = \dfrac{e^x + e^{-x}}{2}$

27 $y = \dfrac{e^x - e^{-x}}{e^x + e^{-x}}$

28 $y = \dfrac{e^x + e^{-x}}{e^x - e^{-x}}$

29 The current I in a certain electrical circuit at time t is given by

$$I = \frac{E}{R}(1 - e^{-Rt/L})$$

where E, R, and L denote the electromotive force, the resistance, and the inductance, respectively (see figure).

Use natural logarithms to solve for t in terms of the remaining symbols.

FIGURE FOR EXERCISE 29

30 Solve the Compound Interest Formula

$$A = P\left(1 + \frac{r}{n}\right)^{nt}$$

for t in terms of the other symbols by using natural logarithms.

31 Refer to Example 5. The most important zone in the sea from the viewpoint of marine biology is the *photic zone*, the zone in which photosynthesis can take place. That zone must end at the depth where about 1% of the surface light penetrates. In very clear waters in the Caribbean, 50% of the light at the surface reaches a depth of about 13 meters. Estimate the depth of the photic zone.

32 In sharp contrast to the situation in Exercise 31, in parts of New York harbor, 50% of the surface light does not reach a depth of 10 cm! Estimate the depth of the photic zone.

33 A drug is eliminated from the body through urine. The initial dose is 10 mg and the amount $A(t)$ in the body t hours later is given by $A(t) = 10(0.8)^t$. In order for the drug to be effective, at least 2 mg must be in the body.

(a) Determine when only 2 mg are left.

(b) What is the half-life of the drug?

34 Radioactive Iodine, ^{131}I, is frequently used in tracer studies involving the thyroid gland. The substance decays according to the formula $A(t) = A_0 a^{-t}$ where A_0 is the initial dose and t is the time in days. Find a assuming the half-life of ^{131}I is eight days.

4.7 Review

Define or discuss each of the following.

1 The exponential function with base a

2 The natural exponential function

3 The logarithmic function with base a

4 Laws of Logarithms

5 Common logarithms

6 The natural logarithmic function

Exercises 4.7

Find the numbers in Exercises 1–10 without the aid of a calculator or table.

1 $\log_2 \frac{1}{16}$

2 $\log_5 \sqrt[3]{5}$

3 $6^{\log_6 4}$

4 $10^{3 \log 2}$

5 $\log 1{,}000{,}000$

6 $\ln e$

7 $\log_4 2$

8 $\log_\pi 1$

9 $e^{\ln 5}$

10 $\log \log 10^{10}$

In Exercises 11–24 sketch the graph of f.

11 $f(x) = 3^{x+2}$

12 $f(x) = (\frac{3}{5})^x$

13 $f(x) = (\frac{3}{2})^{-x}$

14 $f(x) = 3^{-2x}$

15 $f(x) = 3^{-x^2}$

16 $f(x) = 1 - 3^{-x}$

17 $f(x) = \log_6 x$

18 $f(x) = \log_3 (x^2)$

19 $f(x) = 2 \log_3 x$

20 $f(x) = \log_2 (x + 4)$

21 $f(x) = e^{x/2}$

22 $f(x) = \frac{1}{2}e^x$

23 $f(x) = e^{x-2}$

24 $f(x) = e^{2-x}$

Find the solutions of the equations in Exercises 25–34 without using a calculator or table.

25 $\log_8 (x - 5) = \frac{2}{3}$

26 $\log_4 (x + 1) = 2 + \log_4 (3x - 2)$

27 $2 \log_3 (x + 3) - \log_3 (x + 1) = 3 \log_3 2$

28 $\log \sqrt[4]{x + 1} = \frac{1}{2}$

29 $2^{5-x} = 6$

30 $3^{x^2} = 7$

31 $2^{5x+3} = 3^{2x+1}$

32 $e^{\ln (x+1)} = 3$

33 $x^2(-2xe^{-x^2}) + 2xe^{-x^2} = 0$

34 $\ln x = 1 + \ln (x + 1)$

35 Express $\log x^4 \sqrt[3]{y^2/z}$ in terms of logarithms of x, y, and z.

36 Express $\log (x^2/y^3) + 4 \log y - 6 \log \sqrt{xy}$ as one logarithm.

Solve the equations in Exercises 37 and 38 for x in terms of y.

37 $y = \dfrac{10^x + 10^{-x}}{10^x - 10^{-x}}$

38 $y = \dfrac{1}{10^x + 10^{-x}}$

In Exercises 39–44 approximate x using (a) a calculator; (b) tables.

39 $\log x = 1.8938$

40 $\log x = -2.4260$

41 $\ln x = 1.8$

42 $\ln x = -0.75$

43 $x = \ln 6.6$

44 $x = \log 8.4$

45 The number of bacteria in a certain culture at time t is given by $Q(t) = 2(3^t)$, where t is measured in hours and $Q(t)$ in thousands. What is the initial number of bacteria? What is the number after 10 minutes? after 30 minutes? after 1 hour?

46 If \$1000 is invested at a rate of 12% compounded four times per year, what is the principal after one year?

47 Radioactive iodine, ^{131}I, is frequently used in tracer studies in the human body (and notably in tests of the thyroid gland). The substance decays according to the formula $N = N_0(0.5)^{t/8}$ where N_0 is the initial dose and t is the time in days.

(a) Sketch the graph of the equation if $N_0 = 64$.

(b) Show that the half-life of ^{131}I is eight days.

48 One thousand young trout are put in a fishing pond. Three months later, the owner estimates that there are about 600 left. Find an exponential formula $N = N_0a^t$ that fits this information and use it to estimate the number of trout left after one year.

49 Ten thousand dollars is invested in a savings fund in which interest is compounded continuously at the rate of 11% per year. The amount A in the account t years later is given by $A = 10{,}000e^{0.11t}$.

(a) When will the account contain \$35,000?

(b) How long does it take money to double in the account?

50 The current $I(t)$ at time t in a certain electrical circuit is given by $I(t) = I_0e^{-Rt/L}$ where R and L denote the resistance and inductance, respectively, and I_0 is the current at the time $t = 0$. At what time is the current $\frac{1}{100}I_0$?

51 The sound intensity level formula is $\alpha = 10 \log (I/I_0)$.

(a) Solve for I in terms of α and I_0.

(b) Show that a one-decibel rise in the intensity level α corresponds to a 26% increase in the intensity I.

52 Stars are classified into categories of brightness called *magnitudes*. The faintest stars (with light flux L_0) are assigned a magnitude of 6. Brighter stars are assigned magnitudes according to the formula

$$m = 6 - (2.5) \log (L/L_0)$$

where L is the light flux from the star.

(a) Find m if $L = 10^{0.4}L_0$.

(b) Solve the formula for L in terms of m and L_0.

53 Solve the von-Bertalanffy growth formula $y = a(1 - be^{-kt})$ for t in terms of y, a, b, and k. The resulting formula can be used to estimate the age of a fish from a length measurement.

54 For a population of female African elephants, the weight W (in kg) at age t (in years) is given by

$$W = 2600(1 - 0.51e^{-0.075t})^3.$$

(a) What is the weight of a newborn?

(b) Assuming an adult female weighs 1800 kg, estimate her age.

55 If the pollution of Lake Erie were suddenly stopped, it has been estimated that the level y of pollutants would decrease according to the formula $y = y_0e^{-0.3821t}$ where t is in years and y_0 is the pollutant level at which further pollution ceased. How many years would it take to clear 50% of the pollutants?

56 Radioactive Strontium 90, ^{90}Sr, has been deposited into a large field by acid rain. If sufficient amounts make their way through the food chain to man, bone cancer can result. It has been determined that the radioactivity level in the field is 2.5 times the safe level S. For how many years will this field be contaminated? ^{90}Sr decays according to the formula $A(t) = A_0e^{-0.0239t}$ where A_0 is the present amount in the field and t is the time in years.

CHAPTER 5

The Trigonometric Functions

Trigonometry was invented over 2000 years ago by the Greeks, who needed precise methods for measuring angles and sides of triangles. ■ In Section 5.1 we shall discuss angles from a modern point of view. ■ Trigonometric functions of angles and of real numbers are discussed in Section 5.2–5.4. ■ Graphs involving trigonometric functions are considered in Sections 5.5–5.7. ■ The chapter concludes with some applications of trigonometry to problems about right triangles and harmonic motion.

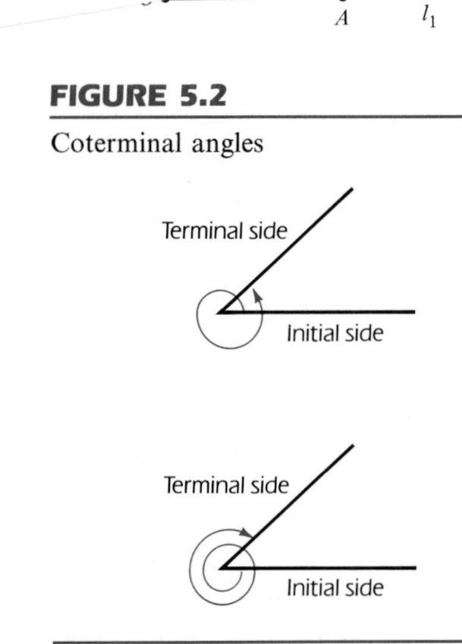

FIGURE 5.2

Coterminal angles

Angles

In geometry an angle is determined by two rays, or half-lines, l_1 and l_2, having the same initial point O. If A and B are points on l_1 and l_2, respectively (see Figure 5.1), then we refer to **angle AOB.** An angle can also be considered as two finite line segments with a common endpoint.

In trigonometry we often describe angles in a different manner: start with a fixed ray l_1 having endpoint O, and rotate it about O, in a plane, to a position specified by ray l_2. We call l_1 the **initial side,** l_2 the **terminal side,** and O the **vertex** of angle AOB. The amount or direction of rotation is not restricted in any way. Thus we might let l_1 make several revolutions in either direction about O before coming to the position l_2, as illustrated by the curved arrows in Figure 5.2. It is important to observe that many different angles have the same initial and terminal sides. Any two such angles are called **coterminal.**

If we introduce a rectangular coordinate system, then the **standard position** of an angle is obtained by taking the vertex at the origin and letting the initial side l_1 coincide with the positive x-axis. If l_1 is rotated in a counterclockwise direction to the terminal position l_2, then the angle is considered **positive,** whereas if l_1 is rotated in a clockwise direction, the angle is **negative.** We often denote angles by lowercase Greek letters and specify the direction of rotation by means of a circular arc or spiral with an arrowhead attached. Figure 5.3 contains sketches of two positive angles, α and β, and a negative angle γ. If the terminal side of an angle in standard position is in a certain quadrant, we speak of the *angle* as being in that quadrant. In Figure 5.3, α is in quadrant III, β is in quadrant I, and γ is in quadrant II. If the terminal side coincides with a coordinate axis, then the angle is referred to as a **quadrantal angle.**

FIGURE 5.3

Standard position of an angle

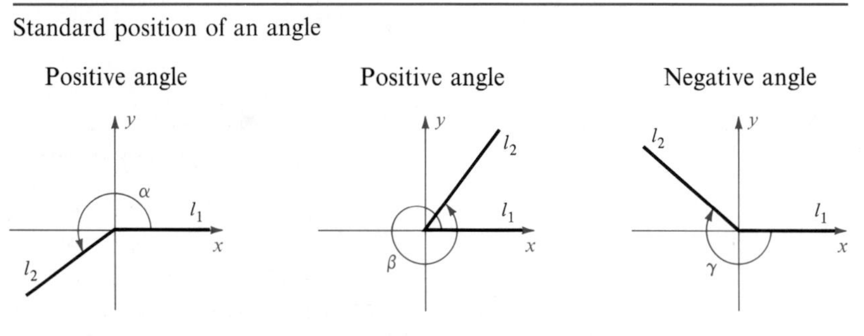

One of the units of measurement for angles is the **degree.** When placed in standard position on a rectangular coordinate system, an angle of 1 degree is by definition the measure of the angle formed by $\frac{1}{360}$ of a complete revolution in the counterclockwise direction. The symbol ° is used to denote the degrees in the measure of an angle. In Figure 5.4 several angles measured in degrees are shown in standard position on a rectangular coordinate system. In the figure, the notation $\theta = 60°$ specifies an angle θ whose measure is 60°. We also use phrases such as "an angle of 60°," or "a 60° angle" instead of the more precise (but cumbersome) phrase, "an angle having degree measure 60°."

FIGURE 5.4

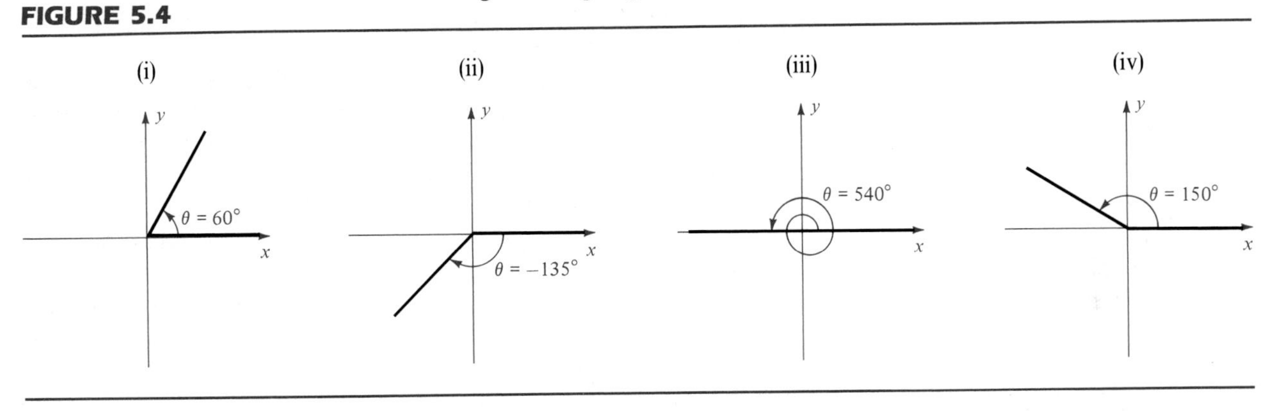

A 90° angle is called a **right angle.** An angle θ is **acute** if $0° < \theta < 90°$ or **obtuse** if $90° < \theta < 180°$. Two acute angles are **complementary** if their sum is 90°. For example, 20° and 70° are complementary angles. In general, if θ is acute, then θ and $90° - \theta$ are complementary. Two positive angles are **supplementary** if their sum is 180°.

If smaller measurements than the degree are required, we can use tenths, hundredths, or thousandths of degrees. Another method is to divide each degree into 60 equal parts, called **minutes** (denoted by ′), and each minute into 60 equal parts, called **seconds** (denoted by ″). Thus 1′ is $\frac{1}{60}$ of 1°, and 1″ is $\frac{1}{60}$ of 1′ or $\frac{1}{3600}$ of 1°. A notation such as $\theta = 73°56'18''$ refers to an angle θ of measure 73 degrees, 56 minutes, and 18 seconds.

EXAMPLE 1 If $\theta = 60°$, find two positive angles and two negative angles that are coterminal with θ.

SOLUTION The angle θ is shown in standard position in Figure 5.4(i). To find positive coterminal angles, we may add 360° or 720° (or any other multiple of 360°) to θ, since this amounts to additional complete revolutions in the positive direction. This gives us

$$60° + 360° = 420° \quad \text{and} \quad 60° + 720° = 780°.$$

To find negative coterminal angles we may add −360° or −720° (or any multiple of −360°), since this corresponds to additional revolutions in the negative direction. This gives us

$$60^\circ + (-360^\circ) = -300^\circ \quad \text{and} \quad 60^\circ + (-720^\circ) = -660^\circ. \qquad \blacksquare$$

EXAMPLE 2 Find the angle that is complementary to each angle θ.
(a) $\theta = 25^\circ 43' 37''$ (b) $\theta = 73.26^\circ$

SOLUTION We wish to find $90^\circ - \theta$. Let us arrange our work as follows:

$$\text{(a)} \quad \begin{array}{rl} 90^\circ & = 89^\circ 59' 60'' \\ \theta & = 25^\circ 43' 37'' \\ \hline 90^\circ - \theta & = 64^\circ 16' 23'' \end{array} \qquad \text{(b)} \quad \begin{array}{rl} 90^\circ & = 90.00^\circ \\ \theta & = 73.26^\circ \\ \hline 90^\circ - \theta & = 16.74^\circ \end{array} \qquad \blacksquare$$

Degree measure for angles is used in applied areas such as surveying, navigation, and the design of mechanical equipment. In scientific applications that require calculus it is customary to employ *radian measure.* A *radian* may be defined as follows.

DEFINITION

An angle has a measure of 1 **radian** if, when its vertex is placed at the center of a circle, the length of the arc intercepted by the angle on the circle equals the radius of the circle.

If we consider a circle of radius r, then an angle α whose measure is 1 radian intercepts an arc $\overset{\frown}{AP}$ of length r, as illustrated in Figure 5.5(i). We also say that the arc $\overset{\frown}{AP}$ **subtends** α or that α is *subtended by* the arc $\overset{\frown}{AP}$.

The angle β in Figure 5.5(ii) has radian measure 2, since it is subtended by an arc of length $2r$. Similarly, γ in (iii) of the figure has radian measure

FIGURE 5.5

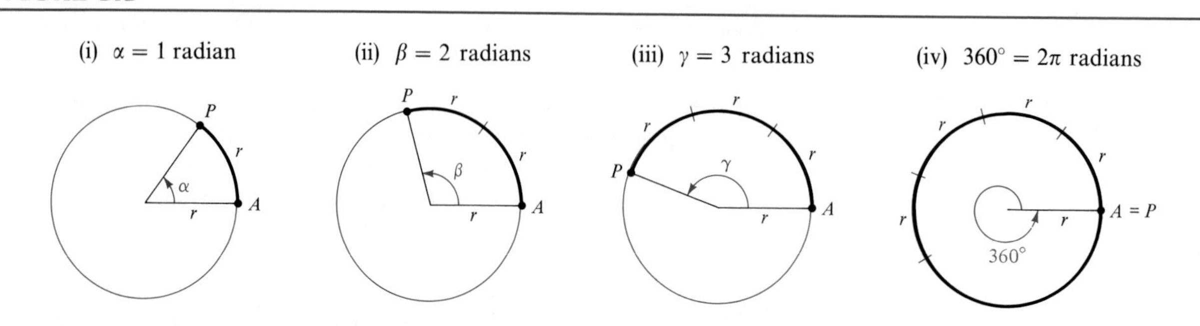

3. To find the radian measure corresponding to 360° it is necessary to find the number of times that a circular arc of length r can be laid off along the circumference. This number is not an integer or even a rational number. Indeed, since the circumference of the circle is $2\pi r$, the number of times r units can be laid off is 2π. Thus an angle of measure 2π radians corresponds to the degree measure 360°. This gives us the following important relationships.

RELATIONSHIPS BETWEEN DEGREES AND RADIANS

$$180^\circ = \pi \text{ radians}, \qquad 1^\circ = \frac{\pi}{180} \text{ radians}, \qquad 1 \text{ radian} = \left(\frac{180}{\pi}\right)^\circ.$$

If we use a calculator to approximate $\pi/180$ and $180/\pi$, we obtain

$$1^\circ \approx 0.0174533 \text{ radians} \quad \text{and} \quad 1 \text{ radian} \approx 57.29578^\circ.$$

The following theorem is a consequence of the preceding formulas.

THEOREM

(i) To change radian measure to degrees, multiply by $180/\pi$.

(ii) To change degree measure to radians, multiply by $\pi/180$.

When radian measure of an angle is used, no units will be indicated. Thus, if an angle has radian measure 5, we write $\theta = 5$ instead of $\theta = 5$ *radians.* There should be no confusion as to whether radian or degree measure is being used, since if θ has degree measure 5°, we write $\theta = 5^\circ$, and *not* $\theta = 5$.

EXAMPLE 3

(a) Find the radian measure of θ if $\theta = 150^\circ$ and if $\theta = 225^\circ$.

(b) Find the degree measure of θ if $\theta = 7\pi/4$ and if $\theta = \pi/3$.

SOLUTION

(a) By (ii) of the preceding theorem, we can find the number of radians in 150° by multiplying 150 by $\pi/180$. Thus,

$$150^\circ = 150\left(\frac{\pi}{180}\right) = \frac{5\pi}{6}.$$

Similarly, $$225^\circ = 225\left(\frac{\pi}{180}\right) = \frac{5\pi}{4}.$$

(b) By (i) of the theorem, we find the number of degrees in $7\pi/4$ radians by multiplying by $180/\pi$, obtaining

$$\frac{7\pi}{4} = \frac{7\pi}{4}\left(\frac{180}{\pi}\right) = 315^\circ.$$

Similarly,
$$\frac{\pi}{3} = \frac{\pi}{3}\left(\frac{180}{\pi}\right) = 60^\circ.$$
■

The following table displays the relationship between the radian and degree measures of several common angles. The entries may be checked by using the last theorem.

Radians	0	$\pi/6$	$\pi/4$	$\pi/3$	$\pi/2$	$2\pi/3$	$3\pi/4$	$5\pi/6$	π
Degrees	0°	30°	45°	60°	90°	120°	135°	150°	180°

Several angles in radian measure are shown in standard position in Figure 5.6.

FIGURE 5.6

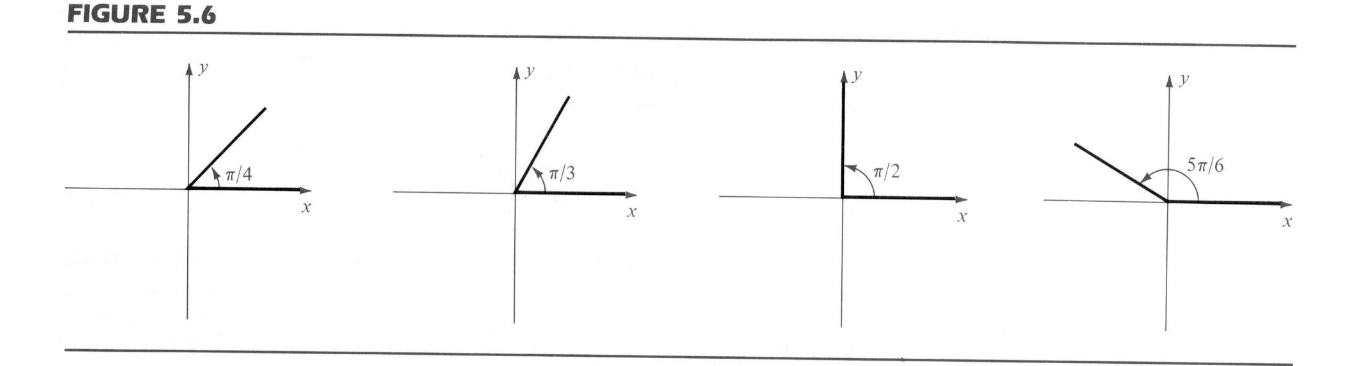

EXAMPLE 4 If the measure of an angle θ is 3 radians, approximate θ in terms of degrees, minutes, and seconds.

SOLUTION Using a calculator,

$$3 \text{ radians} \approx 3\left(\frac{180}{\pi}\right)^\circ \approx 171.88734^\circ = 171^\circ + (0.88734)^\circ.$$

Since $1^\circ = 60'$,

$$(0.88734)^\circ = (0.88734)(1^\circ) = (0.88734)(60') = 53.2404' = 53' + 0.2404'.$$

Since $1' = 60''$,

$$0.2404' = (0.2404)(1') = (0.2404)(60'') = 14.424'' \approx 14''.$$

Hence $$3 \text{ radians} \approx 171°53'14''.$$ ■

Some calculators have keys that can be used to convert the radian measure of an angle to degrees, and vice versa. Calculators may also have a [DMS] key for converting decimal degree measure to degrees, minutes, and seconds, and vice versa. Refer to the user's manual of a particular calculator for specific information. Since entries are made in terms of decimals, angles that are expressed in terms of degrees, minutes, and seconds must be changed to decimal form. If a [DMS] key is not available, we may follow the procedure given in the next example.

EXAMPLE 5 Express 19°47′23″ as a decimal, to the nearest ten thousandths of a degree.

SOLUTION Since $1' = \frac{1}{60}°$ and $1'' = \frac{1}{3600}°$,

$$\begin{aligned} 19°47'23'' &= 19° + \left(\frac{47}{60}\right)^{\circ} + \left(\frac{23}{3600}\right)^{\circ} \\ &\approx 19° + 0.7833° + 0.0064° \\ &= 19.7897° \end{aligned}$$ ■

FIGURE 5.7

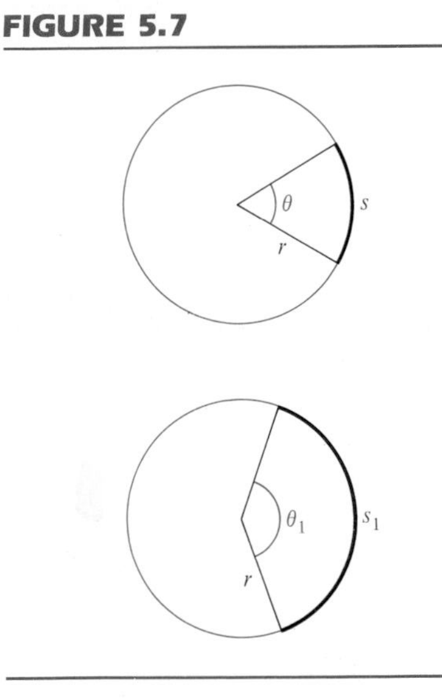

The terminology **central angle** of a circle refers to an angle whose vertex is at the center of the circle. It is shown in plane geometry that if two central angles θ and θ_1 of a circle of radius r are subtended by arcs of lengths s and s_1, respectively (see Figure 5.7), and *if θ and θ_1 are measured in radians*, then

$$\frac{\theta}{\theta_1} = \frac{s}{s_1}.$$

If we consider the special case where θ_1 has radian measure 1, then $s_1 = r$ and we obtain

$$\frac{\theta}{1} = \frac{s}{r}.$$

This gives us the following result.

THEOREM

If a central angle θ of a circle of radius r is subtended by an arc of length s, then the radian measure of θ is

$$\theta = \frac{s}{r}.$$

EXAMPLE 6 A central angle θ is subtended by an arc 10 cm long on a circle of radius 4 cm. Find (a) the radian measure of θ and (b) the degree measure of θ.

SOLUTION

(a) Substituting in the formula $\theta = s/r$ gives us the radian measure:

$$\theta = \frac{10}{4} = 2.5$$

(b) To change the radian measure found in (a) to degrees, we multiply by $180/\pi$. Thus

$$\theta = (2.5)\left(\frac{180}{\pi}\right) = \left(\frac{450}{\pi}\right)^{\circ} \approx 143.24^{\circ}.$$ ■

The formula $\theta = s/r$ for radian measure of an angle is independent of the size of the circle. For example, if the radius of the circle is $r = 4$ cm and an arc of length 8 cm subtends a central angle θ, then the radian measure of θ is

$$\theta = \frac{8 \text{ cm}}{4 \text{ cm}} = 2.$$

If the radius of the circle is 5 km and the arc is 10 km, then

$$\theta = \frac{10 \text{ km}}{5 \text{ km}} = 2.$$

These calculations indicate that the radian measure of an angle is dimensionless and hence may be regarded as a real number. It is for this reason that we employ the notation $\theta = t$ instead of $\theta = t$ radians.

The formula $\theta = s/r$ can be used to find the length of the arc that subtends a central angle θ of a circle. For problems of this type it is convenient to use the equivalent formula

$$s = r\theta.$$

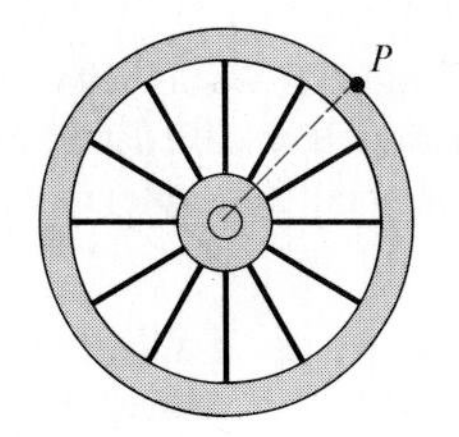

FIGURE 5.8

Rotating wheel

EXAMPLE 7 The **angular speed** of a wheel that is rotating at a constant rate is the angle generated in one unit of time by a line segment from the center of the wheel to a point P on the circumference (see Figure 5.8). Suppose that a machine contains a wheel of diameter 3 feet, rotating at a rate of 1600 rpm (revolutions per minute).

(a) Find the angular speed of the wheel.

(b) Find the speed at which a point P on the circumference of the wheel is moving.

SOLUTION

(a) Let O denote the center of the wheel and let P be a point on the circumference. Since the number of revolutions per minute is 1600, and since each revolution generates an angle of 2π radians, the angle generated by the segment OP in one minute has radian measure $(1600)(2\pi)$; that is,

$$\text{Angular speed} = (1600)(2\pi) = 3200\pi \text{ radians per minute.}$$

Note that the diameter of the wheel is irrelevant in finding the angular speed.

(b) The speed at which the point P moves is the distance it travels per minute. (This is sometimes called the **linear speed** of P.) This distance may be found by using the formula $s = r\theta$, with $r = \frac{3}{2}$ ft and $\theta = 3200\pi$ radians per minute. Thus

$$s = r\theta = \tfrac{3}{2}(3200\pi) = 4800\pi \text{ ft/min}.$$

To the nearest integer, this is approximately 15,080 ft/min. Unlike angular speed, the linear speed *is* dependent on the diameter of the wheel. ■

Exercises 5.1

In Exercises 1–12 place the angle with the indicated measure in standard position on a rectangular coordinate system, and find the measure of two positive angles and two negative angles that are coterminal with the given angle.

1 120° **2** 240° **3** 135°

4 315° **5** -30° **6** -150°

7 620° **8** 570° **9** $5\pi/6$

10 $2\pi/3$ **11** $-\pi/4$ **12** $-5\pi/4$

In Exercises 13–24 find the exact radian measure that corresponds to the degree measure.

13 150° **14** 120° **15** -60°

16 -135° **17** 225° **18** 210°

19 450° **20** 630° **21** 72°

22 54° **23** 100° **24** 95°

In Exercises 25–36 find the exact degree measure that corresponds to the radian measure.

25 $2\pi/3$ **26** $5\pi/6$ **27** $11\pi/6$

28 $4\pi/3$ **29** $3\pi/4$ **30** $11\pi/4$

31 $-7\pi/2$ **32** $-5\pi/2$ **33** 7π

34 9π **35** $\pi/9$ **36** $\pi/16$

In Exercises 37–40 find an approximate measure of θ in terms of degrees, minutes, and seconds.

37 $\theta = 2$ **38** $\theta = 1.5$

39 $\theta = 5$ **40** $\theta = 4$

In Exercises 41–44 express the degree measure as a decimal, to the nearest ten thousandths of a degree.

41 $37°41'$ **42** $83°17'$

43 $115°26'27''$ **44** $258°39'52''$

In Exercises 45–48 express the angle in terms of degrees, minutes, and seconds, rounding off to the nearest second.

45 $63.169°$ **46** $12.864°$

47 $310.6215°$ **48** $81.7238°$

49 A central angle θ is subtended by an arc 7 cm long on a circle of radius 4 cm. Approximate the measure of θ in (a) radians; (b) degrees.

50 A central angle θ is subtended by an arc 3 feet long on a circle of radius 20 inches. Approximate the measure of θ in (a) radians; (b) degrees.

51 Approximate the length of an arc that subtends a central angle of measure 50° on a circle of diameter 16 meters.

52 Approximate the length of an arc that subtends a central angle of 2.2 radians on a circle of diameter 120 cm.

53 If a central angle θ of measure 20° is subtended by a circular arc of length 3 km, find the radius of the circle.

54 If a central angle θ of radian measure 4 is subtended by a circular arc 10 cm long, find the radius of the circle.

55 The distance between two points A and B on the earth is measured along a circle having center C at the center of the earth and radius equal to the distance from C to the surface. If the diameter of the earth is approximately 8000 miles, approximate the distance between A and B if angle ACB has measure (a) 60°, (b) 45°, (c) 30°, (d) 10°, and (e) 1°.

56 Refer to Exercise 55. If angle ACB has measure 1′, then the distance between A and B is called a **nautical mile.** Approximate the number of land **(statute)** miles in a nautical mile.

57 Refer to Exercise 55. If two points A and B are 500 miles apart, find angle ACB in both degree measure and radian measure.

58 Refer to Example 7. If a wheel of diameter 18 inches is rotating at a rate of 2400 rpm, find (a) the angular speed of the wheel, and (b) the speed of a point on the circumference of the wheel in ft/min.

59 A pendulum in a grandfather clock is 4 feet long and swings back and forth along a 6-inch arc. Approximate the angle (in degrees) through which the pendulum passes during one swing.

60 A large winch of diameter 3 feet is used to hoist cargo into railroad cars (see figure).

(a) Find the distance the cargo is lifted if the winch rotates through an angle of radian measure $7\pi/4$.

(b) Find the angle (in radians) through which the winch must rotate in order to lift the cargo d feet.

FIGURE FOR EXERCISE 60

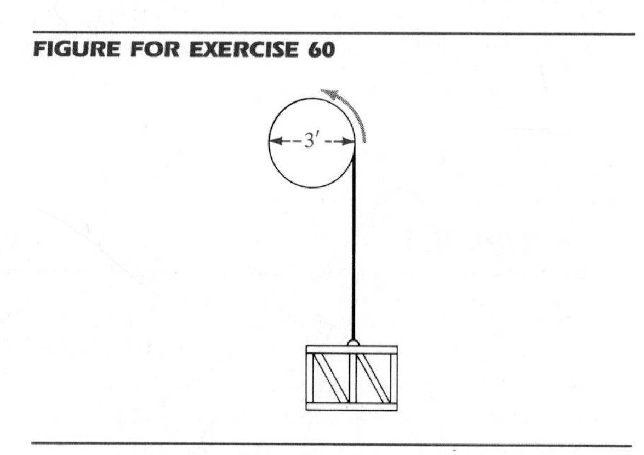

61 The sprocket assembly for a bicycle is shown in the figure on page 203. If the sprocket of radius r_1 rotates through an angle of θ_1 radians, find the corresponding angle of rotation for the sprocket of radius r_2.

FIGURE FOR EXERCISE 61

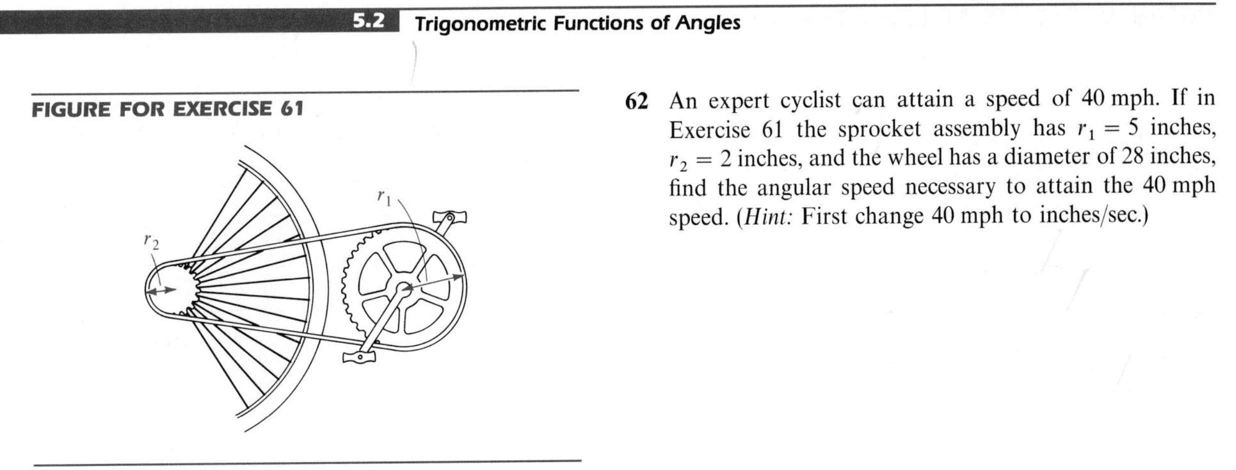

62 An expert cyclist can attain a speed of 40 mph. If in Exercise 61 the sprocket assembly has $r_1 = 5$ inches, $r_2 = 2$ inches, and the wheel has a diameter of 28 inches, find the angular speed necessary to attain the 40 mph speed. (*Hint:* First change 40 mph to inches/sec.)

5.2 Trigonometric Functions of Angles

We shall introduce the trigonometric functions in the manner that they arose historically, as ratios of sides of a right triangle. This will lead, in a natural way, to the definition in terms of arbitrary angles. We will discuss trigonometric functions of real numbers in Section 5.3.

FIGURE 5.9

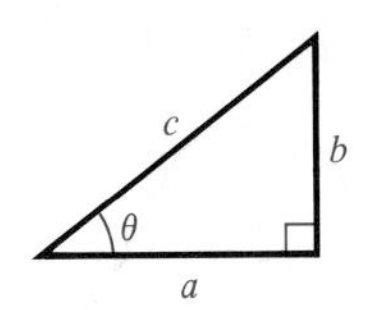

A triangle is called a **right triangle** if one of its angles is a right angle. If θ is any acute angle, we may consider a right triangle having θ as one of its angles, as illustrated in Figure 5.9, where the symbol ⌐ is used to specify the angle that has measure 90°. The following six ratios can be obtained using the lengths a, b, c of the sides of the triangle:

$$\frac{b}{c}, \quad \frac{a}{c}, \quad \frac{b}{a}, \quad \frac{c}{b}, \quad \frac{c}{a}, \quad \frac{a}{b}.$$

FIGURE 5.10

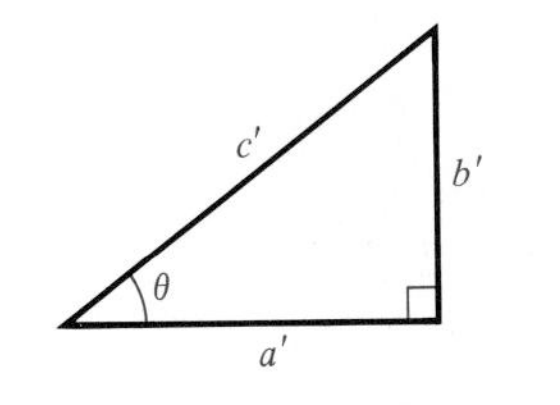

These ratios depend only on θ, and not on the size of the triangle, for suppose that θ is also an angle of a different right triangle, as shown in Figure 5.10. Since the two triangles have equal angles, they are similar, and therefore ratios of corresponding sides are proportional. For example,

$$\frac{b}{c} = \frac{b'}{c'}, \quad \frac{a}{c} = \frac{a'}{c'}, \quad \text{and} \quad \frac{b}{a} = \frac{b'}{a'}.$$

Thus, for each θ, the six ratios are uniquely determined and hence are functions of θ. They are called **trigonometric functions.** It is customary to designate them as the **sine, cosine, tangent, cosecant, secant,** and **cotangent** functions, abbreviated **sin, cos, tan, csc, sec,** and **cot,** respectively. The symbol sin (θ), or sin θ, is used for the ratio b/c, which the sine function associates with θ. Values of the other five functions are denoted in like manner. To summarize, if θ is an acute angle of a right triangle, as in Figure 5.10, then by definition

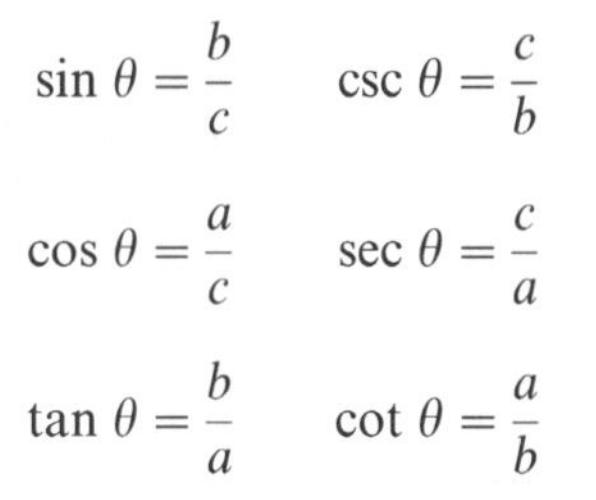

$$\sin\theta = \frac{b}{c} \qquad \csc\theta = \frac{c}{b}$$

$$\cos\theta = \frac{a}{c} \qquad \sec\theta = \frac{c}{a}$$

$$\tan\theta = \frac{b}{a} \qquad \cot\theta = \frac{a}{b}$$

Each of the six trigonometric functions has for its domain the set of all acute angles. Later in the text we will extend the domains to larger sets of angles and to real numbers.

FIGURE 5.11

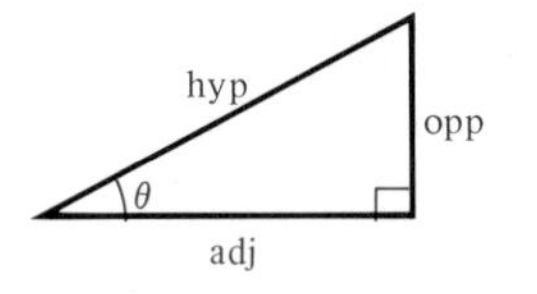

If θ is the angle in Figure 5.9, we refer to the sides of the triangle of lengths a, b, and c as the **adjacent side, opposite side,** and **hypotenuse,** respectively. For convenience we shall use **adj, opp,** and **hyp** to denote the lengths of these sides. We may then represent the triangle as in Figure 5.11. Using this notation, the trigonometric functions may be expressed as follows.

TRIGONOMETRIC FUNCTIONS OF ACUTE ANGLES

$$\sin\theta = \frac{\text{opp}}{\text{hyp}} \qquad \csc\theta = \frac{\text{hyp}}{\text{opp}}$$

$$\cos\theta = \frac{\text{adj}}{\text{hyp}} \qquad \sec\theta = \frac{\text{hyp}}{\text{adj}}$$

$$\tan\theta = \frac{\text{opp}}{\text{adj}} \qquad \cot\theta = \frac{\text{adj}}{\text{opp}}$$

It is better to memorize these formulas than those in the preceding definition, since these can be applied to any right triangle without attaching the labels a, b, c to the sides.

Since the hypotenuse is always greater than the adjacent or opposite sides, we see that $\sin\theta < 1$, $\cos\theta < 1$, $\csc\theta > 1$, and $\sec\theta > 1$ for every acute angle θ. Also note that

$$\csc\theta = \frac{\text{hyp}}{\text{opp}} = \frac{1}{(\text{opp}/\text{hyp})} = \frac{1}{\sin\theta}.$$

This gives us the first formula in the next box. The other two formulas are proved in similar fashion.

RECIPROCAL RELATIONSHIPS

$$\csc\theta = \frac{1}{\sin\theta}, \qquad \sec\theta = \frac{1}{\cos\theta}, \qquad \cot\theta = \frac{1}{\tan\theta}$$

EXAMPLE 1 Find the values of the trigonometric functions that correspond to each θ:

(a) $\theta = 60°$ (b) $\theta = 30°$ (c) $\theta = 45°$

FIGURE 5.12

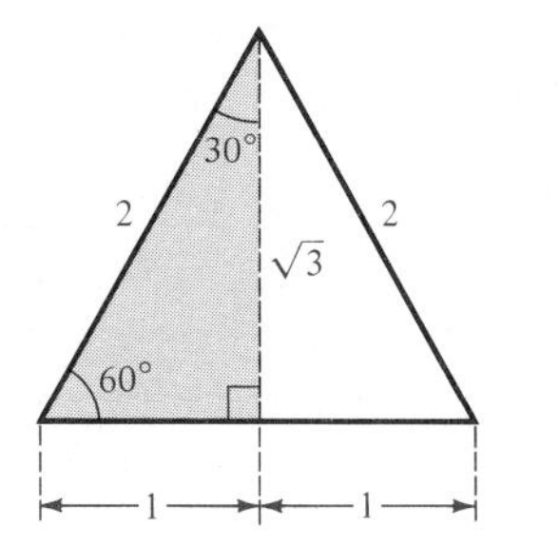

SOLUTION Let us consider an equilateral triangle having sides of length 2. The median from one vertex to the opposite side bisects the angle at that vertex, as illustrated by the dashes in Figure 5.12. By the Pythagorean Theorem, the side opposite 60° in the shaded right triangle has length $\sqrt{3}$. Using the definition of trigonometric functions, we obtain the values at 60° and 30° as follows:

(a)

$$\sin 60° = \frac{\sqrt{3}}{2} \qquad \csc 60° = \frac{2}{\sqrt{3}} = \frac{2\sqrt{3}}{3}$$

$$\cos 60° = \frac{1}{2} \qquad \sec 60° = \frac{2}{1} = 2$$

$$\tan 60° = \frac{\sqrt{3}}{1} = \sqrt{3} \qquad \cot 60° = \frac{1}{\sqrt{3}} = \frac{\sqrt{3}}{3}$$

(b)

$$\sin 30° = \frac{1}{2} \qquad \csc 30° = \frac{2}{1} = 2$$

$$\cos 30° = \frac{\sqrt{3}}{2} \qquad \sec 30° = \frac{2}{\sqrt{3}} = \frac{2\sqrt{3}}{3}$$

$$\tan 30° = \frac{1}{\sqrt{3}} = \frac{\sqrt{3}}{3} \qquad \cot 30° = \frac{\sqrt{3}}{1} = \sqrt{3}$$

FIGURE 5.13

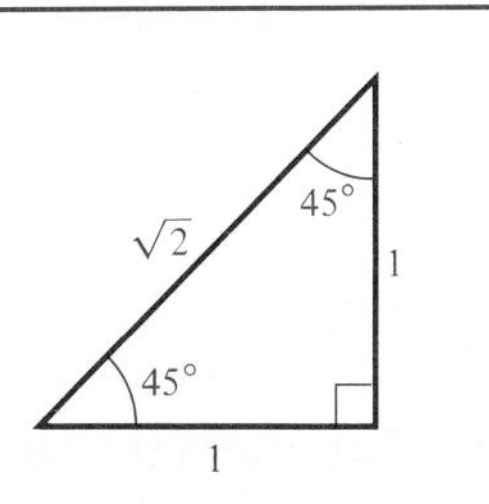

(c) To find the values for $\theta = 45°$, let us consider an isosceles right triangle whose two equal sides have length 1, as illustrated in Figure 5.13. By the Pythagorean Theorem, the length of the hypotenuse is $\sqrt{2}$. Hence,

$$\sin 45° = \frac{1}{\sqrt{2}} = \frac{\sqrt{2}}{2} = \cos 45° \qquad \tan 45° = \frac{1}{1} = 1$$

$$\csc 45° = \frac{\sqrt{2}}{1} = \sqrt{2} = \sec 45° \qquad \cot 45° = \frac{1}{1} = 1$$

■

For reference, the values found in Example 1 are listed in the following table, together with the radian measures of the angles. Two reasons for stressing these values are that (1) they are exact and (2) they occur frequently in work involving trigonometry. Because of the importance of these special values, it is a good idea either to memorize the table or to be able to find the values quickly by using triangles as in Example 1.

SPECIAL VALUES OF THE TRIGONOMETRIC FUNCTIONS

θ (radians)	θ (degrees)	$\sin\theta$	$\cos\theta$	$\tan\theta$	$\cot\theta$	$\sec\theta$	$\csc\theta$
$\frac{\pi}{6}$	30°	$\frac{1}{2}$	$\frac{\sqrt{3}}{2}$	$\frac{\sqrt{3}}{3}$	$\sqrt{3}$	$\frac{2\sqrt{3}}{3}$	2
$\frac{\pi}{4}$	45°	$\frac{\sqrt{2}}{2}$	$\frac{\sqrt{2}}{2}$	1	1	$\sqrt{2}$	$\sqrt{2}$
$\frac{\pi}{3}$	60°	$\frac{\sqrt{3}}{2}$	$\frac{1}{2}$	$\sqrt{3}$	$\frac{\sqrt{3}}{3}$	2	$\frac{2\sqrt{3}}{3}$

By employing advanced techniques it is possible to approximate, to any degree of accuracy, the values of the trigonometric functions for any acute angle. Table 4 in Appendix II gives four-decimal-place approximations to many values. The use of Table 4 is explained in Appendix I. Table 5 can be used to find function values if θ is a two-decimal-place approximation of the radian measure of an angle and if $0 < \theta < 1.57$. (Note that $1.57 \approx \pi/2$.)

Scientific calculators have keys labeled [sin], [cos], and [tan] that can be used to approximate values of these functions. The values of csc, sec, and cot may then be found by means of the reciprocal key. *Before using a calculator to find function values that correspond to the radian measure of an angle, be sure that the calculator is in the radian mode. For values corresponding to degree measure, the degree mode should be selected.*

As an illustration, to find sin 30°, place the calculator in the degree mode, enter the number 30 and press the [sin] key, obtaining $\sin 30° = 0.5$, which is the exact value. Using the same procedure for 60° we obtain, on a typical calculator,

$$\sin 60° \approx 0.8660254,$$

which is a seven-decimal-place approximation to the exact value, $\sqrt{3}/2$.

Similarly, to find a value such as cos (1.3), where 1.3 is the radian measure of an acute angle, we place the calculator in radian mode, enter 1.3, and press [cos], obtaining

$$\cos(1.3) \approx 0.9997426$$

If minutes are involved, as in finding tan 67°29′, then before using the [tan] key, *the degree measure of the angle should be expressed in decimal form.* Since

$$29' = \left(\frac{29}{60}\right)^{\circ} \approx 0.4833333,$$

we may write

$$\tan 67°29' \approx \tan 67.4833333° \approx 2.4122286$$

EXAMPLE 2 Approximate cot 54.8°.

SOLUTION Place the calculator in degree mode and use the reciprocal relationship $\cot\theta = 1/\tan\theta$ as follows:

Enter:	54.8	(degree value of θ)
Press [tan]:	1.4175904	(value of $\tan\theta$)
Press [1/x]:	0.7054224	(value of $1/\tan\theta$)

Hence cot 54.8° ≈ 0.7054224. ■

Inverse functions that correspond to trigonometric functions are defined in Section 6.6, where it is shown that there is a function denoted by $\sin^{-1}$ such that

$$\sin^{-1}(\sin\theta) = \theta$$

provided $$-\pi/2 \le \theta \le \pi/2 \quad \text{or} \quad -90° \le \theta \le 90°.$$

This fact can be used to find θ when given $\sin\theta$. For our present purposes we need not be concerned about the restriction $-\pi/2 \le \theta \le \pi/2$, or $-90° \le \theta \le 90°$, since we are only interested in acute angles. The reason for this restriction on θ is discussed in Section 6.6, where *inverse trigonometric functions* are treated in detail.

To use $\sin^{-1}$ with some calculators, we first press [INV] (for inverse) and then [sin]. This technique is analogous to that in Section 4.5, where given $\log x$, we pressed [INV] followed by [log] to approximate x. Some calculators have a key labeled $\boxed{\sin^{-1}}$ or $\boxed{\text{arcsin}}$ for finding values of inverse trigonometric functions directly.

EXAMPLE 3 For an acute angle θ with $\sin\theta = 0.5$, use a calculator to approximate (a) the degree measure of θ; (b) the radian measure of θ.

SOLUTION

(a) We place the calculator in degree mode and proceed as follows:

Enter:	0.5	(the value of $\sin\theta$)
Press $\boxed{\text{INV}}$ $\boxed{\sin}$:	30	(the degree measure of θ)

(b) We place the calculator in radian mode and do the following:

Enter:	0.5	(the value of $\sin\theta$)
Press $\boxed{\text{INV}}$ $\boxed{\sin}$:	0.5235988	(the radian measure of θ)

The last number is a decimal approximation for an angle of $\pi/6$ radians. ■

In Section 6.6 we will also define functions denoted by $\cos^{-1}$ and $\tan^{-1}$ such that

$$\cos^{-1}(\cos\theta) = \theta \quad \text{if} \quad 0 \le \theta \le \pi \quad \text{or} \quad 0^\circ \le \theta \le 180^\circ$$

and

$$\tan^{-1}(\tan\theta) = \theta \quad \text{if} \quad -\pi/2 < \theta < \pi/2 \quad \text{or} \quad -90^\circ < \theta < 90^\circ$$

These may be used in the same manner that $\sin^{-1}$ was used in Example 3.

EXAMPLE 4 For an acute angle θ with $\cos\theta = 0.4271$, approximate the degree measure of θ to the nearest (a) $(0.01)^\circ$; (b) $1'$.

SOLUTION

(a) We place the calculator in degree mode and then:

Enter:	0.4271	(the value of $\cos\theta$)
Press $\boxed{\text{INV}}$ $\boxed{\cos}$:	64.7163341	(the degree measure of θ)

Thus $$\theta \approx 64.72^\circ.$$

(b) To find θ to the nearest $1'$ we proceed as in Example 4 of the preceding section, obtaining

$$(0.716341)^\circ = (0.716341)(1^\circ) = (0.716341)(60') = 42.98046' \approx 43'$$

and hence $$\theta \approx 64^\circ 43'.$$

Table 4, together with the method of interpolation (see Appendix I), can also be used to approximate θ. ■

Many relationships exist among the trigonometric functions. The formulas listed in the next box are, without doubt, the most important identities in trigonometry, because they may be used to simplify and to unify many different aspects of the subject. Since they are part of the foundation for work in trigonometry, they are called the *Fundamental Identities.*

Three of the fundamental identities involve squares such as $(\sin\theta)^2$ and $(\cos\theta)^2$. In general, if n is an integer and $n \neq -1$, then powers such as $(\cos\theta)^n$ are written in the form $\cos^n\theta$. The symbols $\sin^{-1}$ and $\cos^{-1}$ are reserved for inverse trigonometric functions, which will be discussed in Section 6.6. With this agreement on notation we have, for example,

$$\cos^2\theta = (\cos\theta)^2 = (\cos\theta)(\cos\theta)$$
$$\tan^3\theta = (\tan\theta)^3 = (\tan\theta)(\tan\theta)(\tan\theta)$$
$$\sec^4\theta = (\sec\theta)^4 = (\sec\theta)(\sec\theta)(\sec\theta)(\sec\theta).$$

Let us first list all the fundamental identities and then discuss the proofs. The following formulas are true for every acute angle θ. We shall see later that they are also true for much larger sets of angles and for real numbers.

THE FUNDAMENTAL IDENTITIES

$$\csc\theta = \frac{1}{\sin\theta} \qquad \tan\theta = \frac{\sin\theta}{\cos\theta} \qquad \sin^2\theta + \cos^2\theta = 1$$

$$\sec\theta = \frac{1}{\cos\theta} \qquad \cot\theta = \frac{\cos\theta}{\sin\theta} \qquad 1 + \tan^2\theta = \sec^2\theta$$

$$\cot\theta = \frac{1}{\tan\theta} \qquad 1 + \cot^2\theta = \csc^2\theta$$

FIGURE 5.14

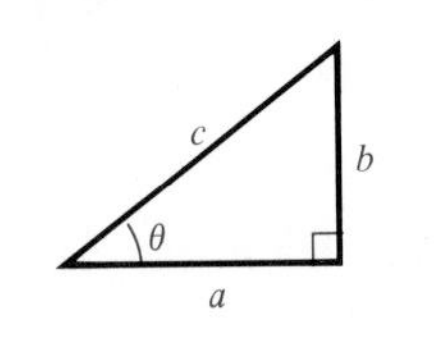

The first three formulas in the left-hand column are the reciprocal relationships discussed in the preceding section. The others can be proved by referring to the right triangle in Figure 5.14. Thus

$$\tan \theta = \frac{b}{a} = \frac{b/c}{a/c} = \frac{\sin \theta}{\cos \theta}.$$

Then, also,

$$\cot \theta = \frac{1}{\tan \theta} = \frac{\cos \theta}{\sin \theta}.$$

If we apply the Pythagorean Theorem to the triangle in Figure 5.14, we obtain

$$b^2 + a^2 = c^2.$$

Dividing both sides by c^2 leads to

$$\left(\frac{b}{c}\right)^2 + \left(\frac{a}{c}\right)^2 = 1, \quad \text{that is,} \quad \sin^2 \theta + \cos^2 \theta = 1.$$

Dividing both sides of the last equation by $\cos^2 \theta$ gives us

$$\frac{\sin^2 \theta}{\cos^2 \theta} + \frac{\cos^2 \theta}{\cos^2 \theta} = \frac{1}{\cos^2 \theta} \quad \text{or} \quad \left(\frac{\sin \theta}{\cos \theta}\right)^2 + 1 = \left(\frac{1}{\cos \theta}\right)^2.$$

Since $\sin \theta/\cos \theta = \tan \theta$ and $1/\cos \theta = \sec \theta$, this implies that

$$\tan^2 \theta + 1 = \sec^2 \theta.$$

The proof that $1 + \cot^2 \theta = \csc^2 \theta$ is left as an exercise.

The fundamental identity $\sin^2 \theta + \cos^2 \theta = 1$ can be used to express $\sin \theta$ in terms of $\cos \theta$, or vice versa. For example, since

$$\sin^2 \theta = 1 - \cos^2 \theta$$

and $\sin \theta$ is positive,

$$\sin \theta = \sqrt{1 - \cos^2 \theta}.$$

Similarly,

$$\cos \theta = \sqrt{1 - \sin^2 \theta}.$$

The formulas $1 + \tan^2 \theta = \sec^2 \theta$ and $1 + \cot^2 \theta = \csc^2 \theta$ can be used in like manner.

Each trigonometric function can be expressed in terms of any other trigonometric function. The next example gives one illustration. (See also Exercises 29–36.)

EXAMPLE 5 Express $\tan\theta$ in terms of $\cos\theta$, where θ is an acute angle.

SOLUTION Consider $\tan\theta = \sin\theta/\cos\theta$. Since $\sin^2\theta = 1 - \cos^2\theta$ and θ is acute, we may write

$$\sin\theta = \sqrt{1 - \cos^2\theta}.$$

Hence,
$$\tan\theta = \frac{\sin\theta}{\cos\theta} = \frac{\sqrt{1-\cos^2\theta}}{\cos\theta}.$$ ■

EXAMPLE 6 For an acute angle θ with $\sec\theta = \frac{5}{3}$, use fundamental identities to find (a) $\tan\theta$; (b) $\sin\theta$.

SOLUTION

(a) Since $1 + \tan^2\theta = \sec^2\theta$, we see that

$$\tan^2\theta = \sec^2\theta - 1$$

and hence

$$\tan\theta = \sqrt{\sec^2\theta - 1} = \sqrt{\left(\frac{5}{3}\right)^2 - 1} = \sqrt{\frac{25}{9} - 1}$$
$$= \sqrt{\frac{16}{9}} = \frac{4}{3}.$$

(b) First we note that $\cos\theta = 1/\sec\theta = \frac{3}{5}$. Hence

$$\sin\theta = \sqrt{1 - \cos^2\theta} = \sqrt{1 - \left(\frac{3}{5}\right)^2} = \sqrt{\frac{16}{25}} = \frac{4}{5}.$$ ■

In Examples 5 and 6 square roots were positive. Later, when we enlarge the domains, the trigonometric functions may take on negative values, and we will have to write

$$\sin\theta = \pm\sqrt{1 - \cos^2\theta} \quad \text{and} \quad \tan\theta = \pm\sqrt{\sec^2\theta - 1},$$

and choose either + or −, depending on the nature of θ.

Many applied problems require the use of angles that are not acute. To employ trigonometry as a tool for solving such problems, it is necessary to extend the definition of the trigonometric functions. This extension may be made by using the standard position of an angle θ on a rectangular coordinate system. If θ is acute, then we have a situation similar to that

FIGURE 5.15

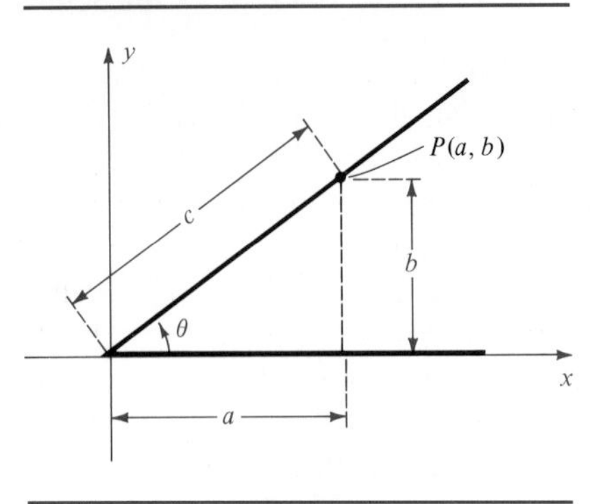

illustrated in Figure 5.15, where we have chosen a point $P(a, b)$ on the terminal side of θ and where $c = \sqrt{a^2 + b^2}$. Evidently,

$$\sin\theta = \frac{b}{c}, \qquad \cos\theta = \frac{a}{c}, \qquad \tan\theta = \frac{b}{a}.$$

We now wish to consider angles of the type illustrated in Figure 5.16 (or any *other* angle, either positive, negative, or zero).

Our objective is to define the trigonometric functions so that their values agree with those given previously whenever the angle is acute. The next definition accomplishes this in a simple manner.

FIGURE 5.16

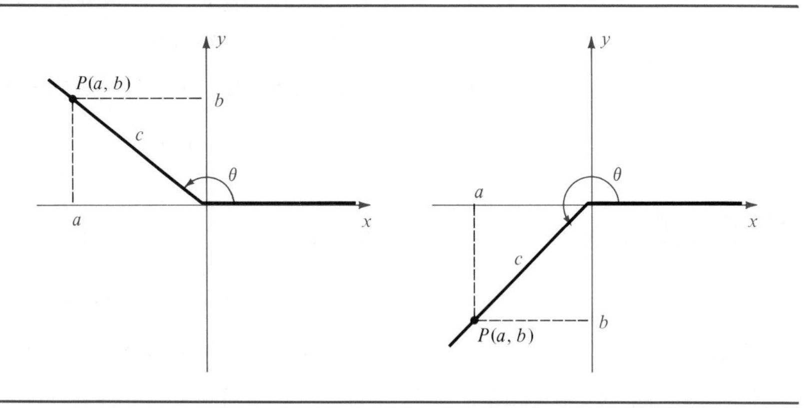

TRIGONOMETRIC FUNCTIONS OF ANY ANGLE

Let θ be an angle in standard position on a rectangular coordinate system and let $P(a, b)$ be any point other than O on the terminal side of θ. If $d(O,P) = c = \sqrt{a^2 + b^2}$, then

$$\sin\theta = \frac{b}{c} \qquad\qquad \csc\theta = \frac{c}{b} \quad (\text{if } b \neq 0)$$

$$\cos\theta = \frac{a}{c} \qquad\qquad \sec\theta = \frac{c}{a} \quad (\text{if } a \neq 0)$$

$$\tan\theta = \frac{b}{a} \quad (\text{if } a \neq 0) \qquad\qquad \cot\theta = \frac{a}{b} \quad (\text{if } b \neq 0)$$

It can be shown, using similar triangles, that the formulas in this definition are independent of the point $P(a, b)$ that is chosen on the terminal side of θ. Moreover, the numbers a and b may be positive, negative, or zero, depending on the quadrant containing θ, or on whether the terminal side is on a coordinate axis.

The domains of the sine and cosine functions consist of all angles θ. Note, however, that $\tan\theta$ and $\sec\theta$ are undefined if $a = 0$, that is, if the terminal side of θ is on the y-axis. Thus the domain of both the tangent and secant functions consists of all angles *except* those having radian measure $(\pi/2) + n\pi$ where n is an integer. Some special cases are $\pm\pi/2$, $\pm 3\pi/2$, and $\pm 5\pi/2$. The corresponding degree measures are $\pm 90°$, $\pm 270°$, and $\pm 450°$. The domains of both the cotangent and cosecant functions consist of all angles except those which lead to $b = 0$, that is, all angles except those having terminal sides on the x-axis. These are the angles of radian measure $n\pi$ (or degree measure $n \cdot 180°$) where n is an integer.

For all points $P(a, b)$ in the preceding definition we have $|a| \le c$ and $|b| \le c$. This implies that

$$|\sin\theta| \le 1, \qquad |\cos\theta| \le 1, \qquad |\csc\theta| \ge 1, \qquad |\sec\theta| \ge 1$$

for every θ in the domains of these functions.

FIGURE 5.17

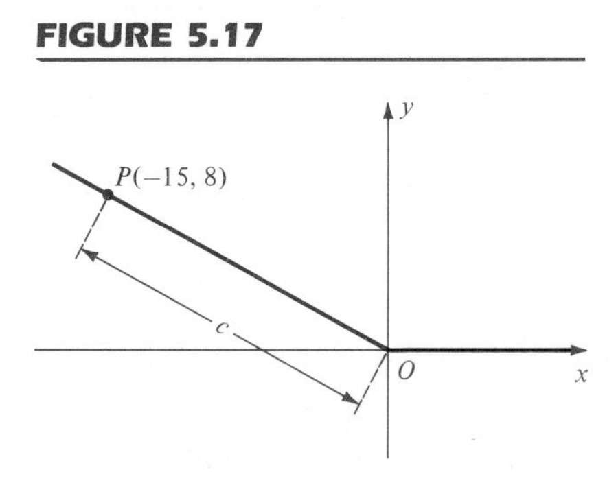

EXAMPLE 7 If θ is an angle in standard position on a rectangular coordinate system and if the point $P(-15, 8)$ is on the terminal side of θ, find the values of the trigonometric functions of θ.

SOLUTION The point $P(-15, 8)$ is shown in Figure 5.17. By the Distance Formula,

$$c = d(O, P) = \sqrt{(-15)^2 + 8^2} = \sqrt{225 + 64} = \sqrt{289} = 17.$$

Applying the definition with $a = -15$, $b = 8$, and $c = 17$,

$$\sin\theta = \tfrac{8}{17} \qquad \csc\theta = \tfrac{17}{8}$$
$$\cos\theta = -\tfrac{15}{17} \qquad \sec\theta = -\tfrac{17}{15}$$
$$\tan\theta = -\tfrac{8}{15} \qquad \cot\theta = -\tfrac{15}{8}$$

■

FIGURE 5.18

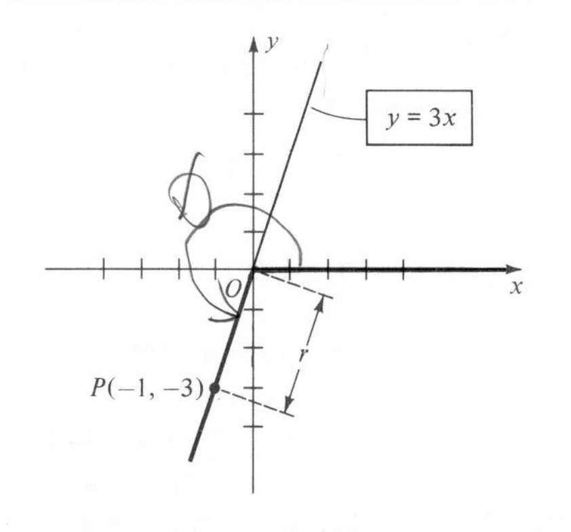

EXAMPLE 8 An angle θ is in standard position on a rectangular coordinate system, and its terminal side is in quadrant III and lies on the line $y = 3x$. Find the values of the trigonometric functions of θ.

SOLUTION The graph of $y = 3x$ is sketched in Figure 5.18, together with the (black) initial and terminal sides of θ.

We begin by choosing a convenient third quadrant point, say $P(-1, -3)$, on the terminal side. The distance c from the origin to P is

$$c = d(O, P) = \sqrt{(-1)^2 + (-3)^2} = \sqrt{10}.$$

Applying the definition with $a = -1$, $b = -3$, and $c = \sqrt{10}$ gives us

$$\sin\theta = \frac{-3}{\sqrt{10}} = -\frac{3\sqrt{10}}{10} \qquad \csc\theta = \frac{\sqrt{10}}{-3} = -\frac{\sqrt{10}}{3}$$

$$\cos\theta = \frac{-1}{\sqrt{10}} = -\frac{\sqrt{10}}{10} \qquad \sec\theta = \frac{\sqrt{10}}{-1} = -\sqrt{10}$$

$$\tan\theta = \frac{-3}{-1} = 3 \qquad \cot\theta = \frac{-1}{-3} = \frac{1}{3}$$ ■

The definition of the trigonometric functions of any angle may be applied if the terminal side of θ is on a coordinate axis, as illustrated in the next example.

EXAMPLE 9 Find the values of the trigonometric functions of θ if $\theta = 3\pi/2$.

FIGURE 5.19

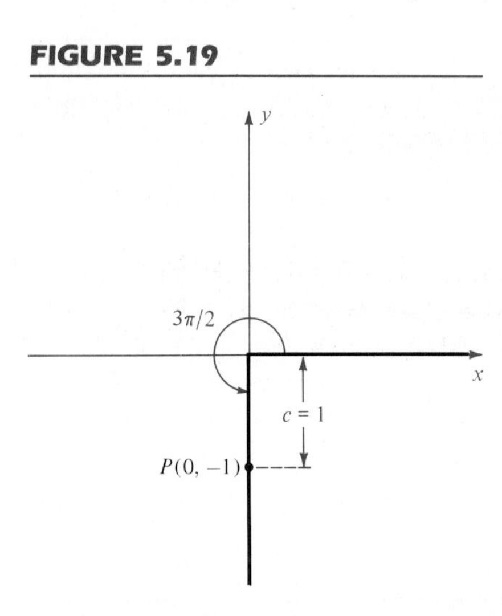

SOLUTION Note that $3\pi/2 = 270°$. Placing θ in standard position, the terminal side of θ coincides with the negative y-axis, as shown in Figure 5.19. To use the definition we may choose any point P on the terminal side of θ. For simplicity we consider $P(0, -1)$. In this case $c = 1$, $a = 0$, $b = -1$, and hence

$$\sin\frac{3\pi}{2} = \frac{-1}{1} = -1 \qquad \csc\frac{3\pi}{2} = \frac{1}{-1} = -1$$

$$\cos\frac{3\pi}{2} = \frac{0}{1} = 0 \qquad \cot\frac{3\pi}{2} = \frac{0}{-1} = 0$$

The tangent and secant functions are undefined, since the meaningless expressions $\tan\theta = (-1)/0$ and $\sec\theta = 1/0$ occur when we substitute in the appropriate formulas. ■

Let us determine the signs associated with the trigonometric functions. If θ is in quadrant II and $P(a, b)$ is a point on the terminal side, then a is negative, b is positive and hence, by definition, $\sin\theta$ and $\csc\theta$ are positive, whereas the other four functions are negative. The reader should check the remaining quadrants. The following table indicates the signs in all four quadrants.

SIGNS OF THE TRIGONOMETRIC FUNCTIONS

Quadrant containing θ	Positive functions	Negative functions
I	All	None
II	sin, csc	cos, sec, tan, cot
III	tan, cot	sin, csc, cos, sec
IV	cos, sec	sin, csc, tan, cot

EXAMPLE 10 Find the quadrant containing θ if both $\sin\theta < 0$ and $\cos\theta > 0$.

SOLUTION Referring to the table of signs, we see that $\sin\theta < 0$ if θ is in quadrants III or IV, and $\cos\theta > 0$ if θ is in quadrants I or IV. Hence, for both conditions to be satisfied, θ must be in quadrant IV. ■

The Fundamental Identities, which were established for acute angles, are also true for trigonometric functions of any angle.

Exercises 5.2

In Exercises 1–10 find the values of the trigonometric functions of the angle θ in the pictured right triangle.

1

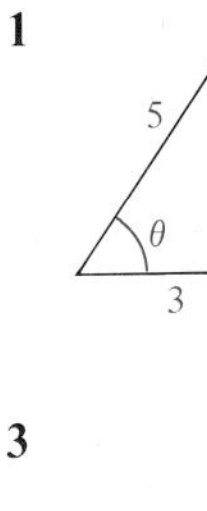

2

4

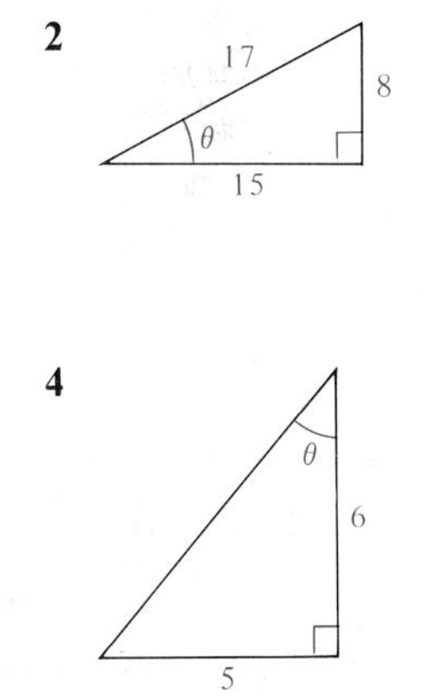

3

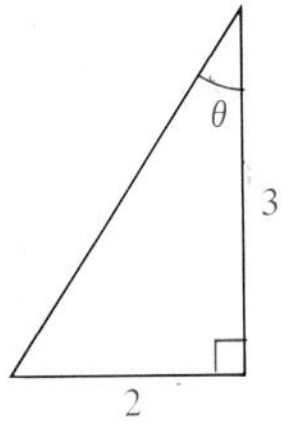

5

7

9

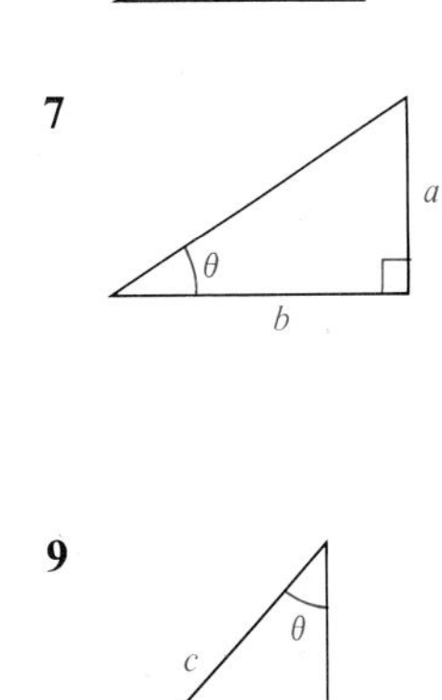

6

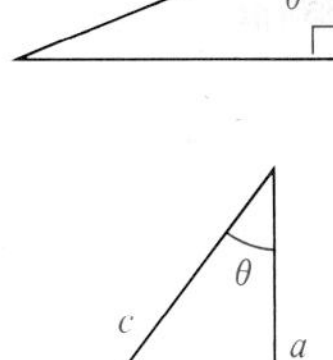

8

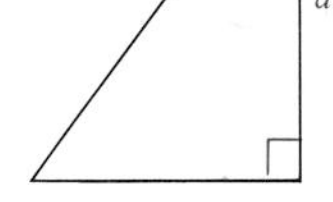

10

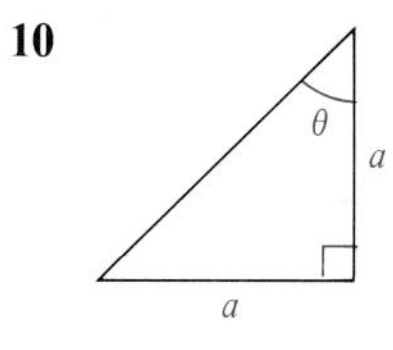

In Exercises 11–16 find the values of all the trigonometric functions for the acute angle θ.

11 $\sin \theta = \frac{3}{5}$ **12** $\cos \theta = \frac{8}{17}$

13 $\sec \theta = \frac{6}{5}$ **14** $\csc \theta = 4$

15 $\cos \theta = \sqrt{2}/2$ **16** $\sin \theta = \sqrt{3}/2$

In Exercises 17–22 approximate the function values to four significant figures using (a) Table 4; (b) a calculator.

17 $\cos 38°30'$ **18** $\sin 73°20'$ **19** $\cot 9°10'$

20 $\csc 43°40'$ **21** $\sec 67°50'$ **22** $\tan 21°10'$

In Exercises 23–28 approximate the function values of the indicated radian measures to four significant figures using (a) Table 5; (b) a calculator.

23 $\cos (0.68)$ **24** $\sin (1.48)$ **25** $\cot (1.13)$

26 $\csc (0.32)$ **27** $\sec (0.26)$ **28** $\tan (0.75)$

In Exercises 29–36 use fundamental identities to write the first expression in terms of the second.

29 $\cot \theta, \sin \theta$ **30** $\tan \theta, \sin \theta$

31 $\sec \theta, \sin \theta$ **32** $\csc \theta, \cos \theta$

33 $\tan \theta, \sec \theta$ **34** $\cot \theta, \csc \theta$

35 $\sin \theta, \sec \theta$ **36** $\cos \theta, \cot \theta$

Prove the identities in Exercises 37 and 38.

37 (a) $\sin \theta = \dfrac{1}{\csc \theta}$

(b) $\cos \theta = \dfrac{1}{\sec \theta}$

(c) $\tan \theta = \dfrac{1}{\cot \theta}$

38 $1 + \cot^2 \theta = \csc^2 \theta$

In Exercises 39–42 use fundamental identities to find the values of the remaining trigonometric functions for the acute angle θ.

39 $\cos \theta = \frac{1}{2}$ **40** $\sin \theta = \frac{2}{5}$

41 $\csc \theta = \frac{3}{2}$ **42** $\cot \theta = \frac{4}{3}$

In Exercises 43–60 find the values of the six trigonometric functions of θ when θ is in standard position and satisfies the given conditions.

43 The point $P(4, -3)$ is on the terminal side of θ.

44 The point $P(-8, -15)$ is on the terminal side of θ.

45 The point $P(-2, -5)$ is on the terminal side of θ.

46 The point $P(-1, 2)$ is on the terminal side of θ.

47 The terminal side of θ is in quadrant II and lies on the line $y = -4x$.

48 The terminal side of θ is in quadrant IV and lies on the line $3y + 5x = 0$.

49 The terminal side of θ is in quadrant III and is parallel to the line $2y - 7x + 2 = 0$.

50 The terminal side of θ is in quadrant II and is parallel to the line through the points $A(1, 4)$ and $B(3, -2)$.

51 The terminal side of θ is in quadrant I and lies on a line having slope $\frac{4}{3}$.

52 The terminal side of θ bisects the third quadrant.

53 $\theta = 90°$ **54** $\theta = 0°$

55 $\theta = 180°$ **56** $\theta = -270°$

57 $\theta = 2\pi$ **58** $\theta = 5\pi/2$

59 $\theta = 7\pi/2$ **60** $\theta = 3\pi$

In Exercises 61–70 find the quadrant containing θ if the given conditions are true.

61 $\cos \theta > 0$ and $\sin \theta < 0$

62 $\tan \theta < 0$ and $\cos \theta > 0$

63 $\sin \theta < 0$ and $\cot \theta > 0$

64 $\sec \theta > 0$ and $\tan \theta < 0$

65 $\csc \theta > 0$ and $\sec \theta < 0$

66 $\csc \theta > 0$ and $\cot \theta < 0$

67 $\sec \theta < 0$ and $\tan \theta > 0$

68 $\sin \theta < 0$ and $\sec \theta > 0$

69 $\cos \theta > 0$ and $\tan \theta > 0$

70 $\cos \theta < 0$ and $\csc \theta < 0$

5.3 The Circular Functions

In Section 2.1 a function was defined as a correspondence that assigns to each element of a set X (the domain) a unique element of a set Y (the range). For each trigonometric function discussed in the preceding section, the domain is a set of angles and the range is a set of real numbers. For example, given the angle 45°, the tangent function assigns the real number tan 45°, or 1. In calculus and in many applications, the domains of functions consist of real numbers. It is therefore necessary to take a different point of view, and to regard the domains of trigonometric functions as subsets of $\mathbb{R}$. Perhaps the simplest way to accomplish this is to use the following definition.

TRIGONOMETRIC FUNCTIONS OF REAL NUMBERS

> If t is a real number, then the value of any trigonometric function at t is its value at an angle of t radians.

According to our definition, a notation such as sin 2 may be interpreted as *either* the sine of the real number 2 *or* the sine of an angle of 2 radians. As in Section 5.2, if degree measure is used we shall write sin 2°. Of course, $\sin 2 \neq \sin 2^\circ$.

To find values of trigonometric functions of real numbers with a calculator, all that is necessary is to use the radian mode. When using the tables in Appendix II, the column labeled "t" should be consulted. For example,

$$\cos(0.6720) \approx 0.7826 \qquad \text{(from Table 4 or a calculator)}$$

$$\sin(1.40) \approx 0.9854 \qquad \text{(from Table 5 or a calculator)}$$

FIGURE 5.20

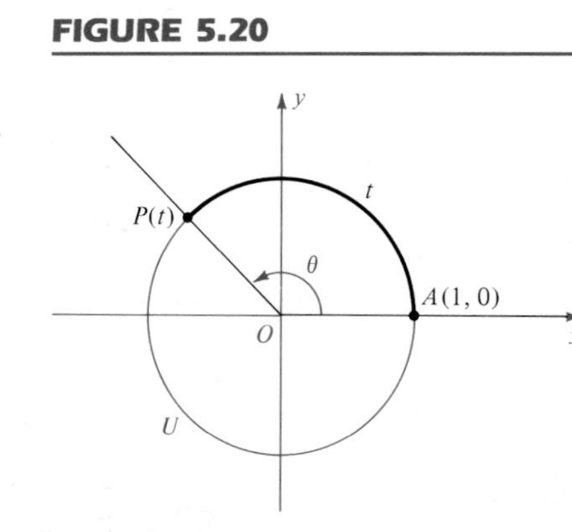

The concept of trigonometric functions as functions of real numbers may be interpreted geometrically by referring to a unit circle U with center at the origin of a coordinate plane. Thus U is the graph of the equation $x^2 + y^2 = 1$. Given any real number t, let θ denote the angle (in standard position) of radian measure t. One possibility, with $0 < \theta < 2\pi$, is illustrated in Figure 5.20, where $P(t)$ denotes the point of intersection of the terminal side of θ and the unit circle U. Using the formula $s = r\theta$ (from Section 5.1) with $r = 1$, we see that the arc $\overset{\frown}{AP}$ that subtends θ has length $s = t$. Thus, *the real number t can be regarded as either the radian measure of the angle θ or as the length of arc $\overset{\frown}{AP}$ on U.*

For any $t > 0$ we may think of the angle θ as having been generated by rotating the positive x-axis about O in the counterclockwise direction. In this case, t is the distance along U that the point $P(t)$ travels before reaching its final position. If $t < 0$, then $|t|$ is the distance traveled by $P(t)$.

The preceding discussion indicates how we may associate with each real number t, a unique point $P(t)$ on U. We shall call $P(t)$ **the point on the unit circle U that corresponds to t.** The rectangular coordinates (x, y) of $P(t)$ may be used to find the six trigonometric functions of t. Thus, by the definition of trigonometric functions of real numbers, together with that for any angle (see page 212), we may write

$$\sin t = \sin \theta = \frac{y}{1} = y.$$

The same procedure for the remaining five functions gives us the following formulas.

TRIGONOMETRIC FUNCTIONS IN TERMS OF A UNIT CIRCLE

If t is a real number and $P(x, y)$ is the point on the unit circle U that corresponds to t, then

$$\sin t = y \qquad \csc t = \frac{1}{y} \quad (\text{if } y \neq 0)$$

$$\cos t = x \qquad \sec t = \frac{1}{x} \quad (\text{if } x \neq 0)$$

$$\tan t = \frac{y}{x} \quad (\text{if } x \neq 0) \qquad \cot t = \frac{x}{y} \quad (\text{if } y \neq 0).$$

Since these formulas express values in terms of the coordinates of a point on a unit circle, the trigonometric functions are sometimes referred to as the **circular functions.**

EXAMPLE 1 Use the unit circle U to find the values of the trigonometric functions at

(a) $t = 0$ (b) $t = \pi/4$ (c) $t = \pi/2$.

SOLUTION The points $P(x, y)$ on the unit circle U that correspond to the given values of t are plotted in Figure 5.21.

(a) If $t = 0$, the point P has coordinates $(1, 0)$ and, accordingly, we take $x = 1$ and $y = 0$ in the definition of the trigonometric functions. This gives us

$$\sin 0 = 0, \qquad \cos 0 = 1, \qquad \tan 0 = \frac{0}{1} = 0, \qquad \sec 0 = \frac{1}{1} = 1.$$

Note that $\csc 0$ and $\cot 0$ are undefined, since $y = 0$ appears in a denominator.

FIGURE 5.21

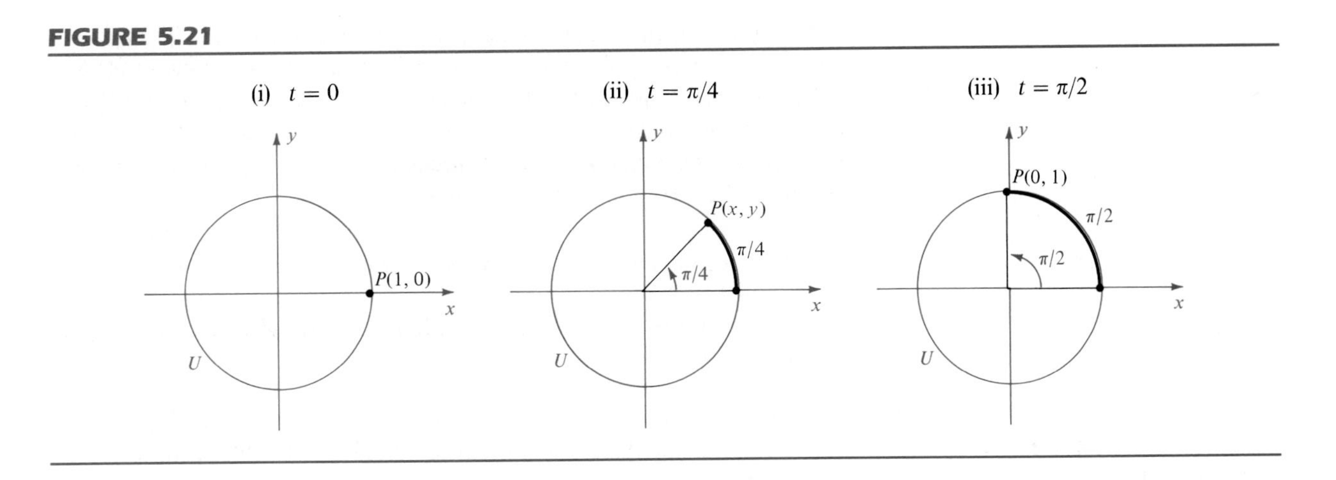

(b) If $t = \pi/4$, then the angle of radian measure $\pi/4$ (or degree measure 45°), shown in Figure 5.21(ii), bisects the first quadrant. It follows that the point P has coordinates of the form (x, x). Since $P(x, x)$ is also on the unit circle $x^2 + y^2 = 1$, we see that

$$x^2 + x^2 = 1, \quad \text{or} \quad 2x^2 = 1.$$

Solving for x gives us $\qquad x = \dfrac{1}{\sqrt{2}} = \dfrac{\sqrt{2}}{2}.$

Thus P is the point $(\sqrt{2}/2, \sqrt{2}/2)$. Letting $x = \sqrt{2}/2$ and $y = \sqrt{2}/2$ in the definition gives us

$$\sin\frac{\pi}{4} = \frac{\sqrt{2}}{2} \qquad \csc\frac{\pi}{4} = \frac{2}{\sqrt{2}} = \sqrt{2}$$

$$\cos\frac{\pi}{4} = \frac{\sqrt{2}}{2} \qquad \sec\frac{\pi}{4} = \frac{2}{\sqrt{2}} = \sqrt{2}$$

$$\tan\frac{\pi}{4} = \frac{\sqrt{2}/2}{\sqrt{2}/2} = 1 \qquad \cot\frac{\pi}{4} = \frac{\sqrt{2}/2}{\sqrt{2}/2} = 1.$$

(c) If $t = \pi/2$, then P has coordinates $(0, 1)$, as shown in Figure 5.21(iii). Letting $x = 0$ and $y = 1$ in the definition, we obtain

$$\sin\frac{\pi}{2} = 1, \qquad \cos\frac{\pi}{2} = 0, \qquad \csc\frac{\pi}{2} = \frac{1}{1} = 1, \qquad \cot\frac{\pi}{2} = \frac{0}{1} = 0.$$

The tangent and cosecant functions are undefined. (Why?) ■

The domains of trigonometric functions of angles were discussed in Section 5.2. If we wish to express the domains in terms of real numbers, we could refer to the radian measures of angles. An alternative method is to use the unit circle approach discussed in this section. Thus, if $P(x, y)$ is the point on U corresponding to the real number t, then since $\sin t = x$ and $\cos t = y$, the sine and cosine functions always exist and, therefore, each has domain $\mathbb{R}$. In the definitions of the tangent and secant functions, x appears in the denominator and hence we must exclude values of t for which x is 0, that is, the values of t corresponding to the points (0, 1) and (0, −1) on the unit circle U. It follows that the domain of the tangent and secant functions consists of all numbers *except* those of the form $(\pi/2) + n\pi$ where n is an integer. In particular, we exclude $\pm\pi/2$, $\pm 3\pi/2$, $\pm 5\pi/2$, etc.

Since $\cot t = x/y$ and $\csc t = 1/y$, the domain of the cotangent and cosecant functions is the set of all real numbers except those numbers t for which the y-coordinate of P is 0. These include the numbers, 0, $\pm\pi$, $\pm 2\pi$, $\pm 3\pi$, and, in general, all numbers of the form $n\pi$ where n is an integer. In the future, when we work with $\tan t$, $\cot t$, $\sec t$, and $\csc t$, we will always assume that t is in the appropriate domain, even though this fact will not always be mentioned explicitly.

FIGURE 5.22

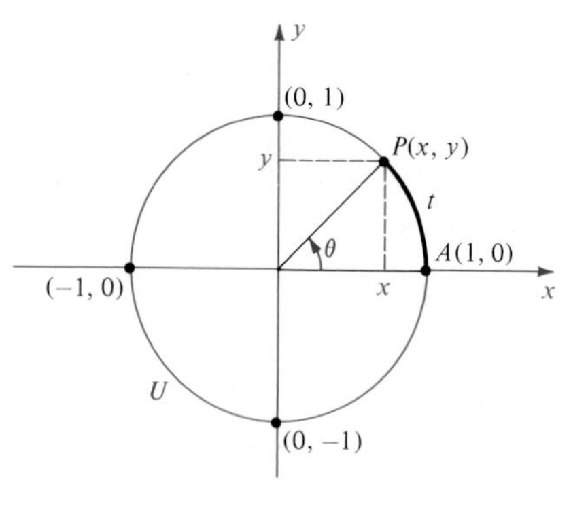

The unit circle approach can be used to make some general observations about the variation of the trigonometric functions. Let us consider the unit circle U shown in Figure 5.22, where the real number t may be interpreted either as the radian measure of the angle θ or as the length of arc $\overset{\frown}{AP}$. If we let t increase from 0 to 2π, then the point $P(x, y)$ that corresponds to t travels around U once in the counterclockwise direction. Since $\sin t = y$, we can study the corresponding values of the sine function by concentrating on the y-coordinate of P. Similarly, since $\cos t = x$, values of the cosine function may be studied by observing the x-coordinate of P. The following table indicates how $\sin t$ and $\cos t$ vary as t changes from 0 to 2π.

t	$P(x, y)$	$\sin t = y$	$\cos t = x$
$0 \to \frac{\pi}{2}$	$(1, 0) \to (0, 1)$	$0 \to 1$	$1 \to 0$
$\frac{\pi}{2} \to \pi$	$(0, 1) \to (-1, 0)$	$1 \to 0$	$0 \to -1$
$\pi \to \frac{3\pi}{2}$	$(-1, 0) \to (0, -1)$	$0 \to -1$	$-1 \to 0$
$\frac{3\pi}{2} \to 2\pi$	$(0, -1) \to (1, 0)$	$-1 \to 0$	$0 \to 1$

The notation $0 \to \pi/2$ means that t increases from 0 to $\pi/2$, and the notation $(1, 0) \to (0, 1)$ denotes the corresponding variation of $P(x, y)$

along U. We see that as t increases from 0 to $\pi/2$, $\sin t$ increases from 0 to 1; this is denoted by $0 \to 1$. Moreover, $\sin t$ takes on every value between 0 and 1. As t increases from $\pi/2$ to π, $\sin t$ decreases from 1 to 0. Other entries in the table may be interpreted in similar fashion.

Let us analyze in more detail the variation of $\sin t$ and $\cos t$. The following table lists several values for $0 \leq t \leq 2\pi$.

t	0	$\frac{\pi}{4}$	$\frac{\pi}{2}$	$\frac{3\pi}{4}$	π	$\frac{5\pi}{4}$	$\frac{3\pi}{2}$	$\frac{7\pi}{4}$	2π
$\sin t$	0	$\frac{\sqrt{2}}{2}$	1	$\frac{\sqrt{2}}{2}$	0	$-\frac{\sqrt{2}}{2}$	-1	$-\frac{\sqrt{2}}{2}$	0
$\cos t$	1	$\frac{\sqrt{2}}{2}$	0	$-\frac{\sqrt{2}}{2}$	-1	$-\frac{\sqrt{2}}{2}$	0	$\frac{\sqrt{2}}{2}$	1

As t increases from 2π to 4π, $P(x, y)$ traces the unit circle U again and the identical patterns for $\sin t$ and $\cos t$ are repeated; that is,

$$\sin(t + 2\pi) = \sin t \quad \text{and} \quad \cos(t + 2\pi) = \cos t$$

for every t in the interval $[0, 2\pi]$. The same is true if t increases from 4π to 6π, from 6π to 8π, and so on. Indeed, for every positive or negative integer n,

$$\sin(t + 2\pi n) = \sin t \quad \text{and} \quad \cos(t + 2\pi n) = \cos t.$$

This repetitive variation is specified by saying that the sine and cosine functions are *periodic*. In general we have the following definition.

DEFINITION

A function f is **periodic** if there exists a positive real number k such that

$$f(t + k) = f(t)$$

for every t in the domain of f. The least such positive real number k, if it exists, is called the **period** of f.

The period of the sine and cosine functions is 2π. (See Exercises 23 and 24.)

The preceding discussion suggests that the graphs of the sine and cosine functions have the shapes illustrated in Figures 5.23 and 5.24.

FIGURE 5.23

Sine function

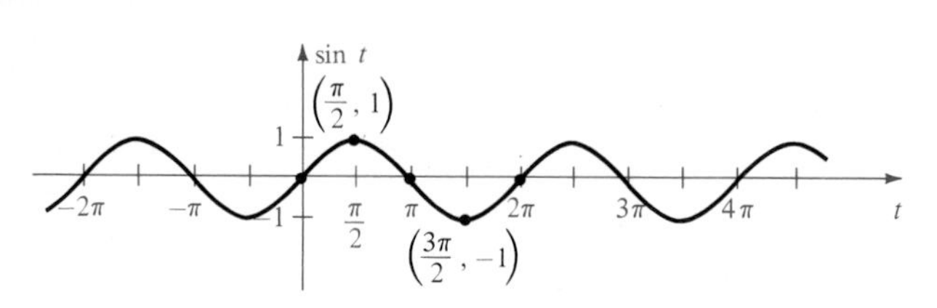

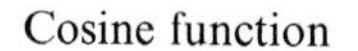

FIGURE 5.24

Cosine function

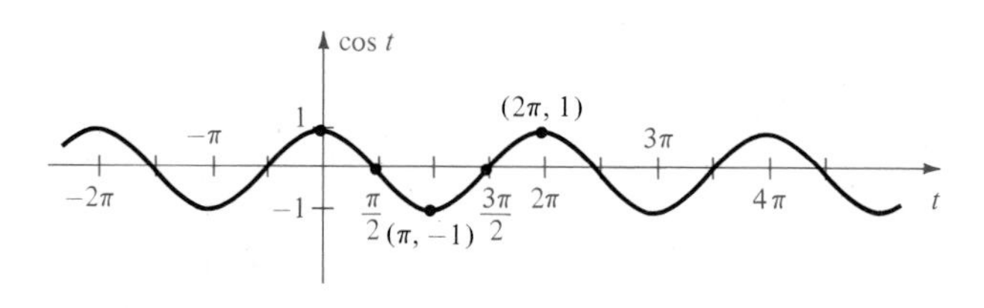

In the figures we have plotted several points of the form $(t, \sin t)$ and $(t, \cos t)$. Since the functions have period 2π, their graphs are repeated every 2π units along the horizontal axes. The sketches are only rough approximations, because we have not yet discussed the use of calculators or tables for obtaining many other points on the graphs.

The secant and cosecant functions are also periodic with period 2π. (See Exercises 25 and 26.) Graphs of these functions will be considered in Section 5.5. At that time we shall also see that the tangent and cotangent functions are periodic with period π.

The range of the sine and cosine functions consists of all real numbers in the closed interval $[-1, 1]$. Since $\csc t = 1/\sin t$ and $\sec t = 1/\cos t$, it follows that the range of the cosecant and secant functions consists of all real numbers having absolute value greater than or equal to 1.

The range of the tangent and cotangent functions consists of all real numbers. (See Exercises 27 and 28.) We will discuss their graphs in Section 5.5.

We shall conclude this section by obtaining formulas for trigonometric functions of $-t$, where t is any real number. Since a minus sign is involved, we shall call them *Formulas for Negatives.*

FORMULAS FOR NEGATIVES

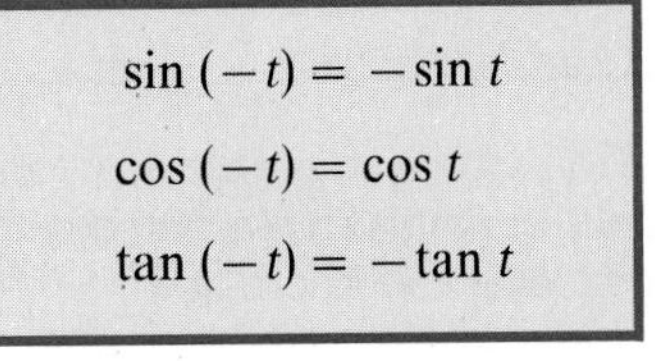

$$\sin(-t) = -\sin t$$
$$\cos(-t) = \cos t$$
$$\tan(-t) = -\tan t$$

FIGURE 5.25

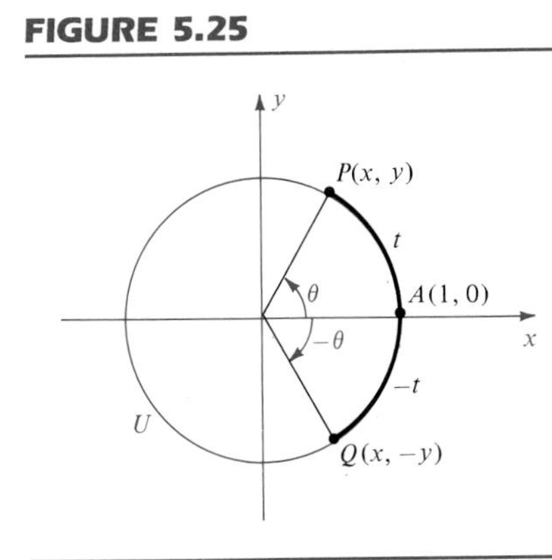

PROOF Consider the unit circle in Figure 5.25. As t increases from 0 to 2π, the point $P(x, y)$ that corresponds to t traces the unit circle U once in the counterclockwise direction, whereas the point $Q(x, -y)$ corresponding to $-t$ traces U once in the clockwise direction. Using the formulas for trigonometric functions in terms of a unit circle,

$$\sin(-t) = -y = -\sin t$$

$$\cos(-t) = x = \cos t$$

$$\tan(-t) = \frac{-y}{x} = -\frac{y}{x} = -\tan t. \qquad \square$$

EXAMPLE 2 Use the formulas for negatives to find the exact values of $\sin(-\pi/4)$, $\cos(-\pi/4)$, and $\tan(-\pi/4)$.

SOLUTION Using the formulas and Example 1,

$$\sin(-\pi/4) = -\sin(\pi/4) = -\sqrt{2}/2$$
$$\cos(-\pi/4) = \cos(\pi/4) = \sqrt{2}/2$$
$$\tan(-\pi/4) = -\tan(\pi/4) = -1 \qquad \blacksquare$$

If $f(t) = \cos t$, then

$$f(-t) = \cos(-t) = \cos t = f(t)$$

and hence the cosine function is even. As we pointed out on page 67, this implies that the graph is symmetric with respect to the y-axis. (See Figure 5.24.)

Similarly, if $f(t) = \sin t$, then

$$f(-t) = \sin(-t) = -\sin t = -f(t).$$

Thus the sine function is odd, and the graph is symmetric with respect to the origin (see Figure 5.23).

Exercises 5.3

In Exercises 1–12 find (a) the coordinates of the point on the unit circle that corresponds to the given real number t, and (b) the values of the trigonometric functions at t.

1 2π **2** -3π **3** $3\pi/2$

4 $5\pi/2$ **5** -7π **6** 98π

7 $-7\pi/2$ **8** $-9\pi/2$ **9** $5\pi/4$

10 $-\pi/4$ **11** $-9\pi/4$ **12** $11\pi/4$

If the point $P(t)$ on U has the rectangular coordinates given in Exercises 13–16, find:

(a) $P(t + \pi)$, (b) $P(t - \pi)$, (c) $P(-t)$, (d) $P(-t - \pi)$.

13 $(\frac{3}{5}, \frac{4}{5})$ **14** $(\frac{4}{5}, -\frac{3}{5})$

15 $(-\frac{8}{17}, \frac{15}{17})$ **16** $(-\frac{15}{17}, \frac{8}{17})$

In Exercises 17–20 use formulas for negatives to find the exact values of $\sin t$ and $\cos t$ for the given value of t.

17 $-3\pi/4$ **18** $-5\pi/4$

19 $-\pi/2$ **20** -2π

21 Prove that (a)–(c) are true for every t in the domain of the function:

(a) $\csc(-t) = -\csc t$ (b) $\sec(-t) = \sec t$

(c) $\cot(-t) = -\cot t$

22 Prove that the tangent function is odd. What does this imply about its graph?

23 Prove that the sine function has period 2π. (*Hint:* Assume that there is a positive real number k less than 2π such that $\sin(t + k) = \sin t$ for all t. Arrive at a contradiction by letting $t = 0$.)

24 Prove that the cosine function has period 2π.

25 Prove that the cosecant function has period 2π.

26 Prove that the secant function has period 2π.

27 Prove that the range of the tangent function is $\mathbb{R}$ by showing that if a is any real number, then there is a point $P(t)$ on U such that $\tan t = a$. (*Hint:* If $P(x, y)$ is on U, consider the equation $\tan t = y/x = a$, where $x^2 + y^2 = 1$.)

28 Prove that the range of the cotangent function is $\mathbb{R}$.

29 Show that for every integer n the point $P(t)$ on the unit circle U that corresponds to $t = n\pi$ is either $(-1, 0)$ or $(1, 0)$, and hence that $\sin n\pi = 0$ and $\cos n\pi = (-1)^n$.

30 Use the accompanying figure to approximate

(a) the numbers $\cos 4$ and $\sin 4$ (to one decimal place);

(b) $\cos(-1.2)$ and $\sin(-1.2)$;

(c) the values of t between 0 and 2π such that $\cos t = -0.6$;

(d) the values of t between 0 and 2π such that $\sin t = 0.5$.

FIGURE FOR EXERCISE 30

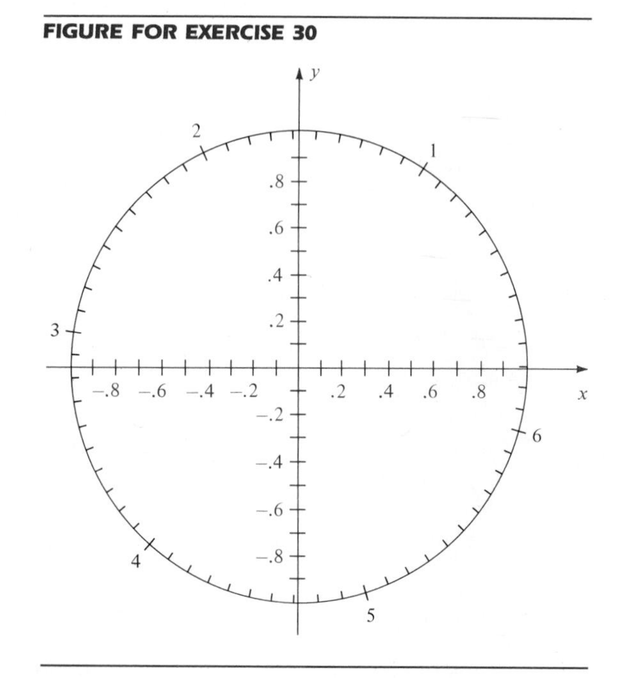

5.4 Values of the Trigonometric Functions

In previous sections we calculated several special values of the trigonometric functions. Let us now consider the problem of finding *all* values. Since the sine function has period 2π, it is sufficient to know the values of $\sin t$ for $0 \le t \le 2\pi$, because these values are repeated in every t-interval of length 2π. The same is true for the other trigonometric functions. As a matter of fact, the values of any trigonometric function can be determined if we know its values for $0 \le t \le \pi/2$. To prove this fact we shall make use of the following concepts.

DEFINITION

Let t be a real number, let θ be the angle in standard position of radian measure t, and let P be the point on the unit circle U that corresponds to t.

(i) The **reference number** associated with t is the shortest arc length t' on U between P and the x-axis.

(ii) The **reference angle** associated with θ is the acute angle θ' that the terminal side of θ makes with the positive or negative x-axis.

Typical reference numbers t' and reference angles θ' are indicated in Figure 5.26. It is important to observe that *the reference angle and reference number are always measured from the x-axis*, and that $0 \le t' \le \pi/2$ and $0 \le \theta' \le \pi/2$ for all values of t and θ.

FIGURE 5.26

Reference numbers t' and reference angles θ'

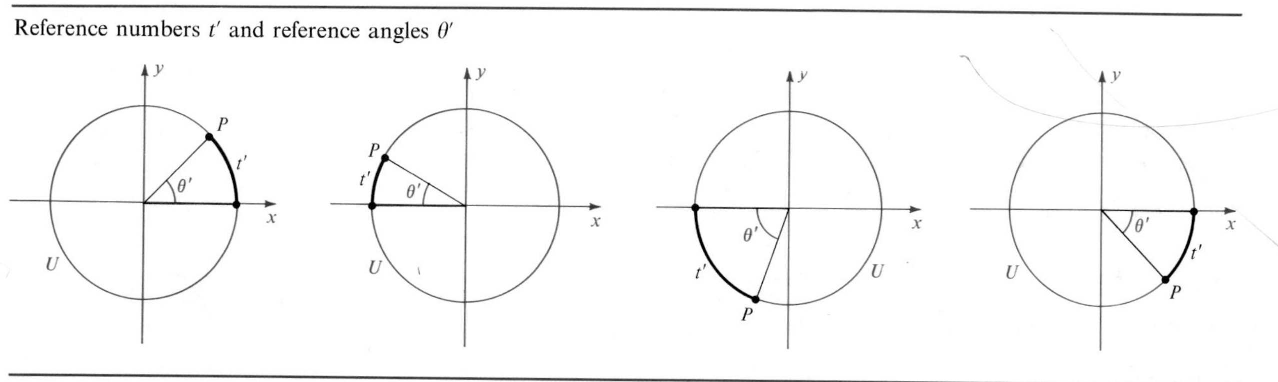

Using the definition of trigonometric functions of real numbers on page 217, we see that

$$\sin \theta' = \sin t', \qquad \cos \theta' = \cos t', \qquad \tan \theta' = \tan t'.$$

Similar formulas are true for the other trigonometric functions. In the following example we shall use the approximation $\pi \approx 3.1416$ for estimating reference numbers.

EXAMPLE 1 Find the reference number t' if

(a) $t = 2$ (b) $t = 4$ (c) $t = 7\pi/4$ (d) $t = -2\pi/3$.

SOLUTION Let $P(t)$ denote the point on the unit circle U that corresponds to t. Each point $P(t)$ in (a)–(d) and the magnitude of its reference number t' are indicated in Figure 5.27.

FIGURE 5.27

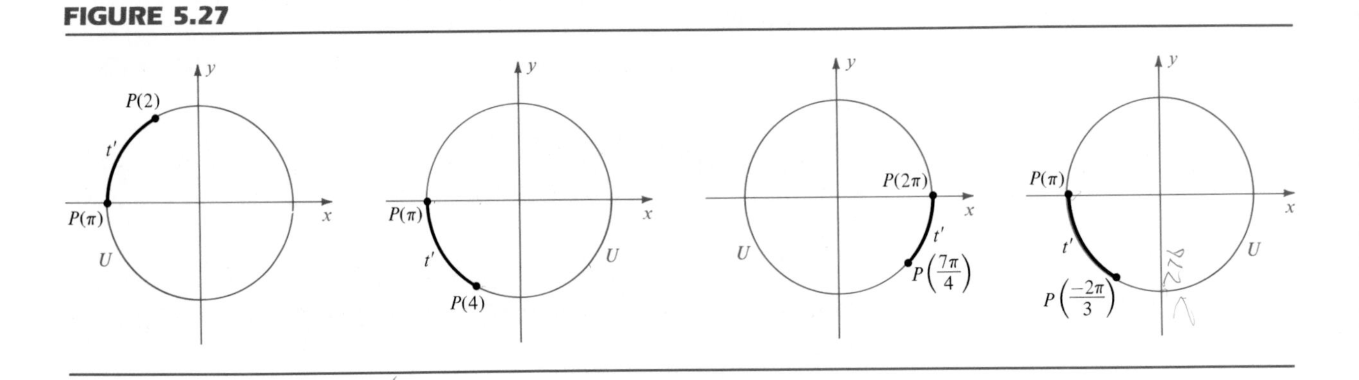

Referring to the figure, we see that:

(a) $t' = \pi - 2 \approx 3.1416 - 2 = 1.1416$

(b) $t' = 4 - \pi \approx 4 - 3.1416 = 0.8584$

(c) $t' = 2\pi - (7\pi/4) = \pi/4$

(d) $t' = \pi - (2\pi/3) = \pi/3$ ■

EXAMPLE 2 Sketch the reference angle θ' and express the measure of θ' in terms of both radians and degrees if

(a) $\theta = 5\pi/6$ (b) $\theta = 315°$ (c) $\theta = -240°$.

SOLUTION Each angle θ and its reference angle θ' are sketched in Figure 5.28.

FIGURE 5.28

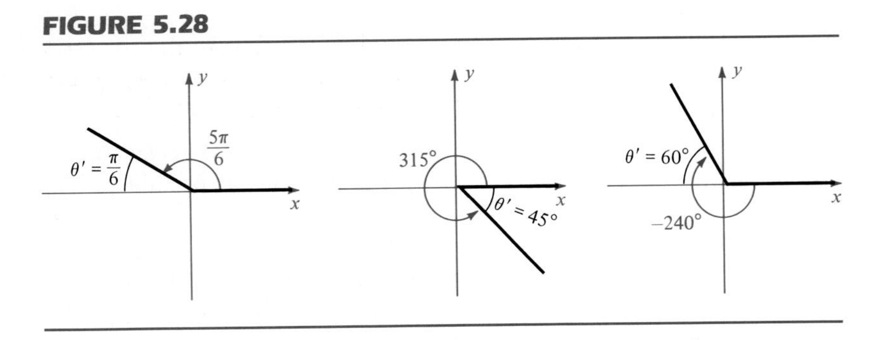

Evidently,

(a) $\theta' = \pi/6 = 30°$ (b) $\theta' = 45° = \pi/4$ (c) $\theta' = 60° = \pi/3$. ■

We shall next show how reference numbers or reference angles can be used to find values of trigonometric functions. Let $P(x, y)$ be the point on the unit circle U that corresponds to t. If P is not on a coordinate axis and t' is the reference number for t, then $0 < t' < \pi/2$. Consider the point $A(1, 0)$ and let $P'(x', y')$ be the point on U in quadrant I such that $\widehat{AP'} = t'$. Illustrations in which $P(x, y)$ lies in quadrants II, III, or IV are given in Figure 5.29.

FIGURE 5.29

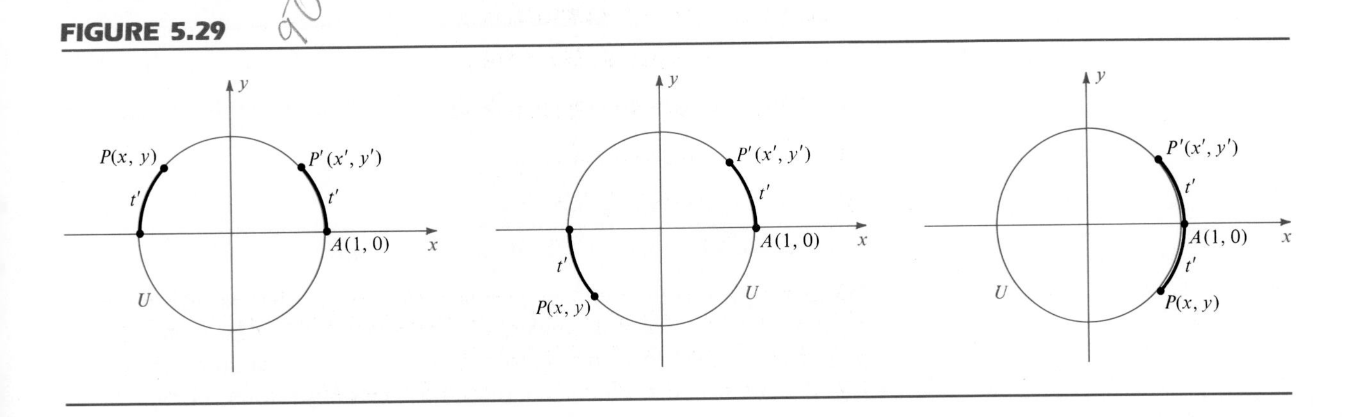

We see that in all cases

$$x' = |x| \quad \text{and} \quad y' = |y|$$

and therefore

$$|\cos t| = |x| = x' = \cos t'$$

$$|\sin t| = |y| = y' = \sin t'.$$

It is easy to show that the absolute value of *every* trigonometric function at t is the same as its value at t'. For example,

$$|\tan t| = \left|\frac{y}{x}\right| = \frac{|y|}{|x|} = \frac{y'}{x'} = \tan t'.$$

In terms of angles, if θ' is the reference angle of θ, then

$$|\sin \theta| = \sin \theta' \quad \text{and} \quad |\cos \theta| = \cos \theta'.$$

Similar formulas are true for the other trigonometric functions. We may state the following rules.

RULES FOR FINDING VALUES OF TRIGONOMETRIC FUNCTIONS

(i) To find the value of a trigonometric function at a real number t, determine its value for the reference number t' associated with t and prefix the appropriate sign.

(ii) To find the value of a trigonometric function at an angle θ, determine its value for the reference angle θ' associated with θ and prefix the appropriate sign.

The "appropriate sign" can be determined from the table of signs on page 215.

EXAMPLE 3 Find $\sin (7\pi/4)$ and $\cos (-7\pi/6)$.

SOLUTION We can interpret $7\pi/4$ and $-7\pi/6$ either as real numbers or as radian measures of angles. If we let $P(t)$ denote the point on the unit circle U that corresponds to t, then we have the situation shown in Figure 5.30, where $\pi/4$ is the reference number (angle) for $7\pi/4$, and $\pi/6$ is the reference number (angle) for $-7\pi/6$.

FIGURE 5.30

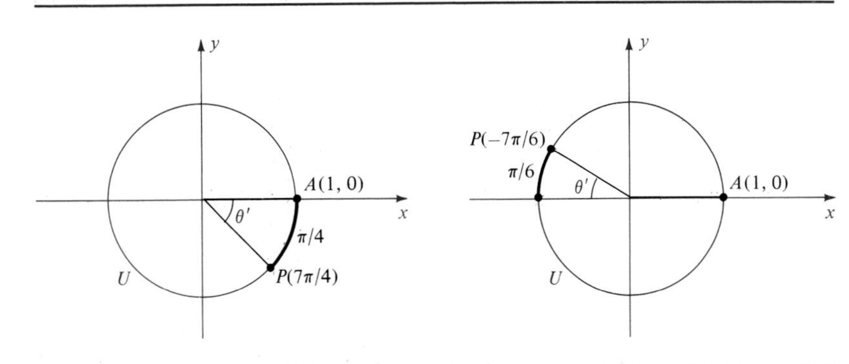

Using the rules preceding this example,

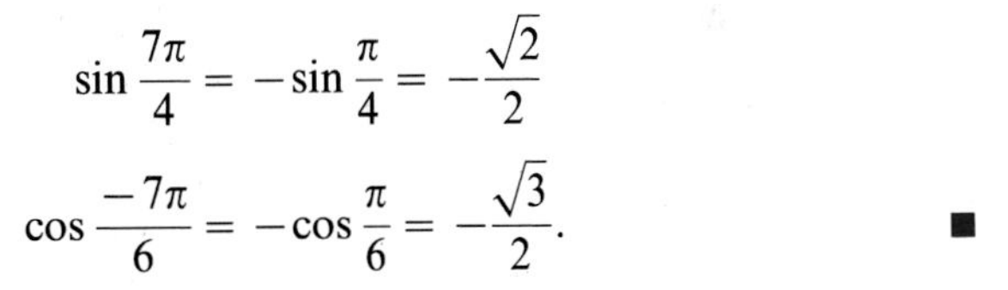

$$\sin \frac{7\pi}{4} = -\sin \frac{\pi}{4} = -\frac{\sqrt{2}}{2}$$

$$\cos \frac{-7\pi}{6} = -\cos \frac{\pi}{6} = -\frac{\sqrt{3}}{2}. \quad \blacksquare$$

EXAMPLE 4 Find the following.

(a) $\sin 150^\circ$ (b) $\tan 315^\circ$ (c) $\sec(-240^\circ)$

SOLUTION The angles and their reference angles are shown in Figure 5.31.

FIGURE 5.31

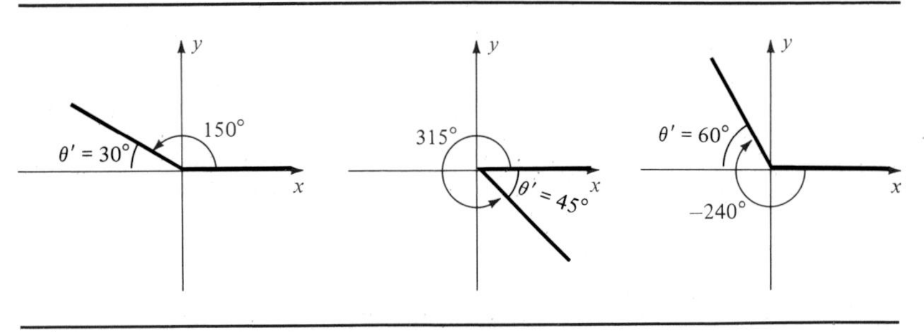

Using the rules for finding function values,

(a) $$\sin 150^\circ = \sin 30^\circ = \tfrac{1}{2}$$

(b) $$\tan 315^\circ = -\tan 45^\circ = -1$$

(c) $$\sec(-240^\circ) = -\sec 60^\circ = -2 \quad \blacksquare$$

If a calculator is used to approximate function values, it is unnecessary to use reference numbers or angles. As an illustration, to find sin 210°, place the calculator in the degree mode, enter the number 210 and press the [sin] key, obtaining $\sin 210° = -0.5$, which is the exact value. Using the same procedure for 240° we obtain, on a typical calculator,

$$\sin 240° \approx -0.8660254,$$

which is a seven-decimal-place approximation. If the *exact* value of sin 240° is desired, a calculator should not be used. In this case we find the reference angle 60° of 240° and use the rules stated on page 228 together with a knowledge of special angles to obtain

$$\sin 240° = -\sin 60° = -\frac{\sqrt{3}}{2}.$$

EXAMPLE 5 Approximate cot 837.4°.

SOLUTION Place the calculator in degree mode and use the reciprocal relationship $\cot \theta = 1/\tan \theta$ as follows:

Enter:	837.4	(degree value of θ)
Press [tan]:	−1.9291956	(value of $\tan \theta$)
Press [1/x]:	−0.5183508	(value of $1/\tan \theta$)

Thus $\cot 837.4° \approx -0.5183508$. ■

Earlier in this chapter we discussed how the [INV] key on a calculator can be used in conjunction with [sin], [cos], or [tan] to find values associated with *acute* angles (see Exercises 3 and 4 of Section 5.2). The next example illustrates how to proceed for nonacute angles.

EXAMPLE 6 If $\tan \theta = -0.4623$, and $0° \le \theta \le 360°$, find θ to the nearest minute.

SOLUTION There are two possible values of θ that satisfy the given conditions—one in quadrant II and the other in quadrant IV. If θ' is the reference angle, then

$$\tan \theta' = 0.4623.$$

We use the degree mode on a calculator and proceed as follows:

Enter: 0.4623 (the value of $\tan\theta'$)

Press [INV] [tan]: 24.811101 (a degree approximation to θ')

To change 0.811101° to minutes we write

$$(0.811101)(60') = 48.66606' \approx 49'$$

Hence $\theta' \approx 24°49'$ and

$$\theta = 180° - \theta' \approx 180° - 24°49' = 155°11'$$

or

$$\theta = 360° - \theta' \approx 360° - 24°49' = 335°11'. \blacksquare$$

Exercises 5.4

In Exercises 1–6 find the reference number t' if t has the given value.

1 (a) $3\pi/4$ (b) $4\pi/3$ (c) $-\pi/6$

2 (a) $5\pi/6$ (b) $2\pi/3$ (c) $-3\pi/4$

3 (a) $9\pi/4$ (b) $7\pi/6$ (c) $-2\pi/3$

4 (a) $8\pi/3$ (b) $7\pi/4$ (c) $-7\pi/6$

5 (a) 1.5 (b) 5

6 (a) 3.5 (b) -4

In Exercises 7–10 find the reference angle θ' if θ has the given measure.

7 (a) 240° (b) 340° (c) $-110°$

8 (a) 165° (b) 275° (c) $-202°$

9 (a) 130°40′ (b) $-405°$ (c) $-260°35'$

10 (a) 335°20′ (b) $-620°$ (c) $-185°40'$

In Exercises 11–16 find the exact values without the use of tables or a calculator.

11 (a) $\sin(2\pi/3)$ (b) $\sin(4\pi/3)$

12 (a) $\cos(5\pi/6)$ (b) $\cos(7\pi/6)$

13 (a) $\tan(-5\pi/4)$ (b) $\cot 315°$

14 (a) $\sin 210°$ (b) $\csc(-150°)$

15 (a) $\csc 300°$ (b) $\sec(-120°)$

16 (a) $\tan(-135°)$ (b) $\sec 225°$

In Exercises 17–24 use Table 4 to approximate the numbers to four significant figures and compare your answers with those found by means of a calculator.

17 $\cos 141°30'$

18 $\sin 286°40'$

19 $\cot 189°10'$

20 $\csc 136°20'$

21 $\sec 168°50'$

22 $\tan 207°10'$

23 $\sin 342°20'$

24 $\cos 432°40'$

Calculator Exercises 5.4

In Exercises 1–8, if $0^\circ \leq \theta \leq 360^\circ$, find θ to the nearest minute.

1 $\cos\theta = 0.8620$ **2** $\sin\theta = 0.6612$

3 $\tan\theta = 3.7$ **4** $\cos\theta = 0.8$

5 $\sin\theta = -0.4213$ **6** $\tan\theta = -2.164$

7 $\tan\theta = -1.456$ **8** $\cos\theta = -0.7869$

9 The fifth-degree polynomial $P(x) = x - \frac{1}{6}x^3 + \frac{1}{120}x^5$ is an excellent approximation to $\sin x$ for $0 \leq x \leq \pi/4$. In fact, it gives four-decimal-place accuracy. Use $P(x)$ to estimate $\sin \frac{1}{2}$ and compare the result to that obtained using the [sin] key on a calculator.

10 Shown in the figure is a circular race track of diameter 2 km. All races begin at S and proceed in the counterclockwise direction. Approximate the coordinates of the point at which the following races end relative to a rectangular coordinate system with origin at the center of the track:

(a) An endurance race of length 500 km;

(b) A drag race of length 2 km.

FIGURE FOR EXERCISE 10

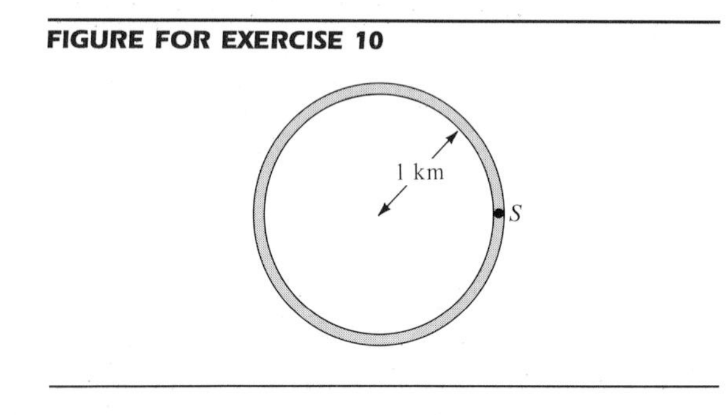

11 Answer (a) and (b) of Exercise 10 for the race track shown in the following figure.

FIGURE FOR EXERCISE 11

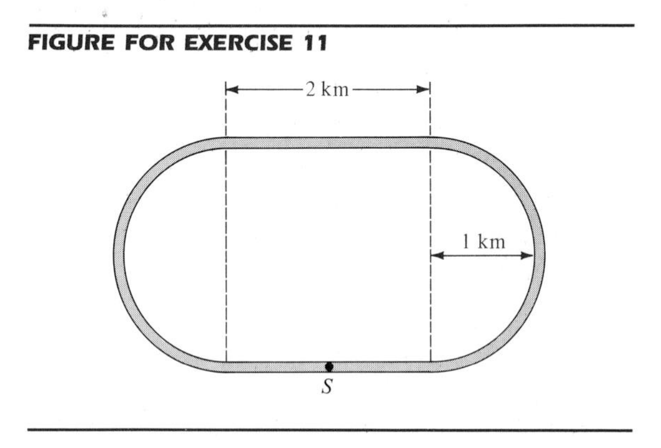

12 Scientific calculators have sin, cos, and tan keys in addition to $1/x$ and x^2 keys. After entering x, how can you calculate (a) $\cos^2 x$; (b) $\csc x$; (c) $\cot^4 x$?

5.5 Graphs of the Trigonometric Functions

In preceding sections we used $\sin t$ or $\sin\theta$ to denote values of the sine function at t or θ, respectively. Since we now wish to sketch the graph of the sine function on an xy-coordinate system, we shall consider equations of the form $y = \sin x$, where x denotes a real number (or the radian measure of an angle). *The variable x used here should not be confused with that employed in Section 5.3*, where x denoted the x-coordinate of a point P on the unit circle U. Equivalent remarks hold for other graphs involving the trigonometric functions.

In Section 5.3 we obtained rough sketches of the graphs of the sine and cosine functions by observing the coordinates of a point P as P traveled around the unit circle U. If we use the methods discussed in the preceding section, we can find coordinates of many additional points on the graphs to any degree of accuracy. A table or calculator could also be used to locate points. Portions of the graphs of $y = \sin x$ and $y = \cos x$ are sketched in Figure 5.32. Note that the graph of $y = \cos x$ can be obtained by shifting the graph of $y = \sin x$ to the left a distance $\pi/2$. The part of the graphs corresponding to the interval $[0, 2\pi]$ is sometimes referred to as one **cycle.** For the sine function we also refer to a cycle as a **sine wave.**

FIGURE 5.32

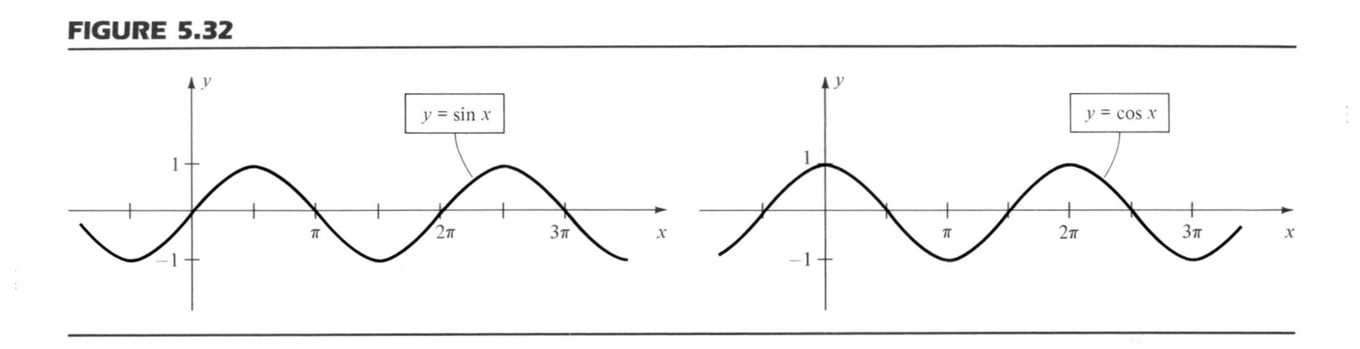

Several values of the tangent function are listed in the following table.

x	$-\frac{\pi}{3}$	$-\frac{\pi}{4}$	$-\frac{\pi}{6}$	0	$\frac{\pi}{6}$	$\frac{\pi}{4}$	$\frac{\pi}{3}$
$\tan x$	$-\sqrt{3} \approx -1.7$	-1	$\frac{-\sqrt{3}}{3} \approx -0.6$	0	$\frac{\sqrt{3}}{3} \approx 0.6$	1	$\sqrt{3} \approx 1.7$

FIGURE 5.33

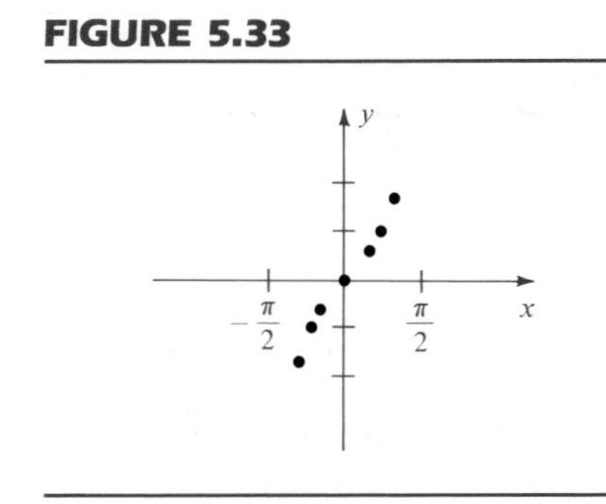

The corresponding points are plotted in Figure 5.33. The values of $\tan x$ near $x = \pi/2$ require special consideration. Let us begin by noting that $\tan x = \sin x/\cos x$. As x increases toward $\pi/2$, the numerator $\sin x$ approaches 1, whereas the denominator $\cos x$ approaches 0, and consequently $\tan x$ takes on large positive values. Recall that $\pi/2 \approx \frac{1}{2}(3.1416) = 1.5708$. The following values of $\tan x$ for x close to $\pi/2$ were obtained using a calculator:

$$\tan(1.5700) \approx 1{,}255.8$$
$$\tan(1.5703) \approx 2{,}014.8$$
$$\tan(1.5706) \approx 5{,}093.5$$
$$\tan(1.5707) \approx 10{,}381.3$$
$$\tan(1.57079) \approx 158{,}057.9$$

Notice how rapidly $\tan x$ increases as x approaches $\pi/2$. We say that $\tan x$ *increases without bound as x approaches $\pi/2$ through values less than $\pi/2$.* Similarly, if x approaches $-\pi/2$ through values greater than $-\pi/2$, then $\tan x$ *decreases without bound.* This variation of $\tan x$ in the open interval $(-\pi/2, \pi/2)$ is illustrated in Figure 5.34. The lines $x = \pi/2$ and $x = -\pi/2$ are vertical asymptotes for the graph. It is not difficult to show that the same pattern is repeated in the open intervals $(\pi/2, 3\pi/2)$, $(3\pi/2, 5\pi/2)$, and in similar intervals of length π, as shown in the figure. Thus, *the tangent function has period π.*

FIGURE 5.34

$y = \tan x$

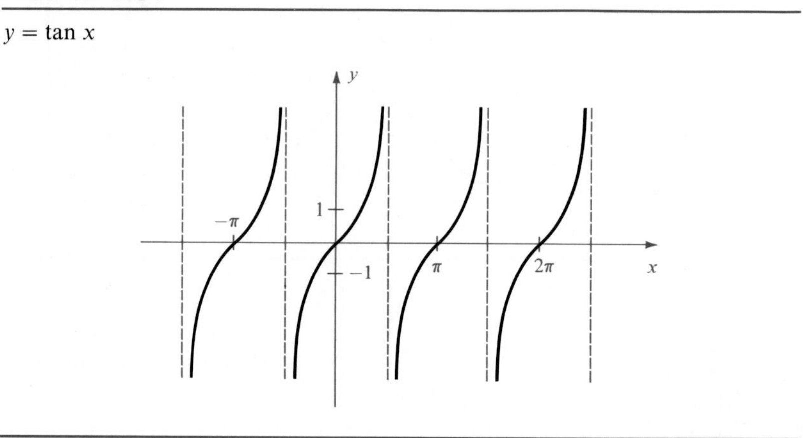

From the Formulas for Negatives, $\tan(-x) = -\tan x$. This implies that the tangent function is odd. Hence the graph is symmetric with respect to the origin.

The graphs of the remaining three trigonometric functions can now be easily obtained. For example, since $\csc x = 1/\sin x$, we may find the y-coordinate of a point on the graph of the cosecant function by taking the reciprocal of the corresponding y-coordinate on the sine graph. This is possible except for $x = n\pi$, where n is an integer, for in this case $\sin x = 0$ and hence $1/\sin x$ is undefined. As an aid to sketching the graph of the cosecant function it is convenient to sketch the graph of the sine function (shown in color in Figure 5.35) and then take reciprocals of y-coordinates

FIGURE 5.35

$y = \csc x$

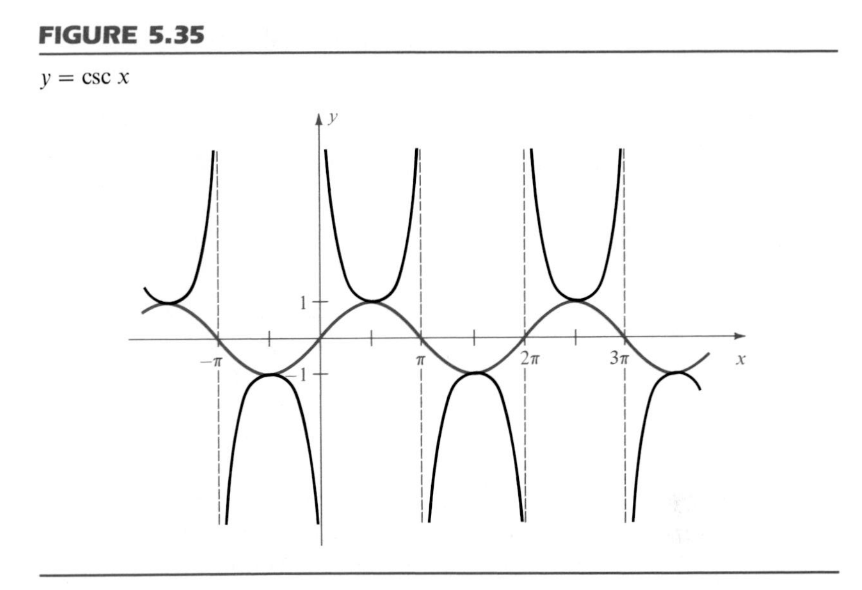

to obtain points on the cosecant graph. Notice the manner in which the cosecant function increases or decreases without bound as x approaches $n\pi$, where n is an integer. The graph has vertical asymptotes as indicated in the figure.

Since $\sec x = 1/\cos x$ and $\cot x = 1/\tan x$, the graphs of the secant and cotangent functions may be obtained in similar fashion (see Figures 5.36 and 5.37). Their verifications are left as exercises.

FIGURE 5.36

$y = \sec x$

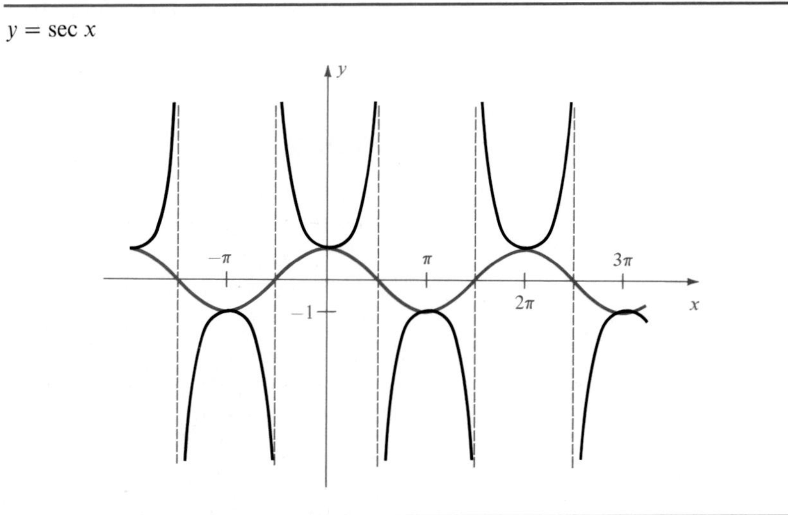

FIGURE 5.37

$y = \cot x$

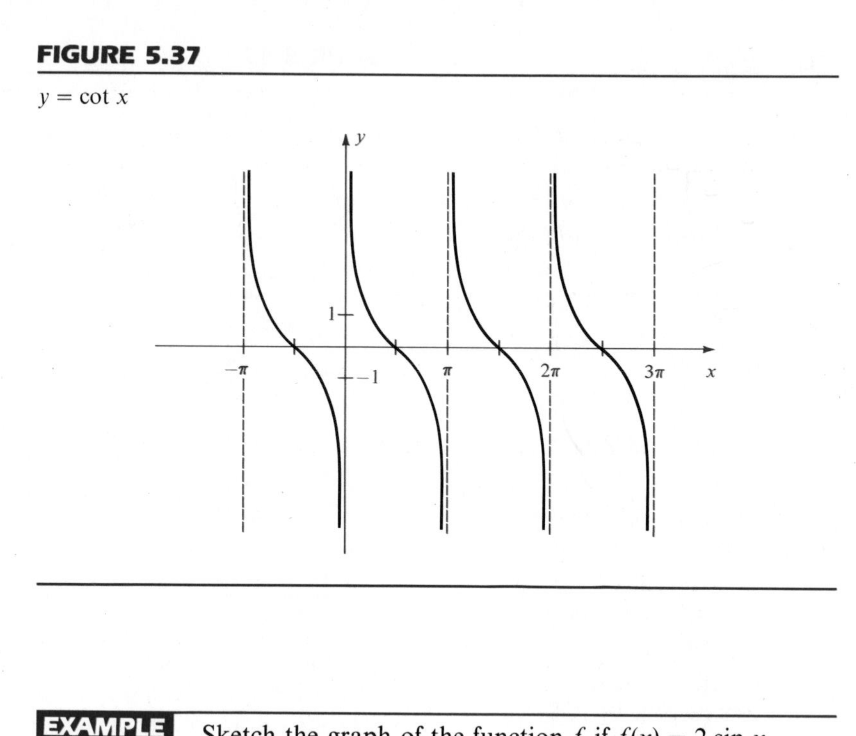

EXAMPLE Sketch the graph of the function f if $f(x) = 2 \sin x$.

SOLUTION Although the graph could be sketched by plotting points, note that for each x-coordinate x_1, the y-coordinate $f(x_1)$ is always twice that of the corresponding y-coordinate on the sine graph. A simple graphical technique is to sketch the graph of $y = \sin x$ (shown in color) and then double each y-coordinate to find points on the graph of $y = 2 \sin x$ (see Figure 5.38). This amounts to stretching the graph of a function, as discussed in Section 2.2.

FIGURE 5.38

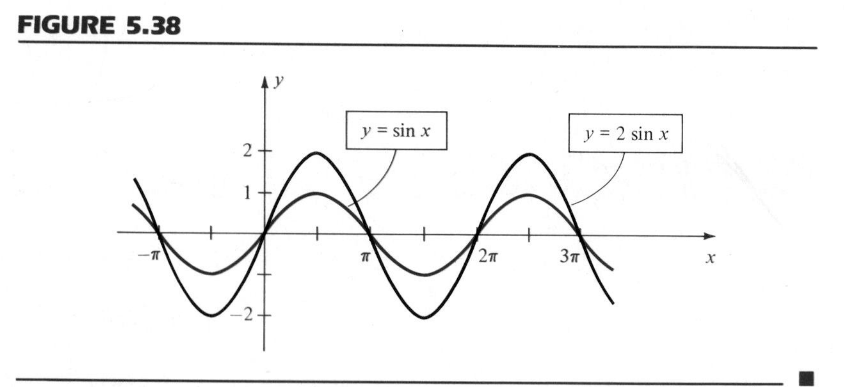

■

Exercises 5.5

1 (a) Describe the intervals between -2π and 2π in which the secant function is increasing.

(b) Describe the intervals between -2π and 2π in which the secant function is decreasing.

(c) With the aid of Table 4 or a calculator, discuss the variation of $\sec x$ as x approaches $\pi/2$ through values less than $\pi/2$ and through values greater than $\pi/2$.

(d) Discuss the symmetry of the graph of the secant function.

2 (a) In what intervals is the cotangent function increasing?

(b) In what intervals is the cotangent function decreasing?

(c) With the aid of Table 4 or a calculator, discuss the variation of $\cot x$ as x approaches 0 through values greater than 0 and through values less than 0.

(d) Discuss the symmetry of the graph of the cotangent function.

3 Practice sketching graphs of the sine, cosine, and tangent functions, taking different units of length on the horizontal and vertical axes. Continue this practice until you reach the stage at which, if you were awakened from a sound sleep in the middle of the night and asked to sketch one of these graphs, you could do so in less than thirty seconds.

4 Repeat Exercise 3 for the cosecant, secant, and cotangent functions.

In Exercises 5–14 use the method illustrated in Example 1 to sketch the graph of f.

5 $f(x) = 4 \sin x$ **6** $f(x) = 3 \sin x$

7 $f(x) = \frac{1}{2} \sin x$ **8** $f(x) = \frac{1}{4} \sin x$

9 $f(x) = 2 \cos x$ **10** $f(x) = 4 \cos x$

11 $f(x) = \frac{1}{3} \cos x$ **12** $f(x) = \frac{1}{2} \cos x$

13 $f(x) = -\sin x$ **14** $f(x) = -\cos x$

Calculator Exercises 5.5

If f is a function, the notation

$$f(x) \to L \quad \text{as} \quad x \to a^+$$

is sometimes used to express the fact that $f(x)$ gets very close to L as x gets close to a (through values greater than a). Use a calculator to support Exercises 1–6 by substituting the following values for x: $x = 0.1$, $x = 0.01$, $x = 0.001$, $x = 0.0001$.

1 $\dfrac{\sin x}{x} \to 1 \quad \text{as} \quad x \to 0^+$ **2** $\dfrac{\tan x}{x} \to 1 \quad \text{as} \quad x \to 0^+$

3 $\dfrac{1 - \cos x}{x} \to 0 \quad \text{as} \quad x \to 0^+$

4 $\dfrac{\sin x}{1 + \cos x} \to 0 \quad \text{as} \quad x \to 0^+$

5 $x \cot x \to 1 \quad \text{as} \quad x \to 0^+$

6 $\dfrac{x + \tan x}{\sin x} \to 2 \quad \text{as} \quad x \to 0^+$

5.6 Trigonometric Graphs

In this section we will consider graphs of the equations

$$y = a \sin (bx + c) \quad \text{and} \quad y = a \cos (bx + c)$$

for real numbers a, b, and c. Our goal is to sketch a graph without plotting many points. To do so we shall use facts about the graphs of $y = \sin x$ and $y = \cos x$, which we discussed in previous sections.

Let us begin by considering the special case $c = 0$ and $b = 1$; that is,

$$y = a \sin x \quad \text{and} \quad y = a \cos x.$$

We can find y-coordinates of points on the graphs by multiplying y-coordinates of points on the graphs of $y = \sin x$ and $y = \cos x$ by a. Thus, if $y = 2 \sin x$, we multiply by 2; if $y = \frac{1}{2} \sin x$, we multiply by $\frac{1}{2}$, and so on. The graph of $y = 2 \sin x$ is sketched in Figure 5.38. The graph of $y = \frac{1}{2} \sin x$ is sketched in Figure 5.39, where for comparison we have also shown the graph of $y = \sin x$ in color.

FIGURE 5.39

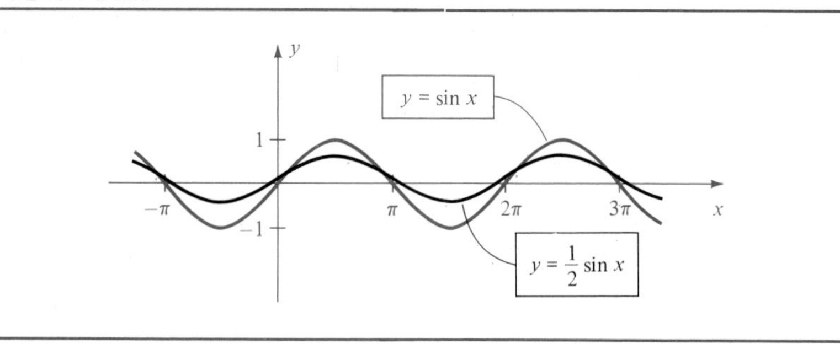

The graph of $y = a \sin x$ always has the general appearance of one of the graphs illustrated in Figure 5.40. The largest value $|a|$ of y is called the **amplitude** of the graph, or the amplitude of the function given by

FIGURE 5.40

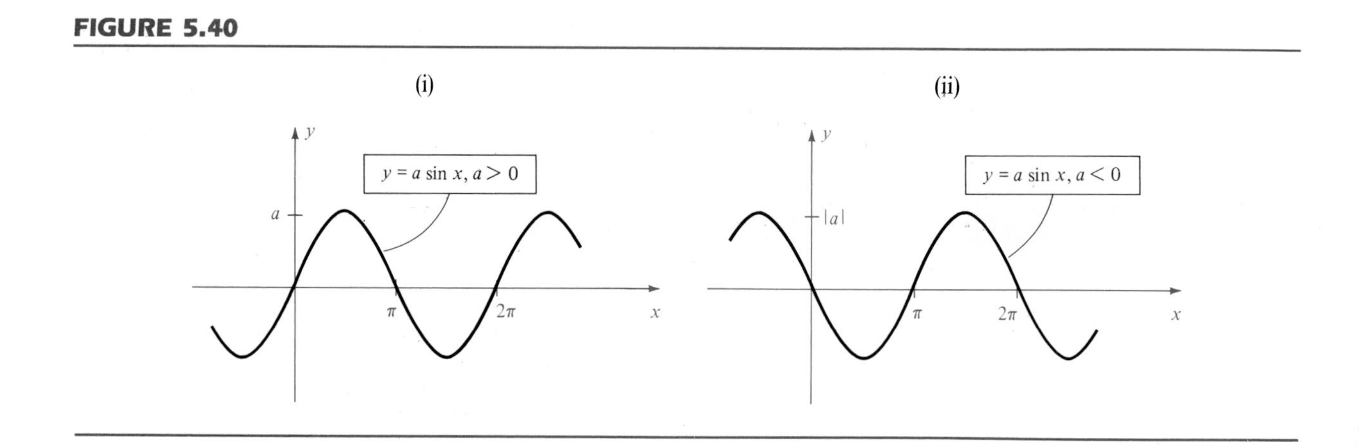

$f(x) = a \sin x$. For example, the amplitude in Figure 5.39 is $\frac{1}{2}$, whereas that in Figure 5.38 is 2.

If $a < 0$, then the y-coordinates of points on the graph of $y = a \sin x$ are the negatives of the corresponding y-coordinates on the graph of $y = |a| \sin x$. Thus the graph in Figure 5.40(ii) is a reflection, through the x-axis, of the graph in (i) of the figure.

FIGURE 5.41

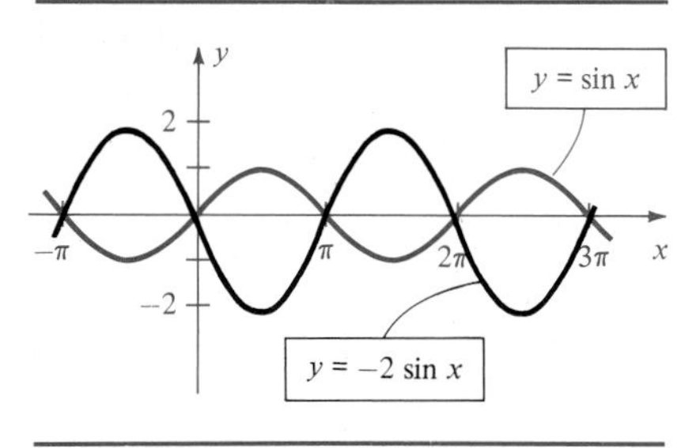

EXAMPLE 1 Find the amplitude and sketch the graph of the equation $y = -2 \sin x$.

SOLUTION Since $a = -2 < 0$, the graph has the general shape shown in Figure 5.40(ii). The amplitude is $|-2| = 2$.

We first sketch the graph of $y = \sin x$ in color and then we multiply each y-coordinate by -2. This gives us the sketch in Figure 5.41. An alternative method is to reflect the graph of $y = 2 \sin x$ (see Figure 5.38) through the x-axis. ■

Similar techniques may be used for $y = a \cos x$.

FIGURE 5.42

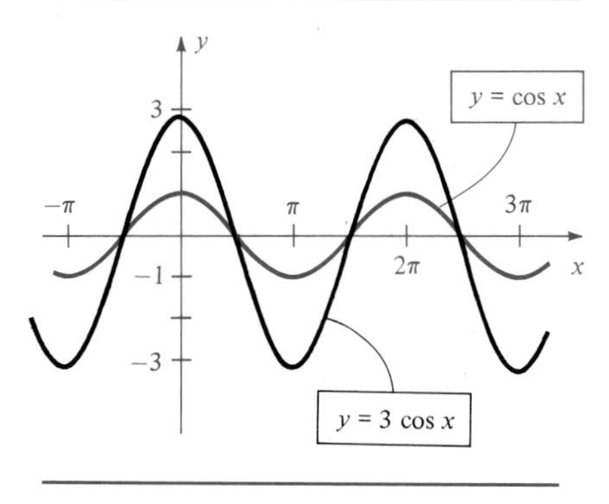

EXAMPLE 2 Sketch the graph of $y = 3 \cos x$.

SOLUTION As indicated in Figure 5.42, we first sketch the graph of $y = \cos x$ in color and then multiply y-coordinates by 3. The amplitude in this case is 3. ■

Let us next consider $y = a \sin bx$ and $y = a \cos bx$ for any $b \neq 0$. As before, the amplitude is $|a|$. The following theorem describes the periods for these graphs.

THEOREM

> If $f(x) = a \sin bx$ or $f(x) = a \cos bx$, where $a \neq 0$ and $b \neq 0$, then the period of f is $2\pi/|b|$.

PROOF If we let bx increase from 0 to 2π we obtain exactly one cycle for f. For $b > 0$, we note that

$$0 \leq bx \leq 2\pi \quad \text{if and only if} \quad 0 \leq x \leq 2\pi/b.$$

For $b < 0$, we have

$$0 \le bx \le 2\pi \quad \text{if and only if} \quad 2\pi/b \le x \le 0.$$

In either case, exactly one cycle occurs in an interval of length $2\pi/|b|$. □

FIGURE 5.43

FIGURE 5.44

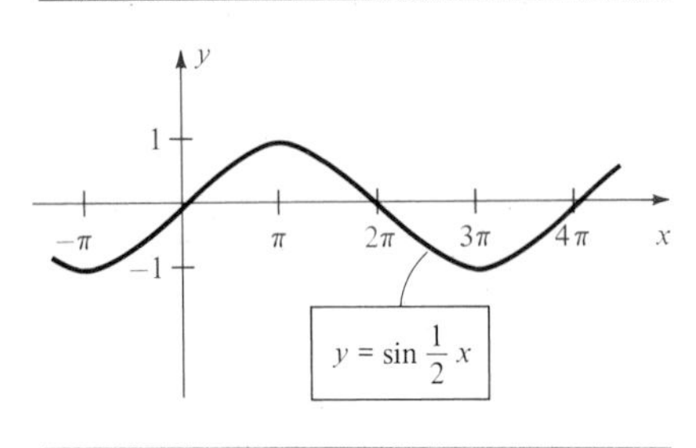

EXAMPLE 3 Find the amplitude, the period, and sketch the graph of f if

(a) $f(x) = \sin 2x$ (b) $f(x) = \sin \frac{1}{2}x$.

SOLUTION Since each expression has the form $f(x) = a \sin bx$ with $a = 1$, the amplitude is 1 for each graph.

In (a) we have $b = 2$, and hence the period $2\pi/|b|$ is $2\pi/2 = \pi$. This means that there is exactly one sine wave (of amplitude 1) on the interval $[0, \pi]$. The graph is sketched in Figure 5.43, where we have used different scales on the x- and y-axes.

In (b), we see that $b = \frac{1}{2}$, and hence the period $2\pi/|b|$ is $2\pi/(\frac{1}{2}) = 4\pi$. Thus there is one sine wave (of amplitude 1) on the interval $[0, 4\pi]$ as shown in Figure 5.44. ■

If $f(x) = a \sin bx$ and if b is a large positive number, then the period $2\pi/b$ is small and the sine waves are close together. As a matter of fact, there are b sine waves on the interval $[0, 2\pi]$. For example, in Figure 5.43 we have $b = 2$ and there are two sine waves on $[0, 2\pi]$. If b is small, then the period $2\pi/b$ is large and the waves are far apart. To illustrate, if $y = \sin \frac{1}{10}x$, then $\frac{1}{10}$ of a sine wave occurs on $[0, 2\pi]$, and an interval 20π units long is required for one complete cycle. (See also Figure 5.44, where $y = \sin \frac{1}{2}x$ and one-half of a sine wave occurs in $[0, 2\pi]$.)

If $b < 0$ we can use the fact that $\sin(-x) = -\sin x$ to obtain the graph of $y = a \sin bx$. To illustrate, the graph of $y = \sin(-2x)$ is the same as the graph of $y = -\sin 2x$.

FIGURE 5.45

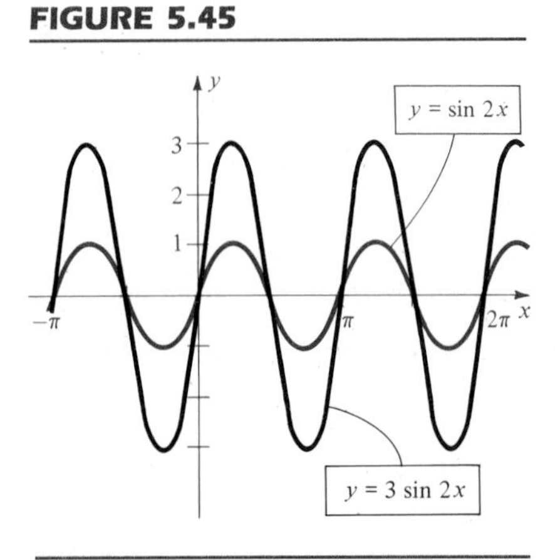

EXAMPLE 4 Find the amplitude, the period, and sketch the graph of f if $f(x) = 3 \sin 2x$.

SOLUTION The amplitude of f is 3 and the period is $2\pi/2 = \pi$. The graph is readily obtained by first sketching the graph of $y = \sin 2x$ in color, and then multiplying the y-coordinate of each point by 3. This gives us the sketch in Figure 5.45. ■

FIGURE 5.46

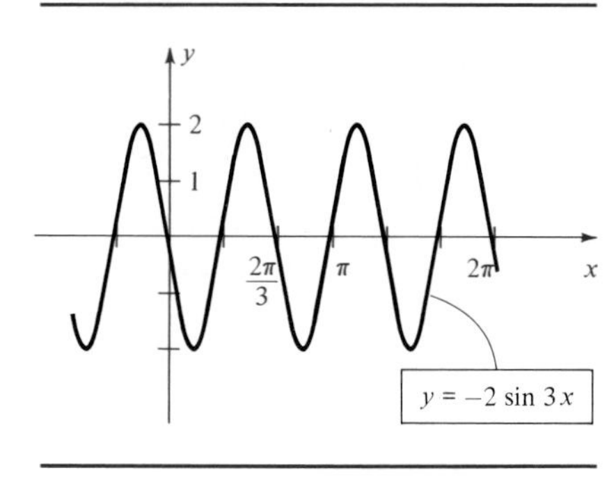

EXAMPLE 5 Find the amplitude, the period, and sketch the graph of f if $f(x) = 2\sin(-3x)$.

SOLUTION Since $f(x) = -2\sin 3x$, we see that the amplitude is $|-2| = 2$ and the period is $2\pi/3$. Thus, there is one cycle on an interval of length $2\pi/3$. The minus sign indicates a reflection through the x-axis. If we consider the interval $[0, 2\pi/3]$ and sketch a sine wave of amplitude 2 (reflected through the x-axis), the shape of the graph is apparent. The part of the graph in the interval $[0, 2\pi/3]$ is repeated periodically, as illustrated in Figure 5.46. ■

Finally, let us consider the graph of

$$f(x) = a\sin(bx + c).$$

As before, the amplitude is $|a|$. One sine wave is obtained if

$$0 \le bx + c \le 2\pi$$

or

$$-c \le bx \le 2\pi - c.$$

Thus if $b > 0$, then one sine wave occurs when x varies through the interval

$$\left[-\frac{c}{b}, \frac{2\pi - c}{b}\right] \quad \text{or} \quad \left[-\frac{c}{b}, -\frac{c}{b} + \frac{2\pi}{b}\right].$$

If $-c/b < 0$, this amounts to shifting the graph of $y = a\sin bx$ to the left a distance $|-c/b|$. If $-c/b > 0$, the shift is to the right. Thus the period of f is $2\pi/b$. The number $-c/b$ is sometimes called the **phase shift** associated with the function. A similar discussion can be given if $b < 0$ or if $f(x) = a\cos(bx + c)$. It is unnecessary to memorize a formula for finding the phase shift. In any specific problem the interval that contains exactly one cycle can be found by solving the two equations

$$bx + c = 0 \quad \text{and} \quad bx + c = 2\pi$$

for x, as illustrated in the next two examples.

EXAMPLE 6 Sketch the graph of f if

$$f(x) = 3\sin\left(2x + \frac{\pi}{2}\right)$$

and find the phase shift.

SOLUTION The equation is of the form discussed in this section, with $a = 3$, $b = 2$, and $c = \pi/2$. Thus the graph has the sine wave pattern with amplitude 3 and period $2\pi/2 = \pi$. To obtain an interval containing exactly one sine wave, we let $2x + (\pi/2)$ vary from 0 to 2π. The endpoints of this interval can be found by solving the two equations

$$2x + \frac{\pi}{2} = 0 \quad \text{and} \quad 2x + \frac{\pi}{2} = 2\pi.$$

This gives us

$$x = -\frac{\pi}{4} \quad \text{and} \quad x = \frac{3\pi}{4}.$$

and hence one sine wave of amplitude 3 occurs on the interval $[-\pi/4, 3\pi/4]$. Sketching that wave and then repeating it to the right and left gives us the graph in Figure 5.47. We see from the graph that the phase shift is $-\pi/4$. ■

FIGURE 5.47

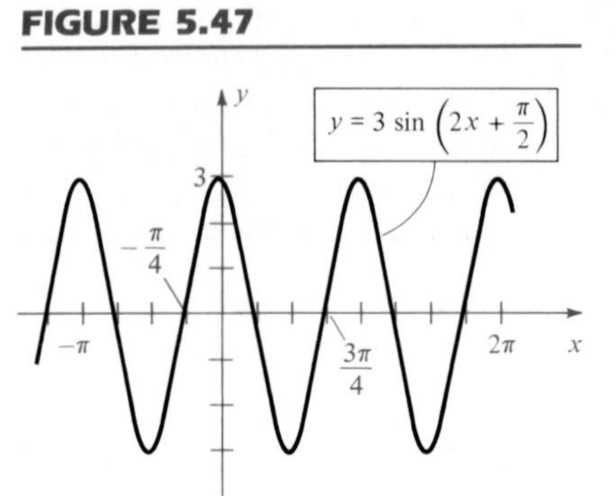

EXAMPLE 7 Sketch the graph of f if $f(x) = 2\cos(3x - \pi)$. Find the phase shift.

SOLUTION The function is given by $f(x) = a\cos(bx + c)$ with $a = 2$, $b = 3$, and $c = -\pi$. Thus the amplitude is 2 and the period is $2\pi/3$. To obtain an interval containing one (cosine) cycle we let $3x - \pi$ vary from 0 to 2π. The endpoints of this interval are the solutions of the equations

$$3x - \pi = 0 \quad \text{and} \quad 3x - \pi = 2\pi.$$

This gives us

$$x = \frac{\pi}{3} \quad \text{and} \quad x = \frac{3\pi}{3} = \pi.$$

Hence one cosine-type curve of amplitude 2 occurs on the interval $[\pi/3, \pi]$. Sketching that part of the graph and then repeating it to the right and left gives us the sketch in Figure 5.48. We see from the graph that the phase shift is $\pi/3$. ■

FIGURE 5.48

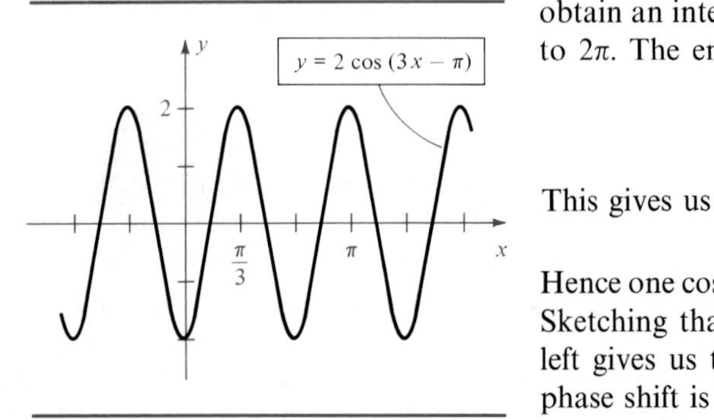

If $f(x) = a\tan(bx + c)$, the period is $\pi/|b|$, since the tangent function has period π. There is no largest y-coordinate for points on the graph of the tangent function, and hence we do not refer to its amplitude; however, we may still use the process of multiplying y-coordinates by a to obtain points on the graph of $y = a\tan(bx + c)$. We shall not refer to phase shifts for the tangent function. Similar remarks may be made for the cotangent, secant, and cosecant functions.

Many phenomena that occur in nature vary in a cyclic or rhythmic manner. It is sometimes possible to represent such behavior by means of trigonometric functions, as illustrated in the next example.

EXAMPLE 8 The rhythmic process of breathing consists of alternating periods of inhaling and exhaling. One complete cycle normally takes place every 5 seconds. If F denotes the air flow rate at time t (in liters/second) and if the maximum flow rate is 0.6 liters/second, find a formula $F = a \sin bt$ that fits this information.

SOLUTION The air flow rate F is a function of t, that is, $F = f(t)$ for some f. If $f(t) = a \sin bt$ for some $b > 0$, then the period of f is $2\pi/b$. In this application the period is 5 seconds, and hence

$$\frac{2\pi}{b} = 5, \quad \text{or} \quad b = \frac{2\pi}{5}.$$

Since the maximum flow rate corresponds to the amplitude a of f, we let $a = 0.6$. This gives us the formula

$$F = 0.6 \sin\left(\tfrac{2}{5}\pi t\right). \qquad \blacksquare$$

Exercises 5.6

1 Sketch the graph and determine the amplitude and period of each function f.

(a) $f(x) = 4 \sin x$ (b) $f(x) = \sin 4x$
(c) $f(x) = \frac{1}{4} \sin x$ (d) $f(x) = \sin \frac{1}{4}x$
(e) $f(x) = 2 \sin \frac{1}{4}x$ (f) $f(x) = \frac{1}{2} \sin 4x$
(g) $f(x) = -4 \sin x$ (h) $f(x) = \sin(-4x)$

2 Sketch the graphs of the functions that involve the cosine and are analogous to those in Exercise 1.

3 Sketch the graph of f and determine the amplitude and period in each case:

(a) $f(x) = 3 \cos x$ (b) $f(x) = \cos 3x$
(c) $f(x) = \frac{1}{3} \cos x$ (d) $f(x) = \cos \frac{1}{3}x$
(e) $f(x) = 2 \cos \frac{1}{3}x$ (f) $f(x) = \frac{1}{3} \cos 2x$
(g) $f(x) = -3 \cos x$ (h) $f(x) = \cos(-3x)$

4 Sketch the graphs of the functions that involve the sine and are analogous to those in Exercise 3.

Sketch the graphs of the equations in Exercises 5–44.

5 $y = \sin\left(x - \dfrac{\pi}{2}\right)$ **6** $y = \cos\left(x + \dfrac{\pi}{4}\right)$

7 $y = 3 \sin\left(x - \dfrac{\pi}{2}\right)$ **8** $y = 4 \cos\left(x + \dfrac{\pi}{4}\right)$

9 $y = \cos\left(x + \dfrac{\pi}{3}\right)$ **10** $y = \sin\left(x - \dfrac{\pi}{3}\right)$

11 $y = 4 \cos\left(x + \dfrac{\pi}{3}\right)$ **12** $y = 5 \sin\left(x - \dfrac{\pi}{3}\right)$

13 $y = \sin(3x + \pi)$ **14** $y = \cos(2x + \pi)$

15 $y = -2 \sin(3x + \pi)$ **16** $y = 3 \cos(3x - \pi)$

17 $y = 5 \sin\left(3x - \dfrac{\pi}{2}\right)$ **18** $y = -4 \cos\left(2x + \dfrac{\pi}{3}\right)$

19 $y = 6 \sin \pi x$ **20** $y = 3 \cos \frac{1}{2}\pi x$

21 $y = 2 \cos \dfrac{\pi}{2}x$ **22** $y = 4 \sin 3\pi x$

23 $y = \frac{1}{2} \sin 2\pi x$ **24** $y = \frac{1}{2} \cos \dfrac{\pi}{2}x$

25 $y = \frac{1}{2} \sec x$ **26** $y = \frac{1}{4} \csc x$

27 $y = -2 \csc x$ **28** $y = -3 \sec x$

29 $y = \sec 2x$ **30** $y = \csc 3x$

31 $y = \csc\left(x - \dfrac{\pi}{4}\right)$

32 $y = \sec\left(x + \dfrac{\pi}{3}\right)$

33 $y = \tan \frac{1}{2}x$

34 $y = \cot 2x$

35 $y = \tan\left(x + \dfrac{\pi}{2}\right)$

36 $y = \cot\left(x - \dfrac{\pi}{4}\right)$

37 $y = \frac{1}{2} \cot x$

38 $y = -2 \tan x$

39 $y = \tan(-x)$

40 $y = -\cot x$

41 $y = 2 \sin\left(x - \dfrac{\pi}{2}\right)$

42 $y = \cos\left(x - \dfrac{\pi}{2}\right)$

43 $y = \sin\left(x + \dfrac{\pi}{2}\right)$

44 $y = \cos\left(x + \dfrac{\pi}{2}\right)$

For the trigonometric functions whose computer-generated graphs are shown in Exercises 45–48 find (a) the amplitude and period; (b) an equation of the graph in the form $y = a \sin(bx + c)$; and (c) the phase shift.

45

46

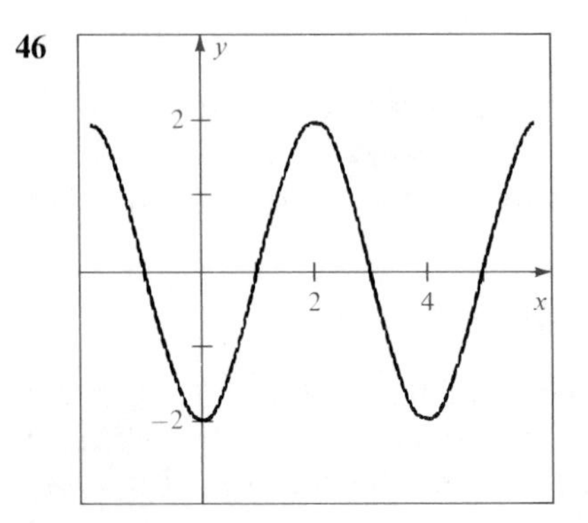

47

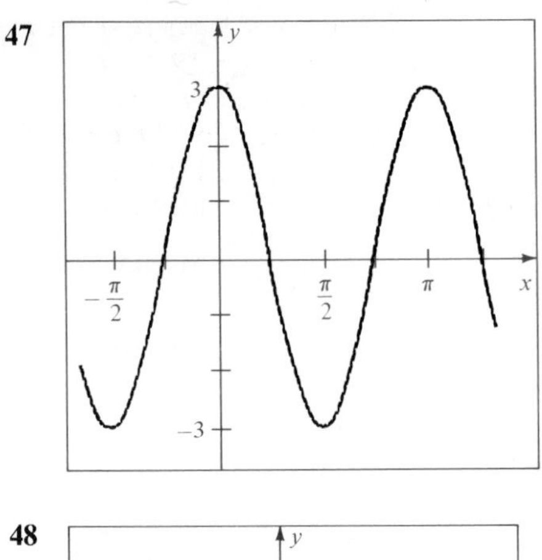

48

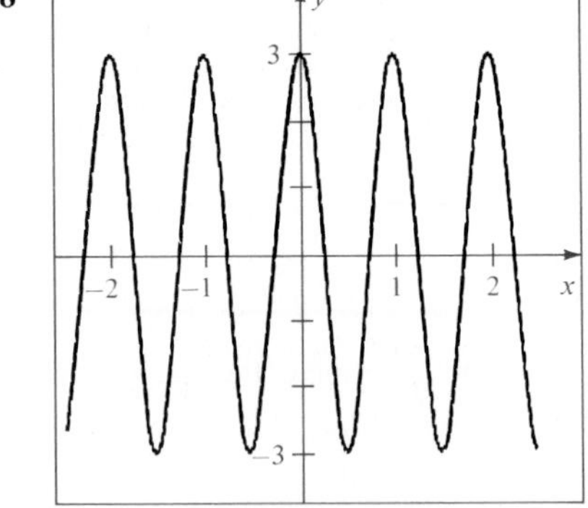

49 Shown in the figure is an electroencephalogram of human brain waves during deep sleep. If we use $W = a \sin(bt + c)$ to represent these waves, what is the value of b?

FIGURE FOR EXERCISE 49

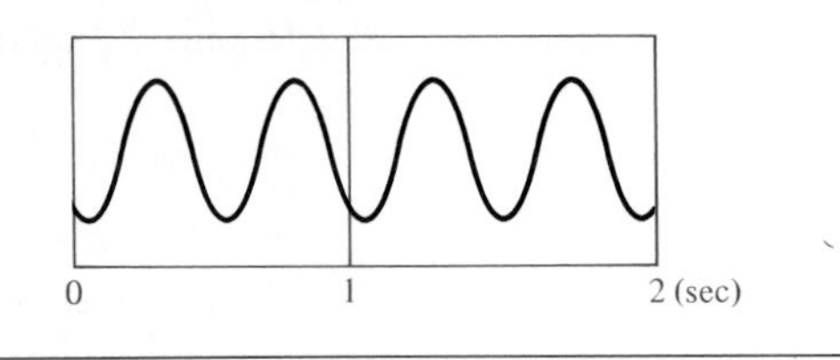

50 The light intensity I on a certain spring day takes on its largest value of 510 calories/cm^2 at midday. If $t = 0$ corresponds to sunrise and if there are 13 hours of daylight, find a formula $I = a \sin bt$ that fits this information.

51 The pumping action of the heart consists of a *systolic phase* in which blood rushes from the left ventricle into the aorta and the *diastolic phase* during which the heart muscle relaxes. The function whose graph is shown in the figure is often used to model one complete cycle of this process. For a particular individual the systolic phase lasts $\frac{1}{4}$ second and has a maximum flow rate of 8 liters/minute. Find a and b.

FIGURE FOR EXERCISE 51

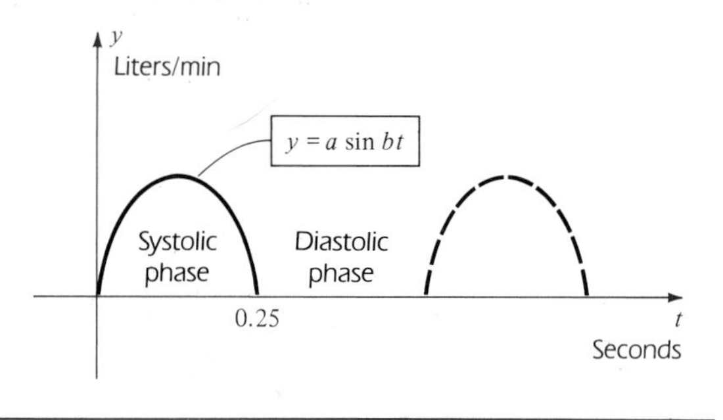

52 The popular biorhythm theory uses the graphs of three simple sine functions to make predictions about an individual's physical, emotional, and intellectual potential for a particular day. The graphs are given by $y = a \sin bt$, where $t = 0$ corresponds to birth and where $a = 1$ is used to denote 100% potential.

(a) Find the value of b for the physical cycle having a period of 23 days; for the emotional cycle (period 28 days); and for the intellectual cycle (period 33 days).

(b) What does the biorhythm chart tell a person who has just become 21 years of age and is exactly 7660 days old?

53 The height of the tide at a particular point on shore can be predicted by using seven trigonometric functions (called *tidal components*) of the form $f(t) = a \cos (bt + c)$. The *principal lunar component* may be approximated by

$$f(t) = a \cos \left(\frac{\pi}{6} t - \frac{11}{12} \pi \right)$$

where t is in hours and $t = 0$ corresponds to midnight. Sketch the graph of f if $a = 0.5$ meter.

54 Refer to Exercise 53. The *principal solar diurnal component* may be approximated by

$$f(t) = a \cos \left(\frac{\pi}{12} t - \frac{7\pi}{12} \right).$$

Sketch the graph of f if $a = 0.2$ meter.

5.7 Additional Graphical Techniques

Mathematical applications often involve functions that are defined in terms of sums and products, such as

$$f(x) = \sin 2x + \cos x \quad \text{or} \quad f(x) = 2^{-x} \sin x.$$

For sums, the graphical technique known as **addition of *y*-coordinates** is useful. The method applies not only to trigonometric expressions, but to arbitrary expressions as well. If f is a sum of two functions g and h that have the same domain D, then

$$f(x) = g(x) + h(x)$$

for every x in D. The graph of f may be obtained from the graphs of g and h. We begin by sketching the graphs of $y = g(x)$ and $y = h(x)$ on the same coordinate axes, as illustrated by the color curves in Figure 5.49.

FIGURE 5.49

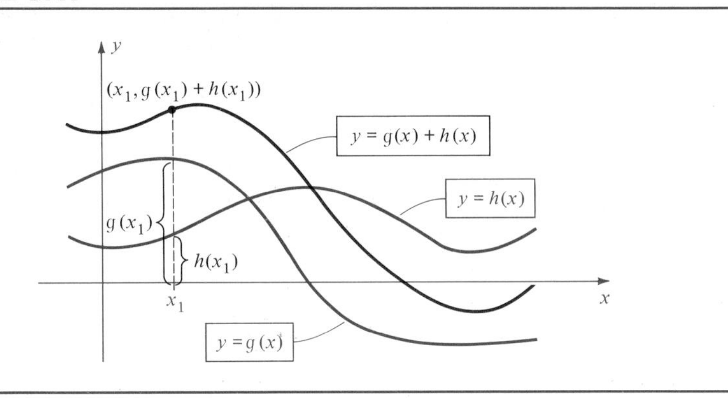

Since $f(x_1) = g(x_1) + h(x_1)$ for every x_1 in D, the y-coordinate of the point on the graph of $y = g(x) + h(x)$ with x-coordinate x_1 is the *sum* of the corresponding y-coordinates of points on the graphs of g and h. If we draw a vertical line at the point $(x_1, 0)$, then the y-coordinates $g(x_1)$ and $h(x_1)$ may be added geometrically by means of a compass or ruler, as is illustrated in Figure 5.49. If either $g(x_1)$ or $h(x_1)$ is negative, then a *subtraction* of y-coordinates may be employed. By using this technique a sufficient number of times, we can obtain a sketch of the graph of f.

FIGURE 5.50

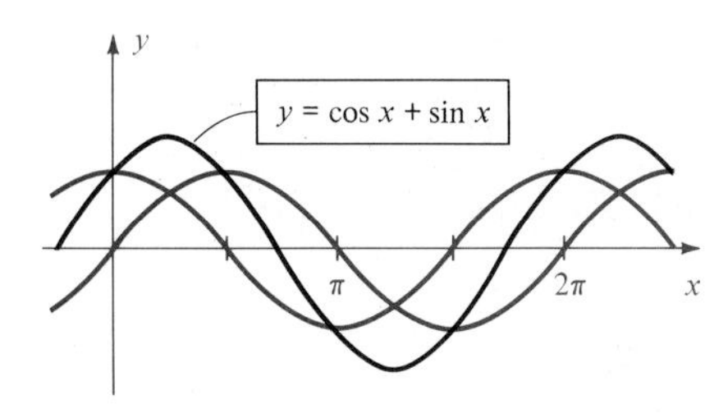

EXAMPLE 1 If $f(x) = \cos x + \sin x$, use addition of y-coordinates to sketch the graph of f.

SOLUTION We begin by sketching (in color) the graphs of the equations $y = \cos x$ and $y = \sin x$. Next, for various numbers x_1, we add y-coordinates geometrically. After a sufficient number of y-coordinates are added and a pattern emerges, we draw a curve through the points, as indicated by the sketch in Figure 5.50. As a check, it is worthwhile to plot some points on the graph by substituting numbers for x; however, we shall leave such verifications to the reader. It can be seen from the graph that the function f is periodic with period 2π. ■

FIGURE 5.51

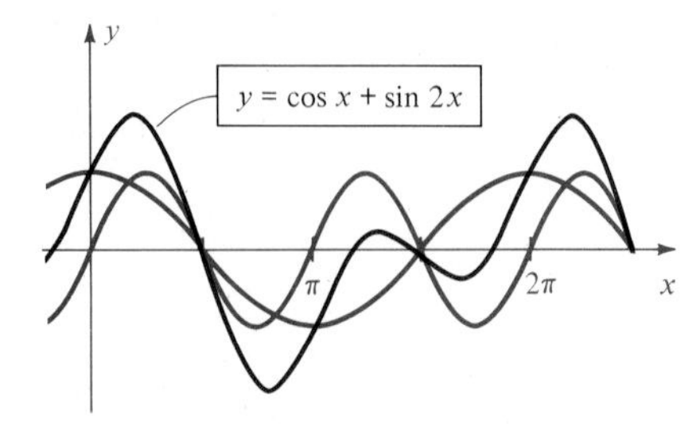

EXAMPLE 2 Sketch the graph of the equation $y = \cos x + \sin 2x$.

SOLUTION We sketch (in color) the graphs of the equations $y = \cos x$ and $y = \sin 2x$ on the same coordinate axes and use the method of addition of y-coordinates. The graph is illustrated in Figure 5.51. Evidently, f is periodic with period 2π. ■

EXAMPLE 3 The number of hours of daylight $D(t)$ at a particular time of the year can be approximated by

$$D(t) = \frac{K}{2} \sin \frac{2\pi}{365}(t - 79) + 12$$

where t is in days and $t = 0$ corresponds to January 1. The constant K determines the total variation in day length and depends on the latitude of the locale.

(a) For Boston, MA, $K \approx 6$. Sketch the graph of D for $0 \leq t \leq 365$.

(b) When is the day length the longest? the shortest?

SOLUTION (a) If $K = 6$, then $K/2 = 3$, and we may write $D(t)$ in the form

$$D(t) = f(t) + 12$$

where

$$f(t) = 3 \sin \frac{2\pi}{365}(t - 79).$$

We could regard D as the sum of f and the constant function g, where $g(t) = 12$; however, it is simpler to sketch the graph of f and then apply a vertical shift through a distance 12 (see page 76). The expression

$$f(t) = 3 \sin \left(\frac{2\pi}{365} t - \frac{158}{365} \pi \right)$$

has the form $f(t) = a \sin(bt + c)$, with $a = 3$, $b = 2\pi/365$, and $c = -158\pi/365$. From our work in the preceding section, the amplitude is 3, and the period of f is

$$\frac{2\pi}{b} = \frac{2\pi}{2\pi/365} = 365 \text{ days}.$$

Referring to the form of $f(t)$ we see that exactly one sine wave occurs if

$$0 \leq \frac{2\pi}{365}(t - 79) \leq 2\pi,$$

or

$$0 \leq t - 79 \leq 365,$$

or

$$79 \leq t \leq 444.$$

Dividing the time interval $[79, 444]$ into four equal parts we obtain the following table, which indicates the familiar sine wave pattern of amplitude 3.

t	79	170.25	261.5	352.75	444
$f(t)$	0	3	0	-3	0

To find other values of $f(t)$ we may use a calculator or Table 5. For example, if $t = 0$,

$$f(0) = 3 \sin \frac{2\pi}{365}(-79) \approx 3 \sin(-1.36) \approx -2.9.$$

Since the period of f is 365, this implies that $f(365) = -2.9$.

The graph of f for the interval [0, 444] is sketched in Figure 5.52, where we have used different scales on the axes, and where t has been rounded off to the nearest day.

FIGURE 5.52

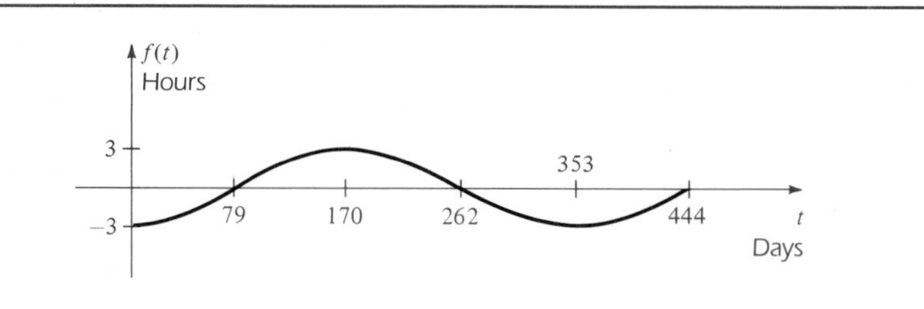

Applying a vertical shift of 12 units gives us the graph of D for $0 \leq t \leq 365$ shown in Figure 5.53.

FIGURE 5.53

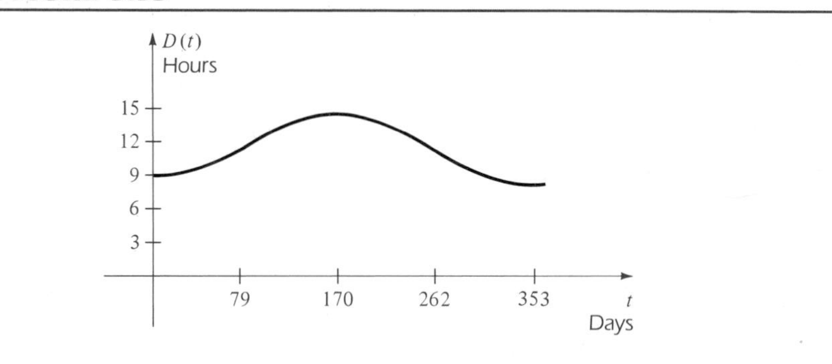

(b) The longest day—that is, the largest value of $D(t)$—occurs 170 days after January 1. Except for leap year, this corresponds to June 20. The shortest day occurs 353 days after January 1, or December 20. ■

The next example illustrates the fact that if f is the product of two functions g and h, then the graphs of g and h may be used to supply information about the graph of f.

EXAMPLE 4 Sketch the graph of f if $f(x) = 2^{-x} \sin x$.

FIGURE 5.54

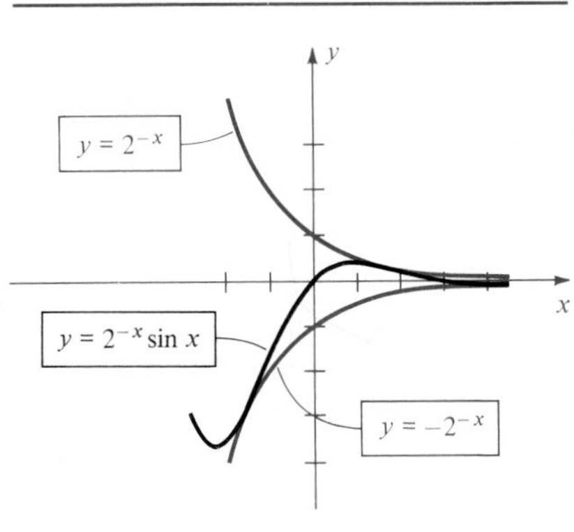

SOLUTION The function f is the product of two functions g and h, where $g(x) = 2^{-x}$ and $h(x) = \sin x$. Using properties of absolute values, we see that $|f(x)| = |2^{-x}||\sin x|$. Since $|\sin x| \leq 1$ and $2^{-x} > 0$, it follows that $|f(x)| \leq 2^{-x}$. Hence

$$-2^{-x} \leq f(x) \leq 2^{-x},$$

which implies that the graph of f lies between the graphs of the equations $y = -2^{-x}$ and $y = 2^{-x}$. The graph of f will coincide with one of these graphs if $|\sin x| = 1$, that is, if $x = (\pi/2) + n\pi$ for some integer n. Since $2^{-x} > 0$, the x-intercepts on the graph of f occur at $\sin x = 0$, that is, at $x = n\pi$. With this information we obtain the sketch shown in Figure 5.54. ■

The graph of $y = 2^{-x} \sin x$, sketched in Figure 5.54, is called a **damped sine wave,** and 2^{-x} is called the **damping factor.** By using different damping factors, we may obtain other compressed or expanded variations of sine waves. The analysis of such graphs is important in physics and engineering.

Exercises 5.7

Use addition of y-coordinates to sketch the graphs of the equations in Exercises 1–20.

1 $y = \cos x + 3 \sin x$

2 $y = \sin x + 3 \cos x$

3 $y = 2 \cos x + 3 \sin x$

4 $y = 2 \sin x + 3 \cos x$

5 $y = \sin x + \cos 2x$

6 $y = 2 \cos x + \sin 2x$

7 $y = \cos x - \sin x$

8 $y = 2 \sin x - \cos x$

9 $y = 2 \cos x - \frac{1}{2} \sin 2x$

10 $y = 2 \cos x + \cos \frac{1}{2}x$

11 $y = 1 + \sin x$

12 $y = 2 + \tan x$

13 $y = \frac{1}{2}x + \sin x$

14 $y = 1 + \cos x$

15 $y = 2^x + \sin x$

16 $y = \csc x - 1$

17 $y = 2 + \sec x$

18 $y = x - \sin x$

19 $y = \cos x - 1$

20 $y = |x + 1| + \cos 2x$

In Exercises 21–30 sketch the graph of f.

21 $f(x) = 2^{-x} \cos x$

22 $f(x) = 2^x \sin x$

23 $f(x) = |x| \sin x$

24 $f(x) = (1 + x^2) \cos x$

25 $f(x) = \frac{1}{2}x^2 \sin x$

26 $f(x) = \cot x \tan x$

27 $f(x) = |\sin x|$

28 $f(x) = |\cos x| + 1$

29 $f(x) = \sin |x|$

30 $f(x) = \cos |x + 1|$

31 If the formula for $D(t)$ in Example 3 is used for Fairbanks, Alaska, then $K \approx 12$. Sketch the graph of D in this case for $0 \le t \le 365$.

32 Based on years of weather data, the expected low temperature T (in °F) in Fairbanks, Alaska, can be approximated by

$$T = 36 \sin \frac{2\pi}{365}(t - 101) + 14$$

where t is in days and $t = 0$ corresponds to January 1.

(a) Sketch the graph of T for $0 \le t \le 365$.

(b) Predict when the coldest day of the year will occur.

Scientists frequently use the formula $f(t) = a \sin (bt + c) + d$ to simulate temperature variations during the day. In Exercises 33–36, $t = 0$ corresponds to midnight, time t is measured in hours, and $f(t)$ is measured in °C. For each exercise, (a) determine values of a, b, c, and d that fit the information, and (b) sketch the graph of f for $0 \le t \le 24$.

33 The high temperature is 10 °C and the low temperature of −10 °C occurs at 4 A.M.

34 The temperature at midnight is 15 °C and the high and low temperatures are 20 °C and 10 °C.

35 The temperature varies between 10 °C and 30 °C, and the average temperature of 20 °C first occurs at 9 A.M.

36 The high temperature of 28 °C occurs at 2 P.M. and the average temperature of 20°C occurs 6 hours later.

37 A dramatic example of the phenomenon of *resonance* occurs when a singer adjusts the pitch of her voice to shatter a wine glass. Functions given by $f(x) = ax \cos bx$ occur in the mathematical analysis of such vibrations. If $a = 1$ and $b = 2$, sketch the graph of f for $x \ge 0$.

5.8 Applications Involving Right Triangles

Trigonometry was developed to help solve problems involving angles and lengths of sides of triangles. Problems of that type are no longer the most important applications; however, questions about triangles still arise in physical situations. We shall restrict our discussion in this section to right triangles. Triangles that do not contain a right angle will be considered in Chapter 7.

We shall often use the following notation. The vertices of a triangle will be denoted by A, B, and C. The angles at A, B, and C will be denoted by α, β, and γ, respectively, and the lengths of the sides opposite these angles by a, b, and c, respectively. The triangle itself will be referred to as *triangle ABC*. If a triangle is a right triangle and if one of the acute angles and a side are known, or if two sides are given, then the formulas in Section 5.2 that express the trigonometric functions as ratios of sides of a triangle may be used to find the remaining parts.

EXAMPLE 1 If, in triangle ABC, $\gamma = 90°$, $\alpha = 34°$, and $b = 10.5$, approximate the remaining parts of the triangle.

FIGURE 5.55

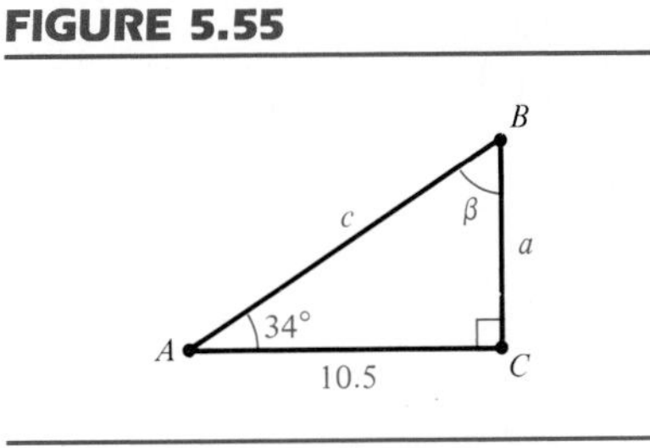

SOLUTION Since the sum of the angles is 180° and $\alpha + \gamma = 124°$, it follows that $\beta = 56°$. Referring to Figure 5.55 and using $\tan \alpha = \text{opp/adj}$,

$$\tan 34° = \frac{a}{10.5} \quad \text{or} \quad a = (10.5) \tan 34°.$$

Using a calculator or Table 4,

$$a \approx (10.5)(0.6745) \approx 7.1.$$

Side c can be found using either the cosine or secant functions. Since $\cos \alpha = \text{adj/hyp}$,

$$\cos 34° = \frac{10.5}{c} \quad \text{or} \quad c = \frac{10.5}{\cos 34°}.$$

Using Table 4 or a calculator,

$$c \approx \frac{10.5}{0.8290} \approx 12.7.$$

We could also have found c as follows:

$$\sec 34° = \frac{c}{10.5} \quad \text{or} \quad c = (10.5) \sec 34°.$$

Using Table 4 or a calculator,

$$c \approx (10.5)(1.2062) = 12.6651 \approx 12.7.$$ ■

As illustrated in Example 1, when working with triangles, we usually round off answers. One reason for doing so is that in most applications, the lengths of sides of triangles and measures of angles are found by mechanical devices, and hence are only approximations to the exact values. Consequently, a number such as 10.5 in Example 1 is assumed to have been rounded off to the nearest tenth. We cannot expect more accuracy in the calculated values for the remaining sides and, therefore, they should also be rounded off to the nearest tenth.

In some problems a number with many digits, such as 36.4635, may be given for the side of a triangle. If a calculator is used, the number may be entered as usual; however, if Table 4 is employed, a number of this type should be rounded off before beginning any calculations. If a number x is written in the scientific form $x = c \cdot 10^k$, where $1 \le c < 10$, then before using Table 4, c should be rounded off to three decimal places. Another way of saying this is that x should be rounded off to four *significant figures*. Some examples will help to clarify the procedure. If $x = 36.4635$, we round off to 36.46. The number 684,279 should be rounded off to 684,300. For a decimal such as 0.096202 we use 0.09620. The reason for rounding off is that the values of the trigonometric functions in Table 4 have been rounded off to four significant figures, and hence we cannot expect more than four-figure accuracy in our computations.

When finding angles, answers should be rounded off as indicated in the following table.

Number of significant figures for sides	Round off degree measure of angles to the nearest
2	1°
3	$10'$ or 0.1°
4	$1'$ or 0.01°

Justification of this table requires a careful analysis of problems that involve approximate data.

EXAMPLE 2 If, in triangle ABC, $\gamma = 90^\circ$, $a = 12.3$, and $b = 31.6$, find the remaining parts.

FIGURE 5.56

SOLUTION Referring to the triangle illustrated in Figure 5.56,

$$\tan \alpha = \frac{12.3}{31.6} \approx 0.3892.$$

Since the sides are given to three significant figures, the rule stated in the preceding table tells us that α should be rounded off to the nearest multiple of $10'$. If a calculator is used to approximate α, we use the degree mode and proceed as follows:

Enter:	0.3892	(the value of $\tan \alpha$)
Press [INV] [tan]:	$21.265988 \approx 21^\circ 16' \approx 21^\circ 20'$	(the approximation to α)

Table 4 can also be used to find α.
Since $\alpha \approx 21^\circ 20'$,

$$\beta \approx 90^\circ - 21^\circ 20' = 68^\circ 40'.$$

Again referring to Figure 5.56,

$$\sec \alpha = \frac{c}{31.6} \quad \text{or} \quad c = (31.6) \sec \alpha$$

and

$$c \approx (31.6) \sec 21^\circ 20' \approx (31.6)(1.0736) \approx 33.9.$$

Side c can also be found using other trigonometric functions. Thus from Figure 5.56 we see that

$$\cos \alpha = \frac{31.6}{c}$$

and hence
$$c = \frac{31.6}{\cos 21°20'} \approx \frac{31.6}{0.9315} \approx 33.9.$$

Still another way to find c is to use the Pythagorean Theorem as follows:

$$c = \sqrt{(31.6)^2 + (12.3)^2} = \sqrt{1149.85} \approx 33.9$$ ■

As illustrated in Figure 5.57, if an observer sights an object, then the angle that the line of sight makes with the horizontal line l is called the **angle of elevation** or the **angle of depression** of the object, depending on whether the object is above or below the horizontal line. The terminology is used in the next two examples.

FIGURE 5.57

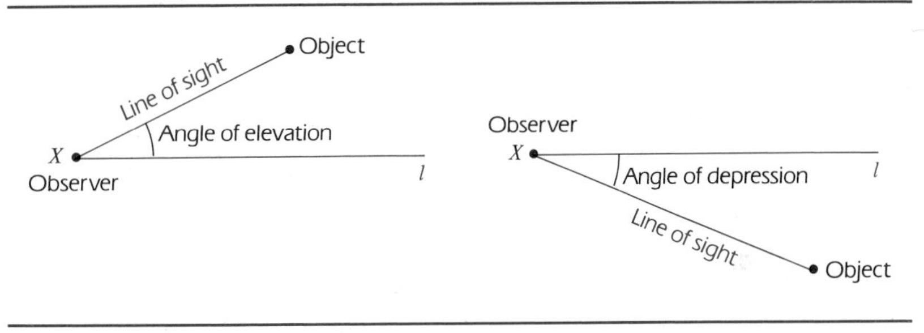

FIGURE 5.58

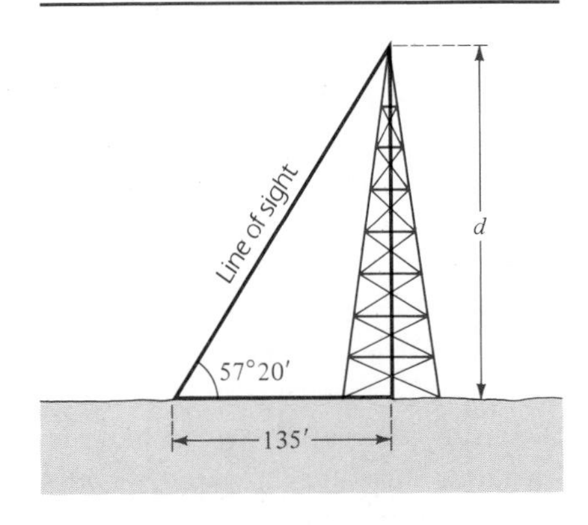

EXAMPLE 3 From a point on level ground 135 feet from the base of a tower, the angle of elevation of the top of the tower is 57°20′. Approximate the height of the tower.

SOLUTION If we let d denote the height of the tower, then the given facts are represented by the triangle in Figure 5.58. Referring to the figure, we see that

$$\tan 57°20' = \frac{d}{135} \quad \text{or} \quad d = (135) \tan 57°20'.$$

Using a table or calculator,

$$d \approx (135)(1.560) \approx 210.6 \approx 211 \text{ feet}.$$ ■

FIGURE 5.59

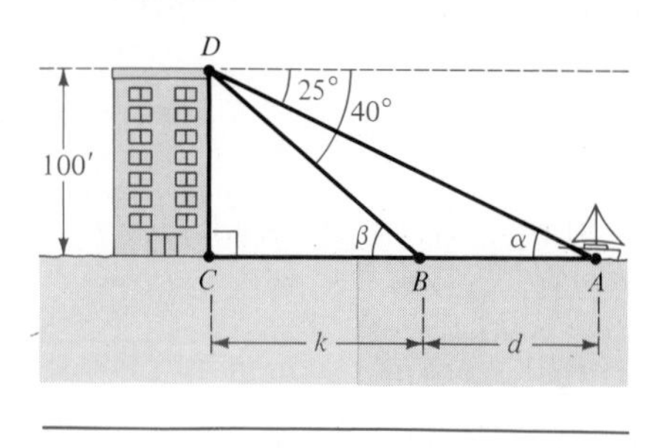

EXAMPLE 4 From the top of a building that overlooks an ocean, a man observes a boat sailing directly toward him. If the man is 100 feet above sea level and if the angle of depression of the boat changes from 25° to 40° during the period of observation, approximate the distance that the boat travels.

SOLUTION As illustrated in Figure 5.59, let A and B be the positions of the boat that correspond to the 25° and 40° angles, respectively. Suppose that the man is at point D, and C is the point 100 feet directly below him. Let d denote the distance the boat travels and let k denote the distance from B to C. If α and β denote angles DAC and DBC, respectively, then it follows from geometry that $\alpha = 25°$ and $\beta = 40°$.

From triangle BCD,

$$\cot \beta = \cot 40° = \frac{k}{100} \quad \text{or} \quad k = 100 \cot 40°.$$

From triangle DAC,

$$\cot \alpha = \cot 25° = \frac{d+k}{100} \quad \text{or} \quad d + k = 100 \cot 25°.$$

Consequently,

$$\begin{aligned} d &= 100 \cot 25° - k = 100 \cot 25° - 100 \cot 40° \\ &= 100(\cot 25° - \cot 40°) \\ &\approx 100(2.145 - 1.192) = 100(0.953) = 95.3. \end{aligned}$$

Hence, $d \approx 95$ feet. ■

FIGURE 5.60

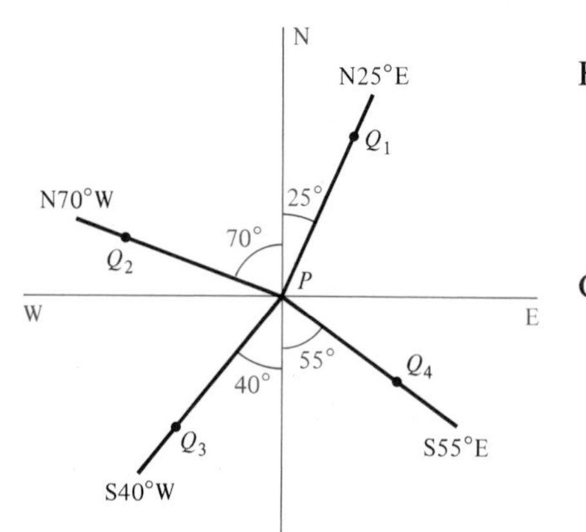

FIGURE 5.61

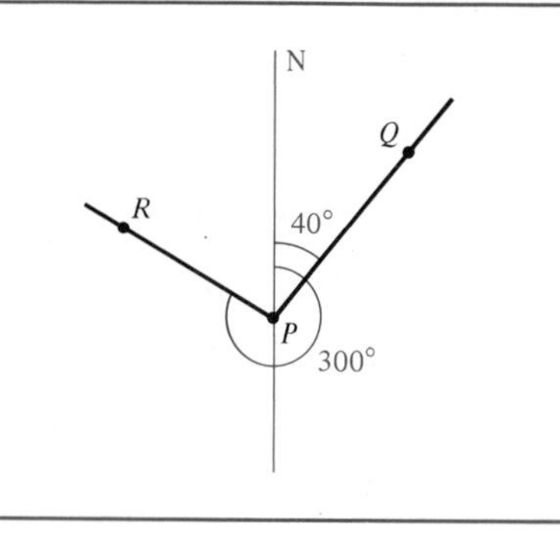

In certain navigation or surveying problems, the **direction,** or **bearing,** from a point P to a point Q is specified by stating the acute angle that the half-line from P through Q varies to the east or west of the north-south line. Figure 5.60 illustrates four such lines. The north-south and east-west lines are labeled NS and WE, respectively. The bearing from P to Q_1 is 25° east of north and is denoted by N25°E. We also refer to the **direction** N25°E, meaning the direction from P to Q_1. The bearings from P to Q_2, Q_3, and Q_4 are represented in a similar manner in the figure.

In air navigation, directions and bearings are specified by measuring from the north in a *clockwise* direction. In this case, a positive measure is assigned to the angle instead of the negative measure to which we are accustomed for clockwise rotations. Thus, referring to Figure 5.61, we see that the direction of PQ is 40°, whereas the direction of PR is 300°.

We shall use these notations in Exercises 35–37.

EXAMPLE 5 Two ships A and B leave port at the same time, ship A sailing in the direction N23°E at a speed of 11 mph, and ship B sailing in the direction S67°E at 15 mph. Approximate the bearing from ship B to ship A one hour later.

FIGURE 5.62

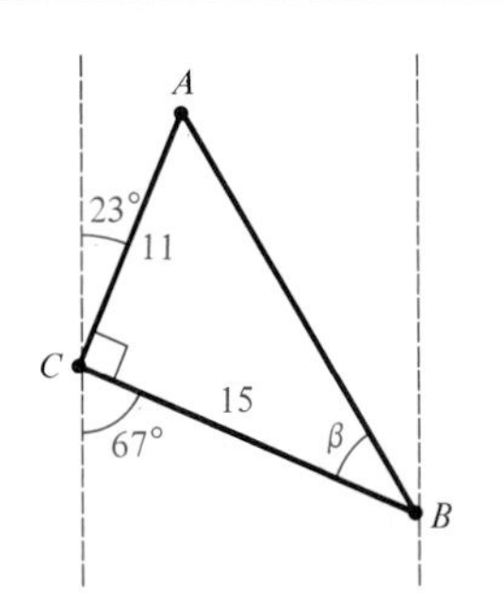

SOLUTION The sketch in Figure 5.62 indicates the positions of the ships at points A and B after one hour, where point C represents the port. Note that

$$\angle ACB = 180° - (23° + 67°) = 90°$$

and hence triangle ABC is a right triangle. Thus

$$\tan \beta = \frac{11}{15} \approx 0.7333.$$

FIGURE 5.63

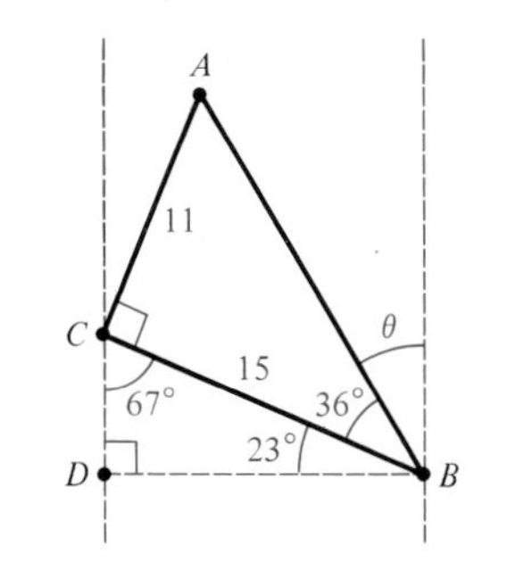

Using Table 4 or a calculator, we obtain

$$\beta \approx 36°.$$

Referring to Figure 5.63 we see that

$$\angle CBD = 90° - 67° = 23° \quad \text{and} \quad \angle ABD \approx 36° + 23° = 59°.$$

Hence

$$\theta \approx 90° - 59° = 31°,$$

which implies that the bearing from B to A is approximately N31°W. ■

Exercises 5.8

Given the indicated parts of triangle ABC with $\gamma = 90°$, approximate the remaining parts of the triangles in Exercises 1–16.

1 $\alpha = 30°,\ b = 20$

2 $\beta = 45°,\ b = 35$

3 $\beta = 52°,\ a = 15$

4 $\alpha = 37°,\ b = 24$

5 $\alpha = 17°40',\ a = 4.50$

6 $\beta = 64°20',\ a = 20.1$

7 $\beta = 71°51',\ b = 240.0$

8 $\alpha = 31°10',\ a = 510$

9 $a = 25,\ b = 45$

10 $a = 31,\ b = 9.0$

11 $c = 5.8,\ b = 2.1$

12 $a = 0.42,\ c = 0.68$

13 $\alpha = 37°46',\ b = 512.0$

14 $\beta = 10°17',\ b = 68.40$

15 $a = 614,\ c = 806$

16 $c = 37.4,\ b = 21.6$

17 Approximate the angle of elevation of the sun if a boy 5.0 feet tall casts a shadow 4.0 feet long on level ground (see figure).

FIGURE FOR EXERCISE 17

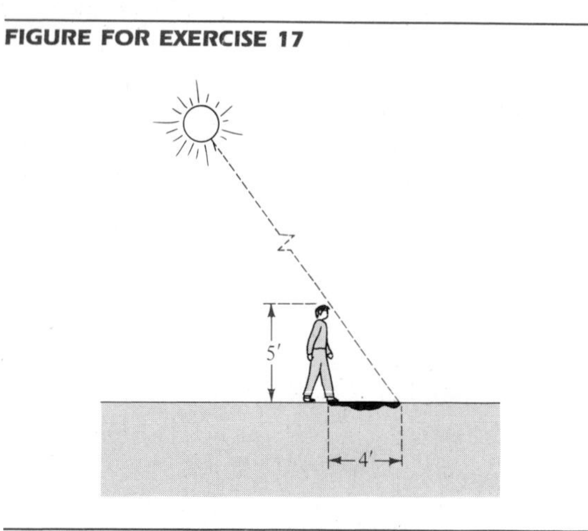

18 From a point 15 meters above level ground an observer measures the angle of depression of an object on the ground as 68°. Approximate the distance from the object to the point on the ground directly beneath the observer.

19 A boy flying a kite holds the string 1.50 meters above ground level. The string of the kite is taut and makes an angle of 54°20′ with the horizontal (see figure). Find the approximate height of the kite above level ground if 85.0 meters of string are out.

FIGURE FOR EXERCISE 19

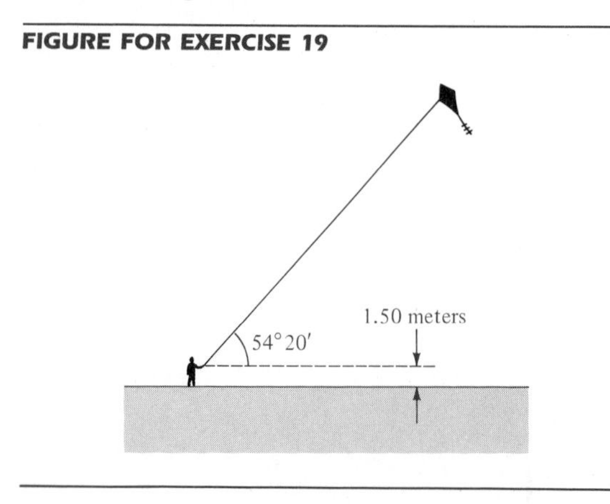

20 The side of a regular pentagon is 24.0 cm long. Approximate the radius of the circumscribed circle.

21 From a point P on level ground the angle of elevation of the top of a tower is 26°50′. From a point 25.0 meters closer to the tower and on the same line with P and the base of the tower, the angle of elevation of the top is 53°30′. Approximate the height of the tower.

22 A ladder 20 feet long leans against the side of a building. If the angle between the ladder and the building is 22°, approximate the distance from the bottom of the ladder to the building. If the distance from the bottom of the ladder to the building is increased by 3.0 feet, approximately how far does the top of the ladder move down the building?

23 As a hot-air balloon rises vertically, its angle of elevation from a point P on level ground 110 km from the point Q directly underneath the balloon changes from 19°20′ to 31°50′ (see figure). Approximately how far does the balloon rise during this period?

FIGURE FOR EXERCISE 23

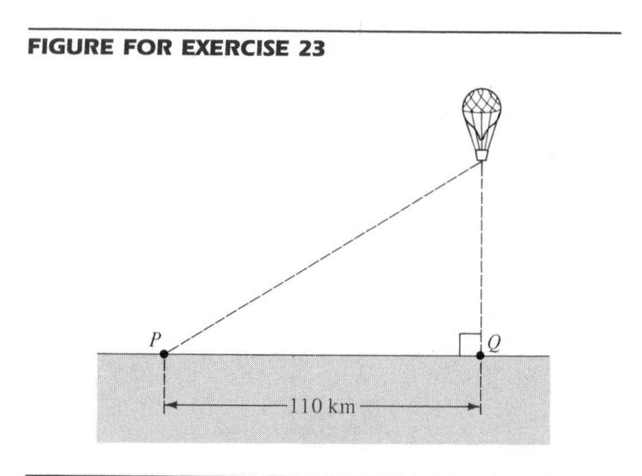

24 From a point A that is 8.20 meters above level ground, the angle of elevation of the top of a building is 31°20′ and the angle of depression of the base of the building is 12°50′. Approximate the height of the building.

25 To find the distance d between two points P and Q on opposite shores of a lake, a surveyor locates a point R that is 50.0 meters from P such that RP is perpendicular to PQ. Next, using a transit, the surveyor measures angle PRQ as 72°40′. What is d?

26 A guy wire is attached to the top of a radio antenna and to a point on horizontal ground that is 40.0 meters from the base of the antenna. If the wire makes an angle of

58°20′ with the ground, approximate the length of the wire.

27 A regular octagon is inscribed in a circle of radius 12.0 cm. Approximate the perimeter of the octagon.

28 A builder wishes to construct a ramp 24 feet long that rises to a height of 5.0 feet above level ground. Approximate the angle that the ramp should make with the horizontal.

29 A rocket is fired at sea level and climbs at a constant angle of 75° through a distance of 10,000 feet. Approximate its altitude to the nearest foot.

30 A CB antenna is located on the top of a garage that is 16 feet tall. From a point on level ground that is 100 feet from a point directly below the antenna, the antenna subtends an angle of 12°. Approximate the length of the antenna.

31 An airplane flying at an altitude of 10,000 feet passes directly over a fixed object on the ground. One minute later the angle of depression of the object is 42°. Approximate the speed of the airplane to the nearest mph.

32 A motorist, traveling along a level highway at a speed of 60 km/h directly toward a distant mountain, observes that between 1:00 P.M. and 1:10 P.M. the angle of elevation of the top of the mountain changes from 10° to 70°. Approximate the height of the mountain.

33 An airplane pilot wishes to make his approach to an airstrip at an angle of 10° with the horizontal. If he is flying at an altitude of 5000 feet, approximately how far from the airstrip should he begin his descent? (Give answer to the nearest 100 feet.)

34 In order to measure the height h of a cloud cover, a spotlight is directed vertically upward from the ground. From a point on level ground that is d meters from the spotlight, the angle of elevation θ of the light image on the clouds is then measured. Find a formula that expresses h in terms of d and θ. As a special case, approximate h if $d = 1000$ meters and $\theta = 59°$.

35 A ship leaves port at 1:00 P.M. and sails in the direction N34°W at a rate of 24 mph. Another ship leaves port at 1:30 P.M. and sails in the direction N56°E at a rate of 18 mph. Approximately how far apart are the ships at 3:00 P.M.? What is the bearing (to the nearest degree) from the first ship to the second?

36 From an observation point A a forest ranger sights a fire in the direction S35°50′W. From a point B, 5 miles due west of A, another ranger sights the same fire in the direction S54°10′E. Approximate, to the nearest tenth of a mile, the distance of the fire from A.

37 An airplane flying at a speed of 360 mph flies from a point A in the direction 137° for 30 minutes and then flies in the direction 227° for 45 minutes. Approximate, to the nearest mile, the distance from the airplane to A.

38 A jet takes off at a 10° angle and travels at the rate of 250 ft/sec. Approximately how long does it take to reach an altitude of 15,000 feet?

39 A draw bridge is 150 feet long when stretched across a river. As shown in the figure, the two sections of the bridge can be rotated upward through an angle of 35°.

(a) If the water level is 15 feet below the bridge, find the distance d between the end of a section and the water level when the bridge is fully open.

(b) Approximately how far apart are the ends of the two sections when the bridge is fully opened, as shown?

FIGURE FOR EXERCISE 39

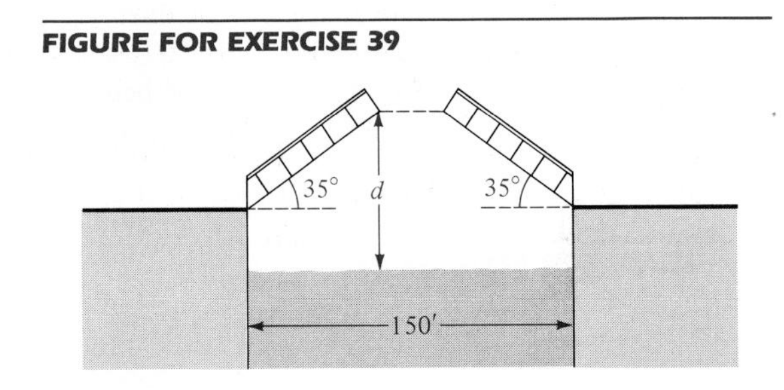

40 Shown in the figure are design plans for a water slide. Find the total length of the slide to the nearest foot.

FIGURE FOR EXERCISE 40

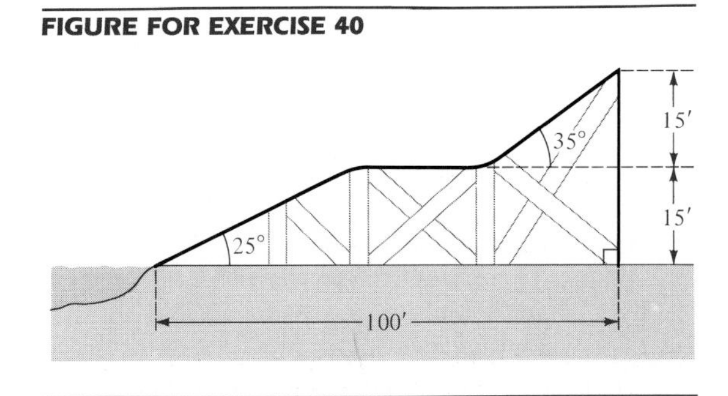

41 Shown in the figure is the screen for a simple video arcade game in which ducks travel from A to B at the rate of 10 cm/sec. Bullets fired from point O travel 20 cm/sec. If a player shoots as soon as a duck appears at A, at what angle ϕ should the gun be aimed to score a direct hit?

FIGURE FOR EXERCISE 41

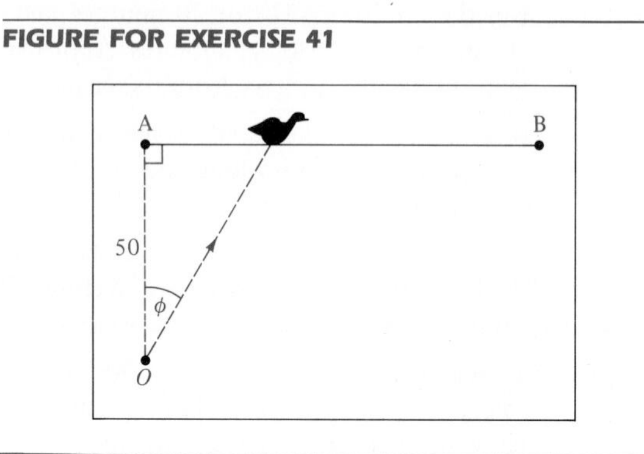

42 A conveyor belt 9 meters long can be hydraulically lifted up to 40° to unload cargo from airplanes (see figure).

(a) Find, to the nearest degree, the angle through which the conveyor belt should be rotated to reach a door that is 4 meters above the level of the belt.

(b) Approximate the maximum height that the belt can reach.

FIGURE FOR EXERCISE 42

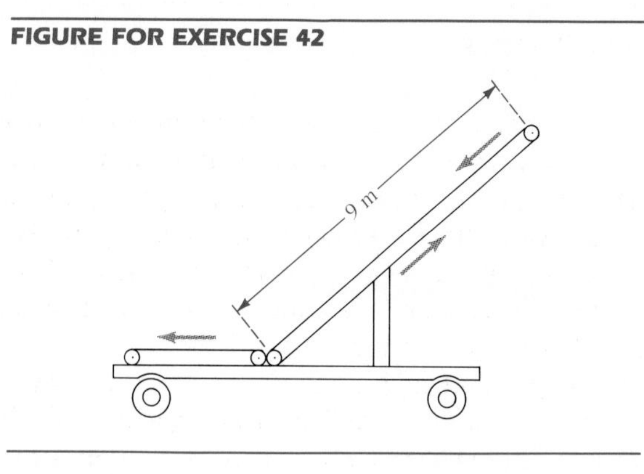

43 A rectangular box has dimensions 8″ × 6″ × 4″. Approximate, to the nearest tenth of a degree, the angle θ formed by a diagonal of the base and the diagonal of the box, as shown in the figure.

FIGURE FOR EXERCISE 43

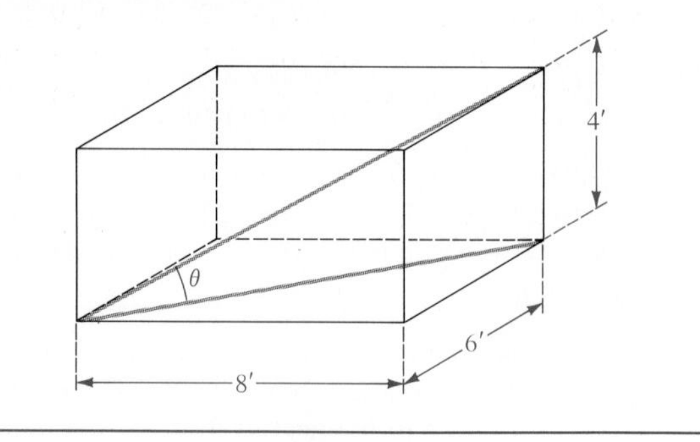

44 A conical paper cup has a radius of 2 inches. Approximate, to the nearest degree, the angle β (see figure) so that the cone will have a volume of 20 in^3.

FIGURE FOR EXERCISE 44

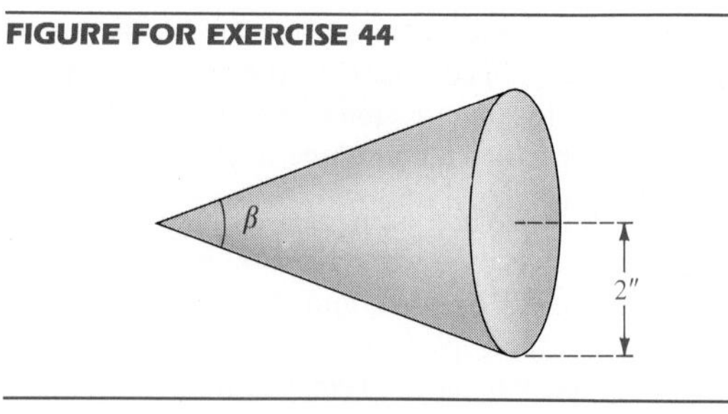

45 A spacelab circles the earth at an altitude of 380 miles. When an astronaut views the horizon of the earth, the angle θ shown in the figure is 65.8°. Use this information to estimate the radius r of the earth.

FIGURE FOR EXERCISE 45

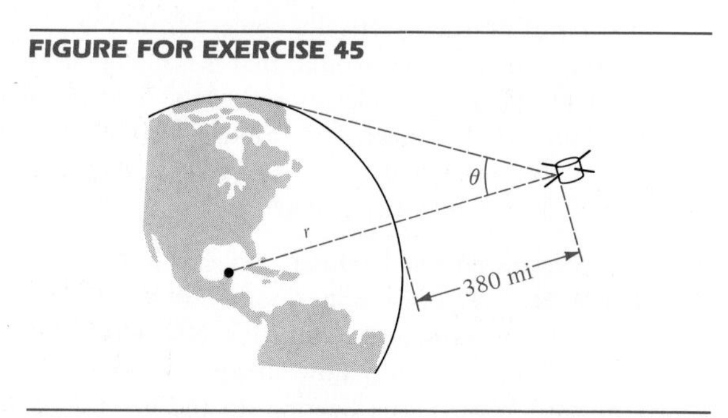

46. When viewed from a point P on the earth, the sun's surface subtends an angle of $32'$ (see figure). Using 92,900,000 miles for the distance between P and the center of the sun, estimate the diameter of the sun.

FIGURE FOR EXERCISE 46

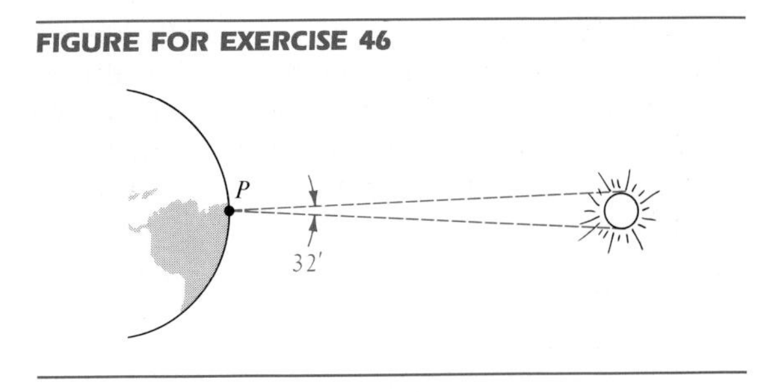

Calculator Exercises 5.8

Given the indicated parts of triangle ABC with $\gamma = 90°$, approximate the remaining parts of each triangle in Exercises 1–10. Round off answers to four significant figures.

1 $\alpha = 41.27°,\ a = 314.6$

2 $\beta = 24.96°,\ b = 209.3$

3 $\beta = 37.06°,\ a = 0.4613$

4 $\alpha = 17.69°,\ b = 1.307$

5 $\beta = 2.71°,\ b = 7149$

6 $\alpha = 84.07°,\ a = 0.1024$

7 $a = 46.87,\ b = 13.12$

8 $a = 6.948,\ b = 8.371$

9 $b = 2462,\ c = 5074$

10 $a = 88.12,\ c = 94.06$

5.9 Harmonic Motion

FIGURE 5.64

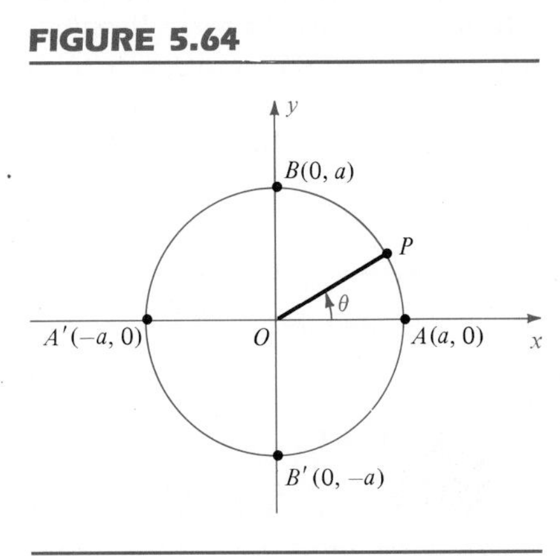

Trigonometric functions are important in the investigation of vibratory or oscillatory motion, such as the motion of a particle in a vibrating guitar string or in a spring that has been compressed or elongated, and then released to oscillate back and forth. The fundamental type of particle displacement inherent in these illustrations is termed *harmonic motion.* To help introduce this concept, let us consider a point P moving at a constant rate around the circle of radius a shown in Figure 5.64.

Suppose that the initial position of P is $A(a, 0)$ and θ is the angle generated by the ray OP after t units of time. The **angular speed** ω of OP is, by definition, the rate at which the measure of θ changes per unit time. To state that P moves around the circle at a constant rate is equivalent to stating that the angular speed ω is constant. If ω is constant, then $\theta = \omega t$. To illustrate, if $\omega = \pi/6$ radians per second, then $\theta = (\pi/6)t$. In this case, at $t = 1$ second, $\theta = (\pi/6)(1) = \pi/6$, and P is one-third of the way from

$A(a, 0)$ to $B(0, a)$. At $t = 2$ seconds, $\theta = (\pi/6)(2) = \pi/3$, and P is two-thirds of the way from A to B. At $t = 6$ seconds, $\theta = (\pi/6)(6) = \pi$, and P is at $A'(-a, 0)$, and so on.

If the coordinates of P are (x, y) as shown in Figure 5.65, then $\cos\theta = x/a$ and $\sin\theta = y/a$.

FIGURE 5.65

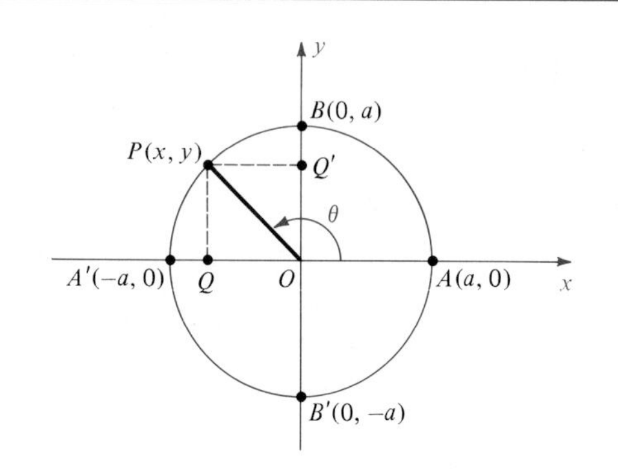

Hence

$$x = a\cos\theta, \qquad y = a\sin\theta.$$

Using the fact that $\theta = \omega t$ gives us

$$x = a\cos\omega t, \qquad y = a\sin\omega t.$$

The last two equations specify the position (x, y) of P at any time t. Let us next consider the point $Q(x, 0)$, which is called the **projection of P on the x-axis.** This point is shown in Figure 5.65. The position of Q is given by $x = a\cos\omega t$. As P moves around the circle several times, the point Q oscillates back and forth between $A(a, 0)$ and $A'(-a, 0)$. Similarly, the point $Q'(0, y)$ is called the **projection of P on the y-axis,** and its position is given by $y = a\sin\omega t$. As P moves around the circle, Q' oscillates between $B'(0, -a)$ and $B(0, a)$. The motions of Q and Q' are of the type described in the next definition.

DEFINITION

A point that moves on a coordinate line such that its distance d from the origin at time t is given by either

$$d = a\cos\omega t \quad \text{or} \quad d = a\sin\omega t$$

for some real numbers a and ω, is said to be in **simple harmonic motion.**

FIGURE 5.66

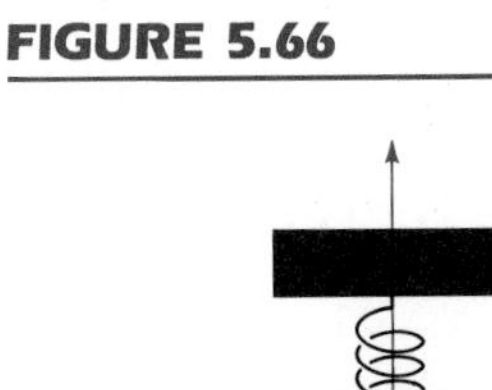

In the preceding definition, the **amplitude** of the motion is the maximum displacement $|a|$ of the point from the origin. The **period** is the time $2\pi/\omega$ required for one complete oscillation. The **frequency** $\omega/2\pi$ is the number of oscillations per unit of time.

A physical interpretation of simple harmonic motion can be obtained by considering a spring with an attached weight that is oscillating vertically relative to a coordinate line, as illustrated in Figure 5.66. The number d represents the coordinate of a fixed point Q in the weight, and we assume that the amplitude a of the motion is constant. In this case no frictional force is retarding the motion. If friction is present, then the amplitude decreases with time, and the motion is said to be *damped*.

EXAMPLE Suppose the weight shown in Figure 5.66 is oscillating according to the law

$$d = 10 \cos\left(\frac{\pi}{6}t\right)$$

where t is measured in seconds and d in centimeters. Discuss the motion of the weight.

SOLUTION By definition, the motion is simple harmonic. The amplitude is $a = 10$ cm. Since $\omega = \pi/6$, the period is $2\pi/\omega = 2\pi/(\pi/6) = 12$. Thus, one complete oscillation takes 12 seconds. The frequency is $\omega/2\pi = (\pi/6)/2\pi = \frac{1}{12}$; that is, $\frac{1}{12}$ of an oscillation takes place each second. The following table indicates the position of Q at various times.

t	0	1	2	3	4	5	6
$(\pi/6)t$	0	$\pi/6$	$\pi/3$	$\pi/2$	$2\pi/3$	$5\pi/6$	π
d	10	$5\sqrt{3}$ ≈ 8.7	5	0	-5	$-5\sqrt{3}$ ≈ -8.7	-10

Note that the initial position of Q is 10 cm above the origin O. It then moves downward, gaining speed until it reaches O. In particular, Q travels approximately 1.3 cm during the first second, 3.7 cm during the next second, and 5 cm during the third second. It then slows down until it reaches a point 10 cm below O at the end of 6 seconds. The direction of motion is then reversed, and the weight moves upward, gaining speed until it reaches O, after which it slows down until it returns to its original position at the end of 12 seconds. The direction of motion is then reversed again, and the same pattern is repeated indefinitely. ■

The preceding example is typical of simple harmonic motion. If the initial angle AOP in Figure 5.65 is ϕ at $t = 0$, then the position (x, y) of P on the circle is given by

$$x = a \cos(\omega t + \phi), \qquad y = a \sin(\omega t + \phi).$$

Points that vary on a coordinate line according to either of these formulas are also said to be in simple harmonic motion.

FIGURE 5.67

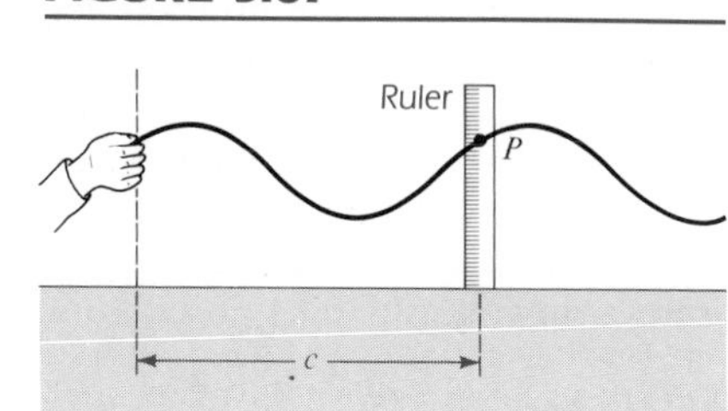

Simple harmonic motion takes place in many different types of wave motion, such as water waves, sound waves, radio waves, light waves, and distortional waves that are present in vibrating bodies. As a specific example, consider the waves made by holding one end of a long rope, as in Figure 5.67, and then causing it to vibrate by raising and lowering the hand in simple harmonic motion.

Waves appear to move along the rope, traveling away from the hand. A ruler, positioned vertically a fixed distance c from the end, as shown in the figure, may be used to study the upward and downward motion of a specific particle P in the rope. If a suitable horizontal axis is chosen, it can be shown that P is in simple harmonic motion, as if it were attached to a spring. A similar motion takes place in water ripples on a large pond.

The definition of simple harmonic motion is usually extended to include situations where d is *any* mathematical or physical quantity (not necessarily a distance). In the next illustration d is itself an angle.

FIGURE 5.68

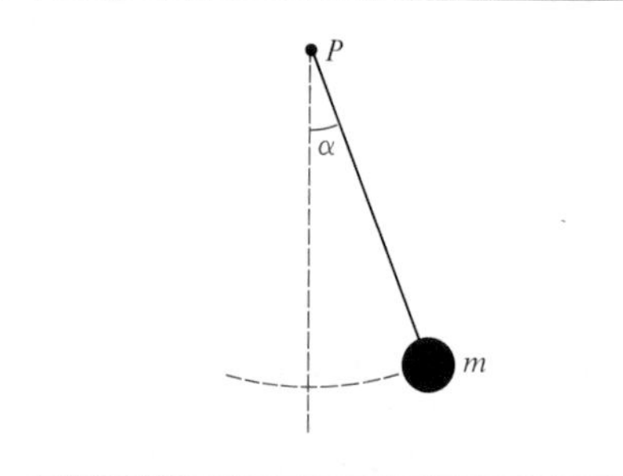

Historically, the first type of harmonic motion to be scientifically investigated involved pendulums. Figure 5.68 illustrates a *simple pendulum* consisting of a bob of mass m attached to one end of a string, with the other end of the string attached to a fixed point P. In the ideal case it is assumed that the string is weightless, that there is no air resistance, and that the only force acting on the bob is gravity. If the bob is displaced sideways and released, the pendulum oscillates back and forth in a vertical plane. Let α denote the *angular displacement* at time t (see Figure 5.68). If the bob moves through a small arc (say, $|\alpha| < 5°$), then it can be shown, by using physical laws, that

$$\alpha = \beta \cos(\omega t + \alpha_0)$$

where α_0 is the initial displacement, ω is the frequency of oscillation of the angle α, and β is the (angular) amplitude of oscillation. This means that the *angle* α is in simple harmonic motion, and hence we refer to the motion as *angular* simple harmonic motion.

As a final illustration, in electric circuits an alternating electromotive force (emf) and the current may vary harmonically. For example, the emf E (measured in volts) at time t may be given by

$$E = E_M \sin \omega t$$

FIGURE 5.69

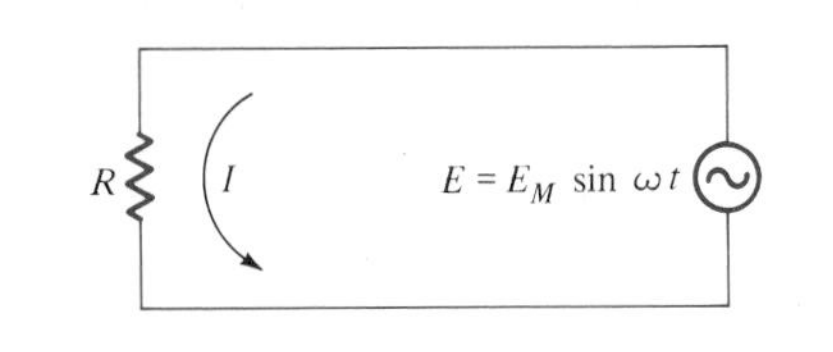

where E_M is the maximum voltage. If an emf of this type is impressed on a circuit containing only a resistor of R ohms, then by Ohm's Law, the current I at time t is

$$I = \frac{E}{R} = \frac{E_M}{R} \sin \omega t = I_M \sin \omega t$$

where $I_M = E_M/R$ is the maximum current. A schematic drawing of an electric circuit of this type is illustrated in Figure 5.69.

In this case the maximum current I_M occurs at the same time as the maximum voltage E_M. In other situations these maximum values may occur at different times, in which case we say there is a **phase difference** between E and I. For example, if $E = E_M \sin \omega t$, we could have either

$$\text{(a)} \quad I = I_M \sin(\omega t - \phi) \quad \text{or} \quad \text{(b)} \quad I = I_M \sin(\omega t + \phi)$$

for some $\phi > 0$. In case (a), the current is said to **lag** the emf by an amount ϕ/ω, and the graph of I can be obtained by shifting the graph of $I = I_M \sin \omega t$ an amount ϕ/ω to the *right*. In case (b) we shift the graph to the *left*, and the current is said to **lead** the emf by an amount ϕ/ω. This is analogous to our discussion of phase shifts in Section 5.6.

Numerous other physical illustrations involve simple harmonic motion. Interested students may find further information in books on physics and engineering.

Exercises 5.9

1 A wheel of diameter 40 cm is rotating about an axle at a rate of 100 revolutions per minute. If a coordinate system is introduced as in Figure 5.65 where P is a point on the rim of the wheel, find:

(a) the angular speed of OP.

(b) the position (x, y) of P after t minutes.

2 A wheel of radius 2 feet is rotating about an axle, and the angular speed of a ray from the center of the wheel to a point P on the rim is $5\pi/6$ radians per second.

(a) How many revolutions does the wheel make in 10 minutes?

(b) If a coordinate system is introduced as in Figure 5.65 find the position (x, y) of P after t seconds.

In Exercises 3–6 the given formula specifies the position of a point P that is moving harmonically on a vertical axis, where t is in seconds and d is in centimeters. Determine the amplitude, period, and frequency, and describe the motion of the point during one complete oscillation (starting at $t = 0$).

3 $d = 10 \sin 6\pi t$

4 $d = \frac{1}{3} \cos \frac{\pi}{4} t$

5 $d = 4 \cos \frac{3\pi}{2} t$

6 $d = 6 \sin \frac{2\pi}{3} t$

7 A point P in simple harmonic motion has a period of 3 seconds and an amplitude of 5 cm. Express the motion of P by means of an equation of the form $d = a \cos \omega t$.

8 A point P in simple harmonic motion has a frequency of $\frac{1}{2}$ oscillation per minute and amplitude of 4 feet. Express the motion of P by means of an equation of the form $d = a \sin \omega t$.

9 The electromotive force E and current I in a certain alternating current circuit are given by

$$E = 220 \sin 360\pi t,$$
$$I = 20 \sin (360\pi t - \tfrac{1}{4}\pi).$$

Sketch the graphs of E and I on the same coordinate axes and determine the lag or lead.

10 Rework Exercise 9 if

$$E = 110 \sin 120\pi t,$$
$$I = 15 \sin (120\pi t + \tfrac{1}{3}\pi).$$

11 A large ferris wheel is 100 feet in diameter and rises 110 feet off the ground as illustrated in the figure. Each revolution of the wheel takes 30 seconds.

(a) Find the angular speed of the wheel.

(b) Express the height h of a passenger off the ground as a function of time t (in seconds) if $t = 0$ corresponds to a time when a passenger is at the bottom.

FIGURE FOR EXERCISE 11

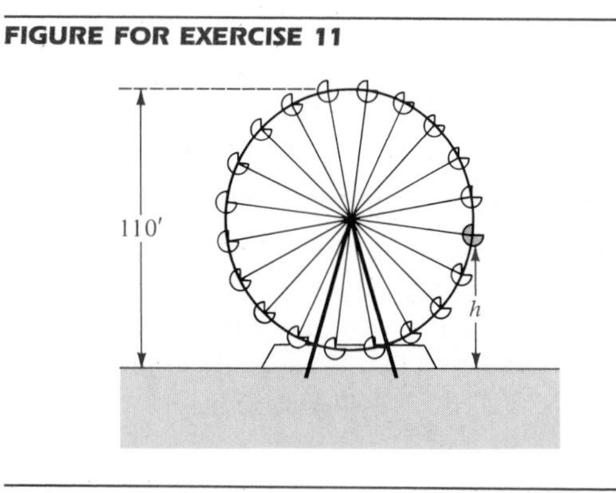

12 A *tsunami wave* is a tidal wave caused by an earthquake beneath the sea. These waves can be more than a hundred feet in height and can travel at great speeds. Engineers represent tsunami waves by trigonometric expressions of the form $y = a \cos bt$ and use these representations to estimate the effectiveness of sea walls. Suppose that a wave has height $h = 50$ feet, period 30 minutes, and is traveling at the rate of 180 feet/second.

(a) Let (x, y) be a point on the wave represented in the figure. Express y as a function of t if $y = 25$ feet when $t = 0$.

(b) The *wave length* L is the distance between two successive crests of the wave. Approximate L in feet and in miles.

FIGURE FOR EXERCISE 12

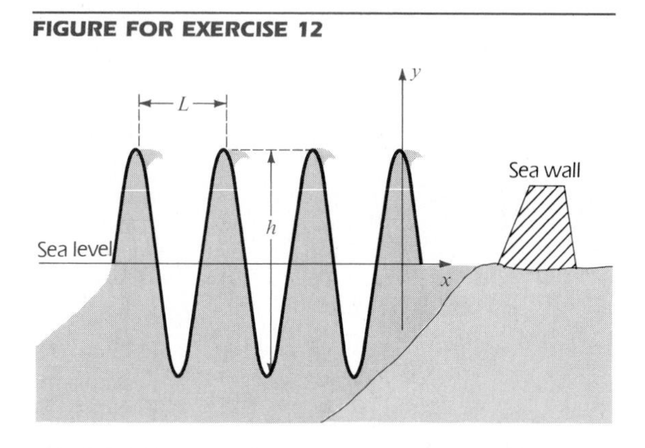

13 The graph in the figure shows the rise and fall of the water level in Boston Harbor during a particular 24-hour period. Approximate the water level y by means of an expression of the form $y = a \sin (bt + c) + d$, where $t = 0$ corresponds to midnight.

FIGURE FOR EXERCISE 13

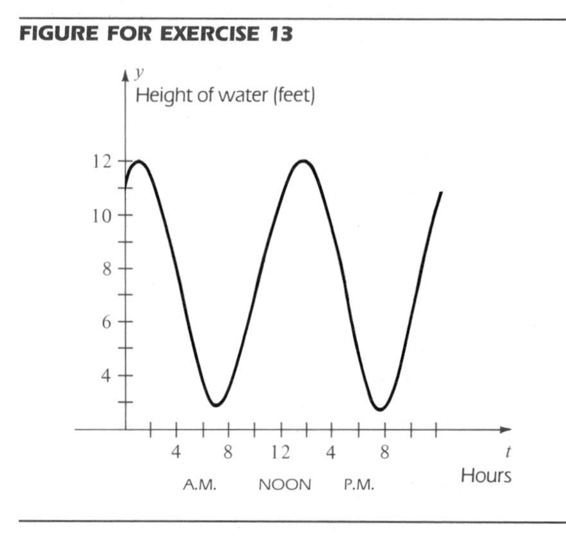

14 A piston is attached to a crankshaft as shown in the figure. The connecting rod AB has length 6 inches and the radius of the crankshaft is 2 inches.

(a) If the crankshaft rotates counterclockwise 2 times per second, find formulas for the position of point A at t seconds after A has coordinates $(2, 0)$.

(b) Find a formula for the position of point B at time t.

FIGURE FOR EXERCISE 14

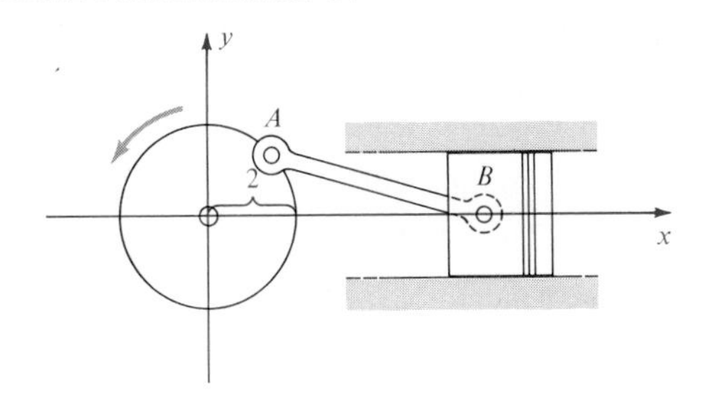

5.10 Review

Define or discuss each of the following.

1 Angle
2 Initial and terminal sides of an angle
3 Coterminal sides of angles
4 Standard position of an angle
5 Positive and negative angles
6 Degree measure
7 Acute and obtuse angles
8 Radian measure
9 The relationship between radians and degrees
10 Trigonometric functions of acute angles
11 The Fundamental Identities
12 Trigonometric functions of any angle
13 Signs of the trigonometric functions
14 Trigonometric functions of real numbers
15 Trigonometric functions in terms of a unit circle
16 Domains and ranges of the trigonometric functions
17 Periodic function; period; periods of the trigonometric functions
18 Formulas for Negatives
19 Reference numbers and angles
20 Finding values of trigonometric functions using reference numbers and angles
21 Graphs of the trigonometric functions
22 The graphs of $f(x) = a \sin(bx + c)$ and $f(x) = a \cos(bx + c)$
23 Trigonometric solutions of right triangles
24 Harmonic motion

Exercises 5.10

1 Find the radian measures that correspond to the following degree measures: $330°$, $405°$, $-150°$, $240°$, $36°$.

2 Find the degree measures that correspond to the following radian measures: $9\pi/2$, $-2\pi/3$, $7\pi/4$, 5π, $\pi/5$.

3 A central angle θ is subtended by an arc 20 cm long on a circle of radius 2 meters. What is the radian measure of θ?

4 Find the length of arc that subtends an angle of measure $70°$ on a circle of diameter 15 cm.

In Exercises 5–7 $P(t)$ denotes the point on the unit circle U that corresponds to the real number t.

5 Find the rectangular coordinates of $P(7\pi)$, $P(-5\pi/2)$, $P(9\pi/2)$, $P(-3\pi/4)$, $P(18\pi)$, and $P(\pi/6)$.

6 If $P(t)$ has coordinates $(-\frac{3}{5}, -\frac{4}{5})$, find the coordinates of $P(t+3\pi)$, $P(t-\pi)$, $P(-t)$, and $P(2\pi - t)$.

7 Find the quadrant containing $P(t)$ if:

(a) $\sec t < 0$ and $\sin t > 0$.

(b) $\cot t > 0$ and $\csc t < 0$.

(c) $\cos t > 0$ and $\tan t < 0$.

8 Use fundamental identities to find the values of the remaining trigonometric functions if:

(a) $\sin \theta = -\frac{4}{5}$ and $\cos \theta = \frac{3}{5}$.

(b) $\csc \theta = \sqrt{13}/2$ and $\cot \theta = -\frac{3}{2}$.

9 Without the use of tables or a calculator, find the values of the trigonometric functions corresponding to each of the following real numbers.

(a) $9\pi/2$ (b) $-5\pi/4$

(c) 0 (d) $11\pi/6$

10 Find the reference number t' if t equals: $5\pi/4$, $-5\pi/6$, $-9\pi/8$.

11 Find the reference angle for each of the following: $245°$, $137°10'$, $892°$, $5\pi/4$, $-\pi/6$, $9\pi/8$.

12 Find the following without the use of tables or calculators.

(a) $\cos 225°$ (b) $\tan 150°$

(c) $\sin(-\pi/6)$ (d) $\sec(4\pi/3)$

(e) $\cot(7\pi/4)$ (f) $\csc(300°)$

13 Find the values of the six trigonometric functions of θ when θ is in standard position and satisfies the stated condition.

(a) The point $(30, -40)$ is on the terminal side of θ.

(b) The terminal side of θ is in quadrant II and is parallel to the line $2x + 3y + 6 = 0$.

(c) $\theta = -90°$.

14 If θ is an acute angle of a right triangle and if the adjacent side and hypotenuse have lengths 4 and 7, respectively, find the values of the six trigonometric functions of θ.

In Exercises 15–22 sketch the graph of f. Find the amplitude and period when appropriate.

15 $f(x) = 5\cos x$ **16** $f(x) = \frac{2}{3}\sin x$

17 $f(x) = -\tan x$ **18** $f(x) = 4\sin 3x$

19 $f(x) = 3\cos 4x$ **20** $f(x) = -4\sin\frac{1}{3}x$

21 $f(x) = 2\sec\frac{1}{2}x$ **22** $f(x) = \frac{1}{2}\csc 2x$

Sketch the graphs of the equations in Exercises 23–29.

23 $y = 2\sin\left(x - \dfrac{2\pi}{3}\right)$ **24** $y = -4\cos\left(x + \dfrac{\pi}{6}\right)$

25 $y = 5\cos\left(2x + \dfrac{\pi}{2}\right)$ **26** $y = \tan\left(x - \dfrac{\pi}{4}\right)$

27 $y = 2\sin x + \sin 2x$ **28** $y = 1 + x + \cos x$

29 $y = 2^{-x}\sin 2x$

Parts of triangle ABC are given in Exercises 30–32 with $\gamma = 90°$. Approximate the remaining parts.

30 $\beta = 60°$, $b = 40$ **31** $\alpha = 54°40'$, $b = 220$

32 $a = 62$, $b = 25$

33 A conical paper cup is formed by cutting a sector AOB from a circle of radius 5 inches and attaching edge OA to OB (see figure). Find angle AOB so that the cup has a depth of 4 inches.

FIGURE FOR EXERCISE 33

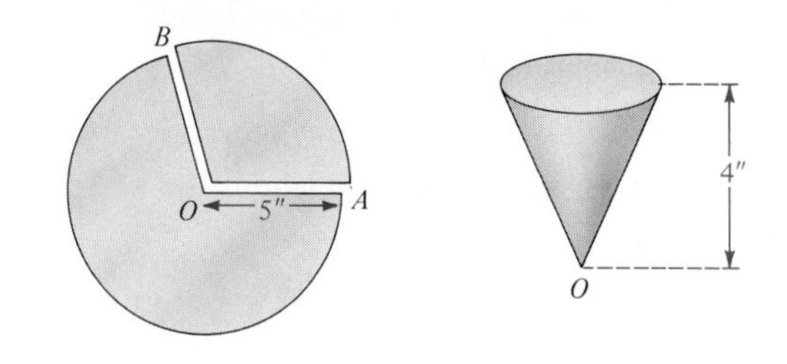

34 The largest airplane propeller ever used had a diameter of 22′7.5″. The plane was powered by four engines that turned the propeller at 545 rpm.

(a) Find the angular speed of the propeller in radians per second.

(b) Approximately how fast (in mph) does the tip of the propeller travel along the circle it generates?

35 In order to simulate the response of a structure to an earthquake, a shape must be chosen for the initial displacement of the beams in the building. When the beam is of length L feet and the maximum displacement is a feet, the equation $y = a - a \cos(\pi x/2L)$ has been used by engineers for the displacement y (see figure). If $a = 1$ and $L = 10$, sketch the graph of the equation for $0 \le x \le 10$.

FIGURE FOR EXERCISE 35

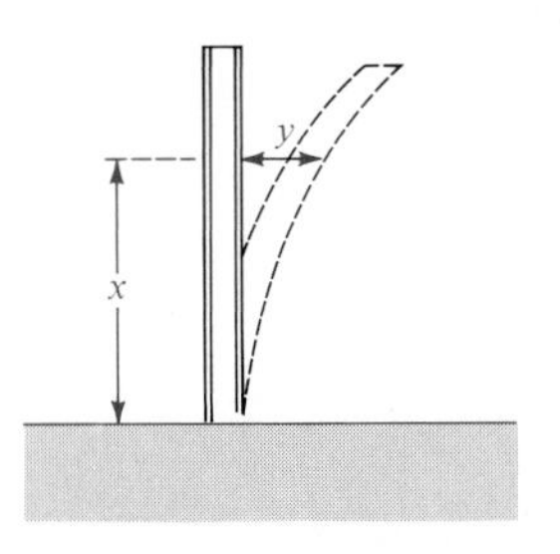

36 The variation in body temperature is an example of a *circadian rhythm,* a biological process that repeats itself approximately every 24 hours. Body temperature is highest around 5 P.M. and lowest at 5 A.M. Let y denote the body temperature (in °F) and let $t = 0$ correspond to midnight. If the low and high body temperatures are 98.3 and 98.9, respectively, find a formula $y = 98.6 + a \sin(bt + c)$ that fits this information.

37 The Great Pyramid of Egypt is 147 meters high with a square base of side 230 meters (see figure). Find, to the nearest degree, the angle φ formed when an observer stands at the midpoint of one of the sides and views the apex of the pyramid.

FIGURE FOR EXERCISE 37

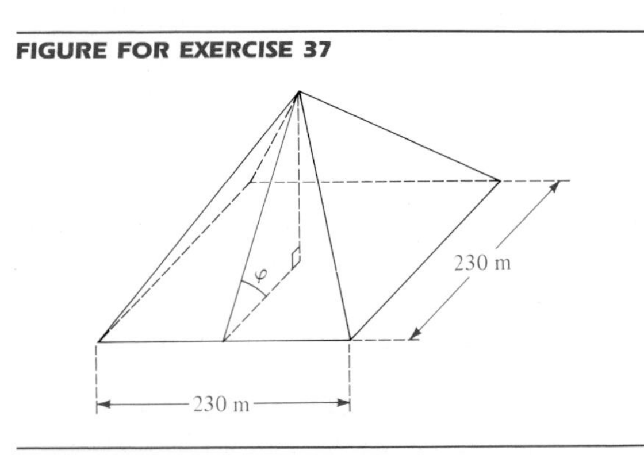

38 When the top of the Eiffel Tower in Paris, France, is viewed at a distance of 200 feet from the base, the angle of elevation is 79.2°. Estimate the height of the tower.

39 A tunnel for a new highway is to be cut through a mountain that is 260 feet high. At a distance of 200 feet from the base of the mountain, the angle of elevation is 36° (see figure). From a distance of 150 feet on the other side, the angle of elevation measures 47°. Find the length of the tunnel to the nearest foot.

FIGURE FOR EXERCISE 39

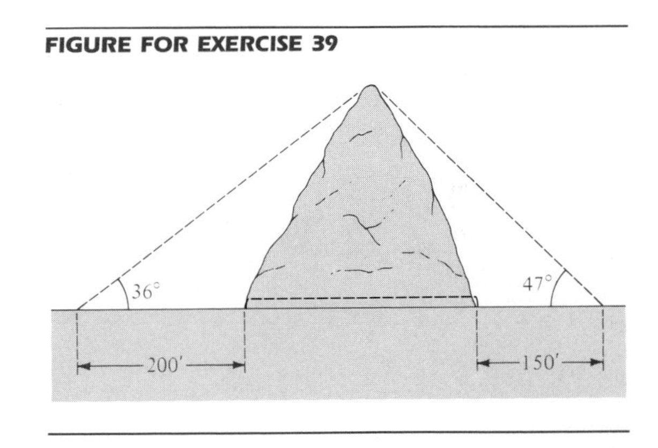

40 When a certain skyscraper is viewed from the top of a building 50 feet tall, the angle of elevation is 59° (see figure). When viewed from the street next to the shorter building, the angle of elevation is 62°.

(a) Approximate the height of the skyscraper to the nearest tenth of a foot.

(b) Approximately how far apart are the two structures?

FIGURE FOR EXERCISE 40

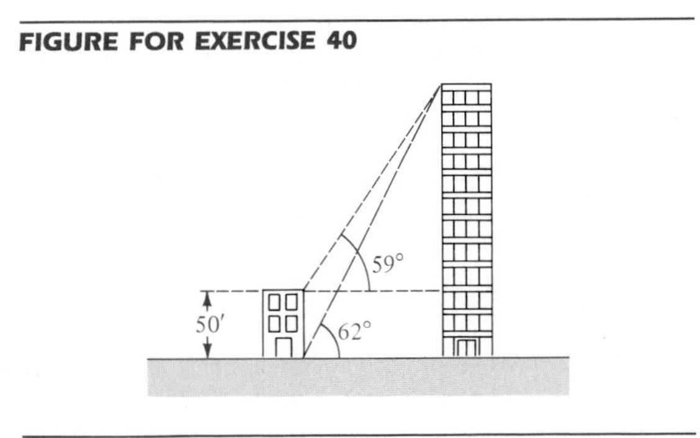

41 When viewed from the earth over a period of time, the planet Venus appears to move back and forth along a line segment with the sun at its midpoint (see figure). At the maximum apparent distance from the sun, angle SEV is approximately 47°. Using $ES = 92{,}900{,}000$ miles, estimate the distance of Venus from the sun.

FIGURE FOR EXERCISE 41

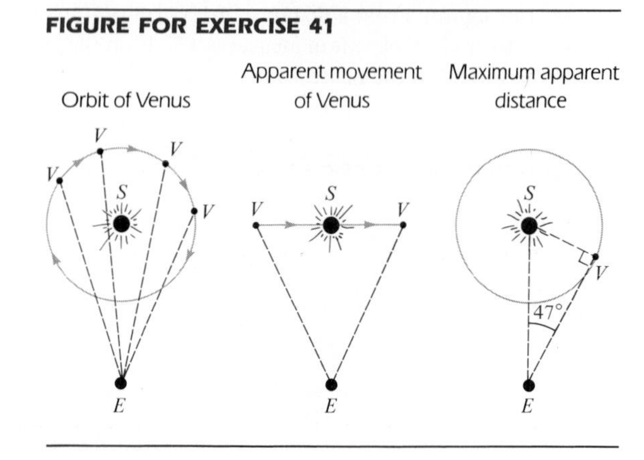

42 A cork bobs up and down in a pail of water such that the distance from the bottom of the pail to the center of the cork at time $t \geq 0$ is given by $s(t) = 12 + \cos(\pi t)$, where $s(t)$ is in inches and t is in seconds (see figure).

(a) Describe the motion of the cork for $0 \leq t \leq 2$.

(b) During what time intervals is the cork rising? When is it falling?

FIGURE FOR EXERCISE 42

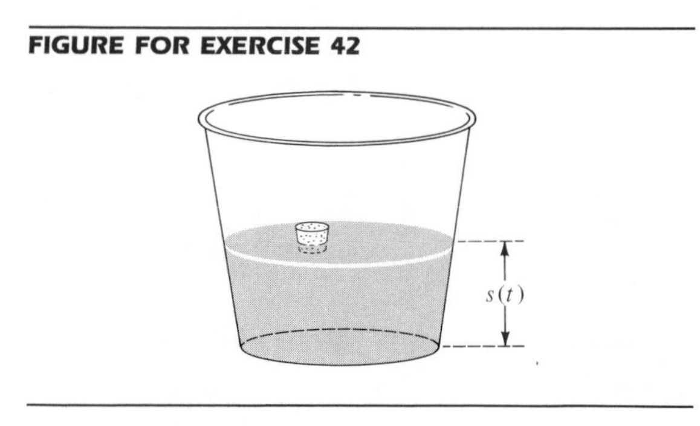

CHAPTER 6

Analytic Trigonometry

As we proceed through this chapter we shall derive many important trigonometric formulas; for reference, they are listed on the inside of the covers of the text. ■ In addition to formal techniques involving trigonometric expressions, we shall also consider numerous applications of trigonometry.

6.1 Trigonometric Identities

A mathematical expression that contains symbols such as $\sin x$, $\cos \beta$, $\tan v$, and so on, where the letters x, β, and v are variables, is referred to as a **trigonometric expression.** The following are examples of trigonometric expressions:

$$x + \sin x \qquad \frac{\sqrt{\theta} + 2^{\sin \theta}}{\cot \theta} \qquad \frac{\cos(3y + 1)}{x^2 + \tan^2(z - y^2)}$$

As for algebraic expressions, we assume that the domain of each variable is the set of real numbers (or angles) for which the expressions are meaningful.

The Fundamental Identities that were introduced in Section 5.2 may be used to help simplify complicated trigonometric expressions. Let us begin by restating these important formulas and working an example. The variable t in each fundamental identity may represent either a real number or the measure of an angle.

THE FUNDAMENTAL IDENTITIES

$$\csc t = \frac{1}{\sin t} \qquad \tan t = \frac{\sin t}{\cos t} \qquad \sin^2 t + \cos^2 t = 1$$

$$\sec t = \frac{1}{\cos t} \qquad \cot t = \frac{\cos t}{\sin t} \qquad 1 + \tan^2 t = \sec^2 t$$

$$\cot t = \frac{1}{\tan t} \qquad 1 + \cot^2 t = \csc^2 t$$

EXAMPLE 1 Simplify the expression $(\sec \theta + \tan \theta)(1 - \sin \theta)$.

SOLUTION Supply reasons for the following steps:

$$\begin{aligned}(\sec \theta + \tan \theta)(1 - \sin \theta) &= \left(\frac{1}{\cos \theta} + \frac{\sin \theta}{\cos \theta}\right)(1 - \sin \theta) \\ &= \left(\frac{1 + \sin \theta}{\cos \theta}\right)(1 - \sin \theta) \\ &= \frac{1 - \sin^2 \theta}{\cos \theta} = \frac{\cos^2 \theta}{\cos \theta} \\ &= \cos \theta.\end{aligned}$$

■

There are other ways to simplify the expression in Example 1. We could first multiply the two factors, and then simplify and combine terms. The method we employed—of changing all expressions to expressions that involve only sines and cosines—is often worthwhile. However, that technique does not always lead to the shortest possible simplification.

Most of the identities to be considered in the remainder of this section are unimportant in their own right. The important thing is the manipulative practice that we will gain. The ability to carry out trigonometric manipulations is essential for solving problems that are encountered in advanced courses in mathematics and science.

We often use the phrase "verify an identity" instead of "prove that an equation is an identity." When verifying an identity, we shall use fundamental identities and algebraic manipulations to change the form of trigonometric expressions in a manner similar to the solution of Example 1. The preferred method of showing that an equation is an identity is to transform one side into the other, as illustrated in the next three examples.

EXAMPLE 2 Verify the identity

$$\frac{\tan t + \cos t}{\sin t} = \sec t + \cot t.$$

SOLUTION We shall transform the left side into the right side. Thus,

$$\frac{\tan t + \cos t}{\sin t} = \frac{\tan t}{\sin t} + \frac{\cos t}{\sin t} = \frac{\left(\dfrac{\sin t}{\cos t}\right)}{\sin t} + \cot t$$

$$= \frac{1}{\cos t} + \cot t = \sec t + \cot t.$$

■

EXAMPLE 3 Verify the identity $\sec \alpha - \cos \alpha = \sin \alpha \tan \alpha$.

SOLUTION We may transform the left side into the right side as follows:

$$\sec \alpha - \cos \alpha = \frac{1}{\cos \alpha} - \cos \alpha = \frac{1 - \cos^2 \alpha}{\cos \alpha}$$

$$= \frac{\sin^2 \alpha}{\cos \alpha} = \sin \alpha \left(\frac{\sin \alpha}{\cos \alpha}\right)$$

$$= \sin \alpha \tan \alpha.$$

■

EXAMPLE 4 Verify the identity

$$\frac{\cos x}{1 - \sin x} = \frac{1 + \sin x}{\cos x}.$$

SOLUTION We begin by multiplying the numerator and denominator of the fraction on the left by $1 + \sin x$. Thus

$$\begin{aligned}\frac{\cos x}{1 - \sin x} &= \frac{\cos x}{1 - \sin x} \cdot \frac{1 + \sin x}{1 + \sin x} \\ &= \frac{\cos x\,(1 + \sin x)}{1 - \sin^2 x} \\ &= \frac{\cos x\,(1 + \sin x)}{\cos^2 x} \\ &= \frac{1 + \sin x}{\cos x}.\end{aligned}$$

■

Another technique for showing that an equation $p = q$ is an identity is to begin by transforming the left side p into another expression s, making sure that each step is *reversible* in the sense that it is possible to transform s back into p by reversing the procedure that has been used. In this case the equation $p = s$ is an identity. Next, as a *separate* exercise, we must show that the right side q can also be transformed to the expression s by means of reversible steps and hence that $q = s$ is an identity. It then follows that $p = q$ is an identity. This method is illustrated in the next example.

EXAMPLE 5 Verify the identity

$$(\tan \theta - \sec \theta)^2 = \frac{1 - \sin \theta}{1 + \sin \theta}.$$

SOLUTION We shall verify the identity by showing that each side of the equation can be transformed into the same expression. Starting with the left side, we may write

$$\begin{aligned}(\tan \theta - \sec \theta)^2 &= \tan^2 \theta - 2 \tan \theta \sec \theta + \sec^2 \theta \\ &= \left(\frac{\sin \theta}{\cos \theta}\right)^2 - 2\left(\frac{\sin \theta}{\cos \theta}\right)\left(\frac{1}{\cos \theta}\right) + \left(\frac{1}{\cos \theta}\right)^2 \\ &= \frac{\sin^2 \theta}{\cos^2 \theta} - \frac{2 \sin \theta}{\cos^2 \theta} + \frac{1}{\cos^2 \theta} = \frac{\sin^2 \theta - 2 \sin \theta + 1}{\cos^2 \theta}.\end{aligned}$$

The right side of the given equation may be changed by multiplying numerator and denominator by $1 - \sin\theta$. Thus,

$$\frac{1-\sin\theta}{1+\sin\theta} = \frac{1-\sin\theta}{1+\sin\theta}\cdot\frac{1-\sin\theta}{1-\sin\theta}$$
$$= \frac{1-2\sin\theta+\sin^2\theta}{1-\sin^2\theta}$$
$$= \frac{1-2\sin\theta+\sin^2\theta}{\cos^2\theta}.$$

The last expression is the same as that obtained for $(\tan\theta - \sec\theta)^2$. Since all steps are reversible, it follows that the given equation is an identity. ■

In calculus it is sometimes convenient to change the forms of certain algebraic expressions by making a **trigonometric substitution,** as illustrated in the following example.

EXAMPLE 6

Express $\sqrt{a^2 - x^2}$ as a trigonometric function of θ, free of radicals, by making the substitution $x = a\sin\theta$ where $a > 0$ and $-\pi/2 \le \theta \le \pi/2$.

SOLUTION Letting $x = a\sin\theta$,

$$\sqrt{a^2 - x^2} = \sqrt{a^2 - (a\sin\theta)^2}$$
$$= \sqrt{a^2 - a^2\sin^2\theta}$$
$$= \sqrt{a^2(1-\sin^2\theta)}$$
$$= \sqrt{a^2\cos^2\theta}$$
$$= a\cos\theta.$$

The last equality is true because (i) $\sqrt{a^2} = a$ if $a > 0$ and (ii) if $-\pi/2 \le \theta \le \pi/2$, then $\cos\theta \ge 0$ and hence $\sqrt{\cos^2\theta} = \cos\theta$. ■

Exercises 6.1

Verify the identities in Exercises 1–88.

1 $\cos\theta\sec\theta = 1$

2 $\tan\alpha\cot\alpha = 1$

3 $\sin\theta\sec\theta = \tan\theta$

4 $\sin\alpha\cot\alpha = \cos\alpha$

5 $\dfrac{\csc x}{\sec x} = \cot x$

6 $\cot\beta\sec\beta = \csc\beta$

7 $(1 + \cos\alpha)(1 - \cos\alpha) = \sin^2\alpha$

8 $\cos^2 x\,(\sec^2 x - 1) = \sin^2 x$

9 $\cos^2 t - \sin^2 t = 2\cos^2 t - 1$

10 $(\tan\theta + \cot\theta)\tan\theta = \sec^2\theta$

11 $\dfrac{\sin t}{\csc t} + \dfrac{\cos t}{\sec t} = 1$

12 $1 - 2\sin^2 x = 2\cos^2 x - 1$

13 $(1 + \sin\alpha)(1 - \sin\alpha) = \dfrac{1}{\sec^2\alpha}$

14 $(1 - \sin^2 t)(1 + \tan^2 t) = 1$

15 $\sec\beta - \cos\beta = \tan\beta\sin\beta$

16 $\dfrac{\sin w + \cos w}{\cos w} = 1 + \tan w$

17 $\dfrac{\csc^2\theta}{1 + \tan^2\theta} = \cot^2\theta$

18 $\sin x + \cos x\cot x = \csc x$

19 $\sin t\,(\csc t - \sin t) = \cos^2 t$

20 $\cot t + \tan t = \csc t\sec t$

21 $\csc\theta - \sin\theta = \cot\theta\cos\theta$

22 $\cos\theta\,(\tan\theta + \cot\theta) = \csc\theta$

23 $\dfrac{\sec^2 u - 1}{\sec^2 u} = \sin^2 u$

24 $(\tan u + \cot u)(\cos u + \sin u) = \sec u + \csc u$

25 $(\cos^2 x - 1)(\tan^2 x + 1) = 1 - \sec^2 x$

26 $(\cot\alpha + \csc\alpha)(\tan\alpha - \sin\alpha) = \sec\alpha - \cos\alpha$

27 $\sec t\csc t + \cot t = \tan t + 2\cos t\csc t$

28 $\dfrac{1 + \cos^2 y}{\sin^2 y} = 2\csc^2 y - 1$

29 $\sec^2\theta\csc^2\theta = \sec^2\theta + \csc^2\theta$

30 $\dfrac{\sec x - \cos x}{\tan x} = \dfrac{\tan x}{\sec x}$

31 $\dfrac{1 + \cos t}{\sin t} + \dfrac{\sin t}{1 + \cos t} = 2\csc t$

32 $\tan^2\alpha - \sin^2\alpha = \tan^2\alpha\sin^2\alpha$

33 $\dfrac{1 + \tan^2 v}{\tan^2 v} = \csc^2 v$

34 $\dfrac{\sec\theta + \csc\theta}{\sec\theta - \csc\theta} = \dfrac{\sin\theta + \cos\theta}{\sin\theta - \cos\theta}$

35 $\dfrac{1 + \sin x}{1 - \sin x} - \dfrac{1 - \sin x}{1 + \sin x} = 4\tan x\sec x$

36 $\dfrac{1}{1 - \cos\gamma} + \dfrac{1}{1 + \cos\gamma} = 2\csc^2\gamma$

37 $\dfrac{1 + \csc\beta}{\sec\beta} - \cot\beta = \cos\beta$

38 $\dfrac{\cos x\cot x}{\cot x - \cos x} = \dfrac{\cot x + \cos x}{\cos x\cot x}$

39 $(\sec u - \tan u)(\csc u + 1) = \cot u$

40 $\dfrac{\cot\theta - \tan\theta}{\sin\theta + \cos\theta} = \csc\theta - \sec\theta$

41 $\dfrac{\cot\alpha - 1}{1 - \tan\alpha} = \cot\alpha$

42 $\dfrac{1 + \sec\beta}{\tan\beta + \sin\beta} = \csc\beta$

43 $\csc^4 t - \cot^4 t = \cot^2 t + \csc^2 t$

44 $\cos^4\theta + \sin^2\theta = \sin^4\theta + \cos^2\theta$

45 $\dfrac{\cos\beta}{1 - \sin\beta} = \sec\beta + \tan\beta$

46 $\dfrac{1}{\csc y - \cot y} = \csc y + \cot y$

47 $\dfrac{\tan^2 x}{\sec x + 1} = \dfrac{1 - \cos x}{\cos x}$

48 $\dfrac{\cot x}{\csc x + 1} = \dfrac{\csc x - 1}{\cot x}$

49 $\dfrac{\cot u - 1}{\cot u + 1} = \dfrac{1 - \tan u}{1 + \tan u}$

50 $\dfrac{1 + \sec x}{\sin x + \tan x} = \csc x$

51 $\sin^4 r - \cos^4 r = \sin^2 r - \cos^2 r$

52 $\sin^4 \theta + 2\sin^2 \theta \cos^2 \theta + \cos^4 \theta = 1$

53 $\tan^4 k - \sec^4 k = 1 - 2\sec^2 k$

54 $\sec^4 u - \sec^2 u = \tan^4 u + \tan^2 u$

55 $(\sec t + \tan t)^2 = \dfrac{1 + \sin t}{1 - \sin t}$

56 $\sec^2 \gamma + \tan^2 \gamma = (1 - \sin^4 \gamma)\sec^4 \gamma$

57 $(\sin^2 \theta + \cos^2 \theta)^3 = 1$

58 $\dfrac{\sin t}{1 - \cos t} = \csc t + \cot t$

59 $\dfrac{1 + \csc \beta}{\cot \beta + \cos \beta} = \sec \beta$

60 $\dfrac{\sin z \tan z}{\tan z - \sin z} = \dfrac{\tan z + \sin z}{\sin z \tan z}$

61 $\left(\dfrac{\sin^2 x}{\tan^4 x}\right)^3 \left(\dfrac{\csc^3 x}{\cot^6 x}\right)^2 = 1$

62 $\dfrac{\cos^3 x - \sin^3 x}{\cos x - \sin x} = 1 + \sin x \cos x$

63 $\dfrac{\sin \theta + \cos \theta}{\tan^2 \theta - 1} = \dfrac{\cos^2 \theta}{\sin \theta - \cos \theta}$

64 $(\csc t - \cot t)^4(\csc t + \cot t)^4 = 1$

65 $(a \cos t - b \sin t)^2 + (a \sin t + b \cos t)^2 = a^2 + b^2$

66 $\sin^6 v + \cos^6 v = 1 - 3\sin^2 v \cos^2 v$

67 $\dfrac{\sin \alpha \cos \beta + \cos \alpha \sin \beta}{\cos \alpha \cos \beta - \sin \alpha \sin \beta} = \dfrac{\tan \alpha + \tan \beta}{1 - \tan \alpha \tan \beta}$

68 $\dfrac{\tan u - \tan v}{1 + \tan u \tan v} = \dfrac{\cot v - \cot u}{1 + \cot u \cot v}$

69 $\sqrt{\dfrac{1 - \cos t}{1 + \cos t}} = \dfrac{1 - \cos t}{|\sin t|}$

70 $\sqrt{\dfrac{1 - \sin \theta}{1 + \sin \theta}} = \dfrac{|\cos \theta|}{1 + \sin \theta}$

71 $\dfrac{\sin \alpha}{1 + \cos \alpha} + \dfrac{1 + \cos \alpha}{\sin \alpha} = 2 \csc \alpha$

72 $\dfrac{\csc x}{1 + \csc x} - \dfrac{\csc x}{1 - \csc x} = 2\sec^2 x$

73 $\dfrac{1}{\tan \beta + \cot \beta} = \sin \beta \cos \beta$

74 $\dfrac{\cot y - \tan y}{\sin y \cos y} = \csc^2 y - \sec^2 y$

75 $\sec \theta + \csc \theta - \cos \theta - \sin \theta = \sin \theta \tan \theta + \cos \theta \cot \theta$

76 $\sin^3 t + \cos^3 t = (1 - \sin t \cos t)(\sin t + \cos t)$

77 $(1 - \tan^2 \phi)^2 = \sec^4 \phi - 4\tan^2 \phi$

78 $\cos^4 w + 1 - \sin^4 w = 2\cos^2 w$

79 $\dfrac{\tan x}{1 - \cot x} + \dfrac{\cot x}{1 - \tan x} = 1 + \sec x \csc x$

80 $\dfrac{\cos \gamma}{1 - \tan \gamma} + \dfrac{\sin \gamma}{1 - \cot \gamma} = \cos \gamma + \sin \gamma$

81 $\sin(-t)\sec(-t) = -\tan t$

82 $\dfrac{\cot(-v)}{\csc(-v)} = \cos v$

83 $\log 10^{\tan t} = \tan t$

84 $10^{\log|\sin t|} = |\sin t|$

85 $\ln \cot x = -\ln \tan x$

86 $\ln \sec \theta = -\ln \cos \theta$

87 $-\ln|\sec \theta - \tan \theta| = \ln|\sec \theta + \tan \theta|$

88 $\ln|\csc x - \cot x| = -\ln|\csc x + \cot x|$

Show that the equations in Exercises 89–100 are not identities. (*Hint:* Find one number in the domain of t or θ for which the equation is false.)

89 $\cos t = \sqrt{1 - \sin^2 t}$

90 $\sqrt{\sin^2 t + \cos^2 t} = \sin t + \cos t$

91 $\sqrt{\sin^2 t} = \sin t$

92 $\sec t = \sqrt{\tan^2 t + 1}$

93 $(\sin \theta + \cos \theta)^2 = \sin^2 \theta + \cos^2 \theta$

94 $\log(1/\sin t) = 1/(\log \sin t)$

95 $\cos(-t) = -\cos t$

96 $\sin(t + \pi) = \sin t$

97 $\cos(\sec t) = 1$

98 $\cot(\tan \theta) = 1$

99 $\sin^2 t - 4 \sin t - 5 = 0$

100 $3 \cos^2 \theta + \cos \theta - 2 = 0$

In Exercises 101–104 make the trigonometric substitution $x = a \sin \theta$ for $-\pi/2 < \theta < \pi/2$, and use fundamental identities to simplify the resulting expression. (Compare Example 6.)

101 $(a^2 - x^2)^{3/2}$

102 $\dfrac{\sqrt{a^2 - x^2}}{x}$

103 $\dfrac{x^2}{\sqrt{a^2 - x^2}}$

104 $\dfrac{1}{x\sqrt{a^2 - x^2}}$

In Exercises 105–108 make the trigonometric substitution $x = a \tan \theta$ for $-\pi/2 < \theta < \pi/2$ and simplify the resulting expression.

105 $\sqrt{a^2 + x^2}$

106 $\dfrac{1}{\sqrt{a^2 + x^2}}$

107 $\dfrac{1}{x^2 + a^2}$

108 $\dfrac{(x^2 + a^2)^{3/2}}{x}$

In Exercises 109–112 make the trigonometric substitution $x = a \sec \theta$ for $0 < \theta < \pi/2$ and simplify the resulting expression.

109 $\sqrt{x^2 - a^2}$

110 $\dfrac{1}{x^2\sqrt{x^2 - a^2}}$

111 $x^3\sqrt{x^2 - a^2}$

112 $\dfrac{\sqrt{x^2 - a^2}}{x^2}$

6.2 Trigonometric Equations

A **trigonometric equation** is an equation that contains trigonometric expressions. Solutions of trigonometric equations may be expressed in terms of either real numbers or angles. If a trigonometric equation is not an identity, then techniques similar to those used for algebraic equations may be employed to find solutions. The main difference is that we usually solve for $\sin x$, $\cos \theta$, and so on, and then find x and θ, as illustrated in the following examples.

EXAMPLE 1 Solve the equation $\sin \theta \tan \theta = \sin \theta$.

SOLUTION Each of the following is equivalent to the given equation:

$$\sin \theta \tan \theta - \sin \theta = 0$$

$$\sin \theta (\tan \theta - 1) = 0$$

To find the solutions we set each factor on the left equal to zero, obtaining

$$\sin \theta = 0 \quad \text{or} \quad \tan \theta = 1.$$

The solutions of the equation $\sin \theta = 0$ are $0, \pm\pi, \pm 2\pi, \ldots$; that is, $\theta = n\pi$ for every integer n.

Since the tangent function has period π, it is sufficient to find the solutions of the equation $\tan\theta = 1$ that are in the interval $[0, \pi)$, for once these are known, we can obtain all the others by adding multiples of π. The only solution of $\tan\theta = 1$ in $[0, \pi)$ is $\pi/4$, and hence every solution has the form

$$\theta = \frac{\pi}{4} + n\pi$$

for some integer n.

It follows that the solutions of the given equation consist of all numbers of the form

$$n\pi \quad \text{or} \quad \frac{\pi}{4} + n\pi$$

where n is an integer. Some particular solutions are 0, $\pm\pi$, $\pm 2\pi$, $\pm 3\pi$, $\pi/4$, $5\pi/4$, $-3\pi/4$, and $-7\pi/4$. ■

Note that in Example 1 it would have been incorrect to begin by dividing both sides by $\sin\theta$, for this manipulation would lose the solutions of $\sin\theta = 0$.

EXAMPLE 2 Solve the equation $2\sin^2 t - \cos t - 1 = 0$.

SOLUTION We first change the equation to an equation that involves only $\cos t$, and then factor as follows:

$$\begin{aligned} 2(1 - \cos^2 t) - \cos t - 1 &= 0 \\ 2 - 2\cos^2 t - \cos t - 1 &= 0 \\ -2\cos^2 t - \cos t + 1 &= 0 \\ 2\cos^2 t + \cos t - 1 &= 0 \\ (2\cos t - 1)(\cos t + 1) &= 0 \end{aligned}$$

The solutions of the last equation are the solutions of

$$2\cos t - 1 = 0 \quad \text{or} \quad \cos t + 1 = 0$$

or equivalently, $\quad \cos t = \frac{1}{2} \quad \text{or} \quad \cos t = -1.$

It is sufficient to find the solutions that are in the interval $[0, 2\pi)$, for once these are known, all solutions may be obtained by adding multiples of 2π.

If $\cos t = \frac{1}{2}$, then the reference number (or reference angle) is $\pi/3$ (or $60°$). Since $\cos t$ is positive, the angle of radian measure t is in quadrant I or quadrant IV. Hence, in the interval $[0, 2\pi)$,

$$t = \frac{\pi}{3} \quad \text{or} \quad t = 2\pi - \frac{\pi}{3} = \frac{5\pi}{3}.$$

If $\cos t = -1$, then $t = \pi$.

Thus, the solutions of the given equation are of the form

$$\frac{\pi}{3} + 2\pi n, \qquad \frac{5\pi}{3} + 2\pi n, \qquad \pi + 2\pi n$$

where n is an integer. If we wish to express the solutions in terms of degrees, we may write

$$60° + 360°n, \qquad 300° + 360°n, \qquad 180° + 360°n.$$ ■

EXAMPLE 3 Find the solutions of $4 \sin^2 x \tan x - \tan x = 0$ that are in the interval $[0, 2\pi)$.

SOLUTION Factoring the left side of the equation, we obtain

$$\tan x\,(4 \sin^2 x - 1) = 0$$

Thus

$$\tan x = 0 \quad \text{or} \quad \sin^2 x = \tfrac{1}{4}.$$

This implies that

$$\tan x = 0, \quad \sin x = \tfrac{1}{2}, \quad \text{or} \quad \sin x = -\tfrac{1}{2}.$$

The equation $\tan x = 0$ has solutions 0 and π in the interval $[0, 2\pi)$. The equation $\sin x = \frac{1}{2}$ has solutions $\pi/6$ and $5\pi/6$. The equation $\sin x = -\frac{1}{2}$ leads to the numbers between π and 2π that have reference number $\pi/6$. These are

$$\pi + \frac{\pi}{6} = \frac{7\pi}{6} \quad \text{and} \quad 2\pi - \frac{\pi}{6} = \frac{11\pi}{6}.$$

Hence, the solutions of the given equation in the interval $[0, 2\pi)$ are

$$0, \quad \pi, \quad \frac{\pi}{6}, \quad \frac{5\pi}{6}, \quad \frac{7\pi}{6}, \quad \frac{11\pi}{6}.$$ ■

EXAMPLE 4 Find the solutions of the equation $\csc^4 2u - 4 = 0$.

SOLUTION Factoring the left side, we obtain

$$(\csc^2 2u - 2)(\csc^2 2u + 2) = 0.$$

It follows that

$$\csc^2 2u = 2 \quad \text{or} \quad \csc^2 2u = -2.$$

The equation $\csc^2 2u = -2$ has no real solutions. The solutions of $\csc^2 2u = 2$ consist of the solutions of

$$\csc 2u = \sqrt{2} \quad \text{or} \quad \csc 2u = -\sqrt{2}.$$

If $\csc 2u = \sqrt{2}$, then

$$2u = \frac{\pi}{4} + 2\pi n \quad \text{or} \quad 2u = \frac{3\pi}{4} + 2\pi n$$

for some integer n. Dividing both sides of the last two equations by 2 gives us

$$u = \frac{\pi}{8} + \pi n \quad \text{or} \quad u = \frac{3\pi}{8} + \pi n.$$

Similarly, from $\csc 2u = -\sqrt{2}$ we obtain

$$2u = \frac{5\pi}{4} + 2\pi n \quad \text{or} \quad 2u = \frac{7\pi}{4} + 2\pi n$$

for some integer n. Dividing both sides of the last two equations by 2 gives us

$$u = \frac{5\pi}{8} + \pi n \quad \text{or} \quad u = \frac{7\pi}{8} + \pi n.$$

Collecting the preceding information, we see that all solutions can be written in the form

$$u = \frac{\pi}{8} + \frac{\pi}{4} n \quad \text{where } n \text{ is an integer.}$$ ■

The next example illustrates the use of a calculator in solving a trigonometric equation.

EXAMPLE 5 Approximate, to the nearest degree, the solutions of the equation

$$5 \sin \theta \tan \theta - 10 \tan \theta + 3 \sin \theta - 6 = 0$$

in the degree interval $[0°, 360°)$.

SOLUTION The equation may be factored by grouping terms as follows:

$$5 \tan \theta (\sin \theta - 2) + 3 (\sin \theta - 2) = 0$$
$$(5 \tan \theta + 3)(\sin \theta - 2) = 0$$

The equation $\sin \theta = 2$ has no solutions since $\sin \theta \leq 1$ for all θ. Thus the solutions of the given equation are the same as those of

$$\tan \theta = -\tfrac{3}{5} = -0.6000.$$

Let us begin by approximating the reference angle, that is, the acute angle θ' such that $\tan \theta' = \frac{3}{5} = 0.6$. Using a calculator in degree mode, we obtain the following:

Enter:	0.6	(the value of $\tan \theta'$)
Press [INV] [tan]:	30.963757	(the degree measure of θ')

Hence $\theta' \approx 31°$. Since θ is in quadrant II or quadrant IV, this implies that

$$\theta \approx 180° - 31° = 149° \quad \text{or} \quad \theta \approx 360° - 31° = 329°.$$ ■

EXAMPLE 6 In Boston, MA, the number of hours of daylight $D(t)$ at a particular time of the year may be approximated by

$$D(t) = 3 \sin \frac{2\pi}{365}(t - 79) + 12$$

where t is in days and $t = 0$ corresponds to January 1. How many days of the year have more than 10.5 hours of daylight?

SOLUTION The graph of D was sketched in Example 3 of Section 5.7 (see Figure 5.53). If we can find two numbers a and b such that $D(a) = 10.5$, $D(b) = 10.5$, and $0 < a < b < 365$, then there will be more than 10.5 hours of daylight in the tth day of the year if $a < t < b$. Let us solve the equation

$D(t) = 10.5$; that is,

$$3 \sin \frac{2\pi}{365}(t - 79) + 12 = 10.5.$$

This is equivalent to

$$3 \sin \frac{2\pi}{365}(t - 79) = -1.5$$

or

$$\sin \frac{2\pi}{365}(t - 79) = -0.5 = -\frac{1}{2}.$$

If $\sin \theta = -\frac{1}{2}$, then the reference angle is $\pi/6$ and the angle θ is either in quadrant III or IV. Thus we can find the numbers a and b by solving the equations

$$\frac{2\pi}{365}(t - 79) = \frac{7\pi}{6} \quad \text{and} \quad \frac{2\pi}{365}(t - 79) = \frac{11\pi}{6}.$$

From the first of these equations we obtain

$$t - 79 = \frac{7\pi}{6} \cdot \frac{365}{2\pi} = \frac{2555}{12} \approx 213$$

or

$$t \approx 213 + 79 = 292.$$

In similar fashion, the second equation gives us $t \approx 414$. Since the period of the function D is 365 days (see Figure 5.53), we also obtain

$$t \approx 414 - 365 = 49.$$

It follows that there will be at least 10.5 hours of daylight from $t = 49$ to $t = 292$, that is, for 243 days of the year. ■

Exercises 6.2

In Exercises 1–16 find all solutions of the equations.

1 $2 \cos t + 1 = 0$

2 $\cot \theta + 1 = 0$

3 $\tan^2 x = 1$

4 $4 \cos \theta - 2 = 0$

5 $(\cos \theta - 1)(\sin \theta + 1) = 0$

6 $2 \cos x = \sqrt{3}$

7 $\sec^2 \alpha - 4 = 0$

8 $3 - \tan^2 \beta = 0$

9 $\sqrt{3} + 2 \sin \beta = 0$

10 $4 \sin^2 x - 3 = 0$

11 $\cot^2 x - 3 = 0$

12 $(\sin t - 1) \cos t = 0$

13 $(2 \sin \theta + 1)(2 \cos \theta + 3) = 0$

14 $(2 \sin u - 1)(\cos u - \sqrt{2}) = 0$

15 $\sin 2x (\csc 2x - 2) = 0$

16 $\tan \alpha + \tan^2 \alpha = 0$

In Exercises 17–36 find the solutions of the equations in the interval $[0, 2\pi)$, and also find the degree measure of each solution.

17 $2 - 8 \cos^2 t = 0$

18 $\cot^2 \theta - \cot \theta = 0$

19 $2 \sin^2 u = 1 - \sin u$

20 $2 \cos^2 t + 3 \cos t + 1 = 0$

21 $\tan^2 x \sin x = \sin x$

22 $\sec \beta \csc \beta = 2 \csc \beta$

23 $2 \cos^2 \gamma + \cos \gamma = 0$

24 $\sin x - \cos x = 0$

25 $\sin^2 \theta + \sin \theta - 6 = 0$

26 $2 \sin^2 u + \sin u - 6 = 0$

27 $1 - \sin t = \sqrt{3} \cos t$

28 $\cos \theta - \sin \theta = 1$

29 $\cos \alpha + \sin \alpha = 1$

30 $2 \tan t - \sec^2 t = 0$

31 $\tan \theta + \sec \theta = 1$

32 $\cot \alpha + \tan \alpha = \csc \alpha \sec \alpha$

33 $2 \sin^3 x + \sin^2 x - 2 \sin x - 1 = 0$

34 $\sec^5 \theta = 4 \sec \theta$

35 $2 \tan t \csc t + 2 \csc t + \tan t + 1 = 0$

36 $2 \sin v \csc v - \csc v = 4 \sin v - 2$

In Exercises 37–40 use a calculator to approximate, to the nearest multiple of ten minutes, the solutions of the equations in the interval $[0°, 360°)$.

37 $\sin^2 t - 4 \sin t + 1 = 0$

38 $\tan^2 \theta + 3 \tan \theta + 2 = 0$

39 $12 \sin^2 u - 5 \sin u - 2 = 0$

40 $5 \cos^2 \alpha + 3 \cos \alpha - 2 = 0$

41 A tidal wave of height 50 feet and period 30 minutes is approaching a sea wall that is 20 feet above sea level. From a particular point on shore, the distance y from sea level to the top of the wave is given by $y = 25 \cos (\pi/15)t$ where t is in minutes. For approximately how many minutes of each 30-minute period is the top of the wave above the sea wall?

42 The expected low temperature T (in °F) in Fairbanks, Alaska, can be approximated by

$$T = 36 \sin \frac{2\pi}{365}(t - 101) + 14$$

where t is in days and $t = 0$ corresponds to January 1. (See Exercise 32 of Section 5.7.) For how many days during the year is the low temperature expected to be below -4°F?

43 On a clear day with D hours of daylight, it is known from empirical studies that the intensity of sunlight I (in calories/cm^2) is well described by $I = I_M \sin^3 (\pi t/D)$ for $0 \le t \le D$, where $t = 0$ corresponds to sunrise and I_M is the maximum intensity. At approximately what times of the day is $I = I_M/2$?

44 Refer to Exercise 43. On cloudy days, the sun intensity I is better represented by $I = I_M \sin^2 (\pi t/D)$. At what times of the day is $I = I_M/2$?

45 Refer to Exercises 43 and 44. A skin specialist recommends protection from the sun when the intensity I exceeds 75% of the maximum intensity. If $D = 12$ hours, approximate the number of hours that protection is required on (a) a clear day; (b) a cloudy day.

46 In the study of frost penetration problems in highway engineering, the temperature T at time t hours and depth x feet is represented by

$$T = T_0 e^{-\lambda x} \sin (\omega t - \lambda x)$$

where T_0, ω, and λ are constants and the period of $\sin (\omega t - \lambda x)$ is 24 hours.

(a) Find a formula for the temperature at the surface. At what times is the surface temperature at a minimum?

(b) For $\lambda = 2.5$, find the times when the temperature is a minimum at a depth of 1 foot.

(c) At what depth is the temperature one-half of the surface temperature?

(*Note:* The expression for T is a solution to the one-dimensional heat equation in physics. For more information, see *The Frost Penetration Problem in Highway Engineering*, A. R. Junikis, Rutgers University Press (1955), pp. 43–61.)

47 Shown in the figure is a computer-generated graph of the equation $y = \frac{1}{2}x + \sin x$ for $-2\pi \le t \le 2\pi$. Using calculus it can be shown that the x-coordinates of the peaks and valleys on the graph are solutions of the equation $\frac{1}{2} + \cos x = 0$. Determine the coordinates of the points A, B, C, and D.

FIGURE FOR EXERCISE 47

48 *Damped oscillations* are oscillations of decreasing magnitude that occur when frictional forces are considered. Shown in the figure is the graph of the damped oscillation given by the equation

$$y = e^{-x/2} \sin 2x.$$

Using calculus it can be shown that the peaks and valleys on the graph are solutions of $2 \cos 2x - \frac{1}{2} \sin 2x = 0$. Approximate the x-coordinates of these peaks and valleys for $x > 0$.

FIGURE FOR EXERCISE 48

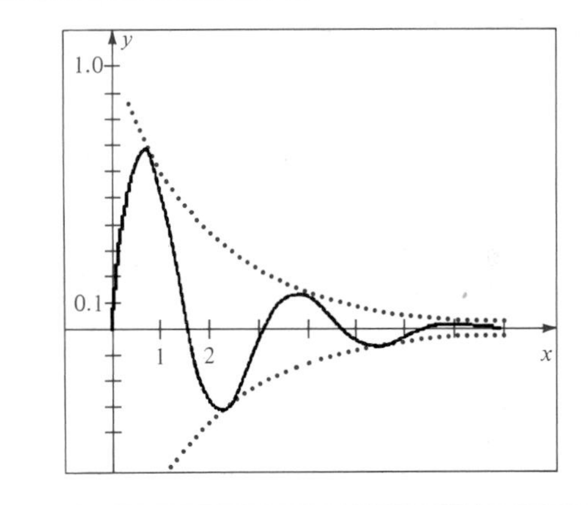

6.3 The Addition and Subtraction Formulas

In this section we derive formulas involving trigonometric functions of $u + v$ or $u - v$, where u and v represent any real numbers or angles. These formulas are known as *addition* or *subtraction formulas*, respectively. The first formula we shall discuss may be stated as follows.

SUBTRACTION FORMULA FOR COSINE

$$\cos(u - v) = \cos u \cos v + \sin u \sin v$$

To prove the formula, let u and v be any real numbers, and consider the angles of radian measure u and v, respectively. Let $w = u - v$. Figure 6.1(i) illustrates one possibility with the angles in standard position on a rectangular coordinate system. For convenience we have assumed that both u and v are positive and that $0 \leq u - v < v$.

FIGURE 6.1

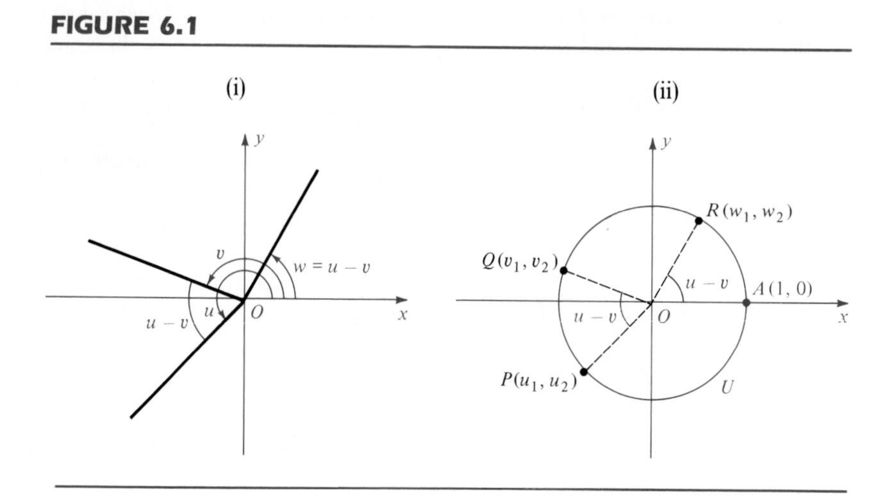

As in (ii) of the figure, let $P(u_1, u_2)$, $Q(v_1, v_2)$, and $R(w_1, w_2)$ be the points on the terminal sides of the indicated angles that are each a distance 1 from the origin. In this case P, Q, and R are on the unit circle U with center at the origin. From the definition of trigonometric functions of any angle (with $c = 1$),

$$(*) \quad \begin{array}{lll} \cos u = u_1, & \cos v = v_1, & \cos(u - v) = w_1 \\ \sin u = u_2, & \sin v = v_2, & \sin(u - v) = w_2. \end{array}$$

The symbol $(*)$ has been used for later reference to these formulas.

We next observe that the distance between $A(1, 0)$ and R must equal the distance between P and Q, because angles AOR and POQ have the same measure $u - v$. Using the distance formula,

$$\sqrt{(w_1 - 1)^2 + (w_2 - 0)^2} = \sqrt{(u_1 - v_1)^2 + (u_2 - v_2)^2}.$$

Squaring both sides and expanding the terms under the radicals gives us

$$w_1^2 - 2w_1 + 1 + w_2^2 = u_1^2 - 2u_1v_1 + v_1^2 + u_2^2 - 2u_2v_2 + v_2^2.$$

Since the points (u_1, u_2), (v_1, v_2), and (w_1, w_2) are on the unit circle U, and since an equation for U is $x^2 + y^2 = 1$, we may substitute 1 for each of

$u_1^2 + u_2^2$, $v_1^2 + v_2^2$ and $w_1^2 + w_2^2$. Doing this and simplifying, we obtain

$$2 - 2w_1 = 2 - 2u_1v_1 - 2u_2v_2,$$

which reduces to

$$w_1 = u_1v_1 + u_2v_2.$$

Substituting from the formulas stated in (*) gives us

$$\cos(u - v) = \cos u \cos v + \sin u \sin v,$$

which is what we wished to prove. It is possible to extend our discussion to all values of u and v, not only those pictured in Figure 6.1.

The next example demonstrates the use of the subtraction formula in finding the exact value of $\cos 15°$. Of course, if only an approximation is desired we could use a calculator or Table 4.

EXAMPLE 1 Use the fact that $15° = 60° - 45°$ to find the exact value of $\cos 15°$.

SOLUTION Using the subtraction formula with $u = 60°$ and $v = 45°$, we obtain

$$\begin{aligned} \cos 15° &= \cos(60° - 45°) \\ &= \cos 60° \cos 45° + \sin 60° \sin 45° \\ &= \frac{1}{2}\frac{\sqrt{2}}{2} + \frac{\sqrt{3}}{2}\frac{\sqrt{2}}{2} \\ &= \frac{\sqrt{2} + \sqrt{6}}{4}. \end{aligned}$$

■

It is easy to obtain a formula for $\cos(u + v)$. We first write $u + v = u - (-v)$ and then employ the subtraction formula. Thus,

$$\begin{aligned} \cos(u + v) &= \cos[u - (-v)] \\ &= \cos u \cos(-v) + \sin u \sin(-v). \end{aligned}$$

Using formulas for negatives (see page 223, $\cos(-v) = \cos v$ and $\sin(-v) = -\sin v$ for all v, and hence we can state the following.

ADDITION FORMULA FOR COSINE

$$\cos(u + v) = \cos u \cos v - \sin u \sin v$$

EXAMPLE 2 Find $\cos(7\pi/12)$ by using $7\pi/12 = (\pi/3) + (\pi/4)$.

SOLUTION Applying the formula for $\cos(u + v)$,

$$\begin{aligned}\cos\frac{7\pi}{12} &= \cos\left(\frac{\pi}{3} + \frac{\pi}{4}\right)\\ &= \cos\frac{\pi}{3}\cos\frac{\pi}{4} - \sin\frac{\pi}{3}\sin\frac{\pi}{4}\\ &= \frac{1}{2}\frac{\sqrt{2}}{2} - \frac{\sqrt{3}}{2}\frac{\sqrt{2}}{2}\\ &= \frac{\sqrt{2}-\sqrt{6}}{4}.\end{aligned}$$

■

It is customary to refer to the sine and cosine functions as **cofunctions** of one another. Similarly, the tangent and cotangent functions are co-functions, as are the secant and cosecant. The subtraction formula for cosine can be used to obtain the following three facts about cofunctions, where u is either a real number or the radian measure of an angle.

COFUNCTION FORMULAS

$$\cos\left(\frac{\pi}{2} - u\right) = \sin u, \qquad \sin\left(\frac{\pi}{2} - u\right) = \cos u,$$

$$\tan\left(\frac{\pi}{2} - u\right) = \cot u$$

The first formula may be proved as follows:

$$\begin{aligned}\cos\left(\frac{\pi}{2} - u\right) &= \cos\frac{\pi}{2}\cos u + \sin\frac{\pi}{2}\sin u\\ &= (0)\cos u + (1)\sin u = \sin u.\end{aligned}$$

To obtain the second formula we substitute $(\pi/2) - v$ for u in the first, obtaining

$$\cos\left[\frac{\pi}{2} - \left(\frac{\pi}{2} - v\right)\right] = \sin\left(\frac{\pi}{2} - v\right)$$

or

$$\cos v = \sin\left(\frac{\pi}{2} - v\right).$$

Since the symbol v is arbitrary, this is equivalent to $\cos u = \sin[(\pi/2) - u]$. Finally,

$$\tan\left(\frac{\pi}{2} - u\right) = \frac{\sin\left(\frac{\pi}{2} - u\right)}{\cos\left(\frac{\pi}{2} - u\right)} = \frac{\cos u}{\sin u} = \cot u.$$

If θ denotes the degree measure of an angle, then

$$\cos(90^\circ - \theta) = \sin\theta, \qquad \sin(90^\circ - \theta) = \cos\theta, \qquad \tan(90^\circ - \theta) = \cot\theta.$$

If θ is acute, then θ and $90^\circ - \theta$ are complementary, since their sum is 90°. The last three formulas constitute a partial description of the fact that *any function value of θ equals the cofunction of the complementary angle $90^\circ - \theta$.*

The following identities may now be established.

ADDITION AND SUBTRACTION FORMULAS FOR SINE AND TANGENT

$$\sin(u + v) = \sin u \cos v + \cos u \sin v$$

$$\sin(u - v) = \sin u \cos v - \cos u \sin v$$

$$\tan(u + v) = \frac{\tan u + \tan v}{1 - \tan u \tan v}$$

$$\tan(u - v) = \frac{\tan u - \tan v}{1 + \tan u \tan v}$$

PROOF We shall prove the first and third and leave the proofs of the remaining two as exercises. Supply reasons for each of the following steps:

$$\begin{aligned}
\sin(u + v) &= \cos\left[\frac{\pi}{2} - (u + v)\right] \\
&= \cos\left[\left(\frac{\pi}{2} - u\right) - v\right] \\
&= \cos\left(\frac{\pi}{2} - u\right)\cos v + \sin\left(\frac{\pi}{2} - u\right)\sin v \\
&= \sin u \cos v + \cos u \sin v.
\end{aligned}$$

To verify the formula for $\tan(u+v)$ we begin as follows:

$$\tan(u+v) = \frac{\sin(u+v)}{\cos(u+v)}$$

$$= \frac{\sin u \cos v + \cos u \sin v}{\cos u \cos v - \sin u \sin v}.$$

If $\cos u \cos v \neq 0$, then we may divide numerator and denominator by $\cos u \cos v$, obtaining

$$\tan(u+v) = \frac{\left(\dfrac{\sin u}{\cos u}\right)\left(\dfrac{\cos v}{\cos v}\right) + \left(\dfrac{\cos u}{\cos u}\right)\left(\dfrac{\sin v}{\cos v}\right)}{\left(\dfrac{\cos u}{\cos u}\right)\left(\dfrac{\cos v}{\cos v}\right) - \left(\dfrac{\sin u}{\cos u}\right)\left(\dfrac{\sin v}{\cos v}\right)}$$

$$= \frac{\tan u + \tan v}{1 - \tan u \tan v}.$$

If $\cos u \cos v = 0$, then either $\cos u = 0$ or $\cos v = 0$. In this case either $\tan u$ or $\tan v$ is undefined and the formula is invalid. □

EXAMPLE 3 If $\sin \alpha = \frac{4}{5}$ for some angle α in quadrant I, and $\cos \beta = -\frac{12}{13}$ for some β in quadrant II, find the exact values of $\sin(\alpha+\beta)$, $\tan(\alpha+\beta)$, and the quadrant containing $\alpha+\beta$.

FIGURE 6.2

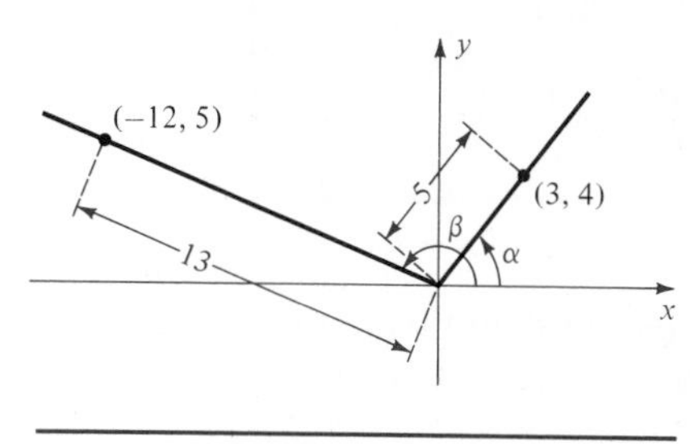

SOLUTION The angles α and β are illustrated in Figure 6.2. There is no loss of generality in picturing α and β as positive angles between 0 and 2π as we have done in the figure. Since $\sin \alpha = \frac{4}{5}$, we may choose the point $(3, 4)$ on the terminal side of α. Similarly, since $\cos \beta = -\frac{12}{13}$, the point $(-12, 5)$ is on the terminal side of β. Referring to Figure 6.2 and using the definition of trigonometric functions of any angle (page 212),

$$\cos \alpha = \tfrac{3}{5}, \quad \tan \alpha = \tfrac{4}{3}, \quad \sin \beta = \tfrac{5}{13}, \quad \text{and} \quad \tan \beta = -\tfrac{5}{12}.$$

Using addition formulas:

$$\sin(\alpha+\beta) = \sin \alpha \cos \beta + \cos \alpha \sin \beta$$

$$= \left(\frac{4}{5}\right)\left(-\frac{12}{13}\right) + \left(\frac{3}{5}\right)\left(\frac{5}{13}\right) = -\frac{33}{65}.$$

$$\tan(\alpha+\beta) = \frac{\tan \alpha + \tan \beta}{1 - \tan \alpha \tan \beta}$$

$$= \frac{\frac{4}{3} + (-\frac{5}{12})}{1 - (\frac{4}{3})(-\frac{5}{12})} = \frac{33}{56}.$$

Since $\sin(\alpha + \beta)$ is negative and $\tan(\alpha + \beta)$ is positive, it follows that $\alpha + \beta$ lies in quadrant III. ■

The type of simplification illustrated in the next example is important in calculus.

EXAMPLE 4 If $f(x) = \sin x$ and $h \neq 0$, prove that

$$\frac{f(x+h) - f(x)}{h} = \sin x \left(\frac{\cos h - 1}{h}\right) + \cos x \left(\frac{\sin h}{h}\right).$$

SOLUTION Using the definition of f and the addition formula for sine,

$$\begin{aligned}
\frac{f(x+h) - f(x)}{h} &= \frac{\sin(x+h) - \sin x}{h} \\
&= \frac{\sin x \cos h + \cos x \sin h - \sin x}{h} \\
&= \frac{\sin x(\cos h - 1) + \cos x \sin h}{h} \\
&= \sin x \left(\frac{\cos h - 1}{h}\right) + \cos x \left(\frac{\sin h}{h}\right).
\end{aligned}$$

■

EXAMPLE 5 Prove that for every x,

$$a \cos Bx + b \sin Bx = A \cos(Bx - C)$$

where $A = \sqrt{a^2 + b^2}$ and $\tan C = b/a$.

SOLUTION Using the formula for $\cos(u - v)$ with $u = Bx$ and $v = C$,

$$A \cos(Bx - C) = A(\cos Bx \cos C + \sin Bx \sin C).$$

We shall complete the proof by determining conditions on a and b such that

$$a \cos Bx + b \sin Bx = A \cos Bx \cos C + A \sin Bx \sin C.$$

The last formula is true for every x if and only if

$$a = A \cos C \quad \text{and} \quad b = A \sin C$$

(to verify this let $x = 0$). Consequently,

$$a^2 + b^2 = A^2 \cos^2 C + A^2 \sin^2 C$$
$$= A^2(\cos^2 C + \sin^2 C) = A^2$$

and we may choose $$A = \sqrt{a^2 + b^2}.$$

Finally, note that

$$\frac{b}{a} = \frac{A \sin C}{A \cos C} = \frac{\sin C}{\cos C} = \tan C.$$

This establishes the identity. ■

EXAMPLE 6 The graph of $f(x) = \cos x + \sin x$ was sketched in Example 1 of Section 5.7 by adding y-coordinates (see Figure 5.50). Use the formula proved in Example 5 to find the amplitude, period, and phase shift of f.

SOLUTION Letting $a = 1$, $b = 1$, and $B = 1$ in Example 5, we obtain

$$A = \sqrt{a^2 + b^2} = \sqrt{1 + 1} = \sqrt{2} \quad \text{and} \quad \tan C = b/a = 1/1 = 1.$$

Since $\tan C = 1$, we may choose $C = \pi/4$. Substitution in the formula gives us

$$\cos x + \sin x = \sqrt{2} \cos\left(x - \frac{\pi}{4}\right).$$

It follows from our work in Section 5.6 that the amplitude is $\sqrt{2}$, the period is 2π, and the phase shift is $\pi/4$. These facts may be checked by referring to Figure 5.50. ■

Exercises 6.3

Write the expressions in Exercises 1–4 in terms of cofunctions of complementary angles.

1 (a) $\sin 46°37'$
(b) $\cos 73°12'$
(c) $\tan(\pi/6)$
(d) $\sec 17.28°$

2 (a) $\tan 24°12'$
(b) $\sin 89°41'$
(c) $\cos(\pi/3)$
(d) $\cot 61.87°$

3 (a) $\cos(7\pi/20)$
(b) $\sin \frac{1}{4}$
(c) $\tan 1$
(d) $\csc(0.53)$

4 (a) $\sin(\pi/12)$
(b) $\cos(0.64)$
(c) $\tan \sqrt{2}$
(d) $\sec(1.2)$

Find the exact values in Exercises 5–10.

5 (a) $\cos(\pi/4) + \cos(\pi/6)$

(b) $\cos(5\pi/12)$

6 (a) $\sin(2\pi/3) + \sin(\pi/4)$

(b) $\sin(11\pi/12)$

7 (a) $\tan 60^\circ + \tan 225^\circ$

(b) $\tan 285^\circ$

8 (a) $\cos 135^\circ - \cos 60^\circ$

(b) $\cos 75^\circ$

9 (a) $\sin(3\pi/4) - \sin(\pi/6)$

(b) $\sin(7\pi/12)$

10 (a) $\tan(3\pi/4) - \tan(\pi/6)$

(b) $\tan(7\pi/12)$

Write the expressions in Exercises 11–16 in terms of one function of one angle.

11 $\cos 48^\circ \cos 23^\circ + \sin 48^\circ \sin 23^\circ$

12 $\cos 13^\circ \cos 50^\circ - \sin 13^\circ \sin 50^\circ$

13 $\cos 10^\circ \sin 5^\circ - \sin 10^\circ \cos 5^\circ$

14 $\sin 57^\circ \cos 4^\circ + \cos 57^\circ \sin 4^\circ$

15 $\cos 3 \sin(-2) - \cos 2 \sin 3$

16 $\sin(-5) \cos 2 + \cos 5 \sin(-2)$

17 If α and β are acute angles such that $\cos\alpha = \frac{4}{5}$ and $\tan\beta = \frac{8}{15}$, find $\cos(\alpha+\beta)$, $\sin(\alpha+\beta)$, and the quadrant containing $\alpha+\beta$.

18 If $\sin\alpha = -\frac{4}{5}$ and $\sec\beta = \frac{5}{3}$, where α is a third-quadrant angle and β is a first-quadrant angle, find $\sin(\alpha+\beta)$, $\tan(\alpha+\beta)$, and the quadrant containing $\alpha+\beta$.

19 If $\tan\alpha = -\frac{7}{24}$ and $\cot\beta = \frac{3}{4}$, where α is in the second quadrant and β is in the third quadrant, find $\sin(\alpha+\beta)$, $\cos(\alpha+\beta)$, $\tan(\alpha+\beta)$, $\sin(\alpha-\beta)$, $\cos(\alpha-\beta)$, and $\tan(\alpha-\beta)$.

20 For the points $P(u)$ and $P(v)$ in quadrant III with $\cos u = -\frac{2}{5}$ and $\cos v = -\frac{3}{5}$, find $\sin(u-v)$, $\cos(u-v)$, and the quadrant containing $P(u-v)$.

Verify the identities in Exercises 21–40.

21 $\sin\left(x + \frac{\pi}{2}\right) = \cos x$

22 $\cos\left(x + \frac{\pi}{2}\right) = -\sin x$

23 $\cos\left(\theta + \frac{3\pi}{2}\right) = \sin\theta$

24 $\sin\left(\alpha - \frac{3\pi}{2}\right) = \cos\alpha$

25 $\sin\left(\theta + \frac{\pi}{4}\right) = \left(\frac{\sqrt{2}}{2}\right)(\sin\theta + \cos\theta)$

26 $\cos\left(\theta + \frac{\pi}{4}\right) = \left(\frac{\sqrt{2}}{2}\right)(\cos\theta - \sin\theta)$

27 $\tan\left(u + \frac{\pi}{4}\right) = \frac{1 + \tan u}{1 - \tan u}$

28 $\tan\left(x - \frac{\pi}{4}\right) = \frac{\tan x - 1}{\tan x + 1}$

29 $\tan\left(u + \frac{\pi}{2}\right) = -\cot u$

30 $\cot\left(t - \frac{\pi}{3}\right) = \frac{\sqrt{3}\tan t + 1}{\tan t - \sqrt{3}}$

31 $\sin(u+v) \cdot \sin(u-v) = \sin^2 u - \sin^2 v$

32 $\cos(u+v) \cdot \cos(u-v) = \cos^2 u - \sin^2 v$

33 $\cos(u+v) + \cos(u-v) = 2\cos u \cos v$

34 $\sin(u+v) + \sin(u-v) = 2\sin u \cos v$

35 $\sin 2u = 2\sin u \cos u$ (*Hint:* $2u = u + u$)

36 $\cos 2u = \cos^2 u - \sin^2 u$

37 $\frac{\sin(u+v)}{\cos(u-v)} = \frac{\tan u + \tan v}{1 + \tan u \tan v}$

38 $\frac{\cos(u+v)}{\cos(u-v)} = \frac{1 - \tan u \tan v}{1 + \tan u \tan v}$

39 $\frac{\sin(u+v)}{\sin(u-v)} = \frac{\tan u + \tan v}{\tan u - \tan v}$

40 $\tan u + \tan v = \dfrac{\sin(u+v)}{\cos u \cos v}$

41 Express $\sin(u + v + w)$ in terms of functions of u, v, and w. (*Hint:* Write $\sin(u + v + w) = \sin[(u + v) + w]$ and use an addition formula.)

42 Express $\tan(u + v + w)$ in terms of functions of u, v, and w.

43 Derive the formula $\cot(u + v) = \dfrac{\cot u \cot v - 1}{\cot u + \cot v}$.

44 If α and β are complementary angles, prove that

$$\sin^2 \alpha + \sin^2 \beta = 1.$$

45 Derive the subtraction formulas for the sine and tangent functions.

46 Prove each of the following:

(a) $\sec\left(\dfrac{\pi}{2} - u\right) = \csc u$

(b) $\csc\left(\dfrac{\pi}{2} - u\right) = \sec u$

(c) $\cot\left(\dfrac{\pi}{2} - u\right) = \tan u$

47 If $f(x) = \cos x$, prove that

$$\frac{f(x+h) - f(x)}{h} = \cos x\left(\frac{\cos h - 1}{h}\right) - \sin x\left(\frac{\sin h}{h}\right).$$

48 If $f(x) = \tan x$, prove that

$$\frac{f(x+h) - f(x)}{h} = \sec^2 x\left(\frac{\sin h}{h}\right)\frac{1}{\cos h - \sin h \tan x}.$$

In Exercises 49 and 50 find the solutions of the equation that are in the interval $[0, 2\pi)$, and find the degree measure of each solution.

49 $\sin 4t \cos t = \sin t \cos 4t$

50 $\cos 5t \cos 3t = 2^{-1} + \sin(-5t) \sin 3t$

In Exercises 51–54 use the formula from Example 5 to express f in terms of the cosine function, and determine the amplitude, period, and phase shift.

51 $f(x) = \sqrt{3} \cos 2x + \sin 2x$

52 $f(x) = \cos 4x + \sqrt{3} \sin 4x$

53 $f(x) = 2 \cos 3x - 2 \sin 3x$

54 $f(x) = -5 \cos 10x + 5 \sin 10x$

55 If a mass that is attached to a spring is raised y_0 feet and given an initial vertical velocity of v_0 feet/second, then the subsequent position y of the mass is given by

$$y = y_0 \cos \omega t + \frac{v_0}{\omega} \sin \omega t$$

where t is the time in seconds and ω is a positive constant.

(a) If $\omega = 1$, $y_0 = 2$, and $v_0 = 3$ feet/second, find the amplitude and period of the resulting motion by first writing y in the form $A \cos(Bx - C)$. (*Hint:* Use the formula in Example 5.)

(b) Determine the times when $y = 0$; that is, determine when the mass passes through the equilibrium position.

56 Refer to Exercise 55. If $y_0 = 1$ and $\omega = 2$, find the initial velocities that result in an amplitude of 4 feet.

For many applications in electrical engineering it is necessary to write the sum of several sinusoidal signals (e.g., voltage signals or radio waves) of the same frequency in the compact form $y = A \cos(Bt - C)$. Express each of the following signals in this form. (See Example 5.)

57 $y = 50 \sin 60\pi t + 40 \cos 60\pi t$

58 $y = 10 \sin(120\pi t - \frac{1}{2}\pi) + 5 \sin 120\pi t$

59 If a mountain top is viewed from a point P due south of the mountain, the angle of elevation is α (see figure). If viewed from a point Q that is d miles east of P, the angle of elevation is β. (a) Show that the height h of the

FIGURE FOR EXERCISE 59

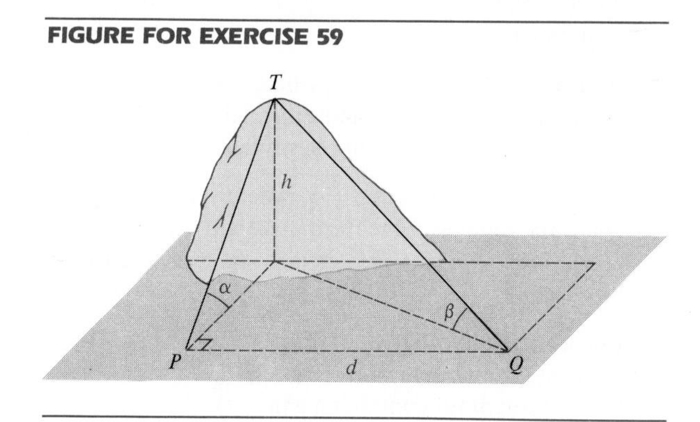

mountain is given by

$$h = \frac{d \sin \alpha \sin \beta}{\sqrt{\sin (\alpha + \beta) \sin (\alpha - \beta)}}.$$

(b) If $\alpha = 30°$, $\beta = 20°$, and $d = 10$ miles, approximate h to the nearest hundredth of a mile.

6.4 Multiple-Angle Formulas

The formulas considered in this section are referred to as **multiple-angle formulas.** In particular, the following identities are called the **double-angle formulas,** because they contain the expression $2u$.

DOUBLE-ANGLE FORMULAS

$$\sin 2u = 2 \sin u \cos u$$

$$\cos 2u = \cos^2 u - \sin^2 u = 1 - 2 \sin^2 u = 2 \cos^2 u - 1$$

$$\tan 2u = \frac{2 \tan u}{1 - \tan^2 u}$$

PROOF The identities may be proved by letting $u = v$ in the appropriate addition formulas. If we use the formula for $\sin (u + v)$, then

$$\begin{aligned} \sin 2u &= \sin (u + u) \\ &= \sin u \cos u + \cos u \sin u \\ &= 2 \sin u \cos u. \end{aligned}$$

Similarly, using the formula for $\cos (u + v)$,

$$\begin{aligned} \cos 2u &= \cos (u + u) \\ &= \cos u \cos u - \sin u \sin u \\ &= \cos^2 u - \sin^2 u. \end{aligned}$$

To obtain the other two forms for $\cos 2u$ we use the fundamental identity $\sin^2 u + \cos^2 u = 1$. Thus,

$$\begin{aligned} \cos 2u &= \cos^2 u - \sin^2 u \\ &= (1 - \sin^2 u) - \sin^2 u \\ &= 1 - 2 \sin^2 u. \end{aligned}$$

Similarly, if we substitute for $\sin^2 u$ instead of $\cos^2 u$, we obtain

$$\begin{aligned}\cos 2u &= \cos^2 u - (1 - \cos^2 u)\\ &= 2\cos^2 u - 1.\end{aligned}$$

The formula for $\tan 2u$ may be obtained by letting $u = v$ in the formula for $\tan (u + v)$. □

EXAMPLE 1 Find $\sin 2\alpha$ and $\cos 2\alpha$ if $\sin \alpha = \frac{4}{5}$ and α is in quadrant I.

SOLUTION As in Example 3 of the preceding section, $\cos \alpha = \frac{3}{5}$. Substitution in double-angle formulas:

$$\sin 2\alpha = 2 \sin \alpha \cos \alpha = 2(\tfrac{4}{5})(\tfrac{3}{5}) = \tfrac{24}{25}$$

$$\cos 2\alpha = \cos^2 \alpha - \sin^2 \alpha = \tfrac{9}{25} - \tfrac{16}{25} = -\tfrac{7}{25}.$$ ■

EXAMPLE 2 Express $\cos 3\theta$ in terms of $\cos \theta$.

SOLUTION

$$\begin{aligned}\cos 3\theta &= \cos (2\theta + \theta)\\ &= \cos 2\theta \cos \theta - \sin 2\theta \sin \theta\\ &= (2\cos^2 \theta - 1)\cos \theta - (2 \sin \theta \cos \theta) \sin \theta\\ &= 2\cos^3 \theta - \cos \theta - 2\cos \theta \sin^2 \theta\\ &= 2\cos^3 \theta - \cos \theta - 2\cos \theta\,(1 - \cos^2 \theta)\\ &= 2\cos^3 \theta - \cos \theta - 2\cos \theta + 2\cos^3 \theta\\ &= 4\cos^3 \theta - 3\cos \theta.\end{aligned}$$ ■

The next three identities are useful for simplifying expressions involving powers of trigonometric functions.

$$\sin^2 u = \frac{1 - \cos 2u}{2}, \qquad \cos^2 u = \frac{1 + \cos 2u}{2}, \qquad \tan^2 u = \frac{1 - \cos 2u}{1 + \cos 2u}.$$

The first and second of these identities may be verified by solving the equations

$$\cos 2u = 1 - 2\sin^2 u \quad \text{and} \quad \cos 2u = 2\cos^2 u - 1$$

for $\sin^2 u$ and $\cos^2 u$, respectively. The third identity may be obtained from the first two by using the fact that $\tan^2 u = \sin^2 u/\cos^2 u$.

EXAMPLE 3 Verify the identity $\sin^2 x \cos^2 x = \frac{1}{8}(1 - \cos 4x)$.

SOLUTION Using the identities for $\sin^2 u$ and $\cos^2 u$ in the preceding box with $u = x$, we obtain

$$\begin{aligned} \sin^2 x \cos^2 x &= \left(\frac{1 - \cos 2x}{2}\right)\left(\frac{1 + \cos 2x}{2}\right) \\ &= \tfrac{1}{4}(1 - \cos^2 2x) \\ &= \tfrac{1}{4}\sin^2 2x. \end{aligned}$$

Finally, using the formula $\sin^2 u = (1 - \cos 2u)/2$ with $u = 2x$,

$$\begin{aligned} \sin^2 x \cos^2 x &= \frac{1}{4}\left(\frac{1 - \cos 4x}{2}\right) \\ &= \tfrac{1}{8}(1 - \cos 4x). \end{aligned}$$

Another method of proof is to use the fact that $\sin 2x = 2 \sin x \cos x$ and hence, that

$$\sin x \cos x = \tfrac{1}{2}\sin 2x.$$

Squaring both sides, we have

$$\sin^2 x \cos^2 x = \tfrac{1}{4}\sin^2 2x.$$

The remainder of the solution is the same as the first proof. ■

EXAMPLE 4 Express $\cos^4 t$ in terms of values of the cosine function with exponent 1.

SOLUTION We begin by writing

$$\begin{aligned} \cos^4 t = (\cos^2 t)^2 &= \left(\frac{1 + \cos 2t}{2}\right)^2 \\ &= \frac{1}{4}(1 + 2\cos 2t + \cos^2 2t). \end{aligned}$$

Next, using the formula $\cos^2 u = (1 + \cos 2u)/2$ with $u = 2t$, we obtain

$$\cos^4 t = \frac{1}{4}\left(1 + 2\cos 2t + \frac{1 + \cos 4t}{2}\right).$$

This simplifies to

$$\cos^4 t = \frac{3}{8} + \frac{1}{2}\cos 2t + \frac{1}{8}\cos 4t.$$

■

Substituting $v/2$ for u in the three formulas for $\sin^2 u$, $\cos^2 u$, and $\tan^2 u$ on page 294 gives us

$$\sin^2 \frac{v}{2} = \frac{1 - \cos v}{2}, \qquad \cos^2 \frac{v}{2} = \frac{1 + \cos v}{2}, \qquad \tan^2 \frac{v}{2} = \frac{1 - \cos v}{1 + \cos v}.$$

If we take the square root of both sides of each of the last three equations and use the fact that $\sqrt{a^2} = |a|$ for every real number a, the following identities result. They are called **half-angle formulas** because of the expression $v/2$.

HALF-ANGLE FORMULAS

$$\left|\sin \frac{v}{2}\right| = \sqrt{\frac{1 - \cos v}{2}}, \qquad \left|\cos \frac{v}{2}\right| = \sqrt{\frac{1 + \cos v}{2}},$$

$$\left|\tan \frac{v}{2}\right| = \sqrt{\frac{1 - \cos v}{1 + \cos v}}.$$

The absolute value signs may be eliminated in the preceding formulas if more information is known about $v/2$. For example, if the angle determined by $v/2$ is in either quadrant I or II, then $\sin (v/2)$ is positive, and

$$\sin \frac{v}{2} = \sqrt{\frac{1 - \cos v}{2}}.$$

However, if $v/2$ leads to an angle in either quadrant III or IV, then $\sin (v/2)$ is negative, and

$$\sin \frac{v}{2} = -\sqrt{\frac{1 - \cos v}{2}}.$$

Similar remarks are true for the other formulas.

An alternative form for $\tan(v/2)$ can be obtained. Multiplying numerator and denominator of the radicand in the third half-angle formula by $1 - \cos v$ gives us

$$\left|\tan\frac{v}{2}\right| = \sqrt{\frac{1-\cos v}{1+\cos v}\cdot\frac{1-\cos v}{1-\cos v}}$$
$$= \sqrt{\frac{(1-\cos v)^2}{\sin^2 v}}$$
$$= \frac{1-\cos v}{|\sin v|}.$$

The absolute value sign is unnecessary in the numerator since $1 - \cos v$ is never negative. It can be shown that $\tan(v/2)$ and $\sin v$ always have the same sign. For example, if $0 < v < \pi$, then $0 < v/2 < \pi/2$, and hence both $\sin v$ and $\tan(v/2)$ are positive. If $\pi < v < 2\pi$, then $\pi/2 < v/2 < \pi$, and hence both $\sin v$ and $\tan(v/2)$ are negative. It is possible to generalize these remarks to all values of v for which the expressions $\tan(v/2)$ and $(1 - \cos v)/|\sin v|$ have meaning. This gives us the first of the next two identities. The second identity for $\tan(v/2)$ may be obtained by multiplying numerator and denominator of the radicand in the third half-angle formula by $1 + \cos v$.

HALF-ANGLE FORMULAS FOR TANGENT

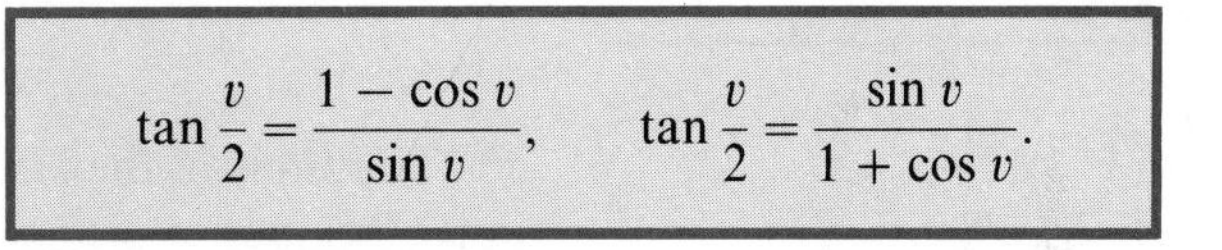

$$\tan\frac{v}{2} = \frac{1-\cos v}{\sin v}, \qquad \tan\frac{v}{2} = \frac{\sin v}{1+\cos v}.$$

EXAMPLE 5 Find the exact values of $\sin 22.5°$ and $\cos 22.5°$.

SOLUTION Using the formula for $\sin(v/2)$ and the fact that $22.5°$ is in quadrant I,

$$\sin 22.5° = \sin\frac{45°}{2} = \sqrt{\frac{1-\cos 45°}{2}}$$
$$= \sqrt{\frac{1-\sqrt{2}/2}{2}} = \frac{\sqrt{2-\sqrt{2}}}{2}$$

$$\cos 22.5° = \sqrt{\frac{1+\cos 45°}{2}}$$
$$= \sqrt{\frac{1+\sqrt{2}/2}{2}} = \frac{\sqrt{2+\sqrt{2}}}{2}$$

■

FIGURE 6.3

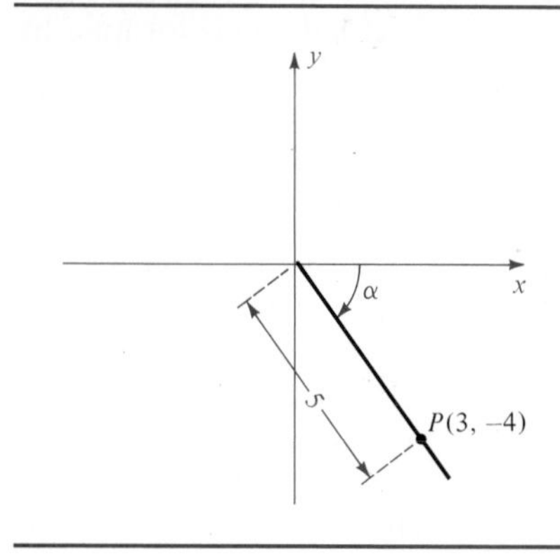

EXAMPLE 6 If $\tan \alpha = -\frac{4}{3}$ and α is in quadrant IV, find $\tan (\alpha/2)$.

SOLUTION If we choose the point $(3, -4)$ on the terminal side of α, as illustrated in Figure 6.3, then $\sin \alpha = -\frac{4}{5}$ and $\cos \alpha = \frac{3}{5}$. Applying a half-angle formula,

$$\tan \frac{\alpha}{2} = \frac{1 - \cos \alpha}{\sin \alpha} = \frac{1 - \frac{3}{5}}{-\frac{4}{5}} = -\frac{1}{2}.$$ ■

EXAMPLE 7 Find the solutions of the equation $\cos 2x + \cos x = 0$ that are in the interval $[0, 2\pi)$. Express the solutions both in radian measure and degree measure.

SOLUTION We first use a double-angle formula to write the equation in terms of $\cos x$, and then solve by factoring as follows:

$$\cos 2x + \cos x = 0$$
$$(2 \cos^2 x - 1) + \cos x = 0$$
$$2 \cos^2 x + \cos x - 1 = 0$$
$$(2 \cos x - 1)(\cos x + 1) = 0$$

Setting each factor equal to zero we obtain

$$\cos x = \tfrac{1}{2} \quad \text{and} \quad \cos x = -1.$$

The solutions of the last two equations (and hence of the given equation) in $[0, 2\pi)$ are

$$\frac{\pi}{3}, \quad \frac{5\pi}{3}, \quad \text{and} \quad \pi.$$

The corresponding degree measures are

$$60°, \quad 300°, \quad \text{and} \quad 180°.$$ ■

FIGURE 6.4

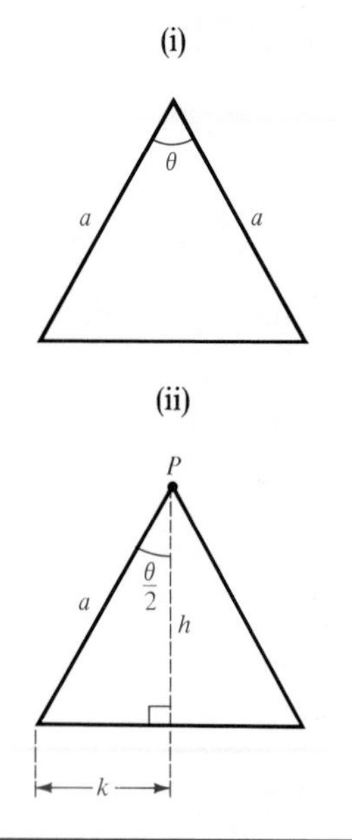

EXAMPLE 8 Suppose that an isosceles triangle has two equal sides of length a and that the angle between them is θ (see Figure 6.4(i)). Express the area A of the triangle in terms of a and θ.

SOLUTION From Figure 6.4(ii) we see that the altitude from point P bisects θ and that $A = hk$. Referring to the right triangle in the figure,

$$\sin \frac{\theta}{2} = \frac{k}{a} \quad \text{and} \quad \cos \frac{\theta}{2} = \frac{h}{a}.$$

Hence $$k = a \sin \frac{\theta}{2} \quad \text{and} \quad h = a \cos \frac{\theta}{2},$$

and $$A = hk = a^2 \sin \frac{\theta}{2} \cos \frac{\theta}{2}.$$

Using half-angle formulas gives us

$$\begin{aligned} A &= a^2 \sqrt{\frac{1 - \cos \theta}{2}} \sqrt{\frac{1 + \cos \theta}{2}} \\ &= a^2 \sqrt{\frac{1 - \cos^2 \theta}{4}} \\ &= a^2 \sqrt{\frac{\sin^2 \theta}{4}}. \end{aligned}$$

Thus $$A = \tfrac{1}{2} a^2 \sin \theta.$$ ■

Exercises 6.4

In Exercises 1–4 find the exact values of $\sin 2\theta$, $\cos 2\theta$, and $\tan 2\theta$ for the given conditions.

1 $\cos \theta = \frac{3}{5}$ and θ acute

2 $\cot \theta = \frac{4}{3}$ and $180^\circ < \theta < 270^\circ$

3 $\sec \theta = -3$ and $90^\circ < \theta < 180^\circ$

4 $\sin \theta = -\frac{4}{5}$ and $270^\circ < \theta < 360^\circ$

In Exercises 5–8 find the exact values of $\sin(\theta/2)$, $\cos(\theta/2)$, and $\tan(\theta/2)$ for the given conditions.

5 $\sec \theta = \frac{5}{4}$ and θ acute

6 $\csc \theta = -\frac{5}{3}$ and $-90^\circ < \theta < 0^\circ$

7 $\tan \theta = 1$ and $-180^\circ < \theta < -90^\circ$

8 $\sec \theta = -4$ and $180^\circ < \theta < 270^\circ$

In Exercises 9 and 10, use half-angle formulas to find exact values.

9 (a) $\cos 67^\circ 30'$
(b) $\sin 15^\circ$
(c) $\tan(3\pi/8)$

10 (a) $\cos 165^\circ$
(b) $\sin 157^\circ 30'$
(c) $\tan(\pi/8)$

Verify the identities stated in Exercises 11–28.

11 $\sin 10\theta = 2 \sin 5\theta \cos 5\theta$

12 $\cos^2 3x - \sin^2 3x = \cos 6x$

13 $4 \sin \frac{x}{2} \cos \frac{x}{2} = 2 \sin x$

14 $\frac{\sin^2 2\alpha}{\sin^2 \alpha} = 4 - 4 \sin^2 \alpha$

15 $(\sin t + \cos t)^2 = 1 + \sin 2t$

16 $\csc 2u = \frac{1}{2} \sec u \csc u$

17 $\sin 3u = \sin u (3 - 4 \sin^2 u)$

18 $\sin 4t = 4 \cos t \sin t (1 - 2 \sin^2 t)$

19 $\cos 4\theta = 8 \cos^4 \theta - 8 \cos^2 \theta + 1$

20 $\cos 6t = 32 \cos^6 t - 48 \cos^4 t + 18 \cos^2 t - 1$

21 $\sin^4 t = \frac{3}{8} - \frac{1}{2} \cos 2t + \frac{1}{8} \cos 4t$

22 $\cos^4 x - \sin^4 x = \cos 2x$

23 $\sec 2\theta = \frac{\sec^2 \theta}{2 - \sec^2 \theta}$

24 $\cot 2u = \dfrac{\cot^2 u - 1}{2 \cot u}$

25 $2 \sin^2 2t + \cos 4t = 1$

26 $\tan \theta + \cot \theta = 2 \csc 2\theta$

27 $\tan 3u = \dfrac{(3 - \tan^2 u)\tan u}{1 - 3 \tan^2 u}$

28 $\dfrac{1 + \sin 2v + \cos 2v}{1 + \sin 2v - \cos 2v} = \cot v$

Write the expressions in Exercises 29 and 30 in terms of values of cosine with exponent 1.

29 $\cos^4 (\theta/2)$ **30** $\sin^4 2x$

In Exercises 31–38 find all solutions of the equations in the interval $[0, 2\pi)$. Express the solutions both in radian measure and degree measure.

31 $\sin 2t + \sin t = 0$

32 $\cos t - \sin 2t = 0$

33 $\cos u + \cos 2u = 0$

34 $\cos 2\theta - \tan \theta = 1$

35 $\tan 2x = \tan x$

36 $\tan 2t - 2 \cos t = 0$

37 $\sin \frac{1}{2}u + \cos u = 1$

38 $2 - \cos^2 x = 4 \sin^2 \frac{1}{2}x$

39 If $a > 0$, $b > 0$, and $0 < u < \pi/2$, prove that

$$a \sin u + b \cos u = \sqrt{a^2 + b^2} \sin (u + v)$$

where $0 < v < \pi/2$, $\sin v = b/\sqrt{a^2 + b^2}$, and $\cos v = a/\sqrt{a^2 + b^2}$.

40 Use Exercise 39 to express $8 \sin u + 15 \cos u$ in the form $c \sin (u + v)$.

41 Shown in the figure is a computer-generated graph of the equation $y = \cos 2x + 2 \cos x$ for $0 \le x \le 2\pi$.

(a) Approximate the x-intercepts to two decimal places.

(b) It can be shown, using calculus, that the turning points P, Q, and R on the graph are solutions of the equation $-2 \sin 2x - 2 \sin x = 0$. Find the coordinates of P, Q, and R.

FIGURE FOR EXERCISE 41

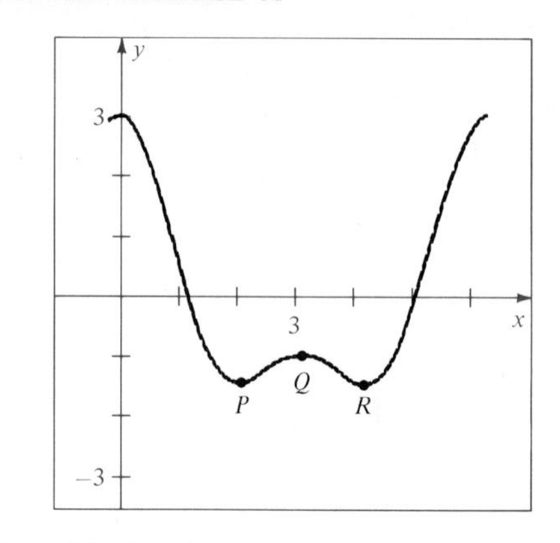

42 A computer-generated graph of the equation $y = \cos x - \sin 2x$ for $-2\pi \le x \le 2\pi$ is shown in the figure.

(a) Find the x-intercepts.

(b) It can be shown, using calculus, that the eight turning points on the graph are solutions of the equation $-\sin x - 2 \cos 2x = 0$. Approximate the x-coordinates of these points to two decimal places.

FIGURE FOR EXERCISE 42

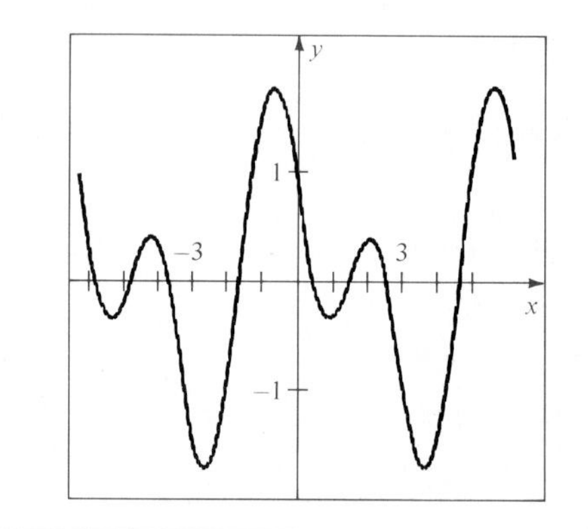

43 A computer-generated graph of $y = \cos 3x - 3 \cos x$ for $-2\pi \le x \le 2\pi$ is shown in the figure.

(a) Find the x-intercepts. (*Hint:* Use the formula for $\cos 3x$ given in Example 2.)

(b) The thirteen turning points on the graph are solutions of the equation $-3 \sin 3x + 3 \sin x = 0$. Find the x-coordinates of these points. (*Hint:* Use the formula for $\sin 3x$ in Exercise 17.)

FIGURE FOR EXERCISE 43

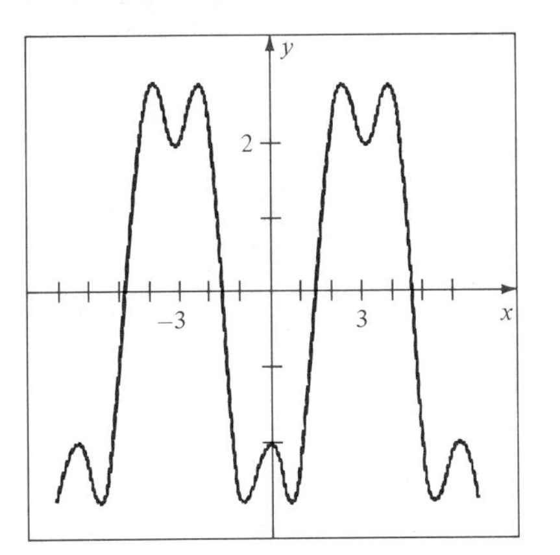

44 A computer-generated graph of $y = \sin 4x - 4 \sin x$ for $-2\pi \leq x \leq 2\pi$ is shown in the figure. Find the x-intercepts. (*Hint:* Use Exercise 18.)

FIGURE FOR EXERCISE 44

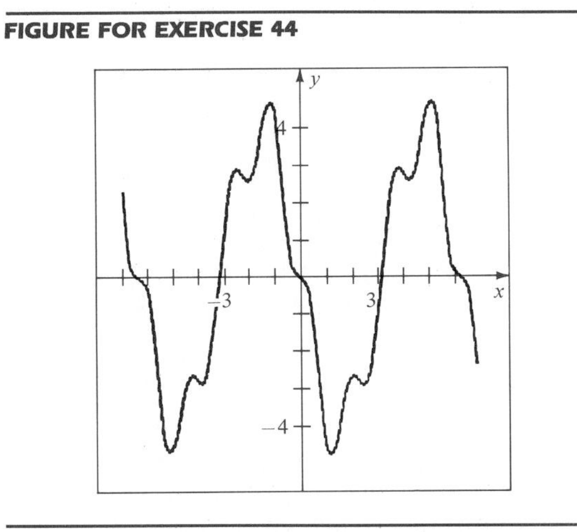

45 Shown in the figure is a railroad route, from town A to town C, that branches out from B toward C at an angle θ.

(a) Show that the total distance d from A to C is given by $d = 20 \tan(\theta/2) + 40$.

(b) Because of mountains between A and C, the branching point B must be at least 20 miles from A. Is there a route that avoids the mountains and measures exactly 50 miles?

FIGURE FOR EXERCISE 45

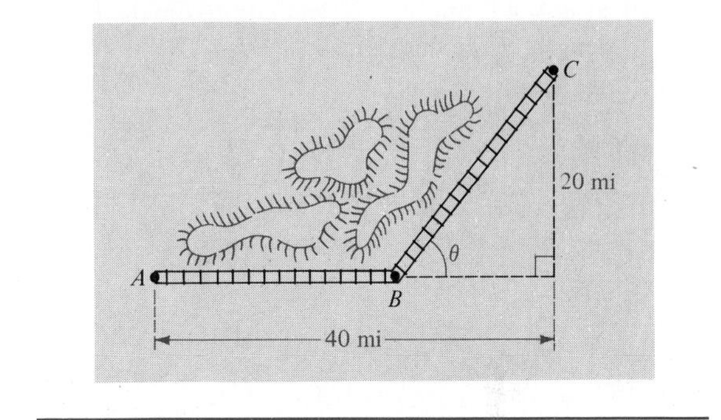

46 Shown in the figure is a design for a rain gutter. Express the volume V as a function of θ and then approximate the value of θ that results in a volume of 2 ft^3. (*Hint:* See Example 8.)

FIGURE FOR EXERCISE 46

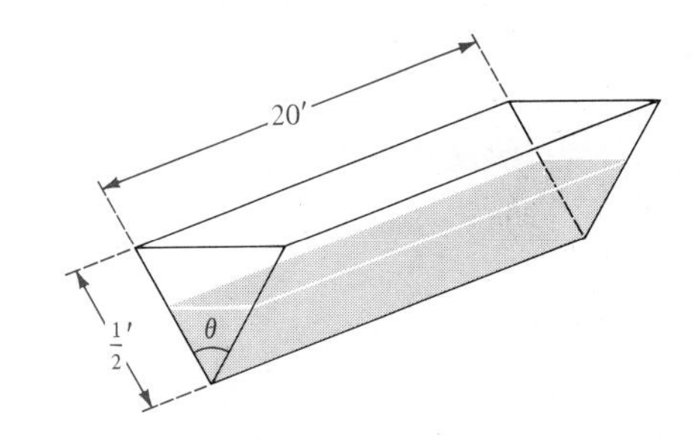

47 If a projectile is fired from ground level with an initial velocity of v feet/second and at an angle of θ degrees with the horizontal, the range R of the projectile is given by

$$R = \frac{v^2}{16} \sin \theta \cos \theta.$$

If $v = 80$ feet/second, approximate the angle(s) that result in a range of 150 feet.

48 A highway engineer is designing the street curbing at an intersection where two highways come together at an angle ϕ (see figure). The curbing joining points A and B

is to be constructed using a circle that is tangent to the highway at these two points.

(a) Show that the relationship between the radius R of the circle and the distance d in the figure is given by $d = R \tan(\phi/2)$.

(b) If $\phi = 45^\circ$ and $d = 20$ feet, approximate R and the length of the curbing.

FIGURE FOR EXERCISE 48

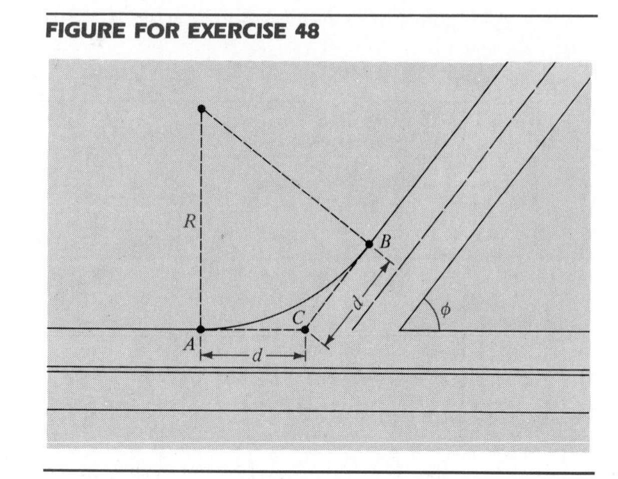

6.5 Product and Factoring Formulas

The following identities may be used to change the form of certain trigonometric expressions.

PRODUCT FORMULAS

$$\sin(u + v) + \sin(u - v) = 2 \sin u \cos v$$

$$\sin(u + v) - \sin(u - v) = 2 \cos u \sin v$$

$$\cos(u + v) + \cos(u - v) = 2 \cos u \cos v$$

$$\cos(u - v) - \cos(u + v) = 2 \sin u \sin v$$

Each of these identities is a consequence of our work in Section 6.3. For example, to verify the first product formula, we merely add the left- and right-hand sides of the identities we obtained for $\sin(u + v)$ and $\sin(u - v)$. The remaining product formulas are obtained in similar fashion.

By starting with the expression on the *right* side of each product formula, we can express certain products as sums, as illustrated in the next example.

EXAMPLE 1 Express each of the following as a sum:

(a) $\sin 4\theta \cos 3\theta$ (b) $\sin 3x \sin x$

SOLUTION

(a) Using the first product formula with $u = 4\theta$ and $v = 3\theta$,

$$2 \sin 4\theta \cos 3\theta = \sin (4\theta + 3\theta) + \sin (4\theta - 3\theta)$$

or

$$\sin 4\theta \cos 3\theta = \tfrac{1}{2} \sin 7\theta + \tfrac{1}{2} \sin \theta.$$

This relationship could also have been obtained by using the second product formula.

(b) Using the fourth product formula with $u = 3x$ and $v = x$,

$$2 \sin 3x \sin x = \cos (3x - x) - \cos (3x + x)$$

or

$$\sin 3x \sin x = \tfrac{1}{2} \cos 2x - \tfrac{1}{2} \cos 4x.$$ ■

Product formulas may also be employed to express a sum as a product. To obtain a form that can be applied more easily, we shall change the notation as follows. If we let

$$u + v = a \quad \text{and} \quad u - v = b,$$

then $(u + v) + (u - v) = a + b$, which simplifies to

$$u = \frac{a + b}{2}.$$

Similarly, since $(u + v) - (u - v) = a - b$,

$$v = \frac{a - b}{2}.$$

If we now substitute for $u + v$ and $u - v$ on the left sides of the product formulas and for u and v on the right sides, we obtain the following.

FACTORING FORMULAS

$$\sin a + \sin b = 2 \sin \frac{a + b}{2} \cos \frac{a - b}{2}$$

$$\sin a - \sin b = 2 \cos \frac{a + b}{2} \sin \frac{a - b}{2}$$

$$\cos a + \cos b = 2 \cos \frac{a + b}{2} \cos \frac{a - b}{2}$$

$$\cos b - \cos a = 2 \sin \frac{a + b}{2} \sin \frac{a - b}{2}$$

EXAMPLE 2 Express $\sin 5x - \sin 3x$ as a product.

SOLUTION Use the second factoring formula with $a = 5x$ and $b = 3x$,

$$\sin 5x - \sin 3x = 2 \cos \frac{5x + 3x}{2} \sin \frac{5x - 3x}{2}$$

$$= 2 \cos 4x \sin x.$$

■

EXAMPLE 3 Verify the identity

$$\frac{\sin 3t + \sin 5t}{\cos 3t - \cos 5t} = \cot t.$$

SOLUTION Using the first and fourth factoring formulas,

$$\frac{\sin 3t + \sin 5t}{\cos 3t - \cos 5t} = \frac{2 \sin \frac{3t + 5t}{2} \cos \frac{3t - 5t}{2}}{2 \sin \frac{5t + 3t}{2} \sin \frac{5t - 3t}{2}}$$

$$= \frac{2 \sin 4t \cos (-t)}{2 \sin 4t \sin t} = \frac{\cos (-t)}{\sin t}$$

$$= \frac{\cos t}{\sin t} = \cot t.$$

■

EXAMPLE 4 Find the solutions of the equation

$$\cos t - \sin 2t - \cos 3t = 0.$$

SOLUTION Using the fourth factoring formula,

$$\cos t - \cos 3t = 2 \sin \frac{3t + t}{2} \sin \frac{3t - t}{2}$$

$$= 2 \sin 2t \sin t.$$

Hence, the given equation is equivalent to

$$2 \sin 2t \sin t - \sin 2t = 0$$

or

$$\sin 2t\, (2 \sin t - 1) = 0.$$

Thus the solutions of the given equation are the solutions of

$$\sin 2t = 0 \quad \text{or} \quad \sin t = \tfrac{1}{2}.$$

The first of these equations has solutions

$$2t = n\pi \quad \text{or} \quad t = \frac{\pi}{2}n$$

where n is any integer. The second equation has solutions

$$t = \frac{\pi}{6} + 2n\pi \quad \text{or} \quad t = \frac{5\pi}{6} + 2n\pi$$

where n is any integer. ■

Addition formulas may also be employed to derive **reduction formulas.** Reduction formulas are used to change expressions such as

$$\sin\left(\theta + \frac{n\pi}{2}\right) \quad \text{and} \quad \cos\left(\theta + \frac{n\pi}{2}\right),$$

where n is an integer, to expressions involving only $\sin\theta$ or $\cos\theta$. Similar formulas are true for the other trigonometric functions. Instead of deriving general reduction formulas, we shall illustrate several special cases in the next example.

EXAMPLE 5 Express $\sin(\theta - 3\pi/2)$ and $\cos(\theta + \pi)$ in terms of a function of θ.

SOLUTION Using addition formulas, we obtain

$$\begin{aligned}\sin\left(\theta - \frac{3\pi}{2}\right) &= \sin\theta\cos\frac{3\pi}{2} - \cos\theta\sin\frac{3\pi}{2}\\ &= \sin\theta\cdot(0) - \cos\theta\cdot(-1) = \cos\theta\end{aligned}$$

$$\begin{aligned}\cos(\theta + \pi) &= \cos\theta\cos\pi - \sin\theta\sin\pi\\ &= \cos\theta\cdot(-1) - \sin\theta\cdot(0) = -\cos\theta\end{aligned}$$

■

We have derived many important identities in this chapter. For reference, these identities are listed on the endpapers of the text.

Exercises 6.5

In Exercises 1–8 express the product as a sum or difference.

1 $2\sin 9\theta\cos 3\theta$

2 $2\cos 5\theta\sin 5\theta$

3 $\sin 7t\sin 3t$

4 $\sin(-4x)\cos 8x$

5 $\cos 6u\cos(-4u)$

6 $\sin 4t\sin 6t$

7 $3\cos x\sin 2x$

8 $5\cos u\sin 5u$

In Exercises 9–16 write the expression as a product.

9 $\sin 6\theta + \sin 2\theta$

10 $\sin 4\theta - \sin 8\theta$

11 $\cos 5x - \cos 3x$

12 $\cos 5t + \cos 6t$

13 $\sin 3t - \sin 7t$

14 $\cos\theta - \cos 5\theta$

15 $\cos x + \cos 2x$

16 $\sin 8t + \sin 2t$

Verify the identities in Exercises 17–24.

17 $\dfrac{\sin 4t + \sin 6t}{\cos 4t - \cos 6t} = \cot t$

18 $\dfrac{\sin \theta + \sin 3\theta}{\cos \theta + \cos 3\theta} = \tan 2\theta$

19 $\dfrac{\sin u + \sin v}{\cos u + \cos v} = \tan \dfrac{u + v}{2}$

20 $\dfrac{\sin u - \sin v}{\cos u - \cos v} = -\cot \dfrac{u + v}{2}$

21 $\dfrac{\sin u - \sin v}{\sin u + \sin v} = \dfrac{\tan \frac{1}{2}(u - v)}{\tan \frac{1}{2}(u + v)}$

22 $\dfrac{\cos u - \cos v}{\cos u + \cos v} = \tan \frac{1}{2}(u + v) \tan \frac{1}{2}(u - v)$

23 $\sin 2x + \sin 4x + \sin 6x = 4 \cos x \cos 2x \sin 3x$

24 $\dfrac{\cos t + \cos 4t + \cos 7t}{\sin t + \sin 4t + \sin 7t} = \cot 4t$

25 Express $(\sin ax)(\cos bx)$ as a sum.

26 Express $(\cos mu)(\cos nu)$ as a sum.

In Exercises 27–30 find the solutions of the equations.

27 $\sin 5t + \sin 3t = 0$

28 $\sin t + \sin 3t = \sin 2t$

29 $\cos x = \cos 3x$

30 $\cos 4x - \cos 3x = 0$

In Exercises 31–36 verify the reduction formulas by using addition formulas.

31 $\sin (\theta + \pi) = -\sin \theta$

32 $\sin \left(\theta + \dfrac{3\pi}{2}\right) = -\cos \theta$

33 $\cos \left(\theta - \dfrac{5\pi}{2}\right) = \sin \theta$

34 $\cos (\theta - 3\pi) = -\cos \theta$

35 $\tan (\pi - \theta) = -\tan \theta$

$\left(Hint\text{: } \tan (\pi - \theta) = \dfrac{\sin (\pi - \theta)}{\cos (\pi - \theta)}.\right)$

36 $\tan \left(\theta + \dfrac{\pi}{2}\right) = -\cot \theta$

Shown in the figures are computer-generated graphs of the functions defined in Exercises 37 and 38, for $0 \le x \le 2\pi$. In each case use a factoring formula to find the x-intercepts.

37 $f(x) = \cos x + \cos 3x$

FIGURE FOR EXERCISE 37

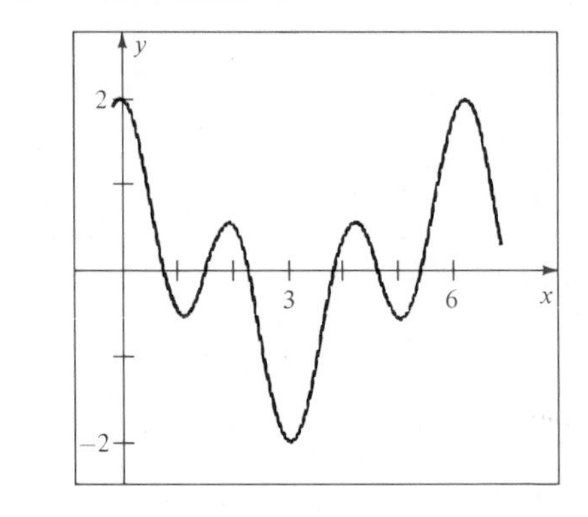

38 $f(x) = \sin 4x - \sin x$

FIGURE FOR EXERCISE 38

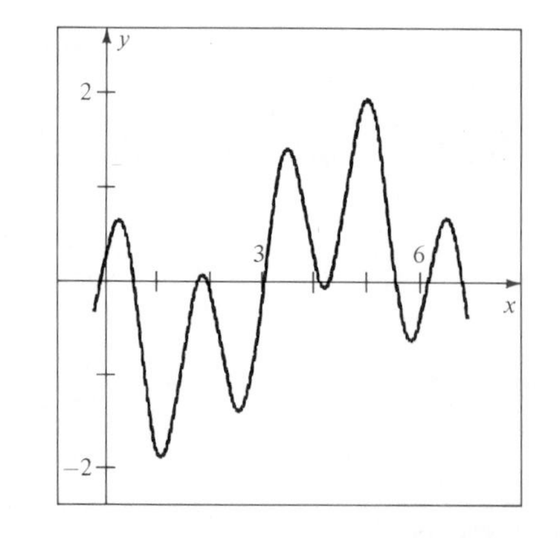

39 The graph of $y = \cos 3x - 3 \cos x$ has 13 turning points between -2π and 2π. Using calculus, it can be shown that the x-coordinates of these points satisfy the equation $\sin 3x - \sin x = 0$. Find these points using a factoring formula. (See Exercise 43 of Section 6.4.)

40 The x-coordinates of the turning points on the graph of $y = \sin 4x - 4 \sin x$ are solutions of the equation

$\cos 4x - \cos x = 0$. Use a factoring formula to locate the turning points for $-2\pi \leq x \leq 2\pi$. (See Exercise 44 of Section 6.4.)

41 The formula $f(x) = \sin(n\pi x/L)\cos(n\pi kt/L)$ arises in the mathematical analysis of a vibrating violin string, where L is the length of the string. Express f as a sum of two sine functions.

42 If two tuning forks are struck with the same force and are held at the same distance from the eardrum, the pressure on the outside of the eardrum at time t can be represented by $p(t) = A\cos\omega_1 t + A\cos\omega_2 t$. If ω_1 is close to ω_2, that is, if the tuning forks have nearly equal frequencies, a tone is heard that alternates between loudness and virtual silence. This phenomenon is known as *beats.*

(a) Use factoring formulas to write $p(t)$ as a product.

(b) Show that $p(t)$ may be considered as a cosine wave with approximate period $2\pi/\omega_1$ and variable amplitude $A(t) = 2A\cos\frac{1}{2}(\omega_1 - \omega_2)t$. What is the maximum amplitude?

(c) Shown in the figure is the graph of the equation $p(t) = \cos 4.5t + \cos 3.5t$. Near silence occurs at points A and B. Find the coordinates of these points and determine how frequently this near silence occurs.

(d) Show that p has period 4π. Conclude that the maximum amplitude of 2 occurs every 4π units of time.

FIGURE FOR EXERCISE 42

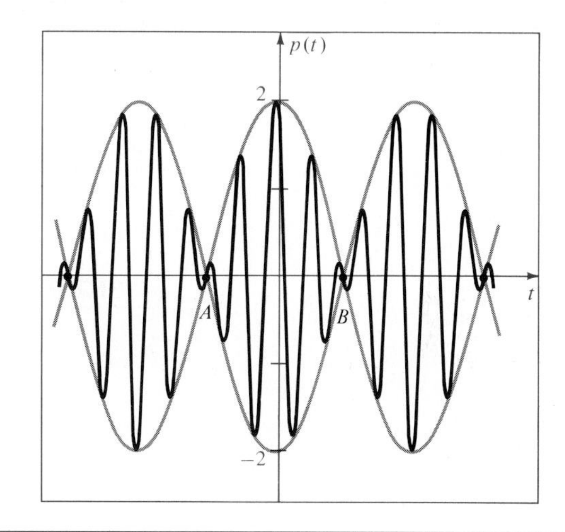

6.6 The Inverse Trigonometric Functions

The concept of inverse function was introduced in Section 2.5. Recall that if f is a one-to-one function with domain D and range E, then, as illustrated in Figure 6.5(i), for each number u in E there is exactly one number v in D such that $f(v) = u$. We may then define the function f^{-1} from E to D by letting $f^{-1}(u) = v$ (see Figure 6.5(ii)).

FIGURE 6.5

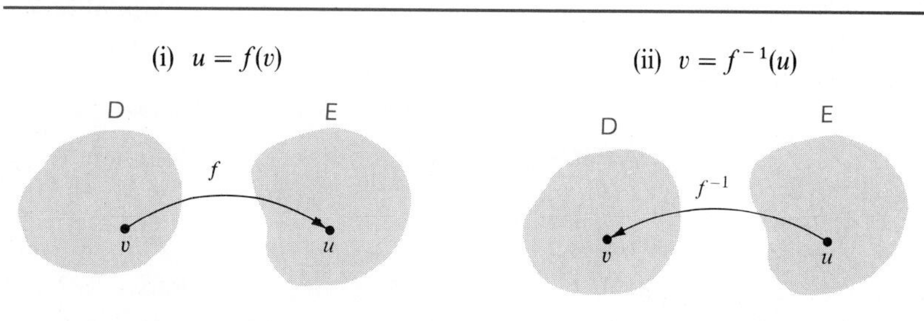

We call f^{-1} the **inverse function** of f. The function f^{-1} *reverses* the correspondence given by f, that is,

$$v = f^{-1}(u) \quad \text{if and only if} \quad f(v) = u$$

for every u in E and v in D.

Since the trigonometric functions are not one-to-one, they do not have inverse functions. However, by restricting the domains, it is possible to obtain functions that have the same values as the trigonometric functions (over the smaller domains) and that *do* possess inverse functions. Let us first consider the sine function. If we restrict the domain to $[-\pi/2, \pi/2]$, then as illustrated by the black portion of the curve in Figure 6.6, we obtain an increasing function that takes on all the values of the sine function once and only once.

FIGURE 6.6

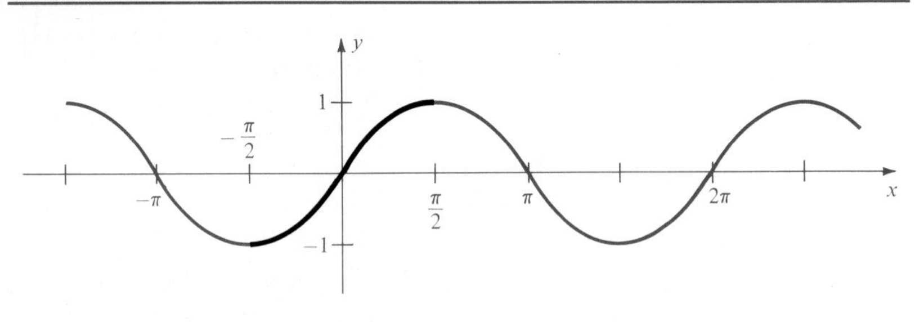

This new function with domain $[-\pi/2, \pi/2]$ and range $[-1, 1]$ is sometimes denoted by Sin (with a capital S). The inverse function Sin^{-1} has domain $[-1, 1]$ and range $[-\pi/2, \pi/2]$ and is called the *inverse sine function.* By definition,

$$v = \text{Sin}^{-1}\, u \quad \text{if and only if} \quad \text{Sin}\, v = u$$

for $-1 \le u \le 1$ and $-\pi/2 \le v \le \pi/2$. Since the letters used for the variables are immaterial, we may rephrase this in terms of the customary symbols x and y. At the same time, let us follow tradition and use a lower case letter in place of the capital S. We can then state the following definition.

DEFINITION

The **inverse sine function,** denoted by $\sin^{-1}$, is defined by

$$y = \sin^{-1} x \quad \text{if and only if} \quad \sin y = x$$

for $-1 \le x \le 1$ and $-\pi/2 \le y \le \pi/2$.

We also refer to $\sin^{-1}$ as the **arcsine function** and use the notation $\arcsin x$ in place of $\sin^{-1} x$. Arcsin x is used because if $t = \arcsin x$, then $\sin t = x$, and t may be interpreted as an *arc*length on the unit circle U. Since both notations, $\sin^{-1}$ and arcsin, are commonly used in mathematics and its applications, we shall employ both of them in our work. Note that, by definition,

$$-\frac{\pi}{2} \le \sin^{-1} x \le \frac{\pi}{2}$$

or equivalently,

$$-\frac{\pi}{2} \le \arcsin x \le \frac{\pi}{2}.$$

FIGURE 6.7

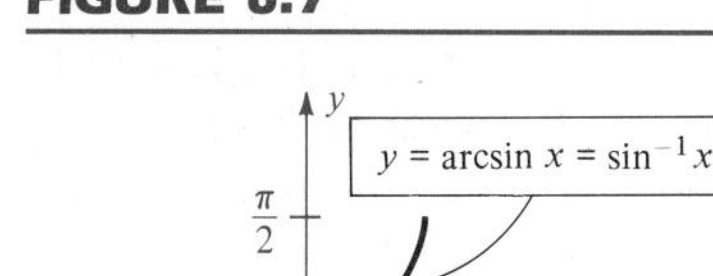

As in the discussion of inverse functions in Section 2.5, the graph of $\sin^{-1}$ can be found by reflecting the black portion of Figure 6.6 through the line $y = x$. This gives us Figure 6.7. We could also use the equation $\sin y = x$ to find points on the graph.

In Section 5.2 we introduced a method for finding values of $\sin^{-1}$ with a calculator. The next example illustrates the use of the definition in finding several special values of the inverse sine function.

EXAMPLE 1 Find (a) $\sin^{-1}(\sqrt{2}/2)$; (b) $\arcsin(-\frac{1}{2})$; (c) $\sin^{-1} 1$.

SOLUTION

(a) By definition,

$$y = \sin^{-1}(\sqrt{2}/2) \quad \text{if and only if} \quad \sin y = \sqrt{2}/2$$

for $-\pi/2 \le y \le \pi/2$. The only number y in the interval $[-\pi/2, \pi/2]$ that satisfies $\sin y = \sqrt{2}/2$ is $y = \pi/4$. Hence

$$\sin^{-1}(\sqrt{2}/2) = \pi/4.$$

Note that it is *essential* to choose y in the interval $[-\pi/2, \pi/2]$. A number such as $3\pi/4$ is incorrect, even though $\sin(3\pi/4) = \sqrt{2}/2$.

(b) By definition,

$$y = \arcsin(-\tfrac{1}{2}) \quad \text{if and only if} \quad \sin y = -\tfrac{1}{2}$$

provided that y is in the interval $[-\pi/2, \pi/2]$. It follows that $y = -\pi/6$. Thus

$$\arcsin(-\tfrac{1}{2}) = -\pi/6.$$

(c) As in parts (a) and (b),

$$y = \sin^{-1} 1 \quad \text{if and only if} \quad \sin y = 1$$

for $-\pi/2 \leq y \leq \pi/2$. Consequently, $y = \pi/2$, and

$$\sin^{-1} 1 = \pi/2.$$

■

The relationships $f^{-1}(f(y)) = y$ and $f(f^{-1}(x)) = x$ that hold for any inverse function f^{-1} (see Section 2.5) give us the following important identities:

$$\sin^{-1}(\sin y) = y \quad \text{if} \quad -\frac{\pi}{2} \leq y \leq \frac{\pi}{2}$$

$$\sin(\sin^{-1} x) = x \quad \text{if} \quad -1 \leq x \leq 1.$$

These may also be written in the form

$$\arcsin(\sin y) = y \quad \text{and} \quad \sin(\arcsin x) = x$$

provided that y and x are suitably restricted.

EXAMPLE 2 Find $\sin^{-1}(\tan 3\pi/4)$.

SOLUTION If we let

$$y = \sin^{-1}\left(\tan\frac{3\pi}{4}\right) = \sin^{-1}(-1)$$

then, by definition, $\sin y = -1.$

Since y must be chosen in the interval $[-\pi/2, \pi/2]$, it follows that $y = -\pi/2$.

■

The other trigonometric functions may also be used to introduce inverse functions. The procedure is first to determine a convenient subset of the domain so that a one-to-one function is obtained. If the domain of the cosine function is restricted to the interval $[0, \pi]$, as illustrated by the black portion of the curve in Figure 6.8, we obtain a decreasing function that takes on all the values of the cosine function once and only once.

FIGURE 6.8

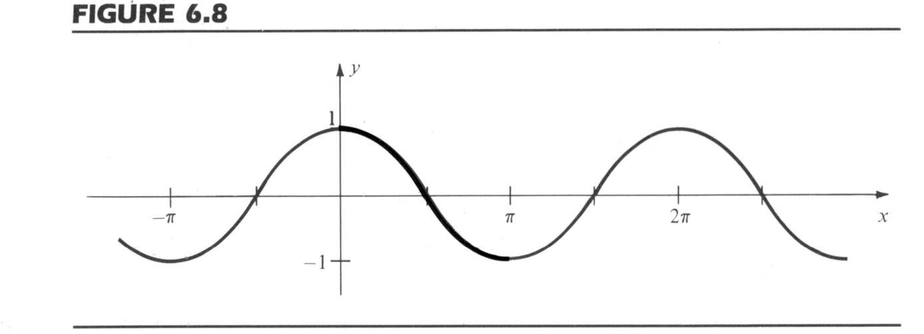

This new function has an inverse called the *inverse cosine function*, defined as follows.

DEFINITION

The **inverse cosine function,** denoted by $\cos^{-1}$, is defined by

$$y = \cos^{-1} x \quad \text{if and only if} \quad \cos y = x$$

for $-1 \leq x \leq 1$ and $0 \leq y \leq \pi$.

The inverse cosine function is also referred to as the **arccosine function,** and the notation arccos x is used interchangeably with $\cos^{-1} x$. Using general properties of inverse functions we obtain

$$\cos(\cos^{-1} x) = \cos(\arccos x) = x$$

and

$$\cos^{-1}(\cos y) = \arccos(\cos y) = y$$

for $-1 \leq x \leq 1$ and $0 \leq y \leq \pi$.

The graph of the inverse cosine function can be found by reflecting the black portion of Figure 6.8 through the line $y = x$. This gives us the sketch in Figure 6.9. We could also use the equation $\cos y = x$ to find points on the graph.

In the next example, several values of the inverse cosine function are determined by means of the definition. Others may be obtained by using a calculator or a table.

FIGURE 6.9

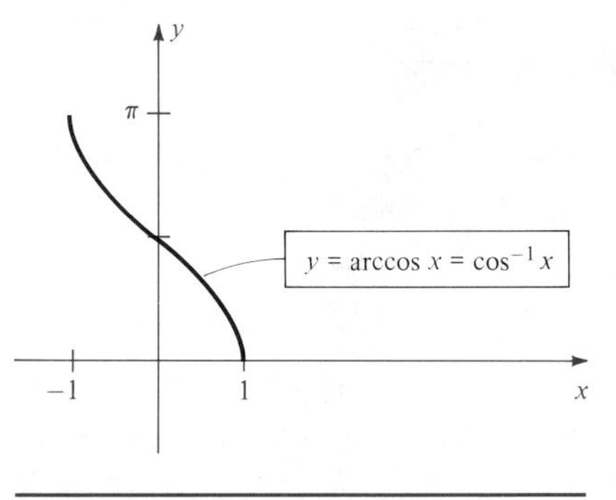

EXAMPLE 3 Find (a) $\cos^{-1}(\frac{1}{2})$; (b) $\cos^{-1}(-\sqrt{3}/2)$; (c) arccos 0.

SOLUTION

(a) By definition,

$$y = \cos^{-1}(\tfrac{1}{2}) \quad \text{if and only if} \quad \cos y = \tfrac{1}{2}$$

for $0 \le y \le \pi$. The only number y in the interval $[0, \pi]$ that satisfies $\cos y = \frac{1}{2}$ is $y = \pi/3$. Hence

$$\cos^{-1}(\tfrac{1}{2}) = \pi/3.$$

(b) By definition,

$$y = \cos^{-1}(-\sqrt{3}/2) \quad \text{if and only if} \quad \cos y = -\sqrt{3}/2$$

for $0 \le y \le \pi$. The reference angle for y is $\pi/6$. Since y must be chosen in the interval $[0, \pi]$ we take $y = \pi - (\pi/6) = 5\pi/6$. Thus

$$\cos^{-1}(-\sqrt{3}/2) = 5\pi/6.$$

(c) As in parts (a) and (b),

$$y = \arccos 0 \quad \text{if and only if} \quad \cos y = 0$$

for $0 \le y \le \pi$. Consequently, $y = \pi/2$ and

$$\arccos 0 = \pi/2. \qquad \blacksquare$$

FIGURE 6.10

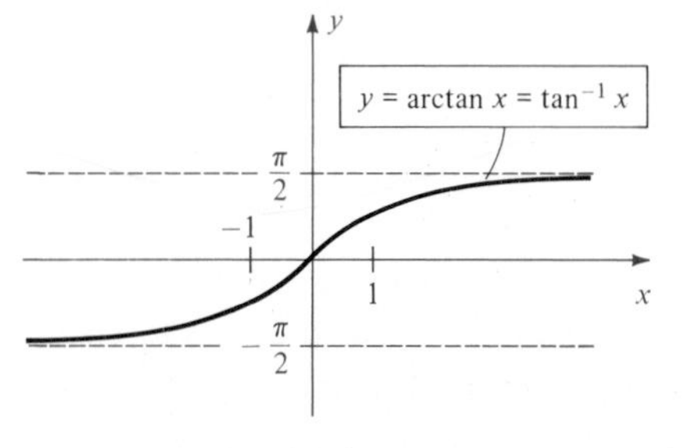

If we restrict the domain of the tangent function to the open interval $(-\pi/2, \pi/2)$, then a one-to-one function is obtained. We may therefore introduce the following definition.

DEFINITION

The **inverse tangent function** or **arctangent function,** denoted by $\tan^{-1}$ or arctan, is defined by

$$y = \tan^{-1} x = \arctan x \quad \text{if and only if} \quad \tan y = x$$

where x is any real number and $-\pi/2 < y < \pi/2$.

Note that the domain of the arctangent function is $\mathbb{R}$ and the range is the open interval $(-\pi/2, \pi/2)$. In Exercise 42 you are asked to verify the graph of the inverse tangent function sketched in Figure 6.10.

An analogous procedure may be used for the remaining trigonometric functions. There is no general acceptance about domains of the inverse secant, cosecant, or cotangent functions. Some typical domains are indicated in Exercises 39–41.

EXAMPLE 4 Without using tables or calculators find sec (arctan $\frac{2}{3}$).

SOLUTION If we let $y = \arctan \frac{2}{3}$, then $\tan y = \frac{2}{3}$. We wish to find sec y. Since $\sec^2 y = 1 + \tan^2 y$ and since $0 < y < \pi/2$,

$$\sec y = \sqrt{1 + \tan^2 y} = \sqrt{1 + \left(\frac{2}{3}\right)^2}$$
$$= \sqrt{1 + \frac{4}{9}} = \sqrt{\frac{13}{9}} = \frac{\sqrt{13}}{3}.$$

Hence, $\sec(\arctan \frac{2}{3}) = \sec y = \sqrt{13}/3$. ∎

The next three examples illustrate some of the manipulations that can be carried out with the inverse trigonometric functions.

EXAMPLE 5 Evaluate $\sin(\arctan \frac{1}{2} - \arccos \frac{4}{5})$.

SOLUTION If we let

$$u = \arctan \tfrac{1}{2} \quad \text{and} \quad v = \arccos \tfrac{4}{5},$$

then

$$\tan u = \tfrac{1}{2} \quad \text{and} \quad \cos v = \tfrac{4}{5}.$$

We wish to find $\sin(u - v)$. Since u and v are in the interval $(0, \pi/2)$, they can be considered as the radian measures of positive acute angles, and we may refer to the right triangles in Figure 6.11. This gives us

FIGURE 6.11

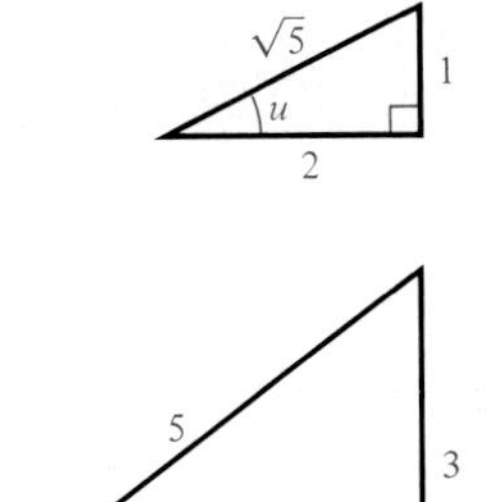

$$\sin u = \frac{1}{\sqrt{5}}, \qquad \cos u = \frac{2}{\sqrt{5}}, \qquad \sin v = \frac{3}{5}.$$

Consequently,

$$\sin(u - v) = \sin u \cos v - \cos u \sin v$$
$$= \frac{1}{\sqrt{5}}\frac{4}{5} - \frac{2}{\sqrt{5}}\frac{3}{5}$$
$$= \frac{-2}{5\sqrt{5}} = \frac{-2\sqrt{5}}{25}.$$ ∎

EXAMPLE 6 Write $\cos(\sin^{-1} x)$ as an algebraic expression in x.

SOLUTION To simplify our work, let

$$y = \sin^{-1} x.$$

Thus

$$\sin y = x.$$

We wish to find an algebraic expression for $\cos(\sin^{-1} x)$, that is, for $\cos y$. Since $-\pi/2 \le y \le \pi/2$, it follows that $\cos y \ge 0$, and hence

$$\cos y = \sqrt{1 - \sin^2 y} = \sqrt{1 - x^2}.$$

Consequently,

$$\cos(\sin^{-1} x) = \sqrt{1 - x^2}.$$

■

EXAMPLE 7 Verify the identity

$$\frac{1}{2}\cos^{-1} x = \tan^{-1}\sqrt{\frac{1 - x}{1 + x}}, \quad \text{for } |x| < 1.$$

SOLUTION Let $y = \cos^{-1} x$. We wish to show that

$$\frac{1}{2} y = \tan^{-1}\sqrt{\frac{1 - x}{1 + x}}.$$

By a half-angle formula (see Section 6.4),

$$\left|\tan\frac{y}{2}\right| = \sqrt{\frac{1 - \cos y}{1 + \cos y}}.$$

Since $y = \cos^{-1} x$ and $|x| < 1$, it follows that $0 < y < \pi$, or $0 < y/2 < \pi/2$. Consequently, $\tan(y/2) > 0$ and we may drop the absolute value sign, obtaining

$$\tan\frac{y}{2} = \sqrt{\frac{1 - \cos y}{1 + \cos y}}.$$

Since $\cos y = x$, this may be written

$$\tan\frac{y}{2} = \sqrt{\frac{1 - x}{1 + x}}.$$

The last equation is equivalent to

$$\frac{y}{2} = \tan^{-1}\sqrt{\frac{1 - x}{1 + x}},$$

which is what we wished to show.

■

Most of the trigonometric equations considered in Section 6.2 had solutions that were rational multiples of π, such as $\pi/3$, $3\pi/4$, π, and so on. If solutions are not of that type we can sometimes use inverse functions to express them in exact form, as illustrated in the next example.

EXAMPLE 8 Find the solutions of the equation

$$5\sin^2 t + 3\sin t - 1 = 0$$

that are in the interval $[-\pi/2, \pi/2]$.

SOLUTION The equation may be regarded as a quadratic equation in $\sin t$. Applying the Quadratic Formula,

$$\sin t = \frac{-3 \pm \sqrt{9+20}}{10} = \frac{-3 \pm \sqrt{29}}{10}.$$

Using the definition of the inverse sine function, we obtain the solutions

$$t = \sin^{-1} \tfrac{1}{10}(-3 + \sqrt{29})$$

and

$$t = \sin^{-1} \tfrac{1}{10}(-3 - \sqrt{29}).$$

If approximations are desired, we place a calculator in radian mode and proceed as follows:

Enter: $\frac{1}{10}(-3 + \sqrt{29}) \approx 0.2385165$

Press [INV] [sin]: 0.240838

Hence, $t \approx 0.2408$.

Similarly,

$$t = \sin^{-1} \tfrac{1}{10}(-3 - \sqrt{29}) \approx \sin^{-1}(-0.8385165) \approx -0.9946.$$ ■

Exercises 6.6

Find the numbers in Exercises 1–8 without the use of tables or calculators.

1 (a) $\sin^{-1} \frac{1}{2}$
(b) $\cos^{-1}(\sqrt{3}/2)$

2 (a) $\sin^{-1}(\sqrt{3}/2)$
(b) $\sin^{-1}(-\sqrt{3}/2)$

3 (a) $\arcsin 0$
(b) $\arccos 1$

4 (a) $\cos^{-1}(\sqrt{2}/2)$
(b) $\cos^{-1}(-\sqrt{2}/2)$

5 (a) $\arcsin(-1)$
(b) $\arccos(-1)$

6 (a) $\tan^{-1}(-1)$
(b) $\arccos(-\frac{1}{2})$

7 (a) $\tan^{-1}\sqrt{3}$
(b) $\arctan(-\sqrt{3})$

8 (a) $\tan^{-1} 0$
(b) $\arctan 1$

Approximate the numbers in Exercises 9–14 by using (a) Table 4; (b) a calculator.

9 $\cos^{-1}(0.5616)$

10 $\sin^{-1}(0.1994)$

11 $\sin^{-1}(-0.6494)$

12 $\cos^{-1}(-0.9112)$

13 $\arctan(2.1775)$

14 $\arcsin(0.8004)$

Determine the numbers in Exercises 15–26 without the use of tables or calculators.

15 $\sin(\cos^{-1}\frac{1}{2})$

16 $\cos(\sin^{-1}0)$

17 $\sin(\arccos\frac{4}{5})$

18 $\tan(\tan^{-1}5)$

19 $\arcsin(\sin 5\pi/4)$

20 $\cos^{-1}(\cos 3\pi/4)$

21 $\cos(\sin^{-1}\frac{4}{5} + \tan^{-1}\frac{3}{4})$

22 $\sin(\arcsin\frac{1}{2} + \arccos 0)$

23 $\tan(\arctan\frac{4}{3} + \arccos\frac{8}{17})$

24 $\cos(2\sin^{-1}\frac{15}{17})$

25 $\sin[2\arccos(-\frac{3}{5})]$

26 $\cos(\frac{1}{2}\tan^{-1}\frac{8}{15})$

In Exercises 27–30 rewrite the expression as an algebraic expression in x. (See Example 6.)

27 $\sin(\tan^{-1}x)$

28 $\tan(\arccos x)$

29 $\cos(\frac{1}{2}\arccos x)$

30 $\cos(2\tan^{-1}x)$

Verify the identities in Exercises 31–38.

31 $\sin^{-1}x + \cos^{-1}x = \pi/2$
(*Hint:* let $\alpha = \sin^{-1}x$, $\beta = \cos^{-1}x$, and consider $\sin(\alpha + \beta)$.)

32 $\arctan x + \arctan(1/x) = \pi/2,\ x > 0$

33 $\arcsin\dfrac{2x}{1 + x^2} = 2\arctan x,\ |x| < 1$

34 $2\cos^{-1}x = \cos^{-1}(2x^2 - 1),\ 0 \le x \le 1$

35 $\arcsin(-x) = -\arcsin x$

36 $\arccos(-x) = \pi - \arccos x$

37 $\sin^{-1}x = \tan^{-1}\dfrac{x}{\sqrt{1 - x^2}}$

38 $\tan^{-1}u + \tan^{-1}v = \tan^{-1}\dfrac{u + v}{1 - uv},\ |u| < 1 \text{ and } |v| < 1$

39 Define $\cot^{-1}$ by restricting the domain of the cotangent function to the interval $(0, \pi)$.

40 Define $\sec^{-1}$ by restricting the domain of the secant function to $[0, \pi/2) \cup [\pi, 3\pi/2)$.

41 Define $\csc^{-1}$ by restricting the domain of the cosecant function to $[-\pi/2, 0) \cup (0, \pi/2]$.

42 Verify the sketch in Figure 6.10.

Sketch the graphs of the equations in Exercises 43–52.

43 $y = \sin^{-1}2x$

44 $y = \cos^{-1}(x/2)$

45 $y = \frac{1}{2}\sin^{-1}x$

46 $y = 2\cos^{-1}x$

47 $y = 2\tan^{-1}x$

48 $y = \tan^{-1}2x$

49 $y = 2 + \tan^{-1}x$

50 $y = \sin^{-1}(x + 1)$

51 $y = \sin(\arccos x)$

52 $y = \sin(\sin^{-1}x)$

Prove that the equations in Exercises 53 and 54 are not identities.

53 $\tan^{-1}x = \dfrac{1}{\tan x}$

54 $(\arcsin x)^2 + (\arccos x)^2 = 1$

In Exercises 55–60, (a) use inverse trigonometric functions to find the solutions of the equations in the given intervals and (b) use a calculator to approximate the solutions in (a) to four decimal places.

55 $2\tan^2 t + 9\tan t + 3 = 0;\ (-\pi/2, \pi/2)$

56 $3\sin^2 t + 7\sin t + 3 = 0;\ [-\pi/2, \pi/2]$

57 $15\cos^4 x - 14\cos^2 x + 3 = 0;\ [0, \pi]$

58 $3\tan^4\theta - 19\tan^2\theta + 2 = 0;\ (-\pi/2, \pi/2)$

59 $6\sin^3\theta + 18\sin^2\theta - 5\sin\theta - 15 = 0;\ (-\pi/2, \pi/2)$

60 $6\sin 2x - 8\cos x + 9\sin x - 6 = 0;\ (-\pi/2, \pi/2)$

Calculator Exercises 6.6

1 Use a calculator to show that $\sin^{-1}(\sin 2) \neq 2$. Explain why 2 is not the answer. Is $\cos^{-1}(\cos 2) = 2$? Why?

2 Use a calculator to show that $\cos^{-1}(\cos(-1)) \neq -1$. Explain why -1 is not the answer. Is $\sin^{-1}(\sin(-1)) = -1$? Why?

3 If a number x between -1 and 1 is entered into a calculator, is it always possible, by pressing INV SIN twice, to calculate $\sin^{-1}(\sin^{-1} x)$? If not, decide which values of x will avoid an error message.

4 For any real number x, will an error message ever result when INV TAN is pressed three times in succession? Explain.

5 The function $\tan^{-1}$ is the only inverse trigonometric function available in many computer languages. (For example, in BASIC the function is denoted by ATN(X).) Express $\sin^{-1} x$, $\cos^{-1} x$, and $\cot^{-1} x$ in terms of $\tan^{-1}$. (See Exercise 37.) Are your formulas valid over the entire domain of the inverse function?

Exercises 6 and 7 employ the notation introduced in the Calculator Exercises of Section 5.5. Use a calculator to support the given statements by substituting the following values for x: 0.1, 0.01, 0.001, 0.0001.

6 $\dfrac{\arcsin 2x}{\arcsin x} \to 2 \quad \text{as} \quad x \to 0^+$

7 $\csc x \tan^{-1} x \to 1 \quad \text{as} \quad x \to 0^+$

6.7 Review

Define or discuss each of the following.

1 The Fundamental Identities
2 Verifying identities
3 Trigonometric equations
4 Addition and subtraction formulas
5 Double-angle formulas
6 Half-angle formulas
7 Product formulas
8 Factoring formulas
9 Reduction formulas
10 Inverse trigonometric functions

Exercises 6.7

Verify the identities in Exercises 1–16.

1 $(\cot^2 x + 1)(1 - \cos^2 x) = 1$

2 $\cos\theta + \sin\theta\tan\theta = \sec\theta$

3 $\dfrac{(\sec^2\theta - 1)\cot\theta}{\tan\theta\sin\theta + \cos\theta} = \sin\theta$

4 $(\tan x + \cot x)^2 = \sec^2 x \csc^2 x$

5 $\dfrac{1}{1 + \sin t} = (\sec t - \tan t)\sec t$

6 $\dfrac{\sin(\alpha - \beta)}{\cos(\alpha + \beta)} = \dfrac{\tan\alpha - \tan\beta}{1 - \tan\alpha\tan\beta}$

7 $\dfrac{2\cot u}{\csc^2 u - 2} = \tan 2u$

8 $\cos^2 \dfrac{v}{2} = \dfrac{1 + \sec v}{2 \sec v}$

9 $\dfrac{\tan^3 \phi - \cot^3 \phi}{\tan^2 \phi + \csc^2 \phi} = \tan \phi - \cot \phi$

10 $\dfrac{\sin u + \sin v}{\csc u + \csc v} = \dfrac{1 - \sin u \sin v}{-1 + \csc u \csc v}$

11 $\cos\left(x - \dfrac{5\pi}{2}\right) = \sin x$

12 $\tan\left(x + \dfrac{3\pi}{4}\right) = \dfrac{\tan x - 1}{\tan x + 1}$

13 $\frac{1}{4} \sin 4\beta = \sin \beta \cos^3 \beta - \cos \beta \sin^3 \beta$

14 $\tan \frac{1}{2}\theta = \csc \theta - \cot \theta$

15 $\sin 8\theta = 8 \sin \theta \cos \theta\, (1 - 2 \sin^2 \theta)(1 - 8 \sin^2 \theta \cos^2 \theta)$

16 $\arctan x = \frac{1}{2} \arctan \dfrac{2x}{1 - x^2},\ |x| \leq 1$

In Exercises 17–28 find the solutions of the equation that are in the interval $[0, 2\pi)$, and also find the degree measure of each solution.

17 $2 \cos^3 \theta - \cos \theta = 0$

18 $2 \cos \alpha + \tan \alpha = \sec \alpha$

19 $\sin \theta = \tan \theta$

20 $\csc^5 \theta - 4 \csc \theta = 0$

21 $2 \cos^3 t + \cos^2 t - 2 \cos t - 1 = 0$

22 $\cos x \cot^2 x = \cos x$

23 $\sin \beta + 2 \cos^2 \beta = 1$

24 $\cos 2x + 3 \cos x + 2 = 0$

25 $2 \sec u \sin u + 2 = 4 \sin u + \sec u$

26 $\sin 2u = \sin u$

27 $2 \cos^2 \frac{1}{2}\theta - 3 \cos \theta = 0$

28 $\sec 2x \csc 2x = 2 \csc 2x$

In Exercises 29–32 find the exact values without the use of tables or calculators.

29 $\cos 75°$ **30** $\tan 285°$

31 $\sin 195°$ **32** $\csc \pi/8$

For acute angles θ and ϕ such that $\csc \theta = \frac{5}{3}$ and $\cos \phi = \frac{8}{17}$, find the numbers in Exercises 33–41 without the use of tables or calculators.

33 $\sin(\theta + \phi)$ **34** $\cos(\theta + \phi)$ **35** $\tan(\theta - \phi)$

36 $\sin(\phi - \theta)$ **37** $\sin 2\phi$ **38** $\cos 2\phi$

39 $\tan 2\theta$ **40** $\sin(\theta/2)$ **41** $\tan(\theta/2)$

42 Express $\cos(\alpha + \beta + \gamma)$ in terms of functions of α, β, and γ.

43 Express each product as a sum or difference.

(a) $\sin 7t \sin 4t$

(b) $\cos(u/4) \cos(-u/6)$

(c) $6 \cos 5x \sin 3x$

44 Express each of the following as a product.

(a) $\sin 8u + \sin 2u$

(b) $\cos 3\theta - \cos 8\theta$

(c) $\sin(t/4) - \sin(t/5)$

Find the numbers in Exercises 45–53 without the use of tables or calculators.

45 $\cos^{-1}\left(\dfrac{-\sqrt{3}}{2}\right)$ **46** $\sin^{-1}\left(-\dfrac{\sqrt{2}}{2}\right)$

47 $\arccos\left(\tan \dfrac{3\pi}{4}\right)$ **48** $\arctan\left(\dfrac{-\sqrt{3}}{3}\right)$

49 $\sin \arccos\left(\dfrac{-\sqrt{3}}{2}\right)$

50 $\cos(\sin^{-1} \frac{15}{17} - \sin^{-1} \frac{8}{17})$

51 $\cos(2 \sin^{-1} \frac{4}{5})$ **52** $\sin(\sin^{-1} \frac{2}{3})$

53 $\cos^{-1}(\sin 0)$

Sketch the graphs of the equations in Exercises 54–56.

54 $y = \cos^{-1} 3x$ **55** $y = 4 \sin^{-1} x$

56 $y = 1 - \sin^{-1} x$

57 When an individual is walking, the magnitude F of the vertical force of one foot on the ground (see figure) can be described by $F = A(\cos bt - a \cos 3bt)$, where t is the time in seconds, $A > 0$, $b > 0$, and $0 < a < 1$.

(a) Show that $F = 0$ when $t = -\pi/2b$ and $\pi/2b$. (The time $t = -\pi/2b$ corresponds to the moment when the foot first touches the ground and the weight of the body is being supported by the other foot.)

(b) Using calculus, it can be shown that the maximum force occurs when $3a \sin 3bt = \sin bt$. If $a = \frac{1}{3}$, find the solutions of this equation for $-\pi/2b < t < \pi/2b$. (*Hint:* Express $\sin 3bt$ in terms of $\sin bt$.)

(c) If $a = \frac{1}{3}$, express the maximum force in terms of A.

FIGURE FOR EXERCISE 57

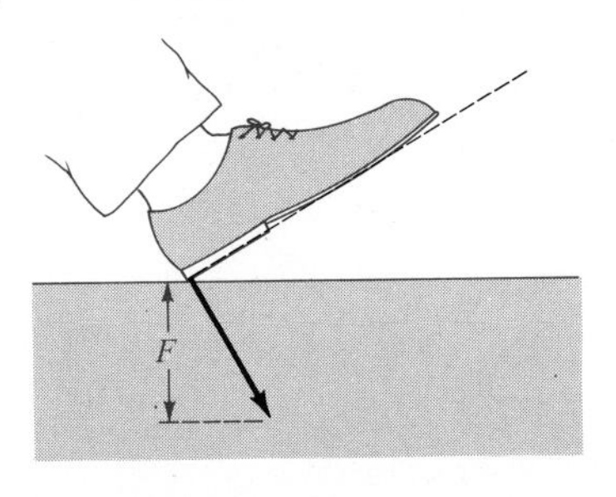

58 Shown in the figure is a computer-generated graph of $f(x) = \sin x - \frac{1}{2} \sin 2x + \frac{1}{3} \sin 3x$. The x-coordinates of peaks and valleys on the graph satisfy the equation $\cos x - \cos 2x + \cos 3x = 0$. Use factoring formulas to find these x-coordinates.

FIGURE FOR EXERCISE 58

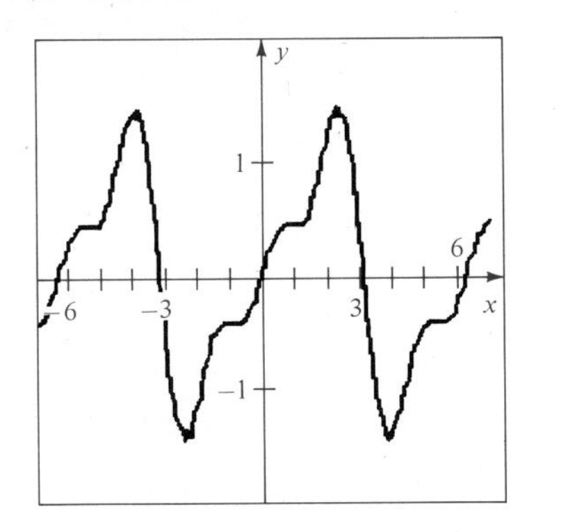

59 The human eye can distinguish clearly between two distant points P and Q provided the angle of resolution θ is not too small. Suppose P and Q are x units apart and are d units from the eye, as illustrated in the figure.

(a) Express x as a function of d and θ.

(b) For a person with normal vision, the smallest angle of resolution is about 0.0005 radians. A pen 6 inches long is viewed by such an individual at a distance of d feet. For what values of d will the endpoints of the pen be clearly distinguishable?

FIGURE FOR EXERCISE 59

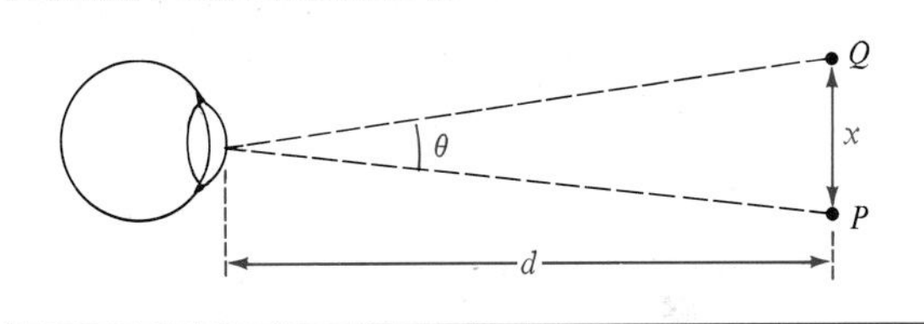

60 A satellite S circles a planet at a distance d miles from the planet's surface. The portion of the planet's surface that is visible from the satellite is determined by the angle θ shown in the figure.

(a) Assuming that the planet is spherical in shape, express d as a function of θ and the radius r of the planet.

(b) If we use $r = 4000$ miles for the radius of the earth, approximate θ for a satellite 300 miles from the surface.

FIGURE FOR EXERCISE 60

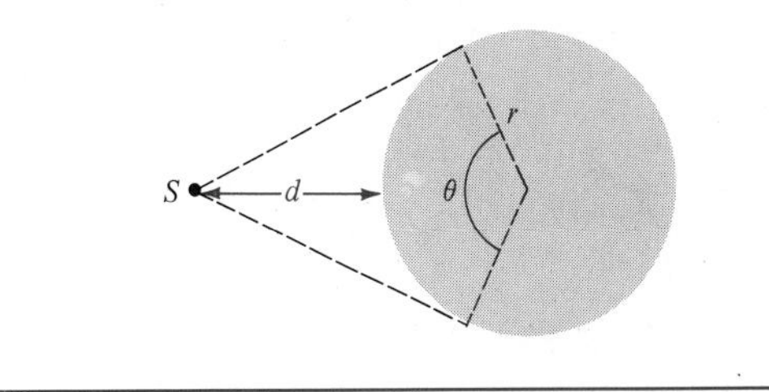

CHAPTER 7

Applications of Trigonometry

This chapter contains some applications of trigonometry to geometry and algebra. ■ First we consider methods for solving oblique triangles. ■ We next introduce the trigonometric form for complex numbers and obtain an important result about nth roots. ■ The chapter concludes with sections on polar coordinates and vectors.

7.1 The Law of Sines

A triangle that does not contain a right angle is called an **oblique triangle.** If two angles and a side are known, or if two sides and an angle opposite one of them are known, then the remaining parts of an oblique triangle may be found by means of the formulas discussed in this section. We shall use the letters A, B, C, a, b, c, α, β, and γ for parts of triangles as they were used in Chapter 5. Given triangle ABC, let us place angle α in standard position on a rectangular coordinate system so that B is on the positive x-axis. The case in which α is obtuse is illustrated in Figure 7.1; however, the following discussion is also valid if α is acute.

FIGURE 7.1

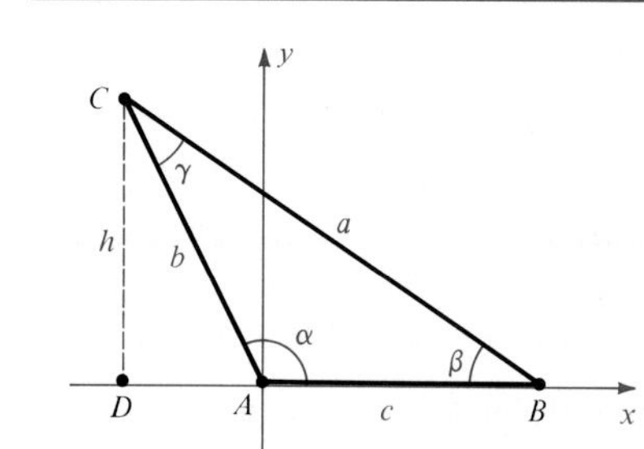

Consider the line through C parallel to the y-axis and intersecting the x-axis at point D. Suppose that $d(C, D) = h$, so that the y-coordinate of C is h. It follows that

$$\sin\alpha = \frac{h}{b} \quad \text{or} \quad h = b\sin\alpha.$$

Referring to triangle BDC, we see that

$$\sin\beta = \frac{h}{a} \quad \text{or} \quad h = a\sin\beta.$$

Consequently, $$b\sin\alpha = a\sin\beta,$$

which may be written $$\frac{\sin\alpha}{a} = \frac{\sin\beta}{b}.$$

If α is taken in standard position, so that C is on the positive x-axis, then by the same reasoning,

$$\frac{\sin\alpha}{a} = \frac{\sin\gamma}{c}.$$

The last two equalities give us the following result.

THE LAW OF SINES

If ABC is an oblique triangle labeled in the usual manner, then

$$\frac{\sin\alpha}{a} = \frac{\sin\beta}{b} = \frac{\sin\gamma}{c}.$$

The Law of Sines can also be written in the form

$$\frac{a}{\sin \alpha} = \frac{b}{\sin \beta} = \frac{c}{\sin \gamma}.$$

The next example illustrates a method of applying the Law of Sines to the case in which two angles and a side of a triangle are known. We shall use the rules for rounding off answers stated in Section 5.8.

EXAMPLE 1 Given triangle ABC with $\alpha = 48°20'$, $\gamma = 57°30'$, and $b = 47.3$, approximate the remaining parts.

FIGURE 7.2

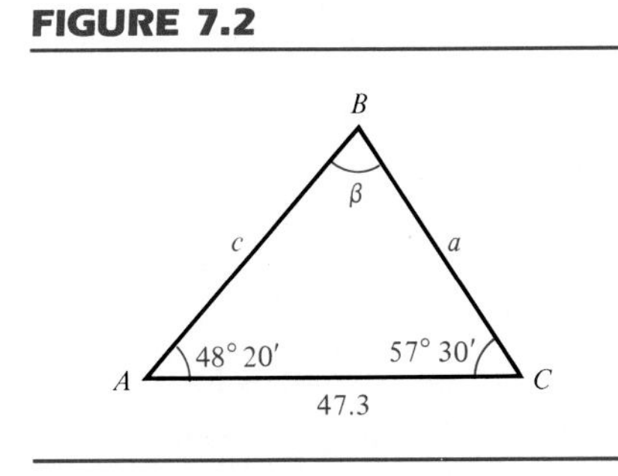

SOLUTION The triangle is sketched in Figure 7.2. Since the sum of the angles is 180°,

$$\beta = 180° - (57°30' + 48°20') = 74°10'.$$

Since

$$\frac{a}{\sin \alpha} = \frac{b}{\sin \beta}$$

we obtain

$$a = \frac{b \sin \alpha}{\sin \beta} = \frac{(47.3) \sin 48°20'}{\sin 74°10'}.$$

Using Table 4 or a calculator,

$$a \approx \frac{(47.3)(0.7470)}{(0.9621)} \approx 36.7.$$

Similarly,

$$c = \frac{b \sin \gamma}{\sin \beta} = \frac{(47.3) \sin 57°30'}{\sin 74°10'}$$

$$\approx \frac{(47.3)(0.8434)}{(0.9621)} \approx 41.5.$$

■

FIGURE 7.3

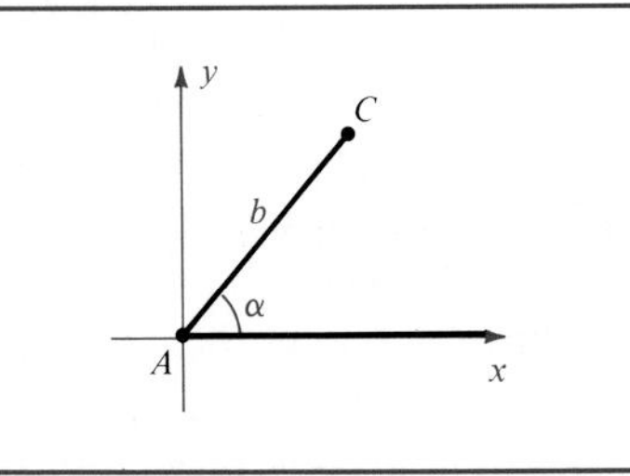

Data such as that in Example 1 always produces a unique triangle ABC. However, if two sides and an angle *opposite* one of them are given, a unique triangle is not always determined. To illustrate, suppose that a and b are to be lengths of sides of triangle ABC and that a specific angle α is to be opposite the side of length a. Let us consider the case in which α is acute. Place α in standard position on a rectangular coordinate system and consider the line segment AC of length b on the terminal side of α, as shown in Figure 7.3. The third vertex, B, should be somewhere on the x-axis. Since the length a of the side opposite α is given, B may be found by striking off a circular arc of length a with center at C. There are four possible outcomes for the construction, as illustrated in Figure 7.4, where the coordinate axes have been deleted.

FIGURE 7.4

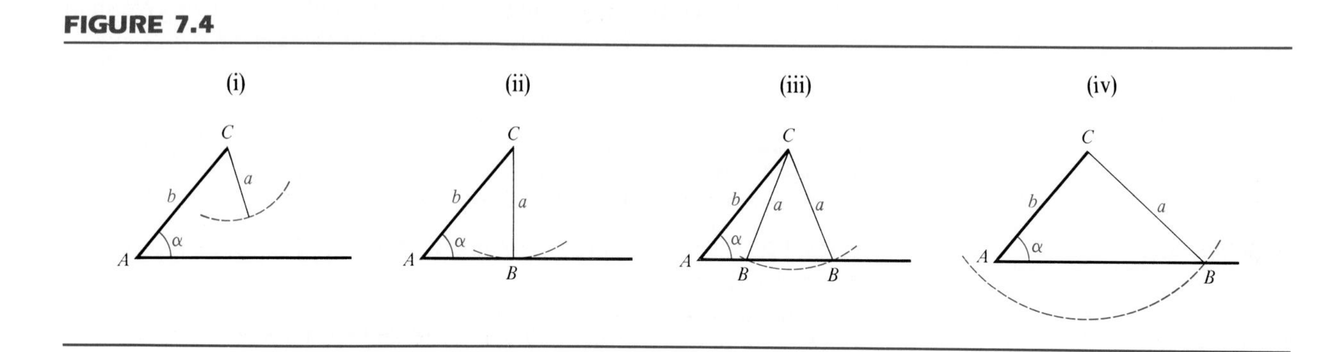

The four possibilities may be listed as follows:

(i) The arc does not intersect the x-axis and no triangle is formed.

(ii) The arc is tangent to the x-axis and a right triangle is formed.

(iii) The arc intersects the positive x-axis in two distinct points and two triangles are formed.

(iv) The arc intersects both the positive and nonpositive parts of the x-axis and one triangle is formed.

FIGURE 7.5

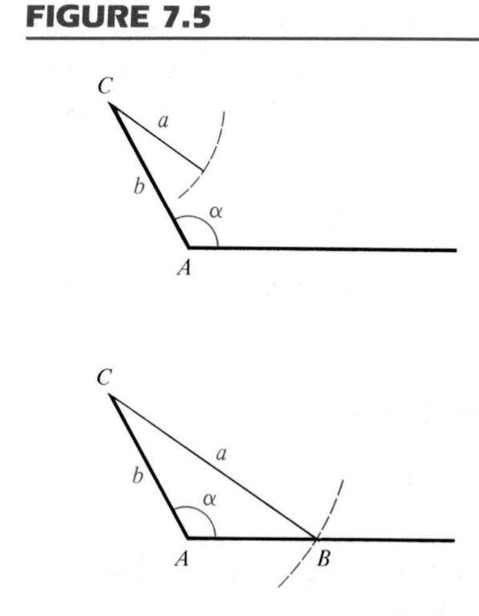

The case that occurs in a specific problem will become evident when the solution is attempted. For example, when solving the equation

$$\frac{\sin \alpha}{a} = \frac{\sin \beta}{b}$$

suppose that we obtain $\sin \beta > 1$. This will indicate that no triangle exists. If we obtain $\sin \beta = 1$, then $\beta = 90^\circ$ and hence case (ii) occurs. If $\sin \beta < 1$, then there are two possible choices for the angle β. By checking both possibilities, it will become apparent whether (iii) or (iv) occurs.

If the measure of α is greater than 90°, then, as shown in Figure 7.5, a triangle exists if and only if $a > b$. Since different possibilities may arise, the case in which two sides and an angle opposite one of them are given is sometimes called the **ambiguous case.**

EXAMPLE 2 Find the remaining parts of triangle ABC if $\alpha = 67°$, $c = 125$, and $a = 100$.

SOLUTION Using $\sin\gamma = \dfrac{c \sin\alpha}{a}$,

$$\sin\gamma = \frac{(125)\sin 67°}{100}$$

$$\approx \frac{(125)(0.9205)}{100} \approx 1.1506.$$

Since $\sin\gamma > 1$, no triangle can be constructed with the given parts. ■

EXAMPLE 3 Approximate the remaining parts of triangle ABC if $a = 12.4$, $b = 8.7$, and $\beta = 36°40'$.

SOLUTION Using the fact that $\sin\alpha = \dfrac{a \sin\beta}{b}$,

$$\sin\alpha = \frac{(12.4)\sin 36°40'}{8.7}$$

$$\approx \frac{(12.4)(0.5972)}{8.7} \approx 0.8512.$$

There are two possible angles α between 0° and 180° such that $\sin\alpha \approx 0.8512$. If we let α' denote the reference angle for α, then using Table 4 or a calculator, we obtain $\alpha' \approx 58°20'$. Consequently, the two possibilities for α are

$$\alpha_1 \approx 58°20' \quad \text{and} \quad \alpha_2 \approx 121°40'.$$

If we let γ_1 and γ_2 denote the third angles of the triangles corresponding to the angles α_1 and α_2, respectively, then

$$\gamma_1 \approx 180° - (36°40' + 58°20') = 85°$$

and

$$\gamma_2 \approx 180° - (36°40' + 121°40') = 21°40'.$$

Thus, there are two possible triangles that have the given parts. They are the triangles A_1BC and A_2BC shown in Figure 7.6.

FIGURE 7.6

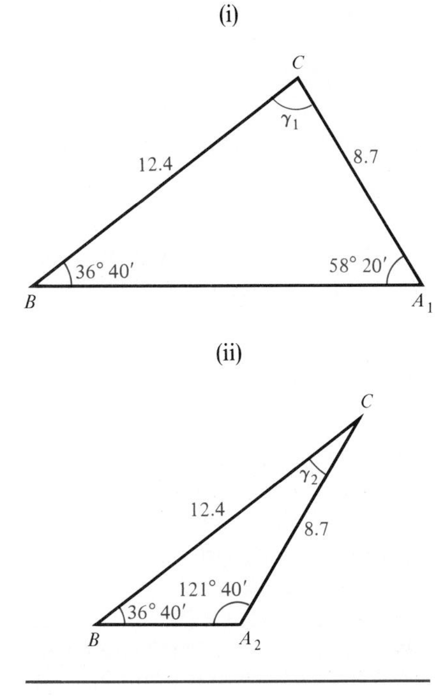

If c_1 is the side opposite γ_1 in triangle A_1BC, then

$$c_1 = \frac{a \sin \gamma_1}{\sin \alpha_1} \approx \frac{(12.4) \sin 85^\circ}{\sin 58^\circ 20'}$$

$$\approx \frac{(12.4)(0.9962)}{0.8511} \approx 14.5.$$

Hence, the remaining parts of triangle A_1BC are $\alpha_1 \approx 58^\circ 20'$, $\gamma_1 \approx 85^\circ$, and $c_1 \approx 14.5$.

If c_2 is the side opposite angle γ_2 in triangle A_2BC, we have

$$c_2 = \frac{a \sin \gamma_2}{\sin \alpha_2} \approx \frac{(12.4) \sin 21^\circ 40'}{\sin 121^\circ 40'} \approx \frac{(12.4)(0.3692)}{0.8511} \approx 5.4.$$

Consequently, the remaining parts of triangle A_2BC are $\alpha_2 \approx 121^\circ 40'$, $\gamma_2 \approx 21^\circ 40'$, and $c_2 \approx 5.4$. ■

FIGURE 7.7

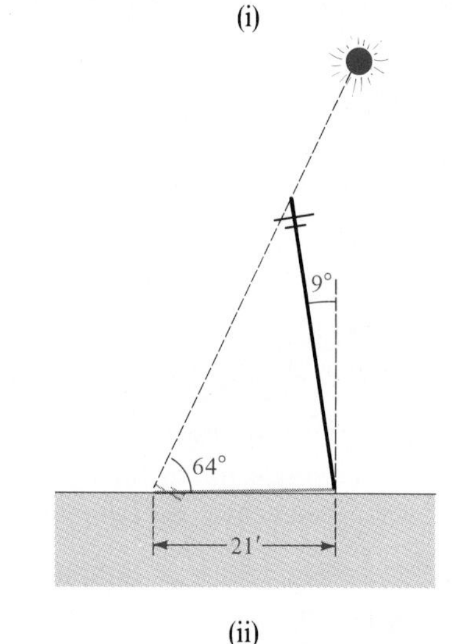

EXAMPLE 4 When the angle of elevation of the sun is 64°, a telephone pole that is tilted at an angle of 9° directly away from the sun casts a shadow 21 feet long on level ground. Approximate the length of the pole.

SOLUTION In applied problems it is important to sketch a diagram and label it appropriately, as illustrated in Figure 7.7(i) (not drawn to scale). In (ii) of the figure we also consider a triangle ABC that displays the given facts. Either of these two drawings is sufficient for our purposes. Note that in Figure 7.7(ii) we have calculated $\beta = 90^\circ - 9^\circ = 81^\circ$. Hence

$$\gamma = 180^\circ - (64^\circ + 81^\circ) = 180^\circ - 145^\circ = 35^\circ.$$

The length of the pole is side a of triangle ABC. Applying the Law of Sines,

$$\frac{a}{\sin 64^\circ} = \frac{21}{\sin \gamma}.$$

Since $\gamma = 35^\circ$, this gives us

$$a = \frac{(21) \sin 64^\circ}{\sin 35^\circ} = \frac{(21)(0.8988)}{0.5736} \approx 33.$$

Thus the telephone pole is approximately 33 feet in length. ■

EXAMPLE 5 A point P on level ground is 3.0 km due north of a point Q. A runner proceeds in the direction N25°E from Q to a point R, and then from R to P in the direction S70°W. Approximate the distance run.

FIGURE 7.8

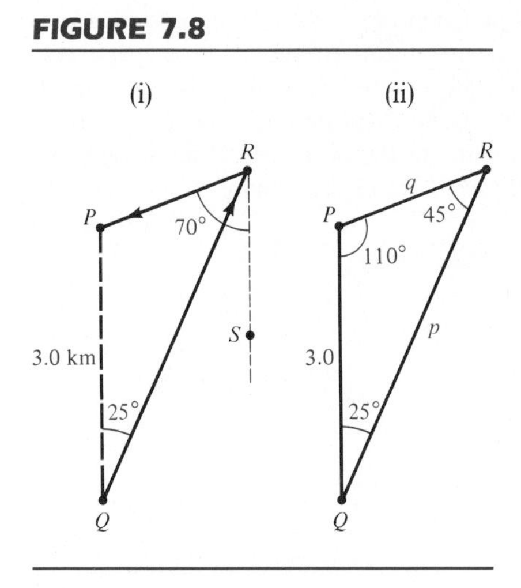

SOLUTION The notation used to specify directions was introduced in Section 5.8. Figure 7.8(i) shows the path of the runner, together with a north-south (dashed) line from R to another point S.

Since the lines through PQ and RS are parallel, it follows from geometry that the alternate interior angles $\angle PQR$ and $\angle QRS$ both have measure 25°. Hence

$$\angle PRQ = 70^\circ - 25^\circ = 45^\circ.$$

These observations give us triangle PQR in Figure 7.8(ii), where

$$\angle QPR = 180^\circ - (25^\circ + 45^\circ) = 180^\circ - 70^\circ = 110^\circ.$$

Applying the Law of Sines twice gives us

$$\frac{q}{\sin 25^\circ} = \frac{3.0}{\sin 45^\circ} \quad \text{and} \quad \frac{p}{\sin 110^\circ} = \frac{3.0}{\sin 45^\circ}.$$

Hence

$$q = \frac{(3.0)\sin 25^\circ}{\sin 45^\circ} \approx \frac{(3.0)(0.4226)}{0.7071} \approx 1.8$$

and

$$p = \frac{(3.0)\sin 110^\circ}{\sin 45^\circ} \approx \frac{(3.0)(0.9397)}{0.7071} \approx 4.0$$

The distance run, to the nearest kilometer, is $p + q = 5.8$ km. ■

Exercises 7.1

In Exercises 1–12 approximate the remaining parts of triangle ABC.

1 $\alpha = 41^\circ,\ \gamma = 77^\circ,\ a = 10.5$

2 $\beta = 20^\circ,\ \gamma = 31^\circ,\ b = 210$

3 $\alpha = 27^\circ 40',\ \beta = 52^\circ 10',\ a = 32.4$

4 $\alpha = 42^\circ 10',\ \gamma = 61^\circ 20',\ b = 19.7$

5 $\beta = 50^\circ 50',\ \gamma = 70^\circ 30',\ c = 537$

6 $\alpha = 7^\circ 10',\ \beta = 11^\circ 40',\ a = 2.19$

7 $\alpha = 65^\circ 10',\ a = 21.3,\ b = 18.9$

8 $\beta = 30^\circ,\ b = 17.9,\ a = 35.8$

9 $\gamma = 53^\circ 20',\ a = 140,\ c = 115$

10 $\alpha = 27^\circ 30',\ c = 52.8,\ a = 28.1$

11 $\beta = 113^\circ 10',\ b = 248,\ c = 195$

12 $\gamma = 81^\circ,\ c = 11,\ b = 12$

13 An engineer wishes to find the distance between two points A and B that lie on opposite banks of a river. A line segment AC of length 240 yards is laid off and the measures of angles BAC and ACB are found to be $63^\circ 20'$

and 54°10′, respectively (see figure). Approximate the distance from A to B.

FIGURE FOR EXERCISE 13

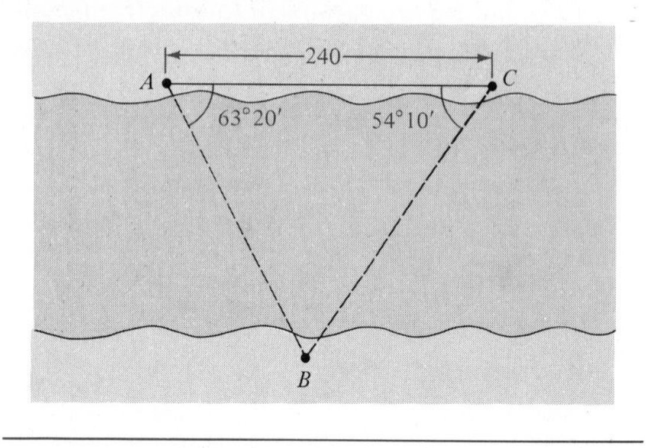

14 In order to determine the distance between two points A and B, a surveyor chooses a point C that is 375 yards from A and 530 yards from B. If $\angle BAC$ has measure 49°30′, approximate the distance between A and B.

15 A cable car carries passengers from point A, which is 1.2 miles from the base of a mountain, to Lookout Peak at point P. As shown in the figure, the angle of elevation of P from A is 21°, while the angle of elevation from point B at the base of the mountain is 65°.

(a) How far does the cable car travel from A to P?

(b) What is the difference in elevation between points A and P?

FIGURE FOR EXERCISE 15

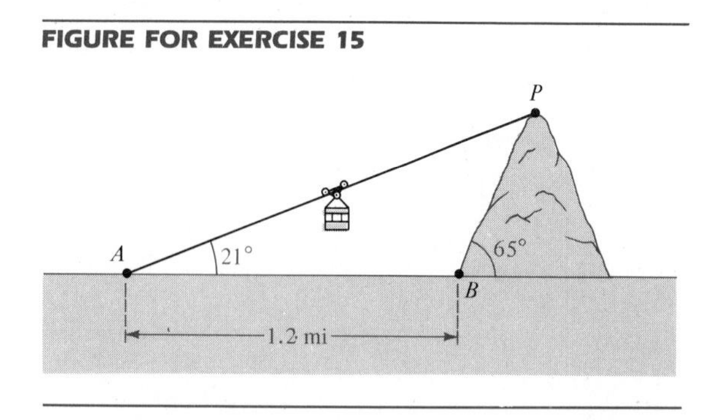

16 A straight road makes an angle of 15° with the horizontal. When the angle of elevation of the sun is 57°, a vertical pole at the side of the road casts a shadow 75 feet long directly down the road. Approximate the length of the pole.

17 The angles of elevation of a balloon from two points A and B on level ground are 24°10′ and 47°40′, respectively. As shown in the figure, points A and B are 8.4 miles apart and the balloon is between the points, in the same vertical plane. Approximate, to the nearest tenth of a mile, the height of the balloon above the ground.

FIGURE FOR EXERCISE 17

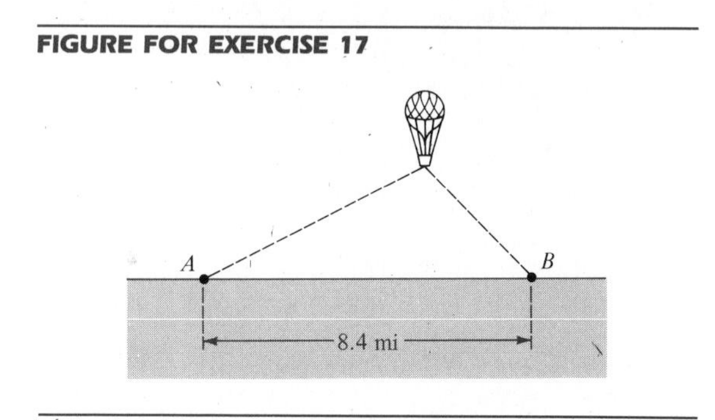

18 Shown in the figure is a solar panel that is 10 feet in width and is to be attached to a roof that makes an angle of 25° with the horizontal. Approximate the length d of the brace that is needed if the panel must make an angle of 45° with the horizontal.

FIGURE FOR EXERCISE 18

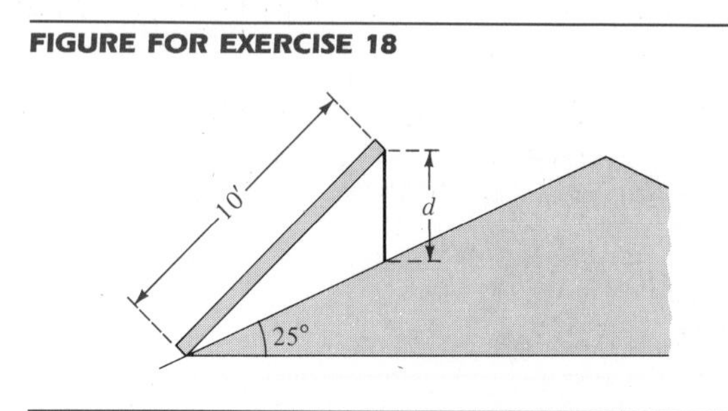

19 A forest ranger at an observation point A sights a fire in the direction N27°10′E. Another ranger at an observation point B, 6.0 miles due east of A, sights the same fire at N52°40′W. Approximate, to the nearest tenth of a mile, the distance from each of the observation points to the fire.

20 A surveyor notes that the direction from point A to point B is S63°W and the direction from A to C is S38°W. The distance from A to B is 239 yards and the distance from B to C is 374 yards. Approximate the distance from A to C.

21 A straight road makes an angle of 22° with the horizontal. From a certain point P on the road the angle of elevation of an airplane at point A is 57°. At the same instant, from another point Q, 100 meters farther up the road, the angle of elevation is 63°. As indicated in the figure, the points P, Q, and A lie in the same vertical plane. Approximate, to the nearest meter, the distance from P to the airplane.

FIGURE FOR EXERCISE 21

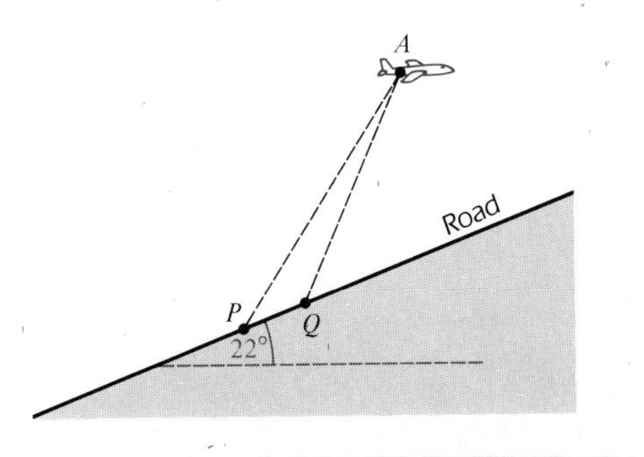

22 The leaning tower of Pisa was originally 179 feet tall, but now, due to sinking into the earth, it leans at a certain angle θ from the perpendicular (see figure). When the top of the tower is viewed from a point 150 feet from its base, the angle of elevation is 53.3°.

(a) Find the angle θ.

(b) Approximate the distance d the top of the tower has moved from the perpendicular position.

FIGURE FOR EXERCISE 22

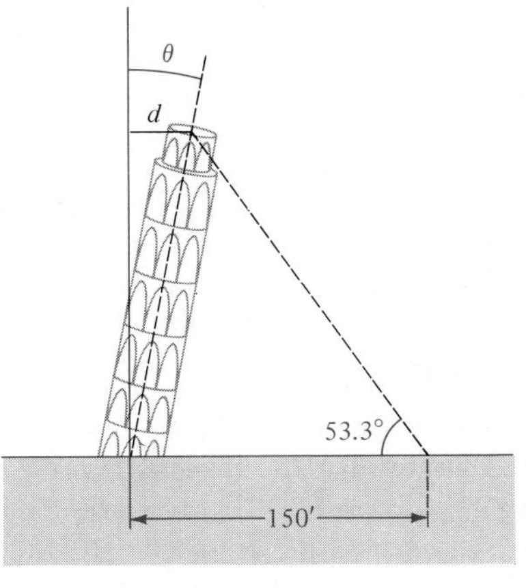

23 A cathedral is located at the top of a hill. When the top of the spire is viewed from the base of the hill, the angle of elevation is 48°. When viewed at a distance of 200 feet from the base of the hill, the angle of elevation is 41°. The hill rises at a 32° angle. Approximate the height of the cathedral.

FIGURE FOR EXERCISE 23

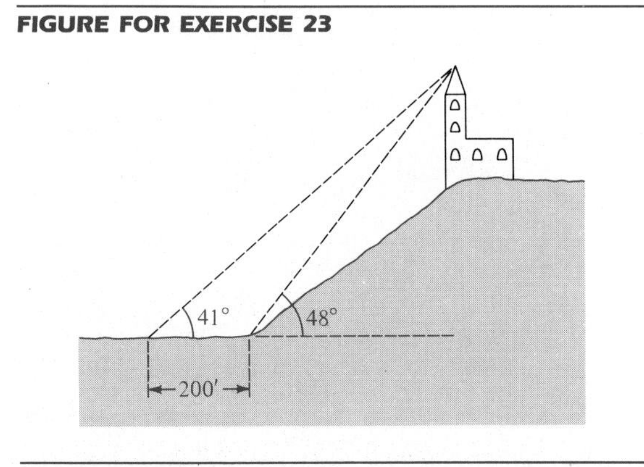

24 A helicopter hovers at an altitude that is 1000 feet above a mountain peak of altitude 5210 feet. A second, taller, peak is viewed from both the mountain top and the helicopter. From the helicopter, the angle of depression is 43°. From the mountain top, the angle of elevation is 18°. Approximate the following to the nearest foot:

(a) The distance from peak to peak.

(b) The altitude of the taller mountain peak.

FIGURE FOR EXERCISE 24

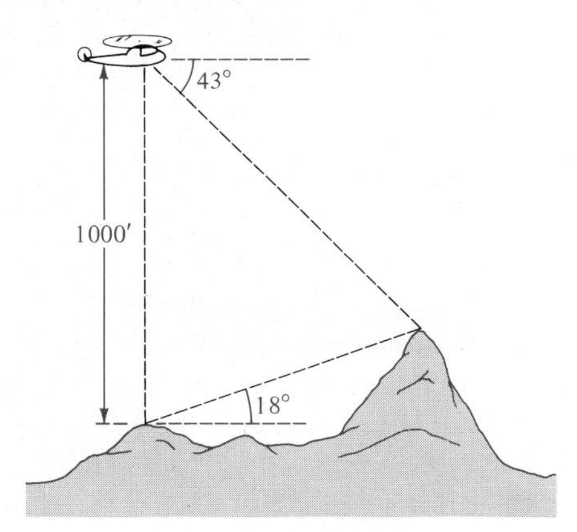

25 Shown in the figure is a right triangular prism of height h. Approximate h and the volume V of the prism.

FIGURE FOR EXERCISE 25

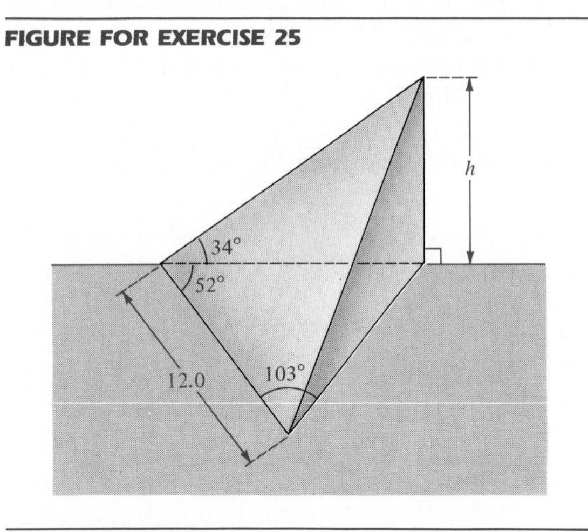

26 Shown in the figure is a top view of a Delta Dart jet with its unusual wing design. Data regarding the wing is indicated in the sketch.

(a) Approximate angle ϕ.

(b) If the fuselage is 4.80 feet wide, approximate the wing span of the plane.

(c) Estimate the area of a wing.

FIGURE FOR EXERCISE 26

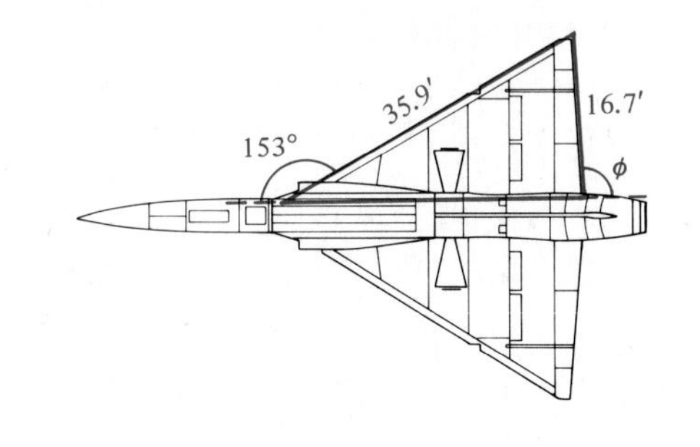

Calculator Exercises 7.1

In Exercises 1–6 approximate the remaining parts of triangle ABC.

1 $\beta = 25.6°$, $\gamma = 34.7°$, $b = 184.8$

2 $\alpha = 6.24°$, $\beta = 14.08°$, $a = 4.56$

3 $\alpha = 103.45°$, $\gamma = 27.19°$, $b = 38.84$

4 $\gamma = 47.74°$, $a = 131.08$, $c = 97.84$

5 $\beta = 121.624°$, $b = 0.283$, $c = 0.178$

6 $\alpha = 32.32°$, $c = 574.3$, $a = 263.6$

7.2 The Law of Cosines

If two sides and the included angle or if the three sides of a triangle are known, we cannot apply the Law of Sines directly to find the remaining parts. We may, however, use the following result.

THE LAW OF COSINES

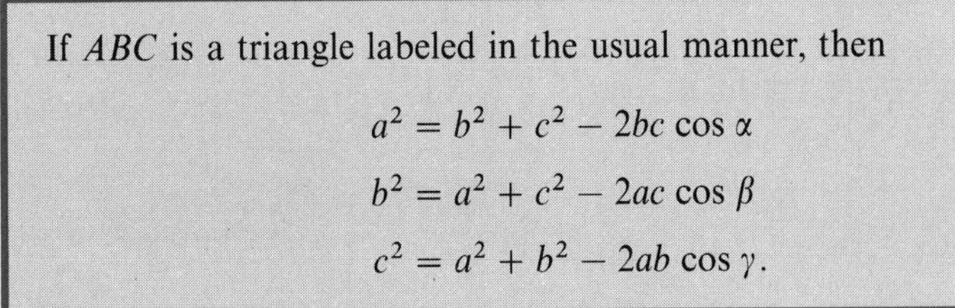

If ABC is a triangle labeled in the usual manner, then

$$a^2 = b^2 + c^2 - 2bc \cos \alpha$$

$$b^2 = a^2 + c^2 - 2ac \cos \beta$$

$$c^2 = a^2 + b^2 - 2ab \cos \gamma.$$

PROOF Let us prove the first of the three formulas. Given triangle ABC, place α in standard position on a rectangular coordinate system as illustrated in Figure 7.9. Although α is pictured as obtuse, our discussion is also valid if α is acute. Consider the line through C, parallel to the y-axis and intersecting the x-axis at the point $K(k, 0)$. If we let $d(C, k) = h$, then C has coordinates (k, h). It follows that

FIGURE 7.9

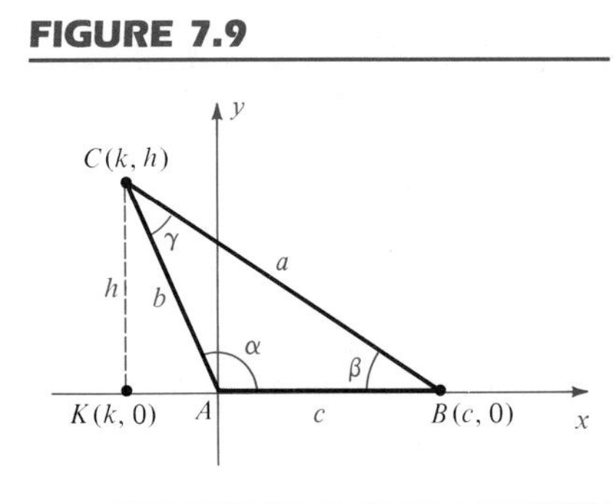

$$\cos \alpha = \frac{k}{b} \quad \text{and} \quad \sin \alpha = \frac{h}{b}$$

and hence $\qquad k = b \cos \alpha \quad \text{and} \quad h = b \sin \alpha.$

Since the segment AB has length c, the coordinates of B are $(c, 0)$. Using the Distance Formula,

$$a^2 = [d(B, C)]^2 = (k - c)^2 + (h - 0)^2.$$

Substituting for k and h gives us

$$\begin{aligned} a^2 &= (b \cos \alpha - c)^2 + (b \sin \alpha)^2 \\ &= b^2 \cos^2 \alpha - 2bc \cos \alpha + c^2 + b^2 \sin^2 \alpha \\ &= b^2 (\cos^2 \alpha + \sin^2 \alpha) + c^2 - 2bc \cos \alpha \\ &= b^2 + c^2 - 2bc \cos \alpha. \end{aligned}$$

This is the first formula stated in the Law of Cosines. The second and third formulas may be obtained by placing β and γ, respectively, in standard position on a rectangular coordinate system and using a similar procedure. □

Note that if $\alpha = 90°$ in Figure 7.9, then $\cos \alpha = 0$ and the Law of Cosines gives us $a^2 = b^2 + c^2$. This shows that the Pythagorean Theorem is a special case of the Law of Cosines.

Instead of memorizing each of the three formulas of the Law of Cosines, it is more convenient to remember the following statement, which takes all of them into account.

THE LAW OF COSINES (ALTERNATIVE FORM)

> The square of the length of any side of a triangle equals the sum of the squares of the lengths of the other two sides minus twice the product of the lengths of the other two sides and the cosine of the angle between them.

EXAMPLE 1 Approximate all parts of triangle ABC for $a = 5.0$, $c = 8.0$, and $\beta = 77°10'$.

FIGURE 7.10

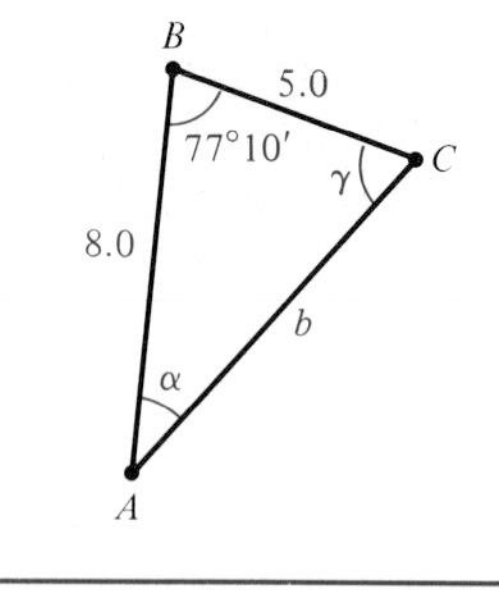

SOLUTION The triangle is sketched in Figure 7.10. Using the Law of Cosines,

$$\begin{aligned} b^2 &= (5.0)^2 + (8.0)^2 - 2(5.0)(8.0)\cos 77°10' \\ &\approx 25 + 64 - (80)(0.2221) \approx 71.2. \end{aligned}$$

Consequently, $b \approx \sqrt{71.2} \approx 8.44 \approx 8.4$.

We next use the Law of Sines to find α. Thus,

$$\sin\alpha = \frac{a\sin\beta}{b} \approx \frac{(5.0)\sin 77°10'}{8.44} \approx \frac{(5.0)(0.9750)}{8.44} \approx 0.5776.$$

Consulting Table 4 or using a calculator, we see that $\alpha \approx 35°20'$, Hence,

$$\gamma \approx 180° - (77°10' + 35°20') = 67°30'.$$ ■

After we found the length of the third side in Example 1, we used the Law of Sines to find a second angle of the triangle. Whenever this procedure is followed, it is best to find the angle opposite the shortest side (as we did), since that angle is always acute. Of course, the Law of Cosines can also be used to find the remaining angles.

EXAMPLE 2 For triangle ABC with sides $a = 90$, $b = 70$, and $c = 40$, approximate angles α, β, and γ.

SOLUTION Using the first formula of the Law of Cosines,

$$\begin{aligned} \cos\alpha &= \frac{b^2 + c^2 - a^2}{2bc} \\ &= \frac{4900 + 1600 - 8100}{5600} \approx -0.2857. \end{aligned}$$

Using Table 4 or a calculator, we see that the reference angle for α is approximately $73°20'$. Hence, $\alpha \approx 180° - 73°20' = 106°40'$.

Similarly, from the second formula of the Law of Cosines,

$$\cos\beta = \frac{a^2 + c^2 - b^2}{2ac}$$

$$= \frac{8100 + 1600 - 4900}{7200} \approx 0.6667$$

and hence $\beta \approx 48°10'$. Finally,

$$\gamma \approx 180° - (106°40' + 48°10') = 25°10'. \qquad \blacksquare$$

After finding the first angle in the preceding solution, we could have used the Law of Sines to find a second angle. Whenever this procedure is followed, it is best to use the Law of Cosines to find the angle opposite the longer side, that is, the largest angle of the triangle (as we did). This will guarantee that the remaining two angles are acute.

EXAMPLE 3 A parallelogram has sides of lengths 30 cm and 70 cm and one of the angles has measure 65°. Approximate the length of each diagonal.

FIGURE 7.11

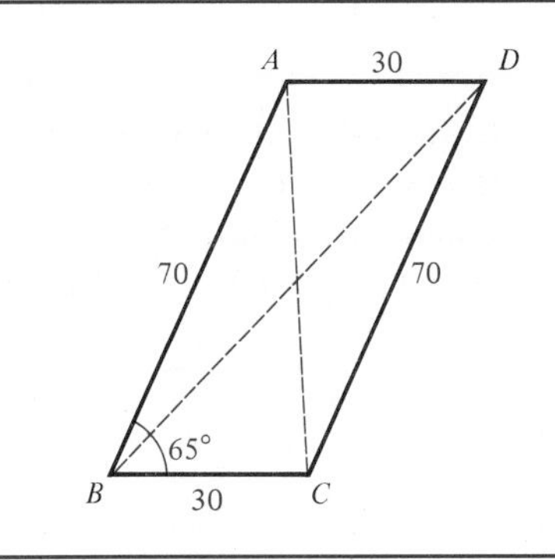

SOLUTION The parallelogram and its diagonals AC and BD are shown in Figure 7.11. Since opposite angles are equal, the measure of $\angle BAD$ is $180° - 65° = 115°$. Applying the Law of Cosines to triangle ABC,

$$(AC)^2 = (30)^2 + (70)^2 - 2(30)(70)\cos 65°$$

$$\approx 900 + 4900 - 4200(0.4226) \approx 4025.1$$

and therefore $AC \approx \sqrt{4025.1} \approx 63.44 \approx 63$.

Similarly, using triangle BAD,

$$(BD)^2 = (30)^2 + (70)^2 - 2(30)(70)\cos 115°$$

$$\approx 900 + 4900 - 4200(-0.4226) \approx 7574.9$$

Hence $BD \approx \sqrt{7574.9} \approx 87$. $\blacksquare$

EXAMPLE 4 A vertical pole 40 feet tall stands on a hillside that makes an angle of 17° with the horizontal. Approximate the minimal length of cable that will reach from the top of the pole to a point directly down the hill 72 feet from the base of the pole.

FIGURE 7.12

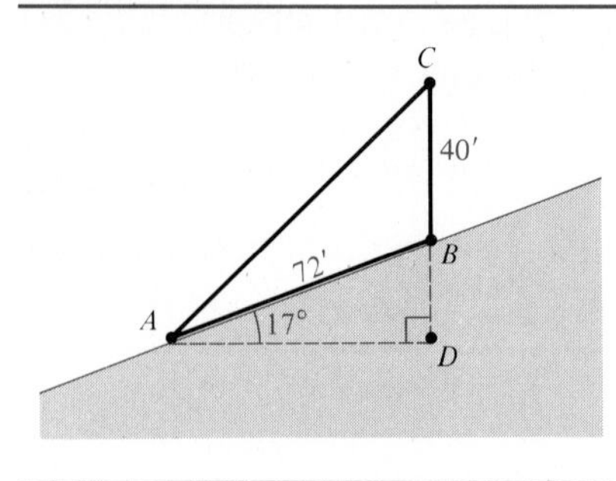

SOLUTION The sketch in Figure 7.12 displays the given data. We wish to find AC. Referring to the figure, we see that

$$\angle ABD = 90^\circ - 17^\circ = 73^\circ \quad \text{and} \quad \angle ABC = 180^\circ - 73^\circ = 107^\circ.$$

Applying the Law of Cosines to triangle ABC,

$$\begin{aligned}(AC)^2 &= (72)^2 + (40)^2 - 2(72)(40)\cos 107^\circ \\ &\approx 5184 + 1600 - 5760(-0.2924) \approx 8468.\end{aligned}$$

Hence $$AC \approx \sqrt{8468} \approx 92 \text{ feet}.$$ ■

The Law of Cosines can be used to derive an interesting formula for the area of a triangle. Let us first prove a preliminary result.

Given triangle ABC, place angle α in standard position, as in Figure 7.9. As shown in the proof of the Law of Cosines, the altitude h from vertex C is $h = b \sin \alpha$. Since the area $\mathscr{A}$ of the triangle is given by $\mathscr{A} = \frac{1}{2}ch$, we see that

$$\mathscr{A} = \tfrac{1}{2}bc \sin \alpha.$$

Our argument is independent of the specific angle that is placed in standard position. By taking β and γ in standard position, the formulas $\mathscr{A} = \frac{1}{2}ac \sin \beta$ and $\mathscr{A} = \frac{1}{2}ab \sin \gamma$, respectively, are obtained. All three formulas are included in the following result.

AREA OF A TRIANGLE

> The area of a triangle equals one-half the product of the lengths of any two sides and the sine of the included angle.

EXAMPLE 5 Approximate the area of triangle ABC with $a = 2.20$ cm, $b = 1.30$ cm, and $\gamma = 43^\circ 10'$.

SOLUTION By the preceding discussion $\mathscr{A} = \frac{1}{2}ab \sin \gamma$. Hence

$$\begin{aligned}\mathscr{A} &= \tfrac{1}{2}(2.2)(1.3)\sin 43^\circ 10' \\ &\approx (1.43)(0.6841) \\ &\approx 0.98 \text{ cm}^2.\end{aligned}$$ ■

The next result, known as *Heron's Formula*, expresses the area of a triangle in terms of the lengths of its sides.

HERON'S FORMULA

The area $\mathscr{A}$ of a triangle with sides a, b, and c is given by

$$\mathscr{A} = \sqrt{s(s-a)(s-b)(s-c)}$$

where $s = \frac{1}{2}(a + b + c)$.

PROOF Using $\mathscr{A} = \frac{1}{2}bc \sin \alpha$ leads to the following equations:

$$\begin{aligned}\mathscr{A} &= \sqrt{\tfrac{1}{4}b^2c^2 \sin^2 \alpha} \\ &= \sqrt{\tfrac{1}{4}b^2c^2(1 - \cos^2 \alpha)} \\ &= \sqrt{\tfrac{1}{2}bc(1 + \cos \alpha) \cdot \tfrac{1}{2}bc(1 - \cos \alpha)}.\end{aligned}$$

We shall obtain Heron's Formula by replacing the expressions under the radical sign by expressions involving only a, b, and c.

Using the Law of Cosines,

$$\begin{aligned}\tfrac{1}{2}bc(1 + \cos \alpha) &= \tfrac{1}{2}bc\left(1 + \frac{b^2 + c^2 - a^2}{2bc}\right) \\ &= \tfrac{1}{2}bc\left(\frac{2bc + b^2 + c^2 - a^2}{2bc}\right) \\ &= \frac{2bc + b^2 + c^2 - a^2}{4} \\ &= \frac{(b + c)^2 - a^2}{4}.\end{aligned}$$

This may be written in the form

$$\tfrac{1}{2}bc(1 + \cos \alpha) = \frac{(b + c) + a}{2} \cdot \frac{(b + c) - a}{2}.$$

By the same type of manipulation,

$$\tfrac{1}{2}bc(1 - \cos \alpha) = \frac{a - b + c}{2} \cdot \frac{a + b - c}{2}.$$

If we now substitute for the expressions under the radical sign, we obtain

$$\mathscr{A} = \sqrt{\frac{b + c + a}{2} \cdot \frac{b + c - a}{2} \cdot \frac{a - b + c}{2} \cdot \frac{a + b - c}{2}}.$$

Letting $s = \frac{1}{2}(a + b + c)$, we see that

$$s - a = \frac{b + c - a}{2}, \qquad s - b = \frac{a - b + c}{2}, \qquad s - c = \frac{a + b - c}{2}.$$

Substitution in the last formula for $\mathscr{A}$ gives us Heron's Formula. □

EXAMPLE 6 A farmer has a triangular field with sides of lengths 125 yards, 160 yards, and 225 yards. Approximate the number of acres in the field. (One acre is equivalent to 4840 square yards.)

SOLUTION We first find the area of the field using Heron's Formula with $a = 125$, $b = 160$, and $c = 225$. Thus

$$s = \tfrac{1}{2}(125 + 160 + 225) = \tfrac{1}{2}(510) = 255$$

$$s - a = 255 - 125 = 130$$

$$s - b = 255 - 160 = 95$$

$$s - c = 255 - 225 = 30$$

and

$$\mathscr{A} = \sqrt{(255)(130)(95)(30)} \approx 9720 \text{ yd}^2.$$

Since there are 4840 square yards in one acre, the number of acres is 9720/4840, or approximately 2. ■

Exercises 7.2

In Exercises 1–10 approximate the remaining parts of triangle ABC.

1 $\alpha = 60°$, $b = 20$, $c = 30$

2 $\gamma = 45°$, $b = 10.0$, $a = 15.0$

3 $\beta = 150°$, $a = 150$, $c = 30.0$

4 $\beta = 73°50'$, $c = 14.0$, $a = 87.0$

5 $\gamma = 115°10'$, $a = 1.10$, $b = 2.10$

6 $\alpha = 23°40'$, $c = 4.30$, $b = 70.0$

7 $a = 2.0$, $b = 3.0$, $c = 4.0$

8 $a = 10$, $b = 15$, $c = 12$

9 $a = 25.0$, $b = 80.0$, $c = 60.0$

10 $a = 20.0$, $b = 20.0$, $c = 10.0$

11 The angle at one corner of a triangular plot of ground has measure 73°40′ and the sides that meet at this corner are 175 feet and 150 feet long. Approximate the length of the third side.

12 To find the distance between two points A and B, a surveyor chooses a point C that is 420 yards from A and 540 yards from B. Angle ACB has measure 63°10′. Approximate the distance AB.

13 Two automobiles leave from the same point and travel along straight highways that differ in direction by 84°. Their speeds are 60 mph and 45 mph, respectively. Approximately how far apart will the cars be at the end of 20 minutes?

14 A triangular plot of land has sides of length 420 feet, 350 feet, and 180 feet. Find the smallest angle between the sides.

15 A ship leaves port at 1:00 P.M. and travels S35°E at the rate of 24 mph. Another ship leaves port at 1:30 P.M. and travels S20°W at 18 mph. Approximately how far apart are the ships at 3:00 P.M.?

16 An airplane flies 165 miles from point A in the direction 130° and then travels in the direction 245° for 80 miles. Approximately how far is the airplane from A?

17 An athlete runs at a constant speed of one mile every eight minutes in the direction S40°E for 20 minutes and then in the direction N20°E for the next 16 minutes. Approximate, to the nearest tenth of a mile, the distance from the endpoint to the starting point of the athlete's course.

18 Two points P and Q on level ground are on opposite sides of a building. In order to find the distance d between the points, a surveyor chooses a point R that is 300 feet from P and 438 feet from Q, and then determines that angle PRQ has measure 37°40′. Approximate d.

19 A motor boat traveled along a triangular course of sides 2 km, 4 km, and 3 km, respectively. The first side was traversed in the direction N20°W and the second in a direction SD°W, where D° is the degree measure of an acute angle. Approximate, to the nearest minute, the direction that the third side was traversed.

20 A rhombus has sides of length 100 cm and the angle at one of the vertices is 70°. Approximate the lengths of the diagonals to the nearest tenth of a centimeter.

21 In major league baseball, the four bases (forming a square) are 90 feet apart and the pitcher's mound is 60.5 feet from home plate. Find the distance from the pitcher's mound to each of the other three bases.

22 The rectangular box shown in the figure has dimensions 8″ × 6″ × 4″. Find the angle θ formed by a diagonal of the base and a diagonal of the 6″ × 4″ side.

FIGURE FOR EXERCISE 22

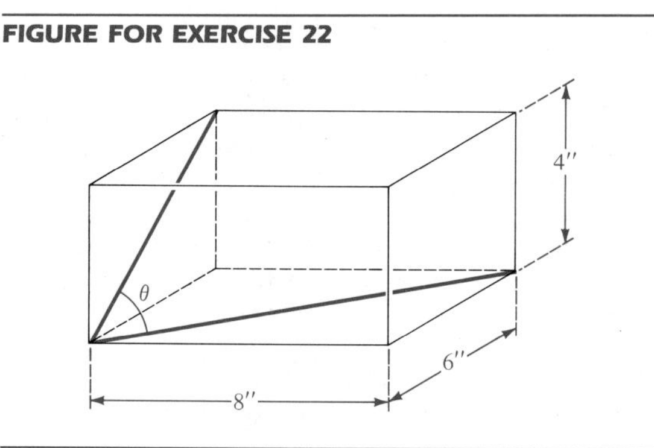

23 A reconaissance jet, J, flying at 10,000 feet, spots a submarine S at an angle of depression of 37° and a tanker T at an angle of depression of 21° (see figure). In addition, $\angle SJT$ is found to be 130°. Find the distance between the submarine and the tanker.

FIGURE FOR EXERCISE 23

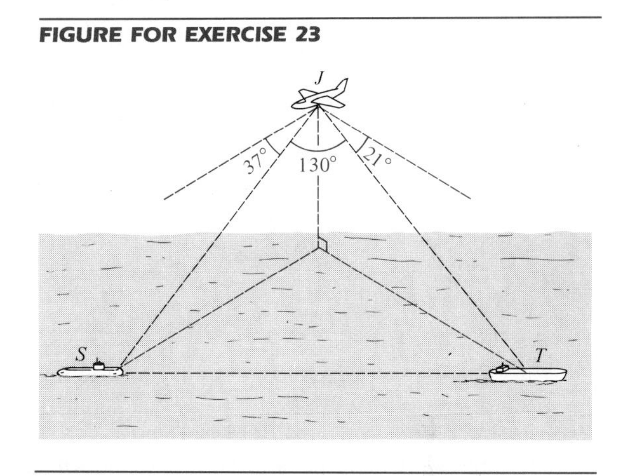

24 A cruise ship sets a course N47°E from an island to a port on the mainland, which is 150 miles away. After moving through strong westerly currents, the captain finds that he is off course at a position P, N33°E and 80 miles from the island.

(a) Approximately how far is the ship from the port?

(b) In what direction should the ship head to correct its course?

FIGURE FOR EXERCISE 24

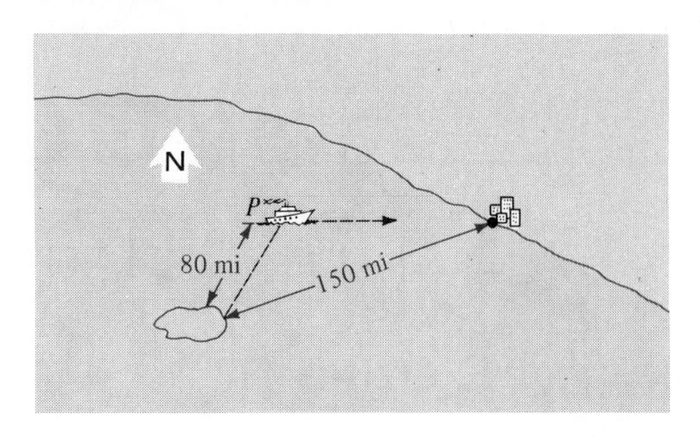

25 Shown in the figure on page 338 is a top view of a jet fighter showing its wing design.

(a) Estimate angle φ and the area of a wing.

(b) For a fuselage width of 5.8 feet, estimate the wing span.

FIGURE FOR EXERCISE 25

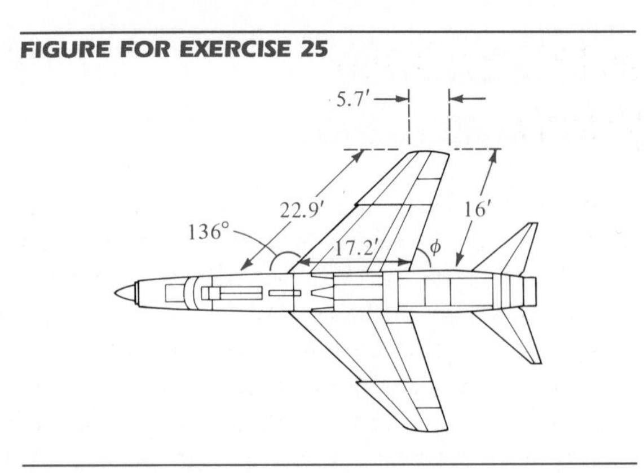

26 The following surveying method can be used to find the distance across a river without measuring angles. Two points B and C on the opposite shore are selected and line segments AB and AC are extended as shown in the figure. Points D and E are chosen as indicated, and the distances BC, BD, BE, CD, and CE are then measured. Suppose that $BC = 184$ feet, $BD = 102$ feet, $BE = 218$ feet, $CE = 118$ feet, and $CD = 236$ feet.

(a) Estimate the distances AB and AC.

(b) Estimate the shortest distance across the river from point A.

FIGURE FOR EXERCISE 26

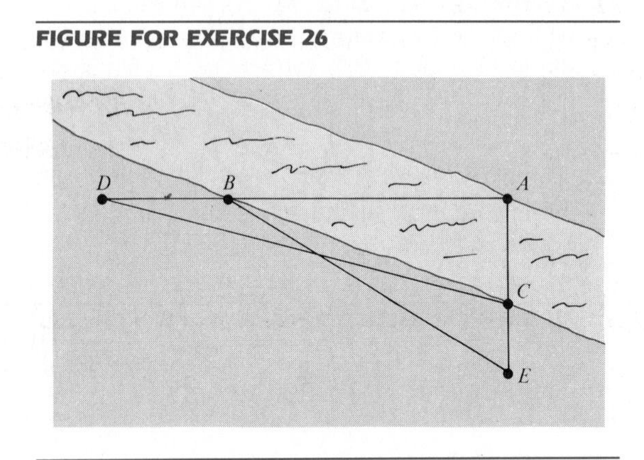

In Exercises 27–34 approximate the area of triangle ABC.

27 $\alpha = 60°,\ b = 20,\ c = 30$

28 $\gamma = 45°,\ b = 10.0,\ a = 15.0$

29 $\beta = 150°,\ a = 150,\ c = 30.0$

30 $\beta = 73°50',\ c = 14.0,\ a = 87.0$

31 $a = 2.0,\ b = 3.0,\ c = 4.0$

32 $a = 10,\ b = 15,\ c = 12$

33 $a = 25.0,\ b = 80.0,\ c = 60.0$

34 $a = 20.0,\ b = 20.0,\ c = 10.0$

35 A triangular field has sides of lengths 115 yards, 140 yards, and 200 yards. Approximate the number of acres in the field. (One acre is equivalent to 4840 square yards.)

36 Find the area of a parallelogram that has sides of lengths 12.0 and 16.0 feet if one angle at a vertex has measure 40°.

Calculator Exercises 7.2

Approximate the remaining parts of triangle ABC in Exercises 1–6.

1 $\alpha = 48.3°,\ b = 24.7,\ c = 52.8$

2 $\beta = 137.8°,\ a = 178.2,\ c = 431.4$

3 $\gamma = 6.85°,\ a = 0.846,\ b = 0.364$

4 $a = 435,\ b = 482,\ c = 78$

5 $a = 5150,\ b = 1814,\ c = 3429$

6 $a = 0.64,\ b = 0.27,\ c = 0.49$

7.3 Trigonometric Form for Complex Numbers

Real numbers may be represented geometrically by points on a coordinate line. We can also obtain geometric representations for complex numbers by using points in a coordinate plane. Specifically, each complex number $a + bi$ determines a unique ordered pair (a, b). The corresponding point $P(a, b)$ in a coordinate plane is called the **geometric representation of** $a + bi$. To emphasize that we are assigning complex numbers to points in a plane, the point $P(a, b)$ may be labeled $a + bi$. A coordinate plane with a complex number assigned to each point is referred to as a **complex plane** instead of an xy-plane. The x-axis is then called the **real axis** and the y-axis the **imaginary axis.** In Figure 7.13 we have indicated geometric representations of several complex numbers. Note that to obtain the point corresponding to the conjugate $a - bi$ of any complex number $a + bi$, we simply reflect through the real axis.

FIGURE 7.13

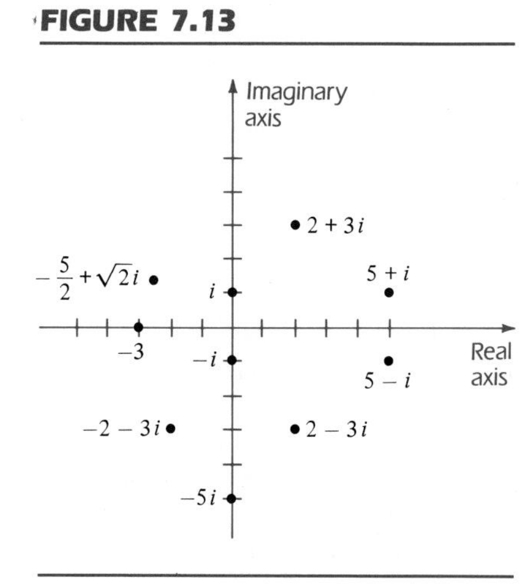

In Chapter 1 we defined the absolute value $|a|$ of a real number a and we noted that, geometrically, $|a|$ is the distance between the origin and the point on a coordinate line that corresponds to a. Thus it is natural to interpret the absolute value $|a + bi|$ of a complex number as the distance $\sqrt{a^2 + b^2}$ between the origin of a complex plane and the point (a, b) that corresponds to $a + bi$.

DEFINITION

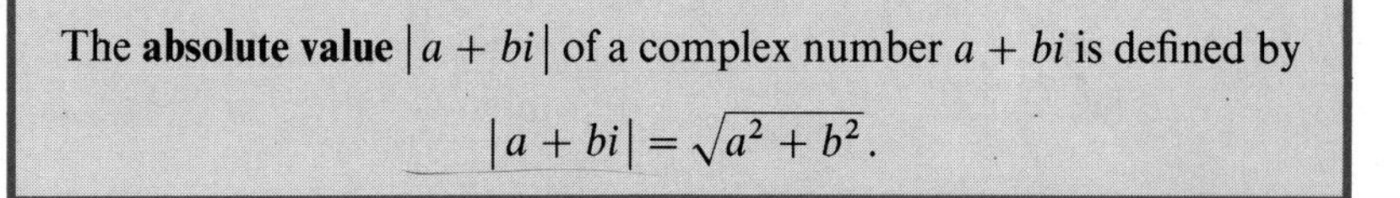
The **absolute value** $|a + bi|$ of a complex number $a + bi$ is defined by

$$|a + bi| = \sqrt{a^2 + b^2}.$$

EXAMPLE 1 Find (a) $|2 - 6i|$; (b) $|3i|$.

SOLUTION Using the definition of absolute value,

(a) $$|2 - 6i| = \sqrt{(2)^2 + (-6)^2} = \sqrt{4 + 36} = \sqrt{40} = 2\sqrt{10}$$

(b) $$|3i| = \sqrt{(0)^2 + (3)^2} = \sqrt{9} = 3$$ ■

It is worth noting that the points corresponding to all complex numbers that have a fixed absolute value lie on a circle with center at the origin

FIGURE 7.14

$z = a + bi = r(\cos\theta + i\sin\theta)$

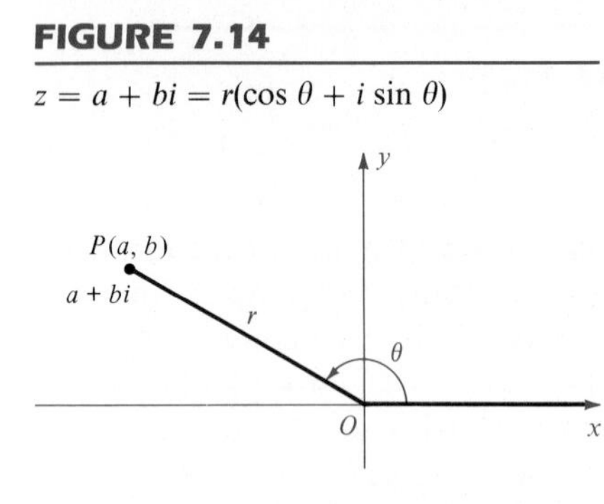

in the complex plane. For example, the points corresponding to the complex numbers z with $|z| = 1$ lie on a unit circle.

Let us consider a nonzero complex number $z = a + bi$ and its geometric representation $P(a, b)$ as illustrated in Figure 7.14. Let θ be any angle in standard position whose terminal side lies on the segment OP and let $r = |z| = \sqrt{a^2 + b^2}$. Since $\cos\theta = a/r$ and $\sin\theta = b/r$, we see that $a = r\cos\theta$ and $b = r\sin\theta$. Substituting for a and b in $z = a + bi$, we obtain

$$z = a + bi = (r\cos\theta) + (r\sin\theta)i,$$

which may be written in the following *trigonometric form.*

TRIGONOMETRIC FORM FOR $a + bi$

$$z = a + bi = r(\cos\theta + i\sin\theta)$$

$$\text{where } r = |z| = \sqrt{a^2 + b^2} \text{ and } \tan\theta = b/a.$$

Sometimes $r(\cos\theta + i\sin\theta)$ is called the **polar form** of the complex number $a + bi$. A common abbreviation is $r \text{ cis } \theta$. Note that θ is not necessarily $\arctan(b/a)$, since θ may not be between $-\pi/2$ and $\pi/2$. Indeed, if $a + bi = 0$, then $r = 0$ and *any* angle θ may be used in the trigonometric form.

The trigonometric form for $z = a + bi$ is not unique, since there are an unlimited number of different choices for the angle θ. When the trigonometric form is used, the absolute value r of z is sometimes referred to as the **modulus** of z and an angle θ associated with z is called an **argument** (or **amplitude**) of z.

EXAMPLE 2 Express the following complex numbers in trigonometric form, where $-\pi \le \theta \le \pi$:

(a) $-4 + 4i$ (b) $2\sqrt{3} - 2i$ (c) $2 + 7i$

SOLUTION The three complex numbers are represented geometrically in Figure 7.15.

Using trigonometric form gives us:

(a) $$-4 + 4i = 4\sqrt{2}\left[\cos\frac{3\pi}{4} + i\sin\frac{3\pi}{4}\right]$$

(b) $$2\sqrt{3} - 2i = 4\left[\cos\left(-\frac{\pi}{6}\right) + i\sin\left(-\frac{\pi}{6}\right)\right]$$

(c) $$2 + 7i = \sqrt{53}\left[\cos\left(\arctan\frac{7}{2}\right) + i\sin\left(\arctan\frac{7}{2}\right)\right]$$

FIGURE 7.15

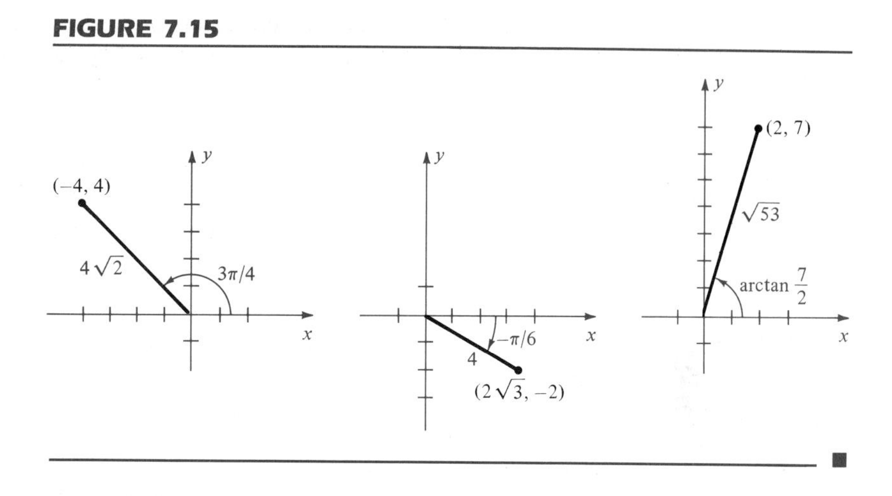

If complex numbers are expressed in trigonometric form, then multiplications and divisions may be carried out as indicated by the next theorem.

THEOREM

If trigonometric forms for two complex numbers z_1 and z_2 are

$$z_1 = r_1(\cos\theta_1 + i\sin\theta_1) \quad \text{and} \quad z_2 = r_2(\cos\theta_2 + i\sin\theta_2)$$

then

(i) $$z_1 z_2 = r_1 r_2[\cos(\theta_1 + \theta_2) + i\sin(\theta_1 + \theta_2)]$$

(ii) $$\frac{z_1}{z_2} = \frac{r_1}{r_2}[\cos(\theta_1 - \theta_2) + i\sin(\theta_1 - \theta_2)], \quad z_2 \neq 0.$$

PROOF We shall prove (i) and leave (ii) as an exercise. Thus,

$$\begin{aligned} z_1 z_2 &= r_1(\cos\theta_1 + i\sin\theta_1)\cdot r_2(\cos\theta_2 + i\sin\theta_2) \\ &= r_1 r_2\{[\cos\theta_1\cos\theta_2 - \sin\theta_1\sin\theta_2] \\ &\qquad + i[\sin\theta_1\cos\theta_2 + \cos\theta_1\sin\theta_2]\}. \end{aligned}$$

Applying the addition formulas for $\cos(\theta_1 + \theta_2)$ and $\sin(\theta_1 + \theta_2)$ gives us (i). □

Part (i) of the preceding theorem states that the modulus of a product of two complex numbers is the product of their moduli, and an argument is the sum of their arguments. An analogous statement can be made for (ii).

EXAMPLE 3 Use trigonometric forms to find z_1z_2 and z_1/z_2 if $z_1 = 2\sqrt{3} - 2i$ and $z_2 = -1 + \sqrt{3}i$. Check by using the methods of Section 3.3.

SOLUTION From Example 2,

$$z_1 = 2\sqrt{3} - 2i = 4\left[\cos\left(-\frac{\pi}{6}\right) + i\sin\left(-\frac{\pi}{6}\right)\right].$$

Similarly, it may be shown that

$$z_2 = -1 + \sqrt{3}i = 2\left[\cos\frac{2\pi}{3} + i\sin\frac{2\pi}{3}\right].$$

Applying (i) of the preceding theorem,

$$\begin{aligned} z_1z_2 &= 8\left[\cos\left(-\frac{\pi}{6} + \frac{2\pi}{3}\right) + i\sin\left(-\frac{\pi}{6} + \frac{2\pi}{3}\right)\right] \\ &= 8\left[\cos\frac{\pi}{2} + i\sin\frac{\pi}{2}\right] = 8i. \end{aligned}$$

As a check, using the methods of Section 3.3,

$$\begin{aligned} z_1z_2 &= (2\sqrt{3} - 2i)(-1 + \sqrt{3}i) \\ &= (-2\sqrt{3} + 2\sqrt{3}) + (2 + 6)i = 8i, \end{aligned}$$

which is in agreement with our previous answer.

Applying (ii) of the theorem,

$$\begin{aligned} \frac{z_1}{z_2} &= 2\left[\cos\left(-\frac{\pi}{6} - \frac{2\pi}{3}\right) + i\sin\left(-\frac{\pi}{6} - \frac{2\pi}{3}\right)\right] \\ &= 2\left[\cos\left(-\frac{5\pi}{6}\right) + i\sin\left(-\frac{5\pi}{6}\right)\right] \\ &= 2\left[-\frac{\sqrt{3}}{2} + i\left(-\frac{1}{2}\right)\right] = -\sqrt{3} - i. \end{aligned}$$

Using the methods of Section 3.3,

$$\begin{aligned} \frac{z_1}{z_2} &= \frac{2\sqrt{3} - 2i}{-1 + \sqrt{3}i} \cdot \frac{-1 - \sqrt{3}i}{-1 - \sqrt{3}i} \\ &= \frac{(-2\sqrt{3} - 2\sqrt{3}) + (2 - 6)i}{4} = -\sqrt{3} - i. \end{aligned}$$

■

Exercises 7.3

Find the absolute values in Exercises 1–10.

1 $|3 - 4i|$

2 $|5 + 8i|$

3 $|-6 - 7i|$

4 $|1 - i|$

5 $|8i|$

6 $|i^7|$

7 $|i^{500}|$

8 $|-15i|$

9 $|0|$

10 $|-15|$

Represent the complex numbers in Exercises 11–20 geometrically.

11 $4 + 2i$

12 $-5 + 3i$

13 $3 - 5i$

14 $-2 - 6i$

15 $-(3 - 6i)$

16 $(1 + 2i)^2$

17 $2i(2 + 3i)$

18 $(-3i)(2 - i)$

19 $(1 + i)^2$

20 $4(-1 + 2i)$

Change the complex numbers in Exercises 21–40 to a trigonometric form in which $0 \leq \theta \leq 2\pi$.

21 $1 - i$

22 $\sqrt{3} + i$

23 $-4\sqrt{3} + 4i$

24 $-2 - 2i$

25 $-20i$

26 15

27 $-5(1 + \sqrt{3}i)$

28 $-6i$

29 -7

30 $2i(1 - \sqrt{3}i)$

31 $2 + i$

32 $4 - 3i$

33 $-4 - 4i$

34 $-10 + 10i$

35 $6i$

36 -5

37 $\sqrt{3} - i$

38 $-5 - 5\sqrt{3}i$

39 12

40 0

In Exercises 41–48 find z_1z_2 and z_1/z_2 by changing to trigonometric form and then using the theorem of this section. Check by using the methods of Section 3.3.

41 $z_1 = -1 + i,\ z_2 = 1 + i$

42 $z_1 = \sqrt{3} - i,\ z_2 = -\sqrt{3} - i$

43 $z_1 = -2 - 2\sqrt{3}i,\ z_2 = 5i$

44 $z_1 = -5 + 5i,\ z_2 = -3i$

45 $z_1 = -10,\ z_2 = -4$

46 $z_1 = 2i,\ z_2 = -3i$

47 $z_1 = 4,\ z_2 = 2 - i$

48 $z_1 = -3,\ z_2 = 5 + 2i$

49 Extend (i) of the theorem in this section to the case of three complex numbers. Generalize to n complex numbers.

50 Prove (ii) of the theorem in this section.

7.4 De Moivre's Theorem and *n*th Roots of Complex Numbers

If z is a complex number and n is a positive integer, then a complex number w is called an **nth root** of z if $w^n = z$. We shall now show that every nonzero complex number has n different nth roots. Since $\mathbb{R}$ is contained in $\mathbb{C}$, it will also follow that every nonzero real number has n distinct nth roots. If a is a positive real number and $n = 2$, then we already know that the roots are $\sqrt{a}$ and $-\sqrt{a}$.

If, in the theorem on page 341, we let both z_1 and z_2 equal the complex number $z = r(\cos\theta + i\sin\theta)$, we obtain

$$z^2 = r^2(\cos 2\theta + i\sin 2\theta).$$

If we next apply the same theorem to z and z^2, then

$$z^2 \cdot z = (r^2 \cdot r)[\cos(2\theta + \theta) + i\sin(2\theta + \theta)]$$

or

$$z^3 = r^3(\cos 3\theta + i\sin 3\theta).$$

Applying the theorem to z^3 and z gives us

$$z^4 = r^4(\cos 4\theta + i\sin 4\theta).$$

In general, we have the following result, named after the French mathematician Abraham De Moivre (1667–1754).

DE MOIVRE'S THEOREM

For every integer n,

$$[r(\cos\theta + i\sin\theta)]^n = r^n[\cos n\theta + i\sin n\theta].$$

EXAMPLE 1 Find $(1 + i)^{20}$.

SOLUTION Introducing trigonometric form, we obtain

$$1 + i = \sqrt{2}\left(\cos\frac{\pi}{4} + i\sin\frac{\pi}{4}\right).$$

Next, by De Moivre's Theorem,

$$\begin{aligned}(1 + i)^{20} &= (2^{1/2})^{20}\left[\cos 20\left(\frac{\pi}{4}\right) + i\sin 20\left(\frac{\pi}{4}\right)\right]\\ &= 2^{10}(\cos 5\pi + i\sin 5\pi) = -1024.\end{aligned}$$

■

If a nonzero complex number z has an nth root w, then $w^n = z$. If trigonometric forms for w and z are

$$w = s(\cos\alpha + i\sin\alpha) \quad \text{and} \quad z = r(\cos\theta + i\sin\theta),$$

then, applying De Moivre's Theorem to $w^n = z$,

$$s^n[\cos n\alpha + i\sin n\alpha] = r(\cos\theta + i\sin\theta).$$

If two complex numbers are equal, then so are their absolute values. Consequently, $s^n = r$, and since s and r are nonnegative, $s = \sqrt[n]{r}$. Substituting s^n for r in the last displayed equation and dividing both sides by s^n, we obtain

$$\cos n\alpha + i \sin n\alpha = \cos \theta + i \sin \theta.$$

It follows that $\quad \cos n\alpha = \cos \theta \quad$ and $\quad \sin n\alpha = \sin \theta.$

Since both the sine and cosine functions have period 2π, the last two equations are true if and only if $n\alpha$ and θ differ by a multiple of 2π. Thus, for some integer k,

$$n\alpha = \theta + 2\pi k$$

and hence

$$\alpha = \frac{\theta + 2\pi k}{n}.$$

Substituting in the trigonometric form for w, we obtain the formula

$$w = \sqrt[n]{r}\left[\cos\left(\frac{\theta + 2\pi k}{n}\right) + i \sin\left(\frac{\theta + 2\pi k}{n}\right)\right].$$

If we substitute $k = 0, 1, \ldots, n - 1$ successively, we obtain n different nth roots of z. No other value of k will produce a new nth root. For example, if $k = n$, we obtain the angle $(\theta + 2\pi n)/n$, or $(\theta/n) + 2\pi$, which gives us the same nth root as $k = 0$. Similarly, $k = n + 1$ yields the same nth root as $k = 1$, and so on. The same is true for negative values of k. We have proved the following theorem.

THEOREM ON *n*th ROOTS

If $z = r(\cos \theta + i \sin \theta)$ is any nonzero complex number and if n is any positive integer, then z has precisely n distinct nth roots $w_0, w_1, w_2, \ldots, w_{n-1}$. Moreover, these roots are given by

$$w_k = \sqrt[n]{r}\left[\cos\left(\frac{\theta + 2\pi k}{n}\right) + i \sin\left(\frac{\theta + 2\pi k}{n}\right)\right]$$

for $k = 0, 1, \ldots, n - 1$.

The nth roots of z all have absolute value $\sqrt[n]{r}$ and hence their geometric representations lie on a circle of radius $\sqrt[n]{r}$ with center at O. Moreover, they are equispaced on this circle, since the difference in the arguments of successive nth roots is $2\pi/n$.

It is sometimes convenient to use degree measure for θ. In this event the formula in the preceding theorem becomes

$$w_k = \sqrt[n]{r}\left[\cos\left(\frac{\theta + k \cdot 360^\circ}{n}\right) + i \sin\left(\frac{\theta + k \cdot 360^\circ}{n}\right)\right]$$

for $k = 0, 1, \ldots, n - 1$.

EXAMPLE 2 Find the four fourth roots of $-8 - (8\sqrt{3})i$.

FIGURE 7.16

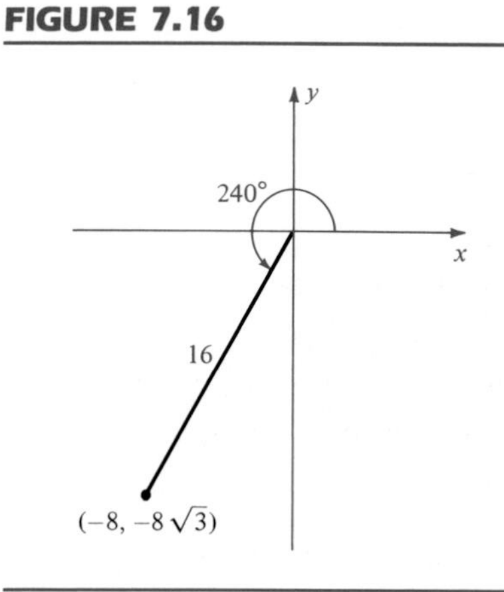

SOLUTION The geometric representation of the complex number is shown in Figure 7.16. Introducing trigonometric form,

$$-8 - (8\sqrt{3})i = 16(\cos 240^\circ + i \sin 240^\circ).$$

Using the preceding theorem with $n = 4$, and noting that $\sqrt[4]{16} = 2$, the fourth roots are

$$w_k = 2\left[\cos\left(\frac{240^\circ + k \cdot 360^\circ}{4}\right) + i \sin\left(\frac{240^\circ + k \cdot 360^\circ}{4}\right)\right]$$

for $k = 0, 1, 2$, and 3. This formula may be written

$$w_k = 2[\cos(60^\circ + k \cdot 90^\circ) + i \sin(60^\circ + k \cdot 90^\circ)].$$

Substituting 0, 1, 2, and 3 for k, we obtain the following fourth roots:

$$w_0 = 2(\cos 60^\circ + i \sin 60^\circ) = 1 + \sqrt{3}i$$
$$w_1 = 2(\cos 150^\circ + i \sin 150^\circ) = -\sqrt{3} + i$$
$$w_2 = 2(\cos 240^\circ + i \sin 240^\circ) = -1 - \sqrt{3}i$$
$$w_3 = 2(\cos 330^\circ + i \sin 330^\circ) = \sqrt{3} - i$$

■

EXAMPLE 3 Find the six sixth roots of -1 and represent them geometrically.

SOLUTION Writing $-1 = 1(\cos \pi + i \sin \pi)$ and using the Theorem on nth Roots with $n = 6$, we find that the sixth roots of -1 are given by

$$w_k = \cos\left(\frac{\pi + 2\pi k}{6}\right) + i \sin\left(\frac{\pi + 2\pi k}{6}\right)$$

or

$$w_k = \cos\left(\frac{\pi}{6} + \frac{\pi}{3}k\right) + i \sin\left(\frac{\pi}{6} + \frac{\pi}{3}k\right).$$

Substituting 0, 1, 2, 3, 4, and 5 for k, we obtain the following sixth roots of -1:

$$w_0 = \cos\frac{\pi}{6} + i\sin\frac{\pi}{6} = \frac{\sqrt{3}}{2} + \frac{1}{2}i$$

$$w_1 = \cos\frac{\pi}{2} + i\sin\frac{\pi}{2} = i$$

$$w_2 = \cos\frac{5\pi}{6} + i\sin\frac{5\pi}{6} = -\frac{\sqrt{3}}{2} + \frac{1}{2}i$$

$$w_3 = \cos\frac{7\pi}{6} + i\sin\frac{7\pi}{6} = -\frac{\sqrt{3}}{2} - \frac{1}{2}i$$

$$w_4 = \cos\frac{3\pi}{2} + i\sin\frac{3\pi}{2} = -i$$

$$w_5 = \cos\frac{11\pi}{6} + i\sin\frac{11\pi}{6} = \frac{\sqrt{3}}{2} - \frac{1}{2}i.$$

FIGURE 7.17

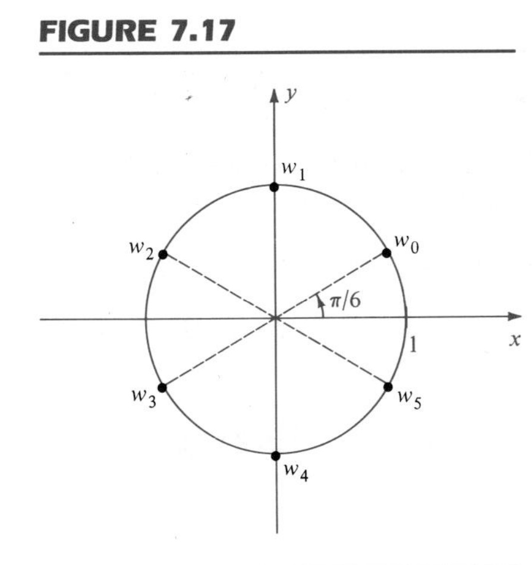

Since $|-1| = 1$, the points that represent the roots of -1 all lie on the unit circle shown in Figure 7.17. As noted earlier, they are equispaced on this circle. ■

The special case in which $z = 1$ is of particular interest. The n distinct nth roots of 1 are called the ***n*th roots of unity.**

EXAMPLE 4 Find the three third roots of unity.

SOLUTION Writing $1 = \cos 0 + i\sin 0$ and using the Theorem on nth Roots with $n = 3$, we obtain the three roots

$$w_k = \cos\frac{2\pi k}{3} + i\sin\frac{2\pi k}{3}$$

for $k = 0$, 1, and 2. Substituting for k gives us

$$w_0 = \cos 0 + i\sin 0 = 1$$

$$w_1 = \cos\frac{2\pi}{3} + i\sin\frac{2\pi}{3} = -\frac{1}{2} + \left(\frac{\sqrt{3}}{2}\right)i$$

$$w_2 = \cos\frac{4\pi}{3} + i\sin\frac{4\pi}{3} = -\frac{1}{2} - \left(\frac{\sqrt{3}}{2}\right)i.$$ ■

Compare the preceding solution with Example 6 of Section 3.3.

Exercises 7.4

In Exercises 1–12 use De Moivre's Theorem to express the numbers in the form $a + bi$, where a and b are real numbers.

1 $(3 + 3i)^5$

2 $(1 + i)^{12}$

3 $(1 - i)^{10}$

4 $(-1 + i)^8$

5 $(1 - \sqrt{3}i)^3$

6 $(1 - \sqrt{3}i)^5$

7 $\left(-\frac{\sqrt{2}}{2} + \frac{\sqrt{2}}{2}i\right)^{15}$

8 $\left(\frac{\sqrt{2}}{2} + \frac{\sqrt{2}}{2}i\right)^{25}$

9 $\left(-\frac{\sqrt{3}}{2} - \frac{1}{2}i\right)^{20}$

10 $\left(-\frac{\sqrt{3}}{2} - \frac{1}{2}i\right)^{50}$

11 $(\sqrt{3} + i)^7$

12 $(-2 - 2i)^{10}$

13 Find the two square roots of $1 + \sqrt{3}i$.

14 Find the two square roots of $-9i$.

15 Find the four fourth roots of $-1 - \sqrt{3}i$.

16 Find the four fourth roots of $-8 + 8\sqrt{3}i$.

17 Find the three cube roots of $-27i$.

18 Find the three cube roots of $64i$.

In Exercises 19–22 find the indicated roots and represent them geometrically.

19 The six sixth roots of unity

20 The eight eighth roots of unity

21 The five fifth roots of $1 + i$

22 The five fifth roots of $-\sqrt{3} - i$

Find the solutions of the equations in Exercises 23–30.

23 $x^4 - 16 = 0$

24 $x^6 - 64 = 0$

25 $x^6 + 64 = 0$

26 $x^5 + 1 = 0$

27 $x^3 + 8i = 0$

28 $x^3 - 64i = 0$

29 $x^5 - 243 = 0$

30 $x^4 + 81 = 0$

7.5 Vectors

In Chapter 1 we assigned (positive) directions to the x- and y-axes. In similar fashion, a **directed line segment** is a line segment to which a direction has been assigned. Another name for a directed line segment is a **vector.** If a vector extends from a point A (called the **initial point**) to a point B (called the **terminal point**), it is customary to place an arrowhead at B and use $\overrightarrow{AB}$ to represent the vector (see Figure 7.18). The length of the directed line segment is called the **magnitude** of the vector $\overrightarrow{AB}$ and is denoted by $|\overrightarrow{AB}|$. The vectors $\overrightarrow{AB}$ and $\overrightarrow{CD}$ are considered **equal,** and we write $\overrightarrow{AB} = \overrightarrow{CD}$, if and only if they have the same magnitude and direction, as illustrated in Figure 7.18. Consequently, vectors may be translated from one position to another, provided neither the magnitude nor direction is changed. Vectors of this type are often referred to as **free vectors.**

FIGURE 7.18

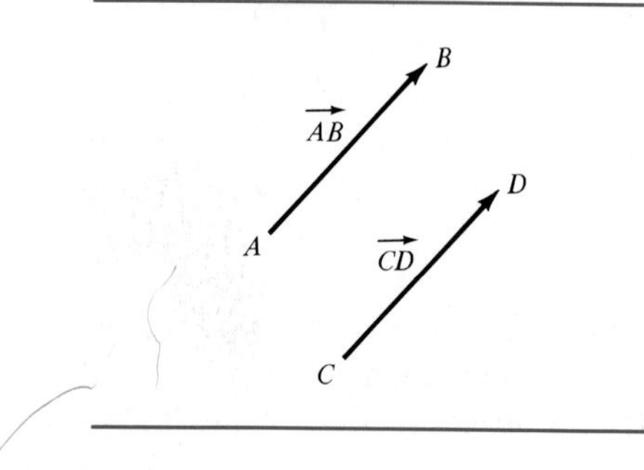

Many physical concepts may be represented by vectors. To illustrate, suppose an airplane is descending at a constant rate of 100 mph and its line of flight makes an angle of 20° with the horizontal. Both of these facts are represented by the vector in Figure 7.19(i). As shown in the figure, we will use a boldface letter such as **v** to denote a vector whose endpoints are not specified. The vector **v** in this illustration is called a **velocity vector.** The magnitude $|\mathbf{v}|$ of **v** is the speed of the airplane.

FIGURE 7.19

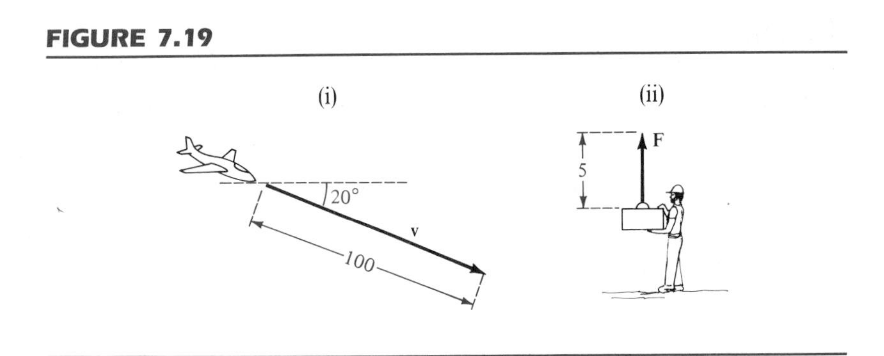

FIGURE 7.20

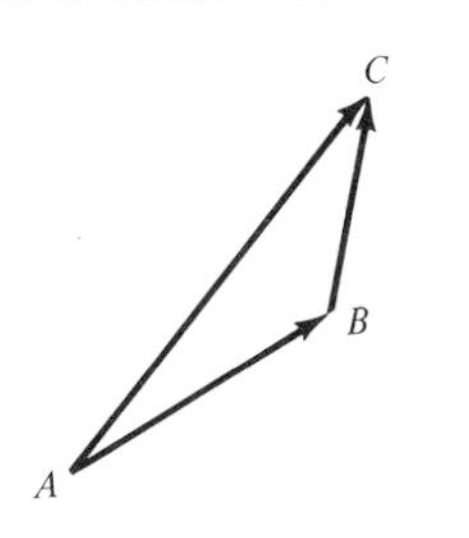

As a second illustration, suppose a person pulls directly upward on an object with a force of 5 kg, as would be the case in lifting a 5-kg weight. We may indicate this fact by the vector **F** in Figure 7.19(ii). A vector that represents a pull or push of some type is called a **force vector.**

Another use for vectors is to let $\overrightarrow{AB}$ represent the path of a point (or some physical particle) as it moves along the line segment from A to B. We then refer to $\overrightarrow{AB}$ as a **displacement** of the point (or particle). As illustrated in Figure 7.20, a displacement $\overrightarrow{AB}$ followed by a displacement $\overrightarrow{BC}$ gives us the same point as the single displacement $\overrightarrow{AC}$. The vector $\overrightarrow{AC}$ is called the **sum** of the first two, written

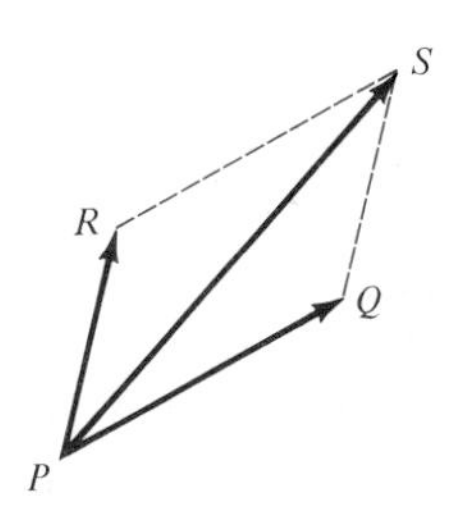

$$\overrightarrow{AC} = \overrightarrow{AB} + \overrightarrow{BC}.$$

Since we are working with free vectors, any two vectors can be added by placing the initial point of one on the terminal point of the other and then finding $\overrightarrow{AC}$ as in Figure 7.20.

Another way to find the sum of two vectors is to consider vectors that are equal to the given ones and have the same initial point, as illustrated by $\overrightarrow{PQ}$ and $\overrightarrow{PR}$ in Figure 7.21. If we construct the parallelogram $RPQS$ with adjacent sides $\overrightarrow{PR}$ and $\overrightarrow{PQ}$, then since $\overrightarrow{PR} = \overrightarrow{QS}$, it follows that

$$\overrightarrow{PS} = \overrightarrow{PQ} + \overrightarrow{PR}.$$

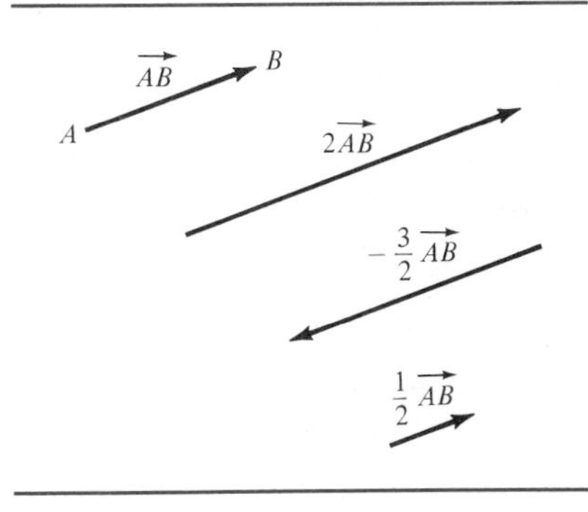

If $\overrightarrow{PQ}$ and $\overrightarrow{PR}$ are two forces acting at P, then it can be shown experimentally that $\overrightarrow{PS}$ is the **resultant force,** that is, the single force that produces the same effect as the two combined forces.

If c is a real number and $\overrightarrow{AB}$ is a vector, then the product $c\overrightarrow{AB}$ is defined as a vector whose magnitude is $|c|$ times the magnitude of $\overrightarrow{AB}$ and whose direction is the same as $\overrightarrow{AB}$ if $c > 0$, and opposite that of $\overrightarrow{AB}$ if $c < 0$. Geometric illustrations are given in Figure 7.22. We refer to c as a **scalar** and $c\overrightarrow{AB}$ as a **scalar multiple** of $\overrightarrow{AB}$.

Let us next introduce a coordinate plane and assume that all vectors under discussion are in that plane. Since we may change the position of a vector, provided that the magnitude and direction are not altered, let

FIGURE 7.23

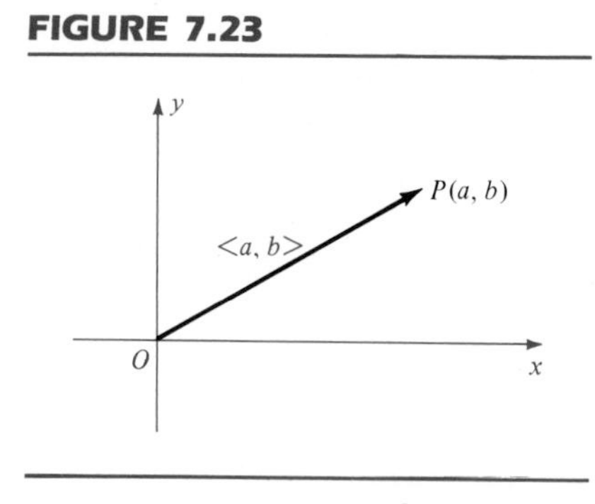

us place the initial point at the origin. The terminal point of a typical vector $\overrightarrow{OP}$ has rectangular coordinates (a, b), as shown in Figure 7.23. Conversely, every ordered pair (a, b), determines the vector $\overrightarrow{OP}$, where P has rectangular coordinates (a, b). This gives us a one-to-one correspondence between vectors and ordered pairs and allows us to regard a vector as an ordered pair of real numbers instead of as a directed line segment.

To avoid confusion with the notation for open intervals or points, we shall use the symbol $\langle a, b\rangle$ for an ordered pair that represents a vector, and we shall refer to $\langle a, b\rangle$ as a vector and denote it by a boldface letter. The numbers a and b are called the **components** of the vector $\langle a, b\rangle$. The magnitude of $\langle a, b\rangle$ is, by definition, the distance from the origin to the point $P(a, b)$. This may also be stated as follows.

DEFINITION

The **magnitude** $|\mathbf{v}|$ of the vector $\mathbf{v} = \langle a, b\rangle$ is

$$|\mathbf{v}| = \sqrt{a^2 + b^2}.$$

FIGURE 7.24

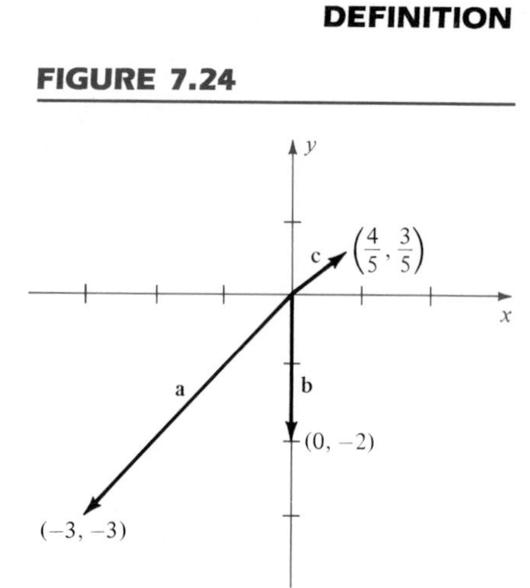

EXAMPLE 1 Sketch the vectors corresponding to each of the following. Then find the magnitude and the smallest positive angle θ from the positive x-axis to each vector.

(a) $\mathbf{a} = \langle -3, -3\rangle$ (b) $\mathbf{b} = \langle 0, -2\rangle$ (c) $\mathbf{c} = \langle \frac{4}{5}, \frac{3}{5}\rangle$

SOLUTION The vectors are sketched in Figure 7.24.

Applying the definition of magnitude, and our knowledge of trigonometry, we obtain:

(a) $|\mathbf{a}| = \sqrt{9 + 9} = 3\sqrt{2}; \quad \theta = 5\pi/4$

(b) $|\mathbf{b}| = \sqrt{0 + 4} = 2; \quad \theta = 3\pi/2$

(c) $|\mathbf{c}| = \sqrt{\frac{16}{25} + \frac{9}{25}} = 1; \quad \theta = \arctan \frac{3}{4}$. ■

FIGURE 7.25

$\langle a, b\rangle + \langle c, d\rangle = \langle a + c, b + d\rangle$

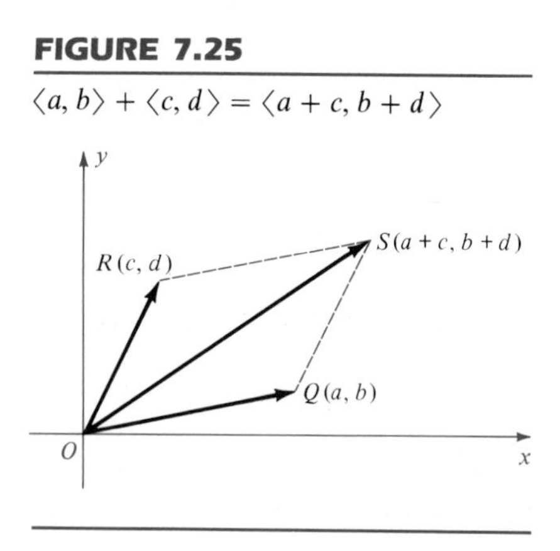

Consider vectors $\overrightarrow{OQ}$ and $\overrightarrow{OR}$ corresponding to $\langle a, b\rangle$ and $\langle c, d\rangle$, respectively, as shown in Figure 7.25. If we let $\overrightarrow{OS}$ be the vector corresponding to $\langle a + c, b + d\rangle$, it can be shown that O, Q, S, and R are vertices of a parallelogram. It follows that

$$\overrightarrow{OQ} + \overrightarrow{OR} = \overrightarrow{OS}.$$

Expressing this fact in terms of ordered pairs gives us the following rule for addition of vectors.

ADDITION OF VECTORS

$$\langle a, b\rangle + \langle c, d\rangle = \langle a + c, b + d\rangle$$

Although we shall not prove it, the rule corresponding to a scalar multiple $k\overrightarrow{OP}$ of a vector is as follows.

SCALAR MULTIPLES OF VECTORS

$$k\langle a, b\rangle = \langle ka, kb\rangle$$

EXAMPLE 2 If $\mathbf{a} = \langle 2, 1\rangle$, find $3\mathbf{a}$ and $-2\mathbf{a}$, and represent all three vectors geometrically.

SOLUTION If $\mathbf{a} = \langle 2, 1\rangle$, then, by the preceding rule for scalar multiples, $3\mathbf{a} = \langle 6, 3\rangle$ and $-2\mathbf{a} = \langle -4, -2\rangle$. The geometric representations are shown in Figure 7.26.

FIGURE 7.26

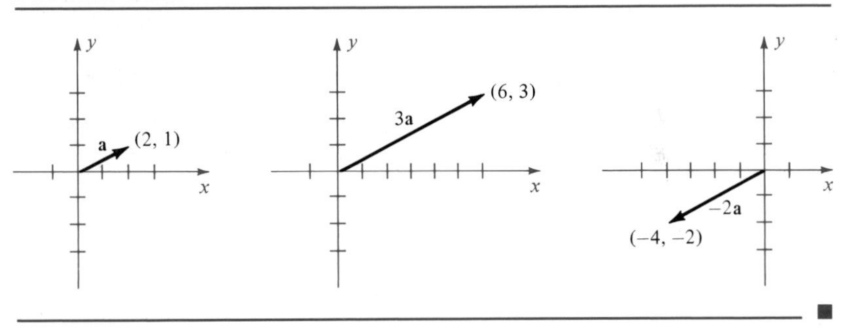

■

EXAMPLE 3 If $\mathbf{a} = \langle 3, -2\rangle$ and $\mathbf{b} = \langle -6, 7\rangle$, find $\mathbf{a} + \mathbf{b}$, $4\mathbf{a}$, and $2\mathbf{a} + 3\mathbf{b}$.

SOLUTION Using the rules for addition and scalar multiples of vectors,

$$\mathbf{a} + \mathbf{b} = \langle 3, -2\rangle + \langle -6, 7\rangle = \langle -3, 5\rangle$$
$$4\mathbf{a} = 4\langle 3, -2\rangle = \langle 12, -8\rangle$$
$$2\mathbf{a} + 3\mathbf{b} = \langle 6, -4\rangle + \langle -18, 21\rangle = \langle -12, 17\rangle$$

■

By definition, the **zero vector 0** corresponds to $\langle 0, 0\rangle$. If $\mathbf{a} = \langle a, b\rangle$, we define $-\mathbf{a} = \langle -a, -b\rangle$. Using these definitions, we may establish the following properties for arbitrary vectors **a**, **b**, and **c**.

PROPERTIES OF ADDITION OF VECTORS

$$\mathbf{a} + \mathbf{b} = \mathbf{b} + \mathbf{a}$$

$$\mathbf{a} + (\mathbf{b} + \mathbf{c}) = (\mathbf{a} + \mathbf{b}) + \mathbf{c}$$

$$\mathbf{a} + \mathbf{0} = \mathbf{a}$$

$$\mathbf{a} + (-\mathbf{a}) = \mathbf{0}$$

The proof of each property follows readily from the rule for addition of vectors and properties of real numbers. For example, if $\mathbf{a} = \langle a_1, a_2\rangle$ and if $\mathbf{b} = \langle b_1, b_2\rangle$, then since $a_1 + b_1 = b_1 + a_1$ and $a_2 + b_2 = b_2 + a_2$,

$$\begin{aligned} \mathbf{a} + \mathbf{b} &= \langle a_1 + b_1, a_2 + b_2\rangle \\ &= \langle b_1 + a_1, b_2 + a_2\rangle \\ &= \mathbf{b} + \mathbf{a}. \end{aligned}$$

The remaining proofs are left as exercises.

Subtraction of vectors (denoted by $-$) is defined by $\mathbf{a} - \mathbf{b} = \mathbf{a} + (-\mathbf{b})$. If we use the ordered pair notation for **a** and **b**, then $-\mathbf{b} = \langle -b_1, -b_2\rangle$, and we obtain the following.

SUBTRACTION OF VECTORS

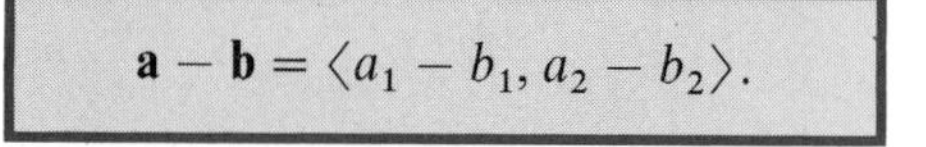

$$\mathbf{a} - \mathbf{b} = \langle a_1 - b_1, a_2 - b_2\rangle.$$

Thus, to find $\mathbf{a} - \mathbf{b}$, we merely subtract the components of **b** from the corresponding components of **a**.

EXAMPLE 4 If $\mathbf{a} = \langle 2, -4\rangle$ and $\mathbf{b} = \langle 6, 7\rangle$, find $3\mathbf{a} - 5\mathbf{b}$.

SOLUTION

$$\begin{aligned} 3\mathbf{a} - 5\mathbf{b} &= 3\langle 2, -4\rangle - 5\langle 6, 7\rangle \\ &= \langle 6, -12\rangle - \langle 30, 35\rangle = \langle -24, -47\rangle \end{aligned}$$

■

The following properties can be proved for any vectors **a**, **b** and real numbers c, d.

PROPERTIES OF SCALAR MULTIPLES OF VECTORS

$$c(\mathbf{a} + \mathbf{b}) = c\mathbf{a} + c\mathbf{b}$$

$$(c + d)\mathbf{a} = c\mathbf{a} + d\mathbf{a}$$

$$(cd)\mathbf{a} = c(d\mathbf{a}) = d(c\mathbf{a})$$

$$1\mathbf{a} = \mathbf{a}$$

$$0\mathbf{a} = \mathbf{0} = c\mathbf{0}$$

We shall prove the first property and leave the remaining proofs as an exercise. Letting $\mathbf{a} = \langle a_1, a_2 \rangle$ and $\mathbf{b} = \langle b_1, b_2 \rangle$,

$$\begin{aligned} c(\mathbf{a} + \mathbf{b}) &= \langle a_1 + b_1, a_2 + b_2 \rangle \\ &= \langle ca_1 + cb_1, ca_2 + cb_2 \rangle \\ &= \langle ca_1, ca_2 \rangle + \langle cb_1, cb_2 \rangle \\ &= c\langle a_1, a_2 \rangle + c\langle b_1, b_2 \rangle \\ &= c\mathbf{a} + c\mathbf{b}. \end{aligned}$$

The special vectors **i** and **j** are defined as follows:

DEFINITION OF i AND j

$$\mathbf{i} = \langle 1, 0 \rangle, \qquad \mathbf{j} = \langle 0, 1 \rangle$$

A **unit vector** is a vector of magnitude 1. The vectors **i** and **j** are unit vectors, as is the vector $\mathbf{c} = \langle \frac{4}{5}, \frac{3}{5} \rangle$ in Example 1 of this section.

The vectors **i** and **j** can be used to obtain an alternative way of denoting vectors. Specifically, if $\mathbf{a} = \langle a_1, a_2 \rangle$, then

$$\mathbf{a} = \langle a_1, 0 \rangle + \langle 0, a_2 \rangle = a_1\langle 1, 0 \rangle + a_2\langle 0, 1 \rangle,$$

that is,

$$\mathbf{a} = \langle a_1, a_2 \rangle = a_1\mathbf{i} + a_2\mathbf{j}$$

Vectors corresponding to **i**, **j**, and **a** are illustrated in Figure 7.27(i). Since **i** and **j** are unit vectors, $a_1\mathbf{i}$ and $a_2\mathbf{j}$ may be represented by horizontal and vertical vectors of magnitudes $|a_1|$ and $|a_2|$, respectively, as illustrated in (ii) of the figure. For this reason a_1 is called the **horizontal component** and a_2 the **vertical component** of the vector **a**.

FIGURE 7.27

$\mathbf{a} = \langle a_1, a_2 \rangle = a_1\mathbf{i} + a_2\mathbf{j}$

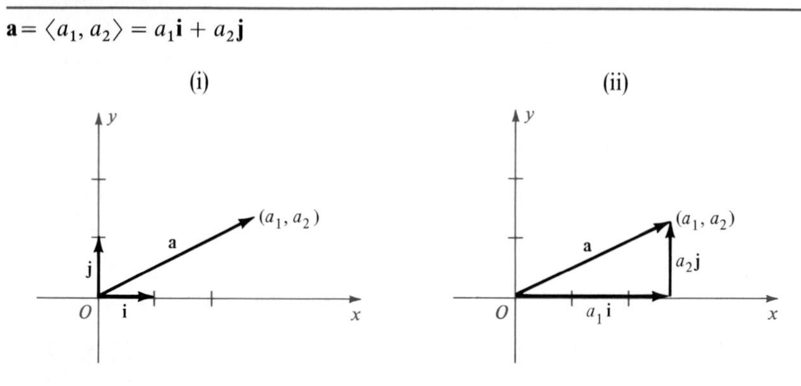

The vector sum $a_1\mathbf{i} + a_2\mathbf{j}$ is called a **linear combination** of **i** and **j**. Rules for addition, subtraction, and multiplication by a scalar may be written as follows, where $\mathbf{b} = \langle b_1, b_2 \rangle = b_1\mathbf{i} + b_2\mathbf{j}$:

$$(a_1\mathbf{i} + a_2\mathbf{j}) + (b_1\mathbf{i} + b_2\mathbf{j}) = (a_1 + b_1)\mathbf{i} + (a_2 + b_2)\mathbf{j}$$

$$(a_1\mathbf{i} + a_2\mathbf{j}) - (b_1\mathbf{i} + b_2\mathbf{j}) = (a_1 - b_1)\mathbf{i} + (a_2 - b_2)\mathbf{j}$$

$$c(a_1\mathbf{i} + a_2\mathbf{j}) = (ca_1)\mathbf{i} + (ca_2)\mathbf{j}.$$

These formulas show that linear combinations of **i** and **j** may be regarded as ordinary algebraic sums.

EXAMPLE 5 If $\mathbf{a} = 5\mathbf{i} + \mathbf{j}$ and $\mathbf{b} = 4\mathbf{i} - 7\mathbf{j}$, express $3\mathbf{a} - 2\mathbf{b}$ as a linear combination of **i** and **j**.

SOLUTION

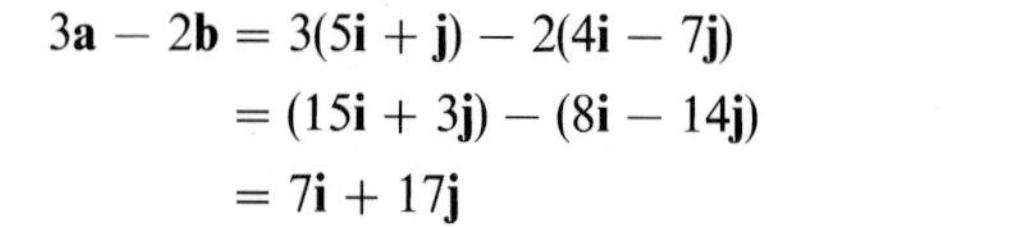

$$\begin{aligned} 3\mathbf{a} - 2\mathbf{b} &= 3(5\mathbf{i} + \mathbf{j}) - 2(4\mathbf{i} - 7\mathbf{j}) \\ &= (15\mathbf{i} + 3\mathbf{j}) - (8\mathbf{i} - 14\mathbf{j}) \\ &= 7\mathbf{i} + 17\mathbf{j} \end{aligned}$$

■

FIGURE 7.28

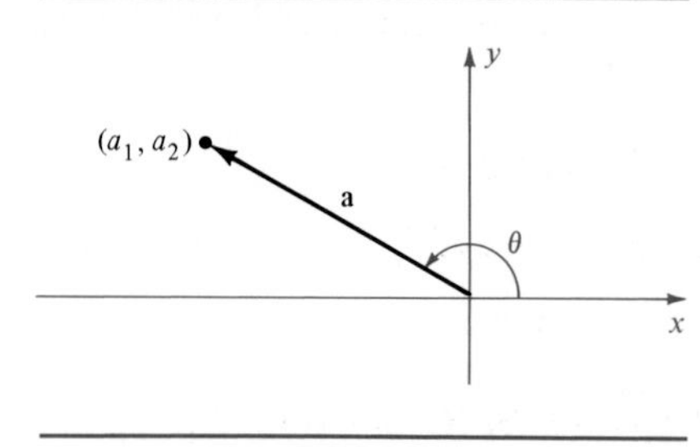

Let θ be an angle in standard position, measured from the positive x-axis to the terminal side of the vector $\mathbf{a} = \langle a_1, a_2 \rangle = a_1\mathbf{i} + a_2\mathbf{j}$ as illustrated in Figure 7.28.

Since $\cos\theta = a_1/|\mathbf{a}|$ and $\sin\theta = a_2/|\mathbf{a}|$ we obtain the following formulas.

HORIZONTAL AND VERTICAL COMPONENTS OF $\mathbf{a} = \langle a_1, a_2 \rangle$

> If $\mathbf{a} = \langle a_1, a_2 \rangle = a_1\mathbf{i} + a_2\mathbf{j}$, then
>
> $$a_1 = |\mathbf{a}|\cos\theta, \qquad a_2 = |\mathbf{a}|\sin\theta,$$
>
> where $\tan\theta = a_2/a_1$.

Using the preceding formulas,

$$\begin{aligned} \mathbf{a} = \langle a_1, a_2 \rangle &= \langle |\mathbf{a}|\cos\theta, |\mathbf{a}|\sin\theta \rangle \\ &= |\mathbf{a}|\cos\theta\mathbf{i} + |\mathbf{a}|\sin\theta\mathbf{j} \end{aligned}$$

FIGURE 7.29

EXAMPLE 6 If a 12 mph wind is blowing in the direction N40°W, express its velocity as a vector **v**.

SOLUTION The vector **v** is represented geometrically in Figure 7.29. Since $\theta = 90° + 40° = 130°$, we have

$$\mathbf{v} = (12\cos 130°)\mathbf{i} + (12\sin 130°)\mathbf{j}.$$

If an approximation is desired, then using Table 4 or a calculator,

$$\mathbf{v} \approx (-7.7)\mathbf{i} + (9.2)\mathbf{j}.$$ ■

FIGURE 7.30

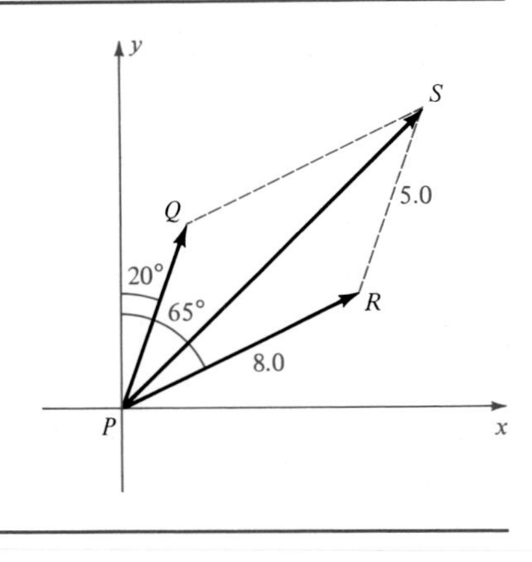

EXAMPLE 7 Two forces $\overrightarrow{PQ}$ and $\overrightarrow{PR}$ of magnitudes 5.0 kg and 8.0 kg, respectively, act at a point P. The direction of $\overrightarrow{PQ}$ is N20°E and the direction of $\overrightarrow{PR}$ is N65°E. Approximate the direction of the resultant vector $\overrightarrow{PS}$ to the nearest degree, and find $|\overrightarrow{PS}|$ to the nearest tenth.

SOLUTION The forces are represented geometrically in Figure 7.30. Note that the angles from the x-axis to $\overrightarrow{PQ}$ and $\overrightarrow{PR}$ have measures 70° and 25°, respectively. Using the formulas for horizontal and vertical components,

$$\overrightarrow{PQ} = (5\cos 70°)\mathbf{i} + (5\sin 70°)\mathbf{j}$$

$$\overrightarrow{PR} = (8\cos 25°)\mathbf{i} + (8\sin 25°)\mathbf{j}.$$

Since $\overrightarrow{PS} = \overrightarrow{PQ} + \overrightarrow{PR}$,

$$\overrightarrow{PS} = (5\cos 70° + 8\cos 25°)\mathbf{i} + (5\sin 70° + 8\sin 25°)\mathbf{j}.$$

Referring to Table 4 or using a calculator gives us the approximation

$$\overrightarrow{PS} \approx (8.9604)\mathbf{i} + (8.0793)\mathbf{j} \approx (9.0)\mathbf{i} + (8.1)\mathbf{j}.$$

Consequently,

$$|\overrightarrow{PS}| \approx \sqrt{(9.0)^2 + (8.1)^2} \approx 12.1$$

We can also find $|\overrightarrow{PS}|$ by using the Law of Cosines. (See Example 3 of Section 7.2.) Since $\angle QPR = 45°$, it follows that $\angle PRS = 135°$ and hence

$$\begin{aligned} |\overrightarrow{PS}|^2 &= (8.0)^2 + (5.0)^2 - 2(8.0)(5.0)\cos 135° \\ &= 64 + 25 - 80(-\sqrt{2}/2) \\ &= 89 + 40\sqrt{2} \approx 145.6 \end{aligned}$$

Thus

$$|\overrightarrow{PS}| \approx \sqrt{145.6} \approx 12.1$$

If θ is the angle from the positive x-axis to the resultant $\overrightarrow{PS} = \langle c_1, c_2 \rangle$, then

$$\tan\theta = \frac{c_2}{c_1} \approx \frac{8.0793}{8.9604} \approx 0.9017$$

Using Table 4 or a calculator, $\theta \approx 42°$ and hence the direction of $\overrightarrow{PS}$ is approximately N48°E. ■

Exercises 7.5

In Exercises 1–10 find $\mathbf{a} + \mathbf{b}$, $\mathbf{a} - \mathbf{b}$, $4\mathbf{a} + 5\mathbf{b}$, and $4\mathbf{a} - 5\mathbf{b}$.

1 $\mathbf{a} = \langle 2, -3 \rangle$, $\mathbf{b} = \langle 1, 4 \rangle$

2 $\mathbf{a} = \langle -2, 6 \rangle$, $\mathbf{b} = \langle 2, 3 \rangle$

3 $\mathbf{a} = -\langle 7, -2 \rangle$, $\mathbf{b} = 4\langle -2, 1 \rangle$

4 $\mathbf{a} = 2\langle 5, -4 \rangle$, $\mathbf{b} = -\langle 6, 0 \rangle$

5 $\mathbf{a} = \mathbf{i} + 2\mathbf{j}$, $\mathbf{b} = 3\mathbf{i} - 5\mathbf{j}$

6 $\mathbf{a} = -3\mathbf{i} + \mathbf{j}$, $\mathbf{b} = -3\mathbf{i} + \mathbf{j}$

7 $\mathbf{a} = -(4\mathbf{i} - \mathbf{j})$, $\mathbf{b} = 2(\mathbf{i} - 3\mathbf{j})$

8 $\mathbf{a} = 8\mathbf{j}$, $\mathbf{b} = (-3)(-2\mathbf{i} + \mathbf{j})$

9 $\mathbf{a} = 2\mathbf{j}$, $\mathbf{b} = -3\mathbf{i}$

10 $\mathbf{a} = \mathbf{0}$, $\mathbf{b} = \mathbf{i} + \mathbf{j}$

In Exercises 11–14 sketch vectors corresponding to $\mathbf{a}$, $\mathbf{b}$, $\mathbf{a} + \mathbf{b}$, $2\mathbf{a}$, and $-3\mathbf{b}$.

11 $\mathbf{a} = 3\mathbf{i} + 2\mathbf{j}$, $\mathbf{b} = -\mathbf{i} + 5\mathbf{j}$

12 $\mathbf{a} = -5\mathbf{i} + 2\mathbf{j}$, $\mathbf{b} = \mathbf{i} - 3\mathbf{j}$

13 $\mathbf{a} = \langle -4, 6 \rangle$, $\mathbf{b} = \langle -2, 3 \rangle$

14 $\mathbf{a} = \langle 2, 0 \rangle$, $\mathbf{b} = \langle -2, 0 \rangle$

In Exercises 15–20 use components to express each sum or difference as a scalar multiple of one of the vectors $\mathbf{a}$, $\mathbf{b}$, $\mathbf{c}$, $\mathbf{d}$, $\mathbf{e}$, or $\mathbf{f}$ shown in the figure.

15 $\mathbf{a} + \mathbf{b}$

16 $\mathbf{c} - \mathbf{d}$

17 $\mathbf{b} + \mathbf{e}$

18 $\mathbf{f} - \mathbf{b}$

19 $\mathbf{b} + \mathbf{d}$

20 $\mathbf{e} + \mathbf{c}$

FIGURE FOR EXERCISES 15–20

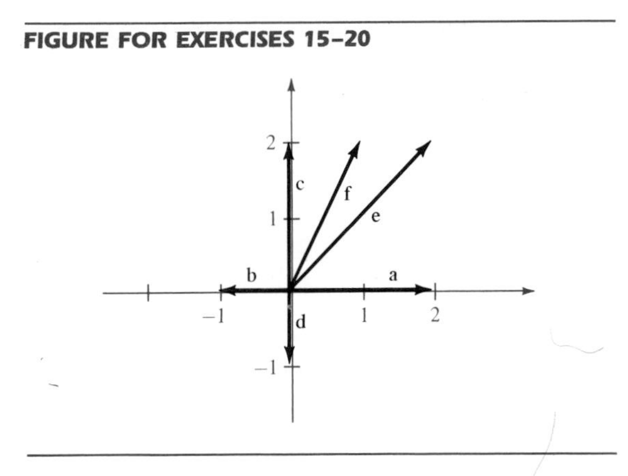

21 Prove the second, third, and fourth properties of addition of vectors listed on page 352.

22 Prove the second through the fifth properties of scalar multiples of vectors listed on page 353.

Prove the properties in Exercises 23–30, where $\mathbf{a} = \langle a_1, a_2 \rangle$, $\mathbf{b} = \langle b_1, b_2 \rangle$, and c is any real number.

23 $(-1)\mathbf{a} = -\mathbf{a}$

24 $(-c)\mathbf{a} = -c\mathbf{a}$

25 $-(\mathbf{a} + \mathbf{b}) = -\mathbf{a} - \mathbf{b}$

26 $c(\mathbf{a} - \mathbf{b}) = c\mathbf{a} - c\mathbf{b}$

27 If $\mathbf{a} + \mathbf{b} = \mathbf{0}$, then $\mathbf{b} = -\mathbf{a}$.

28 If $\mathbf{a} + \mathbf{b} = \mathbf{a}$, then $\mathbf{b} = \mathbf{0}$.

29 If $c\mathbf{a} = \mathbf{0}$ and $c \neq 0$, then $\mathbf{a} = \mathbf{0}$.

30 If $c\mathbf{a} = \mathbf{0}$ and $\mathbf{a} \neq \mathbf{0}$, then $c = 0$.

31 If $\mathbf{v} = \langle a, b \rangle$, prove each of the following.

(a) The magnitude of $2\mathbf{v}$ is twice the magnitude of $\mathbf{v}$.

(b) The magnitude of $\frac{1}{2}\mathbf{v}$ is one-half the magnitude of $\mathbf{v}$.

(c) The magnitude of $-2\mathbf{v}$ is twice the magnitude of $\mathbf{v}$.

(d) If k is any real number, then the magnitude of $k\mathbf{v}$ is $|k|$ times the magnitude of $\mathbf{v}$.

32 If $\mathbf{v} = \langle a, b \rangle$ and $\mathbf{w} = \langle c, d \rangle$, give a geometric interpretation for $\mathbf{v} - \mathbf{w}$.

If $\mathbf{v} = a_1\mathbf{i} + a_2\mathbf{j}$ and $\mathbf{u} = b_1\mathbf{i} + b_2\mathbf{j}$, the **dot product** $\mathbf{v} \cdot \mathbf{u}$ is defined by $\mathbf{v} \cdot \mathbf{u} = a_1b_1 + a_2b_2$. Prove the properties in Exercises 33–36.

33 $\mathbf{v} \cdot \mathbf{u} = \mathbf{u} \cdot \mathbf{v}$

34 $\mathbf{v} \cdot \mathbf{v} = |\mathbf{v}|^2$

35 $(c\mathbf{v}) \cdot \mathbf{u} = c(\mathbf{v} \cdot \mathbf{u})$, for every real number c

36 $\mathbf{v} \cdot (\mathbf{u} + \mathbf{w}) = \mathbf{v} \cdot \mathbf{u} + \mathbf{v} \cdot \mathbf{w}$, where $\mathbf{w} = c_1\mathbf{i} + c_2\mathbf{j}$

In Exercises 37–44 find the magnitude of $\mathbf{a}$ and the smallest positive angle θ from the positive x-axis to the vector $\overrightarrow{OP}$ corresponding to $\mathbf{a}$.

37 $\mathbf{a} = \langle 3, -3 \rangle$

38 $\mathbf{a} = \langle -2, -2\sqrt{3} \rangle$

39 $\mathbf{a} = \langle -5, 0 \rangle$

40 $\mathbf{a} = \langle 0, 10 \rangle$

41 $\mathbf{a} = -4\mathbf{i} + 5\mathbf{j}$

42 $\mathbf{a} = 10\mathbf{i} - 10\mathbf{j}$

43 $\mathbf{a} = -18\mathbf{j}$

44 $\mathbf{a} = 2\mathbf{i} - 3\mathbf{j}$

In Exercises 45–48, (a) and (b) represent the magnitudes and directions of two forces acting at a point P. Approximate the magnitude of the resultant (to two significant figures) and its direction (to the nearest degree).

45 (a) 90 kg, N75°W
(b) 60 kg, S5°E

46 (a) 20 kg, S17°W
(b) 50 kg, N82°W

47 (a) 6.0 lb, 110°
(b) 2.0 lb, 215°

48 (a) 70 lb, 320°
(b) 40 lb, 30°

Approximate the horizontal and vertical components of the vectors that are described in Exercises 49–52.

49 A quarterback releases the football with a velocity of 50 feet/second at an angle of 35° with the horizontal.

50 A girl pulls a sled through the snow by exerting a force of 20 pounds at an angle of 40° with the horizontal.

51 The biceps muscle, in supporting the forearm and a weight held in the hand, exerts a force of 200 pounds. As shown in the figure, the muscle makes an angle of 108° with the forearm.

FIGURE FOR EXERCISE 51

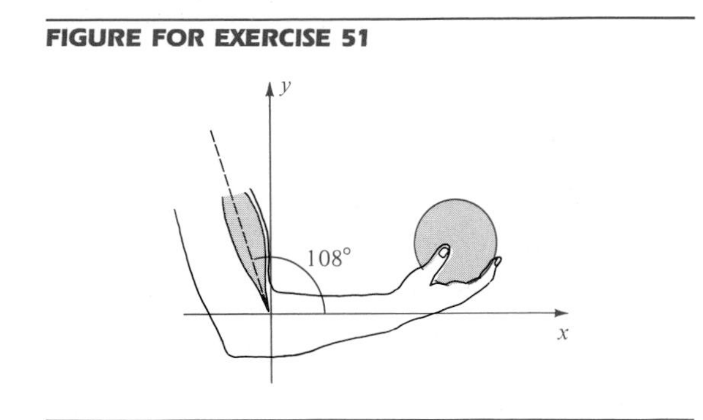

52 A jet airplane approaches a runway at an angle of 7.5° with the horizontal, traveling at a velocity of 160 mph.

When forces $\mathbf{F}_1, \mathbf{F}_2, \mathbf{F}_3, \ldots$ act at a point P, the net (or resultant) force $\mathbf{F}$ is the sum $\mathbf{F}_1 + \mathbf{F}_2 + \mathbf{F}_3 + \cdots$. If $\mathbf{F} = \mathbf{0}$, the system of forces is *in equilibrium.* In Exercises 53–56 the given forces act at the origin O of an xy-plane.

(a) Find the net force $\mathbf{F}$.

(b) Find an additional force $\mathbf{G}$ so that equilibrium occurs.

53 $\mathbf{F}_1 = \langle 4, 3\rangle$, $\mathbf{F}_2 = \langle -2, -3\rangle$, and $\mathbf{F}_3 = \langle 5, 2\rangle$

54 $\mathbf{F}_1 = \langle -3, -1\rangle$, $\mathbf{F}_2 = \langle 0, -3\rangle$, and $\mathbf{F}_3 = \langle 3, 4\rangle$

55 **56**

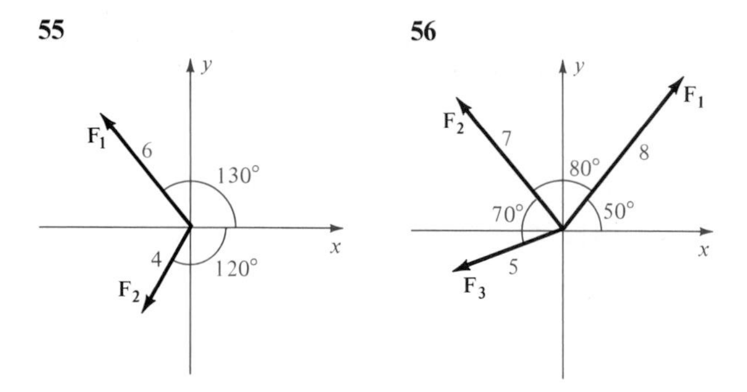

57 Two tugboats are towing a large ship into port, as shown in the figure. The larger tug exerts a force of 4000 pounds on its cable, while a force of 3200 pounds is exerted on the cable by the smaller tug. If the ship is to travel in a straight line from A to B, find the angle θ that the larger tug must make with the line segment AB.

FIGURE FOR EXERCISE 57

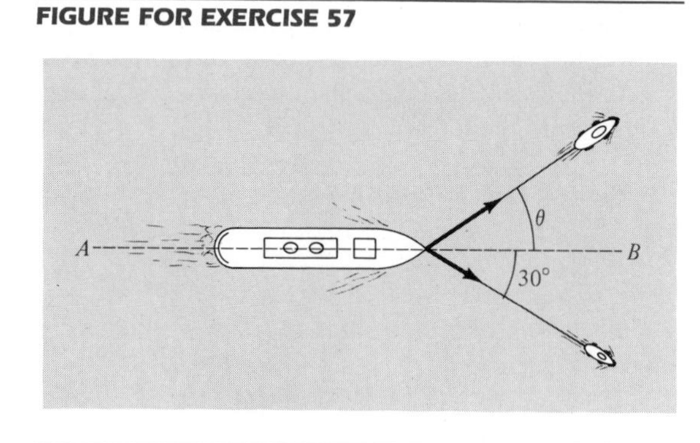

58 Shown in the figure is an apparatus used to simulate gravity conditions on other planets. A rope is attached to an astronaut who maneuvers on an inclined plane that makes an angle of θ degrees with the horizontal.

(a) If the astronaut weighs 160 pounds, find the x and y-components of this downward force. (See figure for axes.)

(b) The y-component in (a) is the weight of the astronaut relative to the inclined plane. The astronaut would weigh 27 pounds on the moon and 60 pounds on Mars. Approximate the angles θ (to the nearest 0.01°) so that the inclined-plane apparatus will simulate walking on these surfaces.

FIGURE FOR EXERCISE 58

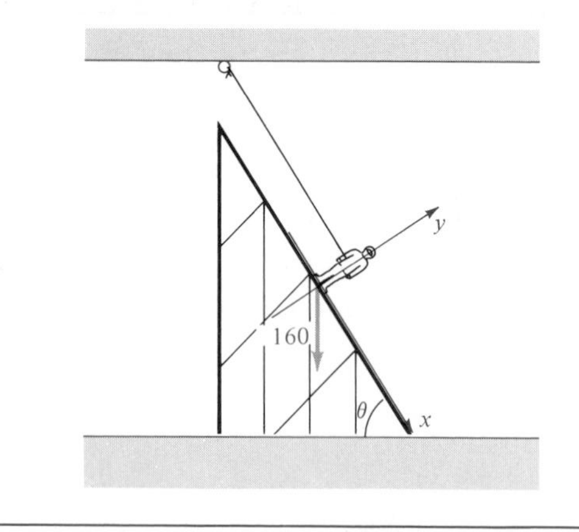

59 An airplane with an airspeed of 200 mph is flying in the direction 50°, and a 40 mph wind is blowing directly from the west. As in the figure, these may be represented by vectors **p** and **w** of magnitudes 200 and 40, respectively.

FIGURE FOR EXERCISE 59

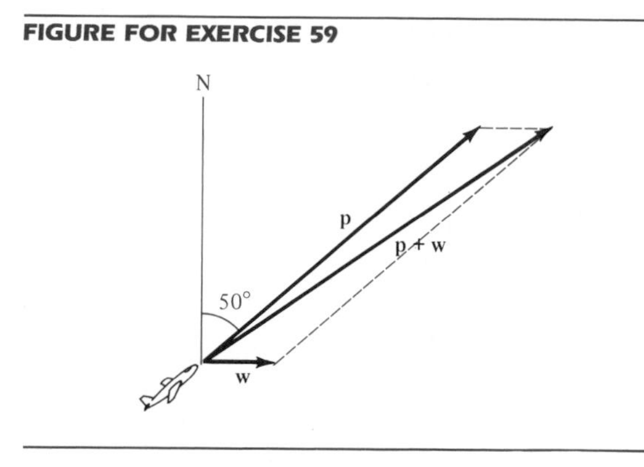

The direction of the resultant $\mathbf{p} + \mathbf{w}$ gives the **true course** of the airplane relative to the ground, and the magnitude $|\mathbf{p} + \mathbf{w}|$ is called the **ground speed** of the airplane. Approximate the true course to the nearest degree and the ground speed to the nearest mph.

60 An airplane is flying in the direction 140° with an airspeed of 500 mph, and a 30 mph wind is blowing in the direction 65°. Approximate the true course (to the nearest degree) and the ground speed (to the nearest mph).

61 An airplane pilot wishes to maintain a true course in the direction 250°, with a ground speed of 400 mph when the wind is blowing directly north at 50 mph. Approximate the required airspeed (to the nearest mph) and compass heading (to the nearest degree).

62 An airplane is flying in the direction 20° with an airspeed of 300 mph. Its ground speed and true course are 350 mph and 30°, respectively. Approximate the direction and speed of the wind.

63 The current in a river flows directly from the west at a rate of 1.5 feet/second. A person who rows a boat at a rate of 4 feet/second in still water wishes to row directly north across the river. Approximate, to the nearest 10′, the direction in which the person should row.

64 In order for a motor boat moving at a speed of 30 mph to travel directly north across a river, it must aim at a point that has a bearing of N15°E. If the current is flowing directly west, approximate the rate at which it flows, to the nearest mph.

7.6 Review

Define or discuss each of the following.

1. The Law of Sines
2. The Law of Cosines
3. Heron's Formula
4. Geometric representation of a complex number
5. The complex plane
6. Absolute value of a complex number
7. De Moivre's Theorem
8. *n*th roots of a complex number
9. Vector
10. Magnitude of a vector
11. Addition of vectors
12. Scalar multiple of a vector
13. Vectors as ordered pairs
14. Components of a vector
15. Subtraction of vectors
16. Unit vector
17. The vectors **i** and **j**
18. Linear combination of **i** and **j**

Exercises 7.6

Without using tables or calculators, find the remaining parts of triangle ABC in Exercises 1–4.

1 $\alpha = 60°,\ \beta = 45°,\ b = 100$

2 $\gamma = 30°,\ a = 2\sqrt{3},\ c = 2$

3 $\alpha = 60°,\ b = 6,\ c = 7$

4 $a = 2,\ b = 3,\ c = 4$

In Exercises 5–8 approximate the remaining parts of triangle ABC.

5 $\beta = 67^\circ$, $\gamma = 75^\circ$, $b = 12$

6 $\alpha = 23^\circ 30'$, $c = 125$, $a = 152$

7 $\beta = 115^\circ$, $a = 4.6$, $c = 7.3$

8 $a = 37$, $b = 55$, $c = 43$

In Exercises 9 and 10 approximate the area of triangle ABC to the nearest 0.1 square unit.

9 $\alpha = 75^\circ$, $b = 20$, $c = 30$

10 $a = 4$, $b = 7$, $c = 10$

Change the complex numbers in Exercises 11–16 to trigonometric form with $0 \le \theta \le 2\pi$.

11 $-10 + 10i$

12 $2 - 2\sqrt{3}i$

13 -17

14 $-12i$

15 $-5\sqrt{3} - 5i$

16 $4 + 5i$

In Exercises 17–20 use De Moivre's Theorem to write the given number in the form $a + bi$, where a and b are real numbers.

17 $(-\sqrt{3} + i)^9$

18 $\left(\frac{\sqrt{2}}{2} - \frac{\sqrt{2}}{2}i\right)^{30}$

19 $(3 - 3i)^5$

20 $(2 + 2\sqrt{3}i)^{10}$

21 Find the three cube roots of -27.

22 Find the solutions of the equation $x^5 - 32 = 0$.

23 For $\mathbf{a} = \langle -4, 5 \rangle$ and $\mathbf{b} = \langle 2, -8 \rangle$, sketch vectors corresponding to $\mathbf{a} + \mathbf{b}$, $\mathbf{a} - \mathbf{b}$, $2\mathbf{a}$, and $-\frac{1}{2}\mathbf{b}$.

24 For $\mathbf{a} = 2\mathbf{i} + 5\mathbf{j}$ and $\mathbf{b} = 4\mathbf{i} - \mathbf{j}$, find the vectors or numbers corresponding to:

(a) $4\mathbf{a} + \mathbf{b}$

(b) $2\mathbf{a} - 3\mathbf{b}$

(c) $|\mathbf{a} - \mathbf{b}|$

(d) $|\mathbf{a}| - |\mathbf{b}|$

25 For $\mathbf{a} = \langle a_1, a_2 \rangle$, $\mathbf{r} = \langle x, y \rangle$, and $c > 0$, describe the set of all points $P(x, y)$ such that $|\mathbf{r} - \mathbf{a}| = c$.

26 For vectors $\mathbf{a}$ and $\mathbf{b}$ with the same initial point and angle θ between them, prove that

$$|\mathbf{a} + \mathbf{b}|^2 = |\mathbf{a}|^2 + |\mathbf{b}|^2 - 2|\mathbf{a}||\mathbf{b}| \cos \theta.$$

27 A ship is sailing at a speed of 14 mph in the direction S50°E. Express its velocity $\mathbf{v}$ as a vector.

28 The magnitudes and directions of two forces are 72 kg, S60°E and 46 kg, N74°E, respectively. Approximate the magnitude and direction of the resultant force.

29 When a mountaintop is viewed from the point P shown in the figure, the angle of elevation is α. From a point Q which is d units closer to the mountain, the angle of elevation increases to β.

(a) Show that the height h of the mountain is

$$h = \frac{\sin \alpha \sin \beta}{\sin (\beta - \alpha)} d.$$

(b) Approximate the height of the mountain if $d = 2$ miles, $\alpha = 15^\circ$, and $\beta = 20^\circ$.

FIGURE FOR EXERCISE 29

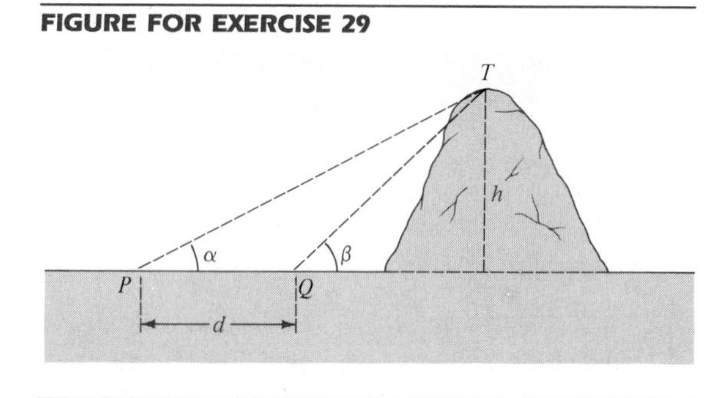

30 A course for a skateboard race consists of a 200-meter downhill run and a 150-meter level portion. When the starting point of the race is spotted from the finishing line, the angle of elevation is 27.4°. What angle does the hill make with the horizontal?

31 If a skyscraper is viewed from the top of a 50-foot building, the angle of elevation is 59°. If viewed from street level, the angle of elevation measures 62°. (See Exercise 40 of Section 5.10.)

(a) Use the Law of Sines to approximate the distance between the tops of the two buildings.

(b) Estimate the height of the skyscraper.

32 The beach communities of San Clemente and Long Beach are 41 miles apart, along a fairly straight stretch of coastline. Shown in the figure is the triangle formed by the two cities and the town of Avalon on the southeast corner of Santa Catalina Island. On a clear day angles ALS and ASL are found to be 66.4° and 47.2°, respectively. Approximate (a) the distance from Avalon to each of the two cities; and (b) the shortest distance from Avalon to the coast.

FIGURE FOR EXERCISE 32

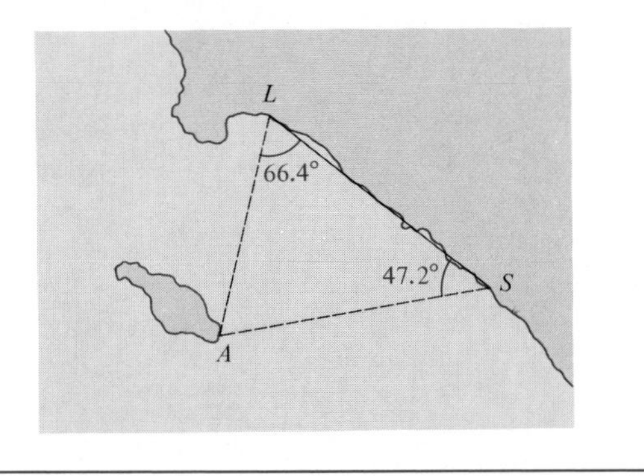

33 The following surveying method can be used to find the distance between two inaccessible points A and B. Two points C and D are selected from which it is possible to view both A and B. The distance CD together with angles ACD, ACB, BDC, and BDA are then measured. If $CD = 120$, $\angle ACD = 115°$, $\angle ACB = 92°$, $\angle BDC = 125°$, and $\angle BDA = 100°$, estimate the distance AB.

FIGURE FOR EXERCISE 33

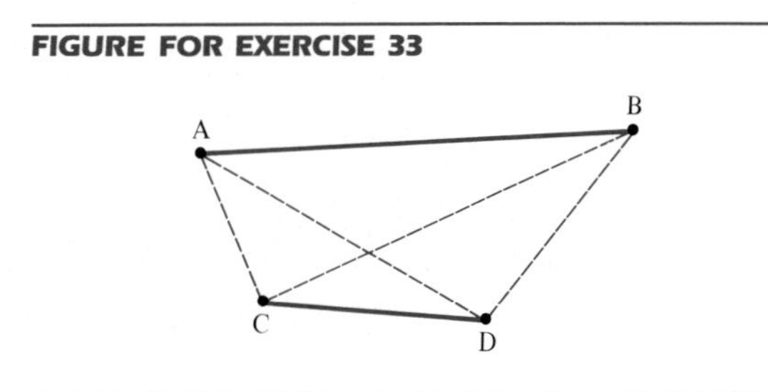

34 Two girls with two-way radios are at the intersection of two country roads that meet at a 105° angle (see figure). One girl begins walking in a northerly direction along one road at the rate of 5 mph, and at the same time the other girl walks east along the other road at the same rate. If the radios have a range of 10 miles, when will they lose contact with one another?

FIGURE FOR EXERCISE 34

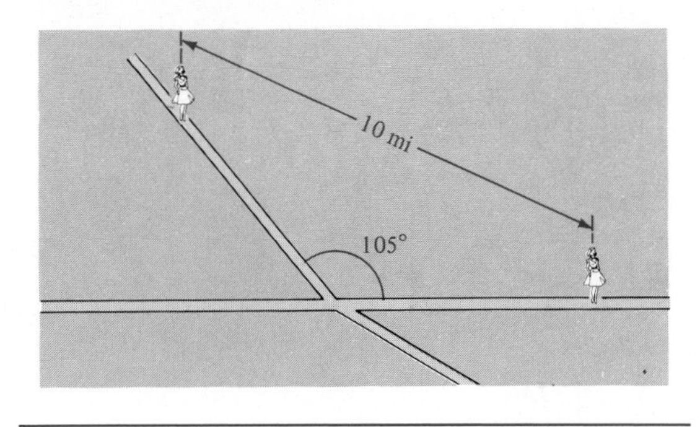

CHAPTER 8

Systems of Equations and Inequalities

It is sometimes necessary to work simultaneously with more than one equation in several variables. ■ We then refer to the equations as a *system of equations.* ■ In this chapter we shall develop methods for finding solutions that are common to all equations in a a system. ■ Of particular importance are the matrix techniques introduced for systems of linear equations. ■ We shall also briefly discuss systems of inequalities and linear programming.

8.1 Systems of Equations

An ordered pair (a, b) is a **solution** of an equation in two variables x and y if a true statement is obtained when a and b are substituted for x and y, respectively. Two equations in x and y are **equivalent** if they have exactly the same solutions. For example, the following equations are equivalent:

$$x^2 - 4y = 3, \qquad x^2 - 3 = 4y$$

As we know, the **graph** of an equation in x and y consists of all points in a coordinate plane that correspond to the solutions of the equation.

A **system** of two equations in x and y is any two equations in those variables. An ordered pair (a, b) is called a **solution of the system** if (a, b) is a solution of both equations. The points that correspond to the solutions are the points at which the graphs of the two equations intersect.

As a concrete example, consider the system

$$\begin{cases} x^2 - y = 0 \\ y - 2x - 3 = 0. \end{cases}$$

The brace is used to indicate that the two equations are to be treated simultaneously. The following table lists some solutions of the equation $x^2 - y = 0$, or equivalently, $y = x^2$.

x	-2	-1	0	1	2	3	4
y	4	1	0	1	4	9	16

FIGURE 8.1

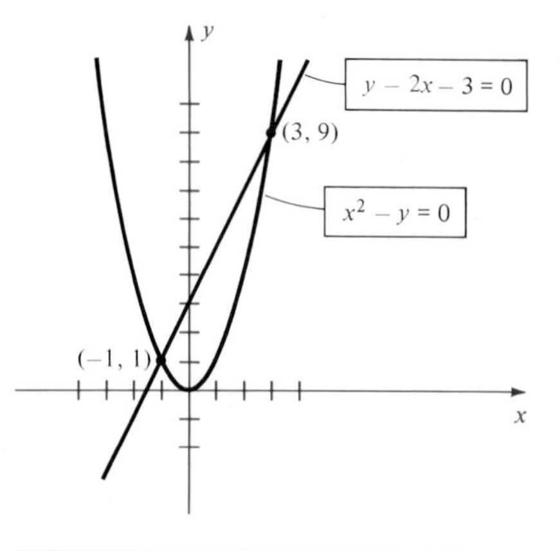

The graph is the parabola sketched in Figure 8.1. Several solutions for $y - 2x - 3 = 0$, or equivalently, $y = 2x + 3$, are listed in the following table.

x	-2	-1	0	1	2	3	4
y	-1	1	3	5	7	9	11

The graph is the line in Figure 8.1. The pairs $(3, 9)$ and $(-1, 1)$ are solutions of both equations and hence are solutions of the system. We shall see later that they are the *only* solutions. Note that the solutions of the system determine the points of intersection of the two graphs.

The solutions of the preceding system can be found algebraically, that is, without reference to graphs. Let us begin by solving the equation $x^2 - y = 0$ for y in terms of x, obtaining $y = x^2$. It follows that if (x, y) is a solution of the system, then it is of the form (x, x^2). In particular, (x, x^2)

is a solution of the equation $y - 2x - 3 = 0$; that is,

$$x^2 - 2x - 3 = 0.$$

Factoring gives us $$(x - 3)(x + 1) = 0$$

and hence either $x = 3$ or $x = -1$. The corresponding values for y (obtained from $y = x^2$) are 9 and 1, respectively. Thus, the only possible solutions of the system are the ordered pairs $(3, 9)$ and $(-1, 1)$. That these are actually solutions can be seen by substitution.

When we found the solutions algebraically in the preceding discussion, we first solved one equation for y in terms of x, and then we substituted for y in the other equation, obtaining an equation in one variable, x. The solutions of the latter equation were the only possible x-values for the solutions of the system. The corresponding y-values were found by means of the equation that expressed y in terms of x. This technique is called the *method of substitution*. Sometimes it is convenient to begin by solving one equation for x in terms of y, and then to substitute for x in the other equation. In general, the steps used in this process may be listed as follows:

METHOD OF SUBSTITUTION

(i) Solve one of the equations for one variable in terms of the other variable.

(ii) Substitute the expression obtained in step (i) into the other equation, obtaining an equation in one variable.

(iii) Find the solutions of the equation obtained in step (ii).

(iv) Use the solutions from step (iii), together with the expression obtained in step (i), to find the solutions of the system.

In the next example we find the solutions of the system considered at the beginning of this section by first solving one equation for x in terms of y.

EXAMPLE 1 Find the solutions of the system

$$\begin{cases} x^2 - y = 0 \\ y - 2x - 3 = 0. \end{cases}$$

SOLUTION We may solve the second equation for x in terms of y as follows:

$$2x = y - 3 \quad \text{or} \quad x = \frac{y - 3}{2}.$$

Substituting for x in the first equation gives us

$$\left(\frac{y-3}{2}\right)^2 - y = 0$$

$$\frac{y^2-6y+9}{4} - y = 0$$

$$y^2 - 6y + 9 - 4y = 0$$

$$y^2 - 10y + 9 = 0$$

$$(y-9)(y-1) = 0.$$

The last equation has solutions $y = 9$ and $y = 1$. If we substitute 9 and 1 for y in the equation $x = (y-3)/2$, we obtain the values $x = 3$ and $x = -1$, respectively. Hence, as before, the solutions of the system are $(3, 9)$ and $(-1, 1)$. ■

EXAMPLE 2 Find the solutions of the following system and then sketch the graph of each equation, showing the points of intersection:

$$\begin{cases} x^2 + y^2 = 25 \\ x^2 + y = 19. \end{cases}$$

SOLUTION Solving the second equation for y, we obtain $y = 19 - x^2$. Substituting for y in the first equation leads to the following list of equivalent equations:

$$x^2 + (19 - x^2)^2 = 25$$

$$x^2 + (361 - 38x^2 + x^4) = 25$$

$$x^4 - 37x^2 + 336 = 0$$

$$(x^2 - 16)(x^2 - 21) = 0.$$

The solutions of the last equation are 4, -4, $\sqrt{21}$, and $-\sqrt{21}$. The corresponding values of y are found by substituting for x in $y = 19 - x^2$. Substitution of 4 or -4 for x gives us $y = 3$, whereas substitution of $\sqrt{21}$ or $-\sqrt{21}$ gives us $y = -2$. Hence, the only possible solutions of the system are

$$(4, 3), \quad (-4, 3), \quad (\sqrt{21}, -2), \quad \text{and} \quad (-\sqrt{21}, -2).$$

It can be seen by substitution in the given equations that all four pairs are solutions.

FIGURE 8.2

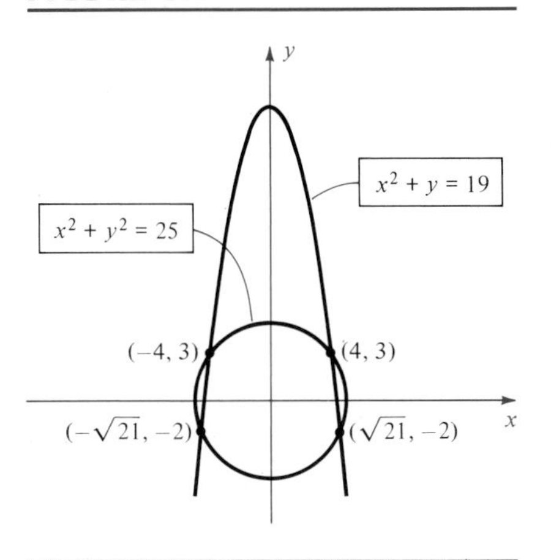

The graph of $x^2 + y^2 = 25$ is a circle of radius 5 with center at the origin, and the graph of $y = 19 - x^2$ is a parabola with a vertical axis. The graphs are illustrated in Figure 8.2. The points of intersection are determined by the solutions of the system. ■

We can also consider equations in three variables x, y, and z, such as

$$x^2y + xz + 3^y = 4z^3.$$

The equation has a **solution** (a, b, c) if substitution of a, b, and c, for x, y, and z, respectively, produces a true statement. We refer to (a, b, c) as an **ordered triple** of real numbers. Equivalent equations are defined as before. A system of equations in three variables and the solutions of such a system are defined as in the two-variable case. In like manner, we can consider systems of *any* number of equations in *any* number of variables.

The method of substitution can be extended to these more complicated systems. For example, given three equations in three variables, suppose that it is possible to solve one of the equations for one variable in terms of the remaining two variables. By substituting that expression in each of the other equations, we obtain a system of two equations in two variables. The solutions of the two-variable system can then be used to find the solutions of the original system, as illustrated in the following example.

EXAMPLE 3 Find the solutions of the system

$$\begin{cases} x - y + z = 2 \\ xyz = 0 \\ 2y + z = 1. \end{cases}$$

SOLUTION Solving the third equation for z,

$$z = 1 - 2y.$$

Substituting for z in the first two equations of the system we obtain the following system of two equations in two variables:

$$\begin{cases} x - y + (1 - 2y) = 2 \\ xy(1 - 2y) = 0. \end{cases}$$

This system is equivalent to

$$\begin{cases} x - 3y - 1 = 0 \\ xy(1 - 2y) = 0. \end{cases}$$

We now find the solutions of the last system. Solving the first equation for x in terms of y gives us

$$x = 3y + 1.$$

Substituting $3y + 1$ for x in the second equation $xy(1 - 2y) = 0$, we obtain

$$(3y + 1)y(1 - 2y) = 0,$$

which has as solutions the numbers $-\frac{1}{3}$, 0, and $\frac{1}{2}$. These are the only possible y-values for the solutions of the system. To obtain the corresponding x-values, we use the equation $x = 3y + 1$, obtaining $x = 0$, 1, and $\frac{5}{2}$, respectively. Finally, substituting in $z = 1 - 2y$ gives us the z-values $\frac{5}{3}$, 1, and 0. It follows that the solutions of the original system consist of the ordered triples

$$(0, -\tfrac{1}{3}, \tfrac{5}{3}), \quad (1, 0, 1), \quad \text{and} \quad (\tfrac{5}{2}, \tfrac{1}{2}, 0).$$ ■

EXAMPLE 4 Is it possible to construct an aquarium with a glass top and two square ends that holds 16 ft^3 of water and requires 40 ft^2 of glass?

FIGURE 8.3

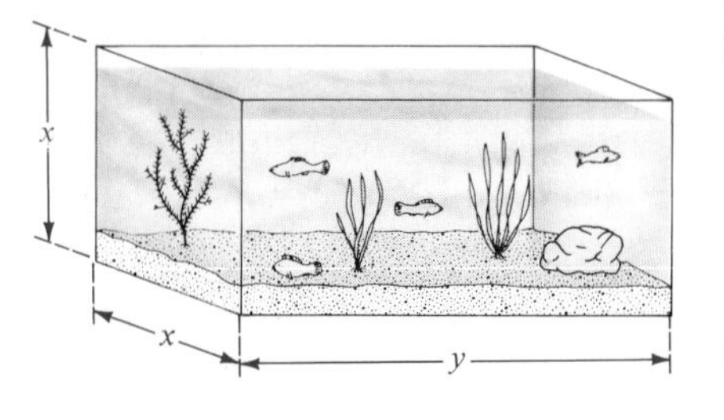

SOLUTION We begin by sketching a typical aquarium and labeling it as in Figure 8.3, in which x and y are in feet. Referring to the figure and using formulas for volume and area, we see that

$$\text{Volume of the aquarium} = x^2y,$$

$$\text{Square feet of glass required} = 2x^2 + 4xy.$$

Since the volume and the glass required are 16 and 40, respectively, we obtain the following system of equations:

$$\begin{cases} x^2y = 16 \\ 2x^2 + 4xy = 40 \end{cases}$$

Solving the first equation for y gives us $y = 16/x^2$. Substituting for y in the second equation leads to

$$2x^2 + 4x\left(\frac{16}{x^2}\right) = 40$$

$$2x^2 + \frac{64}{x} = 40$$

$$2x^3 + 64 = 40x$$

$$x^3 - 20x + 32 = 0$$

We next look for rational solutions of this equation. Dividing the polynomial $x^3 - 20x + 32$ synthetically by $x - 2$ gives us

$$\begin{array}{r|rrrr} 2 & 1 & 0 & -20 & 32 \\ & & 2 & 4 & -32 \\ \hline & 1 & 2 & -16 & 0 \end{array}$$

Thus, one solution of $x^3 - 20x + 32 = 0$ is 2, and the remaining two solutions satisfy

$$x^2 + 2x - 16 = 0.$$

Using the Quadratic Formula, we obtain

$$x = \frac{-2 \pm \sqrt{4 + 64}}{2} = \frac{-2 \pm 2\sqrt{17}}{2} = -1 \pm \sqrt{17}.$$

Since x is positive we may discard $-1 - \sqrt{17}$. Hence,

$$x = -1 + \sqrt{17} \approx 3.12.$$

The corresponding y-values can be determined from $y = 16/x^2$. On the one hand, letting $x = 2$ gives us $y = \frac{16}{4} = 4$. On the other hand, if $x = -1 + \sqrt{17}$, then it can be shown that $y = (9 + \sqrt{17})/8 \approx 1.64$. We have shown that there are two different ways to construct the aquarium. It is likely that most people (and fish) would prefer the dimensions 2 feet by 2 feet by 4 feet. ■

Exercises 8.1

In Exercises 1–30 use the method of substitution to find the solutions of the system of equations.

1 $\begin{cases} y = x^2 - 4 \\ y = 2x - 1 \end{cases}$

2 $\begin{cases} y = x^2 + 1 \\ x + y = 3 \end{cases}$

3 $\begin{cases} y^2 = 1 - x \\ x + 2y = 1 \end{cases}$

4 $\begin{cases} y^2 = x \\ x + 2y + 3 = 0 \end{cases}$

5 $\begin{cases} 2y = x^2 \\ y = 4x^3 \end{cases}$

6 $\begin{cases} x - y^3 = 1 \\ 2x = 9y^2 + 2 \end{cases}$

7 $\begin{cases} x + 2y = -1 \\ 2x - 3y = 12 \end{cases}$

8 $\begin{cases} 3x - 4y + 20 = 0 \\ 3x + 2y + 8 = 0 \end{cases}$

9 $\begin{cases} 2x - 3y = 1 \\ -6x + 9y = 4 \end{cases}$

10 $\begin{cases} 4x - 5y = 2 \\ 8x - 10y = -5 \end{cases}$

11 $\begin{cases} x + 3y = 5 \\ x^2 + y^2 = 25 \end{cases}$

12 $\begin{cases} 3x - 4y = 25 \\ x^2 + y^2 = 25 \end{cases}$

13 $\begin{cases} x^2 + y^2 = 8 \\ y - x = 4 \end{cases}$

14 $\begin{cases} x^2 + y^2 = 25 \\ 3x + 4y = -25 \end{cases}$

15 $\begin{cases} x^2 + y^2 = 9 \\ y - 3x = 2 \end{cases}$

16 $\begin{cases} x^2 + y^2 = 16 \\ y + 2x = -1 \end{cases}$

17 $\begin{cases} x^2 + y^2 = 16 \\ 2y - x = 4 \end{cases}$

18 $\begin{cases} x^2 + y^2 = 1 \\ y + 2x = -3 \end{cases}$

19 $\begin{cases} (x - 1)^2 + (y + 2)^2 = 10 \\ x + y = 1 \end{cases}$

20 $\begin{cases} xy = 2 \\ 3x - y + 5 = 0 \end{cases}$

21 $\begin{cases} y = 20/x^2 \\ y = 9 - x^2 \end{cases}$

22 $\begin{cases} x = y^2 - 4y + 5 \\ x - y = 1 \end{cases}$

23 $\begin{cases} y^2 - 4x^2 = 4 \\ 9y^2 + 16x^2 = 140 \end{cases}$

24 $\begin{cases} 25y^2 - 16x^2 = 400 \\ 9y^2 - 4x^2 = 36 \end{cases}$

25 $\begin{cases} x^2 - y^2 = 4 \\ x^2 + y^2 = 12 \end{cases}$

26 $\begin{cases} 6x^3 - y^3 = 1 \\ 3x^3 + 4y^3 = 5 \end{cases}$

27 $\begin{cases} x + 2y - z = -1 \\ 2x - y + z = 9 \\ x + 3y + 3z = 6 \end{cases}$

28 $\begin{cases} 2x - 3y - z^2 = 0 \\ x - y - z^2 = -1 \\ x^2 - xy = 0 \end{cases}$

29 $\begin{cases} x^2 + z^2 = 5 \\ 2x + y = 1 \\ y + z = 1 \end{cases}$

30 $\begin{cases} x + 2z = 1 \\ 2y - z = 4 \\ xyz = 0 \end{cases}$

31 The perimeter of a rectangle is 40 inches and its area is 96 square inches. Find the length and width.

32 Find the values of b such that the system

$$\begin{cases} x^2 + y^2 = 4 \\ y = x + b \end{cases}$$

has solutions consisting of (a) one real number, (b) two real numbers, (c) no real numbers. Interpret the three cases geometrically.

33 Is there a real number x such that $x = 2^{-x}$? Decide by displaying graphically the system $y - x = 0$ and $y - 2^{-x} = 0$.

34 Is there a real number x such that $x = \log x$? Decide by displaying graphically the system $y - \log x = 0$ and $y - x = 0$.

35 Sections of tin cylindrical tubing are to be made from rectangular sheets that have an area of 200 in^2 (see figure). Is it possible to construct a tube that has a volume of 200 in^3? If so, find the dimensions of the rectangular sheet.

FIGURE FOR EXERCISE 35

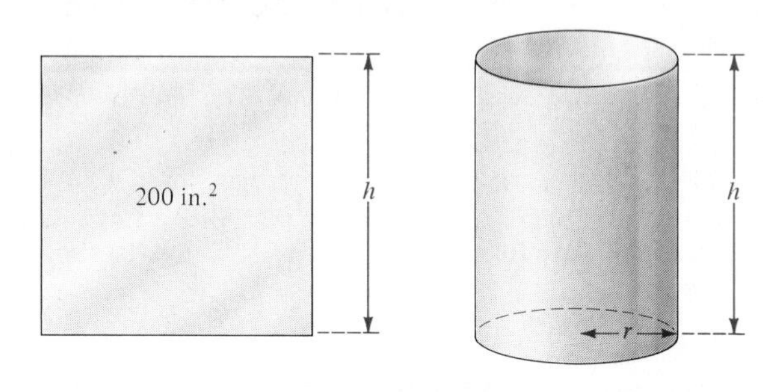

36 Shown in the figure is the graph of $y = x^2$ and a line of slope m that passes through the point (1, 1). Find a value of m such that the line intersects the graph only at (1, 1).

FIGURE FOR EXERCISE 36

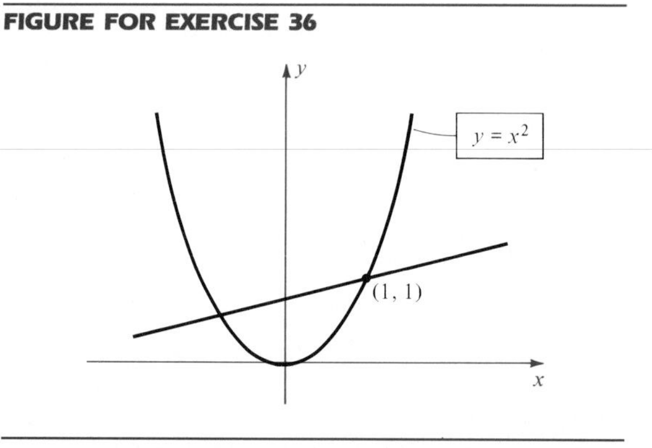

37 In fishery science, spawner-recruit functions are used to predict the number R in next year's breeding population from an estimate S of the number of fish presently spawning. (See Exercise 28 of Section 3.6.)

(a) If $R = aS/(S + b)$ estimate a and b from the data in the following table.

Year	1983	1984	1985
Number spawning	40,000	60,000	72,000

(b) Predict the breeding population for the year 1986.

38 Refer to Exercise 37. Ricker's spawner-recrúit function is given by

$$R = aSe^{-bS}$$

where a and b are positive constants. This relationship predicts low recruitment from very high stocks and has been found to be appropriate for many species, such as arctic cod. Rework Exercise 37 using Ricker's spawner-recruit function.

39 A *competition model* is a collection of equations that specify how two or more species interact in competing for the food resources of an ecosystem. Let x and y denote the numbers (in hundreds) of two competing species and suppose that the respective rates of growth R_1 and R_2 are given by

$$R_1 = 0.01x(50 - x - y),$$

$$R_2 = 0.02y(100 - y - 0.5x).$$

Determine the population levels (x, y) at which both rates of growth are zero. (Such population levels are called *stationary points.*)

40 A rancher has 2420 feet of fence to enclose pastureland that lies along a straight river. If no fence is used along the river, is it possible to enclose 10 acres of pastureland? Recall that 1 acre = 43,560 ft^2.

8.2 Systems of Linear Equations in Two Variables

An equation of the form $ax + by + c = 0$ (or, equivalently, $ax + by = -c$), where a and b are not both zero, is called a *linear equation* in x and y. Similarly, a linear equation in three variables x, y, and z is an equation of the form $ax + by + cz = d$, where the coefficients are real numbers. In general, if we let $x_1, x_2, \ldots, x_n$ denote variables where n is any positive integer, then an equation of the form

$$a_1x_1 + a_2x_2 + \cdots + a_nx_n = a$$

where $a_1, a_2, \ldots, a_n$ and a are real numbers and not every a_j is zero, is called a **linear equation in n variables** (with real coefficients).

The most common systems of equations are those in which all the equations are linear. In this section we shall only consider systems of two linear equations in two variables. Systems involving more than two variables are discussed in Section 8.3.

One method for solving a system of equations in several variables is to replace equations in the system by equivalent equations until we reach a system from which the solutions are easily obtained. Some rules for transforming systems of equations are stated in the next theorem. We shall use the notation $p = 0$ and $q = 0$ for typical equations in a system. Bear in mind that the symbols p and q represent expressions in the variables under consideration. For example, if we let $p = 3x - 2y + 5z - 1$ and $q = 6x + y - 7z - 4$, then the equations $p = 0$ and $q = 0$ have the forms

$$3x - 2y + 5z - 1 = 0, \qquad 6x + y - 7z - 4 = 0.$$

TRANSFORMATIONS THAT LEAD TO EQUIVALENT SYSTEMS OF EQUATIONS

The following transformations do not change the solutions of a system of equations:

(i) Interchanging the position of any two equations;

(ii) Multiplying both sides of an equation in the system by a nonzero real number;

(iii) Replacing an equation $q = 0$ of the system by $kp + q = 0$, where $p = 0$ is any other equation in the system and k is any real number.

PROOF It is easy to show that (i) and (ii) do not change the solutions of the system, and therefore we shall omit the proofs.

To prove (iii), we first note that a solution of the original system is a solution of both equations $p = 0$ and $q = 0$. This means that each of the expressions p and q equals zero if the variables are replaced by appropriate real numbers, and hence the expression $kp + q$ will also equal zero. Since none of the other equations has been changed, this shows that any solution of the original system is also a solution of the transformed system, which is obtained by replacing the equation $q = 0$ by $kp + q = 0$.

Conversely, given a solution of the transformed system, both of the expressions $kp + q$ and p equal zero if the variables are replaced by appropriate numbers. This implies, however, that $(kp + q) - kp$ equals 0. Since $(kp + q) - kp = q$, we see that q must also equal zero when the substitution is made. Thus, a solution of the transformed system is also a solution of the original system. This completes the proof. □

For convenience, we shall describe transformation (iii) by the phrase "add to one equation of the system k times any other equation of the system." Of course, to *add* two equations means to add corresponding sides. To *multiply* an equation by k means to multiply both sides of the equation by k. The process may also be applied if 0 is not on one side of each equation, as illustrated in the next example.

EXAMPLE 1 Find the solutions of the system

$$\begin{cases} x + 3y = -1 \\ 2x - y = 5. \end{cases}$$

SOLUTION By (ii) of the preceding theorem we may multiply the second equation by 3. This gives us the equivalent system

$$\begin{cases} x + 3y = -1 \\ 6x - 3y = 15. \end{cases}$$

Next, by (iii) of the theorem, we may add to the second equation 1 times the first equation, obtaining

$$\begin{cases} x + 3y = -1 \\ 7x \qquad = 14. \end{cases}$$

We see from the last equation that $x = 2$. The corresponding value for y may be found by substituting for x in the equation $x + 3y = -1$. This gives us

$$2 + 3y = -1, \qquad 3y = -3, \qquad y = -1.$$

Thus, the system has one solution, $(2, -1)$.

There are other methods of solution. For example, we could begin by multiplying the first equation by -2, obtaining

$$\begin{cases} -2x - 6y = 2 \\ 2x - y = 5. \end{cases}$$

FIGURE 8.4

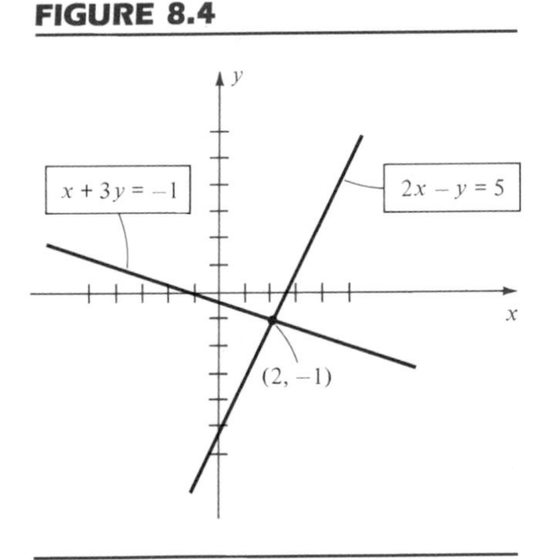

If we next add the first equation to the second, we get

$$\begin{cases} -2x - 6y = 2 \\ \qquad - 7y = 7. \end{cases}$$

The last equation implies that $y = -1$. Substitution for y in the equation $-2x - 6y = 2$ gives us

$$-2x - 6(-1) = 2, \qquad -2x = -4, \qquad x = 2.$$

Again we see that the solution is $(2, -1)$.

The graphs of the two equations, showing the point of intersection $(2, -1)$, are sketched in Figure 8.4. ■

The technique used in Example 1 is called the **method of elimination,** since it involves eliminating a variable from one of the equations. The method of elimination usually leads to solutions in fewer steps than the method of substitution discussed in Section 8.1.

EXAMPLE 2 Find the solutions of the system

$$\begin{cases} 3x + y = 6 \\ 6x + 2y = 12. \end{cases}$$

SOLUTION Multiplying the second equation by $\frac{1}{2}$ gives us

$$\begin{cases} 3x + y = 6 \\ 3x + y = 6. \end{cases}$$

Thus, (a, b) is a solution if and only if $3a + b = 6$; that is, $b = 6 - 3a$. It follows that the solutions consist of all ordered pairs of the form $(a, 6 - 3a)$ where a is a real number. If we wish to find particular solutions we may substitute various values for a. A few solutions are $(0, 6)$, $(1, 3)$, $(3, -3)$, $(-2, 12)$ and $(\sqrt{2}, 6 - 3\sqrt{2})$. Note that the graphs of the two equations are the same line. ■

EXAMPLE 3 Find the solutions of the system

$$\begin{cases} 3x + y = 6 \\ 6x + 2y = 20. \end{cases}$$

SOLUTION If we add to the second equation -2 times the first equation, we obtain the equivalent system

$$\begin{cases} 3x + y = 6 \\ 0 = 8. \end{cases}$$

The last equation can be written $0x + 0y = 8$, which is false for every ordered pair (x, y). Thus, the system has no solutions. This means that the graphs of the given equations do not intersect (see Figure 8.5). ■

FIGURE 8.5

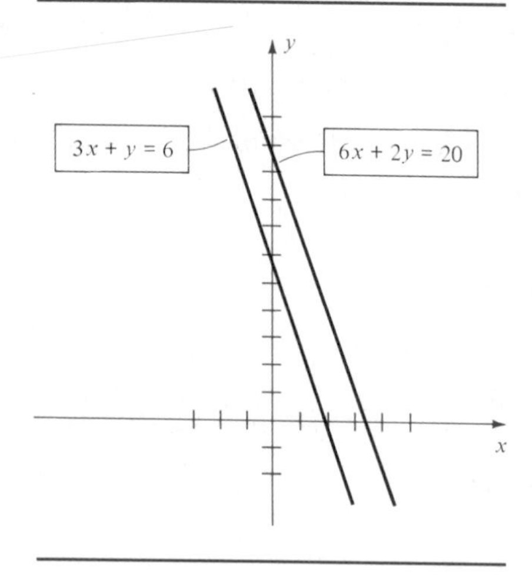

Since the graph of every linear equation $ax + by = c$ is a line, it follows that for every system of two such equations, precisely one of the following three possibilities occurs:

(i) The lines intersect in exactly one point.
(ii) The lines are identical.
(iii) The lines are parallel.

Since the solutions of the system determine the points of intersection of the graphs of the two equations, we may interpret (i)–(iii) in the following way.

THEOREM

For each system of two linear equations in two variables, one and only one of the following statements is true.

(i) The system has exactly one solution;

(ii) The system has an infinite number of solutions;

(iii) The system has no solution.

If (i) occurs the system is said to be **consistent.**
If (ii) occurs the equations are said to be **dependent.**
If (iii) occurs, the system is called **inconsistent.**

In practice, there should be little difficulty in determining which of the three cases occurs. The case of the unique solution (i) will become apparent when suitable transformations are applied to the system, as illustrated in Example 1. In case (ii) the solution is similar to that for Example 2 where one of the equations can be transformed into the other. In case (iii) the lack of a solution is indicated by an absurdity, such as the statement $0 = 8$, which appeared in Example 3.

Certain applied problems can be solved by introducing systems of two linear equations, as illustrated in the next example.

EXAMPLE 4 A motor boat, operating at full throttle, made a trip 4 miles upstream (against a constant current) in 15 minutes. The return trip (with the same current and at full throttle) took 12 minutes. Find the speed of the current and the equivalent speed of the boat in still water.

SOLUTION We shall begin by introducing letters to denote the unknown quantities. Thus, let

$$x = \text{speed of boat (in mph)},$$

$$y = \text{speed of current (in mph)}.$$

We plan to use the formula $d = rt$, where d denotes the distance traveled, r the rate, and t the time. Since the current slows the boat as it travels upstream, but adds to its speed as it travels downstream, we obtain

$$\text{upstream rate} = x - y \qquad \text{(in mph)}$$

$$\text{downstream rate} = x + y \qquad \text{(in mph)}.$$

The time (in hours) traveled in each direction is

$$\text{upstream time} = \tfrac{15}{60} = \tfrac{1}{4}\ \text{hr},$$

$$\text{downstream time} = \tfrac{12}{60} = \tfrac{1}{5}\ \text{hr}.$$

The distance is 4 miles for each trip. Substituting in $d = rt$ gives us the system

$$\begin{cases} 4 = (x - y)(\frac{1}{4}) \\ 4 = (x + y)(\frac{1}{5}) \end{cases}$$

or equivalently,

$$\begin{cases} x - y = 16 \\ x + y = 20. \end{cases}$$

Adding the last two equations, we see that $2x = 36$, or $x = 18$. Consequently, $y = 20 - x = 20 - 18 = 2$. Hence, the speed of the boat in still water is 18 miles per hour and the speed of the current is 2 miles per hour. ■

Exercises 8.2

Find the solutions of the systems in Exercises 1–20.

1 $\begin{cases} 2x + 3y = 2 \\ x - 2y = 8 \end{cases}$

2 $\begin{cases} 4x + 5y = 13 \\ 3x + y = -4 \end{cases}$

3 $\begin{cases} 2x + 5y = 16 \\ 3x - 7y = 24 \end{cases}$

4 $\begin{cases} 7x - 8y = 9 \\ 4x + 3y = -10 \end{cases}$

5 $\begin{cases} 3r + 4s = 3 \\ r - 2s = -4 \end{cases}$

6 $\begin{cases} 9u + 2v = 0 \\ 3u - 5v = 17 \end{cases}$

7 $\begin{cases} 5x - 6y = 4 \\ 3x + 7y = 8 \end{cases}$

8 $\begin{cases} 2x + 8y = 7 \\ 3x - 5y = 4 \end{cases}$

9 $\begin{cases} \frac{1}{3}c + \frac{1}{2}d = 5 \\ c - \frac{2}{3}d = -1 \end{cases}$

10 $\begin{cases} \frac{1}{2}t - \frac{1}{5}v = \frac{3}{2} \\ \frac{2}{3}t + \frac{1}{4}v = \frac{5}{12} \end{cases}$

11 $\begin{cases} \sqrt{3}x - \sqrt{2}y = 2\sqrt{3} \\ 2\sqrt{2}x + \sqrt{3}y = \sqrt{2} \end{cases}$

12 $\begin{cases} 0.11x - 0.03y = 0.25 \\ 0.12x + 0.05y = 0.70 \end{cases}$

13 $\begin{cases} 2x - 3y = 5 \\ -6x + 9y = 12 \end{cases}$

14 $\begin{cases} 3p - q = 7 \\ -12p + 4q = 3 \end{cases}$

15 $\begin{cases} 3m - 4n = 2 \\ -6m + 8n = -4 \end{cases}$

16 $\begin{cases} x - 5y = 2 \\ 3x - 15y = 6 \end{cases}$

17 $\begin{cases} 2y - 5x = 0 \\ 3y + 4x = 0 \end{cases}$

18 $\begin{cases} 3x + 7y = 9 \\ y = 5 \end{cases}$

19 $\begin{cases} \dfrac{2}{x} + \dfrac{3}{y} = -2 \\ \dfrac{4}{x} - \dfrac{5}{y} = 1 \end{cases}$ (*Hint:* Let $x' = 1/x$ and $y' = 1/y$.)

20 $\begin{cases} \dfrac{3}{x-1} + \dfrac{4}{y+2} = 2 \\ \dfrac{6}{x-1} - \dfrac{7}{y+2} = -3 \end{cases}$

21 The price of admission for a high school play was \$1.50 for students and \$2.25 for nonstudents. If 450 tickets were sold for a total of \$777.75, how many of each kind were purchased?

22 New West Airlines flies from Los Angeles to Albuquerque with a stopover in Phoenix. The airfare to Phoenix is \$45 while the fare to Albuquerque is \$60. A total of 185 passengers boarded the plane in Los Angeles and fares totaled \$10,500. How many passengers got off the plane in Phoenix?

23 A crayon is to be 8 cm in length, 1 cm in diameter, and will be made from 5 cm^3 of colored wax. The crayon is to have the shape of a cylinder surmounted by a small conical tip (see figure). Find the length of the cylinder and the height of the cone.

FIGURE FOR EXERCISE 23

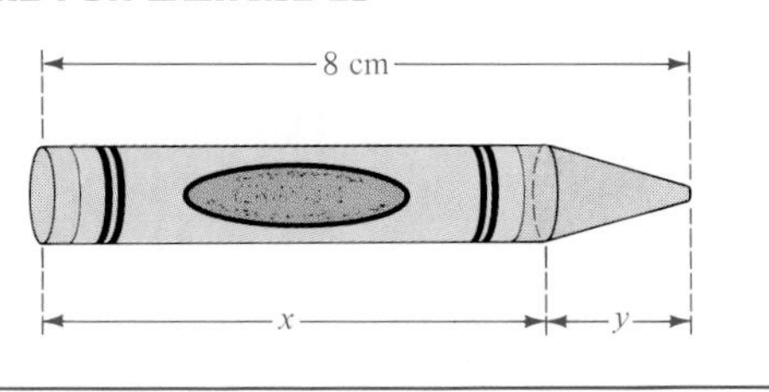

24 A man rows a boat 500 feet upstream against a constant current in 10 minutes. He then rows downstream (with the same current), covering 300 feet in 5 minutes. Find the speed of the current and the equivalent rate at which he can row in still water.

25 A large table for a conference room is to be constructed in the shape of a rectangle with two semicircles at the ends (see figure). Find the length and width of the rectangular portion assuming the table is to have a perimeter of 40 feet and the area of the rectangular portion is to be twice the sum of the areas of the two ends.

FIGURE FOR EXERCISE 25

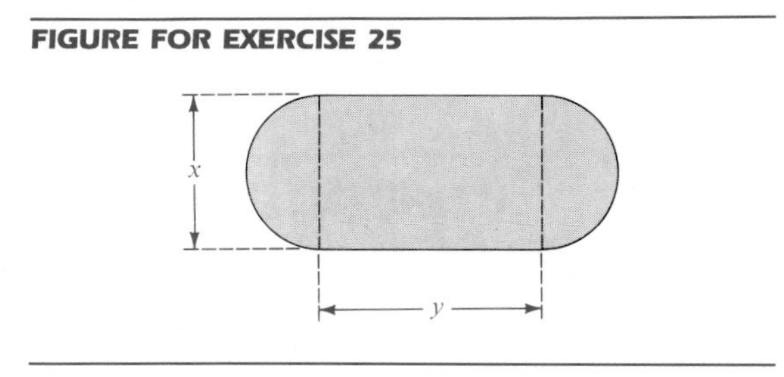

26 A woman has \$10,000 invested in two funds that pay simple interest rates of 8% and 7%, respectively. If she receives yearly interest of \$772, how much is invested in each fund?

27 A man receives income from two investments at simple interest rates of $6\frac{1}{2}\%$ and 8%, respectively. He has twice as much invested at $6\frac{1}{2}\%$ as at 8%. If his annual income from the two investments is \$698.25, find how much is invested at each rate.

28 A 300 gallon water storage tank is filled by a single inlet pipe, and two identical outlet pipes can be used to supply water to the surrounding fields (see figure). It takes 5 hours to fill an empty tank when both outlet pipes are open. When one outlet pipe is closed, it takes 3 hours to fill the tank. Find the flow rates (in gallons per hour) in and out of the tank.

FIGURE FOR EXERCISE 28

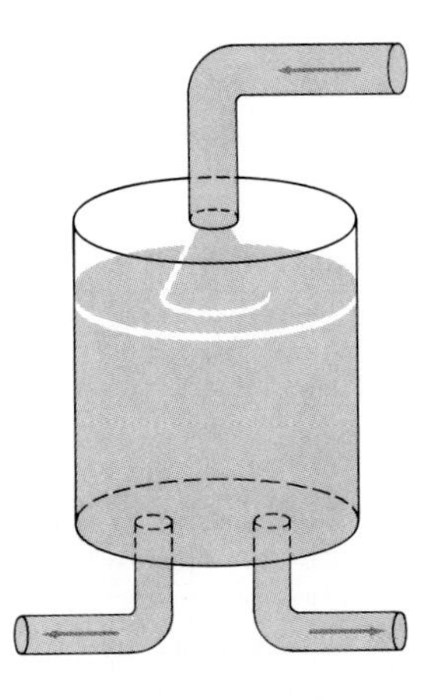

29 A silversmith has two alloys, the first containing 35% silver, and the second 60% silver. How much of each should be melted and combined to obtain 100 grams of an alloy containing 50% silver?

30 A merchant wishes to mix peanuts costing \$2.00 per pound with cashews costing \$3.50 per pound to obtain 60 pounds of a mixture costing \$2.65 per pound. How many pounds of each variety should be mixed?

31 An airplane, flying with a tail wind, travels 1200 miles in 2 hours. The return trip, against the wind, takes $2\frac{1}{2}$ hours. Find the cruising speed of the plane and the speed of the wind (assume that both rates are constant).

32 A stationery company sells two types of notebooks to college bookstores, the first wholesaling for 50¢ and the second for 70¢. The company receives an order for 500 notebooks, together with a check for \$286. If the order fails to specify the number of each type, how should the company fill the order?

33 As a ball rolls down an inclined plane, its velocity $v(t)$ (in cm/second) at time t (in seconds) is given by $v(t) = v_0 + at$, where v_0 is the initial velocity and a is the acceleration (in cm/sec^2). If $v(2) = 16$ and $v(5) = 25$, find v_0 and a.

34 If an object is projected vertically upward from an altitude of s_0 feet with an initial velocity of v_0 feet/second, then its distance $s(t)$ above the ground after t seconds is

$$s(t) = -16t^2 + v_0 t + s_0.$$

If $s(1) = 84$ and $s(2) = 116$, what are v_0 and s_0?

35 A small furniture company manufactures sofas and recliners. Each sofa requires 8 hours of labor and $60 in materials, while a recliner can be built for $35 in 6 hours. The company has 340 hours of labor available each week and can afford to buy $2250 in materials. How many recliners and sofas can be produced if all labor hours and all materials will be used?

36 A rancher is preparing an oat–cornmeal mixture for livestock. Each ounce of oats contain 4 grams of protein and 18 grams of carbohydrates, while an ounce of corn meal has 3 grams of protein and 24 grams of carbohydrates. How many ounces of each can be used to meet the nutritional goals of 200 grams of protein and 1320 grams of carbohydrates per feeding?

37 A plumber and an electrician are each doing repairs on their places of business and agree to swap services. The number of hours spent on each of the projects is shown in the following table.

	Plumber's Business	Electrician's Business
Plumber's hours	6	4
Electrician's hours	5	6

Ordinarily they would call the matter even, but, due to tax laws, they must charge for all work performed. They agree to select hourly wage rates so that the total bill on each place of business will match the income that they would ordinarily receive on the two projects.

(a) If x and y denote the hourly wages of the plumber and electrician, respectively, show that $6x + 5y = 10x$ and $4x + 6y = 11y$. Describe the solutions to this system.

(b) If the plumber ordinarily makes $20 per hour, what should the electrician charge?

8.3 Systems of Linear Equations in More than Two Variables

For systems of linear equations containing more than two variables, we can use either the method of substitution explained in Section 8.1 or the method of elimination developed in Section 8.2. The method of elimination is the shorter and more straightforward technique for finding solutions. In addition, it leads to the matrix technique, discussed in this section.

EXAMPLE 1 Find the solutions of the system

$$\begin{cases} x - 2y + 3z = 4 \\ 2x + y - 4z = 3 \\ -3x + 4y - z = -2. \end{cases}$$

SOLUTION We shall begin by eliminating x from the second and third equations. If we add, to the second equation, -2 times the first equation, we get the equivalent system

$$\begin{cases} x - 2y + 3z = 4 \\ 5y - 10z = -5 \\ -3x + 4y - z = -2. \end{cases}$$

Next we add, to the third equation, 3 times the first equation. This gives us

$$\begin{cases} x - 2y + 3z = 4 \\ 5y - 10z = -5 \\ -2y + 8z = 10. \end{cases}$$

To simplify computations let us multiply the second equation by $\frac{1}{5}$, obtaining

$$\begin{cases} x - 2y + 3z = 4 \\ y - 2z = -1 \\ -2y + 8z = 10. \end{cases}$$

We now eliminate y from the third equation by adding to it 2 times the second equation. This gives us

$$\begin{cases} x - 2y + 3z = 4 \\ y - 2z = -1 \\ 4z = 8. \end{cases}$$

Finally, multiplying the third equation by $\frac{1}{4}$, we obtain

$$\begin{cases} x - 2y + 3z = 4 \\ y - 2z = -1 \\ z = 2. \end{cases}$$

The solutions of the last system are easy to find by a procedure called **back substitution.** Thus, from the third equation we obtain $z = 2$. Substituting 2 for z in the second equation, $y - 2z = -1$, we get

$$y - 2(2) = -1 \quad \text{or} \quad y = 3.$$

Finally, the x-value is found by substituting for y and z in the first equation.

This gives us

$$x - 2(3) + 3(2) = 4 \quad \text{and hence} \quad x = 4.$$

Thus, there is one solution, (4, 3, 2). ■

It can be shown that any system of three linear equations in three variables has either a *unique solution*, an *infinite number of solutions*, or *no solutions*. As with the case of two equations in two variables, the terminology used to describe these cases is *consistent*, *dependent*, or *inconsistent*, respectively.

If we analyze the method of solution in Example 1, we see that the symbols used for the variables are immaterial. The *coefficients* of the variables are the important things to consider. Thus, if different symbols such as r, s, and t are used for the variables, we obtain the system

$$\begin{cases} r - 2s + 3t = 4 \\ 2r + s - 4t = 3 \\ -3r + 4s - t = -2. \end{cases}$$

The method of elimination could then proceed exactly as before. Since this is true, it is possible to simplify the process. Specifically, we introduce a scheme for keeping track of the coefficients in such a way that the variables do not have to be written down. Referring to the preceding system and checking that the variables appear in the same order in each equation, and that terms not involving variables are to the right of the equal signs, we list the numbers that are involved in the equations in the following manner:

$$\begin{bmatrix} 1 & -2 & 3 & 4 \\ 2 & 1 & -4 & 3 \\ -3 & 4 & -1 & -2 \end{bmatrix}$$

If some variable had not appeared in one of the equations, we would have used a zero in the appropriate position. An array of numbers of this type is called a **matrix.** The **rows** of the matrix are the numbers that appear next to one another *horizontally*. Thus, the first row R_1 is 1 −2 3 4, the second row R_2 is 2 1 −4 3, and the third row R_3 is −3 4 −1 −2. The **columns** of the matrix are the numbers that appear *vertically*. For example, the second column consists of the numbers −2, 1, 4 (in that order); the fourth column consists of 4, 3, −2; and so on.

Before discussing a matrix method of solving a system of linear equations, let us state a general definition of matrix. We shall use a **double subscript notation.** Specifically, a_{ij} will denote the number that appears in row i and column j. We call i the **row subscript** and j the **column subscript** of a_{ij}.

DEFINITION

Let m and n be positive integers. An **$m \times n$ matrix** is an array of the form

$$\begin{bmatrix} a_{11} & a_{12} & a_{13} & \cdots & a_{1n} \\ a_{21} & a_{22} & a_{23} & \cdots & a_{2n} \\ a_{31} & a_{32} & a_{33} & \cdots & a_{3n} \\ \vdots & \vdots & \vdots & & \vdots \\ a_{m1} & a_{m2} & a_{m3} & \cdots & a_{mn} \end{bmatrix},$$

where each a_{ij} is a real number.

The notation $m \times n$ in the definition is read "m by n." It is possible to consider matrices in which the symbols a_{ij} represent complex numbers, polynomials, or other mathematical objects; however, we shall not do so in this text. The rows and columns of a matrix are defined as before. Thus, the matrix in the definition has m rows and n columns. Note that a_{23} is in row 2 and column 3, whereas a_{32} is in row 3 and column 2. Each a_{ij} is called an **element of the matrix.** The elements $a_{11}, a_{22}, a_{33}, \ldots$ are called the **main diagonal elements.** If $m \neq n$, the matrix is a **rectangular matrix.** If $m = n$, we refer to the matrix as a **square matrix of order n.**

Let us return to the system of equations considered in Example 1. The 3×4 matrix on page 380 is called the **matrix of the system.** The matrix of the system is also called the **augmented matrix.** In certain cases we may wish to consider only the *coefficients* of the variables. The corresponding matrix is called the **coefficient matrix.** The coefficient matrix in Example 1 is

$$\begin{bmatrix} 1 & -2 & 3 \\ 2 & 1 & -4 \\ -3 & 4 & -1 \end{bmatrix}.$$

After forming the matrix of the system, we work with the rows of the matrix *just as though they were equations.* The only items missing are the symbols for the variables, the addition signs used between terms, and the equal signs. We simply keep in mind that the numbers in the first column are the coefficients of the first variable, the numbers in the second column are the coefficients of the second variable, and so on. The rules for transforming a matrix are formulated so that they always produce a matrix of an equivalent system of equations. The following theorem follows from the theorem on transformations of systems, stated on page 372.

MATRIX ROW TRANSFORMATION THEOREM

Given a matrix of a system of linear equations, each of the following transformations results in a matrix of an equivalent system of linear equations:

(i) Interchanging any two rows;

(ii) Multiplying all of the elements in a row by the same nonzero real number k;

(iii) Adding to the elements in one row, k times the corresponding elements of any other row, for any real number k.

We shall refer to (ii) as "multiplying a row by the number k" and (iii) will be described by "add k times any other row to a row." Rules (i)–(iii) of the preceding theorem are called the **elementary row transformations** of a matrix.

It is convenient to use the following symbols to specify elementary row transformations of a matrix.

Symbol	*Meaning*
R_{ij}	Interchange rows i and j
kR_i	Multiply the ith row by k
$kR_i + R_j$	Add to the jth row, k times the ith row

We shall next rework Example 1 using matrices. It may be enlightening to compare the two solutions, since analogous steps are employed in each case.

EXAMPLE 2 Find the solutions of the system

$$\begin{cases} x - 2y + 3z = 4 \\ 2x + y - 4z = 3 \\ -3x + 4y - z = -2. \end{cases}$$

SOLUTION We begin with the matrix of the system and then change its form by means of elementary row transformations. An arrow, together with the symbols we have introduced, will specify each transformation. Thus, in the first transformation that follows, $-2R_1 + R_2$ indicates that we have added to the second row, -2 times the first row. In the second transformation, $3R_1 + R_3$ indicates that we have added to the third row, 3 times the first row. In the third transformation, $\frac{1}{5}R_2$ denotes the fact that the second row has been multiplied by $\frac{1}{5}$, and so on.

$$\begin{bmatrix} 1 & -2 & 3 & 4 \\ 2 & 1 & -4 & 3 \\ -3 & 4 & -1 & -2 \end{bmatrix} \xrightarrow{-2R_1 + R_2} \begin{bmatrix} 1 & -2 & 3 & 4 \\ 0 & 5 & -10 & -5 \\ -3 & 4 & -1 & -2 \end{bmatrix}$$

$$\xrightarrow{3R_1 + R_3} \begin{bmatrix} 1 & -2 & 3 & 4 \\ 0 & 5 & -10 & -5 \\ 0 & -2 & 8 & 10 \end{bmatrix}$$

$$\xrightarrow{\frac{1}{5}R_2} \begin{bmatrix} 1 & -2 & 3 & 4 \\ 0 & 1 & -2 & -1 \\ 0 & -2 & 8 & 10 \end{bmatrix}$$

$$\xrightarrow{2R_2 + R_3} \begin{bmatrix} 1 & -2 & 3 & 4 \\ 0 & 1 & -2 & -1 \\ 0 & 0 & 4 & 8 \end{bmatrix}$$

$$\xrightarrow{\frac{1}{4}R_3} \begin{bmatrix} 1 & -2 & 3 & 4 \\ 0 & 1 & -2 & -1 \\ 0 & 0 & 1 & 2 \end{bmatrix}$$

We use the final matrix to return to the system of equations

$$\begin{cases} x - 2y + 3z = 4 \\ \quad\;\; y - 2z = -1 \\ \qquad\quad\;\; z = 2, \end{cases}$$

which is equivalent to the original system. The solutions may now be found by back substitution as in Example 1. ■

The final matrix in the solution of Example 2 is said to be in **echelon form.** In general, a matrix is in echelon form if it satisfies the following conditions.

ECHELON FORM OF A MATRIX

(i) The first nonzero number in each row, reading from left to right, is 1.

(ii) The column containing the first nonzero number in any row is to the left of the column containing the first nonzero number in the next row.

(iii) Rows consisting entirely of zeros appear at the bottom of the matrix.

We can use elementary row operations to transform the matrix of any system of linear equations to echelon form. The echelon form can then be used to produce a system of equations that is equivalent to the original system. The solutions of the latter system may be found by back substitution. The next example illustrates this technique for a system of four linear equations.

EXAMPLE 3 Find the solutions of the system

$$\begin{cases} x \phantom{{}+3y} - 2z + 2w = 1 \\ -2x + 3y + 4z \phantom{{}+2w} = -1 \\ \phantom{-2x+{}} y + z - w = 0 \\ 3x + y - 2z - w = 3. \end{cases}$$

SOLUTION We have arranged the equations so that the same variables appear in vertical columns. We begin with the matrix of the system and proceed as follows:

$$\begin{bmatrix} 1 & 0 & -2 & 2 & 1 \\ -2 & 3 & 4 & 0 & -1 \\ 0 & 1 & 1 & -1 & 0 \\ 3 & 1 & -2 & -1 & 3 \end{bmatrix} \xrightarrow{2R_1 + R_2} \begin{bmatrix} 1 & 0 & -2 & 2 & 1 \\ 0 & 3 & 0 & 4 & 1 \\ 0 & 1 & 1 & -1 & 0 \\ 3 & 1 & -2 & -1 & 3 \end{bmatrix}$$

$$\xrightarrow{-3R_1 + R_4} \begin{bmatrix} 1 & 0 & -2 & 2 & 1 \\ 0 & 3 & 0 & 4 & 1 \\ 0 & 1 & 1 & -1 & 0 \\ 0 & 1 & 4 & -7 & 0 \end{bmatrix}$$

$$\xrightarrow{R_{23}} \begin{bmatrix} 1 & 0 & -2 & 2 & 1 \\ 0 & 1 & 1 & -1 & 0 \\ 0 & 3 & 0 & 4 & 1 \\ 0 & 1 & 4 & -7 & 0 \end{bmatrix}$$

$$\xrightarrow{-3R_2 + R_3} \begin{bmatrix} 1 & 0 & -2 & 2 & 1 \\ 0 & 1 & 1 & -1 & 0 \\ 0 & 0 & -3 & 7 & 1 \\ 0 & 1 & 4 & -7 & 0 \end{bmatrix}$$

$$\xrightarrow{(-1)R_2 + R_4} \begin{bmatrix} 1 & 0 & -2 & 2 & 1 \\ 0 & 1 & 1 & -1 & 0 \\ 0 & 0 & -3 & 7 & 1 \\ 0 & 0 & 3 & -6 & 0 \end{bmatrix}$$

$$\xrightarrow{R_3 + R_4} \begin{bmatrix} 1 & 0 & -2 & 2 & 1 \\ 0 & 1 & 1 & -1 & 0 \\ 0 & 0 & -3 & 7 & 1 \\ 0 & 0 & 0 & 1 & 1 \end{bmatrix}$$

$$\xrightarrow{-\frac{1}{3}R_3} \begin{bmatrix} 1 & 0 & -2 & 2 & 1 \\ 0 & 1 & 1 & -1 & 0 \\ 0 & 0 & 1 & -\frac{7}{3} & -\frac{1}{3} \\ 0 & 0 & 0 & 1 & 1 \end{bmatrix}$$

The final matrix is in echelon form and corresponds to the following system of equations

$$\begin{cases} x \quad - 2z + 2w = 1 \\ y + z - w = 0 \\ z - \frac{7}{3}w = -\frac{1}{3} \\ w = 1. \end{cases}$$

We now use back substitution to find the solution. From the last equation we see that $w = 1$. Substituting in the third equation $z - \frac{7}{3}w = -\frac{1}{3}$, we get

$$z - \tfrac{7}{3}(1) = -\tfrac{1}{3} \quad \text{or} \quad z = \tfrac{6}{3} = 2.$$

Substituting $w = 1$ and $z = 2$ in the second equation $y + z - w = 0$, we obtain

$$y + 2 - 1 = 0 \quad \text{or} \quad y = -1.$$

Finally, from the first equation $x - 2z + 2w = 1$, we have

$$x - 2(2) + 2(1) = 1 \quad \text{or} \quad x = 3.$$

Hence, the system has one solution, $x = 3$, $y = -1$, $z = 2$, and $w = 1$. ■

Sometimes it is necessary to consider systems in which the number of equations is not the same as the number of variables. The same matrix techniques are applicable, as illustrated in the next example.

EXAMPLE 4 Find the solutions of the system

$$\begin{cases} 2x + 3y + 4z = 1 \\ 3x + 4y + 5z = 3. \end{cases}$$

SOLUTION When the number of equations is less than the number of unknowns we shall go beyond the echelon form, applying elementary row transformations until there are zeros both below *and* above the first nonzero entry in each row. Thus,

$$\begin{bmatrix} 2 & 3 & 4 & 1 \\ 3 & 4 & 5 & 3 \end{bmatrix} \xrightarrow{-\frac{3}{2}R_1 + R_2} \begin{bmatrix} 2 & 3 & 4 & 1 \\ 0 & -\frac{1}{2} & -1 & \frac{3}{2} \end{bmatrix}$$

$$\xrightarrow{\frac{1}{2}R_1} \begin{bmatrix} 1 & \frac{3}{2} & 2 & \frac{1}{2} \\ 0 & -\frac{1}{2} & -1 & \frac{3}{2} \end{bmatrix}$$

$$\xrightarrow{-2R_2} \begin{bmatrix} 1 & \frac{3}{2} & 2 & \frac{1}{2} \\ 0 & 1 & 2 & -3 \end{bmatrix}$$

$$\xrightarrow{-\frac{3}{2}R_2 + R_1} \begin{bmatrix} 1 & 0 & -1 & 5 \\ 0 & 1 & 2 & -3 \end{bmatrix}$$

The last matrix is the matrix of the system

$$\begin{cases} x \quad - \quad z = 5 \\ \quad y + 2z = -3 \end{cases}$$

or equivalently,

$$\begin{cases} x = z + 5 \\ y = -2z - 3. \end{cases}$$

There are an infinite number of solutions to this system; they can be found by assigning z any value c and then using the last two equations to express x and y in terms of c. This gives us

$$x = c + 5, \qquad y = -2c - 3, \qquad z = c.$$

Thus, the solutions of the system consist of all ordered triples of the form

$$(c + 5, -2c - 3, c)$$

where c is any real number. This may be checked by substituting $c + 5$ for x, $-2c - 3$ for y, and c for z in the two given equations.

We can obtain any number of solutions for the system by substituting specific real numbers for c. For example, if $c = 0$, we obtain $(5, -3, 0)$; if $c = 2$, we have $(7, -7, 2)$; and so on.

There are other ways to specify the general solution. For example, starting with $x = z + 5$ and $y = -2z - 3$, we could let $z = d - 5$, where d is any real number. In this event,

$$x = z + 5 = (d - 5) + 5 = d$$

$$y = -2z - 3 = -2(d - 5) - 3 = -2d + 7$$

and the solutions of the system have the form

$$(d, -2d + 7, d - 5).$$

These triples produce the same solutions as $(c + 5, -2c - 3, c)$. For example, if $d = 5$ we get $(5, -3, 0)$, if $d = 7$ we obtain $(7, -7, 2)$, and so on. ■

A system of linear equations is said to be **homogeneous** if all the terms that do not contain variables are zero. A system of homogeneous equations always has the **trivial solution** obtained by substituting zero for each variable. Nontrivial solutions sometimes exist. The procedure for finding solutions is the same as that used for nonhomogeneous systems.

EXAMPLE 5 Find the solutions of the homogeneous system

$$\begin{cases} x - y + 4z = 0 \\ 2x + y - z = 0 \\ -x - y + 2z = 0. \end{cases}$$

SOLUTION We proceed as follows:

$$\begin{bmatrix} 1 & -1 & 4 & 0 \\ 2 & 1 & -1 & 0 \\ -1 & -1 & 2 & 0 \end{bmatrix} \xrightarrow{-2R_1 + R_2} \begin{bmatrix} 1 & -1 & 4 & 0 \\ 0 & 3 & -9 & 0 \\ -1 & -1 & 2 & 0 \end{bmatrix}$$

$$\xrightarrow{R_1 + R_3} \begin{bmatrix} 1 & -1 & 4 & 0 \\ 0 & 3 & -9 & 0 \\ 0 & -2 & 6 & 0 \end{bmatrix}$$

$$\xrightarrow{\frac{1}{3}R_2} \begin{bmatrix} 1 & -1 & 4 & 0 \\ 0 & 1 & -3 & 0 \\ 0 & -2 & 6 & 0 \end{bmatrix}$$

$$\xrightarrow{2R_2 + R_3} \begin{bmatrix} 1 & -1 & 4 & 0 \\ 0 & 1 & -3 & 0 \\ 0 & 0 & 0 & 0 \end{bmatrix}$$

$$\xrightarrow{R_2 + R_1} \begin{bmatrix} 1 & 0 & 1 & 0 \\ 0 & 1 & -3 & 0 \\ 0 & 0 & 0 & 0 \end{bmatrix}$$

Note that the second from the last matrix is in echelon form. As in Example 4, we have gone one step further, obtaining a zero *above* the first nonzero entry in the second row. The final matrix corresponds to the system

$$\begin{cases} x \quad + \; z = 0 \\ \quad y - 3z = 0 \end{cases} \quad \text{or} \quad \begin{cases} x = -z \\ y = \;\; 3z \end{cases}$$

Assigning any value c to z, we obtain $x = -c$ and $y = 3c$. Thus, the solutions consist of all ordered triples of the form $(-c, 3c, c)$ where c is any real number. ■

EXAMPLE 6 Find the solutions of the system

$$\begin{cases} x + y + z = 0 \\ x - y + z = 0 \\ x - y - z = 0. \end{cases}$$

SOLUTION

$$\begin{bmatrix} 1 & 1 & 1 & 0 \\ 1 & -1 & 1 & 0 \\ 1 & -1 & -1 & 0 \end{bmatrix} \xrightarrow{(-1)R_1 + R_2} \begin{bmatrix} 1 & 1 & 1 & 0 \\ 0 & -2 & 0 & 0 \\ 1 & -1 & -1 & 0 \end{bmatrix}$$

$$\xrightarrow{(-1)R_1 + R_3} \begin{bmatrix} 1 & 1 & 1 & 0 \\ 0 & -2 & 0 & 0 \\ 0 & -2 & -2 & 0 \end{bmatrix}$$

$$\xrightarrow{-\frac{1}{2}R_2} \begin{bmatrix} 1 & 1 & 1 & 0 \\ 0 & 1 & 0 & 0 \\ 0 & -2 & -2 & 0 \end{bmatrix}$$

$$\xrightarrow{2R_2 + R_3} \begin{bmatrix} 1 & 1 & 1 & 0 \\ 0 & 1 & 0 & 0 \\ 0 & 0 & -2 & 0 \end{bmatrix}$$

$$\xrightarrow{-\frac{1}{2}R_3} \begin{bmatrix} 1 & 1 & 1 & 0 \\ 0 & 1 & 0 & 0 \\ 0 & 0 & 1 & 0 \end{bmatrix}$$

The last matrix is in echelon form and is the matrix of the system

$$x + y + z = 0, \qquad y = 0, \qquad z = 0.$$

It follows that the only solution for the given system is the trivial one, (0, 0, 0). ■

The next example is an illustration of an applied problem that can be solved by means of a system of three linear equations.

EXAMPLE 7 A merchant wishes to mix two grades of peanuts costing \$1.50 and \$2.50 per pound, respectively, with cashews costing \$4.00 per pound, to obtain 130 pounds of a mixture costing \$3.00 per pound. If the merchant also wants the amount of cheaper grade peanuts to be twice that of the better grade, how many pounds of each variety should be mixed?

SOLUTION Let us introduce three variables as follows:

$x =$ pounds of peanuts at \$1.50 per pound,

$y =$ pounds of peanuts at \$2.50 per pound,

$z =$ pounds of cashews at \$4.00 per pound.

We refer to the statement of the problem and obtain the following system of equations

$$\begin{cases} x + y + z = 130 \\ 1.50x + 2.50y + 4.00z = 3.00(130) \\ x = 2y. \end{cases}$$

We leave it to the reader to verify that the solution to this system is $x = 40$, $y = 20$, $z = 70$. Thus, the merchant should use 40 pounds of the \$1.50 peanuts, 20 pounds of the \$2.50 peanuts, and 70 pounds of cashews. ■

Exercises 8.3

Use matrices to solve the systems in Exercises 1–26.

1 $\begin{cases} x - 2y - 3z = -1 \\ 2x + y + z = 6 \\ x + 3y - 2z = 13 \end{cases}$

2 $\begin{cases} x + 3y - z = -3 \\ 3x - y + 2z = 1 \\ 2x - y + z = -1 \end{cases}$

3 $\begin{cases} 5x + 2y - z = -7 \\ x - 2y + 2z = 0 \\ 3y + z = 17 \end{cases}$

4 $\begin{cases} 4x - y + 3z = 6 \\ -8x + 3y - 5z = -6 \\ 5x - 4y = -9 \end{cases}$

5 $\begin{cases} 2x + 6y - 4z = 1 \\ x + 3y - 2z = 4 \\ 2x + y - 3z = -7 \end{cases}$

6 $\begin{cases} 2x - y = 5 \\ 5y + 3z = -2 \\ x - 7z = 3 \end{cases}$

7 $\begin{cases} 2x - 3y + 2z = -3 \\ -3x + 2y + z = 1 \\ 4x + y - 3z = 4 \end{cases}$

8 $\begin{cases} 2x - 3y + z = 2 \\ 3x + 2y - z = -5 \\ 5x - 2y + z = 0 \end{cases}$

9 $\begin{cases} x + 3y + z = 0 \\ x + y - z = 0 \\ x - 2y - 4z = 0 \end{cases}$

10 $\begin{cases} 2x - y + z = 0 \\ x - y - 2z = 0 \\ 2x - 3y - z = 0 \end{cases}$

11 $\begin{cases} 2x + y + z = 0 \\ x - 2y - 2z = 0 \\ x + y + z = 0 \end{cases}$

12 $\begin{cases} x + y - 2z = 0 \\ x - y - 4z = 0 \\ y + z = 0 \end{cases}$

13 $\begin{cases} 3x - 2y + 5z = 7 \\ x + 4y - z = -2 \end{cases}$

14 $\begin{cases} 2x - y + 4z = 8 \\ -3x + y - 2z = 5 \end{cases}$

15 $\begin{cases} 4x - 2y + z = 5 \\ 3x + y - 4z = 0 \end{cases}$

16 $\begin{cases} 5x + 2y - z = 10 \\ y + z = -3 \end{cases}$

17 $\begin{cases} x + 2y - z - 3w = 2 \\ 3x + y - 2z - w = 6 \\ x + y + 3z - 2w = -3 \\ 4x - 3y - z - 2w = -8 \end{cases}$

18 $\begin{cases} x - 2y - 5z + w = -1 \\ 2x - y + z + w = 1 \\ 3x - 2y - 4z - 2w = 1 \\ x + y + 3z - 2w = 2 \end{cases}$

19 $\begin{cases} 2x - y - 2z + 2s - 5t = 2 \\ x + 3y - 2z + s - 2t = -5 \\ -x + 4y + 2z - 3s + 8t = -4 \\ 3x - 2y - 4z + s - 3t = -3 \\ 4x - 6y + z - 2s + t = 10 \end{cases}$

20 $\begin{cases} 3x + 2y + z + 3u + v + w = 1 \\ 2x + y - 2z + 3u - v + 4w = 6 \\ 6x + 3y + 4z - u + 2v + w = -6 \\ x + y + z + u - v - w = 8 \\ -2x - 2y + z - 3u + 2v - 3w = -10 \\ x - 3y + 2z + u + 3v + w = -1 \end{cases}$

21 $\begin{cases} 5x + 2z = 1 \\ y - 3z = 2 \\ 2x + y = 3 \end{cases}$

22 $\begin{cases} 2x - 5y = 4 \\ 3y + 2z = -3 \\ 7x - 3z = 1 \end{cases}$

23 $\begin{cases} 4x - 3y = 1 \\ 2x + y = -7 \\ -x + y = -1 \end{cases}$

24 $\begin{cases} 2x + 3y = -2 \\ x + y = 1 \\ x - 2y = 13 \end{cases}$

25 $\begin{cases} 2x + 3y = 5 \\ x - 3y = 4 \\ x + y = -2 \end{cases}$

26 $\begin{cases} 4x - y = 2 \\ 2x + 2y = 1 \\ 4x - 5y = 3 \end{cases}$

27 A chemist has three solutions containing a certain acid. The first contains 10% acid, the second 30%, and the third 50%. He wishes to use all three solutions to obtain a mixture of 50 liters containing 32% acid, using twice as much of the 50% solution as the 30% solution. How many liters of each solution should be used?

28 A swimming pool can be filled by three pipes A, B, and C. Pipe A alone can fill the pool in 8 hours. If pipes A and C are used together, the pool can be filled in 6 hours. If B and C are used together, it takes 10 hours. How long does it take to fill the pool if all three pipes are used?

29 A company has three machines A, B, and C that are each capable of producing a certain item. However, because of a lack of skilled operators, only two of the machines can be used simultaneously. The following table indicates production over a three-day period using various combinations of the machines.

Machines used	Hours used	Items produced
A and B	6	4500
A and C	8	3600
B and C	7	4900

How long would it take each machine, if used alone, to produce 1000 items?

30 In electrical circuits, the formula $1/R = (1/R_1) + (1/R_2)$ is used to find the total resistance R if two resistors R_1 and R_2 are connected in parallel. Given three resistors A, B, and C, suppose that the total resistance is 48 ohms if A and B are connected in parallel; 80 ohms if B and C are connected in parallel, and 60 ohms if A and C are connected in parallel. Find A, B, and C.

31 A supplier of lawn products has three types of grass fertilizer, G_1, G_2, and G_3, having nitrogen contents of 30%, 20%, and 15%, respectively. The supplier plans to mix them, obtaining 600 pounds of fertilizer with a 25% nitrogen content. In addition, the mixture is to contain 100 pounds more of type G_3 than of type G_2. How much of each type should be used?

32 If a particle moves along a coordinate line with a constant acceleration a (cm/sec^2), then at time t (sec) its distance $s(t)$ (cm) from the origin is

$$s(t) = \tfrac{1}{2}at^2 + v_0t + s_0$$

where v_0 and s_0 are the velocity and distance from the origin, respectively, at $t = 0$. If the distances of the particle from the origin at $t = \frac{1}{2}$, $t = 1$ and $t = \frac{3}{2}$ are 7, 11, and 17, respectively, find a, v_0, and s_0.

33 Shown in the figure is an electrical circuit containing three resistors, a 6-volt battery, and a 12-volt battery. It can be shown, using Kirchoff's Laws, that the three currents I_1, I_2, and I_3 satisfy the following system of equations:

$$\begin{cases} I_1 - I_2 + I_3 = 0 \\ R_1I_1 + R_2I_2 = 6 \\ R_2I_2 + R_3I_3 = 12. \end{cases}$$

Find the three currents if

(a) $R_1 = R_2 = R_3 = 3$ ohms.

(b) $R_1 = 4$ ohms, $R_2 = 1$ ohm, and $R_3 = 4$ ohms.

FIGURE FOR EXERCISE 33

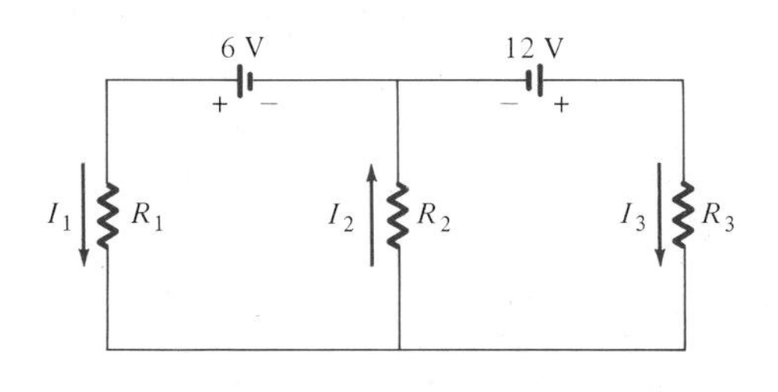

34 A stable population of 35,000 birds lives on three islands. Each year 10% of the population on island A migrates to island B, 20% of the population on island B migrates to island C, and 5% of the population on island C migrates to A. Find the number of birds on each island if the populations on each island do not vary from year to year.

35 A shop specializes in preparing blends of gourmet coffees. The owner wishes to prepare 1-pound bags that will sell for \$8.50 from Columbian, Brazilian, and Kenyan coffees. The cost per pound of these coffees is \$10, \$6, and \$8, respectively. Find the amount of each type of coffee assuming the proprietor decides to use $\frac{1}{8}$ pound of Brazilian coffee.

36 A rancher has 750 head of cattle consisting of 400 adults (aged 2 or more years), 150 yearlings, and 200 calves. The following information is known about this particular species. Each spring an adult female gives birth to a single calf and 75% of these calves will survive the year. The yearly survival percentages for yearlings and adults are 80% and 90% respectively. Finally, the male–female ratio is 1 in all age classes.

(a) Estimate the population of each age class next spring.

(b) Estimate the population of each age class last spring.

37 Find a quadratic function f such that $f(1) = 3$, $f(2) = \frac{5}{2}$, and $f(-1) = 1$.

38 If $f(x) = ax^3 + bx + c$, determine a, b, and c such that the graph of f passes through the points $(-3, -12)$, $(-1, 22)$, and $(2, 13)$.

39 Find an equation of the circle that passes through the three points $P_1(2, 1)$, $P_2(-1, -4)$, and $P_3(3, 0)$. (*Hint:* An equation of the circle has the form $x^2 + y^2 + ax + by + c = 0$.)

40 Determine a, b, and c such that the graph of the equation $y = ax^2 + bx + c$ passes through the points $P_1(3, -1)$, $P_2(1, -7)$, and $P_3(-2, 14)$.

8.4 Partial Fractions

In this section we show how systems of equations can be used to help decompose rational expressions into sums of simpler expressions. This technique is useful in certain parts of advanced mathematics courses, such as calculus.

It is easy to verify that

$$\frac{2}{x^2-1} = \frac{1}{x-1} + \frac{-1}{x+1}.$$

The expression on the right side of this equation is called *the partial fraction decomposition* of $2/(x^2-1)$.

It is theoretically possible to write *any* rational expression as a sum of rational expressions whose denominators involve powers of polynomials of degree not greater than two. Specifically, if $f(x)$ and $g(x)$ are polynomials *and the degree of* $f(x)$ *is less than the degree of* $g(x)$, it can be proved that

$$\frac{f(x)}{g(x)} = F_1 + F_2 + \cdots + F_r$$

where each F_k has one of the forms

$$\frac{A}{(px+q)^m} \quad \text{or} \quad \frac{Cx+D}{(ax^2+bx+c)^n}$$

for some nonnegative integers m and n, and where ax^2+bx+c is **irreducible** in the sense that this quadratic polynomial has no real zeros, that is, $b^2-4ac<0$. The sum $F_1+F_2+\cdots+F_r$ is called the **partial fraction decomposition** of $f(x)/g(x)$ and each F_k is called a **partial fraction.** We shall not prove this result but will, instead, give rules for obtaining the decomposition.

To find the partial fraction decomposition of $f(x)/g(x)$ *it is essential that* $f(x)$ *have lower degree than* $g(x)$. If this is not the case, then long division should be employed to arrive at such an expression. For example, given

$$\frac{x^3-6x^2+5x-3}{x^2-1}$$

we obtain

$$\frac{x^3-6x^2+5x-3}{x^2-1} = x-6+\frac{6x-9}{x^2-1}.$$

The partial fraction decomposition is then found for $(6x-9)/(x^2-1)$.

GUIDELINES: Finding Partial Fraction Decompositions of $\dfrac{f(x)}{g(x)}$

A If the degree of $f(x)$ is not lower than the degree of $g(x)$, use long division to obtain the proper form.

B Express $g(x)$ as a product of linear factors $px + q$ or irreducible quadratic factors $ax^2 + bx + c$, and collect repeated factors so that $g(x)$ is a product of *different* factors of the form $(px + q)^m$ or $(ax^2 + bx + c)^n$, where m and n are nonnegative integers.

C Apply the following rules.

Rule 1. For each factor of the form $(px + q)^m$ where $m \geq 1$, the partial fraction decomposition contains a sum of m partial fractions of the form

$$\frac{A_1}{px + q} + \frac{A_2}{(px + q)^2} + \cdots + \frac{A_m}{(px + q)^m}$$

where each A_k is a real number.

Rule 2. For each factor of the form $(ax^2 + bx + c)^n$ where $n \geq 1$ and $ax^2 + bx + c$ is irreducible, the partial fraction decomposition contains a sum of n partial fractions of the form

$$\frac{A_1x + B_1}{ax^2 + bx + c} + \frac{A_2x + B_2}{(ax^2 + bx + c)^2} + \cdots + \frac{A_nx + B_n}{(ax^2 + bx + c)^n}$$

where each A_k and B_k is a real number. ■■■

EXAMPLE 1 Find the partial fraction decomposition of

$$\frac{4x^2 + 13x - 9}{x^3 + 2x^2 - 3x}.$$

SOLUTION The denominator has the factored form $x(x + 3)(x - 1)$. Each factor has the form stated in Rule 1, with $m = 1$. Thus, for the factor x there corresponds a partial fraction of the form A/x. Similarly, for the factors $x + 3$ and $x - 1$ there correspond partial fractions $B/(x + 3)$ and $C/(x - 1)$, respectively. The partial fraction decomposition has the form

$$\frac{4x^2 + 13x - 9}{x(x + 3)(x - 1)} = \frac{A}{x} + \frac{B}{x + 3} + \frac{C}{x - 1}.$$

Multiplying by the lowest common denominator, $x(x + 3)(x - 1)$, gives us

$$\begin{aligned} 4x^2 + 13x - 9 &= A(x + 3)(x - 1) + Bx(x - 1) + Cx(x + 3) \\ &= A(x^2 + 2x - 3) + B(x^2 - x) + C(x^2 + 3x). \end{aligned}$$

This may also be written

$$4x^2 + 13x - 9 = (A + B + C)x^2 + (2A - B + 3C)x - 3A.$$

If we equate the coefficients of like powers of x on each side of the last equation, we obtain the system of equations

$$\begin{cases} A + B + C = 4 \\ 2A - B + 3C = 13 \\ -3A = -9. \end{cases}$$

It can be shown that the solution is $A = 3$, $B = -1$, $C = 2$. Thus, the partial fraction decomposition is

$$\frac{4x^2 + 13x - 9}{x(x + 3)(x - 1)} = \frac{3}{x} + \frac{-1}{x + 3} + \frac{2}{x - 1}.$$

■

EXAMPLE 2 Find the partial fraction decomposition of

$$\frac{x^2 + 10x - 36}{x(x - 3)^2}.$$

SOLUTION By Rule 1 with $m = 1$, there is a partial fraction A/x corresponding to the factor x. Next, applying Rule 1 with $m = 2$, the factor $(x - 3)^2$ determines a sum of two partial fractions $B/(x - 3) + C/(x - 3)^2$. Thus, the partial fraction decomposition has the form

$$\frac{x^2 + 10x - 36}{x(x - 3)^2} = \frac{A}{x} + \frac{B}{x - 3} + \frac{C}{(x - 3)^2}.$$

Multiplying both sides by $x(x - 3)^2$ gives us

$$\begin{aligned} x^2 + 10x - 36 &= A(x - 3)^2 + Bx(x - 3) + Cx \\ &= A(x^2 - 6x + 9) + B(x^2 - 3x) + Cx \end{aligned}$$

or equivalently,

$$x^2 + 10x - 36 = (A + B)x^2 + (-6A - 3B + C)x + 9A.$$

As in Example 1 we equate the coefficients of like powers of x, obtaining

$$\begin{cases} A + B = 1 \\ -6A - 3B + C = 10 \\ 9A = -36. \end{cases}$$

This system of equations has the solution $A = -4$, $B = 5$, $C = 1$. The partial fraction decomposition is therefore

$$\frac{x^2 + 10x - 36}{x(x-3)^2} = \frac{-4}{x} + \frac{5}{x-3} + \frac{1}{(x-3)^2}. \qquad \blacksquare$$

EXAMPLE 3 Find the partial fraction decomposition of

$$\frac{x^2 - x - 21}{2x^3 - x^2 + 8x - 4}.$$

SOLUTION The denominator may be factored by grouping, as follows:

$$2x^3 - x^2 + 8x - 4 = x^2(2x - 1) + 4(2x - 1) = (x^2 + 4)(2x - 1).$$

Applying Rule 2 to the irreducible quadratic factor $x^2 + 4$, we see that one of the partial fractions has the form $(Ax + B)/(x^2 + 4)$. By Rule 1, there is also a partial fraction $C/(2x - 1)$ corresponding to factor $2x - 1$. Consequently,

$$\frac{x^2 - x - 21}{2x^3 - x^2 + 8x - 4} = \frac{Ax + B}{x^2 + 4} + \frac{C}{2x - 1}.$$

As in previous examples, this leads to

$$\begin{aligned} x^2 - x - 21 &= (Ax + B)(2x - 1) + C(x^2 + 4) \\ &= 2Ax^2 - Ax + 2Bx - B + Cx^2 + 4C \end{aligned}$$

or $$x^2 - x - 21 = (2A + C)x^2 + (-A + 2B)x - B + 4C.$$

This gives us the system

$$\begin{cases} 2A \qquad\quad + C = 1 \\ -A + 2B \qquad = -1 \\ \qquad - B + 4C = -21, \end{cases}$$

which has the solution $A = 3$, $B = 1$, $C = -5$. Thus, the partial fraction decomposition is

$$\frac{x^2 - x - 21}{2x^3 - x^2 + 8x - 4} = \frac{3x + 1}{x^2 + 4} + \frac{-5}{2x - 1}. \qquad \blacksquare$$

EXAMPLE 4 Find the partial fraction decomposition of

$$\frac{5x^3 - 3x^2 + 7x - 3}{(x^2 + 1)^2}.$$

SOLUTION Applying Rule 2 with $n = 2$,

$$\frac{5x^3 - 3x^2 + 7x - 3}{(x^2 + 1)^2} = \frac{Ax + B}{x^2 + 1} + \frac{Cx + D}{(x^2 + 1)^2}.$$

Multiplying both sides by $(x^2 + 1)^2$ gives us

$$5x^3 - 3x^2 + 7x - 3 = (Ax + B)(x^2 + 1) + Cx + D$$

or $$5x^3 - 3x^2 + 7x - 3 = Ax^3 + Bx^2 + (A + C)x + (B + D).$$

Comparing the coefficients of x^3 and x^2, we obtain $A = 5$ and $B = -3$. From the coefficients of x we see that $A + C = 7$, or equivalently, $C = 7 - A = 7 - 5 = 2$. Finally, the constant terms give us $B + D = -3$, or $D = -3 - B = -3 - (-3) = 0$. Therefore,

$$\frac{5x^3 - 3x^2 + 7x - 3}{(x^2 + 1)^2} = \frac{5x - 3}{x^2 + 1} + \frac{2x}{(x^2 + 1)^2}.$$ ■

Exercises 8.4

Find the partial fraction decompositions in Exercises 1–26.

1 $\frac{8x - 1}{(x - 2)(x + 3)}$

2 $\frac{x - 29}{(x - 4)(x + 1)}$

3 $\frac{x + 34}{x^2 - 4x - 12}$

4 $\frac{5x - 12}{x^2 - 4x}$

5 $\frac{4x^2 - 15x - 1}{(x - 1)(x + 2)(x - 3)}$

6 $\frac{x^2 + 19x + 20}{x(x + 2)(x - 5)}$

7 $\frac{4x^2 - 5x - 15}{x^3 - 4x^2 - 5x}$

8 $\frac{37 - 11x}{(x + 1)(x^2 - 5x + 6)}$

9 $\frac{2x + 3}{(x - 1)^2}$

10 $\frac{5x^2 - 4}{x^2(x + 2)}$

11 $\frac{19x^2 + 50x - 25}{3x^3 - 5x^2}$

12 $\frac{10 - x}{x^2 + 10x + 25}$

13 $\frac{x^2 - 6}{(x + 2)^2(2x - 1)}$

14 $\frac{2x^2 + x}{(x - 1)^2(x + 1)^2}$

15 $\frac{3x^3 + 11x^2 + 16x + 5}{x(x + 1)^3}$

16 $\frac{4x^3 + 3x^2 + 5x - 2}{x^3(x + 2)}$

17 $\frac{x^2 + x - 6}{(x^2 + 1)(x - 1)}$

18 $\frac{x^2 - x - 21}{(x^2 + 4)(2x - 1)}$

19 $\frac{9x^2 - 3x + 8}{x^3 + 2x}$

20 $\frac{2x^3 + 2x^2 + 4x - 3}{x^4 + x^2}$

21 $\frac{4x^3 - x^2 + 4x + 2}{(x^2 + 1)^2}$

22 $\frac{3x^3 + 13x - 1}{(x^2 + 4)^2}$

23 $\dfrac{2x^4 - 2x^3 + 6x^2 - 5x + 1}{x^3 - x^2 + x - 1}$

24 $\dfrac{x^3}{x^3 - 3x^2 + 9x - 27}$

25 $\dfrac{4x^3 + 4x^2 - 4x + 2}{2x^2 - x - 1}$

26 $\dfrac{x^5 - 5x^4 + 7x^3 - x^2 - 4x + 12}{x^3 - 3x^2}$

8.5 The Algebra of Matrices

It is possible to develop a comprehensive theory for matrices that has many mathematical and scientific applications. In this section we discuss algebraic properties of matrices that serve as the starting point for that theory.

To conserve space it is sometimes convenient to use the symbol (a_{ij}) to denote an $m \times n$ matrix A of the type displayed in the definition given in Section 8.3. If (b_{ij}) denotes another $m \times n$ matrix B, then we say that A and B are **equal,** and we write

$$A = B \quad \text{if and only if} \quad a_{ij} = b_{ij}$$

for every i and j. For example,

$$\begin{bmatrix} 1 & 0 & 5 \\ \sqrt[3]{8} & 3^2 & -2 \end{bmatrix} = \begin{bmatrix} (-1)^2 & 0 & \sqrt{25} \\ 2 & 9 & -2 \end{bmatrix}.$$

If $A = (a_{ij})$ and $B = (b_{ij})$ are $m \times n$ matrices, then their **sum $A + B$** is defined as the $m \times n$ matrix $C = (c_{ij})$ such that $c_{ij} = a_{ij} + b_{ij}$ for all i and j. Thus, to add two matrices we add the elements that appear in corresponding positions in each matrix. Two matrices can be added only if they have the same number of rows and the same number of columns. Using the parentheses notation, the definition of addition may be expressed as follows:

$$(a_{ij}) + (b_{ij}) = (a_{ij} + b_{ij}).$$

Although we have used the symbol $+$ in two different ways, there is little chance for confusion, since whenever $+$ appears between symbols for matrices it refers to matrix addition, and when $+$ is used between real numbers it denotes their sum. An example of the sum of two 3×2 matrices is

$$\begin{bmatrix} 4 & -5 \\ 0 & 4 \\ -6 & 1 \end{bmatrix} + \begin{bmatrix} 3 & 2 \\ 7 & -4 \\ -2 & 1 \end{bmatrix} = \begin{bmatrix} 7 & -3 \\ 7 & 0 \\ -8 & 2 \end{bmatrix}.$$

It is not difficult to prove that addition of matrices is both commutative and associative, that is,

$$A + B = B + A, \qquad A + (B + C) = (A + B) + C$$

for $m \times n$ matrices A, B, and C.

The ***m* × *n* zero matrix,** denoted by O, is the matrix with m rows and n columns in which every element is 0. It is an identity element relative to addition, since

$$A + O = A$$

for every $m \times n$ matrix A. For example,

$$\begin{bmatrix} a_{11} & a_{12} \\ a_{21} & a_{22} \\ a_{31} & a_{32} \end{bmatrix} + \begin{bmatrix} 0 & 0 \\ 0 & 0 \\ 0 & 0 \end{bmatrix} = \begin{bmatrix} a_{11} & a_{12} \\ a_{21} & a_{22} \\ a_{31} & a_{32} \end{bmatrix}.$$

The **additive inverse** $-A$ of the matrix $A = (a_{ij})$ is, by definition, the matrix $(-a_{ij})$ obtained by changing the sign of each element of A. For example,

$$-\begin{bmatrix} 2 & -3 & 4 \\ -1 & 0 & 5 \end{bmatrix} = \begin{bmatrix} -2 & 3 & -4 \\ 1 & 0 & -5 \end{bmatrix}.$$

It follows that for every $m \times n$ matrix A,

$$A + (-A) = O.$$

Subtraction of two $m \times n$ matrices is defined by

$$A - B = A + (-B).$$

Using the parentheses notation for matrices, this implies that

$$(a_{ij}) - (b_{ij}) = (a_{ij}) + (-b_{ij}) = (a_{ij} - b_{ij}).$$

Thus, to subtract two matrices, we subtract the elements that are in corresponding positions.

The **product** of a real number c and an $m \times n$ matrix $A = (a_{ij})$ is defined by

$$cA = (ca_{ij}).$$

Thus, to find cA, we multiply each element of A by c. For example,

$$3\begin{bmatrix} 4 & -1 \\ 2 & 3 \end{bmatrix} = \begin{bmatrix} 12 & -3 \\ 6 & 9 \end{bmatrix}.$$

The following results may be established, where A and B are $m \times n$ matrices and c and d are any real numbers.

$$c(A + B) = cA + cB$$

$$(c + d)A = cA + dA$$

$$(cd)A = c(dA)$$

The following definition of the product of two matrices may appear unusual to the beginning student; however, there are many applications that justify the form of the definition. To define the product AB of two matrices A and B, *the number of columns of A must be the same as the number of rows of B*. Suppose that $A = (a_{ij})$ is $m \times n$ and $B = (b_{ij})$ is $n \times p$. To determine the element c_{ij} of the product, we single out row i of A and column j of B as follows:

$$\begin{bmatrix} a_{11} & a_{12} & \cdots & a_{1n} \\ \vdots & \vdots & & \vdots \\ a_{i1} & a_{i2} & \cdots & a_{in} \\ \vdots & \vdots & & \vdots \\ a_{m1} & a_{m2} & \cdots & a_{mn} \end{bmatrix} \begin{bmatrix} b_{11} & \cdots & b_{1j} & \cdots & b_{1p} \\ b_{21} & \cdots & b_{2j} & \cdots & b_{2p} \\ \vdots & & \vdots & & \vdots \\ b_{n1} & \cdots & b_{nj} & \cdots & b_{np} \end{bmatrix}$$

Next we multiply pairs of elements and then add them, using the formula

$$c_{ij} = a_{i1}b_{1j} + a_{i2}b_{2j} + \cdots + a_{in}b_{nj}.$$

For example, the element c_{11} in the first row and the first column of AB is

$$c_{11} = a_{11}b_{11} + a_{12}b_{21} + \cdots + a_{1n}b_{n1}.$$

The element c_{12} in the first row and second column of AB is

$$c_{12} = a_{11}b_{12} + a_{12}b_{22} + \cdots + a_{1n}b_{n2}.$$

By definition, the product AB has the same number of rows as A and the same number of columns as B. In particular, if A is $m \times n$ and B is $n \times p$, then AB is $m \times p$. This is illustrated by the following product of a 2×3 matrix and a 3×4 matrix.

$$\begin{bmatrix} 1 & 2 & -3 \\ 4 & 0 & -2 \end{bmatrix} \begin{bmatrix} 5 & -4 & 2 & 0 \\ -1 & 6 & 3 & 1 \\ 7 & 0 & 4 & 8 \end{bmatrix} = \begin{bmatrix} -18 & 8 & -4 & -22 \\ 6 & -16 & 0 & -16 \end{bmatrix}.$$

Here are some typical computations of the elements c_{ij} in the product:

$$c_{11} = (1)(5) + (2)(-1) + (-3)(7) = 5 - 2 - 21 = -18$$

$$c_{13} = (1)(2) + (2)(3) + (-3)(4) = 2 + 6 - 12 = -4$$

$$c_{23} = (4)(2) + (0)(3) + (-2)(4) = 8 + 0 - 8 = 0$$

$$c_{24} = (4)(0) + (0)(1) + (-2)(8) = 0 + 0 - 16 = -16.$$

The reader should calculate the remaining elements.

The product operation for matrices is not commutative. Indeed, if A is 2×3 and B is 3×4, then AB may be found, but BA is undefined, since the number of columns of B is different from the number of rows of A. Even if AB and BA are both defined, it is often true that these products are different. This is illustrated in the next example, along with the fact that the product of two nonzero matrices may equal a zero matrix.

EXAMPLE 1

If $A = \begin{bmatrix} 2 & 2 \\ -1 & -1 \end{bmatrix}$ and $B = \begin{bmatrix} 1 & 2 \\ 1 & 2 \end{bmatrix}$, show that $AB \neq BA$.

SOLUTION Using the definition of product we obtain

$$AB = \begin{bmatrix} 4 & 8 \\ -2 & -4 \end{bmatrix} \quad \text{and} \quad BA = \begin{bmatrix} 0 & 0 \\ 0 & 0 \end{bmatrix}.$$

Hence, $AB \neq BA$. Note that BA is a zero matrix. ■

It can be shown that matrix multiplication is associative. Thus

$$A(BC) = (AB)C$$

provided that the indicated products are defined, which will be the case if A is $m \times n$, B is $n \times p$, and C is $p \times q$.

The Distributive Properties also hold if the matrices involved have the proper number of rows and columns. If A_1 and A_2 are $m \times n$ matrices, and if B_1 and B_2 are $n \times p$ matrices, then

$$A_1(B_1 + B_2) = A_1B_1 + A_1B_2$$

and

$$(A_1 + A_2)B_1 = A_1B_1 + A_2B_1.$$

As a special case, if all matrices are square, of order n, then the Associative and Distributive Properties are true. We shall not prove these properties.

Throughout the remainder of this section we shall concentrate on square matrices. The symbol I_n will denote the square matrix of order n that has 1 in each position on the main diagonal and 0 elsewhere. For example,

$$I_2 = \begin{bmatrix} 1 & 0 \\ 0 & 1 \end{bmatrix}, \qquad I_3 = \begin{bmatrix} 1 & 0 & 0 \\ 0 & 1 & 0 \\ 0 & 0 & 1 \end{bmatrix},$$

and so on. It can be shown that if A is any square matrix of order n, then

$$AI_n = A = I_nA.$$

For that reason I_n is called the **identity matrix of order *n*.** To illustrate, if $A = (a_{ij})$ is of order 2, then a direct calculation shows that

$$\begin{bmatrix} a_{11} & a_{12} \\ a_{21} & a_{22} \end{bmatrix}\begin{bmatrix} 1 & 0 \\ 0 & 1 \end{bmatrix} = \begin{bmatrix} a_{11} & a_{12} \\ a_{21} & a_{22} \end{bmatrix} = \begin{bmatrix} 1 & 0 \\ 0 & 1 \end{bmatrix}\begin{bmatrix} a_{11} & a_{12} \\ a_{21} & a_{22} \end{bmatrix}.$$

Some, but not all, $n \times n$ matrices A have an **inverse** in the sense that there is a matrix B such that $AB = I_n = BA$. If A has an inverse we denote it by A^{-1} and write

$$AA^{-1} = I_n = A^{-1}A.$$

The symbol A^{-1} is read "A inverse." In matrix theory it is *not* acceptable to use the symbol $1/A$ in place of A^{-1}.

There is an interesting technique for finding the inverse of a square matrix A, whenever it exists. We shall not attempt to justify the following procedure, since that would require advanced concepts. Given the $n \times n$ matrix $A = (a_{ij})$, we begin by forming the $n \times 2n$ matrix:

$$\begin{bmatrix} a_{11} & a_{12} & \cdots & a_{1n} & 1 & 0 & \cdots & 0 \\ a_{21} & a_{22} & \cdots & a_{2n} & 0 & 1 & \cdots & 0 \\ \vdots & \vdots & & \vdots & \vdots & \vdots & & \vdots \\ a_{n1} & a_{n2} & \cdots & a_{nn} & 0 & 0 & \cdots & 1 \end{bmatrix}$$

where the $n \times n$ identity matrix I_n appears "to the right" of the matrix A, as indicated. We next apply a succession of elementary row transformations until we arrive at a matrix of the form

$$\begin{bmatrix} 1 & 0 & \cdots & 0 & b_{11} & b_{12} & \cdots & b_{1n} \\ 0 & 1 & \cdots & 0 & b_{21} & b_{22} & \cdots & b_{2n} \\ \vdots & \vdots & & \vdots & \vdots & \vdots & & \vdots \\ 0 & 0 & \cdots & 1 & b_{n1} & b_{n2} & \cdots & b_{nn} \end{bmatrix}$$

where the identity matrix I_n appears "to the left" of the $n \times n$ matrix (b_{ij}). It can be shown that (b_{ij}) is the desired inverse A^{-1}.

EXAMPLE 2 Find A^{-1} if

$$A = \begin{bmatrix} 3 & 5 \\ 1 & 4 \end{bmatrix}.$$

SOLUTION We begin with the matrix $\begin{bmatrix} 3 & 5 & 1 & 0 \\ 1 & 4 & 0 & 1 \end{bmatrix}$ and then perform elementary row transformations until the identity matrix I_2 appears on the left, as follows:

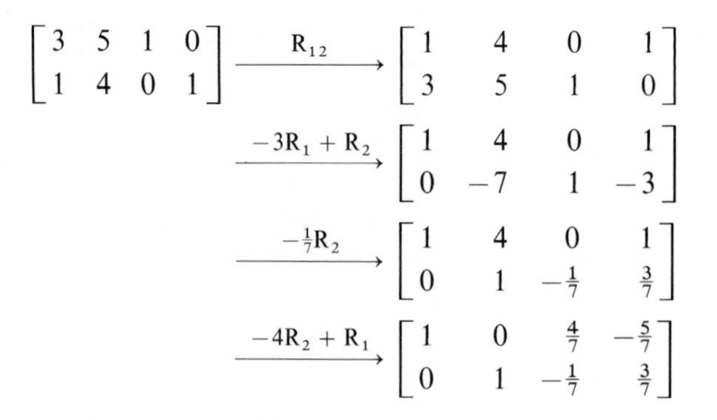

$$\begin{bmatrix} 3 & 5 & 1 & 0 \\ 1 & 4 & 0 & 1 \end{bmatrix} \xrightarrow{R_{12}} \begin{bmatrix} 1 & 4 & 0 & 1 \\ 3 & 5 & 1 & 0 \end{bmatrix}$$

$$\xrightarrow{-3R_1 + R_2} \begin{bmatrix} 1 & 4 & 0 & 1 \\ 0 & -7 & 1 & -3 \end{bmatrix}$$

$$\xrightarrow{-\frac{1}{7}R_2} \begin{bmatrix} 1 & 4 & 0 & 1 \\ 0 & 1 & -\frac{1}{7} & \frac{3}{7} \end{bmatrix}$$

$$\xrightarrow{-4R_2 + R_1} \begin{bmatrix} 1 & 0 & \frac{4}{7} & -\frac{5}{7} \\ 0 & 1 & -\frac{1}{7} & \frac{3}{7} \end{bmatrix}$$

According to the previous discussion,

$$A^{-1} = \begin{bmatrix} \frac{4}{7} & -\frac{5}{7} \\ -\frac{1}{7} & \frac{3}{7} \end{bmatrix}.$$

Checking this fact, we see that

$$\begin{bmatrix} 3 & 5 \\ 1 & 4 \end{bmatrix}\begin{bmatrix} \frac{4}{7} & -\frac{5}{7} \\ -\frac{1}{7} & \frac{3}{7} \end{bmatrix} = \begin{bmatrix} 1 & 0 \\ 0 & 1 \end{bmatrix} = \begin{bmatrix} \frac{4}{7} & -\frac{5}{7} \\ -\frac{1}{7} & \frac{3}{7} \end{bmatrix}\begin{bmatrix} 3 & 5 \\ 1 & 4 \end{bmatrix}.$$ ■

EXAMPLE 3 Find A^{-1} if

$$A = \begin{bmatrix} -1 & 3 & 1 \\ 2 & 5 & 0 \\ 3 & 1 & -2 \end{bmatrix}.$$

SOLUTION We proceed as follows:

$$\begin{bmatrix} -1 & 3 & 1 & 1 & 0 & 0 \\ 2 & 5 & 0 & 0 & 1 & 0 \\ 3 & 1 & -2 & 0 & 0 & 1 \end{bmatrix} \xrightarrow{(-1)R_1} \begin{bmatrix} 1 & -3 & -1 & -1 & 0 & 0 \\ 2 & 5 & 0 & 0 & 1 & 0 \\ 3 & 1 & -2 & 0 & 0 & 1 \end{bmatrix}$$

$$\xrightarrow{-2R_1 + R_2} \begin{bmatrix} 1 & -3 & -1 & -1 & 0 & 0 \\ 0 & 11 & 2 & 2 & 1 & 0 \\ 3 & 1 & -2 & 0 & 0 & 1 \end{bmatrix}$$

$$\xrightarrow{-3R_1 + R_3} \begin{bmatrix} 1 & -3 & -1 & -1 & 0 & 0 \\ 0 & 11 & 2 & 2 & 1 & 0 \\ 0 & 10 & 1 & 3 & 0 & 1 \end{bmatrix}$$

$$\xrightarrow{(-1)R_3 + R_2} \begin{bmatrix} 1 & -3 & -1 & -1 & 0 & 0 \\ 0 & 1 & 1 & -1 & 1 & -1 \\ 0 & 10 & 1 & 3 & 0 & 1 \end{bmatrix}$$

$$\xrightarrow{3R_2 + R_1} \begin{bmatrix} 1 & 0 & 2 & -4 & 3 & -3 \\ 0 & 1 & 1 & -1 & 1 & -1 \\ 0 & 10 & 1 & 3 & 0 & 1 \end{bmatrix}$$

$$\xrightarrow{-10R_2 + R_3} \begin{bmatrix} 1 & 0 & 2 & -4 & 3 & -3 \\ 0 & 1 & 1 & -1 & 1 & -1 \\ 0 & 0 & -9 & 13 & -10 & 11 \end{bmatrix}$$

$$\xrightarrow{-\frac{1}{9}R_3} \begin{bmatrix} 1 & 0 & 2 & -4 & 3 & -3 \\ 0 & 1 & 1 & -1 & 1 & -1 \\ 0 & 0 & 1 & -\frac{13}{9} & \frac{10}{9} & -\frac{11}{9} \end{bmatrix}$$

$$\xrightarrow{-2R_3 + R_1} \begin{bmatrix} 1 & 0 & 0 & -\frac{10}{9} & \frac{7}{9} & -\frac{5}{9} \\ 0 & 1 & 1 & -1 & 1 & -1 \\ 0 & 0 & 1 & -\frac{13}{9} & \frac{10}{9} & -\frac{11}{9} \end{bmatrix}$$

$$\xrightarrow{(-1)R_3 + R_2} \begin{bmatrix} 1 & 0 & 0 & -\frac{10}{9} & \frac{7}{9} & -\frac{5}{9} \\ 0 & 1 & 0 & \frac{4}{9} & -\frac{1}{9} & \frac{2}{9} \\ 0 & 0 & 1 & -\frac{13}{9} & \frac{10}{9} & -\frac{11}{9} \end{bmatrix}$$

Consequently,

$$A^{-1} = \begin{bmatrix} -\frac{10}{9} & \frac{7}{9} & -\frac{5}{9} \\ \frac{4}{9} & -\frac{1}{9} & \frac{2}{9} \\ -\frac{13}{9} & \frac{10}{9} & -\frac{11}{9} \end{bmatrix} = \frac{1}{9}\begin{bmatrix} -10 & 7 & -5 \\ 4 & -1 & 2 \\ -13 & 10 & -11 \end{bmatrix}.$$

It can be shown that

$$AA^{-1} = I_3 = A^{-1}A. \qquad \blacksquare$$

There are many uses for inverses of matrices. One application concerns solutions of systems of linear equations. To illustrate, let us consider the case of two linear equations in two unknowns:

$$\begin{cases} a_{11}x + a_{12}y = k_1 \\ a_{21}x + a_{22}y = k_2. \end{cases}$$

We may express this system in terms of matrices as follows:

$$\begin{bmatrix} a_{11}x + a_{12}y \\ a_{21}x + a_{22}y \end{bmatrix} = \begin{bmatrix} k_1 \\ k_2 \end{bmatrix}.$$

If we let

$$A = \begin{bmatrix} a_{11} & a_{12} \\ a_{21} & a_{22} \end{bmatrix}, \quad X = \begin{bmatrix} x \\ y \end{bmatrix}, \quad \text{and} \quad B = \begin{bmatrix} k_1 \\ k_2 \end{bmatrix},$$

we have

$$AX = B.$$

If A^{-1} exists, then multiplying both sides of the last equation by A^{-1} gives us $A^{-1}AX = A^{-1}B$. Since $A^{-1}A = I_2$ and $I_2X = X$, this leads to

$$X = A^{-1}B$$

from which the solution (x, y) may be found. This technique may be extended to systems of n linear equations in n unknowns.

EXAMPLE 4 Solve the following system of equations:

$$\begin{cases} -x + 3y + z = 1 \\ 2x + 5y = 3 \\ 3x + y - 2z = -2. \end{cases}$$

SOLUTION If we let

$$A = \begin{bmatrix} -1 & 3 & 1 \\ 2 & 5 & 0 \\ 3 & 1 & -2 \end{bmatrix}, \quad X = \begin{bmatrix} x \\ y \\ z \end{bmatrix}, \quad \text{and} \quad B = \begin{bmatrix} 1 \\ 3 \\ -2 \end{bmatrix},$$

then, as in the preceding discussion, the given system may be written in terms of matrices as $AX = B$. This implies that $X = A^{-1}B$. The matrix A^{-1} was found in Example 3. Substituting for X, A^{-1}, and B in the last equation gives us

$$\begin{bmatrix} x \\ y \\ z \end{bmatrix} = \frac{1}{9}\begin{bmatrix} -10 & 7 & -5 \\ 4 & -1 & 2 \\ -13 & 10 & -11 \end{bmatrix}\begin{bmatrix} 1 \\ 3 \\ -2 \end{bmatrix} = \frac{1}{9}\begin{bmatrix} 21 \\ -3 \\ 39 \end{bmatrix} = \begin{bmatrix} \frac{7}{3} \\ -\frac{1}{3} \\ \frac{13}{3} \end{bmatrix}.$$

It follows that $x = \frac{7}{3}$, $y = -\frac{1}{3}$, and $z = \frac{13}{3}$. Hence, the ordered triple $(\frac{7}{3}, -\frac{1}{3}, \frac{13}{3})$ is the solution of the given system. ■

The method of solution employed in Example 4 is beneficial only if A^{-1} is known, or if many systems with the same coefficient matrix are to be considered. The preferred technique for solving an arbitrary system of linear equations is the matrix method discussed in Section 8.3.

Exercises 8.5

In Exercises 1–8 find $A + B$, $A - B$, $2A$, and $-3B$.

1 $A = \begin{bmatrix} 5 & -2 \\ 1 & 3 \end{bmatrix}$, $B = \begin{bmatrix} 4 & 1 \\ -3 & 2 \end{bmatrix}$

2 $A = \begin{bmatrix} 3 & 0 \\ -1 & 2 \end{bmatrix}$, $B = \begin{bmatrix} 3 & -4 \\ 1 & 1 \end{bmatrix}$

3 $A = \begin{bmatrix} 6 & -1 \\ 2 & 0 \\ -3 & 4 \end{bmatrix}$, $B = \begin{bmatrix} 3 & 1 \\ -1 & 5 \\ 6 & 0 \end{bmatrix}$

4 $A = \begin{bmatrix} 0 & -2 & 7 \\ 5 & 4 & -3 \end{bmatrix}$, $B = \begin{bmatrix} 8 & 4 & 0 \\ 0 & 1 & 4 \end{bmatrix}$

5 $A = [4 \quad -3 \quad 2]$, $B = [7 \quad 0 \quad -5]$

6 $A = \begin{bmatrix} 7 \\ -16 \end{bmatrix}$, $B = \begin{bmatrix} -11 \\ 9 \end{bmatrix}$

7 $A = \begin{bmatrix} 0 & 4 & 0 & 3 \\ 1 & 2 & 0 & -5 \end{bmatrix}$,

$B = \begin{bmatrix} -3 & 0 & 1 & 3 \\ 2 & 0 & 7 & -2 \end{bmatrix}$

8 $A = [-7]$, $B = [9]$

In Exercises 9–18 find AB and BA.

9 $A = \begin{bmatrix} 2 & 6 \\ 3 & -4 \end{bmatrix}$, $B = \begin{bmatrix} 5 & -2 \\ 1 & 7 \end{bmatrix}$

10 $A = \begin{bmatrix} 4 & -2 \\ -2 & 1 \end{bmatrix}$, $B = \begin{bmatrix} 2 & 1 \\ 4 & 2 \end{bmatrix}$

11 $A = \begin{bmatrix} 3 & 0 & -1 \\ 0 & 4 & 2 \\ 5 & -3 & 1 \end{bmatrix}$, $B = \begin{bmatrix} 1 & -5 & 0 \\ 4 & 1 & -2 \\ 0 & -1 & 3 \end{bmatrix}$

12 $A = \begin{bmatrix} 5 & 0 & 0 \\ 0 & -3 & 0 \\ 0 & 0 & 2 \end{bmatrix}$, $B = \begin{bmatrix} 3 & 0 & 0 \\ 0 & 4 & 0 \\ 0 & 0 & -2 \end{bmatrix}$

13 $A = \begin{bmatrix} 4 & -3 & 1 \\ -5 & 2 & 2 \end{bmatrix}$, $B = \begin{bmatrix} 2 & 1 \\ 0 & 1 \\ -4 & 7 \end{bmatrix}$

14 $A = \begin{bmatrix} 2 & 1 & -1 & 0 \\ 3 & -2 & 0 & 5 \\ -2 & 1 & 4 & 2 \end{bmatrix}$, $B = \begin{bmatrix} 5 & -3 & 1 \\ 1 & 2 & 0 \\ -1 & 0 & 4 \\ 0 & -2 & 3 \end{bmatrix}$

15 $A = \begin{bmatrix} 1 & 2 & 3 \\ 4 & 5 & 6 \\ 7 & 8 & 9 \end{bmatrix}$, $B = \begin{bmatrix} 1 & 0 & 0 \\ 0 & 1 & 0 \\ 0 & 0 & 1 \end{bmatrix}$

16 $A = \begin{bmatrix} 1 & 2 & 3 \\ 2 & 3 & 1 \\ 3 & 1 & 2 \end{bmatrix}$, $B = \begin{bmatrix} 2 & 0 & 0 \\ 0 & 2 & 0 \\ 0 & 0 & 2 \end{bmatrix}$

17 $A = [-3 \quad 7 \quad 2]$, $B = \begin{bmatrix} 1 \\ 4 \\ -5 \end{bmatrix}$

18 $A = [4 \quad 8]$, $B = \begin{bmatrix} -3 \\ 2 \end{bmatrix}$

In Exercises 19–22 find AB.

19 $A = \begin{bmatrix} 4 & -2 \\ 0 & 3 \\ -7 & 5 \end{bmatrix}$, $B = \begin{bmatrix} 3 \\ 4 \end{bmatrix}$

20 $A = \begin{bmatrix} 4 \\ -3 \\ 2 \end{bmatrix}$, $B = [5 \quad 1]$

21 $A = \begin{bmatrix} 2 & 1 & 0 & -3 \\ -7 & 0 & -2 & 4 \end{bmatrix}$, $B = \begin{bmatrix} 4 & -2 & 0 \\ 1 & 1 & -2 \\ 0 & 0 & 5 \\ -3 & -1 & 0 \end{bmatrix}$

22 $A = \begin{bmatrix} 1 & 2 & -3 \\ 4 & -5 & 6 \end{bmatrix}$, $B = \begin{bmatrix} 1 & -1 & 0 & 2 \\ -2 & 3 & 1 & 0 \\ 0 & 4 & 0 & -3 \end{bmatrix}$

23 If $A = \begin{bmatrix} 1 & 2 \\ 0 & -3 \end{bmatrix}$, and $B = \begin{bmatrix} 2 & -1 \\ 3 & 1 \end{bmatrix}$, show that $(A + B)(A - B) \neq A^2 - B^2$, where $A^2 = AA$ and $B^2 = BB$.

24 If A and B are the matrices of Exercise 23, show that $(A + B)(A + B) \neq A^2 + 2AB + B^2$.

25 If A and B are the matrices of Exercise 23 and $C = \begin{bmatrix} 3 & 1 \\ -2 & 0 \end{bmatrix}$, show that $A(B + C) = AB + AC$.

26 If A, B, and C are the matrices of Exercise 25, show that $A(BC) = (AB)C$.

Prove the identities stated in Exercises 27–30, where

$$A = \begin{bmatrix} a_{11} & a_{12} \\ a_{21} & a_{22} \end{bmatrix}, \quad B = \begin{bmatrix} b_{11} & b_{12} \\ b_{21} & b_{22} \end{bmatrix}, \quad C = \begin{bmatrix} c_{11} & c_{22} \\ c_{21} & c_{22} \end{bmatrix}$$

and c, d are real numbers.

27 $c(A + B) = cA + cB$

28 $(c + d)A = cA + dA$

29 $A(B + C) = AB + AC$

30 $A(BC) = (AB)C$

In Exercises 31–42 find the inverse of the matrix, if it exists.

31 $\begin{bmatrix} 2 & -4 \\ 1 & 3 \end{bmatrix}$

32 $\begin{bmatrix} 3 & 2 \\ 4 & 5 \end{bmatrix}$

33 $\begin{bmatrix} 2 & 4 \\ 4 & 8 \end{bmatrix}$

34 $\begin{bmatrix} 3 & -1 \\ 6 & -2 \end{bmatrix}$

35 $\begin{bmatrix} 3 & -1 & 0 \\ 2 & 2 & 0 \\ 0 & 0 & 4 \end{bmatrix}$

36 $\begin{bmatrix} 3 & 0 & 2 \\ 0 & 1 & 0 \\ -4 & 0 & 2 \end{bmatrix}$

37 $\begin{bmatrix} -2 & 2 & 3 \\ 1 & -1 & 0 \\ 0 & 1 & 4 \end{bmatrix}$

38 $\begin{bmatrix} 1 & 2 & 3 \\ -2 & 1 & 0 \\ 3 & -1 & 1 \end{bmatrix}$

39 $\begin{bmatrix} 2 & 0 & 0 \\ 0 & 4 & 0 \\ 0 & 0 & 6 \end{bmatrix}$

40 $\begin{bmatrix} 1 & 1 & 1 \\ 2 & 2 & 2 \\ 3 & 3 & 3 \end{bmatrix}$

41 $\begin{bmatrix} 1 & -1 & 0 & 1 \\ 0 & 1 & -2 & 0 \\ -1 & 2 & 1 & 2 \\ -2 & 1 & 2 & 0 \end{bmatrix}$

42 $\begin{bmatrix} 1 & 2 & 0 & 1 \\ 0 & -1 & 1 & -2 \\ 0 & 0 & 2 & 0 \\ 0 & 0 & 0 & 1 \end{bmatrix}$

43 State conditions on a and b that guarantee that the matrix $\begin{bmatrix} a & 0 \\ 0 & b \end{bmatrix}$ has an inverse, and find a formula for the inverse, if it exists.

44 If $abc \neq 0$, find the inverse of $\begin{bmatrix} a & 0 & 0 \\ 0 & b & 0 \\ 0 & 0 & c \end{bmatrix}$.

45 If $A = \begin{bmatrix} a_{11} & a_{12} & a_{13} \\ a_{21} & a_{22} & a_{23} \\ a_{31} & a_{32} & a_{33} \end{bmatrix}$, prove that $AI_3 = A = I_3A$.

46 Prove that $AI_4 = A = I_4A$ for every square matrix A of order 4.

Solve the systems in Exercises 47–50 by the method of Example 4. (Refer to inverses of matrices found in Exercises 31, 32, 37, and 38.)

47 $\begin{cases} 2x - 4y = 3 \\ x + 3y = 1 \end{cases}$

48 $\begin{cases} 3x + 2y = -1 \\ 4x + 5y = 1 \end{cases}$

49 $\begin{cases} -2x + 2y + 3z = 1 \\ x - y = 3 \\ y + 4z = -2 \end{cases}$

50 $\begin{cases} x + 2y + 3z = -1 \\ -2x + y = 4 \\ 3x - y + z = 2 \end{cases}$

8.6 Determinants

Throughout this section and the next we will assume that all matrices under discussion are *square* matrices. Associated with each square matrix A is a number called the **determinant of** A, denoted by $|A|$. This notation should not be confused with the symbol for the absolute value of a real number. To avoid any misunderstanding, the expression det A is sometimes used instead of $|A|$. We shall define $|A|$ by beginning with the case in which A has order 1 and then by increasing the order a step at a time.

If A is a square matrix of order 1, then A has only one element. Thus, $A = [a_{11}]$ and we define $|A| = a_{11}$. If A is a square matrix of order 2, then

$$A = \begin{bmatrix} a_{11} & a_{12} \\ a_{21} & a_{22} \end{bmatrix}$$

and the determinant of A is defined by

$$|A| = a_{11}a_{22} - a_{21}a_{12}.$$

Another notation for $|A|$ is obtained by replacing the brackets used for A with vertical bars as follows.

DEFINITION

$$|A| = \begin{vmatrix} a_{11} & a_{12} \\ a_{21} & a_{22} \end{vmatrix} = a_{11}a_{22} - a_{21}a_{12}$$

EXAMPLE 1

Find $|A|$ if $A = \begin{bmatrix} 2 & -1 \\ 4 & -3 \end{bmatrix}$.

SOLUTION By definition,

$$|A| = \begin{vmatrix} 2 & -1 \\ 4 & -3 \end{vmatrix} = (2)(-3) - (4)(-1) = -6 + 4 = -2.$$ ■

For square matrices of order 3 it is convenient to introduce the following terminology.

DEFINITION

Let A be a square matrix of order 3. The **minor** M_{ij} of the element a_{ij} is the determinant of the matrix of order 2 obtained by deleting row i and column j.

To determine the minor of an element, we discard the row and column in which the element appears and then find the determinant of the resulting matrix. To illustrate, given the 3×3 matrix

$$A = \begin{bmatrix} a_{11} & a_{12} & a_{13} \\ a_{21} & a_{22} & a_{23} \\ a_{31} & a_{32} & a_{33} \end{bmatrix}$$

we obtain

$$M_{11} = \begin{vmatrix} a_{22} & a_{23} \\ a_{32} & a_{33} \end{vmatrix} = a_{22}a_{33} - a_{32}a_{23}$$

$$M_{12} = \begin{vmatrix} a_{21} & a_{23} \\ a_{31} & a_{33} \end{vmatrix} = a_{21}a_{33} - a_{31}a_{23}$$

$$M_{13} = \begin{vmatrix} a_{21} & a_{22} \\ a_{31} & a_{32} \end{vmatrix} = a_{21}a_{32} - a_{31}a_{22}$$

$$M_{23} = \begin{vmatrix} a_{11} & a_{12} \\ a_{31} & a_{32} \end{vmatrix} = a_{11}a_{32} - a_{31}a_{12}$$

and likewise for the other minors M_{21}, M_{22}, M_{31}, M_{32}, and M_{33}.

We shall also make use of the following concept.

DEFINITION

The **cofactor** A_{ij} of the element a_{ij} is

$$A_{ij} = (-1)^{i+j}M_{ij}.$$

To obtain the cofactor of a_{ij}, we find the minor and multiply it by 1 or -1, depending on whether the sum of i and j is even or odd, respectively. An easy way to remember the sign $(-1)^{i+j}$ associated with the cofactor A_{ij} is to consider the following "checkerboard" scheme:

$$\begin{bmatrix} + & - & + \\ - & + & - \\ + & - & + \end{bmatrix}$$

EXAMPLE 2 If

$$A = \begin{bmatrix} 1 & -3 & 3 \\ 4 & 2 & 0 \\ -2 & -7 & 5 \end{bmatrix}$$

find M_{11}, M_{21}, M_{22}, A_{11}, A_{21}, and A_{22}.

SOLUTION By definition,

$$M_{11} = \begin{vmatrix} 2 & 0 \\ -7 & 5 \end{vmatrix} = (2)(5) - (-7)(0) = 10$$

$$M_{21} = \begin{vmatrix} -3 & 3 \\ -7 & 5 \end{vmatrix} = (-3)(5) - (-7)(3) = 6$$

$$M_{22} = \begin{vmatrix} 1 & 3 \\ -2 & 5 \end{vmatrix} = (1)(5) - (-2)(3) = 11.$$

To obtain the cofactors, we prefix the corresponding minors with the proper signs. Thus, using the definition of cofactor,

$$A_{11} = (-1)^{1+1}M_{11} = (1)(10) = 10$$

$$A_{21} = (-1)^{2+1}M_{21} = (-1)(6) = -6$$

$$A_{22} = (-1)^{2+2}M_{22} = (1)(11) = 11.$$

The checkerboard scheme can also be used to determine the proper signs. ■

The determinant $|A|$ of a square matrix of order 3 is defined as follows.

DEFINITION

$$|A| = \begin{vmatrix} a_{11} & a_{12} & a_{13} \\ a_{21} & a_{22} & a_{23} \\ a_{31} & a_{32} & a_{33} \end{vmatrix} = a_{11}A_{11} + a_{12}A_{12} + a_{13}A_{13}$$

Since $A_{11} = (-1)^{1+1}M_{11} = M_{11}$, $A_{12} = (-1)^{1+2}M_{12} = -M_{12}$, and $A_{13} = (-1)^{1+3}M_{13} = M_{13}$, the preceding definition may also be written

$$|A| = a_{11}M_{11} - a_{12}M_{12} + a_{13}M_{13}.$$

If we express M_{11}, M_{12}, and M_{13} in terms of elements of A, we obtain the following formula for $|A|$:

$$|A| = a_{11}a_{22}a_{33} - a_{11}a_{32}a_{23} - a_{12}a_{21}a_{33} + a_{12}a_{31}a_{23} + a_{13}a_{21}a_{32} - a_{13}a_{31}a_{22}.$$

The definition of $|A|$ for a square matrix A of order 3 displays a pattern of multiplying each element in row 1 by its cofactor, and then adding to find $|A|$. This is referred to as *expanding* $|A|$ *by the first row.* By actually carrying out the computations, it is not difficult to show that $|A|$ *can be expanded in similar fashion by using any row or column.* As an illustration, the expansion by the second column is

$$\begin{aligned} |A| &= a_{12}A_{12} + a_{22}A_{22} + a_{32}A_{32} \\ &= a_{12}\left(-\begin{vmatrix} a_{21} & a_{23} \\ a_{31} & a_{32} \end{vmatrix}\right) + a_{22}\left(+\begin{vmatrix} a_{11} & a_{13} \\ a_{31} & a_{33} \end{vmatrix}\right) + a_{32}\left(-\begin{vmatrix} a_{11} & a_{13} \\ a_{21} & a_{23} \end{vmatrix}\right). \end{aligned}$$

Applying the definition to the determinants in parentheses, multiplying as indicated, and rearranging the terms in the sum, we could arrive at the formula for $|A|$ in terms of the elements of A. Similarly, the expansion by the third row is

$$|A| = a_{31}A_{31} + a_{32}A_{32} + a_{33}A_{33}.$$

Once again it can be shown that this agrees with previous expansions.

EXAMPLE 3

Find $|A|$ if $A = \begin{bmatrix} -1 & 3 & 1 \\ 2 & 5 & 0 \\ 3 & 1 & -2 \end{bmatrix}$.

SOLUTION Since there is a zero in the second row we shall expand $|A|$ by that row, because then we need only evaluate two cofactors. Thus,

$$|A| = (2)A_{21} + (5)A_{22} + (0)A_{23}.$$

Using the definition of cofactor,

$$A_{21} = (-1)^3 M_{21} = -\begin{vmatrix} 3 & 1 \\ 1 & -2 \end{vmatrix} = -[(3)(-2) - (1)(1)] = 7$$

$$A_{22} = (-1)^4 M_{22} = \begin{vmatrix} -1 & 1 \\ 3 & -2 \end{vmatrix} = [(-1)(-2) - (3)(1)] = -1.$$

Consequently,

$$|A| = (2)(7) + (5)(-1) + (0)A_{23} = 14 - 5 + 0 = 9.$$

■

The definition of the determinant of a matrix of arbitrary order n may be patterned after that used for order 3. Specifically, the **minor** M_{ij} is defined as the determinant of the matrix of order $n - 1$ obtained by deleting row i and column j. The **cofactor** A_{ij} is defined as $(-1)^{i+j}M_{ij}$. The sign $(-1)^{i+j}$ associated with A_{ij} can be remembered by using a checkerboard similar to that used for order 3, extending the rows and columns as far as necessary. We then define the determinant $|A|$ of a matrix A of order n as the expansion by the first row, that is,

$$|A| = a_{11}A_{11} + a_{12}A_{12} + \cdots + a_{1n}A_{1n}$$

or, in terms of minors,

$$|A| = a_{11}M_{11} - a_{12}M_{12} + \cdots + a_{1n}(-1)^{1+n}M_{1n}.$$

The number $|A|$ may be found by using *any* row or column, as stated in the following theorem.

EXPANSION THEOREM FOR DETERMINANTS

If A is a square matrix of order $n > 1$, then the determinant $|A|$ may be found by multiplying the elements of any row (or column) by their respective cofactors, and adding the resulting products.

The proof of this theorem is difficult and may be found in texts on matrix theory. The theorem is quite useful if many zeros appear in a row or column, as illustrated in the following example.

EXAMPLE 4

Find $|A|$ if $A = \begin{bmatrix} 1 & 0 & 2 & 5 \\ -2 & 1 & 5 & 0 \\ 0 & 0 & -3 & 0 \\ 0 & -1 & 0 & 3 \end{bmatrix}$.

SOLUTION Note that all but one of the elements in the third row is zero. Hence if we expand $|A|$ by the third row, there will be at most one nonzero term. Specifically,

$$|A| = (0)A_{31} + (0)A_{32} + (-3)A_{33} + (0)A_{34} = -3A_{33}$$

where $$A_{33} = \begin{vmatrix} 1 & 0 & 5 \\ -2 & 1 & 0 \\ 0 & -1 & 3 \end{vmatrix}.$$

Expanding A_{33} by column 1, we obtain

$$A_{33} = (1)\begin{vmatrix} 1 & 0 \\ -1 & 3 \end{vmatrix} + (-2)\left(-\begin{vmatrix} 0 & 5 \\ -1 & 3 \end{vmatrix}\right) + 0\begin{vmatrix} 0 & 5 \\ 1 & 0 \end{vmatrix} = 3 + 10 + 0 = 13.$$

Thus, $$|A| = -3A_{33} = (-3)(13) = -39.$$ ■

In general, if all but one element a in some row (or column) of A is zero, and if the determinant $|A|$ is expanded by that row (or column), then all terms drop out except the product of the element a with its cofactor. We will make important use of this fact in the next section.

If *every* element in a row (or column) of a matrix A is zero, then upon expanding $|A|$ by that row (or column), we obtain the number 0. This gives us the following result.

THEOREM If every element of a row (or column) of a square matrix A is zero, then $|A| = 0$.

Exercises 8.6

In Exercises 1–4 find all the minors and cofactors of the elements in the matrix.

1 $\begin{bmatrix} 2 & 4 & -1 \\ 0 & 3 & 2 \\ -5 & 7 & 0 \end{bmatrix}$ **2** $\begin{bmatrix} 5 & -2 & 1 \\ 4 & 7 & 0 \\ -3 & 4 & -1 \end{bmatrix}$

3 $\begin{bmatrix} 7 & -1 \\ 5 & 0 \end{bmatrix}$ **4** $\begin{bmatrix} -6 & 4 \\ 3 & 2 \end{bmatrix}$

5–8 Find the determinants of the matrices in Exercises 1–4.

Find the determinants of the matrices in Exercises 9–20.

9 $\begin{bmatrix} -5 & 4 \\ -3 & 2 \end{bmatrix}$ **10** $\begin{bmatrix} 6 & 4 \\ -3 & 2 \end{bmatrix}$

11 $\begin{bmatrix} a & -a \\ b & -b \end{bmatrix}$ **12** $\begin{bmatrix} c & d \\ -d & c \end{bmatrix}$

13 $\begin{bmatrix} 3 & 1 & -2 \\ 4 & 2 & 5 \\ -6 & 3 & -1 \end{bmatrix}$ **14** $\begin{bmatrix} 2 & -5 & 1 \\ -3 & 1 & 6 \\ 4 & -2 & 3 \end{bmatrix}$

15 $\begin{bmatrix} -5 & 4 & 1 \\ 3 & -2 & 7 \\ 2 & 0 & 6 \end{bmatrix}$ **16** $\begin{bmatrix} 2 & 7 & -3 \\ 1 & 0 & 4 \\ 4 & -1 & -2 \end{bmatrix}$

17 $\begin{bmatrix} 3 & -1 & 2 & 0 \\ 4 & 0 & -3 & 5 \\ 0 & 6 & 0 & 0 \\ 1 & 3 & -4 & 2 \end{bmatrix}$

18 $\begin{bmatrix} 2 & 5 & 1 & 0 \\ -4 & 0 & -3 & 0 \\ 3 & -2 & 1 & 6 \\ -1 & 4 & 2 & 0 \end{bmatrix}$

19 $\begin{bmatrix} 0 & b & 0 & 0 \\ 0 & 0 & c & 0 \\ a & 0 & 0 & 0 \\ 0 & 0 & 0 & d \end{bmatrix}$

20 $\begin{bmatrix} a & u & v & w \\ 0 & b & x & y \\ 0 & 0 & c & z \\ 0 & 0 & 0 & d \end{bmatrix}$

Verify the identities in Exercises 21–28 by expanding each determinant.

21 $\begin{vmatrix} a & b \\ c & d \end{vmatrix} = -\begin{vmatrix} c & d \\ a & b \end{vmatrix}$

22 $\begin{vmatrix} a & b \\ c & d \end{vmatrix} = -\begin{vmatrix} b & a \\ d & c \end{vmatrix}$

23 $\begin{vmatrix} a & kb \\ c & kd \end{vmatrix} = k\begin{vmatrix} a & b \\ c & d \end{vmatrix}$

24 $\begin{vmatrix} a & b \\ kc & kd \end{vmatrix} = k\begin{vmatrix} a & b \\ c & d \end{vmatrix}$

25 $\begin{vmatrix} a & b \\ c & d \end{vmatrix} = \begin{vmatrix} a & b \\ ka + c & kb + d \end{vmatrix}$

26 $\begin{vmatrix} a & b \\ c & d \end{vmatrix} = \begin{vmatrix} a & ka + b \\ c & kc + d \end{vmatrix}$

27 $\begin{vmatrix} a & b \\ c & d \end{vmatrix} + \begin{vmatrix} a & e \\ c & f \end{vmatrix} = \begin{vmatrix} a & b + e \\ c & d + f \end{vmatrix}$

28 $\begin{vmatrix} a & b \\ c & d \end{vmatrix} + \begin{vmatrix} a & b \\ e & f \end{vmatrix} = \begin{vmatrix} a & b \\ c + e & d + f \end{vmatrix}$

29 Prove that if a square matrix A of order 2 has two identical rows or columns, then $|A| = 0$.

30 Repeat Exercise 29 for a matrix of order 3.

In Exercises 31–34 let $I = I_2$ be the identity matrix of order 2, and let $f(x) = |A - xI|$. Find (a) the polynomial $f(x)$, and (b) the zeros of $f(x)$. In the study of matrices, $f(x)$ is called the **characteristic polynomial of** A, and the zeros of $f(x)$ are called the **characteristic values,** or **eigenvalues,** of A.

31 $A = \begin{bmatrix} 1 & 2 \\ 3 & 2 \end{bmatrix}$

32 $A = \begin{bmatrix} 3 & 1 \\ 2 & 2 \end{bmatrix}$

33 $A = \begin{bmatrix} -3 & -2 \\ 2 & 2 \end{bmatrix}$

34 $A = \begin{bmatrix} 2 & -4 \\ -3 & 5 \end{bmatrix}$

In Exercises 35–38 let $I = I_3$ and let $f(x) = |A - xI|$. Find (a) the polynomial $f(x)$, and (b) the zeros of $f(x)$.

35 $A = \begin{bmatrix} 1 & 0 & 0 \\ 1 & 0 & -2 \\ -1 & 1 & -3 \end{bmatrix}$

36 $A = \begin{bmatrix} 2 & 1 & 0 \\ -1 & 0 & 0 \\ 1 & 3 & 2 \end{bmatrix}$

37 $A = \begin{bmatrix} 0 & 2 & -2 \\ -1 & 3 & 1 \\ -3 & 3 & 1 \end{bmatrix}$

38 $A = \begin{bmatrix} 3 & 2 & 2 \\ 1 & 0 & 2 \\ -1 & -1 & 0 \end{bmatrix}$

In Exercises 39–42 express the determinants in the form $ai + bj + ck$, where a, b, and c are real numbers.

39 $\begin{vmatrix} i & j & k \\ 2 & -1 & 6 \\ -3 & 5 & 1 \end{vmatrix}$

40 $\begin{vmatrix} i & j & k \\ 1 & -2 & 3 \\ 2 & 1 & -4 \end{vmatrix}$

41 $\begin{vmatrix} i & j & k \\ 5 & -6 & -1 \\ 3 & 0 & 1 \end{vmatrix}$

42 $\begin{vmatrix} i & j & k \\ 4 & -6 & 2 \\ -2 & 3 & -1 \end{vmatrix}$

8.7 Properties of Determinants

Evaluating a determinant by using the Expansion Theorem stated in Section 8.6 is inefficient for matrices of high order. For example, if a determinant of a matrix of order 10 is expanded by any row, a sum of 10 terms is obtained, and each term contains the determinant of a matrix of order 9, which is a cofactor of the original matrix. If any of the latter determinants are expanded by a row (or column), a sum of 9 terms is obtained, each containing the determinant of a matrix of order 8. Hence, at this stage there are 90 determinants of matrices of order 8 to evaluate! The process could be continued until only determinants of matrices of order 2 remain. Unless many elements of the original matrix are zero, it is an enormous task to carry out all of the computations.

In this section we discuss rules that simplify the process of evaluating determinants. The main use for these rules is to introduce zeros into the determinant. They may also be used to change the determinant to **echelon form,** that is, a form in which the elements below the main diagonal elements are all zero. The transformations on rows stated in the next theorem are the same as the elementary row transformations of a matrix introduced in Section 8.3. However, for determinants we may also employ similar transformations on columns.

THEOREM ON ROW AND COLUMN TRANSFORMATIONS OF A DETERMINANT

Let A be a square matrix of order n.

(i) If a matrix B is obtained from A by interchanging two rows (or columns), then $|B| = -|A|$.

(ii) If B is obtained from A by multiplying every element of one row (or column) of A by a real number k, then $|B| = k|A|$.

(iii) If B is obtained from A by adding to any row (or column) of A, k times another row (or column), where k is a real number, then $|B| = |A|$.

We shall not prove this theorem. When using the theorem, we refer to the rows (or columns) of the *determinant* in the obvious way. For example, property (iii) may be phrased: "Adding the product of k times another row (or column) to any row (or column) of a determinant does not affect the value of the determinant."

Row transformations of determinants will be specified by means of the symbols R_{ij}, kR_i, and $kR_i + R_j$, which were introduced in Section 8.3. Analogous symbols are used for column transformations. For example, $kC_i + C_j$ means "add to the jth column, k times the ith column." The following are illustrations of the preceding theorem, with the reason for each equality stated at the right.

$$\begin{vmatrix} 2 & 0 & 1 \\ 6 & 4 & 3 \\ 0 & 3 & 5 \end{vmatrix} = - \begin{vmatrix} 6 & 4 & 3 \\ 2 & 0 & 1 \\ 0 & 3 & 5 \end{vmatrix} \qquad R_{12}$$

$$\begin{vmatrix} 2 & 0 & 1 \\ 6 & 4 & 3 \\ 0 & 3 & 5 \end{vmatrix} = - \begin{vmatrix} 1 & 0 & 2 \\ 3 & 4 & 6 \\ 5 & 3 & 0 \end{vmatrix} \qquad C_{13}$$

$$\begin{vmatrix} 1 & -3 & 4 \\ 2 & -1 & 0 \\ 3 & 1 & 6 \end{vmatrix} = \begin{vmatrix} 1 & -3 & 4 \\ 0 & 5 & -8 \\ 3 & 1 & 6 \end{vmatrix} \qquad -2R_1 + R_2$$

$$\begin{vmatrix} 1 & -3 & 4 \\ 2 & -1 & 0 \\ 3 & 1 & 6 \end{vmatrix} = \begin{vmatrix} -5 & -3 & 4 \\ 0 & -1 & 0 \\ 5 & 1 & 6 \end{vmatrix} \qquad 2C_2 + C_1$$

THEOREM

> If two rows (or columns) of a square matrix A are identical, then $|A| = 0$.

PROOF If B is the matrix obtained from A by interchanging the two identical rows (or columns), then B and A are the same and, consequently, $|B| = |A|$. However, by (i) of the Theorem on Row and Column Transformations of a Determinant, $|B| = -|A|$ and hence $-|A| = |A|$, which implies that $|A| = 0$. □

EXAMPLE 1

Find $|A|$ if $A = \begin{bmatrix} 2 & 3 & 0 & 4 \\ 0 & 5 & -1 & 6 \\ 1 & 0 & -2 & 3 \\ -3 & 2 & 0 & -5 \end{bmatrix}$.

SOLUTION We plan to use (iii) of the Theorem on Row and Column Transformations to introduce many zeros in some row or column. To do this, it is convenient to work with an element of the matrix that equals 1, since this enables us to avoid the use of fractions. If 1 is not an element of the original matrix, it is always possible to introduce the number 1 by using (iii) or (ii) of the theorem. In this example 1 appears in row 3, and we proceed as follows, where the reason for each equality is stated at the right.

$$\begin{vmatrix} 2 & 3 & 0 & 4 \\ 0 & 5 & -1 & 6 \\ 1 & 0 & -2 & 3 \\ -3 & 2 & 0 & -5 \end{vmatrix} = \begin{vmatrix} 0 & 3 & 4 & -2 \\ 0 & 5 & -1 & 6 \\ 1 & 0 & -2 & 3 \\ -3 & 2 & 0 & -5 \end{vmatrix} \qquad -2R_3 + R_1$$

$$= \begin{vmatrix} 0 & 3 & 4 & -2 \\ 0 & 5 & -1 & 6 \\ 1 & 0 & -2 & 3 \\ 0 & 2 & -6 & 4 \end{vmatrix} \qquad 3R_3 + R_4$$

$$= (1) \begin{vmatrix} 3 & 4 & -2 \\ 5 & -1 & 6 \\ 2 & -6 & 4 \end{vmatrix} \qquad \text{Expand by the first column.}$$

$$= \begin{vmatrix} 23 & 4 & -2 \\ 0 & -1 & 6 \\ -28 & -6 & 4 \end{vmatrix} \qquad 5C_2 + C_1$$

$$= \begin{vmatrix} 23 & 4 & 22 \\ 0 & -1 & 0 \\ -28 & -6 & -32 \end{vmatrix} \qquad 6C_2 + C_3$$

$$= (-1) \begin{vmatrix} 23 & 22 \\ -28 & -32 \end{vmatrix} \qquad \text{Expand by the second row.}$$

$$= (-1)[(23)(-32) - (-28)(22)] \qquad \text{Definition of determinant}$$

$$= 120$$

■

Part (ii) of the Theorem on Row and Column Transformations is useful for finding factors of determinants. To illustrate, for a determinant of a matrix of order 3 we have the following:

$$\begin{vmatrix} a_{11} & a_{12} & a_{13} \\ ka_{21} & ka_{22} & ka_{23} \\ a_{31} & a_{32} & a_{33} \end{vmatrix} = k \begin{vmatrix} a_{11} & a_{12} & a_{13} \\ a_{21} & a_{22} & a_{23} \\ a_{31} & a_{32} & a_{33} \end{vmatrix}.$$

Similar formulas hold if k is a common factor of the elements of some other row or column. When referring to this manipulation, we often use the phrase "k is a common factor in the row (or column)."

EXAMPLE 2

Find $|A|$ if $A = \begin{bmatrix} 14 & -6 & 4 \\ 4 & -5 & 12 \\ -21 & 9 & -6 \end{bmatrix}$.

SOLUTION

$$|A| = 2\begin{vmatrix} 7 & -3 & 2 \\ 4 & -5 & 12 \\ -21 & 9 & -6 \end{vmatrix} \qquad \text{2 is a common factor in row 1.}$$

$$= (2)(-3)\begin{vmatrix} 7 & -3 & 2 \\ 4 & -5 & 12 \\ 7 & -3 & 2 \end{vmatrix} \qquad -3 \text{ is a common factor in row 3.}$$

$$= 0 \qquad \text{Two rows are identical.}$$

■

EXAMPLE 3 Without expanding, show that $a - b$ is a factor of

$$\begin{vmatrix} 1 & 1 & 1 \\ a & b & c \\ a^2 & b^2 & c^2 \end{vmatrix}.$$

SOLUTION

$$\begin{vmatrix} 1 & 1 & 1 \\ a & b & c \\ a^2 & b^2 & c^2 \end{vmatrix} = \begin{vmatrix} 0 & 1 & 1 \\ a-b & b & c \\ a^2-b^2 & b^2 & c^2 \end{vmatrix} \qquad (-1)C_2 + C_1$$

$$= (a-b)\begin{vmatrix} 0 & 1 & 1 \\ 1 & b & c \\ a+b & b^2 & c^2 \end{vmatrix} \qquad a - b \text{ is a common factor of column 1.}$$

■

Exercises 8.7

Without expanding, explain why the statements in Exercises 1–14 are true.

1 $\begin{vmatrix} 1 & 0 & 1 \\ 0 & 1 & 1 \\ 1 & 1 & 0 \end{vmatrix} = -\begin{vmatrix} 1 & 0 & 1 \\ 1 & 1 & 0 \\ 0 & 1 & 1 \end{vmatrix}$

2 $\begin{vmatrix} 1 & 0 & 1 \\ 0 & 1 & 1 \\ 1 & 1 & 0 \end{vmatrix} = -\begin{vmatrix} 1 & 1 & 0 \\ 0 & 1 & 1 \\ 1 & 0 & 1 \end{vmatrix}$

3 $\begin{vmatrix} 1 & 0 & 1 \\ 2 & 1 & 0 \\ 1 & 1 & 2 \end{vmatrix} = \begin{vmatrix} 1 & 0 & 1 \\ 2 & 1 & 0 \\ 0 & 1 & 1 \end{vmatrix}$

4 $\begin{vmatrix} 1 & 1 & 2 \\ 1 & 0 & 1 \\ 2 & 1 & 1 \end{vmatrix} = \begin{vmatrix} 0 & 1 & 1 \\ 1 & 0 & 1 \\ 2 & 1 & 1 \end{vmatrix}$

5 $\begin{vmatrix} 2 & 4 & 2 \\ 1 & 2 & 4 \\ 2 & 6 & 4 \end{vmatrix} = 4\begin{vmatrix} 1 & 2 & 1 \\ 1 & 2 & 4 \\ 1 & 3 & 2 \end{vmatrix}$

6 $\begin{vmatrix} 2 & 1 & 6 \\ 4 & 3 & 3 \\ 2 & 1 & 3 \end{vmatrix} = 6\begin{vmatrix} 1 & 1 & 2 \\ 2 & 3 & 1 \\ 1 & 1 & 1 \end{vmatrix}$

7 $\begin{vmatrix} 1 & -1 & 2 \\ 1 & 2 & -1 \\ 1 & -1 & 2 \end{vmatrix} = 0$

8 $\begin{vmatrix} 1 & -1 & 1 \\ 0 & 1 & 0 \\ -1 & 0 & -1 \end{vmatrix} = 0$

9 $\begin{vmatrix} 1 & 5 \\ -3 & 2 \end{vmatrix} = -\begin{vmatrix} 1 & 5 \\ 3 & -2 \end{vmatrix}$

10 $\begin{vmatrix} 2 & -2 \\ 1 & 1 \end{vmatrix} = -\begin{vmatrix} -2 & 2 \\ 1 & 1 \end{vmatrix}$

11 $\begin{vmatrix} 0 & 0 & 1 \\ 1 & 0 & 0 \\ 0 & 0 & 2 \end{vmatrix} = 0$

12 $\begin{vmatrix} 1 & 0 & 1 \\ 0 & 0 & 0 \\ 1 & 1 & 0 \end{vmatrix} = 0$

13 $\begin{vmatrix} 1 & -1 & -2 \\ -1 & 2 & 1 \\ 0 & 1 & 1 \end{vmatrix} = \begin{vmatrix} 1 & -1 & 0 \\ -1 & 2 & -1 \\ 0 & 1 & 1 \end{vmatrix}$

14 $\begin{vmatrix} a & 0 & 0 \\ 0 & b & 0 \\ 0 & 0 & c \end{vmatrix} = -\begin{vmatrix} 0 & 0 & a \\ 0 & b & 0 \\ c & 0 & 0 \end{vmatrix}$

In Exercises 15–24 find the determinant of the matrix after introducing zeros, as in Example 1.

15 $\begin{bmatrix} 3 & 1 & 0 \\ -2 & 0 & 1 \\ 1 & 3 & -1 \end{bmatrix}$

16 $\begin{bmatrix} -3 & 0 & 4 \\ 1 & 2 & 0 \\ 4 & 1 & -1 \end{bmatrix}$

17 $\begin{bmatrix} 5 & 4 & 3 \\ -3 & 2 & 1 \\ 0 & 7 & -2 \end{bmatrix}$

18 $\begin{bmatrix} 0 & 2 & -6 \\ 5 & 1 & -3 \\ 6 & -2 & 5 \end{bmatrix}$

19 $\begin{bmatrix} 2 & 2 & -3 \\ 3 & 6 & 9 \\ -2 & 5 & 4 \end{bmatrix}$

20 $\begin{bmatrix} 3 & 8 & 5 \\ 5 & 3 & -6 \\ 2 & 4 & -2 \end{bmatrix}$

21 $\begin{bmatrix} 3 & 1 & -2 & 2 \\ 2 & 0 & 1 & 4 \\ 0 & 1 & 3 & 5 \\ -1 & 2 & 0 & -3 \end{bmatrix}$

22 $\begin{bmatrix} 3 & 2 & 0 & 4 \\ -2 & 0 & 5 & 0 \\ 4 & -3 & 1 & 6 \\ 2 & -1 & 2 & 0 \end{bmatrix}$

23 $\begin{bmatrix} 2 & -2 & 0 & 0 & -3 \\ 3 & 0 & 3 & 2 & -1 \\ 0 & 1 & -2 & 0 & 2 \\ -1 & 2 & 0 & 3 & 0 \\ 0 & 4 & 1 & 0 & 0 \end{bmatrix}$

24 $\begin{bmatrix} 2 & 0 & -1 & 0 & 2 \\ 1 & 3 & 0 & 0 & 1 \\ 0 & 4 & 3 & 0 & -1 \\ -1 & 2 & 0 & -2 & 0 \\ 0 & 1 & 5 & 0 & -4 \end{bmatrix}$

25 Prove that

$$\begin{vmatrix} 1 & 1 & 1 \\ a & b & c \\ a^2 & b^2 & c^2 \end{vmatrix} = (a-b)(b-c)(c-a).$$

(*Hint:* See Example 3.)

26 Prove that

$$\begin{vmatrix} 1 & 1 & 1 \\ a & b & c \\ a^3 & b^3 & c^3 \end{vmatrix} = (a-b)(b-c)(c-a)(a+b+c).$$

27 If A is a matrix of order 4 of the form

$$A = \begin{bmatrix} a_{11} & a_{12} & a_{13} & a_{14} \\ 0 & a_{22} & a_{23} & a_{24} \\ 0 & 0 & a_{33} & a_{34} \\ 0 & 0 & 0 & a_{44} \end{bmatrix}$$

show that $|A| = a_{11}a_{22}a_{33}a_{44}$.

28 If

$$A = \begin{bmatrix} a & b & 0 & 0 \\ c & d & 0 & 0 \\ 0 & 0 & e & f \\ 0 & 0 & g & h \end{bmatrix}$$

prove that

$$|A| = \begin{vmatrix} a & b \\ c & d \end{vmatrix} \begin{vmatrix} e & f \\ g & h \end{vmatrix}.$$

29 If $A = (a_{ij})$ and $B = (b_{ij})$ are arbitrary square matrices of order 2, prove that $|AB| = |A||B|$.

30 If $A = (a_{ij})$ is a square matrix of order n and k is any real number, prove that $|kA| = k^n|A|$. (*Hint:* Use (ii) of the Theorem on Row and Column Transformations of a Determinant.)

31 Use properties of determinants to show that

$$\begin{vmatrix} x & y & 1 \\ x_1 & y_1 & 1 \\ x_2 & y_2 & 1 \end{vmatrix} = 0$$

is an equation of a line through the points (x_1, y_1) and (x_2, y_2).

32 Use properties of determinants to show that

$$\begin{vmatrix} x^2 + y^2 & x & y & 1 \\ x_1^2 + y_1^2 & x_1 & y_1 & 1 \\ x_2^2 + y_2^2 & x_2 & y_2 & 1 \\ x_3^2 + y_3^2 & x_3 & y_3 & 1 \end{vmatrix} = 0$$

is an equation of a circle through the three points (x_1, y_1), (x_2, y_2), and (x_3, y_3).

8.8 Cramer's Rule

Determinants arise in the study of solutions of systems of linear equations. To illustrate, let us consider the following case of two linear equations in two variables x and y:

$$\begin{cases} a_{11}x + a_{12}y = k_1 \\ a_{21}x + a_{22}y = k_2 \end{cases}$$

where at least one nonzero coefficient appears in each equation. We may as well assume that $a_{11} \neq 0$, for otherwise $a_{12} \neq 0$, and we could regard y as the "first" variable instead of x. We shall use elementary row transformations to obtain the matrix of an equivalent system as follows:

$$\begin{bmatrix} a_{11} & a_{12} & k_1 \\ a_{21} & a_{22} & k_2 \end{bmatrix} \xrightarrow{-\frac{a_{21}}{a_{11}}\text{R}_1 + \text{R}_2} \begin{bmatrix} a_{11} & a_{12} & k_1 \\ 0 & a_{22} - \left(\dfrac{a_{12}a_{21}}{a_{11}}\right) & k_2 - \left(\dfrac{a_{21}k_1}{a_{11}}\right) \end{bmatrix}$$

$$\xrightarrow{a_{11}\text{R}_2} \begin{bmatrix} a_{11} & a_{12} & k_1 \\ 0 & (a_{11}a_{22} - a_{12}a_{21}) & (a_{11}k_2 - a_{21}k_1) \end{bmatrix}$$

Thus, the given system is equivalent to

$$\begin{cases} a_{11}x + a_{12}y = k_1 \\ (a_{11}a_{22} - a_{12}a_{21})y = a_{11}k_2 - a_{21}k_1 \end{cases}$$

which may also be written

$$\begin{cases} a_{11}x + a_{12}y = k_1 \\ \begin{vmatrix} a_{11} & a_{12} \\ a_{21} & a_{22} \end{vmatrix} y = \begin{vmatrix} a_{11} & k_1 \\ a_{21} & k_2 \end{vmatrix}. \end{cases}$$

If $\begin{vmatrix} a_{11} & a_{12} \\ a_{21} & a_{22} \end{vmatrix} \neq 0$, we can solve the second equation for y, obtaining

$$y = \frac{\begin{vmatrix} a_{11} & k_1 \\ a_{21} & k_2 \end{vmatrix}}{\begin{vmatrix} a_{11} & a_{12} \\ a_{21} & a_{22} \end{vmatrix}}$$

The corresponding value for x may be found by substituting for y in the first equation. It can be shown that this leads to

$$x = \frac{\begin{vmatrix} k_1 & a_{12} \\ k_2 & a_{22} \end{vmatrix}}{\begin{vmatrix} a_{11} & a_{12} \\ a_{21} & a_{22} \end{vmatrix}}$$

This proves that *if the determinant of the coefficient matrix of a system of two linear equations in two variables is not zero, then the system has a unique solution.* The last two formulas for x and y as quotients of certain determinants constitute what is known as **Cramer's Rule.**

There is an easy way to remember Cramer's Rule. Let

$$D = \begin{bmatrix} a_{11} & a_{12} \\ a_{21} & a_{22} \end{bmatrix}$$

be the coefficient matrix of the system and let D_x denote the matrix obtained from D by replacing the coefficients a_{11}, a_{21} of x by the numbers k_1, k_2, respectively. Similarly, let D_y denote the matrix obtained from D by replacing the coefficients a_{12}, a_{22} of y by the numbers k_1, k_2, respectively. Thus,

$$D_x = \begin{bmatrix} k_1 & a_{12} \\ k_2 & a_{22} \end{bmatrix}, \qquad D_y = \begin{bmatrix} a_{11} & k_1 \\ a_{21} & k_2 \end{bmatrix}.$$

If $|D| \neq 0$, the solution (x, y) is given by

CRAMER'S RULE

$$x = \frac{|D_x|}{|D|}, \qquad y = \frac{|D_y|}{|D|}.$$

EXAMPLE 1 Use Cramer's Rule to solve the system

$$\begin{cases} 2x - 3y = -4 \\ 5x + 7y = 1. \end{cases}$$

SOLUTION The determinant of the coefficient matrix is

$$|D| = \begin{vmatrix} 2 & -3 \\ 5 & 7 \end{vmatrix} = 29.$$

Using the notation introduced previously,

$$|D_x| = \begin{vmatrix} -4 & -3 \\ 1 & 7 \end{vmatrix} = -25, \qquad |D_y| = \begin{vmatrix} 2 & -4 \\ 5 & 1 \end{vmatrix} = 22.$$

Hence,

$$x = \frac{|D_x|}{|D|} = \frac{-25}{29}, \qquad y = \frac{|D_y|}{|D|} = \frac{22}{29}.$$

Thus, the system has the unique solution $(-\frac{25}{29}, \frac{22}{29})$. ■

Cramer's Rule can be extended to systems of n linear equations in n variables $x_1, x_2, \ldots x_n$, where each equation has the form

$$a_1x_1 + a_2x_2 + \cdots + a_nx_n = a.$$

To solve such a system, let D denote the coefficient matrix and let D_{x_i} denote the matrix obtained by replacing the coefficients of x_i in D by the column of numbers $k_1, \ldots, k_n$ that appears to the right of the equal signs in the system. It can be shown that if $|D| \neq 0$, then the system has the following unique solution.

CRAMER'S RULE (GENERAL FORM)

$$x_1 = \frac{|D_{x_1}|}{|D|}, \qquad x_2 = \frac{|D_{x_2}|}{|D|}, \qquad \ldots, \qquad x_n = \frac{|D_{x_n}|}{|d|}.$$

EXAMPLE 2 Use Cramer's Rule to solve the system

$$\begin{cases} x - 2z = 3 \\ - y + 3z = 1 \\ 2x + 5z = 0. \end{cases}$$

SOLUTION We shall merely list the various determinants, leaving the reader to check the answers:

$$|D| = \begin{vmatrix} 1 & 0 & -2 \\ 0 & -1 & 3 \\ 2 & 0 & 5 \end{vmatrix} = -9, \qquad |D_x| = \begin{vmatrix} 3 & 0 & -2 \\ 1 & -1 & 3 \\ 0 & 0 & 5 \end{vmatrix} = -15,$$

$$|D_y| = \begin{vmatrix} 1 & 3 & -2 \\ 0 & 1 & 3 \\ 2 & 0 & 5 \end{vmatrix} = 27, \qquad |D_z| = \begin{vmatrix} 1 & 0 & 3 \\ 0 & -1 & 1 \\ 2 & 0 & 0 \end{vmatrix} = 6.$$

By Cramer's Rule, the solution is

$$x = \frac{|D_x|}{|D|} = \frac{-15}{-9} = \frac{5}{3}, \quad y = \frac{|D_y|}{|D|} = \frac{27}{-9} = -3, \quad z = \frac{|D_z|}{|D|} = \frac{6}{-9} = -\frac{2}{3}.$$

■

Cramer's Rule is an inefficient method to apply if there are a large number of equations, since many determinants of matrices of high order must be evaluated. Note also that Cramer's Rule cannot be used directly if $|D| = 0$ or if the number of equations is not the same as the number of variables. In general, the matrix method is far superior to Cramer's Rule.

Exercises 8.8

1–18 Use Cramer's Rule to solve the systems in Exercises 1–18 of Section 8.2.

19–26 Use Cramer's Rule to solve the systems in Exercises 1–8 of Section 8.3.

27–30 Use Cramer's Rule to solve the systems in Exercises 17, 18, 21, and 22 of Section 8.3.

8.9 Systems of Inequalities

The discussion of inequalities in Chapter 1 was restricted to inequalities in one variable. Inequalities in several variables can also be considered. For example, expressions of the form

$$3x + y < 5y^2 + 1,$$

$$2x^2 \geq 4 - 3y,$$

and so on are called **inequalities in x and y.** A **solution** of an inequality in x and y is defined as an ordered pair (a, b) that produces a true statement if a and b are substituted for x and y, respectively. The **graph of an inequality** is the graph of all the solutions. Two inequalities are **equivalent** if they have exactly the same solutions. An inequality in x and y can often be simplified by adding an expression in x and y to both sides or by multiplying both sides by some expression (provided care is taken with regard to signs). Similar remarks apply to inequalities in more than two variables. We shall restrict our discussion, however, to the case of inequalities in two variables.

EXAMPLE 1 Find the solutions and sketch the graph of the inequality

$$3x - 3 < 5x - y.$$

SOLUTION The inequalities in the following list are equivalent:

$$3x - 3 < 5x - y$$
$$y + 3x - 3 < 5x$$
$$y < 5x - (3x - 3)$$
$$y < 2x + 3.$$

Hence, the solutions consist of all ordered pairs (x, y) such that $y < 2x + 3$. It is convenient to denote the solutions as follows:

$$\{(x, y): y < 2x + 3\}.$$

FIGURE 8.6

There is a close relationship between the graph of the inequality $y < 2x + 3$ and the graph of the equation $y = 2x + 3$. The graph of the equation is the line sketched in Figure 8.6. For each real number a, the point on the line with x-coordinate a has coordinates $(a, 2a + 3)$. A point $P(a, b)$ belongs to the graph of the *inequality* if and only if $b < 2a + 3$; that is, if and only if the point $P(a, b)$ lies directly below the point with coordinates $(a, 2a + 3)$ as shown in Figure 8.6. It follows that the graph of the inequality $y < 2x + 3$ consists of all points that lie below the line $y = 2x + 3$. In Figure 8.7 we have shaded a portion of the graph of the inequality. Dashes used for the line indicate that it is not part of the graph. ■

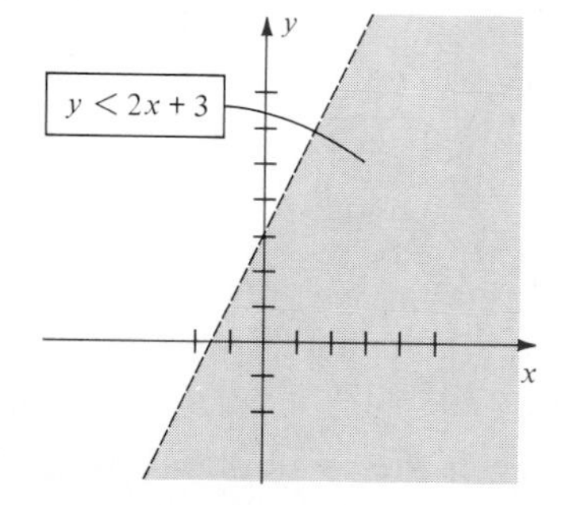

A region of the type shown in Figure 8.7 is called a **half-plane.** More precisely, if the line is *not* included, we refer to the region as an **open half-plane.** If the line *is* included, as would be the case for the graph of the inequality $y \leq 2x + 3$, then the region is called a **closed half-plane.**

By an argument similar to that used in Example 1, it can be shown that the graph of the inequality $y > 2x + 3$ is the open half-plane that lies *above* the line $y = 2x + 3$.

If an inequality involves only polynomials of the first degree in x and y, as was the case in Example 1, it is called a **linear inequality.**

The procedure used in Example 1 can be generalized to inequalities of the form $y < f(x)$ for any function f. Specifically, the following theorem is true.

THEOREM

Let f be a function.

(i) The graph of the inequality $y < f(x)$ is the set of points that lie *below* the graph of the equation $y = f(x)$.

(ii) The graph of $y > f(x)$ is the set of points that lie *above* the graph of $y = f(x)$.

EXAMPLE 2 Find the solutions and sketch the graph of the inequality

$$x(x + 1) - 2y > 3(x - y).$$

FIGURE 8.8

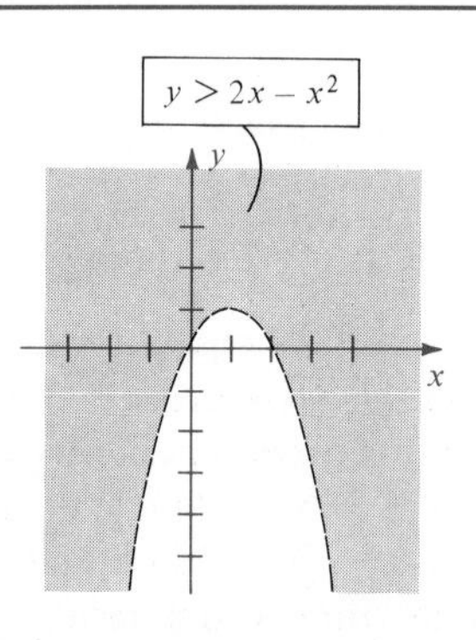

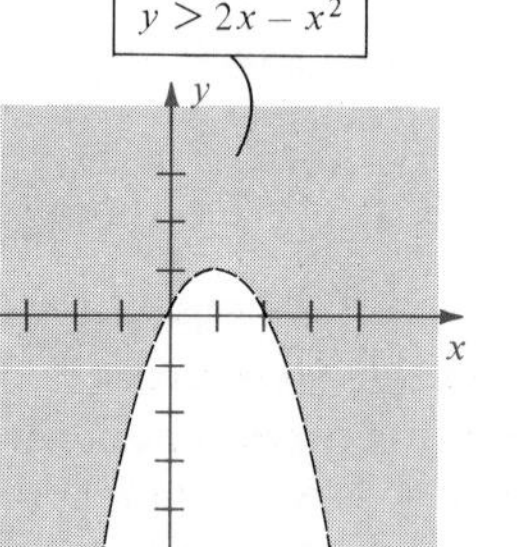

SOLUTION The inequality is equivalent to

$$x^2 + x - 2y > 3x - 3y.$$

Adding $3y - x^2 - x$ to both sides, we obtain

$$y > 2x - x^2.$$

Hence, the solutions are $\{(x, y): y > 2x - x^2\}$.

To find the graph of $y > 2x - x^2$, we begin by sketching the graph of $y = 2x - x^2$ (a parabola) with dashes as illustrated in Figure 8.8. By (ii) of the preceding theorem, the graph is the region above the parabola, as indicated by the shaded portion of the figure. ■

The following can also be proved.

THEOREM

Let g be a function.

(i) The graph of the inequality $x < g(y)$ is the set of points to the *left* of the graph of the equation $x = g(y)$.

(ii) The graph of $x > g(y)$ is the set of points to the *right* of the graph of $x = g(y)$.

FIGURE 8.9

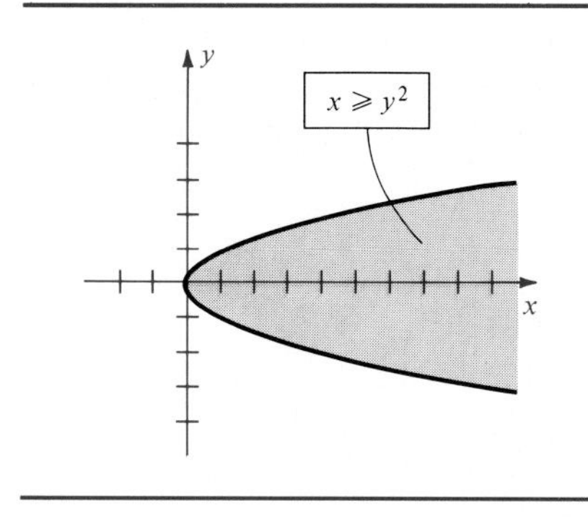

EXAMPLE 3 Sketch the graph of $x \geq y^2$.

SOLUTION The graph of the equation $x = y^2$ is a parabola that is symmetric to the x-axis. By the preceding theorem, the graph of the inequality consists of all points on the parabola, together with the points in the region to the right of the parabola (see Figure 8.9). ■

It is sometimes necessary to work simultaneously with several inequalities in two variables. In this case we refer to the inequalities as a **system of inequalities.** The **solutions of a system** of inequalities are, by definition, the solutions that are common to all inequalities in the system. It should be clear how to define **equivalent systems** and the **graph of a system** of inequalities. The following examples illustrate a method for solving systems of inequalities.

EXAMPLE 4 Find the solutions and sketch the graph of the system

$$\begin{cases} x + y \leq 4 \\ 2x - y \leq 4. \end{cases}$$

FIGURE 8.10

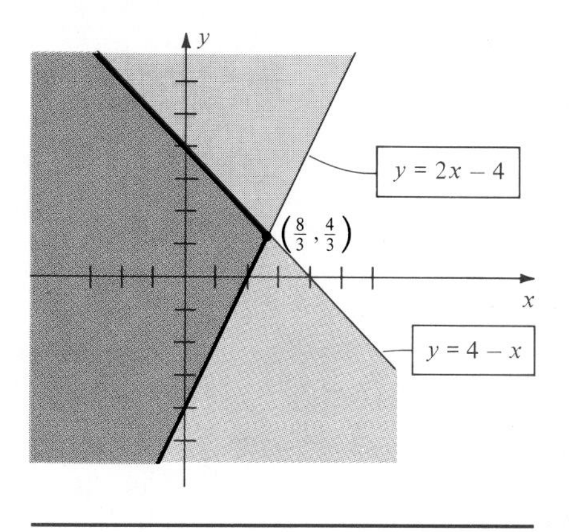

SOLUTION The system is equivalent to

$$\begin{cases} y \leq 4 - x \\ y \geq 2x - 4. \end{cases}$$

We begin by sketching the graphs of the lines $y = 4 - x$ and $y = 2x - 4$. The lines intersect at the point $(\frac{8}{3}, \frac{4}{3})$ shown in Figure 8.10. The graph of $y \leq 4 - x$ includes the points on the graph of $y = 4 - x$ together with the points that lie below this line. The graph of $y \geq 2x - 4$ includes the points on the graph of $y = 2x - 4$ together with the points that lie above this line. A portion of each of these regions is shown in Figure 8.10. The graph of the system consists of the points that are in *both* regions as indicated by the double-shaded portion of the figure. ■

EXAMPLE 5 Sketch the graph of the system

$$\begin{cases} x + y \leq 4 \\ 2x - y \leq 4 \\ x \geq 0 \\ y \geq 0. \end{cases}$$

FIGURE 8.11

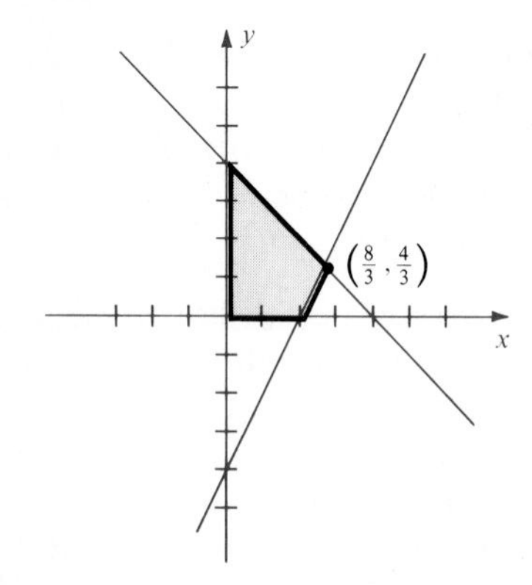

SOLUTION The first two inequalities are the same as those considered in Example 4, and hence the points on the graph of the system must lie within the double-shaded region shown in Figure 8.10. In addition, the third and fourth inequalities in the system tell us that the points must lie in the first quadrant or on its boundaries. This gives us the region shown in Figure 8.11. ■

EXAMPLE 6 Sketch the graph of the system

$$\begin{cases} x^2 + y^2 \le 1 \\ (x-1)^2 + y^2 \le 1. \end{cases}$$

SOLUTION The graph of the equation $x^2 + y^2 = 1$ is a unit circle with center at the origin, and the graph of $(x-1)^2 + y^2 = 1$ is a unit circle with center at the point $C(1, 0)$. To find the points of intersection of the two circles, let us solve the equation $x^2 + y^2 = 1$ for y^2, obtaining $y^2 = 1 - x^2$. Substituting for y^2 in $(x-1)^2 + y^2 = 1$ leads to the following equations:

FIGURE 8.12

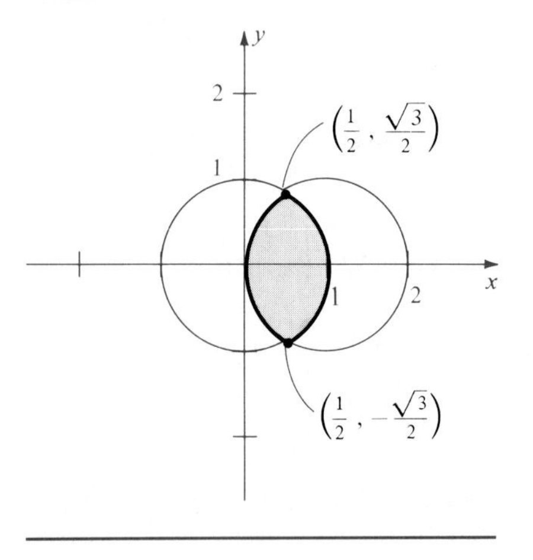

$$\begin{aligned} (x-1)^2 + (1 - x^2) &= 1 \\ x^2 - 2x + 1 + 1 - x^2 &= 1 \\ -2x &= -1 \\ x &= \tfrac{1}{2}. \end{aligned}$$

The corresponding values for y are given by

$$y^2 = 1 - x^2 = 1 - (\tfrac{1}{2})^2 = \tfrac{3}{4}$$

and hence $y = \pm\sqrt{3}/2$. Thus, the points of intersection are $(\frac{1}{2}, \sqrt{3}/2)$ and $(\frac{1}{2}, -\sqrt{3}/2)$ as shown in Figure 8.12. By the Distance Formula, the graphs of the inequalities are the regions within and on the two circles. The graph of the system consists of the points common to both regions, as indicated by the shaded portion of the figure. ■

EXAMPLE 7 The manager of a baseball team wishes to buy bats and balls costing \$12 and \$3 each, respectively. At least five bats and ten balls are required, and the total cost is not to exceed \$180. Find a system of inequalities that describes all possibilities and sketch the graph.

FIGURE 8.13

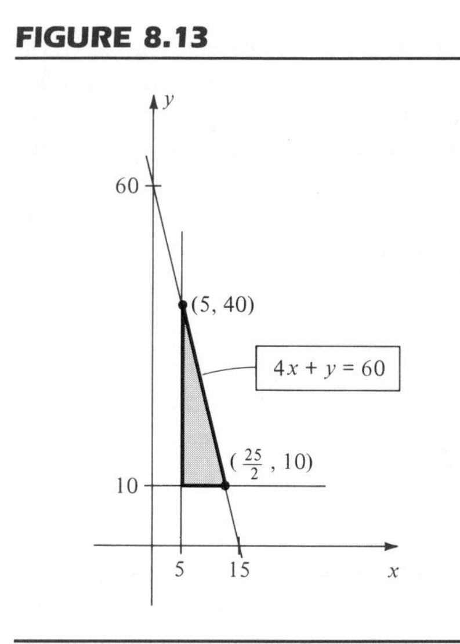

SOLUTION We begin by introducing the following variables:

$$x = \text{number of bats},$$
$$y = \text{number of balls}.$$

Thus, the bats will cost $12x$ dollars and the balls will cost $3y$ dollars. Since the total cost is not to exceed \$180, we must have

$$12x + 3y \leq 180 \quad \text{or} \quad 4x + y \leq 60.$$

The other restrictions are

$$x \geq 5 \quad \text{and} \quad y \geq 10.$$

The graph of the system is sketched in Figure 8.13. ■

Exercises 8.9

In Exercises 1–10 find the solutions and sketch the graph of the inequality.

1 $3x - 2y < 6$

2 $4x + 3y < 12$

3 $2x + 3y \geq 2y + 1$

4 $2x - y > 3$

5 $y + 2 < x^2$

6 $y^2 - x \leq 0$

7 $x^2 + 1 \leq y$

8 $y - x^3 < 1$

9 $yx^2 \geq 1$

10 $x^2 + 4 \geq y$

In Exercises 11–24 sketch the graph of the system of inequalities.

11 $\begin{cases} 3x + y < 3 \\ 4 - y < 2x \end{cases}$

12 $\begin{cases} y + 2 < 2x \\ y - x > 4 \end{cases}$

13 $\begin{cases} y - x < 0 \\ 2x + 5y < 10 \end{cases}$

14 $\begin{cases} 2y - x \leq 4 \\ 3y + 2x < 6 \end{cases}$

15 $\begin{cases} 3x + y \leq 6 \\ y - 2x \geq 1 \\ x \geq -2 \\ y \leq 4 \end{cases}$

16 $\begin{cases} 3x - 4y \geq 12 \\ x - 2y \leq 2 \\ x \geq 9 \\ y \leq 5 \end{cases}$

17 $\begin{cases} x^2 + y^2 \leq 4 \\ x + y \geq 1 \end{cases}$

18 $\begin{cases} x^2 + y^2 > 1 \\ x^2 + y^2 < 4 \end{cases}$

19 $\begin{cases} x^2 \leq 1 - y \\ x \geq 1 + y \end{cases}$

20 $\begin{cases} x - y^2 < 0 \\ x + y^2 > 0 \end{cases}$

21 $\begin{cases} y < 3^x \\ y > 2^x \\ x \geq 0 \end{cases}$

22 $\begin{cases} y \geq \log x \\ y - x \leq 1 \\ x \geq 1 \end{cases}$

23 $\begin{cases} y \leq \log x \\ y + x \geq 1 \\ x \leq 10 \end{cases}$

24 $\begin{cases} y \leq 3^{-x} \\ y \geq 2^{-x} \\ y < 9 \end{cases}$

25 A store sells two brands of television sets. Customer demand indicates that it is necessary to stock at least twice as many sets of brand A as of brand B. It is also necessary to have on hand at least 20 of brand A and 10 of brand B. If there is room for not more than 100 sets in the store, find a system of inequalities that describes all possibilities, and sketch the graph.

26 An auditorium contains 600 seats. For a certain event it is planned to charge \$8.00 for some seats and \$5.00 for others. At least 225 tickets are to be sold for \$5.00, and total sales of more than \$3000 are desired. Find a system of inequalities that describes all possibilities, and sketch the graph.

27 A woman wishes to invest \$15,000 in two different savings accounts. She also wants to have at least \$2000 in each account, with the amount in one account being at least three times that in the other. Find a system of inequalities that describes all possibilities, and sketch the graph.

28 The manager of a college bookstore stocks two types of notebooks, the first wholesaling for 55 cents and the second for 85 cents. If the maximum amount to be spent is \$600 and if an inventory of at least 300 of the 85-cent variety and 400 of the 55-cent variety is desired, find a system of inequalities that describes all possibilities and sketch the graph.

29 An aerosol can is to be constructed in the shape of a circular cylinder with a small cone on the top. The total height of the can is to be no more than 9 inches and the cylinder must contain at least 75% of the total volume. In addition, the height of the conical top must be at least 1 inch. Find and graph the system of inequalities that describes the possible relationships between the height of the cylinder and the height of the cone.

30 A stained-glass window is to be constructed in the form of a rectangle surmounted by a semicircle. The total height of the window can be no more than 6 feet, and the area of the rectangular part must be at least twice the area of the semicircle. In addition, the diameter of the semicircle must be at least 2 feet. Find and graph the system of inequalities that describes the possibilities for the length and width of the rectangle.

31 A nuclear power plant will be constructed to serve the power needs of cities A and B. City B is 100 miles due east of A. The state has promised that the plant will be at least 60 miles from each city. It is not possible, however, to locate the plant south of either city because of rough terrain, and the plant must be within 100 miles of both A and B. Assuming A is at the origin, find and graph the system of inequalities that describes all possible locations for the plant.

32 A man has a rectangular back yard that is 50 feet wide and 60 feet deep. He plans to construct a pool area and a patio area as shown in the figure, and can spend at most \$12,000 on the project. The patio area must be at least as large as the pool area. The pool area will cost \$5 per square foot and the patio will cost \$3 per square foot. Find and graph the system of inequalities that describes the possibilities for the width of the patio and pool areas.

FIGURE FOR EXERCISE 32

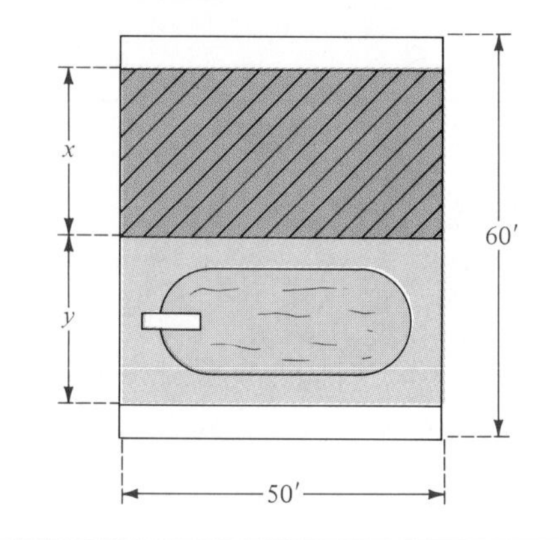

8.10 Linear Programming

Certain applications require finding particular solutions of systems of inequalities. A typical problem consists of finding maximum and minimum values of expressions involving variables that are subject to various constraints. If all the expressions and inequalities are linear in the variables, then a technique called **linear programming** may be used to help solve such problems. This technique has become very important in businesses that require decisions to be made concerning the best use of stock, parts, or manufacturing processes. Usually the objective of management is to maximize profit or to minimize cost. Since there are often many choices, it may be extremely difficult to arrive at a correct decision. A mathematical theory such as that afforded by linear programming can simplify the task considerably. The logical development of the theorems and methods that are needed would take us beyond the objectives of this text. We shall, therefore, limit ourselves to several examples.

EXAMPLE 1 A manufacturer of a certain product has two warehouses, W_1 and W_2. There are 80 units of the product stored at W_1 and 70 units at W_2. Two customers, A and B, order 35 units and 60 units, respectively. The shipping cost from each warehouse to A and B is determined according to the following table. How should the order be filled to minimize the total shipping cost?

Warehouse	Customer	Shipping cost per unit
W_1	A	\$ 8
W_1	B	12
W_2	A	10
W_2	B	13

SOLUTION Let

$$x = \text{number of units sent to } A \text{ from } W_1,$$

$$y = \text{number of units sent to } B \text{ from } W_1.$$

FIGURE 8.14

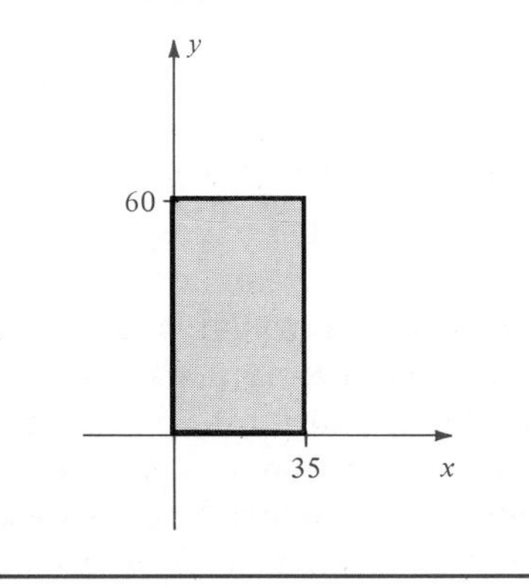

Thus, to fill the orders we must have

$$35 - x = \text{number sent to } A \text{ from } W_2,$$

$$60 - y = \text{number sent to } B \text{ from } W_2.$$

We wish to determine values for x and y that make the total shipping costs minimal. Since x and y are between 35 and 60, respectively, the pair (x, y) must be a solution of the following system of inequalities:

$$\begin{cases} 0 \leq x \leq 35 \\ 0 \leq y \leq 60. \end{cases}$$

The graph of this system is the rectangular region shown in Figure 8.14.

FIGURE 8.15

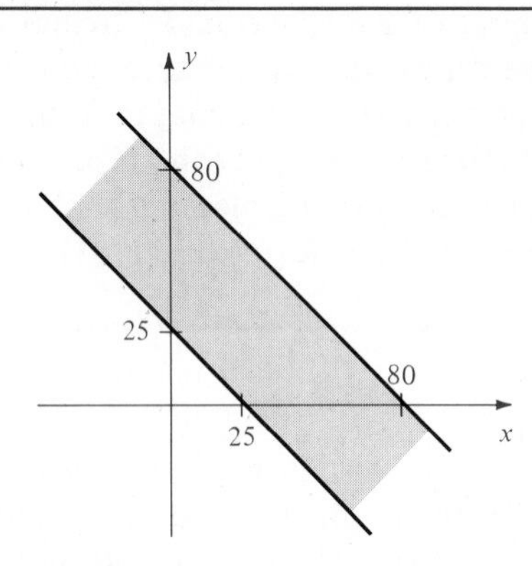

FIGURE 8.16

There are further constraints on x and y that make it possible to reduce the size of the region in Figure 8.14. Since the total number of units shipped from W_1 cannot exceed 80 and the total shipped from W_2 cannot exceed 70, the pair (x, y) must also be a solution of the system

$$\begin{cases} x + y \leq 80 \\ (35 - x) + (60 - y) \leq 70. \end{cases}$$

This system is equivalent to

$$\begin{cases} x + y \leq 80 \\ x + y \geq 25. \end{cases}$$

The graph of this system is the region between the parallel lines $x + y = 80$ and $x + y = 25$ (see Figure 8.15). Since the pair (x, y) that we seek must be a solution of this system, and also of the system $0 \leq x \leq 35$, $0 \leq y \leq 60$, the corresponding point must lie in the region shown in Figure 8.16.

Let C denote the total cost (in dollars) of shipping the merchandise to A and B. We see from the table listing shipping costs that the following are true:

$$\text{Cost of shipping 35 units to } A = 8x + 10(35 - x),$$

$$\text{Cost of shipping 60 units to } B = 12y + 13(60 - y).$$

Hence, the total cost is

$$\begin{aligned} C &= 8x + 10(35 - x) + 12y + 13(60 - y) \\ &= (8x + 350 - 10x) + (12y + 780 - 13y) \end{aligned}$$

or

$$C = 1130 - 2x - y.$$

For each point (x, y) of the region shown in Figure 8.16 there corresponds a value for C. For example, at (20, 40)

$$C = 1130 - 40 - 40 = 1050,$$

and at (10, 50),

$$C = 1130 - 20 - 50 = 1060.$$

Since x and y are integers, there is only a finite number of possible values for C. By checking each possibility, we could find the pair (x, y) that produces the smallest cost. However, since there is a very large number of pairs, the task of checking each one would be very tedious. This is where the theory developed in linear programming is helpful. It can be shown that if we are interested in the value C of a linear expression $ax + by + c$, and if each pair (x, y) is a solution of a system of linear inequalities, and hence corresponds to a point that is common to several half-planes, then C takes on its maximum or minimum value at a point of intersection of the lines that determine the half-planes. This means that to determine the minimum (or maximum) value of C we need only check the points (0, 25), (0, 60), (20, 60), (35, 45), (35, 0), and (25, 0) shown in Figure 8.16. The values are displayed in the following table.

Point	$1130 - 2x - y = C$
(0. 25)	$1130 - 2(0) - 25 = 1105$
(0, 60)	$1130 - 2(0) - 60 = 1070$
(20, 60)	$1130 - 2(20) - 60 = 1030$
(35, 45)	$1130 - 2(35) - 45 = 1015$
(35, 0)	$1130 - 2(35) - 0 = 1060$
(25, 0)	$1130 - 2(25) - 0 = 1080$

According to our remarks, the minimal shipping cost $1015 occurs if $x = 35$ and $y = 45$. This means that the manufacturer should ship all of the units to A from W_1 and none from W_2. In addition, the manufacturer should ship 45 units to B from W_1 and 15 units to B from W_2. Note that the *maximum* shipping cost will occur if $x = 0$ and $y = 25$, that is, if all 35 units are shipped to A from W_2 and if B receives 25 units from W_1 and 35 units from W_2. ■

The preceding example demonstrates how linear programming can be used to minimize the cost in a certain situation. The next example illustrates maximization of profit.

EXAMPLE 2 A firm manufactures two products X and Y. For each product it is necessary to use three different machines, A, B, and C. To manufacture one unit of product X, machine A must be used for 3 hours, machine B for 1 hour, and machine C for 1 hour. To manufacture one unit of product Y requires 2 hours on A, 2 hours on B, and 1 hour on C. The profit on product X is $500 per unit and the profit on product Y is $350 per unit. Machine A is available for a total of 24 hours per day; however, B can only be used for 16 hours and C for 9 hours. Assuming the machines are available when needed (subject to the noted total hour restrictions), determine the number of units of each product that should be manufactured each day in order to maximize the profit.

SOLUTION The following table summarizes the data given in the statement of the problem.

Machine	Hours required for 1 unit of X	Hours required for one unit of Y	Hours available
A	3	2	24
B	1	2	16
C	1	1	9

Let x and y denote the number of units of products X and Y, respectively, to be produced per day. Since each unit of product X requires 3 hours on machine A, x units require $3x$ hours. Similarly, since each unit of product Y requires 2 hours on A, y units require $2y$ hours. Hence, the total number of hours per day that machine A must be used is $3x + 2y$. Since A can be used for at most 24 hours per day,

$$3x + 2y \leq 24.$$

FIGURE 8.17

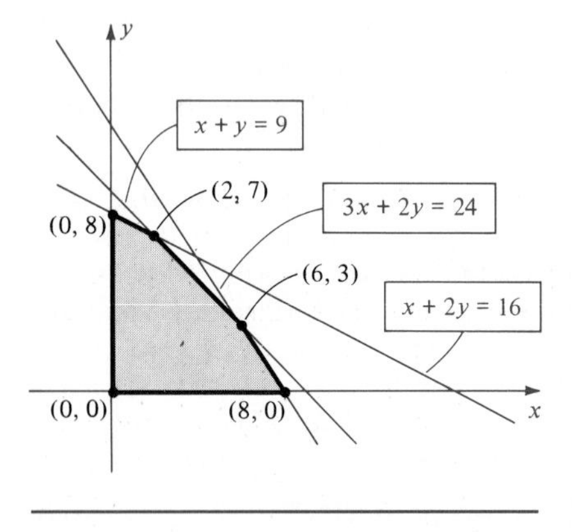

Using the same type of reasoning on rows two and three of the table, we see that

$$x + 2y \leq 16$$

$$x + \ \ y \leq \ 9.$$

This system of three linear inequalities together with the obvious inequalities

$$x \geq 0, \qquad y \geq 0$$

states, in mathematical form, the constraints that occur in the manufacturing process. The graph of the preceding system of five linear inequalities is sketched in Figure 8.17.

The points shown in the figure are found by solving systems of linear equations. Specifically, (6, 3) is the solution of the system $3x + 2y = 24$, $x + y = 9$, and (2, 7) is the solution of the system $x + 2y = 16$, $x + y = 9$.

Since the production of each unit of product X yields a profit of \$500, and each unit of product Y yields a profit of \$350, the profit P obtained by producing x units of X together with y units of Y is

$$P = 500x + 350y.$$

The maximum value of P must occur at one of the points shown in Figure 8.17. The values of P at all the points are shown in the following table.

(x, y)	$500x + 350y = P$
$(0, 8)$	$500(0) + 350(8) = 2800$
$(2, 7)$	$500(2) + 350(7) = 3450$
$(6, 3)$	$500(6) + 350(3) = 4050$
$(8, 0)$	$500(8) + 350(0) = 4000$
$(0, 0)$	$500(0) + 350(0) = 0$

We see from the table that a maximum profit of $4050 occurs for a daily production of 6 units of product X and 3 units of product Y. ■

The two illustrations given in this section are elementary problems in linear programming that can be solved by rather crude methods. The much more complicated problems that occur in practice are usually solved by employing matrix techniques that are adapted for solutions by computers.

Exercises 8.10

1 A manufacturer of tennis rackets makes a profit of $15 on each Set Point racket and $8 on each Double Fault racket. To meet dealer demand, daily production of Double Faults should be between 30 and 80, whereas the number of Set Points produced should be between 10 and 30. To maintain high quality, the total number of rackets produced should not exceed 80 per day. How many of each type should be manufactured daily to maximize the profit?

2 A manufacturer of CB radios makes a profit of $25 on a deluxe model and $30 on a standard model. The company wishes to produce at least 80 deluxe models and at least 100 standard models per day. To maintain high quality, the daily production should not exceed 200 radios. How many of each type should be produced daily in order to maximize the profit?

3 Two substances S and T each contain two types of ingredients I and G. One pound of S contains 2 ounces of I and 4 ounces of G. One pound of T contains 2 ounces of I and 6 ounces of G. A manufacturer plans to combine quantities of the two substances to obtain a mixture that contains at least 9 ounces of I and 20 ounces of G. If the cost of S is $3.00 per pound and the cost of T is $4.00 per pound, how much of each substance should be used to keep the cost to a minimum?

4 A stationery company makes two types of notebooks: a deluxe notebook, with subject dividers, that sells for $1.25, and a regular notebook that sells for $0.90. It costs the company $1.00 to produce each deluxe notebook and $0.75 to produce each regular notebook. The company has the facilities to manufacture between 2000 and 3000 deluxe and between 3000 and 6000 regular, but not more than 7000 altogether. How many notebooks of each type should be manufactured to maximize the difference between the selling prices and the costs of production?

5 Refer to Example 1 of this section. If the shipping costs are $12 per unit from W_1 to A, $10 per unit from W_2 to A, $16 per unit from W_1 to B, and $12 per unit from W_2 to B, determine how the order should be filled to minimize shipping costs.

6 A coffee company purchases mixed lots of coffee beans and then grades them into premium, regular, and unusable beans. The company needs at least 280 tons of premium-grade and 200 tons of regular-grade coffee beans. The company can purchase ungraded coffee from

two suppliers in any amount desired. Samples from the two suppliers contain the following percentages of premium, regular, and unusable beans:

Supplier	Premium	Regular	Unusable
A	20%	50%	30%
B	40%	20%	40%

If supplier *A* charges \$125 per ton and *B* charges \$200 per ton, how much should the company purchase from each supplier to fulfill its needs at minimum cost?

7 A farmer, in the business of growing fodder for livestock, has 100 acres available for planting alfalfa and corn. The cost of seed per acre is \$4 for alfalfa and \$6 for corn. The total cost of labor will amount to \$20 per acre for alfalfa and \$10 per acre for corn. The expected income from alfalfa is \$110 per acre, and from corn, \$150 per acre. If the farmer does not wish to spend more than \$480 for seed and \$1400 for labor, how many acres of each should be planted to obtain the maximum profit?

8 A small firm manufactures bookshelves and desks for microcomputers. For each product it is necessary to use a table saw and a power router. To manufacture each bookshelf, the saw must be used for $\frac{1}{2}$ hour and the router for 1 hour. A desk requires the use of each machine for 2 hours. The profits are \$20 per bookshelf and \$50 per desk. If the saw can be used 8 hours per day, and the router 12 hours per day, how many bookshelves and desks should be manufactured each day to maximize profits?

9 Three substances *X*, *Y*, and *Z* each contain four ingredients *A*, *B*, *C*, and *D*. The percentage of each ingredient and the cost in cents per ounce of each substance are given in the following table.

Substance	*A*	*B*	*C*	*D*	Cost/ ounce
X	20%	10%	25%	45%	25¢
Y	20%	40%	15%	25%	35¢
Z	10%	20%	25%	45%	50¢

If the cost is to be minimum, how many ounces of each substance should be combined to obtain a mixture of 20 ounces containing at least 14% *A*, 16% *B*, and 20% *C*? What combination would make the cost greatest?

10 A man plans to operate a stand at a one-day fair at which he will sell bags of peanuts and bags of candy. He has \$100 available to purchase his stock, which will cost 10¢ per bag of peanuts and 20¢ per bag of candy. He intends to sell the peanuts at 25¢ and the candy at 40¢ per bag. His stand can accommodate up to 500 bags of peanuts and 400 bags of candy. From past experience he knows that he will sell no more than a total of 700 bags. Find the number of bags of each that he should have available in order to maximize his profit. What is the maximum profit?

11 A small community wishes to purchase vans and small buses for its public transportation system. The community can spend no more than \$100,000 for the vehicles and no more than \$500 per month for maintenance. The vans sell for \$10,000 each and average \$100 per month in maintenance costs. The corresponding cost estimates for each bus are \$20,000 and \$75 per month. If each van can carry 10 passengers and each bus can accomodate 15 riders, determine the number of vans and buses that should be purchased to maximize the passenger capacity of the system.

12 Refer to Exercise 11. The monthly fuel costs (based on 5000 miles of service) for each van is \$550 while each bus consumes \$850 in fuel. Find the number of vans and buses that should be purchased to minimize the monthly fuel costs if the passenger capacity of the system must be at least 75.

13 A fish farmer will purchase no more than 5000 young trout and bass from the hatchery and feed them a special diet over the next year. Each trout will consume \$0.50 in food while \$0.75 will be spent per bass. The total amount spent on the special diet is not to exceed \$3000. At the end of the year, a typical trout will weigh 2 pounds and a bass will weigh 3 pounds. How many fish of each type should be stocked in the pond in order to maximize the total number of pounds of fish at the end of the year?

14 A hospital dietician wishes to prepare a corn–squash vegetable dish that will provide at least 3 grams of protein and cost no more than 36 cents per serving. An ounce of creamed corn provides $\frac{1}{2}$ gram of protein and costs 4 cents. An ounce of squash supplies $\frac{1}{4}$ gram of protein and costs 3 cents. For taste, there must be at least 2 ounces of corn and at least as much squash as corn. It is important to keep the total number of ounces in a serving as small as possible. Find the combination of corn and squash that minimizes the size of the dish.

15 A contractor has a large building that he wishes to convert into a series of self-storage spaces. He will construct basic 8×10 foot units and deluxe 12×10 foot units that contain extra shelves and a clothes closet. Market considerations dictate that there be at least twice as many smaller units as larger units and that he rent the smaller units for \$40 per month and the deluxe units for \$75 per month. There is at most 7200 ft^2 for the storage spaces and no more than \$30,000 can be spent on construction. If each small unit will cost \$300 to make while a deluxe unit will cost \$600, how many units of each type should be constructed to maximize monthly revenues?

8.11 Review

Define or discuss each of the following.

1 System of equations
2 Solution of a system of equations
3 Equivalent systems of equations
4 System of linear equations
5 An $m \times n$ matrix
6 A square matrix of order n
7 The coefficient matrix of a system of linear equations; the augmented matrix
8 Elementary row transformations
9 Homogeneous system of linear equations
10 Partial fraction decompositions
11 The sum and product of two matrices
12 Zero matrix
13 Identity matrix
14 Inverse of a matrix
15 Determinant
16 Minor
17 Cofactor
18 Properties of determinants
19 Cramer's Rule
20 System of inequalities
21 Linear programming

Exercises 8.11

Find the solutions of the systems of equations in Exercises 1–16.

1 $\begin{cases} 2x - 3y = 4 \\ 5x + 4y = 1 \end{cases}$

2 $\begin{cases} x - 3y = 4 \\ -2x + 6y = 2 \end{cases}$

3 $\begin{cases} y + 4 = x^2 \\ 2x + y = -1 \end{cases}$

4 $\begin{cases} x^2 + y^2 = 25 \\ x - y = 7 \end{cases}$

5 $\begin{cases} 9x^2 + 16y^2 = 140 \\ x^2 - 4y^2 = 4 \end{cases}$

6 $\begin{cases} 2x = y^2 + 3z \\ x = y^2 + z - 1 \\ x^2 = xz \end{cases}$

7 $\begin{cases} \dfrac{1}{x} + \dfrac{3}{y} = 7 \\ \dfrac{4}{x} - \dfrac{2}{y} = 1 \end{cases}$

8 $\begin{cases} 2^x + 3^{y+1} = 10 \\ 2^{x+1} - 3^y = 5 \end{cases}$

9 $\begin{cases} 3x + y - 2z = -1 \\ 2x - 3y + z = 4 \\ 4x + 5y - z = -2 \end{cases}$

10 $\begin{cases} x + 3y = 0 \\ y - 5z = 3 \\ 2x + z = -1 \end{cases}$

11 $\begin{cases} 4x - 3y - z = 0 \\ x - y - z = 0 \\ 3x - y + 3z = 0 \end{cases}$

12 $\begin{cases} 2x + y - z = 0 \\ x - 2y + z = 0 \\ 3x + 3y + 2z = 0 \end{cases}$

13 $\begin{cases} 4x + 2y - z = 1 \\ 3x + 2y + 4z = 2 \end{cases}$

14 $\begin{cases} 2x + y = 6 \\ x - 3y = 17 \\ 3x + 2y = 7 \end{cases}$

15 $\begin{cases} \frac{4}{x} + \frac{1}{y} + \frac{2}{z} = 4 \\ \frac{2}{x} + \frac{3}{y} - \frac{1}{z} = 1 \\ \frac{1}{x} + \frac{1}{y} + \frac{1}{z} = 4 \end{cases}$

16 $\begin{cases} 2x - y + 3z - w = -3 \\ 3x + 2y - z + w = 13 \\ x - 3y + z - 2w = -4 \\ -x + y + 4z + 3w = 0 \end{cases}$

Find the solutions and sketch the graphs of the systems in Exercises 17–20.

17 $\begin{cases} x^2 + y^2 < 16 \\ y - x^2 > 0 \end{cases}$

18 $\begin{cases} y - x \leq 0 \\ y + x \geq 2 \\ x \leq 5 \end{cases}$

19 $\begin{cases} x - 2y \leq 2 \\ y - 3x \leq 4 \\ 2x + y \leq 4 \end{cases}$

20 $\begin{cases} x^2 - y < 0 \\ y - 2x < 5 \\ xy < 0 \end{cases}$

Find the determinants of the matrices in Exercises 21–30.

21 $[-6]$

22 $\begin{bmatrix} 3 & 4 \\ -6 & -5 \end{bmatrix}$

23 $\begin{bmatrix} 3 & -4 \\ 6 & 8 \end{bmatrix}$

24 $\begin{bmatrix} 0 & 4 & -3 \\ 2 & 0 & 4 \\ -5 & 1 & 0 \end{bmatrix}$

25 $\begin{bmatrix} 2 & -3 & 5 \\ -4 & 1 & 3 \\ 3 & 2 & -1 \end{bmatrix}$

26 $\begin{bmatrix} 3 & 1 & -2 \\ -5 & 2 & -4 \\ 7 & 3 & -6 \end{bmatrix}$

27 $\begin{bmatrix} 5 & 0 & 0 & 0 \\ 6 & -3 & 0 & 0 \\ 1 & 4 & -4 & 0 \\ 7 & 2 & 3 & 2 \end{bmatrix}$

28 $\begin{bmatrix} 1 & 2 & 0 & 3 & 1 \\ -2 & -1 & 4 & 1 & 2 \\ 3 & 0 & -1 & 0 & -1 \\ 2 & -3 & 2 & -4 & 2 \\ -1 & 1 & 0 & 1 & 3 \end{bmatrix}$

29 $\begin{bmatrix} 2 & 0 & 1 & 0 & -1 \\ 0 & 1 & 0 & 1 & 2 \\ 2 & -2 & 1 & -2 & 0 \\ 0 & 0 & -2 & 0 & 1 \\ 1 & -1 & 0 & -1 & 0 \end{bmatrix}$

30 $\begin{bmatrix} 1 & 2 & 0 & 0 & 0 \\ 3 & 4 & 0 & 0 & 0 \\ 0 & 0 & 1 & 2 & 3 \\ 0 & 0 & 2 & -1 & 1 \\ 0 & 0 & 1 & 3 & -1 \end{bmatrix}$

31 Find the determinant of the $n \times n$ matrix (a_{ij}) where $a_{ij} = 0$ if $i \neq j$.

32 Without expanding, show that

$$\begin{vmatrix} 1 & a & b + c \\ 1 & b & a + c \\ 1 & c & a + b \end{vmatrix} = 0.$$

Find the inverses of the matrices in Exercises 33–36.

33 $\begin{bmatrix} 5 & -4 \\ -3 & 2 \end{bmatrix}$

34 $\begin{bmatrix} 2 & -1 & 0 \\ 1 & 4 & 2 \\ 3 & -2 & 1 \end{bmatrix}$

35 $\begin{bmatrix} 3 & -1 & 0 & 0 \\ 1 & 2 & 0 & 0 \\ 0 & 0 & -1 & -2 \\ 0 & 0 & 5 & 3 \end{bmatrix}$

36 $\begin{bmatrix} 2 & 0 & 0 & 0 \\ 0 & 3 & 0 & 0 \\ 0 & 0 & 4 & 0 \\ 0 & 0 & 0 & 5 \end{bmatrix}$

Express each product or sum in Exercises 37–46 as a single matrix.

37 $\begin{bmatrix} 2 & -1 & 0 \\ 3 & 0 & -2 \end{bmatrix} \begin{bmatrix} 2 & -1 & 3 \\ 0 & 3 & 0 \\ 1 & 4 & 2 \end{bmatrix}$

38 $\begin{bmatrix} 4 & 2 \\ 5 & -3 \end{bmatrix}\begin{bmatrix} 3 \\ 7 \end{bmatrix}$

39 $\begin{bmatrix} 2 & 0 \\ 1 & 4 \\ -2 & 3 \end{bmatrix}\begin{bmatrix} 0 & 2 & -3 \\ 4 & 5 & 1 \end{bmatrix}$

40 $\begin{bmatrix} 0 & -2 & 3 \\ 4 & 1 & 2 \end{bmatrix}\begin{bmatrix} 2 & 0 \\ 3 & 8 \\ 2 & -7 \end{bmatrix}$

41 $2\begin{bmatrix} 0 & -1 & -4 \\ 3 & 2 & 1 \end{bmatrix} - 3\begin{bmatrix} 4 & -2 & 1 \\ 0 & 5 & -1 \end{bmatrix}$

42 $\begin{bmatrix} 1 & 3 \\ 2 & 4 \end{bmatrix}\begin{bmatrix} a & 0 \\ 0 & a \end{bmatrix}$

43 $\begin{bmatrix} a & 0 \\ 0 & b \end{bmatrix}\begin{bmatrix} 1 & 3 \\ 2 & 4 \end{bmatrix}$

44 $\begin{bmatrix} 3 & 2 \\ 0 & 0 \end{bmatrix}\begin{bmatrix} -2 & 0 \\ 3 & 0 \end{bmatrix}$

45 $\begin{bmatrix} 1 & 2 \\ 3 & 4 \end{bmatrix}\left\{\begin{bmatrix} 2 & -4 \\ 3 & 7 \end{bmatrix} + \begin{bmatrix} 1 & 5 \\ -2 & -3 \end{bmatrix}\right\}$

46 $\begin{bmatrix} 3 & 2 & 5 \\ -3 & 4 & 7 \\ 6 & 5 & 1 \end{bmatrix}\begin{bmatrix} 3 & 2 & 5 \\ -3 & 4 & 7 \\ 6 & 5 & 1 \end{bmatrix}^{-1}$

Verify Exercises 47 and 48 without expanding the determinants.

47 $\begin{vmatrix} 2 & 4 & -6 \\ 1 & 4 & 3 \\ 2 & 2 & 0 \end{vmatrix} = 12\begin{vmatrix} 1 & 1 & -1 \\ 1 & 2 & 1 \\ 2 & 1 & 0 \end{vmatrix}$

48 $\begin{vmatrix} a & b & c \\ d & e & f \\ g & h & k \end{vmatrix} = \begin{vmatrix} d & e & f \\ g & h & k \\ a & b & c \end{vmatrix}$

In Exercises 49 and 50, find the solutions of the equation $|A - xI| = 0$, where x denotes a real number.

49 $A = \begin{bmatrix} 2 & 3 \\ 1 & -4 \end{bmatrix}$, $I = I_2$

50 $A = \begin{bmatrix} 2 & -1 & 3 \\ 0 & 4 & 0 \\ 1 & 0 & -2 \end{bmatrix}$, $I = I_3$

51 Suppose that $A = (a_{ij})$ is a square matrix of order n such that $a_{ij} = 0$ if $i < j$. Prove that

$$|A| = a_{11}a_{22} \cdots a_{nn}.$$

52 If $A = (a_{ij})$ is any 2×2 matrix such that $|A| \neq 0$, prove that A has an inverse and find a general formula for A^{-1}.

In Exercises 53 and 54 find the partial fraction decompositions.

53 $\dfrac{4x^2 + 54x + 134}{(x + 3)(x^2 + 4x - 5)}$

54 $\dfrac{x^2 + 14x - 13}{x^3 + 5x^2 + 4x + 20}$

55 A rotating sprinkler head with a range of 50 feet is to be placed in the center of a rectangular field (see figure). Find the dimensions of the field assuming it is to contain 4000 ft^2 and the water is to just reach the corners of the field.

FIGURE FOR EXERCISE 55

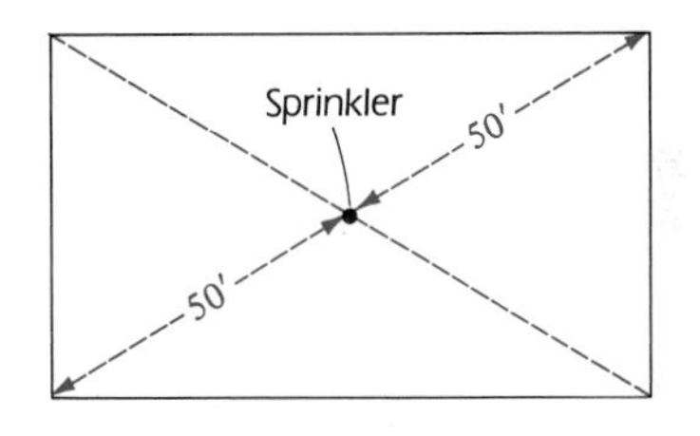

56 Find equations of the two lines that are tangent to the circle $x^2 + y^2 = 1$ and pass through the point $(0, 3)$. (*Hint:* Let $y = mx + 3$ and determine conditions on m that will ensure that the system has only one solution.)

57 An accountant must pay taxes and payroll bonuses to employees from the company's profits of \$50,000. The total tax is 40% of the amount left after bonuses are paid, and the total paid in bonuses is 10% of the amount left after taxes. Find the total tax and the total bonus amount.

58 A circular track is to have a 10-foot wide running lane around the outside, and the inside distance around the track is to be 90% of the outside distance. Find the dimensions of the track.

59 Three inlet pipes A, B, and C can be used to fill a 1000-ft^3 water storage tank. When all three pipes are in operation, the tank can be filled in 10 hours. When only A and B are used, the time increases to 20 hours. Pipes A and C can fill the tank in 12.5 hours. Find the individual flow rates (in cubic feet per hour) for each of the three pipes.

60 A deer spends the day in three basic activities: rest, searching for food, and grazing. At least 6 hours each day must be spent resting, and the number of hours spent searching for food will be at least two times the number of hours spent grazing. Using x as the number of hours spent searching for food and y as the number of hours spent grazing, find and graph the system of inequalities that describes the possible divisions of the day.

61 The Mower-for-the-Money Company manufactures a power lawn mower and a power edger. These two products are of such high quality that the company can sell all the products it makes, but production capability is limited in the areas of machining, welding, and assembly. Each week, the company has available 600 hours for machining, 300 hours for welding, and 550 hours for assembly. The number of hours required for the production of a single item is shown in the following table:

Product	Machining	Welding	Assembly
Lawn mower	6	2	5
Edger	4	3	5

The profits from the sale of a mower and an edger are \$100 and \$80, respectively. How many mowers and edgers should be produced each week in order to maximize profits?

CHAPTER 9

Sequences and Series

The method of *mathematical induction,* considered in the first section of this chapter, is needed to prove that certain statements are true for every positive integer. ■ In particular, in Section 9.2 it is used to prove the *Binomial Theorem.* ■ This is followed by material on *sequences* and *summation notation.* ■ Of special interest are the *arithmetic* and *geometric sequences* discussed in the last two sections.

9.1 Mathematical Induction

If n is a positive integer, let P_n denote the statement

$$(xy)^n = x^n y^n$$

where x and y are real numbers. Thus, P_1 represents $(xy)^1 = x^1 y^1$, P_2 denotes $(xy)^2 = x^2 y^2$, P_3 is $(xy)^3 = x^3 y^3$, and so on. It is easy to show that P_1, P_2, and P_3 are *true* statements. However, since the set of positive integers is infinite, it is impossible to check the validity of P_n for every positive integer n. To give a proof, the method of mathematical induction is required. This method is based on the following fundamental axiom.

AXIOM OF MATHEMATICAL INDUCTION

Suppose a set S of positive integers has the following two properties:

(i) S contains the integer 1.

(ii) Whenever S contains a positive integer k, S also contains $k + 1$.

Then S contains every positive integer.

If S is a set of positive integers that satisfies property (ii), then whenever S contains an arbitrary positive integer k, it must also contain the next positive integer, $k + 1$. If S also satisfies property (i), then S contains 1 and hence by (ii), S contains $1 + 1$, or 2. Applying (ii) again, we see that S contains $2 + 1$, or 3. Once again, S must contain $3 + 1$, or 4. If we continue in this manner, it can be argued that if n is any *specific* positive integer, then n is in S, since we can proceed a step at a time, eventually reaching n. Although this argument does not *prove* the axiom, it certainly makes it plausible.

We shall use the preceding axiom to establish the following fundamental principle.

PRINCIPLE OF MATHEMATICAL INDUCTION

If with each positive integer n there is associated a statement P_n, then all the statements P_n are true, provided the following two conditions are satisfied:

(i) P_1 is true.

(ii) Whenever k is a positive integer such that P_k is true, then P_{k+1} is also true.

PROOF Assume that conditions (i) and (ii) of the Principle hold, and let S denote the set of all positive integers n such that P_n is true. By assumption, P_1 is true, and consequently, 1 is in S. Thus, S satisfies property (i) of the Axiom of Mathematical Induction. Whenever S contains a positive integer k, then by the definition of S, P_k is true and hence from condition (ii) of the Principle, P_{k+1} is also true. This means that S contains $k + 1$. We have shown that whenever S contains a positive integer k, then S also contains $k + 1$. Consequently, property (ii) of the Axiom of Mathematical Induction is true, and hence S contains every positive integer; that is, P_n is true for every positive integer n. □

When applying the Principle of Mathematical Induction always follow these two steps:

Step (i) Prove that P_1 is true.

Step (ii) Assume that P_k is true and prove that P_{k+1} is true.

Step (ii) is usually the most confusing for the beginning student. We do not *prove* that P_k is true (except for $k = 1$). Instead, we show that *if* P_k is true, then the statement P_{k+1} is true. That is all that is necessary according to the Principle of Mathematical Induction. The assumption that P_k is true is referred to as the **induction hypothesis.**

EXAMPLE 1 Prove that for every positive integer n, the sum of the first n positive integers is

$$\frac{n(n+1)}{2}.$$

SOLUTION If n is any positive integer, let P_n denote the statement

$$1 + 2 + 3 + \cdots + n = \frac{n(n+1)}{2}.$$

The following are some special cases of P_n:

If $n = 2$, then P_2 is

$$1 + 2 = \frac{2(2+1)}{2} \quad \text{or} \quad 3 = 3.$$

If $n = 3$, then P_3 is

$$1 + 2 + 3 = \frac{3(3+1)}{2} \quad \text{or} \quad 6 = 6.$$

If $n = 5$, then P_5 is

$$1 + 2 + 3 + 4 + 5 = \frac{5(5+1)}{2} \quad \text{or} \quad 15 = 15.$$

Although it is instructive to check the validity of P_n for several values of n as we have done, it is unnecessary to do so. We need only apply the two-step process outlined prior to this example. Thus, we proceed as follows:

Step (i): If we substitute $n = 1$ in P_n, then the left side contains only the number 1 and the right side is $\frac{1(1+1)}{2}$, which also equals 1. This proves that P_1 is true.

Step (ii): Assume that P_k is true. Thus, the induction hypothesis is

$$1 + 2 + 3 + \cdots + k = \frac{k(k+1)}{2}.$$

Our goal is to prove that P_{k+1} is true; that is,

$$1 + 2 + 3 + \cdots + (k+1) = \frac{(k+1)[(k+1)+1]}{2}.$$

By the induction hypothesis we already have a formula for the sum of the first k positive integers. Hence, a formula for the sum of the first $k + 1$ positive integers may be found simply by adding $(k + 1)$ to both sides. Doing so and simplifying, we obtain

$$\begin{aligned} 1 + 2 + 3 + \cdots + k + (k+1) &= \frac{k(k+1)}{2} + (k+1) \\ &= \frac{k(k+1) + 2(k+1)}{2} \\ &= \frac{k^2 + 3k + 2}{2} \\ &= \frac{(k+1)(k+2)}{2} \\ &= \frac{(k+1)[(k+1)+1]}{2}. \end{aligned}$$

We have shown that P_{k+1} is true and, therefore, the proof by mathematical induction is complete. ■

EXAMPLE 2 Prove that for each positive integer n,

$$1^2 + 3^2 + \cdots + (2n - 1)^2 = \frac{n(2n - 1)(2n + 1)}{3}.$$

SOLUTION For each positive integer n, let P_n denote the given statement. Note that this is a formula for the sum of the squares of the first n odd positive integers. We again follow the two-step procedure.

Step (i): Substituting 1 for n in P_n, we obtain

$$1^2 = \frac{(1)(2 - 1)(2 + 1)}{3} = \frac{3}{3} = 1.$$

This shows that P_1 is true.

Step (ii): Assume that P_k is true. Thus, the induction hypothesis is

$$1^2 + 3^2 + \cdots + (2k - 1)^2 = \frac{k(2k - 1)(2k + 1)}{3}.$$

We wish to prove that P_{k+1} is true, that is,

$$1^2 + 3^2 + \cdots + [2(k + 1) - 1]^2 = \frac{(k + 1)[2(k + 1) - 1][2(k + 1) + 1]}{3}.$$

This equation for P_{k+1} simplifies to

$$1^2 + 3^2 + \cdots + (2k + 1)^2 = \frac{(k + 1)(2k + 1)(2k + 3)}{3}.$$

Observe that the second from the last term on the left-hand side is $(2k - 1)^2$. In a manner similar to the solution of Example 1, we may obtain the left side of P_{k+1} by adding $(2k + 1)^2$ to both sides of the induction hypothesis P_k. This gives us

$$1^2 + 3^2 + \cdots + (2k - 1)^2 + (2k + 1)^2 = \frac{k(2k - 1)(2k + 1)}{3} + (2k + 1)^2.$$

It is left to the reader to show that the right side of the preceding equation may be written in the form of the right side of P_{k+1}. This proves that P_{k+1} is true, and hence P_n is true for every n. ■

Let j be a positive integer and suppose that with each integer $n \geq j$ there is associated a statement P_n. For example, if $j = 6$, then the statements are numbered $P_6, P_7, P_8, \ldots$. The principle of mathematical induction may be extended to cover this situation. Just as before, two steps are used. Specifically, to prove that the statements P_n are true for $n \geq j$, we use the following steps.

EXTENDED PRINCIPLE OF MATHEMATICAL INDUCTION FOR P_k, $k \geq j$

(i′) Prove that P_j is true.

(ii′) Assume that P_k is true for $k \geq j$ and prove that P_{k+1} is true.

EXAMPLE 3 Let a be a nonzero real number such that $a > -1$. Prove that $(1 + a)^n > 1 + na$ for every integer $n \geq 2$.

SOLUTION For each positive integer n, let P_n denote the inequality $(1 + a)^n > 1 + na$. Note that P_1 is *false*, since $(1 + a)^1 = 1 + (1)(a)$. However, we can show that P_n is true for $n \geq 2$ by using the Extended Principle with $j = 2$.

Step (i′): We first note that $(1 + a)^2 = 1 + 2a + a^2$. Since $a \neq 0$, we have $a^2 > 0$ and therefore $1 + 2a + a^2 > 1 + 2a$. This gives us $(1 + a)^2 > 1 + 2a$, and hence P_2 is true.

Step (ii′): Assume that P_k is true. Thus, the induction hypothesis is

$$(1 + a)^k > 1 + ka.$$

We wish to show that P_{k+1} is true, that is,

$$(1 + a)^{k+1} > 1 + (k + 1)a.$$

Since $a > -1$, we have $1 + a > 0$, and hence multiplying both sides of the induction hypothesis by $1 + a$ will not change the inequality sign. Consequently,

$$(1 + a)^k(1 + a) > (1 + ka)(1 + a),$$

which may be rewritten as

$$(1 + a)^{k+1} > 1 + ka + a + ka^2$$

or as

$$(1 + a)^{k+1} > 1 + (k + 1)a + ka^2.$$

Since $ka^2 > 0$,

$$1 + (k + 1)a + ka^2 > 1 + (k + 1)a$$

and therefore,

$$(1 + a)^{k+1} > 1 + (k + 1)a.$$

Thus, P_{k+1} is true and the proof is complete. ■

Exercises 9.1

In Exercises 1–18 prove that the formula is true for every positive integer n.

1 $2 + 4 + 6 + \cdots + 2n = n(n + 1)$

2 $1 + 4 + 7 + \cdots + (3n - 2) = \dfrac{n(3n - 1)}{2}$

3 $1 + 3 + 5 + \cdots + (2n - 1) = n^2$

$3 + 9 + 15 + \cdots + (6n - 3) = 3n^2$

$7 + 12 + \cdots + (5n - 3) = \frac{1}{2}n(5n - 1)$

6 $8 + \cdots + 2 \cdot 3^{n-1} = 3^n - 1$

7 $1 + 2 \cdot 2 + 3 \cdot 2^2 + \cdots + n \cdot 2^{n-1} = 1 + (n - 1) \cdot 2^n$

8 $(-1)^1 + (-1)^2 + (-1)^3 + \cdots + (-1)^n = \dfrac{(-1)^n - 1}{2}$

9 $1^2 + 2^2 + 3^2 + \cdots + n^2 = \dfrac{n(n + 1)(2n + 1)}{6}$

10 $1^3 + 2^3 + 3^3 + \cdots + n^3 = \left[\dfrac{n(n + 1)}{2}\right]^2$

11 $\dfrac{1}{1 \cdot 2} + \dfrac{1}{2 \cdot 3} + \dfrac{1}{3 \cdot 4} + \cdots + \dfrac{1}{n(n + 1)} = \dfrac{n}{n + 1}$

12 $\dfrac{1}{1 \cdot 2 \cdot 3} + \dfrac{1}{2 \cdot 3 \cdot 4} + \dfrac{1}{3 \cdot 4 \cdot 5} + \cdots + \dfrac{1}{n(n + 1)(n + 2)} = \dfrac{n(n + 3)}{4(n + 1)(n + 2)}$

13 $3 + 3^2 + 3^3 + \cdots + 3^n = \frac{3}{2}(3^n - 1)$

14 $1^3 + 3^3 + 5^3 + \cdots + (2n - 1)^3 = n^2(2n^2 - 1)$

15 $n < 2^n$

16 $1 + 2n \leq 3^n$

17 $1 + 2 + 3 + \cdots + n < \frac{1}{8}(2n + 1)^2$

18 If $0 < a < b$, then $\left(\dfrac{a}{b}\right)^{n+1} < \left(\dfrac{a}{b}\right)^n$.

Prove that the statements in Exercises 19–22 are true for every positive integer n.

19 3 is a factor of $n^3 - n + 3$.

20 2 is a factor of $n^2 + n$.

21 4 is a factor of $5^n - 1$.

22 9 is a factor of $10^{n+1} + 3 \cdot 10^n + 5$.

23 Use mathematical induction to prove that if a is any real number greater than 1, then $a^n > 1$ for every positive integer n.

24 Prove that

$$a + ar + ar^2 + \cdots + ar^{n-1} = \frac{a(1 - r^n)}{1 - r}$$

where n is any positive integer and a and r are real numbers with $r \neq 1$.

25 Use mathematical induction to prove that $a - b$ is a factor of $a^n - b^n$ for every positive integer n. (*Hint:* $a^{k+1} - b^{k+1} = a^k(a - b) + (a^k - b^k)b$.)

26 Prove that $a + b$ is a factor of $a^{2n-1} + b^{2n-1}$ for every positive integer n.

27 Use mathematical induction to prove De Moivre's Theorem:

$$[r(\cos\theta + i\sin\theta)]^n = r^n[\cos n\theta + i\sin n\theta]$$

for every positive integer n.

28 Prove that for every positive integer $n \geq 3$, the sum of the interior angles of a polygon of n sides is $(n - 2) \cdot 180°$.

Prove that the formulas in Exercises 29 and 30 are true for every positive integer n.

29 $\sin(\theta + n\pi) = (-1)^n \sin\theta$

30 $\cos(\theta + n\pi) = (-1)^n \cos\theta$

9.2 The Binomial Theorem

A sum $a + b$ is called a **binomial.** Sometimes it is necessary to consider $(a + b)^n$, where n is a large positive integer. A general formula for *expanding* $(a + b)^n$, that is, for expressing it as a sum, is given by the **Binomial Theorem.** Let us begin by considering some special cases. If we actually perform the multiplications, we obtain

$$(a + b)^2 = a^2 + 2ab + b^2$$

$$(a + b)^3 = a^3 + 3a^2b + 3ab^2 + b^3$$

$$(a + b)^4 = a^4 + 4a^3b + 6a^2b^2 + 4ab^3 + b^4$$

$$(a + b)^5 = a^5 + 5a^4b + 10a^3b^2 + 10a^2b^3 + 5ab^4 + b^5.$$

These expansions of $(a + b)^n$, for $n = 2, 3, 4$, and 5, have the following properties:

(i) There are $n + 1$ terms, the first being a^n and the last b^n.

(ii) In going from any term to the next, the power of a decreases by 1 and the power of b increases by 1. Thus, the sum of the exponents of a and b is always n.

(iii) Each term has the form $(c)a^{n-k}b^k$, where the coefficient c is a real number and $k = 0, 1, 2, \ldots, n$.

(iv) The following formula is true for each of the first n terms of the expansion:

$$\frac{(\text{coefficient of the term}) \cdot (\text{exponent of } a)}{\text{number of the term}} = \text{coefficient of the next term}.$$

The next table illustrates property (iv) for the expansion of $(a + b)^5$.

Number of the term	Coefficient of the term	Exponent of a	Coefficient of the next term
1	1	5	$\frac{1 \cdot 5}{1} = 5$
2	5	4	$\frac{5 \cdot 4}{2} = 10$
3	10	3	$\frac{10 \cdot 3}{3} = 10$
4	10	2	$\frac{10 \cdot 2}{4} = 5$
5	5	1	$\frac{5 \cdot 1}{5} = 1$

Thus, as before, we get

$$(a+b)^5 = a^5 + 5a^4b + 10a^3b^2 + 10a^2b^3 + 5ab^4 + b^5.$$

Let us next consider $(a+b)^n$, for an arbitrary positive integer n. The first term is a^n, which has coefficient 1. If we assume that property (iv) is true, we obtain the successive coefficients listed in the following table.

Number of the term	Coefficient of the term	Exponent of a	Coefficient of the next term
1	1	n	$\frac{1\cdot n}{1} = \frac{n}{1}$
2	$\frac{n}{1}$	$n-1$	$\frac{n(n-1)}{1\cdot 2}$
3	$\frac{n(n-1)}{1\cdot 2}$	$n-2$	$\frac{n(n-1)(n-2)}{1\cdot 2\cdot 3}$
4	$\frac{n(n-1)(n-2)}{1\cdot 2\cdot 3}$	$n-3$	$\frac{n(n-1)(n-2)(n-3)}{1\cdot 2\cdot 3\cdot 4}$

The last coefficient that we calculated belongs to the fifth term. In similar fashion, the coefficient of the sixth term is

$$\frac{n(n-1)(n-2)(n-3)(n-4)}{1\cdot 2\cdot 3\cdot 4\cdot 5}.$$

It now appears likely that the coefficient of the $(k+1)$st term in the expansion of $(a+b)^n$ is the following:

$(k+1)$st COEFFICIENT OF $(a+b)^n$

$$\frac{n(n-1)(n-2)(n-3)\cdots(n-k+1)}{1\cdot 2\cdot 3\cdot 4\cdots k}$$

This fraction can be written in a compact form by using **factorial notation.** If n is any positive integer, the symbol $n!$ (read "n factorial") is defined as follows:

DEFINITION OF $n!$ $(n > 0)$

$$n! = n(n-1)(n-2)\cdots 1$$

Thus, $n!$ is the product of the first n positive integers. Special cases are:

$$1! = 1, \qquad 2! = 2 \cdot 1, \qquad 3! = 3 \cdot 2 \cdot 1 = 6,$$
$$4! = 4 \cdot 3 \cdot 2 \cdot 1 = 24, \qquad 5! = 5 \cdot 4 \cdot 3 \cdot 2 \cdot 1 = 120.$$

To ensure that certain formulas will be true for all *nonnegative* integers, we define $0!$ as follows:

DEFINITION OF $0!$

$$0! = 1$$

Note that for $n \geq 1$,

$$n! = n[(n-1)!] = n(n-1)!$$

For larger values of n we can write

$$n! = n(n-1)(n-2)! = n(n-1)(n-2)(n-3)!$$

and so on. The formula for the $(k+1)$st coefficient in the expansion of $(a+b)^n$ can be written as follows:

$$\begin{aligned}\frac{n(n-1)(n-2)\cdots(n-k+1)}{1 \cdot 2 \cdot 3 \cdots k} &= \frac{\dfrac{n(n-1)(n-2)\cdots(n-k+1)(n-k)!}{(n-k)!}}{k!} \\ &= \frac{\dfrac{n!}{(n-k)!}}{k!} \\ &= \frac{n!}{k!(n-k)!}\end{aligned}$$

These numbers are called **binomial coefficients** and are often represented by the following notation.

DEFINITION

$$\binom{n}{k} = \frac{n!}{k!(n-k)!}, \qquad k = 0, 1, 2, \ldots, n$$

The symbol $\binom{n}{k}$, which denotes the $(k+1)$st coefficient in the expansion of $(a+b)^n$, is sometimes read "n choose k".

EXAMPLE 1

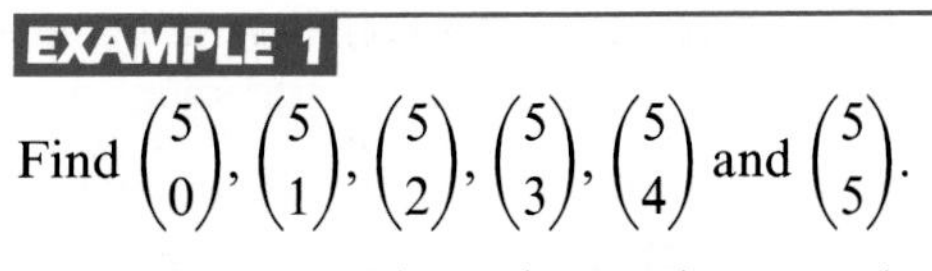

Find $\binom{5}{0}, \binom{5}{1}, \binom{5}{2}, \binom{5}{3}, \binom{5}{4}$ and $\binom{5}{5}$.

SOLUTION These six numbers are the coefficients in the expansion of $(a + b)^5$, which we tabulated earlier in this section. By definition,

$$\binom{5}{0} = \frac{5!}{0!(5-0)!} = \frac{5!}{1 \cdot 5!} = 1$$

$$\binom{5}{1} = \frac{5!}{1!(5-1)!} = \frac{5!}{1 \cdot 4!} = \frac{5 \cdot 4!}{4!} = 5$$

$$\binom{5}{2} = \frac{5!}{2!(5-2)!} = \frac{5!}{2!3!} = \frac{5 \cdot 4 \cdot 3!}{2!3!} = \frac{20}{2} = 10$$

$$\binom{5}{3} = \frac{5!}{3!(5-3)!} = \frac{5!}{3!2!} = \frac{5 \cdot 4 \cdot 3!}{3!2!} = \frac{20}{2} = 10$$

$$\binom{5}{4} = \frac{5!}{4!(5-4)!} = \frac{5!}{4!1!} = \frac{5 \cdot 4!}{4!} = 5$$

$$\binom{5}{5} = \frac{5!}{5!(5-5)!} = \frac{1}{0!} = \frac{1}{1} = 1$$

■

The Binomial Theorem may be stated as follows.

THE BINOMIAL THEOREM

$$(a+b)^n = a^n + \binom{n}{1} a^{n-1}b + \binom{n}{2} a^{n-2}b^2 + \cdots + \binom{n}{k} a^{n-k}b^k + \cdots + \binom{n}{n-1} ab^{n-1} + b^n$$

A statement of the Binomial Theorem without using symbols for binomial coefficients is:

THE BINOMIAL THEOREM (ALTERNATIVE FORM)

$$(a+b)^n = a^n + na^{n-1}b + \frac{n(n-1)}{2!} a^{n-2}b^2 + \cdots + \frac{n(n-1)(n-2)\cdots(n-r+1)}{r!} a^{n-r}b^r + \cdots + nab^{n-1} + b^n$$

PROOF We shall use mathematical induction. For each positive integer n, let P_n denote the statement given in the Binomial Theorem.

Step (i): If $n = 1$, the statement reduces to $(a + b)^1 = a^1 + b^1$. Consequently, P_1 is true.

Step (ii): Assume that P_k is true. Thus, the induction hypothesis is

$$\begin{aligned}(a + b)^k = a^k + ka^{k-1}b + \frac{k(k-1)}{2!}a^{k-2}b^2 + \cdots \\ + \frac{k(k-1)(k-2)\cdots(k-r+2)}{(r-1)!}a^{k-r+1}b^{r-1} \\ + \frac{k(k-1)(k-2)\cdots(k-r+1)}{r!}a^{k-r}b^r \\ + \cdots + kab^{k-1} + b^k.\end{aligned}$$

We have shown both the rth and the $(r + 1)$st terms in the expansion.
Multiplying both sides of the last equation by $(a + b)$, we obtain

$$\begin{aligned}(a + b)^{k+1} = \left[a^{k+1} + ka^kb + \frac{k(k-1)}{2!}a^{k-1}b^2 + \cdots \right. \\ \left. + \frac{k(k-1)\cdots(k-r+1)}{r!}a^{k-r+1}b^r + \cdots + ab^k\right] \\ + \left[a^kb + ka^{k-1}b^2 + \cdots + \frac{k(k-1)\cdots(k-r+2)}{(r-1)!}a^{k-r+1}b^r \right. \\ \left. + \cdots + kab^k + b^{k+1}\right]\end{aligned}$$

where the terms in the first pair of brackets result from multiplying the right side of the induction hypothesis by a, and the terms in the second pair of brackets result from multiplying by b. Rearranging and combining terms,

$$\begin{aligned}(a + b)^{k+1} = a^{k+1} + (k + 1)a^kb + \left[\frac{k(k-1)}{2!} + k\right]a^{k-1}b^2 + \cdots \\ + \left[\frac{k(k-1)\cdots(k-r+1)}{r!} + \frac{k(k-1)\cdots(k-r+2)}{(r-1)!}\right]a^{k-r+1}b^r \\ + \cdots + (1 + k)ab^k + b^{k+1}.\end{aligned}$$

It can be shown that if the coefficients are simplifed we obtain statement P_n with $k + 1$ substituted for n. Thus, P_{k+1} is true and therefore P_n holds for every positive integer n. This completes the proof. □

The following examples may be solved by using either the general formulas for the Binomial Theorem, or by repeated use of property (iv) stated at the beginning of this section.

EXAMPLE 2 Find the binomial expansion of $(2x + 3y^2)^4$.

SOLUTION Using the Binomial Theorem with $a = 2x$, $b = 3y^2$, and $n = 4$,

$$(2x + 3y^2)^4 = (2x)^4 + \binom{4}{1}(2x)^3(3y^2) + \binom{4}{2}(2x)^2(3y^2)^2 + \binom{4}{3}(2x)(3y^2)^3 + (3y^2)^4.$$

This simplifies to

$$\begin{aligned}(2x + 3y^2)^4 &= 16x^4 + 4(8x^3)(3y^2) + 6(4x^2)(9y^4) + 4(2x)(27y^6) + 81y^8 \\ &= 16x^4 + 96x^3y^2 + 216x^2y^4 + 216xy^6 + 81y^8.\end{aligned}$$ ■

The next example illustrates the fact that if either a or b is negative, then the terms of the expansion are alternatively positive and negative.

EXAMPLE 3

Expand $\left(\dfrac{1}{x} - 2\sqrt{x}\right)^5$.

SOLUTION The binomial coefficients for $(a + b)^5$ were calculated in Example 1. Thus, if we let $a = 1/x$, $b = -2\sqrt{x}$, and $n = 5$ in the Binomial Theorem, then

$$\left(\frac{1}{x} - 2\sqrt{x}\right)^5 = \left(\frac{1}{x}\right)^5 + 5\left(\frac{1}{x}\right)^4(-2\sqrt{x}) + 10\left(\frac{1}{x}\right)^3(-2\sqrt{x})^2 + 10\left(\frac{1}{x}\right)^2(-2\sqrt{x})^3 + 5\left(\frac{1}{x}\right)(-2\sqrt{x})^4 + (-2\sqrt{x})^5.$$

This simplifies to

$$\left(\frac{1}{x} - 2\sqrt{x}\right)^5 = \frac{1}{x^5} - \frac{10}{x^{7/2}} + \frac{40}{x^2} - \frac{80}{x^{1/2}} + 80x - 32x^{5/2}.$$ ■

Certain problems only require finding a specific term in the expansion of $(a + b)^n$. To work such problems, it is convenient to first find the exponent k that is to be assigned to b. Notice that, by the Binomial Theorem, *the exponent of b is always one less than the number of the term.* Once k is found, the exponent of a is $n - k$, and the coefficient is $\binom{n}{k}$.

EXAMPLE 4 Find the fifth term in the expansion of $(x^3 + \sqrt{y})^{13}$.

SOLUTION Let $a = x^3$ and $b = \sqrt{y}$. The exponent of b in the fifth term is 4 and hence the exponent of a is 9. From the discussion of the preceding paragraph we obtain

$$\binom{13}{4}(x^3)^9(\sqrt{y})^4 = \frac{13!}{4!(13-4)!}x^{27}y^2 = \frac{13 \cdot 12 \cdot 11 \cdot 10}{4!}x^{27}y^2$$
$$= 715x^{27}y^2.$$ ■

Finally, there is an interesting triangular array of numbers called **Pascal's Triangle,** which can be used to obtain binomial coefficients. The numbers are arranged as follows.

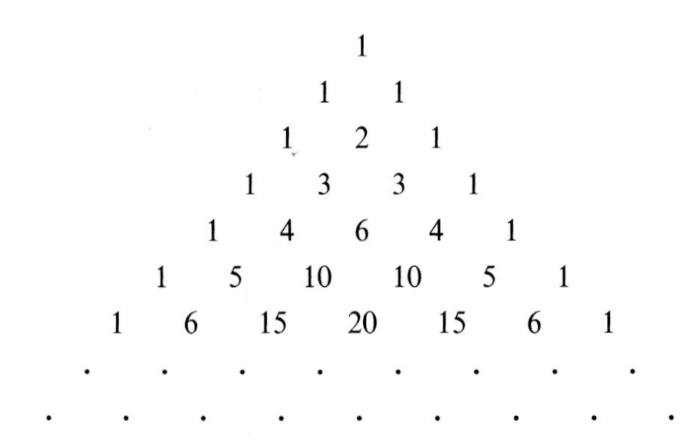

The numbers in the second row are the coefficients in the expansion of $(a + b)^1$; those in the third row are the coefficients determined by $(a + b)^2$; those in the fourth row are obtained from $(a + b)^3$, and so on. Each number in the array that is different from 1 can be found by adding the two numbers in the previous row that appear above and immediately to the left and right of the number.

EXAMPLE 5 Find the eighth row of Pascal's Triangle and use it to expand $(a + b)^7$.

SOLUTION Let us rewrite the seventh row and then use the process described previously. In the following display the arrows indicate which two numbers in row seven are added to obtain the numbers in row eight.

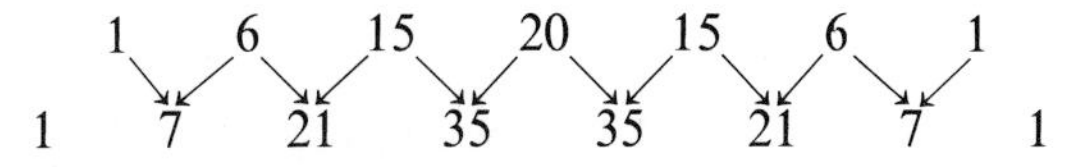

The eighth row gives us the coefficients in the expansion of $(a + b)^7$. Thus,

$$(a + b)^7 = a^7 + 7a^6b + 21a^5b^2 + 35a^4b^3 + 35a^3b^4 + 21a^2b^5 + 7ab^6 + b^7.$$ ■

Pascal's Triangle is useful to expand small powers of $a + b$; however, to expand large powers or to find a specific term, as in Example 4, it is better to refer to the general formula given by the Binomial Theorem.

Exercises 9.2

Find the numbers in Exercises 1–12.

1 $6!$ **2** $5!$ **3** $3!2!$

4 $2!0!$ **5** $\dfrac{8!}{5!}$ **6** $\dfrac{7!}{3!}$

7 $\dbinom{7}{3}$ **8** $\dbinom{6}{2}$ **9** $\dbinom{8}{5}$

10 $\dbinom{4}{4}$ **11** $\dbinom{9}{4}$ **12** $\dbinom{3}{2}$

Expand and simplify the expressions in Exercises 13–24.

13 $(a + b)^6$ **14** $(a - b)^7$

15 $(a - b)^8$ **16** $(a + b)^9$

17 $(3x - 5y)^4$ **18** $(2t - s)^5$

19 $(u^2 + 4v)^5$ **20** $(\frac{1}{2}c + d^3)^4$

21 $(r^{-2} - 2r)^6$ **22** $(x^{1/2} - y^{-1/2})^6$

23 $(1 + x)^{10}$ **24** $(1 - x)^{10}$

25 Find the first four terms in the binomial expansion of $(3c^{2/5} + c^{4/5})^{25}$.

26 Find the first three terms and the last three terms in the binomial expansion of $(x^3 + 5x^{-2})^{20}$.

27 Find the last two terms in the expansion of $(4b^{-1} - 3b)^{15}$.

28 Find the last three terms in the expansion of $(s - 2t^3)^{12}$.

Solve Exercises 29–40 without expanding completely.

29 Find the fifth term in the expansion of $(3a^2 + \sqrt{b})^9$.

30 Find the sixth term in the expansion of $\left(\dfrac{2}{c} + \dfrac{c^2}{3}\right)^7$.

31 Find the seventh term in the expansion of $(\frac{1}{2}u - 2v)^{10}$.

32 Find the fourth term in the expansion of $(2x^3 - y^2)^6$.

33 Find the middle term in the expansion of $(x^{1/3} + y^{1/3})^{12}$.

34 Find the two middle terms in the expansion of $(rs + t)^7$.

35 Find the term that does not contain x in the expansion of $\left(6x - \dfrac{1}{2x}\right)^{10}$.

36 Find the term involving x^8 in the expansion of $(y + 3x^2)^6$.

37 Find the term containing y^6 in the expansion of $(x - 2y^3)^4$.

38 Find the term containing b^9 in the expansion of $(5a + 2b^3)^4$.

39 Find the term containing c^3 in the expansion of $(\sqrt{c} + \sqrt{d})^{10}$.

40 In the expansion of $(xy - 2y^{-3})^8$ find the term that does not contain y.

41 Use the first four terms in the binomial expansion of $(1 + 0.02)^{10}$ to approximate $(1.02)^{10}$. Compare with the answer obtained using a calculator.

42 Use the first four terms in the binomial expansion of $(1 - 0.01)^4$ to approximate $(0.99)^4$. Compare with the answer obtained using a calculator.

9.3 Infinite Sequences and Summation Notation

Recall that a function f from a set D to a set E is a correspondence that associates with each element x of D a unique element $f(x)$ of E. Up to now the domain D has usually been a set of real numbers. In this section we shall consider a different class of functions.

DEFINITION

An **infinite sequence** is a function whose domain is the set of positive integers.

For convenience we sometimes refer to an infinite sequence as a **sequence.** In this book the range of an infinite sequence will be a set of real numbers.

If f is an infinite sequence, then to each positive integer n there corresponds a real number $f(n)$. These numbers in the range of f may be represented by writing

$$f(1), f(2), f(3), \ldots, f(n), \ldots$$

where the dots at the end indicate that the sequence does not terminate. The number $f(1)$ is called the **first term** of the sequence, $f(2)$ the **second**

term and, in general, $f(n)$ the ***n*****th term** of the sequence. It is customary to use a subscript notation instead of the function notation and to write these numbers as

$$a_1, a_2, a_3, \ldots, a_n, \ldots$$

where it is understood that for each positive integer n, the symbol a_n denotes the real number $f(n)$. In this way we obtain an infinite collection of real numbers that is *ordered* in the sense that there is a first number, a second number, a forty-fifth number, and so on. Although sequences are functions, an ordered collection of the type displayed above will also be referred to as an infinite sequence. If we wish to convert the collection to a function f, we let $f(n) = a_n$ for every positive integer n.

From the definition of equality of functions we see that a sequence

$$a_1, a_2, a_3, \ldots, a_n, \ldots$$

is **equal** to a sequence $\quad b_1, b_2, b_3, \ldots, b_n, \ldots$

if and only if $a_k = b_k$ for every positive integer k. Infinite sequences are often defined by stating a formula for the nth term, as in the following example.

EXAMPLE 1 List the first four terms and the tenth term of the sequence whose nth term is as follows:

(a) $a_n = \dfrac{n}{n+1}$ (b) $a_n = 2 + (0.1)^n$

(c) $a_n = (-1)^{n+1}\dfrac{n^2}{3n-1}$ (d) $a_n = 4$

SOLUTION To find the first four terms we substitute, successively, $n = 1, 2, 3$, and 4 in the formula for a_n. The tenth term is found by substituting 10 for n. Doing this and simplifying gives us the following:

	*n*th *term*	*First four terms*	*Tenth term*
(a)	$\dfrac{n}{n+1}$	$\dfrac{1}{2}, \dfrac{2}{3}, \dfrac{3}{4}, \dfrac{4}{5}$	$\dfrac{10}{11}$
(b)	$2 + (0.1)^n$	2.1, 2.01, 2.001, 2.0001	2.0000000001
(c)	$(-1)^{n+1}\dfrac{n^2}{3n-1}$	$\dfrac{1}{2}, -\dfrac{4}{5}, \dfrac{9}{8}, -\dfrac{16}{11}$	$-\dfrac{100}{29}$
(d)	4	4, 4, 4, 4	4

■

Some infinite sequences are described by stating the first term a_1, together with a rule that shows how to obtain any term a_{k+1} from the preceding term a_k whenever $k \geq 1$. A description of this type is called a **recursive definition,** and the sequence is said to be defined **recursively.**

EXAMPLE 2 Find the first four terms and the nth term of the infinite sequence defined as follows:

$$a_1 = 3 \quad \text{and} \quad a_{k+1} = 2a_k \quad \text{for } k \geq 1.$$

SOLUTION The sequence is defined recursively since the first term is given and, moreover, whenever a term a_k is known, then the next term a_{k+1} can be found. Thus,

$$\begin{aligned} a_1 &= 3 \\ a_2 &= 2a_1 = 2 \cdot 3 = 6 \\ a_3 &= 2a_2 = 2 \cdot 2 \cdot 3 = 2^2 \cdot 3 = 12 \\ a_4 &= 2a_3 = 2 \cdot 2 \cdot 2 \cdot 3 = 2^3 \cdot 3 = 24. \end{aligned}$$

We have written the terms as products in order to gain some insight into the nature of the nth term. Continuing, we obtain $a_5 = 2^4 \cdot 3$ and $a_6 = 2^5 \cdot 3$; and it appears that

$$a_n = 2^{n-1} \cdot 3$$

for every positive integer n. We shall prove that this guess is correct by mathematical induction. If we let P_n denote the statement $a_n = 2^{n-1} \cdot 3$, then P_1 is true since $a_1 = 2^0 \cdot 3 = 3$. Next, *assume* that P_k is true, that is, $a_k = 2^{k-1} \cdot 3$. We then have

$$\begin{aligned} a_{k+1} &= 2a_k && \text{(definition of } a_{k+1}) \\ &= 2 \cdot 2^{k-1} \cdot 3 && \text{(induction hypothesis)} \\ &= 2^k \cdot 3 && \text{(a law of exponents)} \\ &= 2^{(k+1)-1} \cdot 3, && \text{(rewriting } 2^k \cdot 3) \end{aligned}$$

which shows that P_{k+1} is true. Hence, $a_n = 2^{n-1} \cdot 3$ for every positive integer n.

It is important to observe that if only the first few terms of an infinite sequence are known, then it is impossible to predict additional terms. For example, if we were given 3, 6, 9, . . . and asked to find the fourth term, we could not proceed without further information. The infinite sequence with nth term

$$a_n = 3n + (1 - n)^3(2 - n)^2(3 - n)$$

has for its first four terms 3, 6, 9, and 120. It is possible to describe sequences in which the first three terms are 3, 6, and 9, and the fourth term is *any* given number. This shows that when we work with an infinite sequence it is essential to have specific information about the nth term or to know a general scheme for obtaining each term from the preceding one.

It is sometimes necessary to find the sum of many terms of an infinite sequence. For ease in expressing such sums we use the following **summation notation.** Given an infinite sequence

$$a_1, a_2, a_3, \ldots, a_n, \ldots$$

the symbol $\sum_{k=1}^{m} a_k$ represents the sum of the first m terms, that is,

SUMMATION NOTATION

$$\sum_{k=1}^{m} a_k = a_1 + a_2 + a_3 + \cdots + a_m.$$

The Greek capital letter $\sum$ (sigma) indicates a sum and the symbol a_k represents the kth term. The letter k is called the **index of summation** or the **summation variable,** and the numbers 1 and m indicate the smallest and largest values of the summation variable.

EXAMPLE 3 Find the sum

$$\sum_{k=1}^{4} k^2(k-3).$$

SOLUTION In this case, $a_k = k^2(k-3)$. To find the sum we merely substitute, in succession, the integers 1, 2, 3, and 4 for k and add the resulting terms.

$$\begin{aligned}\sum_{k=1}^{4} k^2(k-3) &= 1^2(1-3) + 2^2(2-3) + 3^2(3-3) + 4^2(4-3)\\ &= (-2) + (-4) + 0 + 16 = 10.\end{aligned}$$

■

The letter used for the summation variable is arbitrary. To illustrate, if j is the summation variable, then

$$\sum_{j=1}^{m} a_j = a_1 + a_2 + a_3 + \cdots + a_m,$$

which is the same as $\sum_{k=1}^{m} a_k$. Other symbols can also be used. As a numerical example, the sum in Example 3 can be written

$$\sum_{j=1}^{4} j^2(j-3).$$

If n is a positive integer, then the sum of the first n terms of an infinite sequence will be denoted by S_n. For example, given $a_1, a_2, a_3, \ldots, a_n, \ldots,$

$$S_1 = a_1$$
$$S_2 = a_1 + a_2$$
$$S_3 = a_1 + a_2 + a_3$$
$$S_4 = a_1 + a_2 + a_3 + a_4$$

and, in general,

$$S_n = \sum_{k=1}^{n} a_k = a_1 + a_2 + \cdots + a_n.$$

The real number S_n is called the **nth partial sum** of the infinite sequence $a_1, a_2, a_3, \ldots, a_n, \ldots,$ and the sequence

$$S_1, S_2, S_3, \ldots, S_n, \ldots$$

is called a **sequence of partial sums.** Sequences of partial sums are very important in calculus, where the concept of *infinite series* is studied. We shall discuss some special types of infinite series in Section 9.5.

EXAMPLE 4 Find the first four terms and the nth term of the sequence of partial sums associated with the sequence $1, 2, 3, \ldots, n, \ldots$ of positive integers.

SOLUTION The first four terms of the sequence of partial sums are

$$S_1 = 1$$
$$S_2 = 1 + 2 = 3$$
$$S_3 = 1 + 2 + 3 = 6$$
$$S_4 = 1 + 2 + 3 + 4 = 10.$$

From Example 1 of Section 9.1 we see that

$$S_n = 1 + 2 + 3 + \cdots + n = \frac{n(n+1)}{2}.$$

■

If a_k is the same for every positive integer k, say $a_k = c$, where c is a real number, then

$$\sum_{k=1}^{n} a_k = a_1 + a_2 + a_3 + \cdots + a_n$$
$$= c + c + c + \cdots + c = nc.$$

This gives us the following result.

THEOREM

$$\sum_{k=1}^{n} c = nc$$

The domain of the summation variable does not have to begin at 1. For example, the following is self-explanatory:

$$\sum_{k=4}^{8} a_k = a_4 + a_5 + a_6 + a_7 + a_8.$$

As another variation, if the first term of an infinite sequence is a_0, as in

$$a_0, a_1, a_2, \ldots, a_n, \ldots,$$

then sums of the form

$$\sum_{k=0}^{n} a_i = a_0 + a_1 + a_2 + \cdots + a_n$$

may be considered. Note that this is the sum of the first $n + 1$ terms of the sequence.

EXAMPLE 5 Find the sum

$$\sum_{k=0}^{3} \frac{2^k}{(k+1)}.$$

SOLUTION

$$\sum_{k=0}^{3} \frac{2^k}{(k+1)} = \frac{2^0}{(0+1)} + \frac{2^1}{(1+1)} + \frac{2^2}{(2+1)} + \frac{2^3}{(3+1)}$$
$$= 1 + 1 + \frac{4}{3} + 2 = \frac{16}{3}.$$

■

Summation notation can be used to denote polynomials compactly. For example, in place of

$$f(x) = a_0 + a_1x + a_2x^2 + \cdots + a_nx^n$$

we may write
$$f(x) = \sum_{k=0}^{n} a_kx^k.$$

As another illustration, the rather cumbersome formula for the Binomial Theorem (see page 449) can be written

$$(a + b)^n = \sum_{k=0}^{n} \binom{n}{k} a^{n-k}b^k.$$

The following theorem concerning sums has many uses in advanced courses in mathematics.

THEOREM ON SUMS

If $a_1, a_2, \ldots, a_n, \ldots$ and $b_1, b_2, \ldots, b_n, \ldots$ are infinite sequences, then for every positive integer n,

(i) $\sum_{k=1}^{n} (a_k + b_k) = \sum_{k=1}^{n} a_k + \sum_{k=1}^{n} b_k$

(ii) $\sum_{k=1}^{n} (a_k - b_k) = \sum_{k=1}^{n} a_k - \sum_{k=1}^{n} b_k$

(iii) $\sum_{k=1}^{n} ca_k = c\left(\sum_{k=1}^{n} a_k\right)$ for every number c.

PROOF Although the theorem can be proved by mathematical induction, we shall use an argument that makes the truth of the formulas transparent. We begin as follows:

$$\sum_{k=1}^{n} (a_k + b_k) = (a_1 + b_1) + (a_2 + b_2) + (a_3 + b_3) + \cdots + (a_n + b_n).$$

Using commutative and associative properties many times, we may rearrange the terms on the right to produce

$$\sum_{k=1}^{n} (a_k + b_k) = (a_1 + a_2 + a_3 + \cdots + a_n) + (b_1 + b_2 + b_3 + \cdots + b_n).$$

Expressing the right side in summation notation gives us formula (i).

For formula (iii) we have

$$\begin{aligned}\sum_{k=1}^{n} (ca_k) &= ca_1 + ca_2 + ca_3 + \cdots + ca_n \\ &= c(a_1 + a_2 + a_3 + \cdots + a_n) \\ &= c\left(\sum_{k=1}^{n} a_k\right).\end{aligned}$$

The proof of (ii) is left as an exercise. □

Exercises 9.3

In Exercises 1–16 find the first five terms and the eighth term of the sequence that has the given nth term.

1 $a_n = 12 - 3n$

2 $a_n = \dfrac{3}{5n - 2}$

3 $a_n = \dfrac{3n - 2}{n^2 + 1}$

4 $a_n = 10 + \dfrac{1}{n}$

5 $a_n = 9$

6 $a_n = (n - 1)(n - 2)(n - 3)$

7 $a_n = 2 + (-0.1)^n$

8 $a_n = 4 + (0.1)^n$

9 $a_n = (-1)^{n-1} \dfrac{n + 7}{2n}$

10 $a_n = (-1)^n \dfrac{6 - 2n}{\sqrt{n + 1}}$

11 $a_n = 1 + (-1)^{n+1}$

12 $a_n = (-1)^{n+1} + (0.1)^{n-1}$

13 $a_n = \dfrac{2^n}{n^2 + 2}$

14 $a_n = \sqrt{2}$

15 a_n is the number of decimal places in $(0.1)^n$.

16 a_n is the number of positive integers less than n^3.

Find the first five terms of the infinite sequences defined recursively in Exercises 17–24.

17 $a_1 = 2,\ a_{k+1} = 3a_k - 5$

18 $a_1 = 5,\ a_{k+1} = 7 - 2a_k$

19 $a_1 = -3,\ a_{k+1} = a_k^2$

20 $a_1 = 128,\ a_{k+1} = a_k/4$

21 $a_1 = 5,\ a_{k+1} = ka_k$

22 $a_1 = 3,\ a_{k+1} = 1/a_k$

23 $a_1 = 2,\ a_{k+1} = (a_k)^k$

24 $a_1 = 2,\ a_{k+1} = (a_k)^{1/k}$

25 A test question lists the first four terms of a sequence as 2, 4, 6, and 8 and asks for the fifth term. Show that the fifth term can be any real number a by finding the nth term of a sequence that has for its first five terms, 2, 4, 6, 8, and a.

26 The number of bacteria in a certain culture doubles every day. If the initial number of bacteria is 500, how many are present after one day? Two days? Three days? Find a formula for the number of bacteria present after n days.

Find the number given in each of Exercises 27–42.

27 $\displaystyle\sum_{k=1}^{5} (2k - 7)$

28 $\displaystyle\sum_{k=1}^{6} (10 - 3k)$

29 $\displaystyle\sum_{k=1}^{4} (k^2 - 5)$

30 $\displaystyle\sum_{k=1}^{10} [1 + (-1)^k]$

31 $\displaystyle\sum_{k=0}^{5} k(k - 2)$

32 $\displaystyle\sum_{k=0}^{4} (k - 1)(k - 3)$

33 $\displaystyle\sum_{k=3}^{6} \frac{k - 5}{k - 1}$

34 $\displaystyle\sum_{k=1}^{6} \frac{3}{k + 1}$

35 $\displaystyle\sum_{k=1}^{5} (-3)^{k-1}$

36 $\displaystyle\sum_{k=0}^{4} 3(2^k)$

37 $\displaystyle\sum_{k=1}^{100} 100$

38 $\displaystyle\sum_{k=1}^{1000} 5$

39 $\sum_{k=1}^{n} (k^2 + 3k + 5)$ (*Hint:* Use the Theorem on Sums to write the sum as $\sum_{k=1}^{n} k^2 + 3\sum_{k=1}^{n} k + \sum_{k=1}^{n} 5$. Next employ Exercise 9 and Example 1 of Section 9.1, together with the formula for $\sum_{k=1}^{n} c$.)

40 $\sum_{k=1}^{n} (3k^2 - 2k + 1)$

41 $\sum_{k=1}^{n} (2k - 3)^2$

42 $\sum_{k=1}^{n} (k^3 + 2k^2 - k + 4)$

(*Hint:* See Exercise 10 of Section 9.1.)

Express the sums in Exercises 43–52 in terms of summation notation.

43 $1 + 5 + 9 + 13 + 17$

44 $2 + 5 + 8 + 11 + 14$

45 $\frac{1}{2} + \frac{2}{5} + \frac{3}{8} + \frac{4}{11}$

46 $\frac{1}{4} + \frac{2}{9} + \frac{3}{14} + \frac{4}{19}$

47 $1 - \frac{x^2}{2} + \frac{x^4}{4} - \frac{x^6}{6} + \cdots + (-1)^n \frac{x^{2n}}{2n}$

48 $2 - 4 + 8 - 16 + 32 - 64$

49 $1 - \frac{1}{2} + \frac{1}{3} - \frac{1}{4} + \frac{1}{5} - \frac{1}{6} + \frac{1}{7}$

50 $1 + x + \frac{x^2}{2} + \frac{x^3}{3} + \cdots + \frac{x^n}{n}$

51 $\frac{1}{1 \cdot 2} + \frac{1}{2 \cdot 3} + \frac{1}{3 \cdot 4} + \cdots + \frac{1}{99 \cdot 100}$

52 $\frac{1}{1 \cdot 2 \cdot 3} + \frac{1}{2 \cdot 3 \cdot 4} + \frac{1}{3 \cdot 4 \cdot 5} + \cdots + \frac{1}{98 \cdot 99 \cdot 100}$

53 Prove (ii) of the Theorem on Sums.

54 Extend (i) of the Theorem on Sums to $\sum_{k=1}^{n} (a_k + b_k + c_k)$.

55 Prove the Theorem on Sums by mathematical induction.

56 Prove $\sum_{k=1}^{n} c = nc$ by mathematical induction.

Calculator Exercises 9.3

1 Terms of the sequence defined recursively by $a_1 = 5$, $a_{k+1} = \sqrt{a_k}$ may be generated by entering 5 and pressing the square root key repeatedly. Describe what happens to the terms of the sequence as k increases.

2 Terms of the sequence defined recursively by $a_1 = 1$, $a_{k+1} = \cos a_k$ may be generated by entering 1 and pressing the cosine key (in radian mode) repeatedly. Describe what happens to the terms of the sequence as k increases.

3 Approximations to π may be generated from the sequence

$$x_1 = 3, x_{k+1} = x_k - \tan x_k.$$

Find the first five terms of this sequence. What happens to the terms of the sequence when $x_1 = 6$?

4 Approximations to $\sqrt{N}$ may be generated from the sequence

$$x_1 = \frac{N}{2}, x_{k+1} = \frac{1}{2}\left(x_k + \frac{N}{x_k}\right).$$

Approximate x_2, x_3, x_4, x_5, x_6 if $N = 10$.

5 *Bode's Sequence*, defined by the equation $a_1 = \frac{4}{10}$ and $a_k = (3 \cdot 2^{k-2} + 4)/10$ for $k \geq 2$, can be used to approximate distances of the planets from the sun. The third term of the sequence is $a_3 = 1$ *astronomical unit* (or 92,900,000 miles) and corresponds to earth. Approximate the first five terms of the sequence. (The fifth term corresponds to the minor planet Ceres.)

6 The famous *Fibonacci sequence* is defined recursively by $a_{k+1} = a_k + a_{k-1}$ with $a_1 = a_2 = 1$.

(a) Find the first ten terms of the sequence.

(b) The terms of the sequence $r_k = a_{k+1}/a_k$ give progressively better approximations to τ, the *golden ratio.* Approximate the first ten terms of this sequence.

7 The *discrete logistic sequence* is the sequence defined recursively by

$$y_{k+1} = y_k + \frac{r}{K} y_k(K - y_k)$$

where r and K are positive constants. This sequence is used to model the growth of a seasonally breeding animal population in an environment with limited resources.

(a) Find the terms in the sequence if $y_1 = K$.

(b) Approximate the first ten terms in the sequence if $y_1 = 400$, $r = 2$, and $K = 500$. Describe the behavior of this sequence, which predicts the number y_k in the population after k years.

9.4 Arithmetic Sequences

In this section and the next we shall concentrate on two special types of sequences. The first may be defined as follows:

DEFINITION

An **arithmetic sequence** is a sequence such that successive terms differ by the same real number.

Arithmetic sequences are also called **arithmetic progressions.** By definition, a sequence

$$a_1, a_2, a_3, \ldots, a_n, \ldots$$

is arithmetic if and only if there is a real number d such that

$$a_{k+1} - a_k = d$$

for every positive integer k. The number d is called the **common difference** associated with the arithmetic sequence.

EXAMPLE 1 Show that the sequence

$$1, 4, 7, 10, \ldots, 3n - 2, \ldots$$

is arithmetic and find the common difference.

SOLUTION If $a_n = 3n - 2$, then for every positive integer k,

$$\begin{aligned} a_{k+1} - a_k &= [3(k + 1) - 2] - (3k - 2) \\ &= 3k + 3 - 2 - 3k + 2 = 3. \end{aligned}$$

Hence, by definition, the given sequence is arithmetic with common difference 3. ■

Given an arithmetic sequence, we know that

$$a_{k+1} = a_k + d$$

for every positive integer k. This provides a recursive formula for obtaining successive terms. Beginning with any real number a_1, we can obtain an arithmetic sequence with common difference d simply by adding d to a_1, then to $a_1 + d$, and so on, obtaining

$$a_1,\ a_1 + d,\ a_1 + 2d,\ a_1 + 3d,\ a_1 + 4d, \ldots$$

It is evident that the nth term a_n of this sequence is given by the next formula.

nth TERM OF AN ARITHMETIC SEQUENCE

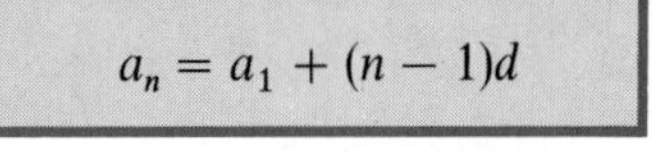

$$a_n = a_1 + (n - 1)d$$

EXAMPLE 2 Find the fifteenth term of the arithmetic sequence whose first three terms are 20, 16.5, and 13.

SOLUTION The common difference is -3.5. Substituting $a_1 = 20$, $d = -3.5$, and $n = 15$ in the formula $a_n = a_1 + (n - 1)d$,

$$a_{15} = 20 + (15 - 1)(-3.5) = 20 - 49 = -29. \qquad \blacksquare$$

EXAMPLE 3 If the fourth term of an arithmetic sequence is 5 and the ninth term is 20, find the sixth term.

SOLUTION Substituting $n = 4$ and $n = 9$ in $a_n = a_1 + (n - 1)d$, and using the fact that $a_4 = 5$ and $a_9 = 20$, we obtain the following system of linear equations in the variables a_1 and d:

$$\begin{cases} 5 = a_1 + (4 - 1)d \\ 20 = a_1 + (9 - 1)d \end{cases} \quad \text{or} \quad \begin{cases} 5 = a_1 + 3d \\ 20 = a_1 + 8d \end{cases}$$

This system has the solution $d = 3$ and $a_1 = -4$. (Verify this fact.) Substitution in the formula $a_n = a_1 + (n - 1)d$ gives us

$$a_6 = (-4) + (6 - 1)(3) = 11. \qquad \blacksquare$$

THEOREM

If $a_1, a_2, \ldots, a_n, \ldots$ is an arithmetic sequence with common difference d, then the nth partial sum S_n is given by both

$$S_n = \frac{n}{2}[2a_1 + (n - 1)d] \quad \text{and} \quad S_n = \frac{n}{2}(a_1 + a_n).$$

PROOF We may write

$$\begin{aligned} S_n &= a_1 + a_2 + a_3 + \cdots + a_n \\ &= a_1 + (a_1 + d) + (a_1 + 2d) + \cdots + [a_1 + (n-1)d]. \end{aligned}$$

Employing commutative and associative properties many times, we obtain

$$S_n = (a_1 + a_1 + a_1 + \cdots + a_1) + [d + 2d + \cdots + (n-1)d]$$

where a_1 appears n times within the first parentheses. It follows that

$$S_n = na_1 + d[1 + 2 + \cdots + (n-1)].$$

The expression within brackets is the sum of the first $n - 1$ positive integers. From Example 1 of Section 9.1 (with $n - 1$ in place of n),

$$1 + 2 + \cdots + (n-1) = \frac{(n-1)n}{2}.$$

Substituting in the last equation for S_n and factoring,

$$S_n = na_1 + d\frac{n(n-1)}{2} = \frac{n}{2}[2a_1 + (n-1)d].$$

Since $a_n = a_1 + (n-1)d$, this is the same as

$$S_n = \frac{n}{2}(a_1 + a_n).$$ □

EXAMPLE 4 Find the sum of all the even integers from 2 through 100.

SOLUTION This problem is equivalent to finding the sum of the first 50 terms of the arithmetic sequence $2, 4, 6, \ldots, 2n, \ldots$ Substituting $n = 50$, $a_1 = 2$, and $a_{50} = 100$ in the second formula of the preceding theorem,

$$S_{50} = \tfrac{50}{2}(2 + 100) = 2550.$$

As a check on our work we may use the first formula of the theorem. Thus,

$$S_{50} = \tfrac{50}{2}[2 \cdot 2 + (50 - 1)2] = 25[4 + 98] = 2550.$$ ■

The **arithmetic mean** of two numbers a and b is defined as $(a + b)/2$. This is also called the **average** of a and b. Note that

$$a, \frac{a+b}{2}, b$$

is an arithmetic sequence. This concept may be generalized as follows: If $c_1, c_2, \ldots, c_k$ are real numbers such that

$$a, c_1, c_2, \ldots, c_k, b$$

is an arithmetic sequence, then $c_1, c_2, \ldots, c_k$ are called the ***k* arithmetic means** of the numbers a and b. The process of determining these numbers is referred to as *inserting* k arithmetic means between a and b.

EXAMPLE 5 Insert three arithmetic means between 2 and 9.

SOLUTION We wish to find three real numbers c_1, c_2, and c_3 such that $2, c_1, c_2, c_3, 9$ is an arithmetic sequence. The common difference d may be found by using the formula $a_n = a_1 + (n - 1)d$ with $n = 5$, $a_5 = 9$, and $a_1 = 2$. This gives us

$$9 = 2 + (5 - 1)d \quad \text{or} \quad d = \tfrac{7}{4}.$$

The three arithmetic means are

$$c_1 = a_1 + d = 2 + \tfrac{7}{4} = \tfrac{15}{4}$$
$$c_2 = c_1 + d = \tfrac{15}{4} + \tfrac{7}{4} = \tfrac{22}{4} = \tfrac{11}{2}$$
$$c_3 = c_2 + d = \tfrac{11}{2} + \tfrac{7}{4} = \tfrac{29}{4}.$$

■

Exercises 9.4

In Exercises 1–8 find the fifth term, the tenth term, and the nth term of the arithmetic sequence.

1 2, 6, 10, 14, . . .

2 16, 13, 10, 7, . . .

3 3, 2.7, 2.4, 2.1, . . .

4 $-6, -4.5, -3, -1.5, \ldots$

5 $-7, -3.9, -0.8, 2.3, \ldots$

6 $x - 8, x - 3, x + 2, x + 7, \ldots$

7 $\ln 3, \ln 9, \ln 27, \ln 81, \ldots$

8 $\log 1000, \log 100, \log 10, 0, \ldots$

9 Find the twelfth term of the arithmetic sequence whose first two terms are 9.1 and 7.5.

10 Find the eleventh term of the arithmetic sequence whose first two terms are $2 + \sqrt{2}$ and 3.

11 The sixth and seventh terms of an arithmetic sequence are 2.7 and 5.2. Find the first term.

12 Given an arithmetic sequence with $a_3 = 7$ and $a_{20} = 43$, find a_{15}.

In Exercises 13–16 find the sum S_n of the arithmetic sequence that satisfies the stated condition.

13 $a_1 = 40,\ d = -3,\ n = 30$

14 $a_1 = 5,\ d = 0.1,\ n = 40$

15 $a_1 = -9,\ a_{10} = 15,\ n = 10$

16 $a_7 = \frac{7}{3},\ d = -\frac{2}{3},\ n = 15$

Find the sums in Exercises 17–20.

17 $\sum_{k=1}^{20} (3k - 5)$

18 $\sum_{k=1}^{12} (7 - 4k)$

19 $\sum_{k=1}^{18} \left(\frac{1}{2}k + 7\right)$

20 $\sum_{k=1}^{10} \left(\frac{1}{4}k + 3\right)$

21 How many integers between 32 and 395 are divisible by 6? Find their sum.

22 How many negative integers greater than -500 are divisible by 33? Find their sum.

23 How many terms are in an arithmetic sequence with first term -2, common difference $\frac{1}{4}$, and sum 21?

24 How many terms are in an arithmetic sequence with sixth term -3, common difference 0.2, and sum -33?

25 Insert five arithmetic means between 2 and 10.

26 Insert three arithmetic means between 3 and -5.

27 A pile of logs has 24 logs in the first layer, 23 in the second, 22 in the third, and so on. The top layer contains 10 logs. Find the total number of logs in the pile.

28 A seating section in a certain athletic stadium has 30 seats in the first row, 32 seats in the second, 34 in the third, and so on, until the tenth row is reached, after which there are 10 more rows, each containing 50 seats. Find the total number of seats in the section.

29 A man wishes to construct a ladder with nine rungs that diminish uniformly from 24 inches at the base to 18 inches at the top. Determine the lengths of the seven intermediate rungs.

30 A boy on a bicycle coasts down a hill, covering 4 feet the first second and in each succeeding second 5 feet more than in the preceding second. If he reaches the bottom of the hill in 11 seconds, find the total distance traveled.

31 A contest will have five cash prizes totaling \$5000, and there will be a \$100 difference between successive prizes. Find the first prize.

32 A company is to distribute \$46,000 in bonuses to its top ten salespeople. The tenth salesperson on the list will receive \$1000, and the difference in bonus money between successively ranked salespeople is to be constant. Find the bonuses for each salesperson.

33 Certain polygons with n sides have the property that the measures of the interior angles can be arranged in an arithmetic sequence. If, in a polygon of this type, the smallest angle is 20° and the largest angle is 160°, determine the number of sides of the polygon. (*Hint:* Find an expression for the sum of the interior angles if the polygon has n sides.)

34 If f is a linear function, show that the sequence with nth term $a_n = f(n)$ is an arithmetic sequence.

35 The sequence defined recursively by $x_{k+1} = x_k/(1 + x_k)$ arises in genetics in the study of the elimination of a deficient gene from a population. Show that the sequence with nth term $1/x_n$ is arithmetic.

36 Find the total length of the curve in the figure if the width of the maze formed by the curve is 16 inches and all halls in the maze have width 1 inch. What is the length if the width of the maze is 32 inches?

FIGURE FOR EXERCISE 36

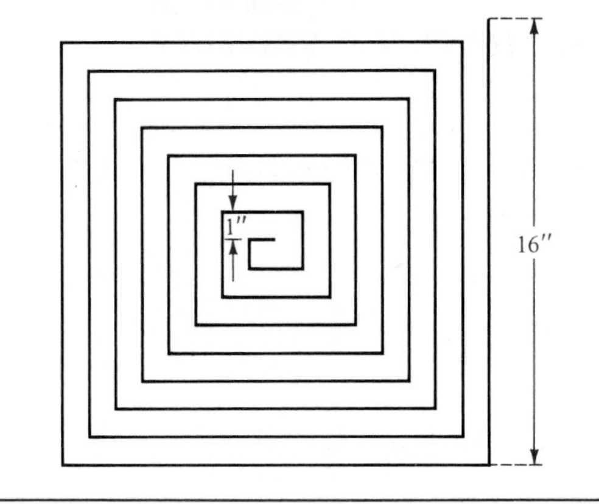

9.5 Geometric Sequences

Another important type of infinite sequence is defined as follows.

DEFINITION

A sequence $a_1, a_2, \ldots, a_n, \ldots$ is a **geometric sequence** if there is a real number $r \neq 0$ such that

$$\frac{a_{k+1}}{a_k} = r$$

for every positive integer k.

Geometric sequences are also called **geometric progressions.** The number r in the definition is called the **common ratio** associated with the geometric sequence. Thus $a_1, a_2, \ldots, a_n, \ldots$ is geometric (with common ratio r) if

$$a_{k+1} = a_k r$$

for every positive integer k. This provides a recursive method for obtaining terms. Beginning with any nonzero real number a_1, we multiply by the number r successively, obtaining

$$a_1, a_1r, a_1r^2, a_1r^3, \ldots$$

The nth term a_n of this sequence is given by the next formula.

***n*th TERM OF A GEOMETRIC SEQUENCE**

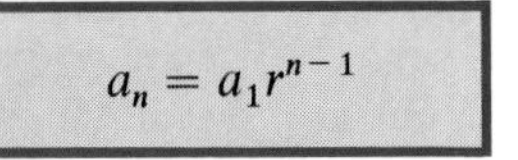

$$a_n = a_1r^{n-1}$$

EXAMPLE 1 Find the first five terms and the tenth term of the geometric sequence having first term 3 and common ratio $-\frac{1}{2}$.

SOLUTION If we let $a_1 = 3$ and $r = -\frac{1}{2}$, then the first five terms are

$$3, -\tfrac{3}{2}, \tfrac{3}{4}, -\tfrac{3}{8}, \tfrac{3}{16}.$$

Using the formula $a_n = a_1r^{n-1}$ with $n = 10$,

$$a_{10} = 3(-\tfrac{1}{2})^9 = -\tfrac{3}{512}.$$

■

EXAMPLE 2 If the third term of a geometric sequence is 5 and the sixth term is -40, find the eighth term.

SOLUTION We are given $a_3 = 5$ and $a_6 = -40$. Substituting $n = 3$ and $n = 6$ in the formula $a_n = a_1 r^{n-1}$ leads to the following system of equations:

$$\begin{cases} 5 = a_1 r^2 \\ -40 = a_1 r^5. \end{cases}$$

Since $r \neq 0$, the first equation is equivalent to $a_1 = 5/r^2$. Substituting for a_1 in the second equation,

$$-40 = \left(\frac{5}{r^2}\right) \cdot r^5 = 5r^3.$$

Hence, $r^3 = -8$ and $r = -2$. If we now substitute -2 for r in the equation $5 = a_1 r^2$, we obtain $a_1 = \frac{5}{4}$. Finally, using $a_n = a_1 r^{n-1}$ with $n = 8$,

$$a_8 = (\tfrac{5}{4})(-2)^7 = -160.$$ ■

Let us find a formula for S_n, the nth partial sum of a geometric sequence. If the first term is a_1 and the common ratio is r,

$$S_n = a_1 + a_1 r + a_1 r^2 + \cdots + a_1 r^{n-2} + a_1 r^{n-1}.$$

If $r = 1$, then $S_n = na_1$. Next suppose that $r \neq 1$. Multiplying both sides of the general formula for S_n by r,

$$rS_n = a_1 r + a_1 r^2 + a_1 r^3 + \cdots + a_1 r^{n-1} + a_1 r^n.$$

If we subtract the preceding equation from that for S_n, then many terms on the right side drop out, leaving

$$S_n - rS_n = a_1 - a_1 r^n \quad \text{or} \quad (1 - r)S_n = a_1(1 - r^n).$$

Since $r \neq 1$, we have $1 - r \neq 0$, and dividing both sides by $1 - r$ gives us

$$S_n = \frac{a_1(1 - r^n)}{1 - r}.$$

We have proved the following theorem.

THEOREM

The nth partial sum of a geometric sequence with first term a_1 and common ratio $r \neq 1$ is

$$S_n = a_1 \frac{(1 - r^n)}{1 - r}.$$

EXAMPLE 3 Find the sum of the first five terms of the geometric sequence that begins as follows: 1, 0.3, 0.09, 0.027, ...

SOLUTION If we let $a_1 = 1$, $r = 0.3$, and $n = 5$ in the formula for S_n, then

$$S_5 = 1\left(\frac{1 - (0.3)^5}{1 - 0.3}\right).$$

This reduces to $S_5 = 1.4251$. ■

EXAMPLE 4 A man wishes to save money by setting aside 1 cent the first day, 2 cents the second day, 4 cents the third day and so on, doubling the amount each day. If this is continued, how much must be set aside on the fifteenth day? Assuming he does not run out of money, what is the total amount saved at the end of 30 days?

SOLUTION The amount (in cents) set aside on successive days forms a geometric sequence

$$1, 2, 4, 8, \ldots$$

with first term 1 and common ratio 2. The amount needed for the fifteenth day is found by using $a_n = a_1 r^{n-1}$ with $a_1 = 1$ and $n = 15$. This gives us $1 \cdot 2^{14}$, or \$163.84. To find the total amount set aside after 30 days, we use the formula for S_n with $n = 30$. Thus,

$$S_{30} = 1\,\frac{(1 - 2^{30})}{1 - 2},$$

which simplifies to \$10,737,418.23. ■

Given the geometric series with first term a_1 and common ratio $r \neq 1$, we may write the formula for S_n of the last theorem in the form

$$S_n = \frac{a_1}{1 - r} - \frac{a_1}{1 - r}\,r^n.$$

Although we shall not do it here, it can be shown that if $|r| < 1$, then r^n *approaches* 0 *as n increases* without bound; that is, we can make r^n as close as we wish to 0 by taking n sufficiently large. It follows that S_n approaches $a_1/(1 - r)$ as n increases without bound. Using the notation introduced in our work with rational functions in Chapter 3, this may be expressed symbolically as

$$S_n \to \frac{a_1}{1 - r} \quad \text{as} \quad n \to \infty.$$

The number $a_1/(1 - r)$ is called the *sum* of the **infinite geometric series**

$$a_1 + a_1r + a_1r^2 + \cdots + a_1r^{n-1} + \cdots$$

This gives us the next result.

THEOREM

> If $|r| < 1$, then the infinite geometric series
>
> $$a_1 + a_1r + a_1r^2 + \cdots + a_1r^{n-1} + \cdots$$
>
> has the sum $a_1/(1 - r)$.

The preceding theorem implies that if we add more and more terms of the indicated infinite geometric series, the sums get closer and closer to $a_1/(1 - r)$. The next example illustrates how the theorem can be used to show that every real number represented by a repeating decimal is rational.

EXAMPLE 5 Find the rational number that corresponds to the infinite repeating decimal $5.4\overline{27}$, where the bar means that the block of digits underneath is repeated indefinitely.

SOLUTION From the decimal expression 5.4272727 . . . , we obtain the infinite series

$$5.4 + 0.027 + 0.00027 + 0.0000027 + \cdots$$

The part of the expression after the first term is

$$0.027 + 0.00027 + 0.0000027 + \cdots,$$

which has the form given in the last theorem with $a_1 = 0.027$ and $r = 0.01$. Hence, the sum of this infinite geometric series is

$$\frac{0.027}{1 - 0.01} = \frac{0.027}{0.99} = \frac{27}{990} = \frac{3}{110}.$$

Thus, it appears that the desired number is $5.4 + \frac{3}{110}$, or $\frac{597}{110}$. A check by division shows that $\frac{597}{110}$ does equal the given repeating decimal. ■

In general, given any infinite sequence $a_1, a_2, \ldots, a_n, \ldots$, the expression

$$a_1 + a_2 + \cdots + a_n + \cdots$$

is called an **infinite series,** or simply a **series.** In summation notation, this series is denoted by

$$\sum_{n=1}^{\infty} a_n.$$

Each number a_k is called a **term** of the series and a_n is called the ***n*th term.** Since only finite sums may be added algebraically, it is necessary to *define* what is meant by an infinite sum. This is done by considering the sequence of partial sums

$$S_1, S_2, \ldots, S_n, \ldots$$

If there is a number S such that $S_n \to S$ as $n \to \infty$, then, as in our discussion of infinite geometric series, we call S the **sum** of the series and write

$$S = a_1 + a_2 + \cdots + a_n + \cdots$$

Thus, in Example 5 we may write

$$\tfrac{597}{110} = 5.4 + 0.027 + 0.00027 + 0.0000027 + \cdots$$

The next example provides an illustration of a nongeometric infinite series. Since a complete discussion is beyond the scope of this text, the solution is presented in an intuitive manner.

EXAMPLE 6 Show that the following infinite series has a sum:

$$\frac{1}{1 \cdot 2} + \frac{1}{2 \cdot 3} + \frac{1}{3 \cdot 4} + \cdots + \frac{1}{n(n+1)} + \cdots$$

SOLUTION The nth term $a_n = 1/n(n+1)$ has the partial fraction decomposition

$$a_n = \frac{1}{n(n+1)} = \frac{1}{n} - \frac{1}{n+1}.$$

Consequently, the nth partial sum of the series may be written

$$\begin{aligned} S_n &= a_1 + a_2 + a_3 + \cdots + a_n \\ &= \left(1 - \frac{1}{2}\right) + \left(\frac{1}{2} - \frac{1}{3}\right) + \left(\frac{1}{3} - \frac{1}{4}\right) + \cdots + \left(\frac{1}{n} - \frac{1}{n+1}\right) \\ &= 1 - \frac{1}{n+1}. \end{aligned}$$

Since $1/(n+1) \to 0$ as $n \to \infty$, it follows that $S_n \to 1$, and we may write

$$1 = \frac{1}{1 \cdot 2} + \frac{1}{2 \cdot 3} + \frac{1}{3 \cdot 4} + \cdots + \frac{1}{n(n+1)} + \cdots$$

This means that, as we add more and more terms of the series, the sums get closer and closer to 1. ■

If the terms of an infinite sequence are alternately positive and negative, and if we consider the expression

$$a_1 + (-a_2) + a_3 + (-a_4) + \cdots + [(-1)^{n+1}a_n] + \cdots$$

where all the a_k are positive real numbers, then this expression is referred to as an **alternating infinite series,** and we write it in the form

$$a_1 - a_2 + a_3 - a_4 + \cdots + (-1)^{n+1}a_n + \cdots$$

Illustrations of alternating infinite series can be obtained by using infinite geometric series with negative common ratio.

Infinite series have many applications in mathematics and the sciences. Texts on calculus contain careful treatments of this important concept.

Exercises 9.5

In Exercises 1–12 find the fifth term, the eighth term, and the nth term of the geometric sequence.

1. $8, 4, 2, 1, \ldots$
2. $4, 1.2, 0.36, 0.108, \ldots$
3. $300, -30, 3, -0.3, \ldots$
4. $1, -\sqrt{3}, 3, -\sqrt{27}, \ldots$
5. $5, 25, 125, 625, \ldots$
6. $2, 6, 18, 54, \ldots$
7. $4, -6, 9, -13.5, \ldots$
8. $162, -54, 18, -6, \ldots$
9. $1, -x^2, x^4, -x^6, \ldots$
10. $1, -\frac{x}{3}, \frac{x^2}{9}, -\frac{x^3}{27}, \ldots$
11. $2, 2^{x+1}, 2^{2x+1}, 2^{3x+1}, \ldots$
12. $10, 10^{2x-1}, 10^{4x-3}, 10^{6x-5}, \ldots$
13. Find the sixth term of the geometric sequence whose first two terms are 4 and 6.
14. Find the seventh term of the geometric sequence that has 2 and $-\sqrt{2}$ for its second and third terms, respectively.
15. In a certain geometric sequence $a_5 = \frac{1}{16}$ and $r = \frac{3}{2}$. Find a_1 and S_5.
16. Given a geometric sequence such that $a_4 = 4$ and $a_7 = 12$, find r and a_{10}.

Find the sums in Exercises 17–20.

17. $\sum_{k=1}^{10} 3^k$
18. $\sum_{k=1}^{9} (-\sqrt{5})^k$
19. $\sum_{k=0}^{9} (-\frac{1}{2})^{k+1}$
20. $\sum_{k=1}^{7} (3^{-k})$
21. A vacuum pump removes one-half of the air in a container at each stroke. After 10 strokes, what percentage of the original amount of air remains in the container?
22. The yearly depreciation of a certain machine is 25% of its value at the beginning of the year. If the original cost of the machine is \$20,000, what is its value after 6 years?

23 A culture of bacteria increases 20% every hour. For the original culture containing 10,000 bacteria find a formula for the number of bacteria present after t hours. How many bacteria are in the culture at the end of 10 hours?

24 If an amount of money P is deposited in a savings account that pays interest at a rate of r percent per year compounded quarterly, and if the principal and accumulated interest are left in the account, find a formula for the total amount in the account after n years.

Find the sums of the infinite geometric series in Exercises 25–30, whenever they exist.

25 $1 - \frac{1}{2} + \frac{1}{4} - \frac{1}{8} + \cdots$

26 $2 + \frac{2}{3} + \frac{2}{9} + \frac{2}{27} + \cdots$

27 $1.5 + 0.015 + 0.00015 + \cdots$

28 $1 - 0.1 + 0.01 - 0.001 + \cdots$

29 $\sqrt{2} - 2 + \sqrt{8} - 4 + \cdots$

30 $250 - 100 + 40 - 16 + \cdots$

In Exercises 31–38 find the rational number represented by the repeating decimal.

31 $0.\overline{23}$

32 $0.0\overline{71}$

33 $2.4\overline{17}$

34 $10.\overline{55}$

35 $5.\overline{146}$

36 $3.2\overline{394}$

37 $1.\overline{6124}$

38 $123.61\overline{83}$

39 A rubber ball is dropped from a height of 10 meters. If it rebounds approximately one-half the distance after each fall, use an infinite geometric series to approximate the total distance the ball travels before coming to rest.

40 The bob of a pendulum swings through an arc 24 cm long on its first swing. If each successive swing is approximately five-sixths the length of the preceding swing, use an infinite geometric series to approximate the total distance it travels before coming to rest.

41 A branch of a clothing manufacturing company has just located in a small community and will pay two million dollars per year in salaries. It has been estimated that 60% of these salaries will be spent in the local area, and of this money spent another 60% will again change hands within the community. This process will be repeated ad infinitum. This is called the Multiplier Effect. Find the total amount of spending that will be generated by company salaries.

42 In a pest eradication program, N sterilized male flies are released into the general population each day, and 90% of these flies will survive a given day.

(a) Show that the number of sterilized flies in the population n days after the program has begun is

$$N + (0.9)N + (0.9)^2 N + \cdots + (0.9)^{n-1} N.$$

(b) If the *long-range* goal of the program is to keep 20,000 sterilized males in the population, how many flies should be released each day?

43 A certain drug has a half-life of about 2 hours in the bloodstream. Doses of D mg will be administered every 4 hours, where D is still to be determined.

(a) Show that the number of milligrams of drug in the bloodstream after the nth dose has been administered is

$$D + \tfrac{1}{4}D + \cdots + (\tfrac{1}{4})^{n-1} D$$

and that this sum is approximately $\frac{4}{3}D$ for large values of n.

(b) If it is considered dangerous to have more than 500 mg of the drug in the bloodstream, find the largest possible dose that can be given repeatedly over a long period of time.

44 Shown in the figure is a family tree displaying 3 prior generations and a total of 12 grandparents. If you were to trace your family history back 10 generations, how many grandparents would you find?

FIGURE FOR EXERCISE 44

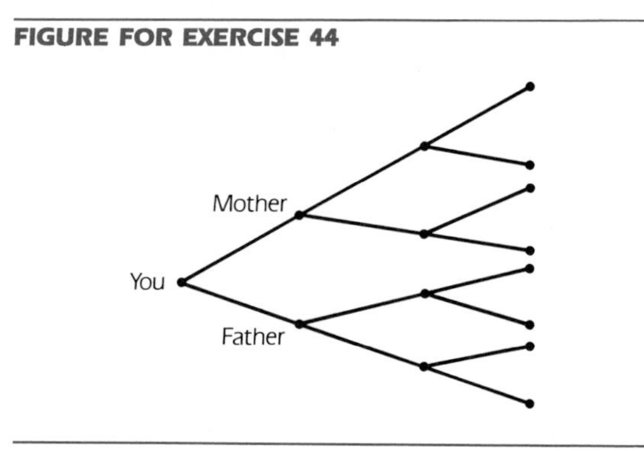

45 Shown in the first figure is a nested sequence of squares $S_1, S_2, \ldots, S_k, \ldots$. Let a_k, A_k, and P_k denote the side, area, and perimeter, respectively, of the square S_k. The square S_{k+1} is constructed from S_k by selecting four points on S_k that are at a distance of $\frac{1}{4}a_k$ from the vertices and connecting them (see figure at right).

(a) Find the relationship between a_{k+1} and a_k.

(b) Find a_n, A_n, and P_n.

(c) Calculate $\sum_{n=1}^{\infty} P_n$.

FIGURES FOR EXERCISE 45

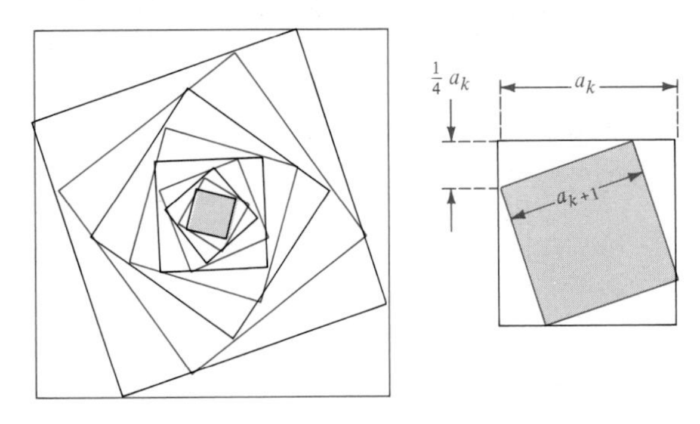

9.6 Review

Define or discuss each of the following.

1 Axiom of Mathematical Induction
2 Principle of Mathematical Induction
3 Factorial notation
4 The Binomial Theorem
5 Binomial coefficients
6 Pascal's Triangle
7 Infinite sequence
8 Summation notation
9 The nth partial sum of an infinite sequence
10 Arithmetic sequence
11 Arithmetic mean of two numbers
12 Geometric sequence
13 Infinite geometric series
14 Infinite series

Exercises 9.6

Prove that the statements in Exercises 1–5 are true for every positive integer n.

1 $2 + 5 + 8 + \cdots + (3n - 1) = \dfrac{n(3n + 1)}{2}$

2 $2^2 + 4^2 + 6^2 + \cdots + (2n)^2 = \dfrac{2n(2n + 1)(n + 1)}{3}$

3 $\dfrac{1}{1 \cdot 3} + \dfrac{1}{3 \cdot 5} + \dfrac{1}{5 \cdot 7} + \cdots + \dfrac{1}{(2n - 1)(2n + 1)} = \dfrac{n}{2n + 1}$

4 $1 \cdot 2 + 2 \cdot 3 + 3 \cdot 4 + \cdots + n(n + 1) = \dfrac{n(n + 1)(n + 2)}{3}$

5 3 is a factor of $n^3 + 2n$.

6 Prove that $2^n > n^2$ for every positive integer $n \geq 5$.

7 Expand and simplify $(x^2 - 3y)^6$.

8 Find the first four terms in the binomial expansion of $(a^{2/5} + 2a^{-3/5})^{20}$.

9 Find the sixth term in the expansion of $(b^3 - \frac{1}{2}c^2)^9$.

10 In the expansion of $(2c^3 + 5c^{-2})^{10}$ find the term that does not contain c.

In Exercises 11–14 find the first four terms and the seventh term of the sequence that has the given nth term.

11 $a_n = \dfrac{5n}{3 - 2n^2}$

12 $a_n = (-1)^{n+1} - (0.1)^n$

13 $a_n = 1 + (-\frac{1}{2})^{n-1}$

14 $a_n = \dfrac{2^n}{(n+1)(n+2)(n+3)}$

Find the first five terms of the infinite sequences defined recursively in Exercises 15–18.

15 $a_1 = 10,\ a_{k+1} = 1 + (1/a_k)$

16 $a_1 = 2,\ a_{k+1} = a_k!$

17 $a_1 = 9,\ a_{k+1} = \sqrt{a_k}$

18 $a_1 = 1,\ a_{k+1} = (1 + a_k)^{-1}$

Find the numbers represented by the sums in Exercises 19–22.

19 $\sum_{k=1}^{5} (k^2 + 4)$

20 $\sum_{k=2}^{6} \dfrac{2k - 8}{k - 1}$

21 $\sum_{k=1}^{100} 10$

22 $\sum_{k=1}^{4} (2^k - 10)$

In Exercises 23–26 use summation notation to represent the sums.

23 $3 + 6 + 9 + 12 + 15$

24 $2 + 4 + 8 + 16 + 32 + 64 + 128$

25 $100 - 95 + 90 - 85 + 80$

26 $a_0 + a_4x^4 + a_8x^8 + \cdots + a_{100}x^{100}$

27 Find the tenth term and the sum of the first ten terms of the arithmetic sequence whose first two terms are $4 + \sqrt{3}$ and 3.

28 Find the sum of the first eight terms of an arithmetic sequence in which the fourth term is 9 and the common difference is -5.

29 The fifth and thirteenth terms of an arithmetic sequence are 5 and 77, respectively. Find the first term and the tenth term.

30 Insert four arithmetic means between 20 and -10.

31 Find the tenth term of the geometric sequence whose first two terms are $\frac{1}{8}$ and $\frac{1}{4}$.

32 If a geometric sequence has 3 and -0.3 as its third and fourth terms, find the eighth term.

33 Find a positive number c such that 4, c, 8 are successive terms of a geometric sequence.

34 In a certain geometric sequence the eighth term is 100 and the common ratio is $-\frac{3}{2}$. Find the first term.

Find the sums in Exercises 35–38.

35 $\sum_{k=1}^{15} (5k - 2)$

36 $\sum_{k=1}^{10} (6 - \frac{1}{2}k)$

37 $\sum_{k=1}^{10} (2^k - \frac{1}{2})$

38 $\sum_{k=1}^{8} (\frac{1}{2} - 2^k)$

39 Find the sum of the infinite geometric series

$$1 - \frac{2}{5} + \frac{4}{25} - \frac{8}{125} + \cdots.$$

40 Find the rational number whose decimal representation is $6.\overline{274}$.

41 Ten-foot lengths of 2×2 lumber are to be cut into five pieces to form children's building blocks, and the lengths of the five blocks are to form an arithmetic sequence.

(a) Show that the difference d in lengths must be less than 1 foot.

(b) If the smallest block is to have a length of 6 inches, find the lengths of the other four pieces.

42 When a ball is dropped from a height of h feet, it takes $\sqrt{h}/4$ seconds to reach the ground. It then takes $\sqrt{d}/4$ seconds for the ball to rebound to a height of d feet. If a rubber ball is dropped from a height of 10 feet and rebounds to three-fourths of its height after each fall, how many seconds elapse before the ball comes to rest?

CHAPTER 10

Topics in Analytic Geometry

Plane geometry includes the study of figures, such as lines, circles, and triangles, that lie in a plane. ■ Theorems are proved by reasoning deductively from certain postulates. ■ In *analytic* geometry, plane geometric figures are investigated by introducing a coordinate system and then using equations and formulas of various types. ■ If the study of analytic geometry were to be summarized by means of one statement, perhaps the following would be appropriate: "Given an equation, find its graph and, conversely, given a graph, find its equation." ■ In this chapter we shall apply coordinate methods to several basic plane figures.

10.1 Conic Sections

Each of the geometric figures discussed in Sections 10.2–10.5 can be obtained by intersecting a double-napped right circular cone with a plane. For this reason they are called **conic sections** or simply **conics.** If, as in Figure 10.1(i), the plane cuts entirely across one nappe of the cone and is not perpendicular to the axis, then the curve of intersection is called an **ellipse.** If the plane is perpendicular to the axis of the cone, a **circle** results. If the plane does not cut across one entire nappe and does not intersect both nappes, as illustrated in Figure 10.1(ii), then the curve of intersection is a **parabola.** If the plane cuts through both nappes of the cone, as in (iii) of the figure, we obtain a **hyperbola.**

FIGURE 10.1

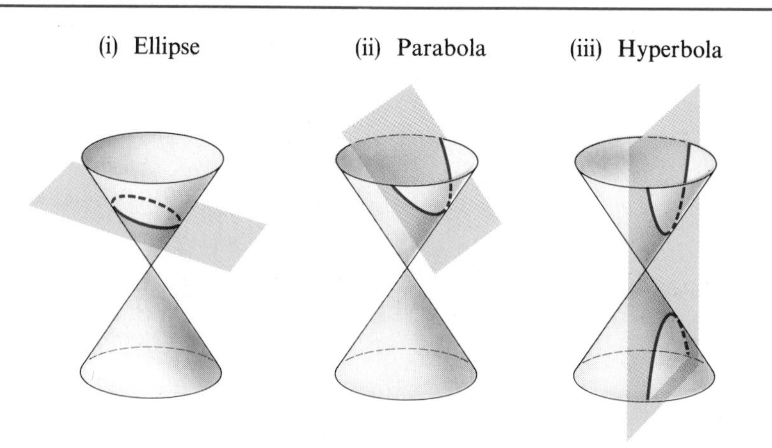

By changing the position of the plane and the shape of the cone, conics can be made to vary considerably. Certain positions of the plane result in **degenerate conics.** For example, if the plane intersects the cone only at the vertex, then the conic consists of one point. If the axis of the cone lies on the plane, then a pair of intersecting lines is obtained. Finally, if we begin with the parabolic case, as in Figure 10.1(ii), and move the plane parallel to its initial position until it coincides with one of the generators of the cone, we obtain a line.

The conic sections were studied extensively by the ancient Greeks, who discovered properties that enable us to define conics in terms of points (foci) and lines (directrices) in the plane of the conic. A remarkable fact about conic sections is that although they were studied thousands of years ago, they are far from obsolete. Indeed, they are important tools for present-day investigations in outer space and for the study of the behavior of atomic particles. It is shown in physics that if a particle moves under the influence of an *inverse square force field*, then its path may be described by means of a conic section. Examples of inverse square fields are gravitational and electromagnetic fields. Planetary orbits are elliptical. If the ellipse is very "flat," the curve resembles the path of a comet. The hyperbola is useful for describing the path of an alpha particle in the electric field of the nucleus of an atom. Parabolic mirrors are sometimes used to collect solar energy. The interested person can find many other applications of conic sections.

10.2 Parabolas

Parabolas were discussed in Section 2.3; however, the definition was not stated at that time. Moreover, we concentrated on parabolas with vertical axes. We shall now define *parabola* and derive equations for parabolas that have either vertical or horizontal axes.

DEFINITION

A **parabola** is the set of all points in a plane equidistant from a fixed point F (the **focus**) and a fixed line l (the **directrix**) in the plane.

We shall assume that F is not on l, for otherwise, a line is obtained. If P is a point in the plane and P' is the point on l determined by a line through P that is perpendicular to l (see Figure 10.2), then by definition, P is on the parabola if and only if $d(P, F) = d(P, P')$. The point P can lie anywhere on the curve in Figure 10.2. The line through F, perpendicular to the directrix, is called the **axis** of the parabola. The point V on the axis, half-way from F to l, is called the **vertex** of the parabola.

FIGURE 10.2

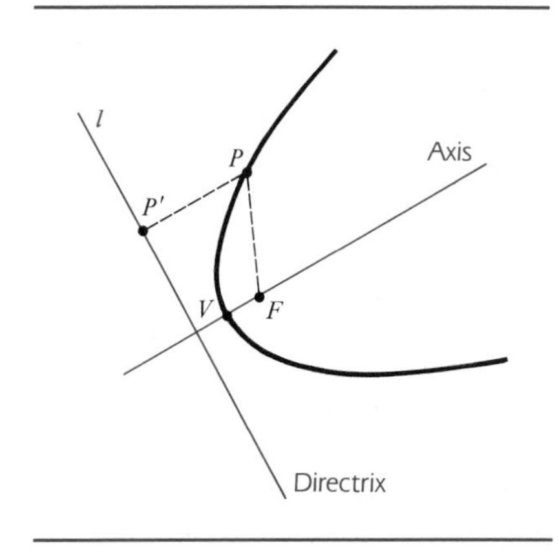

We can obtain a simple equation for a parabola by placing the y-axis along the axis of the parabola, with the origin at the vertex V, as illustrated in Figure 10.3. In this case, the focus F has coordinates $(0, p)$ for some real number $p \neq 0$, and the equation of the directrix is $y = -p$. By the Distance Formula, a point $P(x, y)$ is on the parabola if and only if

$$\sqrt{(x - 0)^2 + (y - p)^2} = \sqrt{(x - x)^2 + (y + p)^2}.$$

Squaring both sides gives us

$$(x - 0)^2 + (y - p)^2 = (y + p)^2,$$

FIGURE 10.3

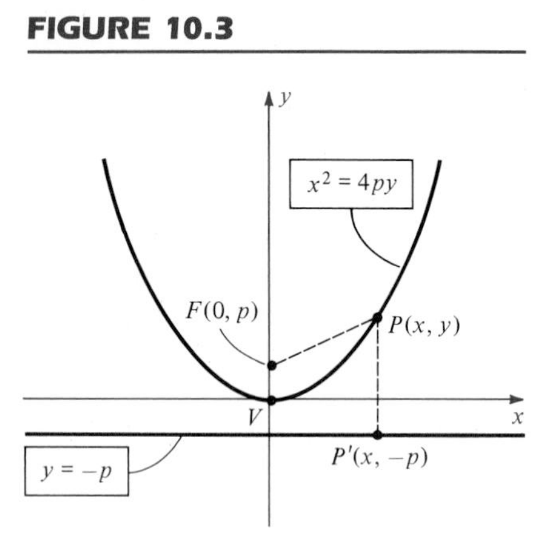

or $$x^2 + y^2 - 2py + p^2 = y^2 + 2py + p^2.$$

This simplifies to $$x^2 = 4py.$$

We have shown that the coordinates of every point (x, y) on the parabola satisfy $x^2 = 4py$. Conversely, if (x, y) is a solution of this equation, then by reversing the previous steps we see that the point (x, y) is on the parabola. If $p > 0$ the parabola opens upward, as in Figure 10.3, whereas if $p < 0$ the parabola opens downward. Note that the graph is symmetric with respect to the y-axis, since the solutions of the equation $x^2 = 4py$ are unchanged if $-x$ is substituted for x.

An analogous situation exists if the axis of the parabola is placed along the x-axis. If the vertex is $V(0, 0)$, the focus is $F(p, 0)$, and the directrix has equation $x = -p$ (see Figure 10.4), then using the same type of argument we obtain the equation $y^2 = 4px$. If $p > 0$ the parabola opens to the right, whereas if $p < 0$ it opens to the left. In this case the graph is symmetric with respect to the x-axis.

FIGURE 10.4

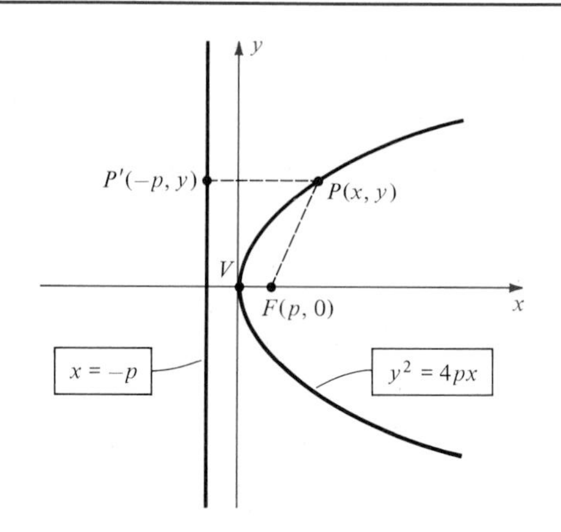

The following theorem summarizes our discussion.

THEOREM

The graph of each of the following equations is a parabola that has its vertex at the origin and has the indicated focus and directrix.

(i) $x^2 = 4py$: focus $F(0, p)$, directrix $y = -p$.

(ii) $y^2 = 4px$: focus $F(p, 0)$, directrix $x = -p$.

FIGURE 10.5

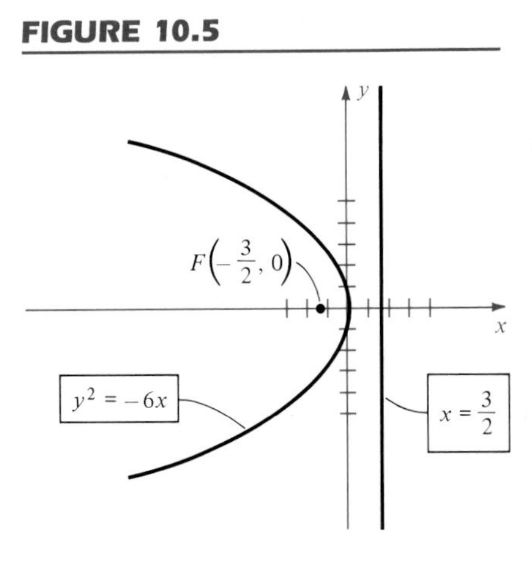

EXAMPLE 1 Find the focus and directrix of the parabola that has the equation $y^2 = -6x$, and sketch its graph.

SOLUTION The equation $y^2 = -6x$ has form (ii) of the preceding theorem with $4p = -6$, and hence $p = -\frac{3}{2}$. Thus the focus is $F(p, 0)$, that is, $F(-\frac{3}{2}, 0)$. The equation of the directrix is $x = -p$, or $x = \frac{3}{2}$. The graph is sketched in Figure 10.5. ■

EXAMPLE 2 Find an equation of the parabola that has its vertex at the origin, opens upward, and passes through the point $P(-3, 7)$.

SOLUTION The general form of the equation is $x^2 = 4py$ (see (i) of the theorem). If $P(-3, 7)$ is on the parabola, then $(-3, 7)$ is a solution of the equation. Hence we must have $(-3)^2 = 4p(7)$, or $p = \frac{9}{28}$. Substituting for p in $x^2 = 4py$ leads to the desired equation $x^2 = \frac{9}{7}y$, or $7x^2 = 9y$. ■

We may use a technique called **translation of axes** to handle the case in which the vertex is not at the origin. Recall that if a and b are the coordinates of two points A and B, respectively, on a coordinate line l, then the distance between A and B is $d(A, B) = |b - a|$. If we wish to take into account the direction of l, then we use the **directed distance** $\overline{AB}$ from A to B, which is, by definition,

$$\overline{AB} = b - a.$$

Since $\overline{BA} = a - b$, we have $\overline{AB} = -\overline{BA}$. If the positive direction on l is to the right, then B is to the right of A if and only if $\overline{AB} > 0$, and B is to the left of A if and only if $\overline{AB} < 0$. If C is any other point on l with coordinate c, it follows that

$$\overline{AC} = \overline{AB} + \overline{BC},$$

since $c - a = (b - a) + (c - b)$. We shall use this formula in the following discussion to develop formulas for translation of axes.

FIGURE 10.6

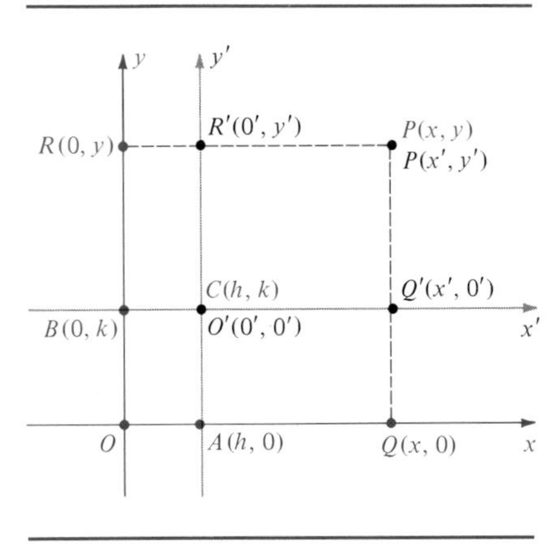

Suppose that $C(h, k)$ is an arbitrary point in an xy-coordinate plane. Let us introduce a new $x'y'$-coordinate system with origin O' at C such that the x'- and y'-axes are parallel to, and have the same unit lengths and positive directions as, the x- and y-axes, respectively. A typical situation is illustrated in Figure 10.6 where, for simplicity, we have placed C in the first quadrant. We shall use primes on letters to denote coordinates of points in the $x'y'$-coordinate system, to distinguish them from coordinates with respect to the xy-coordinate system. Thus the point $P(x, y)$ in the xy-system will be denoted by $P(x', y')$ in the $x'y'$-system. If we label projections of P on the various axes as in Figure 10.6, and let A and B denote projections of C on the x- and y-axes, respectively, then using directed distances we obtain

$$x = \overline{OQ} = \overline{OA} + \overline{AQ} = \overline{OA} + \overline{O'Q'} = h + x'$$
$$y = \overline{OR} = \overline{OB} + \overline{BR} = \overline{OB} + \overline{O'R'} = k + y'.$$

This proves the following result.

TRANSLATION OF AXES FORMULAS

If (x, y) are the coordinates of a point P relative to an xy-coordinate system, and if (x', y') are the coordinates of P relative to an $x'y'$-coordinate system with origin at the point $C(h, k)$ of the xy-system, then

$$\text{(i)} \quad x = x' + h, \quad y = y' + k$$
$$\text{(ii)} \quad x' = x - h, \quad y' = y - k.$$

The Translation of Axes Formulas enable us to go from either coordinate system to the other. Their major use is to change the form of equations of graphs. To be specific, if, in the xy-plane, a certain collection of points is the graph of an equation in x and y, then to find an equation in x' and y' that has the same graph in the $x'y'$-plane, we may substitute $x' + h$ for x and $y' + k$ for y in the given equation. Conversely, if a set of points in the $x'y'$-plane is the graph of an equation in x' and y', then to find the corresponding equation in x and y we substitute $x - h$ for x' and $y - k$ for y'.

As a simple illustration, the equation

$$(x')^2 + (y')^2 = r^2$$

has, for its graph in the $x'y'$-plane, a circle of radius r with center at the origin O'. Using the Translation of Axes Formulas, an equation for this circle in the xy-plane is

$$(x - h)^2 + (y - k)^2 = r^2,$$

which is in agreement with the formula for a circle of radius r with center at $C(h, k)$ in the xy-plane.

As another illustration, we know that

$$(x')^2 = 4py'$$

is an equation of a parabola with vertex at the origin O' of the $x'y'$-plane. Using the Translation of Axes Formulas, we see that

$$(x - h)^2 = 4p(y - k)$$

is an equation of the same parabola in the xy-plane with vertex $V(h, k)$. The focus is $F(h, k + p)$ and the directrix is $y = k - p$. Similarly, starting with $(y')^2 = 4px'$ gives us $(y - k)^2 = 4p(x - h)$. The next theorem summarizes this discussion.

THEOREM

The graph of each of the following equations is a parabola that has vertex $V(h, k)$ and has the indicated focus and directrix.

(i) $(x - h)^2 = 4p(y - k)$: focus $F(h, k + p)$, directrix $y = k - p$.

(ii) $(y - k)^2 = 4p(x - h)$: focus $F(h + p, k)$, directrix $x = h - p$.

In each case the axis of the parabola is parallel to a coordinate axis. The parabola having equation (i) opens upward or downward, whereas the parabola in (ii) opens to the right or left. Typical graphs are sketched in Figure 10.7.

Squaring the left side of the equation in (i) of the theorem and simplifying leads to an equation of the form

$$y = ax^2 + bx + c$$

for real numbers a, b, and c. Conversely, the graph of such an equation is a parabola with a vertical axis. As in Section 2.3, we may complete the square in x to find the vertex (see Examples 3 and 4 of Section 2.3).

The equation in (ii) of the theorem may be written

$$x = ay^2 + by + c$$

for real numbers a, b, and c. Conversely, the last equation can be expressed in form (ii) by completing the square in y as illustrated next, in Example 3. Hence, if $a \neq 0$, the graph of $x = ay^2 + by + c$ is a parabola with a horizontal axis.

FIGURE 10.7

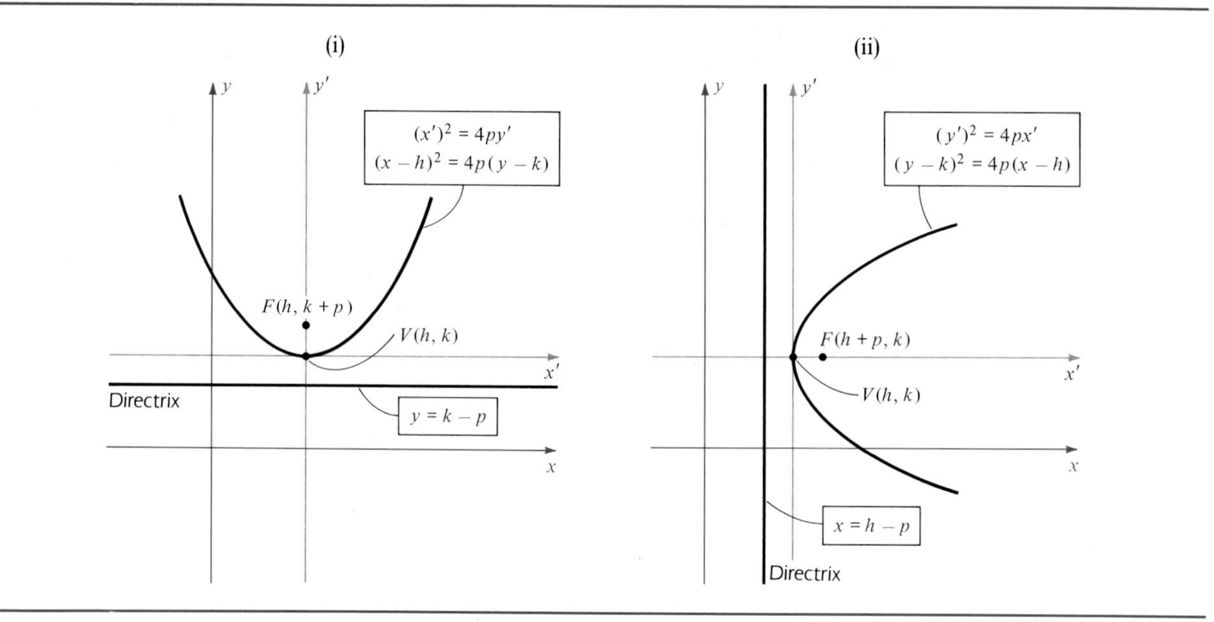

EXAMPLE 3 Discuss and sketch the graph of the equation

$$2x = y^2 + 8y + 22.$$

SOLUTION The graph is a parabola with a horizontal axis. Writing

$$y^2 + 8y = 2x - 22$$

we complete the square on the left by adding 16 to both sides. This gives us

$$y^2 + 8y + 16 = 2x - 6.$$

The last equation may be written

$$(y + 4)^2 = 2(x - 3),$$

which is in form (ii) of the theorem, with $h = 3$, $k = -4$, and $4p = 2$, or $p = \frac{1}{2}$. Hence the vertex is $V(3, -4)$. Since $p = \frac{1}{2} > 0$, the parabola opens to the right with focus at $F(h + p, k)$, that is, $F(\frac{7}{2}, -4)$. The equation of the directrix is $x = h - p$, or $x = \frac{5}{2}$. The parabola is sketched in Figure 10.8. ■

FIGURE 10.8

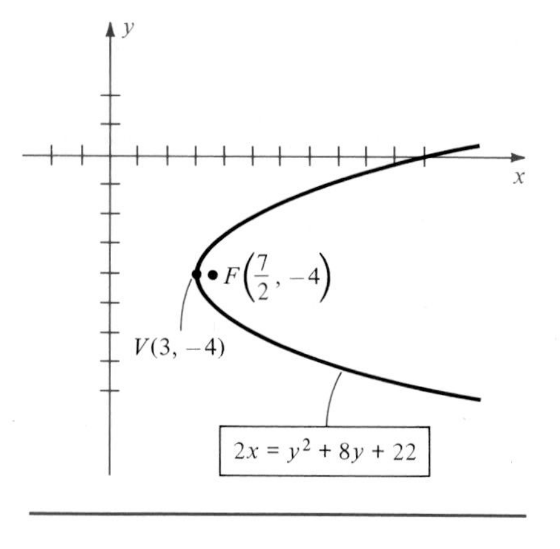

FIGURE 10.9

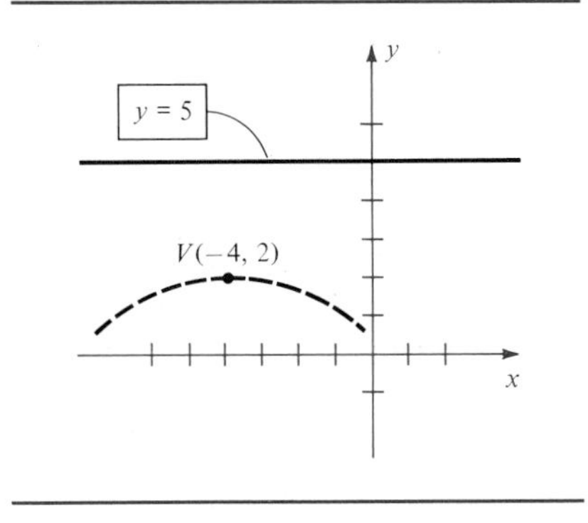

EXAMPLE 4 Find an equation of the parabola with vertex $V(-4, 2)$ and directrix $y = 5$.

SOLUTION The vertex and directrix are shown in Figure 10.9. The dashes indicate a possible position for the parabola. It follows that an equation of the parabola is

$$(x - h)^2 = 4p(y - k)$$

where $h = -4$, $k = 2$, and $p = -3$. This gives us

$$(x + 4)^2 = -12(y - 2).$$

The last equation can be expressed in the form $y = ax^2 + bx + c$ as follows:

$$x^2 + 8x + 16 = -12y + 24$$

$$12y = -x^2 - 8x + 8$$

$$y = -\tfrac{1}{12}x^2 - \tfrac{2}{3}x + \tfrac{2}{3}$$ ■

Exercises 10.2

In Exercises 1–16 find the vertex, focus, and directrix of the parabola with the given equation, and sketch the graph.

1 $x^2 = -12y$

2 $y^2 = \frac{1}{2}x$

3 $2y^2 = -3x$

4 $x^2 = -3y$

5 $8x^2 = y$

6 $y^2 = -100x$

7 $y^2 - 12 = 12x$

8 $y = 40x - 97 - 4x^2$

9 $y = x^2 - 4x + 2$

10 $y = 8x^2 + 16x + 10$

11 $y^2 - 4y - 2x - 4 = 0$

12 $y^2 + 14y + 4x + 45 = 0$

13 $4x^2 + 40x + y + 106 = 0$

14 $y^2 - 20y + 100 = 6x$

15 $x^2 + 20y = 10$

16 $4x^2 + 4x + 4y + 1 = 0$

In Exercises 17–22 find an equation of the parabola that satisfies the given conditions.

17 Focus $(2, 0)$, directrix $x = -2$

18 Focus $(0, -4)$, directrix $y = 4$

19 Focus $(6, 4)$, directrix $y = -2$

20 Focus $(-3, -2)$, directrix $y = 1$

21 Vertex at the origin, symmetric to the y-axis, and passing through the point $A(2, -3)$

22 Vertex $V(-3, 5)$, axis parallel to the x-axis, and passing through $A(5, 9)$

23 A searchlight reflector is designed so that a cross section through its axis is a parabola and the light source is at the focus. Find the focus if the reflector is 3 feet across at the opening and 1 foot deep.

24 Find an equation of a parabola that has a horizontal axis and passes through the points $A(2, 1)$, $B(6, 2)$, and $C(12, -1)$.

10.3 Ellipses

An ellipse may be defined as follows.

DEFINITION

An **ellipse** is the set of all points in a plane, the sum of whose distances from two fixed points in the plane (the **foci**) is constant.

It is known that the orbits of planets in the solar system are elliptical, with the sun at one of the foci. This is only one of many important applications of ellipses.

There is an easy way to construct an ellipse on paper. We begin by inserting two thumbtacks in the paper at points labeled F and F' and fastening the ends of a piece of string to the thumbtacks. If the string is now looped around a pencil and drawn taut at point P, as in Figure 10.10, then moving the pencil and at the same time keeping the string taut, the sum of the distances $d(F, P)$ and $d(F', P)$ is the length of the string, and hence is constant. The pencil will trace out an ellipse with foci at F and F'. By varying the positions of F and F', but keeping the length of string fixed, the shape of the ellipse can be made to change considerably. If F and F' are far apart, in the sense that $d(F, F')$ is almost the same as the length of the string, then the ellipse is quite flat. If $d(F, F')$ is close to zero, the ellipse is almost circular. Indeed, if $F = F'$, a circle is obtained.

FIGURE 10.10

By introducing suitable coordinate systems, we may derive simple equations for ellipses. Let us choose the x-axis as the line through the two foci F and F', with the origin at the midpoint of the segment $F'F$. This point is called the **center** of the ellipse. If F has coordinates $(c, 0)$, with $c > 0$, then, as shown in Figure 10.11, F' has coordinates $(-c, 0)$ and hence the distance between F and F' is $2c$. Let the constant sum of the distances of P from F and F' be denoted by $2a$, where in order to get points that are not on the x-axis we must have $2a > 2c$, that is, $a > c$. By definition, $P(x, y)$ is on the ellipse if and only if

FIGURE 10.11

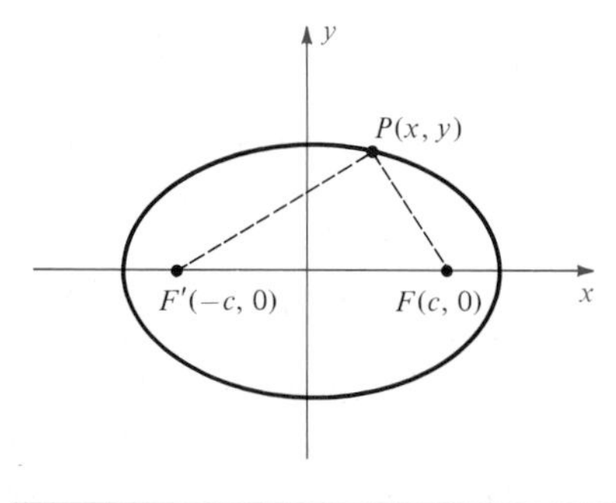

$$d(P, F) + d(P, F') = 2a$$

or, by the Distance Formula,

$$\sqrt{(x - c)^2 + (y - 0)^2} + \sqrt{(x + c)^2 + (y - 0)^2} = 2a.$$

Writing the preceding equation as

$$\sqrt{(x - c)^2 + y^2} = 2a - \sqrt{(x + c)^2 + y^2}$$

and squaring both sides, we obtain

$$x^2 - 2cx + c^2 + y^2 = 4a^2 - 4a\sqrt{(x+c)^2 + y^2} + x^2 + 2cx + c^2 + y^2,$$

which simplifies to

$$a\sqrt{(x+c)^2 + y^2} = a^2 + cx.$$

Squaring both sides gives us

$$a^2(x^2 + 2cx + c^2 + y^2) = a^4 + 2a^2cx + c^2x^2,$$

which may be written in the form

$$x^2(a^2 - c^2) + a^2y^2 = a^2(a^2 - c^2).$$

Dividing both sides by $a^2(a^2 - c^2)$ leads to

$$\frac{x^2}{a^2} + \frac{y^2}{a^2 - c^2} = 1.$$

For convenience, we let

$$b^2 = a^2 - c^2 \qquad \text{where } b > 0$$

in the preceding equation, obtaining

$$\frac{x^2}{a^2} + \frac{y^2}{b^2} = 1.$$

Since $c > 0$ and $b^2 = a^2 - c^2$, it follows that $a^2 > b^2$ and hence $a > b$.

We have shown that the coordinates of every point (x, y) on the ellipse in Figure 10.11 satisfy the equation $(x^2/a^2) + (y^2/b^2) = 1$. Conversely, if (x, y) is a solution of this equation, then by reversing the preceding steps we see that the point (x, y) is on the ellipse.

FIGURE 10.12

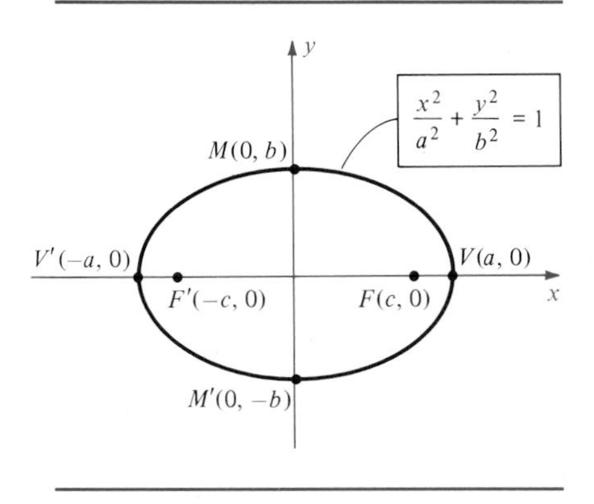

The x-intercepts may be found by setting $y = 0$. Doing so gives us $x^2/a^2 = 1$, or $x^2 = a^2$, and consequently the x-intercepts are a and $-a$. The corresponding points $V(a, 0)$ and $V'(-a, 0)$ on the graph are called the **vertices** of the ellipse. (See Figure 10.12.) The line segment $V'V$ is referred to as the **major axis.** Similarly, letting $x = 0$ in the equation of the ellipse, we obtain $y^2/b^2 = 1$, or $y^2 = b^2$. Hence the y-intercepts are b and $-b$. The segment from $M'(0, -b)$ to $M(0, b)$ is called the **minor axis** of the ellipse. Note that the major axis is longer than the minor axis, since $a > b$.

Applying tests for symmetry we see that the ellipse is symmetric to both the x-axis and the y-axis. It is also symmetric with respect to the origin, since substitution of $-x$ for x and $-y$ for y does not change the equation.

The preceding discussion may be summarized as follows.

THEOREM

The graph of the equation

$$\frac{x^2}{a^2} + \frac{y^2}{b^2} = 1,$$

for $a^2 > b^2$, is an ellipse with vertices $(\pm a, 0)$. The endpoints of the minor axis are $(0, \pm b)$. The foci are $(\pm c, 0)$, where $c^2 = a^2 - b^2$.

EXAMPLE 1 Discuss and sketch the graph of the equation

$$4x^2 + 18y^2 = 36.$$

SOLUTION To obtain the form in the theorem we divide both sides of the equation by 36 and simplify. This leads to

$$\frac{x^2}{9} + \frac{y^2}{2} = 1,$$

which is in the proper form with $a^2 = 9$ and $b^2 = 2$. Thus $a = 3$, $b = \sqrt{2}$, and hence the endpoints of the major axis are $(\pm 3, 0)$ and the endpoints of the minor axis are $(0, \pm\sqrt{2})$. Since

$$c^2 = a^2 - b^2 = 9 - 2 = 7 \quad \text{or} \quad c = \sqrt{7},$$

the foci are $(\pm\sqrt{7}, 0)$. The graph is sketched in Figure 10.13. ■

FIGURE 10.13

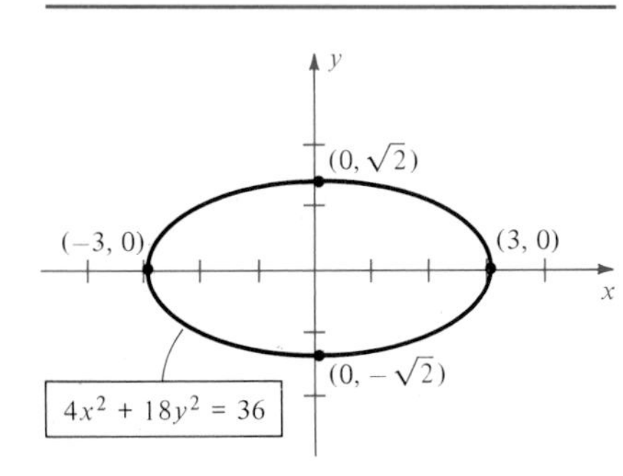

EXAMPLE 2 Find an equation of the ellipse with vertices $(\pm 4, 0)$ and foci $(\pm 2, 0)$.

SOLUTION Using the notation of the theorem, $a = 4$ and $c = 2$. Since $c^2 = a^2 - b^2$, we see that $b^2 = a^2 - c^2 = 16 - 4 = 12$. This gives us

$$\frac{x^2}{16} + \frac{y^2}{12} = 1.$$

Multiplying both sides by 48 leads to $3x^2 + 4y^2 = 48$. ■

It is sometimes convenient to choose the major axis of the ellipse along the y-axis. If the foci are $(0, \pm c)$, then by the same type of argument used previously, we obtain the following.

THEOREM

The graph of the equation

$$\frac{x^2}{b^2} + \frac{y^2}{a^2} = 1,$$

for $a^2 > b^2$, is an ellipse with vertices $(0, \pm a)$. The endpoints of the minor axis are $(\pm b, 0)$. The foci are $(0, \pm c)$, where $c^2 = a^2 - b^2$.

FIGURE 10.14

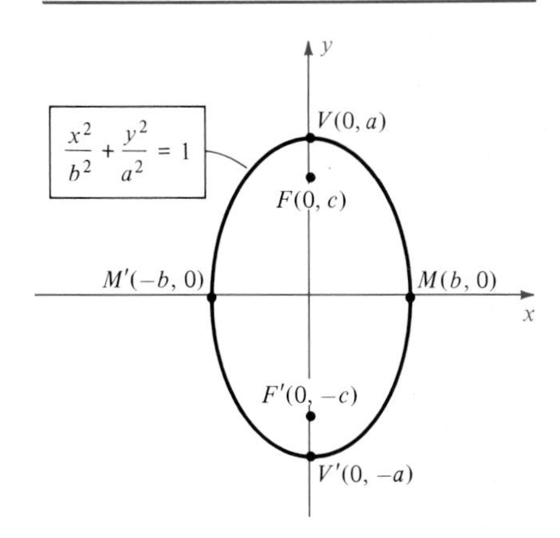

A typical graph is sketched in Figure 10.14.

The preceding discussion shows that an equation of an ellipse with center at the origin and foci on a coordinate axis can always be written in the form

$$\frac{x^2}{p} + \frac{y^2}{q} = 1 \quad \text{or} \quad qx^2 + py^2 = pq$$

where p and q are positive and $p \neq q$. If $p > q$ the major axis lies on the x-axis, whereas if $q > p$ the major axis is on the y-axis. It is unnecessary to memorize these facts, since in any given problem the major axis can be determined by examining the x- and y-intercepts.

FIGURE 10.15

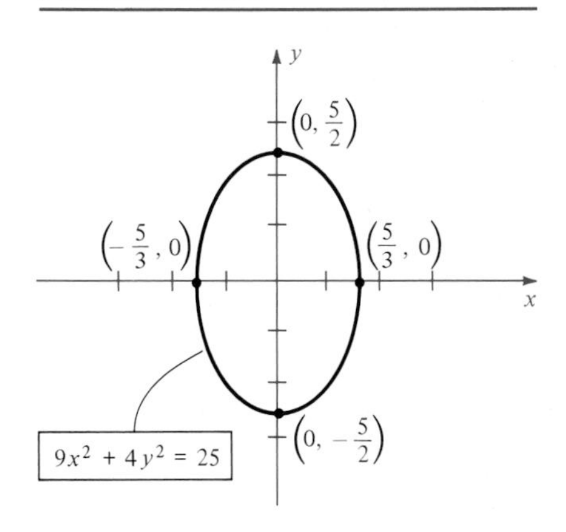

EXAMPLE 3 Sketch the graph of the equation $9x^2 + 4y^2 = 25$.

SOLUTION The graph is an ellipse with center at the origin and foci on one of the coordinate axes. To find the x-intercepts, we let $y = 0$, obtaining $9x^2 = 25$, or $x = \pm\frac{5}{3}$. Similarly, to find the y-intercepts, we let $x = 0$, obtaining $4y^2 = 25$, or $y = \pm\frac{5}{2}$. This enables us to sketch the ellipse (see Figure 10.15). Since $\frac{5}{3} < \frac{5}{2}$, the major axis is on the y-axis. ■

By using the Translation of Axes Formulas we can extend our work to an ellipse with center at any point $C(h, k)$ in the xy-plane. For example, since the graph of

$$\frac{(x')^2}{a^2} + \frac{(y')^2}{b^2} = 1$$

FIGURE 10.16

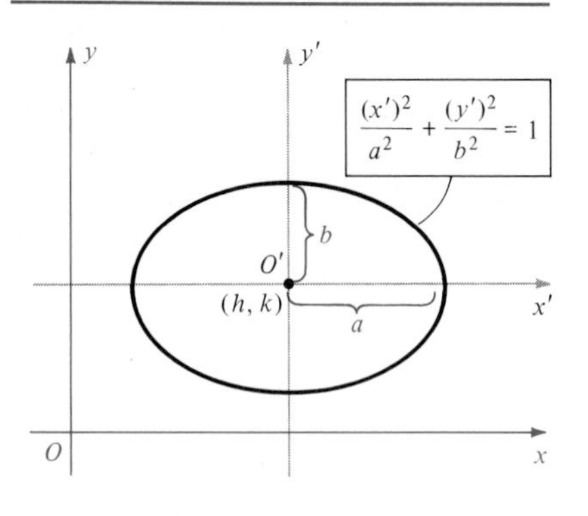

is an ellipse with center at O' in an $x'y'$-plane (see Figure 10.16), its equation relative to the xy-coordinate system is

$$\frac{(x-h)^2}{a^2} + \frac{(y-k)^2}{b^2} = 1.$$

Squaring the indicated terms in the last equation and simplifying gives us an equation of the form

$$Ax^2 + Cy^2 + Dx + Ey + F = 0$$

where the coefficients are real numbers and A and C are both positive. Conversely, if we start with such an equation, then by completing squares we can obtain a form that displays the center of the ellipse and the lengths of the major and minor axes. This technique is illustrated in the next example.

EXAMPLE 4 Discuss and sketch the graph of the equation

$$16x^2 + 9y^2 + 64x - 18y - 71 = 0.$$

SOLUTION We begin by writing the equation in the form

$$16(x^2 + 4x) + 9(y^2 - 2y) = 71.$$

FIGURE 10.17

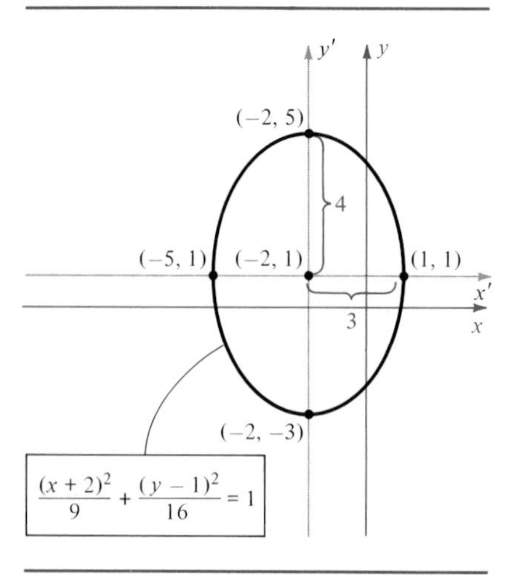

Next, we complete the squares for the expressions within parentheses, obtaining

$$16(x^2 + 4x + 4) + 9(y^2 - 2y + 1) = 71 + 64 + 9.$$

Note that by adding 4 to the expression within the first parentheses we have added 64 to the left side of the equation, and hence must compensate by adding 64 to the right side. Similarly, by adding 1 to the expression within the second parentheses, 9 is added to the left side and, consequently, 9 must also be added to the right side. The last equation may be written

$$16(x + 2)^2 + 9(y - 1)^2 = 144.$$

Dividing by 144, we obtain

$$\frac{(x+2)^2}{9} + \frac{(y-1)^2}{16} = 1,$$

which is of the form $$\frac{(x')^2}{9} + \frac{(y')^2}{16} = 1$$

with $x' = x + 2$ and $y' = y - 1$. This corresponds to letting $h = -2$ and $k = 1$ in the Translation of Axes Formulas. Since the graph of the equation $(x')^2/9 + (y')^2/16 = 1$ is an ellipse with center at the origin O' in the $x'y'$-plane, it follows that the graph of the given equation is an ellipse with center $C(-2, 1)$ in the xy-plane and with axes parallel to the coordinate axes. The graph is sketched in Figure 10.17. ■

Exercises 10.3

In Exercises 1–14 sketch the graph of the equation and find the coordinates of the vertices and foci.

1 $\dfrac{x^2}{9} + \dfrac{y^2}{4} = 1$

2 $\dfrac{x^2}{25} + \dfrac{y^2}{16} = 1$

3 $4x^2 + y^2 = 16$

4 $y^2 + 9x^2 = 9$

5 $5x^2 + 2y^2 = 10$

6 $\frac{1}{2}x^2 + 2y^2 = 8$

7 $4x^2 + 25y^2 = 1$

8 $10y^2 + x^2 = 5$

9 $4x^2 + 9y^2 - 32x - 36y + 64 = 0$

10 $x^2 + 2y^2 + 2x - 20y + 43 = 0$

11 $9x^2 + 16y^2 + 54x - 32y - 47 = 0$

12 $4x^2 + 9y^2 + 24x + 18y + 9 = 0$

13 $25x^2 + 4y^2 - 250x - 16y + 541 = 0$

14 $4x^2 + y^2 = 2y$

In Exercises 15–20 find an equation for the ellipse satisfying the given conditions.

15 Vertices $V(\pm 8, 0)$, foci $F(\pm 5, 0)$

16 Vertices $V(0, \pm 7)$, foci $F(0, \pm 2)$

17 Vertices $V(0, \pm 5)$, length of minor axis 3

18 Foci $F(\pm 3, 0)$, length of minor axis 2

19 Vertices $V(0, \pm 6)$, passing through (3, 2)

20 Center at the origin, symmetric with respect to both axes, passing through $A(2, 3)$ and $B(6, 1)$

In Exercises 21 and 22 find the points of intersection of the graphs of the given equations. Sketch both graphs on the same coordinate axes, showing points of intersection.

21 $\begin{cases} x^2 + 4y^2 = 20 \\ x + 2y = 6 \end{cases}$

22 $\begin{cases} x^2 + 4y^2 = 36 \\ x^2 + y^2 = 12 \end{cases}$

23 An arch of a bridge is semi-elliptical with major axis horizontal. The base of the arch is 30 feet across and the highest part of the arch is 10 feet above the horizontal roadway (see figure). Find the height of the arch 6 feet from the center of the base.

FIGURE FOR EXERCISE 23

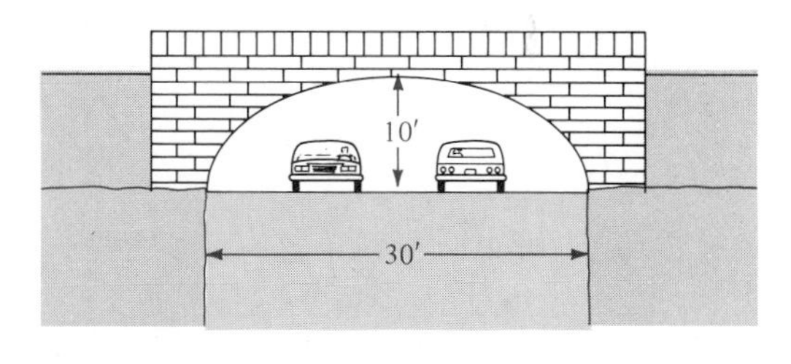

24 The **eccentricity** of an ellipse is defined as the ratio $(\sqrt{a^2 - b^2})/a$. If a is fixed and b varies, describe the general shape of the ellipse when the eccentricity is close to 1 and when it is close to zero.

25 A line segment of length $a + b$ moves with its endpoints A and B attached to the coordinate axes, as illustrated in the figure on page 492. Prove that if $a \neq b$, then the point P traces an ellipse.

FIGURE FOR EXERCISE 25

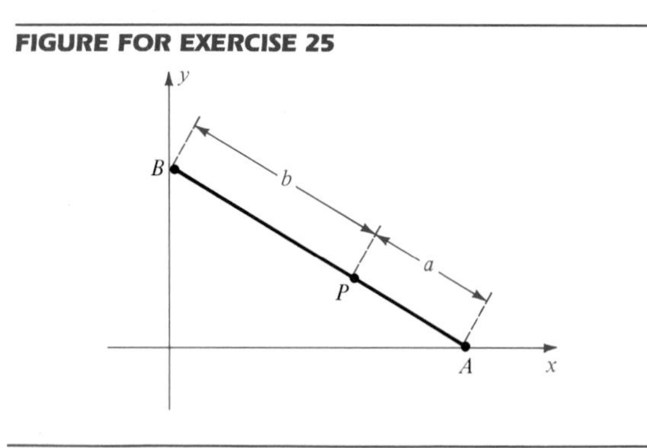

26 Consider the ellipse $px^2 + qy^2 = pq$ where $p > 0$ and $q > 0$. Prove that if m is any real number, there are exactly two lines of slope m that intersect the ellipse at precisely one point, and that equations of the lines are $y = mx \pm \sqrt{p + qm^2}$.

27 A square with sides parallel to the coordinate axes is inscribed in the ellipse with equation $(x^2/a^2) + (y^2/b^2) = 1$. Express the area A of the square in terms of a and b.

28 From a point P on the circle $x^2 + y^2 = 4$, a line segment is drawn perpendicular to the diameter AB, and the midpoint M is found (see figure). Find an equation of the collection of all such midpoints M, and sketch the graph.

FIGURE FOR EXERCISE 28

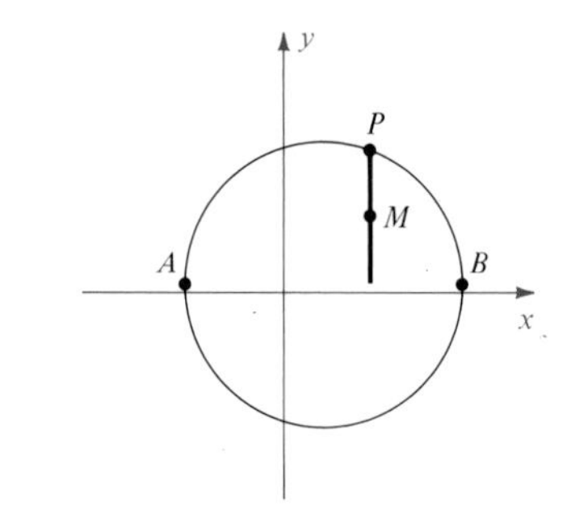

10.4 Hyperbolas

The definition of a hyperbola is similar to that of an ellipse. The only change is that instead of using the *sum* of distances from two fixed points, we use the *difference.*

DEFINITION

A **hyperbola** is the set of all points in a plane, the difference of whose distances from two fixed points in the plane (the **foci**) is a positive constant.

To find a simple equation for a hyperbola, we choose a coordinate system with foci at $F(c, 0)$ and $F'(-c, 0)$, and denote the (constant) distance by $2a$. Referring to Figure 10.18, we see that a point $P(x, y)$ is on the

FIGURE 10.18

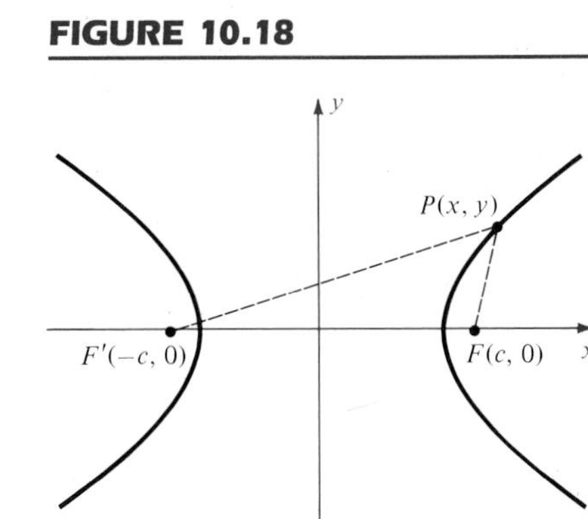

hyperbola if and only if either one of the following is true:

$$d(P, F') - d(P, F) = 2a$$
$$d(P, F) - d(P, F') = 2a.$$

For hyperbolas (unlike ellipses), we need $a < c$ to obtain points on the hyperbola that are not on the x-axis, for if P is such a point, then from Figure 10.18 we see that

$$d(P, F) < d(F', F) + d(P, F'),$$

because the length of one side of a triangle is always less than the sum of the lengths of the other two sides. Similarly,

$$d(P, F') < d(F', F) + d(P, F).$$

Equivalent forms for the previous two inequalities are

$$d(P, F) - d(P, F') < d(F', F)$$
$$d(P, F') - d(P, F) < d(F', F).$$

Since the differences on the left both equal $2a$, and since $d(F', F) = 2c$, the last two inequalities imply that $2a < 2c$, or $a < c$.

The equations $d(P, F') - d(P, F) = 2a$ and $d(P, F) - d(P, F') = 2a$ may be replaced by the single equation

$$|d(P, F) - d(P, F')| = 2a.$$

It follows from the Distance Formula that an equation of the hyperbola is

$$\left|\sqrt{(x - c)^2 + (y - 0)^2} - \sqrt{(x + c)^2 + (y - 0)^2}\right| = 2a.$$

Employing the type of simplification procedure used to derive an equation for an ellipse, we arrive at the equivalent equation

$$\frac{x^2}{a^2} - \frac{y^2}{c^2 - a^2} = 1.$$

For convenience, let

$$b^2 = c^2 - a^2 \qquad \text{where } b > 0$$

in the preceding equation, obtaining

$$\frac{x^2}{a^2} - \frac{y^2}{b^2} = 1.$$

We have shown that the coordinates of every point (x, y) on the hyperbola in Figure 10.18 satisfy the last equation. Conversely, if (x, y) is a solution of that equation, then by reversing the preceding steps we see that the point (x, y) is on the hyperbola.

By the Tests for Symmetry, the hyperbola is symmetric with respect to both axes and the origin. The x-intercepts are $\pm a$. The corresponding points $V(a, 0)$ and $V'(-a, 0)$ are called the **vertices,** and the line segment $V'V$ is known as the **transverse axis** of the hyperbola. The origin is called the **center** of the hyperbola. There are no y-intercepts, since the equation $-y^2/b^2 = 1$ has no solutions.

The preceding discussion may be summarized as follows.

THEOREM

> The graph of the equation
>
> $$\frac{x^2}{a^2} - \frac{y^2}{b^2} = 1$$
>
> is a hyperbola with vertices $(\pm a, 0)$. The foci are $(\pm c, 0)$, where $c^2 = a^2 + b^2$.

If the equation $(x^2/a^2) - (y^2/b^2) = 1$ is solved for y, we obtain

$$y = \pm\frac{b}{a}\sqrt{x^2 - a^2}.$$

There are no points (x, y) on the graph if $x^2 - a^2 < 0$, that is, if $-a < x < a$. However, there *are* points $P(x, y)$ on the graph if $x \geq a$ or $x \leq -a$. If $x \geq a$, we may write the last equation in the form

$$y = \pm\frac{b}{a}\sqrt{x^2\left(1 - \frac{a^2}{x^2}\right)} = \pm\frac{b}{a}x\sqrt{1 - \frac{a^2}{x^2}}.$$

If x is large (in comparison to a), then $1 - (a^2/x^2) \approx 1$, and hence the y-coordinate of the point $P(x, y)$ on the hyperbola is close to either $(b/a)x$

FIGURE 10.19

$$\frac{x^2}{a^2} - \frac{y^2}{b^2} = 1$$

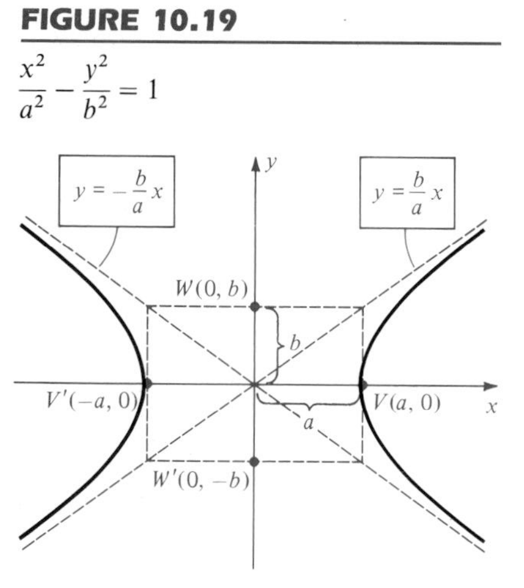

or $-(b/a)x$. Thus the point $P(x, y)$ is close to the line $y = (b/a)x$ when y is positive, or to the line $y = -(b/a)x$ when y is negative. As x increases, we say that the point $P(x, y)$ *approaches* one of these lines. A corresponding situation exists if $x \leq -a$. The lines with equations

$$y = \pm\frac{b}{a}x$$

are called the **asymptotes** of the hyperbola $(x^2/a^2) - (y^2/b^2) = 1$.

The asymptotes serve as excellent guides for sketching the graph. A convenient way to sketch the asymptotes is to first plot the vertices $V(a, 0)$, $V'(-a, 0)$ and the points $W(0, b)$, $W'(0, -b)$ (see Figure 10.19). The line segment $W'W$ of length $2b$ is called the **conjugate axis** of the hyperbola. If horizontal and vertical lines are drawn through the endpoints of the conjugate and transverse axes, respectively, then the diagonals of the resulting rectangle have slopes b/a and $-b/a$. Hence, by extending these diagonals we obtain lines with equations $y = (\pm b/a)x$. The hyperbola is then sketched as in Figure 10.19, using the asymptotes as guides. The two curves that make up the hyperbola are called the **branches** of the hyperbola.

FIGURE 10.20

$9x^2 - 4y^2 = 36$

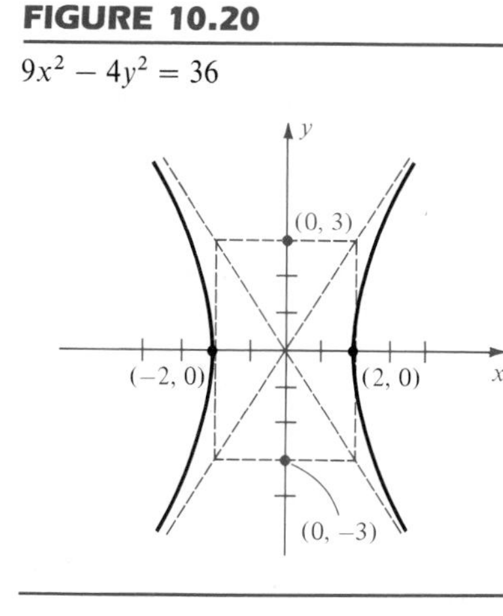

EXAMPLE 1 Discuss and sketch the graph of the equation

$$9x^2 - 4y^2 = 36.$$

SOLUTION Dividing both sides by 36, we have

$$\frac{x^2}{4} - \frac{y^2}{9} = 1,$$

which is of the form stated in the theorem with $a^2 = 4$ and $b^2 = 9$. Hence, $a = 2$ and $b = 3$. The vertices $(\pm 2, 0)$ and the endpoints $(0, \pm 3)$ of the conjugate axis determine a rectangle whose diagonals (extended) give us the asymptotes. The graph of the equation is sketched in Figure 10.20. The equations of the asymptotes, $y = \pm\frac{3}{2}x$, can be found by referring to the graph or to the equations $y = \pm(b/a)x$. Since $c^2 = a^2 + b^2 = 4 + 9 = 13$, the foci are $(\pm\sqrt{13}, 0)$. ■

The preceding example indicates that for hyperbolas it is not always true that $a < b$, as is the case for ellipses. Indeed, we may have $a < b$, $a > b$, or $a = b$.

EXAMPLE 2 Find an equation, the foci, and the asymptotes of a hyperbola that has vertices $(\pm 3, 0)$ and passes through the point $P(5, 2)$.

SOLUTION Substituting $a = 3$ in $(x^2/a^2) - (y^2/b^2) = 1$, we obtain the equation

$$\frac{x^2}{9} - \frac{y^2}{b^2} = 1.$$

If (5, 2) is a solution of this equation, then

$$\frac{25}{9} - \frac{4}{b^2} = 1.$$

This gives us $b^2 = \frac{9}{4}$, and hence the desired equation is

$$\frac{x^2}{9} - \frac{4y^2}{9} = 1,$$

or equivalently, $x^2 - 4y^2 = 9$.

Since $c^2 = a^2 + b^2 = 9 + \frac{9}{4} = \frac{45}{4}$, the foci are $(\pm\frac{3}{2}\sqrt{5}, 0)$. Substituting for b and a in $y = \pm(b/a)x$ and simplifying, we obtain equations $y = \pm\frac{1}{2}x$ for the asymptotes. ■

If the foci of a hyperbola are the points $(0, \pm c)$ on the y-axis, then by the same type of argument used previously, we obtain the following theorem.

THEOREM

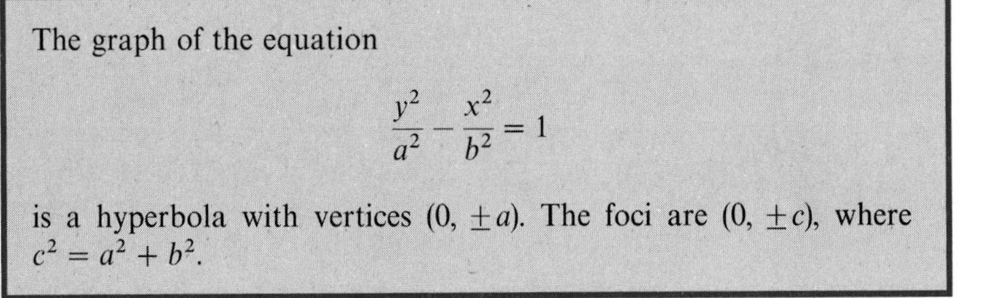

The graph of the equation

$$\frac{y^2}{a^2} - \frac{x^2}{b^2} = 1$$

is a hyperbola with vertices $(0, \pm a)$. The foci are $(0, \pm c)$, where $c^2 = a^2 + b^2$.

In this case the endpoints of the conjugate axis are $W(b, 0)$ and $W'(-b, 0)$. The asymptotes are found, as before, by using the diagonals of the rectangle determined by these points, the vertices, and lines parallel to the coordinate axes. The graph is sketched in Figure 10.21. The equations of the asymptotes are $y = \pm(a/b)x$. Note the difference between these equations and the equations $y = \pm(b/a)x$ for the asymptotes of the hyperbola considered first in this section.

FIGURE 10.21

$$\frac{y^2}{a^2} - \frac{x^2}{b^2} = 1$$

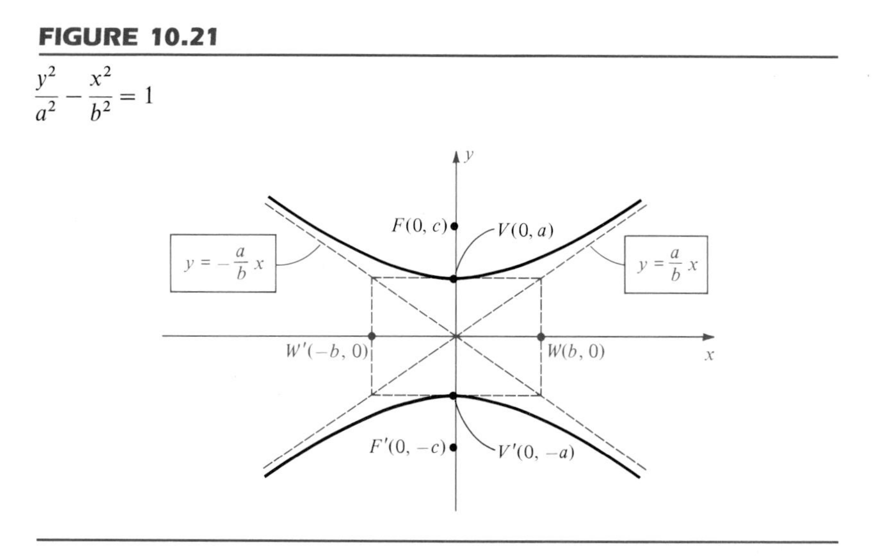

EXAMPLE 3 Discuss and sketch the graph of the equation

$$4y^2 - 2x^2 = 1.$$

SOLUTION The form in the last theorem may be obtained by writing the equation as

$$\frac{y^2}{\frac{1}{4}} - \frac{x^2}{\frac{1}{2}} = 1.$$

Thus $a^2 = \frac{1}{4}$, $b^2 = \frac{1}{2}$, and $c^2 = \frac{1}{4} + \frac{1}{2} = \frac{3}{4}$. Consequently, $a = \frac{1}{2}$, $b = \sqrt{2}/2$, and $c = \sqrt{3}/2$. The vertices are $(0, \pm\frac{1}{2})$ and the foci are $(0, \pm\sqrt{3}/2)$. The graph has the general appearance of the graph shown in Figure 10.21. ■

As was the case for ellipses, we may use translations of axes to generalize our work. The following example illustrates this technique.

FIGURE 10.22

$$\frac{(x-3)^2}{4} - \frac{(y+2)^2}{9} = 1$$

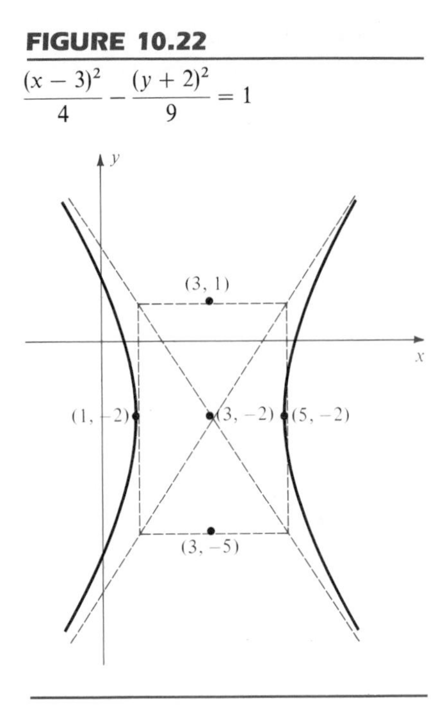

EXAMPLE 4 Discuss and sketch the graph of the equation

$$9x^2 - 4y^2 - 54x - 16y + 29 = 0.$$

SOLUTION We arrange our work as follows:

$$9(x^2 - 6x) - 4(y^2 + 4y) = -29$$

$$9(x^2 - 6x + 9) - 4(y^2 + 4y + 4) = -29 + 81 - 16$$

$$9(x-3)^2 - 4(y+2)^2 = 36$$

$$\frac{(x-3)^2}{4} - \frac{(y+2)^2}{9} = 1,$$

which is of the form $$\frac{(x')^2}{4} - \frac{(y')^2}{9} = 1$$

with $x' = x - 3$ and $y' = y + 2$. By translating the x- and y-axes to the new origin $C(3, -2)$, we obtain the sketch shown in Figure 10.22. ■

The results of the last three sections indicate that the graph of every equation of the form

$$Ax^2 + Cy^2 + Dx + Ey + F = 0$$

is a conic, except for certain degenerate cases in which points, lines, or no graphs are obtained. Although we have only considered special examples, our methods are perfectly general. If A and C are equal and not zero, then the graph, when it exists, is a circle or, in exceptional cases, a point. If A and C are unequal but have the same sign, then by completing squares and properly translating axes, we obtain an equation whose graph, when it exists, is an ellipse (or a point). If A and C have opposite signs, an equation of a hyperbola is obtained, or possibly, in the degenerate case, two intersecting straight lines. Finally, if either A or C (but not both) is zero, the graph is a parabola or, in certain cases, a pair of parallel lines.

Exercises 10.4

In Exercises 1–18 sketch the graph of the equation, find the coordinates of the vertices and foci, and write equations for the asymptotes.

1 $\frac{x^2}{9} - \frac{y^2}{4} = 1$

2 $\frac{y^2}{49} - \frac{x^2}{16} = 1$

3 $\frac{y^2}{9} - \frac{x^2}{4} = 1$

4 $\frac{x^2}{49} - \frac{y^2}{16} = 1$

5 $y^2 - 4x^2 = 16$

6 $x^2 - 2y^2 = 8$

7 $x^2 - y^2 = 1$

8 $y^2 - 16x^2 = 1$

9 $x^2 - 5y^2 = 25$

10 $4y^2 - 4x^2 = 1$

11 $3x^2 - y^2 = -3$

12 $16x^2 - 36y^2 = 1$

13 $25x^2 - 16y^2 + 250x + 32y + 109 = 0$

14 $y^2 - 4x^2 - 12y - 16x + 16 = 0$

15 $4y^2 - x^2 + 40y - 4x + 60 = 0$

16 $25x^2 - 9y^2 - 100x - 54y + 10 = 0$

17 $9y^2 - x^2 - 36y + 12x - 36 = 0$

18 $4x^2 - y^2 + 32x - 8y + 49 = 0$

In Exercises 19–26 find an equation for the hyperbola that satisfies the given conditions.

19 Foci $F(0, \pm 4)$, vertices $V(0, \pm 1)$

20 Foci $F(\pm 8, 0)$, vertices $V(\pm 5, 0)$

21 Foci $F(\pm 5, 0)$, vertices $V(\pm 3, 0)$

22 Foci $F(0, \pm 3)$, vertices $V(0, \pm 2)$

23 Foci $F(0, \pm 5)$, length of conjugate axis 4

24 Vertices $V(\pm 4, 0)$, passing through $P(8, 2)$

25 Vertices $V(\pm 3, 0)$, equations of asymptotes $y = \pm 2x$

26 Foci $F(0, \pm 10)$, equations of asymptotes $y = \pm \frac{1}{3}x$

In Exercises 27 and 28 find the points of intersection of the graphs of the given equations and sketch both graphs on the same coordinate axes, showing points of intersection.

27 $\begin{cases} y^2 - 4x^2 = 16 \\ y - x = 4 \end{cases}$

28 $\begin{cases} x^2 - y^2 = 4 \\ y^2 - 3x = 0 \end{cases}$

29 The graphs of the equations

$$\frac{x^2}{a^2} - \frac{y^2}{b^2} = 1 \quad \text{and} \quad \frac{x^2}{a^2} - \frac{y^2}{b^2} = -1$$

are called **conjugate hyperbolas.** Sketch the graphs of both equations on the same coordinate system with $a = 2$ and $b = 5$. Describe the relationship between the two graphs.

30 Find an equation of the hyperbola with foci $(h \pm c, k)$ and vertices $(h \pm a, k)$, where $0 < a < c$ and $c^2 = a^2 + b^2$.

31 The physicist Ernest Rutherford discovered that when alpha particles are shot toward the nucleus of an atom, they are eventually repulsed away from the nucleus along hyperbolic paths. The figure illustrates the path of a particle that starts toward the origin along the line $y = \frac{1}{2}x$ and comes within 3 units of the nucleus. Find an equation of the path.

FIGURE FOR EXERCISE 31

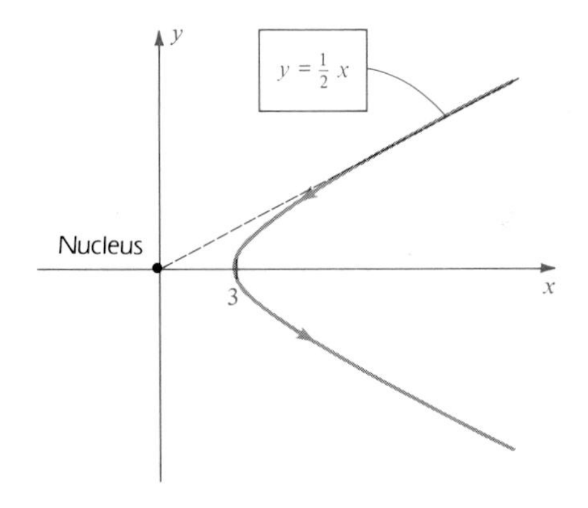

32 An Air Force jet is executing a high-speed maneuver along the path $2y^2 - x^2 = 8$. How close does the jet come to a town located at $(3, 0)$? (*Hint:* Let S denote the square of the distance from a point (x, y) on the path to $(3, 0)$ and find the minimum value of S. What is $\sqrt{S}$?)

FIGURE FOR EXERCISE 32

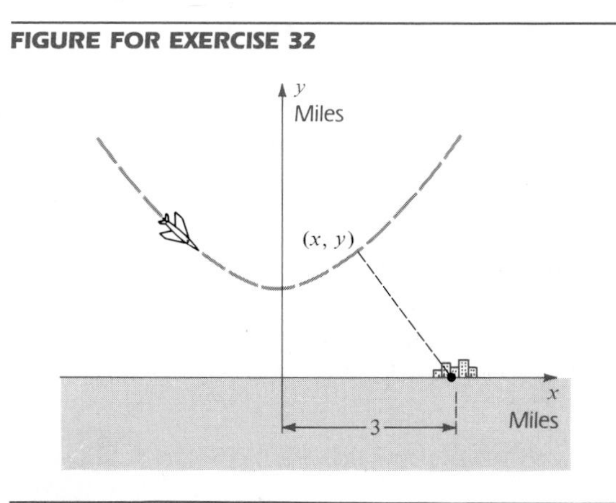

10.5 Rotation of Axes

The $x'y'$-coordinate system used in a translation of axes may be thought of as having been obtained by moving the origin O of the xy-system to a new position $C(h, k)$ while, at the same time, not changing the positive directions of the axes or the units of length. We shall next introduce a new coordinate system by keeping the origin O fixed and rotating the x- and y-axes about O to another position denoted by x' and y'. A transformation of this type will be referred to as a **rotation of axes.**

FIGURE 10.23

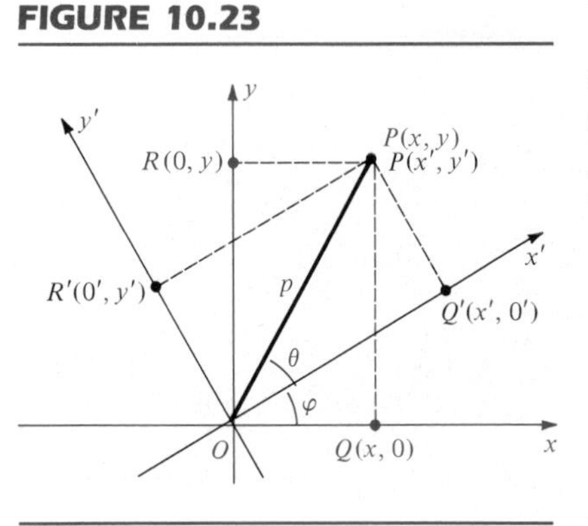

Consider the rotation of axes in Figure 10.23, and let φ denote the angle through which the positive x-axis must be rotated in order to coincide with the positive x'-axis. If (x, y) are the coordinates of a point P relative to the xy-plane, then (x', y') will denote its coordinates relative to the new $x'y'$-coordinate system.

Let the projections of P on the various axes be denoted as in Figure 10.23, and let θ denote angle POQ'. If $p = d(O, P)$, then

$$x' = p\cos\theta, \qquad y' = p\sin\theta,$$

$$x = p\cos(\theta + \varphi), \qquad y = p\sin(\theta + \varphi).$$

Applying the addition formulas for the sine and cosine, we see that

$$x = p\cos\theta\cos\varphi - p\sin\theta\sin\varphi$$

$$y = p\sin\theta\cos\varphi + p\cos\theta\sin\varphi.$$

Using the fact that $x' = p\cos\theta$ and $y' = p\sin\theta$ gives us (i) of the next theorem. The formulas in (ii) may be obtained from (i) by solving for x' and y'.

ROTATION OF AXES FORMULAS

If the x- and y-axes are rotated about the origin O, through an angle φ, then the coordinates (x, y) and (x', y') of a point P in the two systems are related as follows:

(i) $x = x'\cos\varphi - y'\sin\varphi, \qquad y = x'\sin\varphi + y'\cos\varphi$

(ii) $x' = x\cos\varphi + y\sin\varphi, \qquad y' = -x\sin\varphi + y\cos\varphi.$

EXAMPLE 1 The graph of the equation $xy = 1$, or equivalently $y = 1/x$, is sketched in Figure 10.24. If the coordinate axes are rotated through

FIGURE 10.24

$y = 1/x$

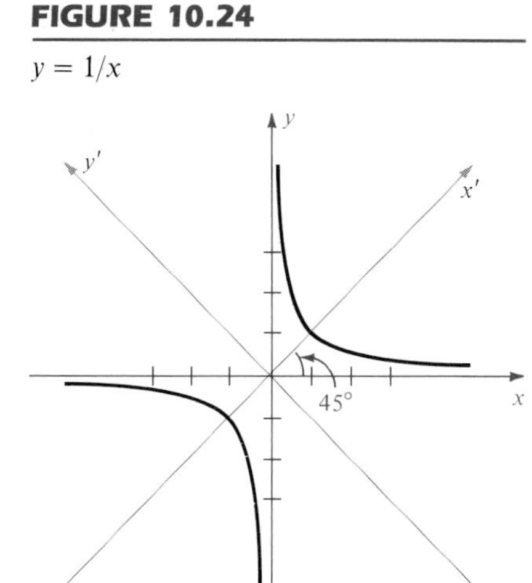

an angle of 45°, find an equation of the graph relative to the new $x'y'$-coordinate system.

SOLUTION Letting $\varphi = 45°$ in (i) of the Rotation of Axes Formulas,

$$x = x'\left(\frac{\sqrt{2}}{2}\right) - y'\left(\frac{\sqrt{2}}{2}\right) = \left(\frac{\sqrt{2}}{2}\right)(x' - y')$$

$$y = x'\left(\frac{\sqrt{2}}{2}\right) + y'\left(\frac{\sqrt{2}}{2}\right) = \left(\frac{\sqrt{2}}{2}\right)(x' + y').$$

Substituting for x and y in the equation $xy = 1$ gives us

$$\left(\frac{\sqrt{2}}{2}\right)(x' - y')\left(\frac{\sqrt{2}}{2}\right)(x' + y') = 1.$$

This reduces to $$\frac{(x')^2}{2} - \frac{(y')^2}{2} = 1,$$

which is an equation of a hyperbola with vertices $(\pm\sqrt{2}, 0)$ on the x'-axis. Note that the asymptotes for the hyperbola have equations $y' = \pm x'$ in the new system. These correspond to the original x- and y-axes. ■

Example 1 illustrates a method for eliminating a term of an equation that contains the product xy. This method can be used to transform any equation of the form

$$Ax^2 + Bxy + Cy^2 + Dx + Ey + F = 0,$$

where $B \neq 0$, into an equation in x' and y' that contains no $x'y'$ term. Let us prove that this may always be done. If we rotate the axes through an angle φ, then using (i) of the Rotation of Axes Formulas to substitute for x and y gives us

$$\begin{aligned} &A(x' \cos \varphi - y' \sin \varphi)^2 \\ &\qquad + B(x' \cos \varphi - y' \sin \varphi)(x' \sin \varphi + y' \cos \varphi) \\ &\qquad + C(x' \sin \varphi + y' \cos \varphi)^2 + D(x' \cos \varphi - y' \sin \varphi) \\ &\qquad + E(x' \sin \varphi + y' \cos \varphi) + F = 0. \end{aligned}$$

This equation may be written in the form

$$A'(x')^2 + B'x'y' + C'(y')^2 + D'x' + E'y' + F = 0$$

where the coefficient B' of $x'y'$ is

$$B' = 2(C - A) \sin \varphi \cos \varphi + B(\cos^2 \varphi - \sin^2 \varphi).$$

To eliminate the $x'y'$ term, we must select φ such that

$$2(C - A) \sin \varphi \cos \varphi + B(\cos^2 \varphi - \sin^2 \varphi) = 0.$$

Using double-angle formulas, the last equation may be written

$$(C - A) \sin 2\varphi + B \cos 2\varphi = 0,$$

which is equivalent to $\cot 2\varphi = \dfrac{A - C}{B}.$

This proves the next result.

THEOREM

To eliminate the xy-term from the equation

$$Ax^2 + Bxy + Cy^2 + Dx + Ey + F = 0,$$

where $B \neq 0$, choose an angle φ such that

$$\cot 2\varphi = \frac{A - C}{B}$$

and use the Rotation of Axes Formulas.

It follows that the graph of any equation in x and y of the type displayed in the preceding theorem is a conic, except for certain degenerate cases.

EXAMPLE 2 Discuss and sketch the graph of the equation

$$41x^2 - 24xy + 34y^2 - 25 = 0.$$

SOLUTION Using the notation of the preceding theorem, we have

$$A = 41, \qquad B = -24, \qquad C = 34,$$

and

$$\cot 2\varphi = \frac{41 - 34}{-24} = -\frac{7}{24}.$$

Since $\cot 2\varphi$ is negative, we may choose 2φ such that $90^\circ < 2\varphi < 180^\circ$, and consequently $\cos 2\varphi = -\frac{7}{25}$. (Why?) We now use the half-angle formulas to obtain

$$\sin\varphi = \sqrt{\frac{1-\cos 2\varphi}{2}} = \sqrt{\frac{1-(-\frac{7}{25})}{2}} = \frac{4}{5}$$

$$\cos\varphi = \sqrt{\frac{1+\cos 2\varphi}{2}} = \sqrt{\frac{1+(-\frac{7}{25})}{2}} = \frac{3}{5}.$$

Thus, the desired rotation formulas are

$$x = \tfrac{3}{5}x' - \tfrac{4}{5}y', \qquad y = \tfrac{4}{5}x' + \tfrac{3}{5}y'.$$

We leave it to the reader to show that after substituting for x and y in the given equation and simplifying, we obtain the equation

$$(x')^2 + 2(y')^2 = 1.$$

Thus the graph is an ellipse with vertices at $(\pm 1, 0)$ on the x'-axis. Since $\tan\varphi = \sin\varphi/\cos\varphi = (\frac{4}{5})/(\frac{3}{5}) = \frac{4}{3}$, we obtain $\varphi = \tan^{-1}(\frac{4}{3})$. To the nearest minute, $\varphi \approx 53°8'$. The graph is sketched in Figure 10.25. ■

FIGURE 10.25

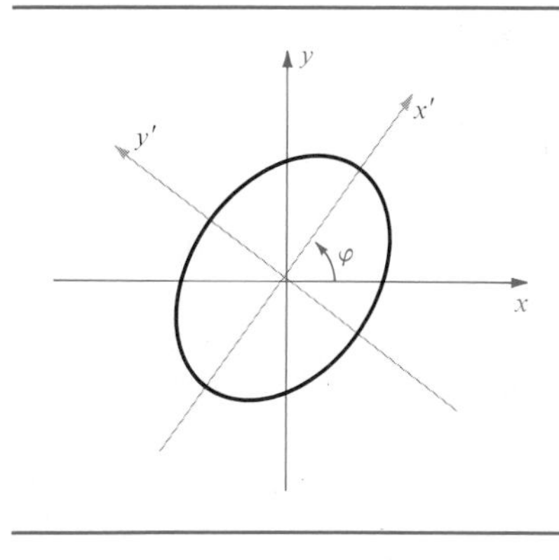

In some cases, after eliminating the xy-term, the resulting equation will contain an x' or y' term. It is then necessary to translate the axes of the $x'y'$-coordinate system to obtain the graph. Several problems of this type are contained in the following exercises.

Exercises 10.5

After a suitable rotation of axes, describe and sketch the graph of each of the equations in Exercises 1–14.

1 $32x^2 - 72xy + 53y^2 = 80$

2 $7x^2 - 48xy - 7y^2 = 225$

3 $11x^2 + 10\sqrt{3}xy + y^2 = 4$

4 $x^2 - xy + y^2 = 3$

5 $5x^2 - 8xy + 5y^2 = 9$

6 $11x^2 - 10\sqrt{3}xy + y^2 = 20$

7 $16x^2 - 24xy + 9y^2 - 60x - 80y + 100 = 0$

8 $x^2 + 2\sqrt{3}xy + 3y^2 + 8\sqrt{3}x - 8y + 32 = 0$

9 $5x^2 + 6\sqrt{3}xy - y^2 + 8x - 8\sqrt{3}y - 12 = 0$

10 $18x^2 - 48xy + 82y^2 + 6\sqrt{10}x + 2\sqrt{10}y - 80 = 0$

11 $x^2 + 4xy + 4y^2 + 6\sqrt{5}x - 18\sqrt{5}y + 45 = 0$

12 $15x^2 + 20xy - 4\sqrt{5}x + 8\sqrt{5}y - 100 = 0$

13 $40x^2 - 36xy + 25y^2 - 8\sqrt{13}x - 12\sqrt{13}y = 0$

14 $64x^2 - 240xy + 225y^2 + 1020x - 544y = 0$

15 Prove that, except for degenerate cases, the graph of $Ax^2 + Bxy + Cy^2 + Dx + Ey + F = 0$ is

(a) a parabola if $B^2 - 4AC = 0$.

(b) an ellipse if $B^2 - 4AC < 0$.

(c) a hyperbola if $B^2 - 4AC > 0$.

16 Use the results of Exercise 15 to determine the nature of the graphs in Exercises 1–14.

17 Let the matrices X, X', and A be defined by

$$X = \begin{bmatrix} x \\ y \end{bmatrix}, \quad X' = \begin{bmatrix} x' \\ y' \end{bmatrix}, \quad A = \begin{bmatrix} \cos\varphi & -\sin\varphi \\ \sin\varphi & \cos\varphi \end{bmatrix}.$$

Show that the Rotation of Axes Formulas may be written

$$\text{(i)}\ X = AX', \qquad \text{(ii)}\ X' = A^{-1}X.$$

10.6 Polar Coordinates

In a rectangular coordinate system the ordered pair (a, b) denotes the point whose directed distances from the x- and y-axes are b and a, respectively. Another method for representing points is to use *polar coordinates.* We begin with a fixed point O (called the **origin,** or **pole**) and a directed half-line (called the **polar axis**) with endpoint O. Next we consider any point P in the plane different from O. If, as illustrated in Figure 10.26, $r = d(O, P)$ and θ denotes the measure of any angle determined by the polar axis and OP, then r and θ are called **polar coordinates** of P and the symbols (r, θ) or $P(r, \theta)$ are used to denote P. As usual, θ is considered positive if the angle is generated by a counterclockwise rotation of the polar axis and negative if the rotation is clockwise. Either radians or degrees may be used for the measure of θ.

FIGURE 10.26

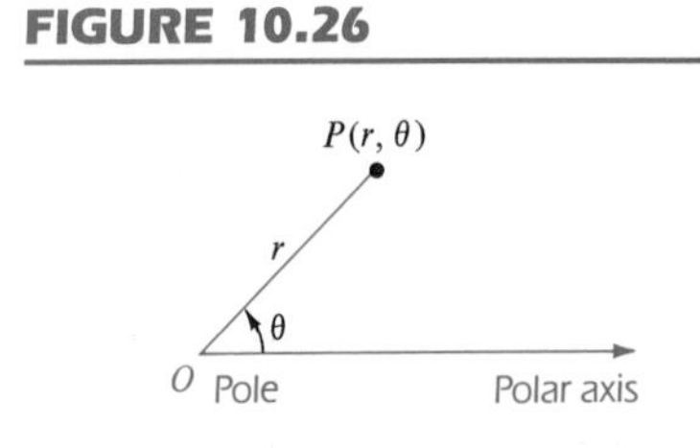

The polar coordinates of a point are not unique. For example $(3, \pi/4)$, $(3, 9\pi/4)$, and $(3, -7\pi/4)$ all represent the same point (see Figure 10.27). We shall also allow r to be negative. In this event, instead of measuring $|r|$ units along the terminal side of the angle θ, we measure along the half-line with endpoint O that has direction *opposite* to that of the terminal side. The points corresponding to the pairs $(-3, 5\pi/4)$ and $(-3, -3\pi/4)$ are also plotted in Figure 10.27.

FIGURE 10.27

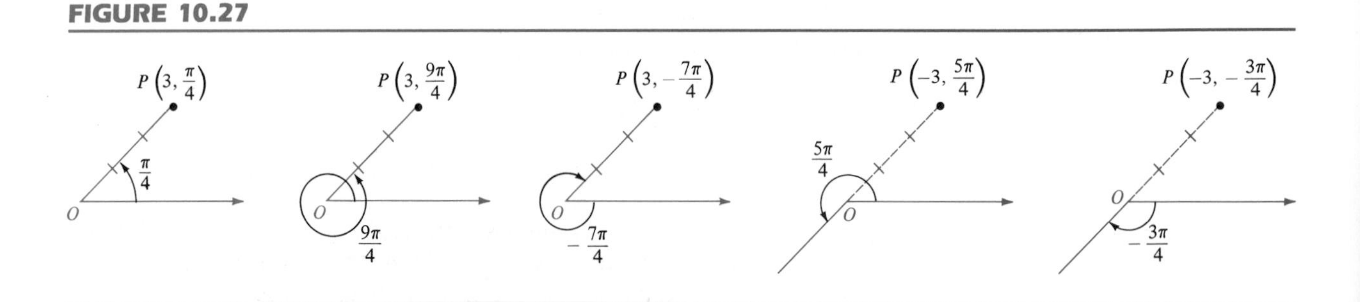

Finally, we agree that the pole O has polar coordinates $(0, \theta)$ for *any* θ. An assignment of ordered pairs of the form (r, θ) to points in a plane will be referred to as a **polar coordinate system** and the plane will be called an **$r\theta$-plane.**

A **polar equation** is an equation in r and θ. A **solution** of a polar equation is an ordered pair (a, b) that leads to equality if a is substituted for r and b for θ. The **graph** of a polar equation is the set of all points (in an $r\theta$-plane) that correspond to the solutions.

EXAMPLE 1 Sketch the graph of the polar equation $r = 4 \sin \theta$.

SOLUTION The following table displays some solutions of the equation. As an aid to plotting points, we have included a third row in the table that gives one-decimal-place approximations to r.

θ	0	$\pi/6$	$\pi/4$	$\pi/3$	$\pi/2$	$2\pi/3$	$3\pi/4$	$5\pi/6$	π
r	0	2	$2\sqrt{2}$	$2\sqrt{3}$	4	$2\sqrt{3}$	$2\sqrt{2}$	2	0
r (approx.)	0	2	2.8	3.4	4	3.4	2.8	2	0

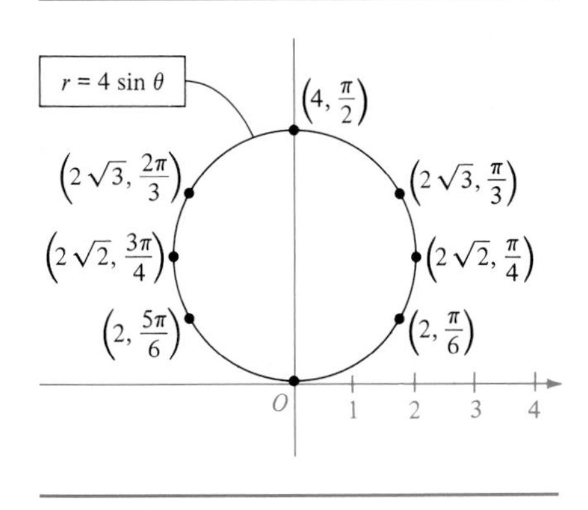

FIGURE 10.28

In rectangular coordinates, the graph of the equation consists of sine waves of amplitude 4 and period 2π. However, if polar coordinates are used, then the points that correspond to the pairs in the table appear to lie on a circle of radius 2 and we draw the graph accordingly (see Figure 10.28). As an aid to plotting points, we have extended the polar axis in the negative direction and introduced a vertical line through the pole.

The proof that the graph of $r = 4 \sin \theta$ is a circle will be given in Example 5. Additional points obtained by letting θ vary from π to 2π lie on the same circle. For example, the solution $(-2, 7\pi/6)$ gives us the same point as $(2, \pi/6)$; the point corresponding to $(-2\sqrt{2}, 5\pi/4)$ is the same as that obtained from $(2\sqrt{2}, \pi/4)$, and so on. If we let θ increase through all real numbers, we obtain the same points again and again because of the periodicity of the sine function. ■

EXAMPLE 2 Sketch the graph of the equation $r = 2 + 2 \cos \theta$.

SOLUTION Since the cosine function decreases from 1 to -1 as θ varies from 0 to π, it follows that r decreases from 4 to 0 in this θ-interval. The following table exhibits some solutions of the equation.

FIGURE 10.29

$r = 2 + 2\cos\theta$

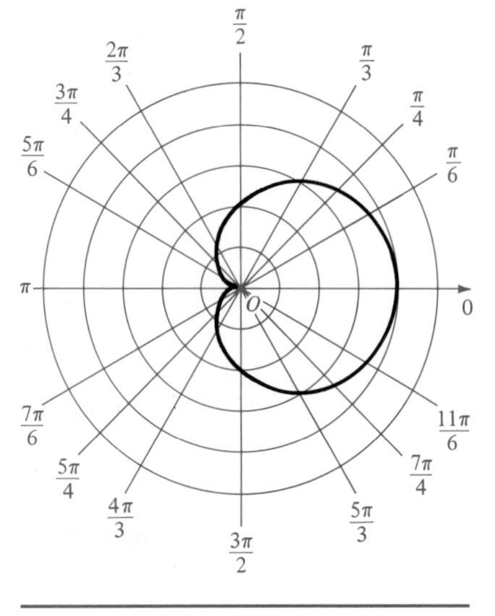

θ	0	$\pi/6$	$\pi/4$	$\pi/3$	$\pi/2$	$2\pi/3$	$3\pi/4$	$5\pi/6$	π
r	4	$2+\sqrt{3}$	$2+\sqrt{2}$	3	2	1	$2-\sqrt{2}$	$2-\sqrt{3}$	0
r (approx.)	4	3.7	3.4	3	2	1	0.6	0.3	0

Plotting points gives us the upper half of the graph sketched in Figure 10.29, where we have used polar coordinate graph paper, which displays lines through O at various angles and concentric circles with centers at the pole.

If θ increases from π to 2π, then $\cos\theta$ increases from -1 to 1 and, consequently, r increases from 0 to 4. Plotting points for $\pi \le \theta \le 2\pi$ would give us the lower half of the graph.

The same graph may be obtained by taking other intervals of length 2π for θ. ■

The heart-shaped graph in Example 2 is called a **cardioid.** In general, the graph of any polar equation of the form

$$r = a(1 + \cos\theta), \qquad r = a(1 + \sin\theta),$$
$$r = a(1 - \cos\theta), \qquad r = a(1 - \sin\theta),$$

where a is a real number, is a cardioid.

The graph of an equation of the form

$$r = a + b\cos\theta \quad \text{or} \quad r = a + b\sin\theta,$$

where $a \neq b$, is called a **limaçon.** The graph is similar in shape to a cardioid; however, there may be an added "loop" as shown in the next example.

FIGURE 10.30

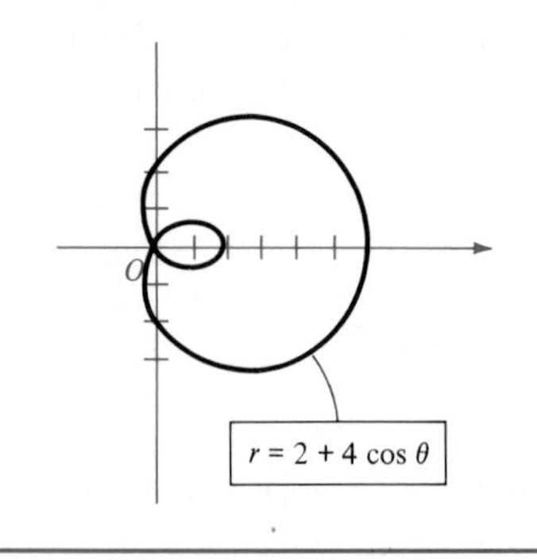

EXAMPLE 3 Sketch the graph of $r = 2 + 4\cos\theta$.

SOLUTION Some points corresponding to $0 \le \theta \le \pi$ are exhibited in the following table.

θ	0	$\pi/6$	$\pi/4$	$\pi/3$	$\pi/2$	$2\pi/3$	$3\pi/4$	$5\pi/6$	π
r	6	$2+2\sqrt{3}$	$2+2\sqrt{2}$	4	2	0	$2-2\sqrt{2}$	$2-2\sqrt{3}$	-2
r (approx.)	6	5.4	4.8	4	2	0	-0.8	-1.4	-2

Note that $r = 0$ at $\theta = 2\pi/3$. The values of r are negative if $2\pi/3 < \theta \le \pi$, and this leads to the lower half of the small loop in Figure 10.30. (Verify this fact.) Letting θ range from π to 2π gives us the upper half of the small loop and the lower half of the large loop. ■

EXAMPLE 4 Sketch the graph of the equation $r = a \sin 2\theta$ for $a > 0$.

SOLUTION Instead of tabulating solutions, let us reason as follows: If θ increases from 0 to $\pi/4$, then 2θ varies from 0 to $\pi/2$ and hence $\sin 2\theta$ increases from 0 to 1. It follows that r increases from 0 to a in the θ-interval $[0, \pi/4]$. If we next let θ increase from $\pi/4$ to $\pi/2$, then 2θ changes from $\pi/2$ to π and hence $\sin 2\theta$ decreases from 1 to 0. Thus r decreases from a to 0 in the θ-interval $[\pi/4, \pi/2]$. The corresponding points on the graph constitute the first-quadrant loop illustrated in Figure 10.31. Note that the point $P(r, \theta)$ traces the loop in a *counterclockwise* direction (indicated by the arrows) as θ increases from 0 to $\pi/2$.

FIGURE 10.31

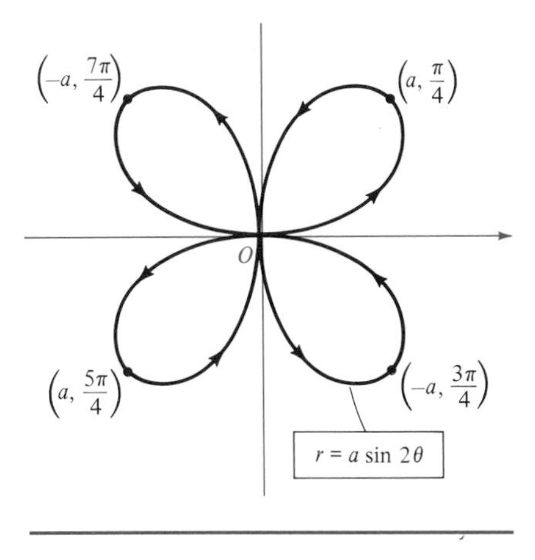

If $\pi/2 \le \theta \le \pi$, then $\pi \le 2\theta \le 2\pi$ and, therefore, $r = a \sin 2\theta \le 0$. Thus, if $\pi/2 < \theta < \pi$, then r is negative and *the points $P(r, \theta)$ are in the fourth quadrant*. If θ increases from $\pi/2$ to π, then we can show, by plotting points, that $P(r, \theta)$ traces (in a counterclockwise direction) the loop shown in the *fourth* quadrant.

Similarly, for $\pi \le \theta \le 3\pi/2$ we get the loop in the third quadrant, whereas for $3\pi/2 \le \theta \le 2\pi$ we get the loop in the second quadrant. Both loops are traced in a counterclockwise direction as θ increases. The reader should verify these facts by plotting some points with, say, $a = 1$. In Figure 10.31 we have plotted only those points on the graph that correspond to the largest numerical values of r. ■

The graph in Example 4 is called a **four-leafed rose.** In general, any equation of the form

$$r = a \sin n\theta \quad \text{or} \quad r = a \cos n\theta,$$

where n is a positive integer greater than 1 and a is a real number, has a graph that consists of a number of loops attached to the origin. If n is even there are $2n$ loops, whereas if n is odd there are n loops. (See, for example, Exercises 9, 10, and 22.)

Many other interesting graphs result from polar equations. Some are included in the exercises at the end of this section. Polar coordinates are useful in applications involving circles with centers at the origin or lines that pass through the origin, since equations that have these graphs may be written in the simple forms $r = k$ or $\theta = k$ for some fixed number k. (Verify this fact.)

Let us next superimpose an xy-plane on an $r\theta$-plane such that the positive x-axis coincides with the polar axis. Any point P in the plane may then be assigned rectangular coordinates (x, y) or polar coordinates (r, θ). If $r > 0$ we have a situation similar to that illustrated in Figure 10.32(i). If $r < 0$ we have that shown in (ii) of the figure where, for later purposes, we have also plotted the point P' having polar coordinates $(|r|, \theta)$ and rectangular coordinates $(-x, -y)$.

FIGURE 10.32

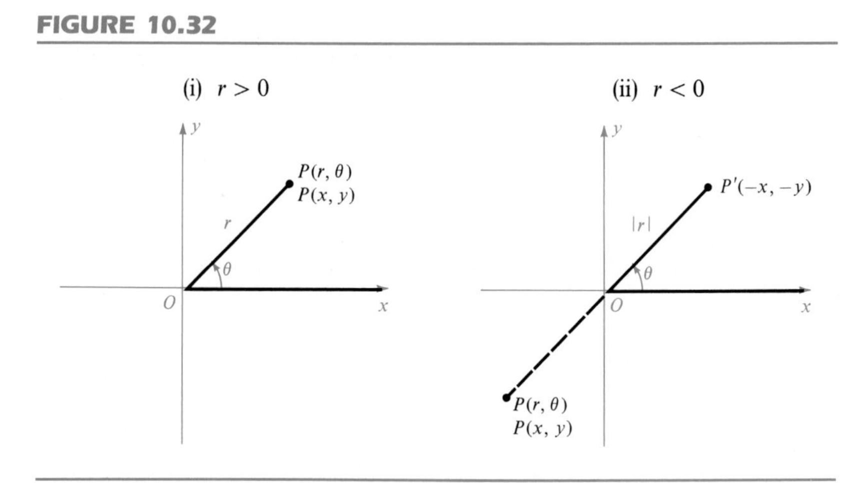

The following theorem specifies relationships between (x, y) and (r, θ). In the statement of the theorem it is assumed that the positive x-axis coincides with the polar axis.

THEOREM

The rectangular coordinates (x, y) and polar coordinates (r, θ) of a point P are related as follows:

$$\text{(i)}\quad x = r\cos\theta, \qquad y = r\sin\theta$$

$$\text{(ii)}\quad \tan\theta = \frac{y}{x}, \qquad r^2 = x^2 + y^2.$$

PROOF Although we have pictured θ as an acute angle in Figure 10.32, the discussion that follows is valid for all angles. On the one hand, if $r > 0$ as in Figure 10.32(i), then $\cos\theta = x/r$, $\sin\theta = y/r$, and hence

$$x = r\cos\theta, \qquad y = r\sin\theta.$$

On the other hand, if $r < 0$ then $|r| = -r$ and from Figure 10.32(ii) we see that

$$\cos\theta = \frac{-x}{|r|} = \frac{-x}{-r} = \frac{x}{r},$$

$$\sin\theta = \frac{-y}{|r|} = \frac{-y}{-r} = \frac{y}{r}.$$

Multiplication by r produces (i) of the theorem and, therefore, these formulas hold whether r is positive or negative. If $r = 0$, then the point is the pole and we again see that the formulas in (i) are true.

The formulas in (ii) follow readily from Figure 10.32. □

We may use the preceding theorem to change from one system of coordinates to the other. A more important use is for transforming a polar equation to an equation in x and y, and vice versa. This is illustrated in the next two examples.

EXAMPLE 5 Find an equation in x and y that has the same graph as the polar equation $r = 4\sin\theta$.

SOLUTION The equation $r = 4\sin\theta$ was considered in Example 1. If we multiply both sides by r, we obtain $r^2 = 4r\sin\theta$. Applying the theorem gives us $x^2 + y^2 = 4y$, which is equivalent to $x^2 + (y - 2)^2 = 4$. Thus, the graph is a circle of radius 2 with center at $(0, 2)$ in the xy-plane. ■

EXAMPLE 6 Find a polar equation of an arbitrary line.

SOLUTION We know that every line in an xy-coordinate system is the graph of a linear equation $ax + by + c = 0$. Using the formulas $x = r\cos\theta$ and $y = r\sin\theta$ gives us the polar equation

$$ar\cos\theta + br\sin\theta + c = 0$$

or

$$r(a\cos\theta + b\sin\theta) + c = 0.$$ ■

If we superimpose an xy-plane on an $r\theta$-plane, then the graph of a polar equation may be symmetric with respect to the x-axis, the y-axis, or the origin. It is left to the reader to show that if a substitution listed in the following table does not change the solutions of a polar equation, then the graph has the indicated symmetry.

TESTS FOR SYMMETRY

Substitution	Symmetry
$-\theta$ for θ	the line $\theta = 0$ (x-axis)
$-r$ for r	the pole (origin)
$\pi + \theta$ for θ	the pole (origin)
$\pi - \theta$ for θ	the line $\theta = \pi/2$ (y-axis)

To illustrate, since $\cos(-\theta) = \cos\theta$, the graph of the equation in Example 2 is symmetric with respect to the x-axis. Since $\sin(\pi - \theta) = \sin\theta$, the graph in Example 1 is symmetric with respect to the y-axis. The graph in Example 4 is symmetric to both axes and to the origin. Other tests for symmetry may be stated; however, those listed above are among the easiest to apply.

Exercises 10.6

Sketch the graphs of the polar equations in Exercises 1–24.

1 $r = 5$ **2** $\theta = \pi/4$

3 $\theta = -\pi/6$ **4** $r = -2$

5 $r = 4\cos\theta$ **6** $r = -2\sin\theta$

7 $r = 4(1 - \sin\theta)$ **8** $r = 1 + 2\cos\theta$

9 $r = 8\cos 3\theta$ **10** $r = 2\sin 4\theta$

11 $r^2 = 4\cos 2\theta$ (lemniscate)

12 $r = 6\sin^2(\frac{1}{2}\theta)$ **13** $r = 4\csc\theta$

14 $r = -3\sec\theta$ **15** $r = 2 - \cos\theta$

16 $r = 2 + 2\sec\theta$ (conchoid)

17 $r = 2^\theta$, $\theta \geq 0$ (spiral)

18 $r\theta = 1$, $\theta > 0$ (spiral)

19 $r = -6(1 + \cos\theta)$

20 $r = e^{2\theta}$ (logarithmic spiral)

21 $r = 2 + 4\sin\theta$ **22** $r = 8\cos 5\theta$

23 $r^2 = -16\sin 2\theta$ **24** $4r = \theta$

In Exercises 25–32 find a polar equation that has the same graph as the given equation.

25 $x = -3$ **26** $y = 2$

27 $x^2 + y^2 = 16$ **28** $x^2 = 8y$

29 $y = 6$ **30** $y = 6x$

31 $x^2 - y^2 = 16$ **32** $9x^2 + 4y^2 = 36$

In Exercises 33–48 find an equation in x and y that has the same graph as the given polar equation, and use it as an aid in sketching the graph in a polar coordinate system.

33 $r\cos\theta = 5$ **34** $r\sin\theta = -2$

35 $r - 6\sin\theta = 0$ **36** $r = 6\cot\theta$

37 $r = 2$ **38** $\theta = \pi/4$

39 $r = \tan\theta$ **40** $r = 4\sec\theta$

41 $r^2(4\sin^2\theta - 9\cos^2\theta) = 36$

42 $r^2(\cos^2\theta + 4\sin^2\theta) = 16$

43 $r^2\cos 2\theta = 1$ **44** $r^2\sin 2\theta = 4$

45 $r(\sin\theta - 2\cos\theta) = 6$

46 $r(\sin\theta + r\cos^2\theta) = 1$

47 $r = 1/(1 + \cos\theta)$ **48** $r = 4/(2 + \sin\theta)$

49 If $P_1(r_1, \theta_1)$ and $P_2(r_2, \theta_2)$ are points in an $r\theta$-plane, use the Law of Cosines to prove that

$$[d(P_1, P_2)]^2 = r_1^2 + r_2^2 - 2r_1r_2\cos(\theta_2 - \theta_1).$$

50 Prove that the graphs of the following polar equations are circles, and find the center and radius in each case.

(a) $r = a\sin\theta,\quad a \neq 0$

(b) $r = b\cos\theta,\quad b \neq 0$

(c) $r = a\sin\theta + b\cos\theta,\quad ab \neq 0$

10.7 Polar Equations of Conics

The following theorem provides another method for describing conic sections.

THEOREM

> Let F be a fixed point and l a fixed line in a plane. The set of all points P in the plane such that the ratio $d(P, F)/d(P, Q)$ is a positive constant e, where $d(P, Q)$ is the distance from P to l, is a conic section. Moreover, the conic is a parabola if $e = 1$, an ellipse if $0 < e < 1$, and a hyperbola if $e > 1$.

The constant e is called the **eccentricity** of the conic. It will be seen that the point F is a focus of the conic. The line l is called a **directrix.** A typical situation is illustrated in Figure 10.33, where the curve indicates possible positions of P.

FIGURE 10.33

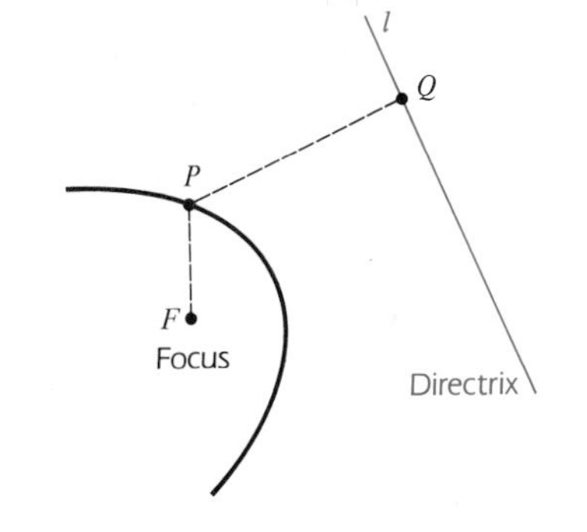

If $e = 1$, then $d(P, F) = d(P, Q)$ and, by definition, we obtain a parabola with focus F and directrix l.

Suppose next that $0 < e < 1$. It is convenient to introduce a polar coordinate system in the plane with F as the pole and with l perpendicular to the polar axis at the point $D(d, 0)$ where $d > 0$. If $P(r, \theta)$ is a point in the plane such that $d(P, F)/d(P, Q) = e < 1$, then from Figure 10.34 we see that P lies to the left of l. Let C be the projection of P on the polar axis. Since

$$d(P, F) = r \quad\text{and}\quad d(P, Q) = \overline{FD} - \overline{FC} = d - r\cos\theta,$$

it follows that P satisfies the condition in the theorem if and only if

$$\frac{r}{d - r\cos\theta} = e$$

FIGURE 10.34

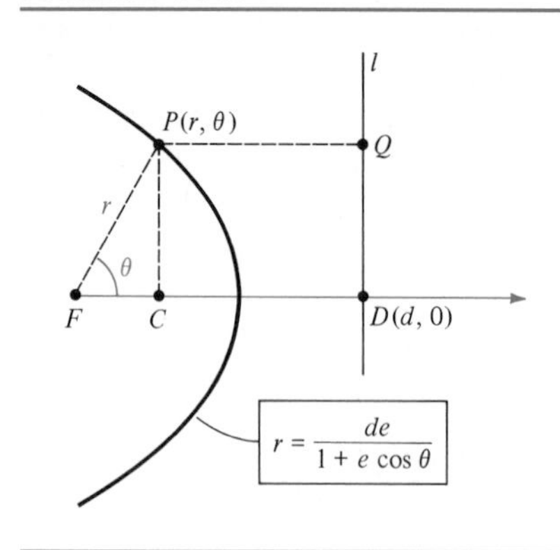

or $$r = de - er\cos\theta.$$

Solving for r gives us $$r = \frac{de}{1 + e\cos\theta},$$

which is a polar equation of the graph. Actually, the same equation is obtained if $e = 1$; however, there is no point (r, θ) on the graph if $1 + \cos\theta = 0$.

The rectangular equation corresponding to $r = de - er\cos\theta$ is

$$\pm\sqrt{x^2 + y^2} = de - ex.$$

Squaring both sides and rearranging terms leads to

$$(1 - e^2)x^2 + 2de^2x + y^2 = d^2e^2.$$

Completing the square in the previous equation and simplifying, we obtain

$$\left(x + \frac{de^2}{1 - e^2}\right)^2 + \frac{y^2}{1 - e^2} = \frac{d^2e^2}{(1 - e^2)^2}.$$

Finally, dividing both sides by $d^2e^2/(1 - e^2)^2$ leads to an equation of the form

$$\frac{(x - h)^2}{a^2} + \frac{y^2}{b^2} = 1$$

where $h = -de^2/(1 - e^2)$. Consequently, the graph is an ellipse with center at the point $(-de^2/(1 - e^2), 0)$ on the x-axis and where

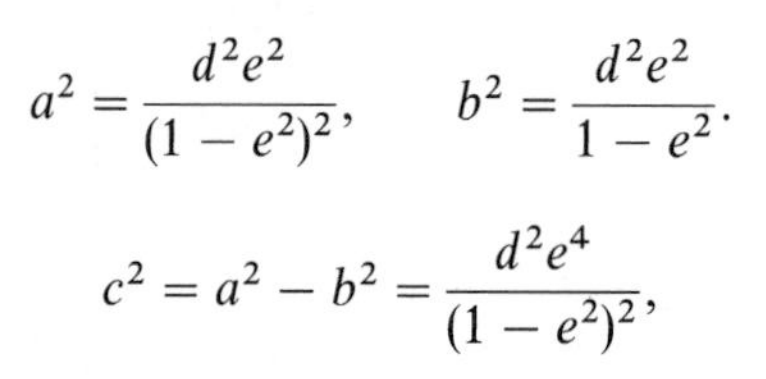

$$a^2 = \frac{d^2e^2}{(1 - e^2)^2}, \qquad b^2 = \frac{d^2e^2}{1 - e^2}.$$

FIGURE 10.35

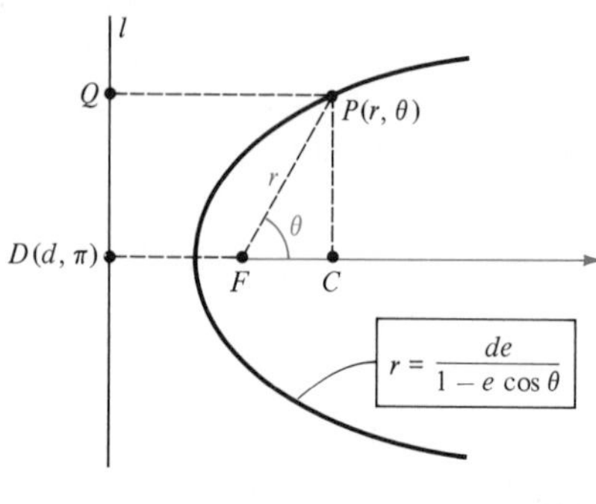

Since $$c^2 = a^2 - b^2 = \frac{d^2e^4}{(1 - e^2)^2},$$

we have $c = de^2/(1 - e^2)$. This proves that F is a focus of the ellipse. It also follows that $e = c/a$. A similar proof may be given for the case $e > 1$.

It can be shown, conversely, that every conic that is not a circle may be described by means of the statement in the theorem. This gives us a formulation of conic sections that is equivalent to the approach used previously. Since the theorem includes all three types of conics, it is sometimes regarded as a definition for the conic sections.

If we had chosen the focus F to the *right* of the directrix, as illustrated in Figure 10.35 (where $d > 0$), then the equation $r = de/(1 - e\cos\theta)$ would have resulted. (Note the minus sign in place of the plus sign.) Other sign changes occur if d is allowed to be negative.

If l is taken *parallel* to the polar axis through one of the points $(d, \pi/2)$ or $(d, 3\pi/2)$, as illustrated in Figure 10.36, then the corresponding equations would contain $\sin\theta$ instead of $\cos\theta$. The proofs of these facts are left to the reader.

FIGURE 10.36

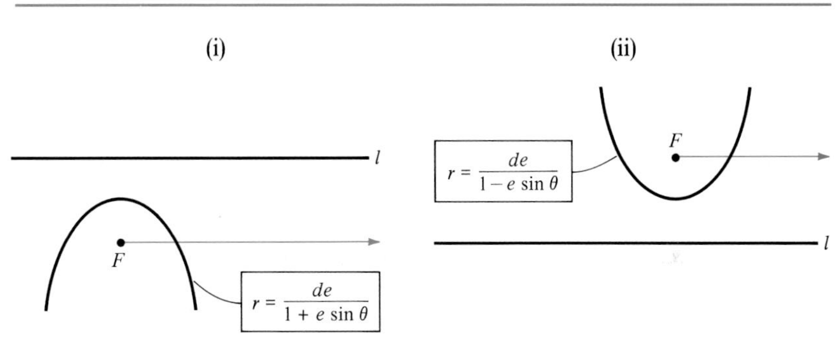

The following theorem summarizes our discussion.

THEOREM

A polar equation that has one of the four forms

$$r = \frac{de}{1 \pm e\cos\theta}, \qquad r = \frac{de}{1 \pm e\sin\theta}$$

is a conic section. Moreover, the conic is a parabola if $e = 1$, an ellipse if $0 < e < 1$, or a hyperbola if $e > 1$.

EXAMPLE 1 Describe and sketch the graph of the equation

$$r = \frac{10}{3 + 2\cos\theta}.$$

SOLUTION Dividing numerator and denominator of the fraction by 3, we obtain

$$r = \frac{\frac{10}{3}}{1 + \frac{2}{3}\cos\theta},$$

which has one of the forms in the theorem with $e = \frac{2}{3}$. Thus the graph is an ellipse with focus F at the pole and major axis along the polar axis. The endpoints of the major axis may be found by setting θ equal to 0 and π. This gives us $V(2, 0)$ and $V'(10, \pi)$. Hence

$$2a = d(V', V) = 12 \quad \text{or} \quad a = 6.$$

The center of the ellipse is the midpoint of the segment $V'V$, namely $(4, \pi)$. Using the fact that $e = c/a$, we obtain

$$c = ae = 6(\tfrac{2}{3}) = 4.$$

Hence,
$$b^2 = a^2 - c^2 = 36 - 16 = 20,$$

that is, the semiminor axis has length $\sqrt{20}$. The graph is sketched in Figure 10.37 where, for reference, we have superimposed a rectangular coordinate system on the polar system. ■

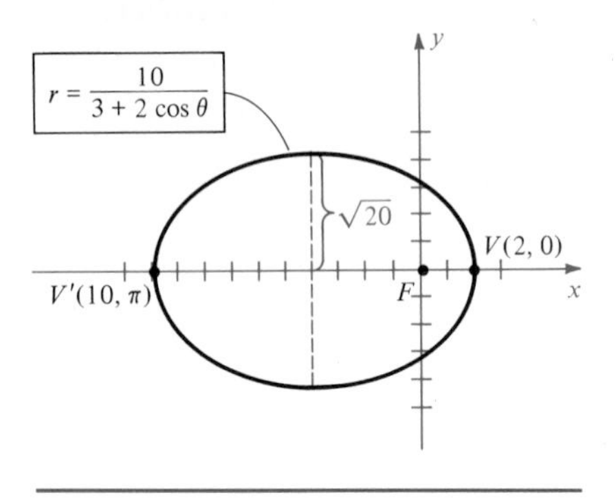

EXAMPLE 2 Describe and sketch the graph of the equation

$$r = \frac{10}{2 + 3\sin\theta}.$$

SOLUTION To express the equation in the proper form we divide numerator and denominator of the fraction by 2, obtaining

$$r = \frac{5}{1 + \frac{3}{2}\sin\theta}.$$

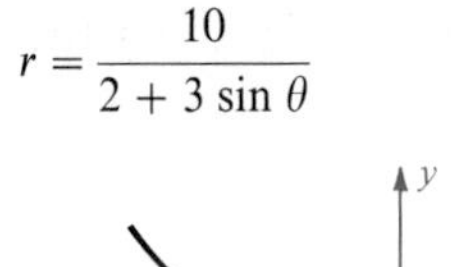

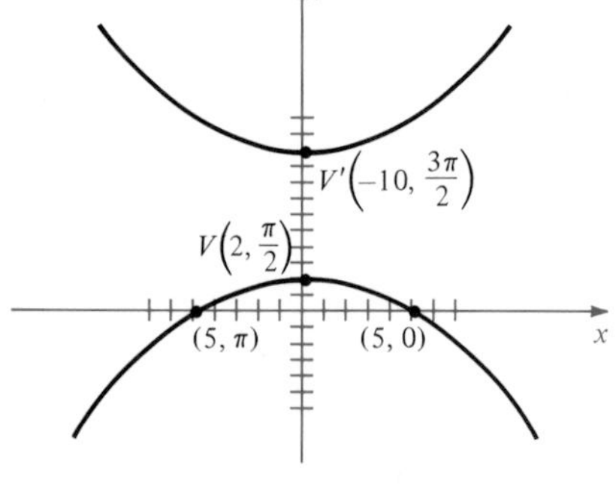

Thus $e = \frac{3}{2}$ and the graph is a hyperbola with a focus at the pole. The expression $\sin\theta$ tells us that the transverse axis of the hyperbola is perpendicular to the polar axis. To find the vertices we let θ equal $\pi/2$ and $3\pi/2$ in the given equation. This gives us the points $V(2, \pi/2)$, $V'(-10, 3\pi/2)$, and hence

$$2a = d(V, V') = 8 \quad \text{or} \quad a = 4.$$

The points $(5, 0)$ and $(5, \pi)$ on the graph can be used to get a rough estimate of the lower branch of the hyperbola. The upper branch is obtained by symmetry, as illustrated in Figure 10.38. If more accuracy or additional information is desired, we may calculate

$$c = ae = 4(\tfrac{3}{2}) = 6$$

and $$b^2 = c^2 - a^2 = 36 - 16 = 20.$$

Asymptotes may then be constructed in the usual way. ■

EXAMPLE 3 Sketch the graph of the equation

$$r = \frac{15}{4 - 4\cos\theta}.$$

SOLUTION To obtain the proper form we divide numerator and denominator by 4, obtaining

$$r = \frac{\frac{15}{4}}{1 - \cos\theta}.$$

Consequently, $e = 1$ and the graph is a parabola with focus at the pole. A rough sketch can be found by plotting the points that correspond to the x- and y- intercepts. These are indicated in the following table.

θ	0	$\pi/2$	π	$3\pi/2$
r	—	15/4	15/8	15/4

FIGURE 10.39

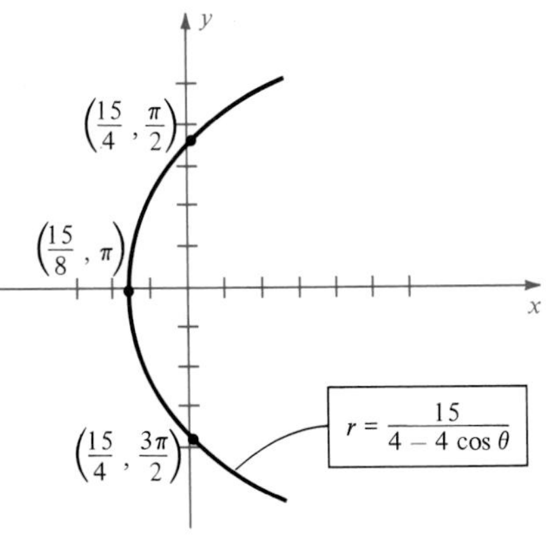

Note that there is no point on the graph corresponding to $\theta = 0$. Plotting the three points and using the fact that the graph is a parabola with focus at the pole gives us the sketch in Figure 10.39. ■

If only a rough sketch of a conic is desired, then the technique employed in Example 3 is recommended. To use this method, we plot (if possible) points corresponding to $\theta = 0$, $\pi/2$, π, and $3\pi/2$. These points, together with the type of conic (obtained from the value of e), readily lead to the sketch.

Exercises 10.7

In Exercises 1–10 identify and sketch the graph of the equation.

1 $r = \dfrac{12}{6 + 2\sin\theta}$

2 $r = \dfrac{12}{6 - 2\sin\theta}$

3 $r = \dfrac{12}{2 - 6\cos\theta}$

4 $r = \dfrac{12}{2 + 6\cos\theta}$

5 $r = \dfrac{3}{2 + 2\cos\theta}$

6 $r = \dfrac{3}{2 - 2\sin\theta}$

7 $r = \dfrac{4}{\cos\theta - 2}$

8 $r = \dfrac{4\sec\theta}{2\sec\theta - 1}$

9 $r = \dfrac{6 \csc \theta}{2 \csc \theta + 3}$

10 $r = \csc \theta (\csc \theta - \cot \theta)$

11–20 Find rectangular equations for the graphs in Exercises 1–10.

In Exercises 21–26 express the equation in polar form and then find the eccentricity and an equation for the directrix.

21 $y^2 = 4 - 4x$

22 $x^2 = 1 - 2y$

23 $3y^2 - 16y - x^2 + 16 = 0$

24 $5x^2 + 9y^2 = 32x + 64$

25 $8x^2 + 9y^2 + 4x = 4$

26 $4x^2 - 5y^2 + 36y - 36 = 0$

In Exercises 27–32 find a polar equation of the conic with focus at the pole and the given eccentricity and equation of directrix.

27 $e = \frac{1}{3},\ r = 2 \sec \theta$

28 $e = \frac{2}{5},\ r = 4 \csc \theta$

29 $e = 4,\ r = -3 \csc \theta$

30 $e = 3,\ r = -4 \sec \theta$

31 $e = 1,\ r \cos \theta = 5$

32 $e = 1,\ r \sin \theta = -2$

33 Find a polar equation of the parabola with focus at the pole and vertex $(4, \pi/2)$.

34 Find a polar equation of the ellipse with eccentricity $\frac{2}{3}$, a vertex at $(1, 3\pi/2)$, and a focus at the pole.

35 Prove the theorem of this section for the case $e > 1$.

36 Derive the formulas $r = de/(1 \pm e \sin \theta)$ discussed in this section.

10.8 Plane Curves and Parametric Equations

The graph of an equation $y = f(x)$, where the domain of the function f is an interval I, is often called a *plane curve*. However, to use this as a definition is unnecessarily restrictive, since it rules out most of the conic sections and many other useful graphs. The following statement is satisfactory for most applications.

DEFINITION

A **plane curve** is a set C of ordered pairs of the form

$$(f(t), g(t))$$

where f and g are functions defined on an interval I.

For simplicity, we shall often refer to a plane curve as a **curve.** The **graph** of the curve C consists of all points $P(t) = (f(t), g(t))$ in a rectangular coordinate system that correspond to the ordered pairs. Each $P(t)$ is referred to as a *point* on the curve. We shall use the term *curve* interchangeably with *graph of a curve.* In some cases it is convenient to imagine that the point $P(t)$ traces the curve C as t varies through the interval I. This is especially true in applications where t represents time and $P(t)$ is the position of a moving particle at time t.

The graphs of several curves are sketched in Figure 10.40 for the case where I is a closed interval $[a, b]$. If, as in (i) of the figure, $P(a) \neq P(b)$, then $P(a)$ and $P(b)$ are called the **endpoints** of C. Note that the curve illustrated in (i) intersects itself in the sense that two different values of t give rise to the same point. If $P(a) = P(b)$, as illustrated in Figure 10.40(ii), then C is called a **closed curve.** If $P(a) = P(b)$ and C does not intersect itself at any other point, as illustrated in (iii), then C is called a **simple closed curve.**

FIGURE 10.40

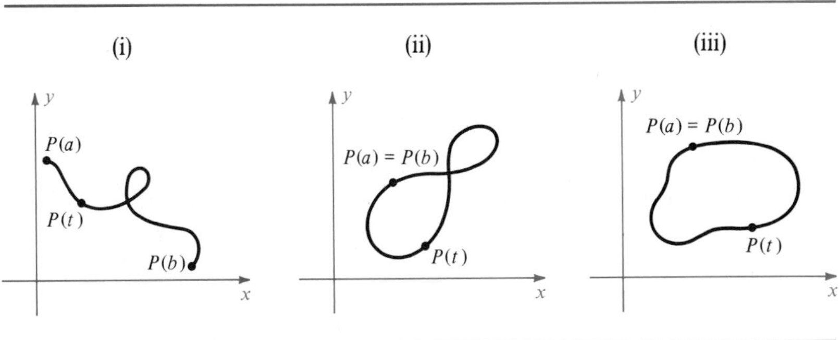

A convenient way to represent curves is given in the next definition.

DEFINITION

> Let C be the curve consisting of all ordered pairs $(f(t), g(t))$ where f and g are defined on an interval I. The equations
>
> $$x = f(t), \qquad y = g(t),$$
>
> for t in I, are called **parametric equations** for C, and t is called a **parameter.**

If we are given parametric equations, then as t varies through I, the point $P(x, y)$ traces the curve. As shown in the next example, it is sometimes possible to eliminate the parameter and obtain an equation for C that involves the variables x and y.

EXAMPLE 1 Sketch and identify the graph of the curve C given parametrically by

$$x = 2t, \quad y = t^2 - 1 \qquad \text{for } -1 \le t \le 2.$$

SOLUTION The parametric equations can be used to tabulate coordinates for points $P(x, y)$ on C as in the following table.

t	-1	$-\frac{1}{2}$	0	$\frac{1}{2}$	1	$\frac{3}{2}$	2
x	-2	-1	0	1	2	3	4
y	0	$-\frac{3}{4}$	-1	$-\frac{3}{4}$	0	$\frac{5}{4}$	3

FIGURE 10.41

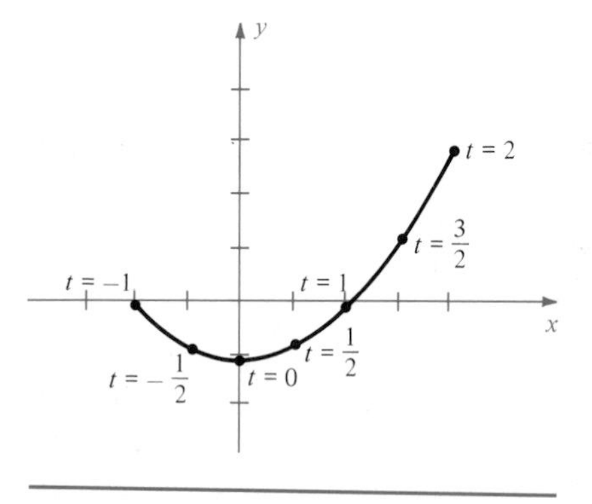

Plotting points leads to the sketch in Figure 10.41.

A precise description of the graph may be obtained by eliminating the parameter. To illustrate, solving the first parametric equation for t, we obtain $t = x/2$. Substituting for t in the second equation gives us

$$y = \left(\frac{x}{2}\right)^2 - 1 \quad \text{or} \quad y + 1 = \frac{1}{4}x^2.$$

The graph of this equation is a parabola with vertical axis and vertex at the point $(0, -1)$. The curve C is that part of the parabola shown in Figure 10.41. ■

Parametric equations of curves are never unique. The reader should verify that the curve C in Example 1 is given by any of the following parametric representations:

$$\begin{aligned} &x = 2t, \quad y = t^2 - 1; && -1 \le t \le 2 \\ &x = t, \quad y = \tfrac{1}{4}t^2 - 1; && -2 \le t \le 4 \\ &x = t^3, \quad y = \tfrac{1}{4}t^6 - 1; && \sqrt[3]{-2} \le t \le \sqrt[3]{4}. \end{aligned}$$

The next example illustrates the fact that it is often useful to eliminate the parameter *before* plotting points.

EXAMPLE 2 Sketch the graph of the curve C that is given parametrically by

$$x = -2 + t^2, \quad y = 1 + 2t^2 \qquad \text{for } t \text{ in } \mathbb{R}.$$

FIGURE 10.42

(i)

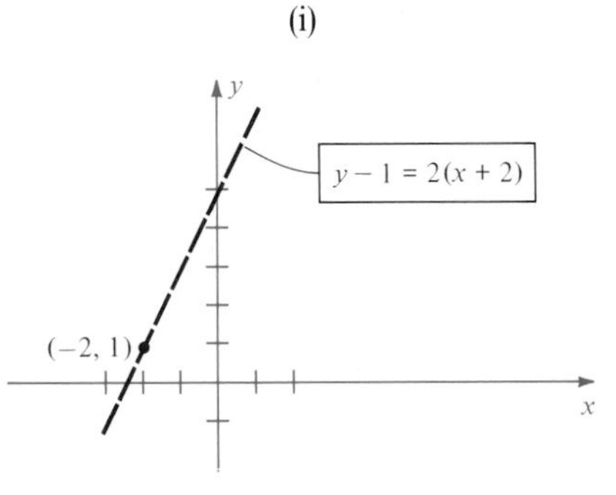

(ii)

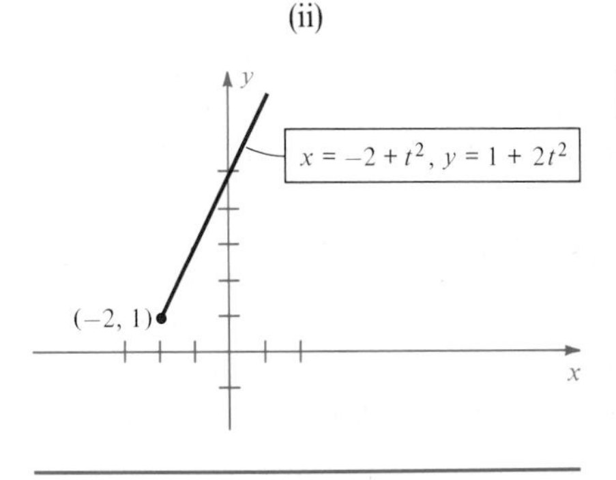

SOLUTION To eliminate the parameter, we note from the first equation that $t^2 = x + 2$. Substituting for t^2 in the second equation gives us

$$y = 1 + 2(x + 2) \quad \text{or} \quad y - 1 = 2(x + 2).$$

This is an equation of the line of slope 2 through the point $(-2, 1)$, as indicated by the dashes in Figure 10.42(i). However, since $t^2 \geq 0$,

$$x = -2 + t^2 \geq -2 \quad \text{and} \quad y = 1 + 2t^2 \geq 1.$$

It follows that the graph of C is that part of the line to the right of $(-2, 1)$, as shown in (ii) of the figure. This fact may also be verified by plotting several points. ■

EXAMPLE 3 Describe the graph of the curve C having parametric equations

$$x = \cos t, \quad y = \sin t \qquad \text{for } 0 \leq t \leq 2\pi.$$

SOLUTION We may use the identity $\cos^2 t + \sin^2 t = 1$ to eliminate the parameter. This gives us $x^2 + y^2 = 1$, and hence points on C are on the unit circle with center at the origin. As t increases from 0 to 2π, $P(t)$ starts at the point $A(1, 0)$ and traverses the circle once in the counterclockwise direction. In this example the parameter t may be interpreted geometrically as the radian measure of angle θ in Figure 5.22. ■

If a curve C is described by means of an equation $y = f(x)$, where f is a function, then an easy way to obtain parametric equations is to let

$$x = t, \qquad y = f(t)$$

where t is in the domain of f. For example, if $y = x^3$, then parametric equations are

$$x = t, \quad y = t^3 \qquad \text{for } t \text{ in } \mathbb{R}.$$

We can use many different substitutions for x, provided that as t varies through some interval, x takes on all values in the domain of f. Thus the graph of $y = x^3$ is also given by

$$x = t^{1/3}, \quad y = t \qquad \text{for } t \text{ in } \mathbb{R}.$$

Note, however, that the parametric equations

$$x = \sin t, \quad y = \sin^3 t \qquad \text{for } t \text{ in } \mathbb{R}$$

give only that part of the graph of $y = x^3$ that lies between the points $(-1, -1)$ and $(1, 1)$.

EXAMPLE 4 Find three different parametric representations for the line of slope m through the point (x_1, y_1).

SOLUTION By the Point-Slope Form, an equation for the line is

$$y - y_1 = m(x - x_1).$$

If we let $x = t$, then $y - y_1 = m(t - x_1)$ and we obtain the parametric equations

$$x = t, \quad y = y_1 + m(t - x_1) \qquad \text{for } t \text{ in } \mathbb{R}.$$

Another pair of parametric equations results if we let $x - x_1 = t$. In this case $y - y_1 = mt$ and hence the line is given parametrically by

$$x = x_1 + t, \quad y = y_1 + mt \qquad \text{for } t \text{ in } \mathbb{R}.$$

As a third illustration, if we let $x - x_1 = \tan t$, then

$$x = x_1 + \tan t, \quad y = y_1 + m \tan t \qquad \text{for } -\frac{\pi}{2} < t < \frac{\pi}{2}.$$

There are many other ways to represent the line parametrically. ■

EXAMPLE 5 The curve traced by a fixed point P on the circumference of a circle as the circle rolls along a line in a plane is called a **cycloid.** Find parametric equations for a cycloid.

SOLUTION Suppose that the circle has radius a and that it rolls along (and above) the x-axis in the positive direction. If one position of P is the origin, then Figure 10.43 displays part of the curve and a possible position of the circle.

Let K denote the center of the circle and T the point of tangency with the x-axis. We introduce a parameter t as the radian measure of angle TKP. Since $\overline{OT}$ is the distance the circle has rolled, $\overline{OT} = at$. Consequently, the coordinates of K are (at, a). If we consider an $x'y'$-coordinate system with origin at $K(at, a)$, and if $P(x', y')$ denotes the point P relative to this

FIGURE 10.43

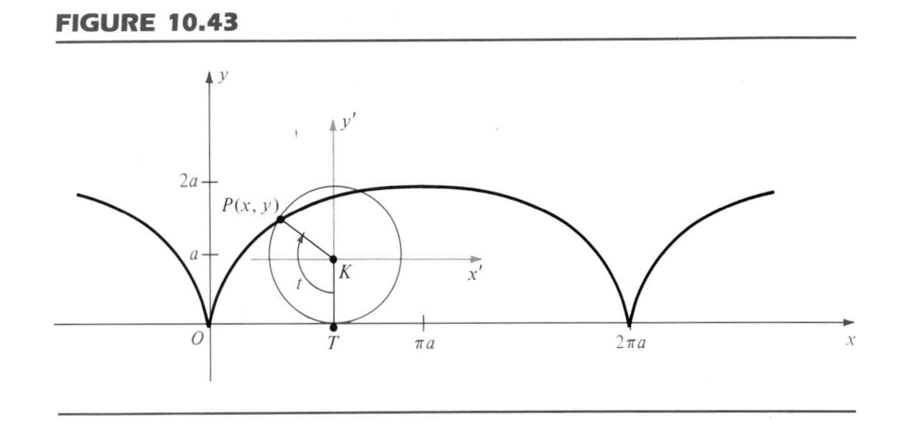

FIGURE 10.44

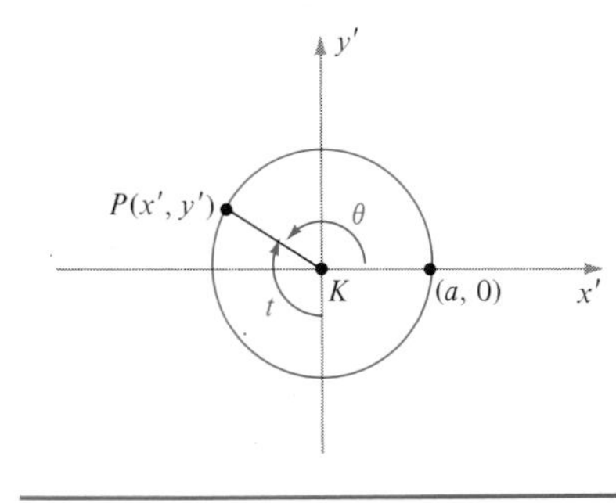

system, then by the Translation of Axes Formulas, with $h = at$ and $k = a$,

$$x = at + x', \qquad y = a + y'.$$

If, as in Figure 10.44, θ denotes an angle in standard position on the $x'y'$-system, then $\theta = (3\pi/2) - t$. Hence

$$x' = a \cos \theta = a \cos \left[(3\pi/2) - t\right] = -a \sin t$$

$$y' = a \sin \theta = a \sin \left[(3\pi/2) - t\right] = -a \cos t,$$

and substitution in $x = at + x'$, $y = a + y'$ gives us parametric equations for the cycloid, namely

$$x = a(t - \sin t), \quad y = a(1 - \cos t) \qquad \text{for } t \text{ in } \mathbb{R}. \quad \blacksquare$$

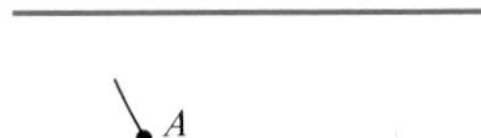

FIGURE 10.45

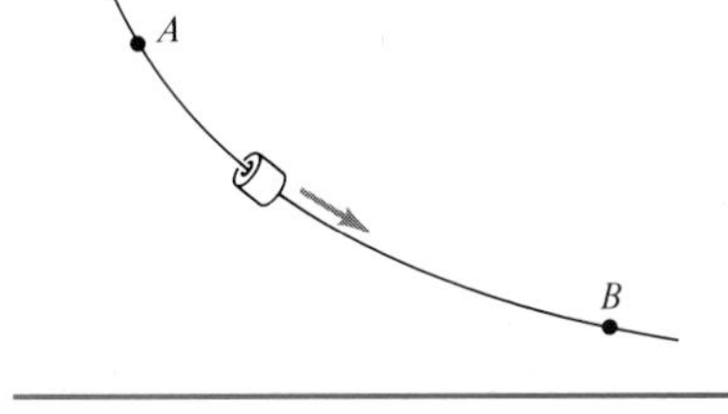

If $a < 0$, then the graph of $x = a(t - \sin t)$, $y = a(1 - \cos t)$ is the inverted cycloid that results if the circle of Example 5 rolls *below* the x-axis. This curve has a number of important physical properties. In particular, suppose a thin wire passes through two fixed points A and B as illustrated in Figure 10.45, and that the shape of the wire can be changed by bending it in any manner. Suppose further, that a bead is allowed to slide along the wire and the only force acting on the bead is gravity. We now ask which of all the possible paths will allow the bead to slide from A to B in the least amount of time. It is natural to conjecture that the desired path is the straight line segment from A to B; however, this is not the correct answer. The path that requires the least time coincides with the graph of the inverted cycloid with A at the origin and B the lowest point on the curve. The proof of this result will not be given in this text.

To cite another interesting property of this curve, suppose that A is the origin and B is the point with x-coordinate $\pi|a|$, that is, the lowest point on the cycloid occurring in the first arc to the right of A. It can be shown that if the bead is released at *any* point between A and B, the time required for it to reach B is always the same!

Variations of the cycloid occur in practical problems. For example, if a motorcycle wheel rolls along a straight road, then the curve traced by a fixed point on one of the spokes is a type of cycloid curve. In this case, the curve does not have sharp corners, nor does it intersect the road (the x-axis) as does the graph of a cycloid. In like manner, if the wheel of a train rolls along a railroad track, then the curve traced by a fixed point on the circumference of the wheel (which extends below the track) contains loops at regular intervals (see Exercise 32). Several other cycloids are defined in Exercises 27 and 28.

Exercises 10.8

In Exercises 1–18, (a) sketch the graph of the curve C having the indicated parametric equations, and (b) find a rectangular equation of a graph that contains the points on C.

1 $x = t - 2, \quad y = 2t + 3; \qquad 0 \le t \le 5$

2 $x = 1 - 2t, \quad y = 1 + t; \qquad -1 \le t \le 4$

3 $x = t^2 + 1, \quad y = t^2 - 1; \qquad -2 \le t \le 2$

4 $x = t^3 + 1, \quad y = t^3 - 1; \qquad -2 \le t \le 2$

5 $x = 4t^2 - 5, \quad y = 2t + 3; \qquad t \text{ in } \mathbb{R}$

6 $x = t^3, \quad y = t^2; \qquad t \text{ in } \mathbb{R}$

7 $x = e^t, \quad y = e^{-2t}; \qquad t \text{ in } \mathbb{R}$

8 $x = \sqrt{t}, \quad y = 3t + 4; \qquad t \ge 0$

9 $x = 2 \sin t, \quad y = 3 \cos t; \qquad 0 \le t \le 2\pi$

10 $x = \cos t - 2, \quad y = \sin t + 3; \qquad 0 \le t \le 2\pi$

11 $x = \sec t, \quad y = \tan t; \qquad -\pi/2 < t < \pi/2$

12 $x = \cos 2t, \quad y = \sin t; \qquad -\pi \le t \le \pi$

13 $x = t^2, \quad y = 2 \ln t; \qquad t > 0$

14 $x = \cos^3 t, \quad y = \sin^3 t; \qquad 0 \le t \le 2\pi$

15 $x = \sin t, \quad y = \csc t; \qquad 0 < t \le \pi/2$

16 $x = e^t, \quad y = e^{-t}; \qquad t \text{ in } \mathbb{R}$

17 $x = t, \quad y = \sqrt{t^2 - 1}; \qquad |t| \ge 1$

18 $x = -2\sqrt{1 - t^2}, \quad y = t; \qquad |t| \le 1$

In Exercises 19–22 sketch the graph of the curve C having the indicated parametric equations.

19 $x = t, \quad y = \sqrt{t^2 - 2t + 1}; \qquad 0 \le t \le 4$

20 $x = 2t, \quad y = 8t^3; \qquad -1 \le t \le 1$

21 $x = (t + 1)^3, \quad y = (t + 2)^2; \qquad 0 \le t \le 2$

22 $x = \tan t, \quad y = 1; \qquad -\pi/2 < t < \pi/2$

In Exercises 23 and 24 curves C_1, C_2, C_3, and C_4 are given parametrically, for t in $\mathbb{R}$. Sketch the graphs of C_1, C_2, C_3, C_4, and discuss their similarities and differences.

23 $C_1\colon x = t^2, \quad y = t$
$C_2\colon x = t^4, \quad y = t^2$
$C_3\colon x = \sin^2 t, \quad y = \sin t$
$C_4\colon x = e^{2t}, \quad y = -e^t$

24 $C_1\colon x = t, \quad y = 1 - t$
$C_2\colon x = 1 - t^2, \quad y = t^2$
$C_3\colon x = \cos^2 t, \quad y = \sin^2 t$
$C_4\colon x = \ln t - t, \quad y = 1 + t - \ln t$

25 If $P_1(x_1, y_1)$ and $P_2(x_2, y_2)$ are distinct points, show that

$$x = (x_2 - x_1)t + x_1, \qquad y = (y_2 - y_1)t + y_1,$$

for t in $\mathbb{R}$, are parametric equations of the line l through P_1 and P_2. Find three other pairs of parametric equations for the line l. Show that there is an infinite number of different pairs of parametric equations for l.

26 Show that

$$x = a \cos t + h, \qquad y = b \sin t + k,$$

for $0 \le t \le 2\pi$, are parametric equations of an ellipse with center at the point (h, k) and semi-axes of lengths a and b.

27 A circle C of radius b rolls on the inside of a second circle having equation $x^2 + y^2 = a^2$, where $b < a$. Let P be a fixed point on C and let the initial position of P be $A(a, 0)$, as shown in the figure. If the parameter t is the angle from the positive x-axis to the line segment from O to the center of C, show that parametric equations for the curve traced by P (called a **hypocycloid**) are

$$x = (a - b) \cos t + b \cos \frac{a - b}{b} t,$$

$$y = (a - b) \sin t - b \sin \frac{a - b}{b} t,$$

for $0 \le t \le 2\pi$. If $b = a/4$, show that

$$x = a \cos^3 t, \qquad y = a \sin^3 t$$

and sketch the graph of the curve.

FIGURE FOR EXERCISE 27

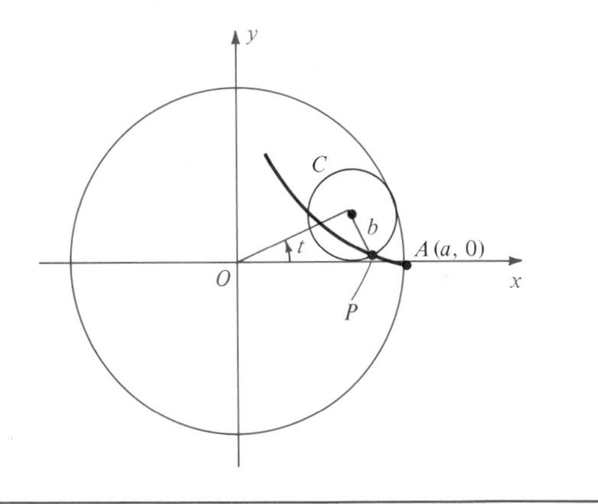

28 If the circle C of Exercise 27 rolls on the outside of the second circle (see figure), then the curve traced by P is called an **epicycloid.** Show that parametric equations for this curve are

$$x = (a + b) \cos t - b \cos \frac{a + b}{b} t,$$

$$y = (a + b) \sin t - b \sin \frac{a + b}{b} t,$$

for $0 \le t \le 2\pi$.

FIGURE FOR EXERCISE 28

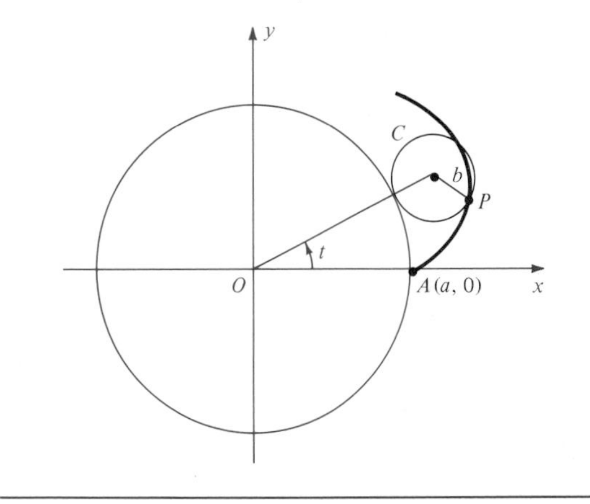

29 For $b = a/3$ in Exercise 28, find parametric equations for the epicycloid and sketch the graph.

30 The radius of circle A is one-third that of circle B. How many revolutions will circle A make as it rolls around circle B until it reaches its starting point? (*Hint:* Use Exercise 29.)

31 If a string is unwound from around a circle and is kept taut in the plane of the circle, then a fixed point P on the string will trace a curve called the **involute of the circle.** If the circle is chosen as in Exercise 27 and the parameter t is the angle from the positive x-axis to the point of tangency of the string, show that parametric equations for the involute are

$$x = a(\cos t + t \sin t), \quad y = a(\sin t - t \cos t) \quad \text{for } t \text{ in } \mathbb{R}.$$

32 Generalize the cycloid of Example 5 to the case where P is any point on a fixed line through the center K of the circle. If $b = d(K, P)$, derive the parametric equations

$$x = at - b \sin t, \qquad y = a - b \cos t \quad \text{for } t \text{ in } \mathbb{R}.$$

Sketch a typical graph if $b < a$ (a **curtate cycloid**) and if $b > a$ (a **prolate cycloid**). The term **trochoid** is sometimes used for either of these curves.

10.9 Review

Define or discuss each of the following.

1 Conic sections
2 Parabola
3 Focus, directrix, vertex, and axis of a parabola
4 Ellipse
5 Major and minor axes of an ellipse
6 Foci and vertices of an ellipse
7 Hyperbola
8 Transverse and conjugate axes of a hyperbola
9 Foci and vertices of a hyperbola
10 Asymptotes of a hyperbola
11 Translation of axes
12 Rotation of axes
13 Polar coordinates of a point
14 The relationship between polar and rectangular coordinates
15 Graphs of polar equations
16 Polar equations of conics
17 Plane curve
18 Parametric equations of a curve

Exercises 10.9

In Exercises 1–10 find the foci and vertices and sketch the graph of the conic that has the given equation.

1 $y^2 = 64x$

2 $y - 1 = 8(x + 2)^2$

3 $9y^2 = 144 - 16x^2$

4 $9y^2 = 144 + 16x^2$

5 $x^2 - y^2 - 4 = 0$

6 $25x^2 + 36y^2 = 1$

7 $25y = 100 - x^2$

8 $3x^2 + 4y^2 - 18x + 8y + 19 = 0$

9 $x^2 - 9y^2 + 8x + 90y - 210 = 0$

10 $x = 2y^2 + 8y + 3$

Find equations for the conics in Exercises 11–18.

11 The hyperbola with vertices $V(0, \pm 7)$ and endpoints of conjugate axes $(\pm 3, 0)$

12 The parabola with focus $F(-4, 0)$ and directrix $x = 4$

13 The parabola with focus $(0, -10)$ and directrix $y = 10$

14 The parabola with vertex at the origin, symmetric to the x-axis, and passing through the point $(5, -1)$

15 The ellipse with vertices $V(0, \pm 10)$ and foci $F(0, \pm 5)$

16 The hyperbola with foci $F(\pm 10, 0)$ and vertices $V(\pm 5, 0)$

17 The hyperbola with vertices $V(0, \pm 6)$ and asymptotes that have equations $y = \pm 9x$

18 The ellipse with foci $F(\pm 2, 0)$ and passing through the point $(2, \sqrt{2})$

Discuss and sketch the graph of each of the equations in Exercises 19–24 after making a suitable translation of axes.

19 $4x^2 + 9y^2 + 24x - 36y + 36 = 0$

20 $4x^2 - y^2 - 40x - 8y + 88 = 0$

21 $y^2 - 8x + 8y + 32 = 0$

22 $4x^2 + y^2 - 24x + 4y + 36 = 0$

23 $y^2 - 2x^2 + 6y + 8x - 3 = 0$

24 $x^2 - 9y^2 + 8x + 7 = 0$

Sketch the graphs of the polar equations in Exercises 25–32.

25 $r = -4 \sin \theta$

26 $r = 3 \sin 5\theta$

27 $r = 6 - 3 \cos \theta$

28 $r^2 = 9 \sin 2\theta$

29 $2r = \theta$

30 $r = \dfrac{8}{1 - 3 \sin \theta}$

31 $r = 6 - r \cos \theta$

32 $r = 8 \sec \theta$

Change the equations in Exercises 33–36 to polar equations.

33 $y^2 = 4x$

34 $x^2 + y^2 - 3x + 4y = 0$

35 $2x - 3y = 8$

36 $x^2 + y^2 = 2xy$

In Exercises 37–40 change each equation to an equation in x and y.

37 $r^2 = \tan\theta$

38 $r = 2\cos\theta + 3\sin\theta$

39 $r^2 = 4\sin 2\theta$

40 $\theta = \sqrt{3}$

In Exercises 41–43 sketch the graph of the curve and find a rectangular equation of a graph that contains the points on the curve.

41 $x = (1/t) + 1, \quad y = (2/t) - t; \qquad 0 < t \le 4$

42 $x = \cos^2 t - 2, \quad y = \sin t + 1; \qquad 0 \le t \le 2\pi$

43 $x = \sqrt{t}, \quad y = 2^{-t}; \qquad t \ge 0$

44 Let the curves C_1, C_2, C_3, and C_4 be given parametrically by

$$C_1\colon x = t, \quad y = \sqrt{t}$$
$$C_2\colon x = t^2, \quad y = t$$
$$C_3\colon x = 1 - \sin^2 t, \quad y = \cos t$$
$$C_4\colon x = e^{2t}, \quad y = -e^t$$

for t in $\mathbb{R}$. Sketch the graphs of C_1, C_2, C_3, and C_4 and discuss their similarities and differences.

After making a suitable rotation of axes, describe and sketch the graphs of the equations in Exercises 45 and 46.

45 $x^2 - 8xy + 16y^2 - 12\sqrt{17}x - 3\sqrt{17}y = 0$

46 $8x^2 + 12xy + 17y^2 - 16\sqrt{5}x - 12\sqrt{5}y = 0$

47 A bridge is to be constructed across a river that is 200 feet wide. The arch of the bridge is to be semi-elliptical and must be constructed so that a ship less than 50 feet wide and 30 feet high can pass safely through (see figure). Find an equation for the arch and calculate the height of the arch in the middle of the bridge.

FIGURE FOR EXERCISE 47

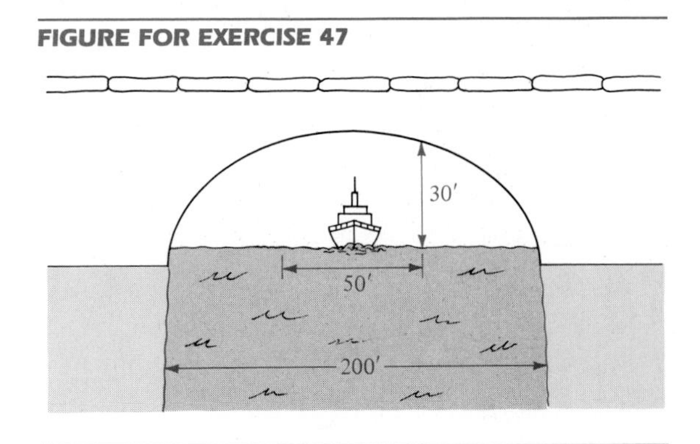

48 As illustrated in the figure, point $P(x, y)$ is the same distance from the point $(4, 0)$ as it is from the circle $x^2 + y^2 = 4$ $(d_1 = d_2)$. Show that the collection of all such points forms a branch of a hyperbola and sketch its graph.

FIGURE FOR EXERCISE 48

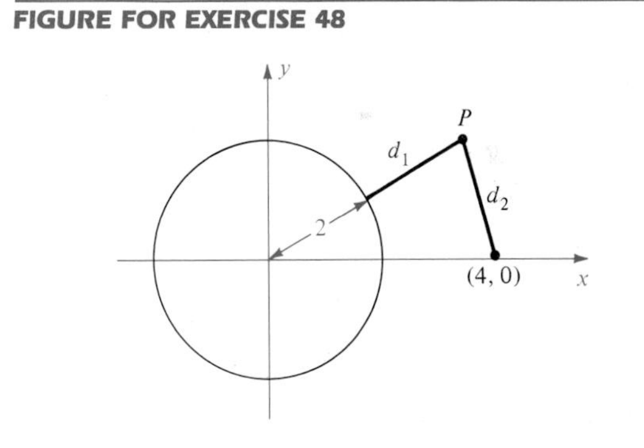

APPENDIX I

Using Logarithmic and Trigonometric Tables

If x is any positive real number and we write

$$x = c \cdot 10^k$$

where $1 \leq c < 10$ and k is an integer, then applying (iii) of the Laws of Logarithms,

$$\log x = \log c + \log 10^k.$$

Since $\log 10^k = k$, we see that

$$\log x = \log c + k.$$

The last equation tells us that to find $\log x$ for any positive real number x it is sufficient to know the logarithms of numbers between 1 and 10. The number $\log c$, for $1 \leq c < 10$, is called the **mantissa,** and the integer k is called the **characteristic** of $\log x$.

If $1 \leq c < 10$, then, since $\log x$ increases as x increases,

$$\log 1 \leq \log c < \log 10,$$

or equivalently,

$$0 \leq \log c < 1.$$

Hence, the mantissa of a logarithm is a number between 0 and 1. In numerical problems it is usually necessary to approximate logarithms. For example, it can be shown that

$$\log 2 = 0.3010299957\ldots$$

where the decimal is nonrepeating and nonterminating. We often round off such logarithms to four decimal places and write

$$\log 2 \approx 0.3010.$$

If a number between 0 and 1 is written as a finite decimal, it is sometimes referred to as a **decimal fraction.** Thus, the equation $\log x = \log c + k$ implies that if x is any positive real number, then $\log x$ *may be approximated by the sum of a positive decimal fraction* (*the mantissa*) *and an integer* k (*the characteristic*). We shall refer to this representation as the **standard form** for $\log x$.

Common logarithms of many of the numbers between 1 and 10 have been calculated. Table 1 of Appendix II contains four-decimal-place approximations for logarithms of numbers between 1.00 and 9.99 at intervals of 0.01. This table can be used to find the common logarithm of any three-digit number to four-decimal-place accuracy. The use of Table 1 is illustrated in the following examples.

EXAMPLE 1 Approximate each of the following:

(a) $\log 43.6$ (b) $\log 43{,}600$ (c) $\log 0.0436$

SOLUTION

(a) Since $43.6 = (4.36)10^1$, the characteristic of $\log 43.6$ is 1. Referring to Table 1, we find that the mantissa of $\log 4.36$ may be approximated by 0.6395. Hence, as in the preceding discussion,

$$\log 43.6 \approx 0.6395 + 1 = 1.6395.$$

(b) Since $43{,}600 = (4.36)10^4$, the mantissa is the same as in part (a); however, the characteristic is 4. Consequently,

$$\log 43{,}600 \approx 0.6395 + 4 = 4.6395.$$

(c) If we write $0.0436 = (4.36)10^{-2}$, then

$$\log 0.0436 = \log 4.36 + (-2).$$

Hence,
$$\log 0.0436 \approx 0.6395 + (-2).$$

We could subtract 2 from 0.6395 and obtain

$$\log 0.0436 \approx -1.3605$$

but this is not standard form, since $-1.3605 = -0.3605 + (-1)$, a number in which the decimal part is *negative.* A common error is to write $0.6395 + (-2)$ as -2.6395. This is incorrect since

$$-2.6395 = -0.6395 + (-2),$$

which is not the same as $0.6395 + (-2)$. ■

If a logarithm has a negative characteristic, we usually either leave it in standard form or rewrite the logarithm, keeping the decimal part positive. To illustrate the latter technique, let us add and subtract 8 on the right side of the equation

$$\log 0.0436 \approx 0.6395 + (-2).$$

This gives us $$\log 0.0436 \approx 0.6395 + (8 - 8) + (-2)$$

or $$\log 0.0436 \approx 8.6395 - 10.$$

We could also write

$$\log 0.0436 \approx 18.6395 - 20 = 43.6395 - 45$$

and so on, as long as the *integral part* of the logarithm is -2.

EXAMPLE 2 Approximate each of the following:

(a) $\log (0.00652)^2$ (b) $\log (0.00652)^{-2}$ (c) $\log (0.00652)^{1/2}$

SOLUTION

(a) By (iii) of the Laws of Logarithms,

$$\log (0.00652)^2 = 2 \log 0.00652.$$

Since $0.00652 = (6.52)10^{-3}$,

$$\log 0.00652 = \log 6.52 + (-3).$$

Referring to Table 1, we see that $\log 6.52$ is approximately 0.8142 and, therefore,

$$\log 0.00652 \approx 0.8142 + (-3).$$

Hence, $$\begin{aligned}\log (0.00652)^2 &= 2 \log 0.00652\\ &\approx 2[0.8142 + (-3)]\\ &= 1.6284 + (-6).\end{aligned}$$

The standard form is $0.6284 + (-5)$.

(b) Again using Law (iii) and the value for log 0.00652 found in part (a),

$$\begin{aligned}\log(0.00652)^{-2} &= -2\log 0.00652\\ &\approx -2[0.8142 + (-3)]\\ &= -1.6284 + 6.\end{aligned}$$

It is important to note that -1.6284 means $-0.6284 + (-1)$ and, consequently, the decimal part is negative. To obtain the standard form, we may write

$$\begin{aligned}-1.6284 + 6 &= 6.0000 - 1.6284\\ &= 4.3716.\end{aligned}$$

This shows that the mantissa is 0.3716 and the characteristic is 4.

(c) By Law (iii),

$$\begin{aligned}\log(0.00652)^{1/2} &= \tfrac{1}{2}\log 0.00652\\ &\approx \tfrac{1}{2}[0.8142 + (-3)].\end{aligned}$$

If we multiply by $\frac{1}{2}$, the standard form is not obtained, since neither number in the resulting sum is the characteristic. In order to avoid this, we may adjust the expression within brackets by adding and subtracting a suitable number. If we use 1 in this way, we obtain

$$\begin{aligned}\log(0.00652)^{1/2} &\approx \tfrac{1}{2}[1.8142 + (-4)]\\ &= 0.9071 + (-2),\end{aligned}$$

which is in standard form. We could also have added and subtracted a number other than 1. For example,

$$\begin{aligned}\tfrac{1}{2}[0.8142 + (-3)] &= \tfrac{1}{2}[17.8142 + (-20)]\\ &= 8.9071 + (-10).\end{aligned}$$

■

Table 1 can be used to find an approximation to x if $\log x$ is given, as illustrated in the following example.

EXAMPLE 3 Find a decimal approximation to x for each $\log x$:

(a) $\log x = 1.7959$ (b) $\log x = -3.5918$

SOLUTION

(a) The mantissa 0.7959 determines the sequence of digits in x and the characteristic determines the position of the decimal point. Referring to the *body* of Table 1, we see that the mantissa 0.7959 is the logarithm of 6.25.

Since the characteristic is 1, x lies between 10 and 100. Consequently, $x \approx 62.5$.

(b) To find x from Table 1, $\log x$ must be written in standard form. To change $\log x = -3.5918$ to standard form, we may add and subtract 4, obtaining

$$\begin{aligned}\log x &= (4 - 3.5918) - 4 \\ &= 0.4082 - 4.\end{aligned}$$

Referring to Table 1, we see that the mantissa 0.4082 is the logarithm of 2.56. Since the characteristic of $\log x$ is -4, it follows that $x \approx 0.000256$. ■

If a calculator with a log key is used to determine common logarithms, then the standard form for $\log x$ is obtained only if $x \geq 1$. For example, to find log 43.6 on a typical calculator, we enter 43.6 and press log, obtaining the standard form

1.6394865

If we find log 0.0436 in similar fashion, then the following number appears on the display panel:

-1.3605135

This is not the standard form for the logarithm, but is similar to that which occurred in the solution to Example 1(c). To find the standard form we could add 2 to the logarithm (using a calculator) and then subtract 2 as follows:

$$\begin{aligned}\log 0.0436 &\approx -1.3605135 \\ &= (-1.3605135 + 2) - 2 \\ &= 0.6394865 - 2 \\ &= 0.6394865 + (-2)\end{aligned}$$

The only common logarithms that can be found *directly* from Table 1 are logarithms of numbers that contain at most three nonzero digits. If *four* nonzero digits are involved, then it is possible to obtain an approximation by using the method of linear interpolation described next. The terminology **linear interpolation** is used because, as we shall see, the method is based upon approximating portions of the graph of $y = \log x$ by line segments.

FIGURE A1.1

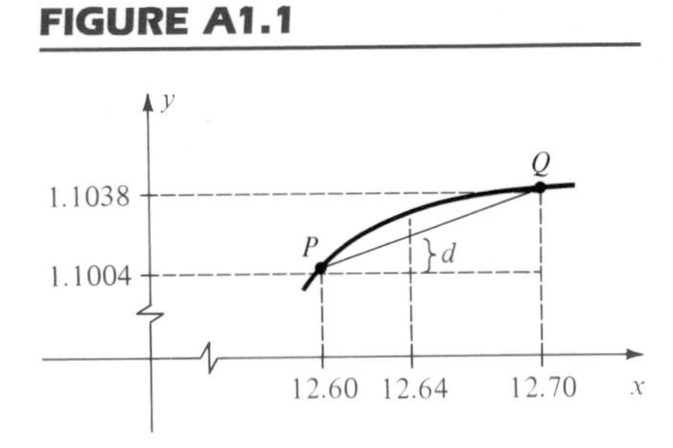

FIGURE A1.2

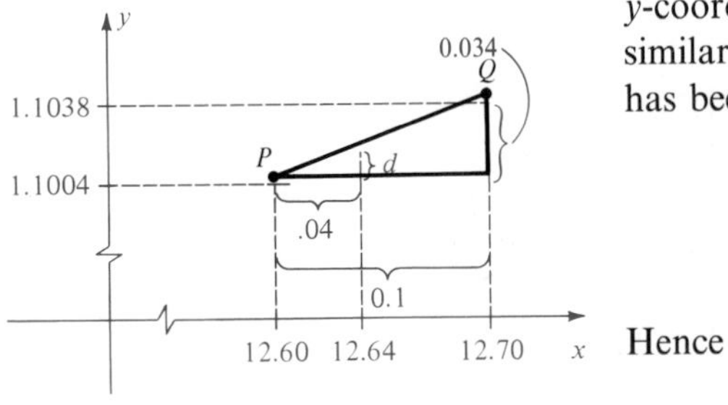

To illustrate the process of linear interpolation, and at the same time give some justification for it, let us consider the specific example log 12.64. Since the logarithmic function with base 10 is increasing, this number lies between $\log 12.60 \approx 1.1004$ and $\log 12.70 \approx 1.1038$. Examining the graph of $y = \log x$, we have the situation shown in Figure A1.1, where we have distorted the units on the x- and y-axes and also the portion of the graph shown. A more accurate drawing would indicate that the graph of $y = \log x$ is much closer to the line segment joining $P(12.60, 1.1004)$ to $Q(12.70, 1.1038)$ than is shown in the figure. Since log 12.64 is the y-coordinate of the point on the graph having x-coordinate 12.64, it can be approximated by the y-coordinate of the point with x-coordinate 12.64 on the *line segment PQ*. Referring to Figure A1.1, we see that the latter y-coordinate is $1.1004 + d$. The number d can be approximated by using similar triangles. Referring to Figure A1.2, where the graph of $y = \log x$ has been deleted, we may form the following proportion:

$$\frac{d}{0.0034} = \frac{0.04}{0.1}.$$

Hence

$$d = \frac{(0.04)(0.0034)}{0.1} = 0.00136.$$

When using this technique, we always round off decimals to the same number of places as appear in the body of the table. Consequently, $d \approx 0.0014$ and

$$\log 12.64 \approx 1.1004 + 0.0014 = 1.1018.$$

Hereafter we shall not sketch a graph when interpolating. Instead we shall use the scheme illustrated in the next example.

EXAMPLE 4 Approximate log 572.6.

SOLUTION It is convenient to arrange our work as follows:

$$1.0\left\{\begin{matrix} 0.6\left\{\begin{matrix}\log 572.0 \approx 2.7574 \\ \log 572.6 = ?\end{matrix}\right\} d \\ \log 573.0 \approx 2.7582 \end{matrix}\right\} 0.0008$$

where differences are indicated by appropriate symbols alongside of the braces. This leads to the proportion

$$\frac{d}{0.0008} = \frac{0.6}{1.0} = \frac{6}{10} \quad \text{or} \quad d = \left(\frac{6}{10}\right)(0.0008) = 0.00048 \approx 0.0005.$$

Hence, $\log 572.6 \approx 2.7574 + 0.0005 = 2.7579.$

Another way of working this type of problem is to reason that since 572.6 is $\frac{6}{10}$ of the way from 572.0 to 573.0, then log 572.6 is (approximately) $\frac{6}{10}$ of the way from 2.7574 to 2.7582. Hence,

$$\log 572.6 \approx 2.7574 + (\tfrac{6}{10})(0.0008) \approx 2.7574 + 0.0005 = 2.7579. \quad \blacksquare$$

EXAMPLE 5 Approximate log 0.003678.

SOLUTION We begin by arranging our work as in the solution of Example 1. Thus,

$$10\left\{\begin{array}{l} 8\left\{\begin{array}{l}\log 0.003670 \approx 0.5647 + (-3)\\ \log 0.003678 = ?\end{array}\right\}d \\ \log 0.003680 \approx 0.5658 + (-3)\end{array}\right\}0.0011$$

Since we are only interested in ratios, we have used the numbers 8 and 10 on the left side because their ratio is the same as the ratio of 0.000008 to 0.000010. This leads to the proportion

$$\frac{d}{0.0011} = \frac{8}{10} = 0.8 \quad \text{or} \quad d = (0.0011)(0.8) = 0.00088 \approx 0.0009.$$

Hence,
$$\begin{aligned}\log 0.003678 &\approx [0.5647 + (-3)] + 0.0009\\ &= 0.5656 + (-3). \quad \blacksquare\end{aligned}$$

If a number x is written in the form $x = c \cdot 10^k$, where $1 \le c < 10$, then before using Table 1 to find $\log x$ by interpolation, c should be rounded off to three decimal places. Another way of saying this is that x should be rounded off to four **significant figures.** Some examples will help to clarify the procedure. If $x = 36.4635$, we round off to 36.46 before approximating $\log x$. The number 684,279 should be rounded off to 684,300. For a decimal such as 0.096202 we use 0.09620. The reason for doing this is that Table 1 does not guarantee more than four-digit accuracy, since the mantissas that appear in it are approximations. This means that if *more* than four-digit accuracy is required in a problem, then Table 1 cannot be used. If, in more extensive tables, the logarithm of a number containing n digits can be found directly, then interpolation is allowed for numbers involving $n + 1$ digits, and numbers should be rounded off accordingly.

The method of interpolation can also be used to find x when we are given $\log x$. If we use Table 1, then x may be found to four significant figures. In this case we are given the *y-coordinate* of a point on the graph of $y = \log x$ and are asked to find the *x-coordinate.* A geometric argument similar to the one given earlier can be used to justify the procedure illustrated in the next example.

EXAMPLE 6 Find x to four significant figures if $\log x = 1.7949$.

SOLUTION The mantissa 0.7949 does not appear in Table 1, but it can be isolated between adjacent entries, namely the mantissas corresponding to 6.230 and 6.240. We shall arrange our work as follows:

$$0.1\left\{\begin{array}{l} r\left\{\begin{array}{ll}\log 62.30 \approx 1.7945 \\ \log x \quad\;\; = 1.7949\end{array}\right\} 0.0004 \\ \log 62.40 \approx 1.7952 \end{array}\right\} 0.0007.$$

This leads to the proportion

$$\frac{r}{0.1} = \frac{0.0004}{0.0007} = \frac{4}{7} \quad \text{or} \quad r = (0.1)\left(\frac{4}{7}\right) \approx 0.06.$$

Hence, $$x \approx 62.30 + 0.06 = 62.36.$$ ■

By employing advanced techniques it is possible to approximate, to any degree of accuracy, all values of the trigonometric functions in the t-interval $[0, \pi/2]$ or, equivalently, in the degree interval $[0°, 90°]$. Table 4, part of

t	t degrees	sin t	cos t	tan t	cot t	sec t	csc t		
.4887	**28°00′**	.4695	.8829	.5317	1.881	1.133	2.130	**62°00′**	1.0821
.4916	10	.4720	.8816	.5354	1.868	1.134	2.118	50	1.0792
.4945	20	.4746	.8802	.5392	1.855	1.136	2.107	40	1.0763
.4974	30	.4772	.8788	.5430	1.842	1.138	2.096	30	1.0734
.5003	40	.4797	.8774	.5467	1.829	1.140	2.085	20	1.0705
.5032	50	.4823	.8760	.5505	1.816	1.142	2.074	10	1.0676
.5061	**29°00′**	.4848	.8746	.5543	1.804	1.143	2.063	**61°00′**	1.0647
.5091	10	.4874	.8732	.5581	1.792	1.145	2.052	50	1.0617
.5120	20	.4899	.8718	.5619	1.780	1.147	2.041	40	1.0588
.5149	30	.4924	.8704	.5658	1.767	1.149	2.031	30	1.0559
.5178	40	.4950	.8689	.5696	1.756	1.151	2.202	20	1.0530
.5207	50	.4975	.8675	.5735	1.744	1.153	2.010	10	1.0501
.5236	**30°00′**	.5000	.8660	.5774	1.732	1.155	2.000	**60°00′**	1.0472
.5265	10	.5025	.8646	.5812	1.720	1.157	1.990	50	1.0443
.5294	20	.5050	.8631	.5851	1.709	1.159	1.980	40	1.0414
.5323	30	.5075	.8616	.5890	1.698	1.161	1.970	30	1.0385
.5352	40	.5100	.8601	.5930	1.686	1.163	1.961	20	1.0356
.5381	50	.5125	.8587	.5969	1.675	1.165	1.951	10	1.0327
		cos t	sin t	cot t	tan t	csc t	sec t	t degrees	t

which is reproduced here, gives four-decimal-place approximations to such values.

In Table 4, $0 \le t \le 1.5708$. The number 1.5708 is a four-decimal-place approximation to $\pi/2$. The table is arranged so that function values corresponding to angles in degree measure may be found directly. Angular measures are given in 10′ intervals from 0° to 90°. The reason that t varies at intervals of approximately 0.0029 is because $10' \approx 0.0029$ radians.

To find values of trigonometric functions if $0 \le t \le 0.7854 \approx \pi/4$ or $0° \le \theta \le 45°$, the labels at the *top* of the columns in Table 4 should be used. For example,

$$\sin(0.5003) \approx 0.4797 \qquad \tan 28°30' \approx 0.5430$$
$$\cos(0.4945) \approx 0.8802 \qquad \sec 29°00' \approx 1.143$$
$$\cot(0.5120) \approx 1.780 \qquad \csc 29°40' \approx 2.020.$$

If $0.7854 \le t \le 1.5708$, or if $45° \le \theta \le 90°$, then the labels at the *bottom* of the columns should be employed. For example,

$$\sin(1.0705) \approx 0.8774 \qquad \csc 62°00' \approx 1.133$$
$$\cos(1.0530) \approx 0.4950 \qquad \cot 61°10' \approx 0.5505$$
$$\tan(1.0821) \approx 1.881 \qquad \sec 60°30' \approx 2.031.$$

The reason that the table can be so arranged follows from the fact that

$$\sin t = \cos\left(\frac{\pi}{2} - t\right), \qquad \cot t = \tan\left(\frac{\pi}{2} - t\right), \qquad \csc t = \sec\left(\frac{\pi}{2} - t\right)$$

or equivalently,

$$\sin\theta = \cos(90° - \theta), \qquad \cot\theta = \tan(90° - \theta), \qquad \csc\theta = \sec(90° - \theta).$$

In particular,

$$\sin 29° = \cos(90° - 29°) = \cos 61°$$
$$\cot 28°20' = \tan(90° - 28°20') = \tan 61°40',$$

as shown in the table.

If it is necessary to find function values when t lies *between* numbers given in the table, the method of linear interpolation may be employed. Similarly, given a value such as $\sin t = 0.6371$, we may refer to the body of Table 4 and use linear interpolation, if necessary, to obtain an approximation to t. If t is measured in degrees we round off to the nearest minute.

The following examples illustrate the use of interpolation in Table 4.

EXAMPLE 7 Approximate $\tan 24°16'$.

SOLUTION We consult Table 4 and interpolate as follows:

$$10' \left\{ \begin{array}{l} 6' \left\{ \begin{array}{l} \tan 24°10' \approx 0.4487 \\ \tan 24°16' = ? \end{array} \right\} d \\ \quad \tan 24°20' \approx 0.4522 \end{array} \right\} 0.0035$$

$$\frac{d}{0.0035} = \frac{6}{10} \quad \text{or} \quad d = \frac{6}{10}(0.0035) \approx 0.0021$$

$$\tan 24°16' \approx 0.4487 + 0.0021 = 0.4508$$

■

EXAMPLE 8 Approximate $\cos(-117°47')$.

SOLUTION The angle is in quadrant III. The reader should check that the reference angle is $62°13'$. Consequently, $\cos(-117°47') = -\cos 62°13'$. Interpolating in Table 4 we have

$$10' \left\{ \begin{array}{l} 3' \left\{ \begin{array}{l} \cos 62°10' \approx 0.4669 \\ \cos 62°13' = ? \end{array} \right\} d \\ \quad \cos 62°20' \approx 0.4643 \end{array} \right\} 0.0026$$

$$\frac{d}{0.0026} = \frac{3}{10} \quad \text{or} \quad d \approx 0.0008$$

Since the cosine function is decreasing,

$$\cos 62°13' \approx 0.4669 - 0.0008 = 0.4661$$

Hence, $\cos(-117°47') \approx -0.4661$. ■

EXAMPLE 9 Approximate the smallest positive real number t such that $\sin t = 0.6635$.

SOLUTION We locate 0.6635 between successive entries in the sine column of Table 4 and interpolate as follows:

$$0.0029 \left\{ \begin{array}{l} d \left\{ \begin{array}{l} \sin(0.7243) \approx 0.6626 \\ \sin t \qquad\; = 0.6635 \end{array} \right\} 0.0009 \\ \quad \sin(0.7272) \approx 0.6648 \end{array} \right\} 0.0022$$

$$\frac{d}{0.0029} = \frac{0.0009}{0.0022} \quad \text{or} \quad d = \frac{9}{22}(0.0029) \approx 0.0012$$

Hence, $t \approx 0.7243 + 0.0012 = 0.7255$ ■

EXAMPLE 10 If $\sin \theta = -0.7963$, approximate the degree measure of all angles θ that are in the interval $[0°, 360°]$.

SOLUTION Let θ' be the reference angle, so that $\sin \theta' = 0.7963$. Interpolating in Table 4,

$$10' \left\{ \begin{array}{l} d \left\{ \begin{array}{l} \sin 52°40' \approx 0.7951 \\ \sin \theta' \quad\;\; = 0.7963 \end{array} \right\} 0.0012 \\ \sin 52°50' \approx 0.7969 \end{array} \right\} 0.0018$$

$$\frac{d}{10} = \frac{0.0012}{0.0018} \quad \text{or} \quad d \approx 7'$$

$$\theta' \approx 52°47'$$

Since $\sin \theta$ is negative, θ lies in quadrant III or IV. Using the reference angle $52°47'$, we have

$$\theta \approx 180° + 52°47' = 232°47'$$

$$\theta \approx 360° - 52°47' = 307°13'.$$ ■

Exercises A.1

In Exercises 1–16 use Table 1 and the Laws of Logarithms to approximate the common logarithms of the numbers.

1 347; 0.00347; 3.47

2 86.2; 8620; 0.862

3 0.54; 540; 540,000

4 208; 2.08; 20,800

5 60.2; 0.0000602; 602

6 5; 0.5; 0.0005

7 $(44.9)^2$; $(44.9)^{1/2}$; $(44.9)^{-2}$

8 $(1810)^4$; $(1810)^{40}$; $(1810)^{1/4}$

9 $(0.943)^3$; $(0.943)^{-3}$; $(0.943)^{1/3}$

10 $(0.017)^{10}$; $10^{0.017}$; $10^{1.43}$

11 $(638)(17.3)$

12 $\dfrac{(2.73)(78.5)}{621}$

13 $\dfrac{(47.4)^3}{(29.5)^2}$

14 $\dfrac{(897)^4}{\sqrt{17.8}}$

15 $\sqrt[3]{20.6}(371)^3$

16 $\dfrac{(0.0048)^{10}}{\sqrt{0.29}}$

In Exercises 17–30 use Table 1 to find a decimal approximation to x.

17 $\log x = 3.6274$

18 $\log x = 1.8965$

19 $\log x = 0.9469$

20 $\log x = 0.5729$

21 $\log x = 5.2095$

22 $\log x = 6.7300 - 10$

23 $\log x = 9.7348 - 10$

24 $\log x = 7.6739 - 10$

25 $\log x = 8.8306 - 10$

26 $\log x = 4.9680$

27 $\log x = 2.2765$ **28** $\log x = 3.0043$

29 $\log x = -1.6253$ **30** $\log x = -2.2118$

Use interpolation in Table 1 to approximate the common logarithms of the numbers in Exercises 31–50.

31 25.48 **32** 421.6 **33** 5363

34 0.3817 **35** 0.001259 **36** 69,450

37 123,400 **38** 0.0212 **39** 0.7786

40 1.203 **41** 384.7 **42** 54.44

43 0.9462 **44** 7259 **45** 66,590

46 0.001428 **47** 0.04321 **48** 300,100

49 3.003 **50** 1.236

In Exercises 51–70 use interpolation in Table 1 to approximate x.

51 $\log x = 1.4437$ **52** $\log x = 3.7455$

53 $\log x = 4.6931$ **54** $\log x = 0.5883$

55 $\log x = 9.1664 - 10$ **56** $\log x = 8.3902 - 10$

57 $\log x = 3.8153 - 6$ **58** $\log x = 5.9306 - 9$

59 $\log x = 2.3705$ **60** $\log x = 4.2867$

61 $\log x = 0.1358$ **62** $\log x = 0.0194$

63 $\log x = 8.9752 - 10$ **64** $\log x = 2.4979 - 5$

65 $\log x = 5.0409$ **66** $\log x = 1.3796$

67 $\log x = -2.8712$ **68** $\log x = -1.8164$

69 $\log x = -0.6123$ **70** $\log x = -3.1426$

In Exercises 71–82 use interpolation in Table 4 to approximate the numbers.

71 $\sin(0.46)$ **72** $\cos(0.82)$ **73** $\tan 3$

74 $\cot 6$ **75** $\sec \frac{1}{4}$ **76** $\csc(1.54)$

77 $\cos 37°43'$ **78** $\sin 22°34'$ **79** $\cot 62°27'$

80 $\tan 57°16'$ **81** $\csc 16°55'$ **82** $\sec 9°12'$

In Exercises 83–88 use interpolation in Table 4 to approximate the smallest positive number t for which the equality is true.

83 $\cos t = 0.8620$ **84** $\sin t = 0.6612$

85 $\tan t = 4.501$ **86** $\sec t = 3.641$

87 $\csc t = 1.436$ **88** $\cot t = 1.165$

In Exercises 89–96 use interpolation in Table 4 to approximate, to the nearest minute, the degree measure of all angles θ that lie in the interval $[0°, 360°]$.

89 $\sin\theta = 0.3672$ **90** $\cos\theta = 0.8426$

91 $\tan\theta = 0.5042$ **92** $\cot\theta = 1.348$

93 $\cos\theta = 0.3465$ **94** $\csc\theta = 1.219$

95 $\sec\theta = 1.385$ **96** $\sin\theta = 0.7534$

APPENDIX II Tables

TABLE 1 Common Logarithms

N	0	1	2	3	4	5	6	7	8	9
1.0	.0000	.0043	.0086	.0128	.0170	.0212	.0253	.0294	.0334	.0374
1.1	.0414	.0453	.0492	.0531	.0569	.0607	.0645	.0682	.0719	.0755
1.2	.0792	.0828	.0864	.0899	.0934	.0969	.1004	.1038	.1072	.1106
1.3	.1139	.1173	.1206	.1239	.1271	.1303	.1335	.1367	.1399	.1430
1.4	.1461	.1492	.1523	.1553	.1584	.1614	.1644	.1673	.1703	.1732
1.5	.1761	.1790	.1818	.1847	.1875	.1903	.1931	.1959	.1987	.2014
1.6	.2041	.2068	.2095	.2122	.2148	.2175	.2201	.2227	.2253	.2279
1.7	.2304	.2330	.2355	.2380	.2405	.2430	.2455	.2480	.2504	.2529
1.8	.2553	.2577	.2601	.2625	.2648	.2672	.2695	.2718	.2742	.2765
1.9	.2788	.2810	.2833	.2856	.2878	.2900	.2923	.2945	.2967	.2989
2.0	.3010	.3032	.3054	.3075	.3096	.3118	.3139	.3160	.3181	.3201
2.1	.3222	.3243	.3263	.3284	.3304	.3324	.3345	.3365	.3385	.3404
2.2	.3424	.3444	.3464	.3483	.3502	.3522	.3541	.3560	.3579	.3598
2.3	.3617	.3636	.3655	.3674	.3692	.3711	.3729	.3747	.3766	.3784
2.4	.3802	.3820	.3838	.3856	.3874	.3892	.3909	.3927	.3945	.3962
2.5	.3979	.3997	.4014	.4031	.4048	.4065	.4082	.4099	.4116	.4133
2.6	.4150	.4166	.4183	.4200	.4216	.4232	.4249	.4265	.4281	.4298
2.7	.4314	.4330	.4346	.4362	.4378	.4393	.4409	.4425	.4440	.4456
2.8	.4472	.4487	.4502	.4518	.4533	.4548	.4564	.4579	.4594	.4609
2.9	.4624	.4639	.4654	.4669	.4683	.4698	.4713	.4728	.4742	.4757
3.0	.4771	.4786	.4800	.4814	.4829	.4843	.4857	.4871	.4886	.4900
3.1	.4914	.4928	.4942	.4955	.4969	.4983	.4997	.5011	.5024	.5038
3.2	.5051	.5065	.5079	.5092	.5105	.5119	.5132	.5145	.5159	.5172
3.3	.5185	.5198	.5211	.5224	.5237	.5250	.5263	.5276	.5289	.5302
3.4	.5315	.5328	.5340	.5353	.5366	.5378	.5391	.5403	.5416	.5428
3.5	.5441	.5453	.5465	.5478	.5490	.5502	.5514	.5527	.5539	.5551
3.6	.5563	.5575	.5587	.5599	.5611	.5623	.5635	.5647	.5658	.5670
3.7	.5682	.5694	.5705	.5717	.5729	.5740	.5752	.5763	.5775	.5786
3.8	.5798	.5809	.5821	.5832	.5843	.5855	.5866	.5877	.5888	.5899
3.9	.5911	.5922	.5933	.5944	.5955	.5966	.5977	.5988	.5999	.6010
4.0	.6021	.6031	.6042	.6053	.6064	.6075	.6085	.6096	.6107	.6117
4.1	.6128	.6138	.6149	.6160	.6170	.6180	.6191	.6201	.6212	.6222
4.2	.6232	.6243	.6253	.6263	.6274	.6284	.6294	.6304	.6314	.6325
4.3	.6335	.6345	.6355	.6365	.6375	.6385	.6395	.6405	.6415	.6425
4.4	.6435	.6444	.6454	.6464	.6474	.6484	.6493	.6503	.6513	.6522
4.5	.6532	.6542	.6551	.6561	.6571	.6580	.6590	.6599	.6609	.6618
4.6	.6628	.6637	.6646	.6656	.6665	.6675	.6684	.6693	.6702	.6712
4.7	.6721	.6730	.6739	.6749	.6758	.6767	.6776	.6785	.6794	.6803
4.8	.6812	.6821	.6830	.6839	.6848	.6857	.6866	.6875	.6884	.6893
4.9	.6902	.6911	.6920	.6928	.6937	.6946	.6955	.6964	.6972	.6981
5.0	.6990	.6998	.7007	.7016	.7024	.7033	.7042	.7050	.7059	.7067
5.1	.7076	.7084	.7093	.7101	.7110	.7118	.7126	.7135	.7143	.7152
5.2	.7160	.7168	.7177	.7185	.7193	.7202	.7210	.7218	.7226	.7235
5.3	.7243	.7251	.7259	.7267	.7275	.7284	.7292	.7300	.7308	.7316
5.4	.7324	.7332	.7340	.7348	.7356	.7364	.7372	.7380	.7388	.7396

N	0	1	2	3	4	5	6	7	8	9
5.5	.7404	.7412	.7419	.7427	.7435	.7443	.7451	.7459	.7466	.7474
5.6	.7482	.7490	.7497	.7505	.7513	.7520	.7528	.7536	.7543	.7551
5.7	.7559	.7566	.7574	.7582	.7589	.7597	.7604	.7612	.7619	.7627
5.8	.7634	.7642	.7649	.7657	.7664	.7672	.7679	.7686	.7694	.7701
5.9	.7709	.7716	.7723	.7731	.7738	.7745	.7752	.7760	.7767	.7774
6.0	.7782	.7789	.7796	.7803	.7810	.7818	.7825	.7832	.7839	.7846
6.1	.7853	.7860	.7868	.7875	.7882	.7889	.7896	.7903	.7910	.7917
6.2	.7924	.7931	.7938	.7945	.7952	.7959	.7966	.7973	.7980	.7987
6.3	.7993	.8000	.8007	.8014	.8021	.8028	.8035	.8041	.8048	.8055
6.4	.8062	.8069	.8075	.8082	.8089	.8096	.8102	.8109	.8116	.8122
6.5	.8129	.8136	.8142	.8149	.8156	.8162	.8169	.8176	.8182	.8189
6.6	.8195	.8202	.8209	.8215	.8222	.8228	.8235	.8241	.8248	.8254
6.7	.8261	.8267	.8274	.8280	.8287	.8293	.8299	.8306	.8312	.8319
6.8	.8325	.8331	.8338	.8344	.8351	.8357	.8363	.8370	.8376	.8382
6.9	.8388	.8395	.8401	.8407	.8414	.8420	.8426	.8432	.8439	.8445
7.0	.8451	.8457	.8463	.8470	.8476	.8482	.8488	.8494	.8500	.8506
7.1	.8513	.8519	.8525	.8531	.8537	.8543	.8549	.8555	.8561	.8567
7.2	.8573	.8579	.8585	.8591	.8597	.8603	.8609	.8615	.8621	.8627
7.3	.8633	.8639	.8645	.8651	.8657	.8663	.8669	.8675	.8681	.8686
7.4	.8692	.8698	.8704	.8710	.8716	.8722	.8727	.8733	.8739	.8745
7.5	.8751	.8756	.8762	.8768	.8774	.8779	.8785	.8791	.8797	.8802
7.6	.8808	.8814	.8820	.8825	.8831	.8837	.8842	.8848	.8854	.8859
7.7	.8865	.8871	.8876	.8882	.8887	.8893	.8899	.8904	.8910	.8915
7.8	.8921	.8927	.8932	.8938	.8943	.8949	.8954	.8960	.8965	.8971
7.9	.8976	.8982	.8987	.8993	.8998	.9004	.9009	.9015	.9020	.9025
8.0	.9031	.9036	.9042	.9047	.9053	.9058	.9063	.9069	.9074	.9079
8.1	.9085	.9090	.9096	.9101	.9106	.9112	.9117	.9122	.9128	.9133
8.2	.9138	.9143	.9149	.9154	.9159	.9165	.9170	.9175	.9180	.9186
8.3	.9191	.9196	.9201	.9206	.9212	.9217	.9222	.9227	.9232	.9238
8.4	.9243	.9248	.9253	.9258	.9263	.9269	.9274	.9279	.9284	.9289
8.5	.9294	.9299	.9304	.9309	.9315	.9320	.9325	.9330	.9335	.9340
8.6	.9345	.9350	.9355	.9360	.9365	.9370	.9375	.9380	.9385	.9390
8.7	.9395	.9400	.9405	.9410	.9415	.9420	.9425	.9430	.9435	.9440
8.8	.9445	.9450	.9455	.9460	.9465	.9469	.9474	.9479	.9484	.9489
8.9	.9494	.9499	.9504	.9509	.9513	.9518	.9523	.9528	.9533	.9538
9.0	.9542	.9547	.9552	.9557	.9562	.9566	.9571	.9576	.9581	.9586
9.1	.9590	.9595	.9600	.9605	.9609	.9614	.9619	.9624	.9628	.9633
9.2	.9638	.9643	.9647	.9652	.9657	.9661	.9666	.9671	.9675	.9680
9.3	.9685	.9689	.9694	.9699	.9703	.9708	.9713	.9717	.9722	.9727
9.4	.9731	.9736	.9741	.9745	.9750	.9754	.9759	.9763	.9768	.9773
9.5	.9777	.9782	.9786	.9791	.9795	.9800	.9805	.9809	.9814	.9818
9.6	.9823	.9827	.9832	.9836	.9841	.9845	.9850	.9854	.9859	.9863
9.7	.9868	.9872	.9877	.9881	.9886	.9890	.9894	.9899	.9903	.9908
9.8	.9912	.9917	.9921	.9926	.9930	.9934	.9939	.9943	.9948	.9952
9.9	.9956	.9961	.9965	.9969	.9974	.9978	.9983	.9987	.9991	.9996

TABLE 2 Natural Exponential Function

x	e^x	e^{-x}	x	e^x	e^{-x}
0.00	1.0000	1.0000	2.50	12.182	0.0821
0.05	1.0513	0.9512	2.60	13.464	0.0743
0.10	1.1052	0.9048	2.70	14.880	0.0672
0.15	1.1618	0.8607	2.80	16.445	0.0608
0.20	1.2214	0.8187	2.90	18.174	0.0550
0.25	1.2840	0.7788	3.00	20.086	0.0498
0.30	1.3499	0.7408	3.10	22.198	0.0450
0.35	1.4191	0.7047	3.20	24.533	0.0408
0.40	1.4918	0.6703	3.30	27.113	0.0369
0.45	1.5683	0.6376	3.40	29.964	0.0334
0.50	1.6487	0.6065	3.50	33.115	0.0302
0.55	1.7333	0.5769	3.60	36.598	0.0273
0.60	1.8221	0.5488	3.70	40.447	0.0247
0.65	1.9155	0.5220	3.80	44.701	0.0224
0.70	2.0138	0.4966	3.90	49.402	0.0202
0.75	2.1170	0.4724	4.00	54.598	0.0183
0.80	2.2255	0.4493	4.10	60.340	0.0166
0.85	2.3396	0.4274	4.20	66.686	0.0150
0.90	2.4596	0.4066	4.30	73.700	0.0136
0.95	2.5857	0.3867	4.40	81.451	0.0123
1.00	2.7183	0.3679	4.50	99.017	0.0111
1.10	3.0042	0.3329	4.60	99.484	0.0101
1.20	3.3201	0.3012	4.70	109.95	0.0091
1.30	3.6693	0.2725	4.80	121.51	0.0082
1.40	4.0552	0.2466	4.90	134.29	0.0074
1.50	4.4817	0.2231	5.00	148.41	0.0067
1.60	4.9530	0.2019	6.00	403.43	0.0025
1.70	5.4739	0.1827	7.00	1096.6	0.0009
1.80	6.0496	0.1653	8.00	2981.0	0.0003
1.90	6.6859	0.1496	9.00	8103.1	0.0001
2.00	7.3891	0.1353	10.00	22026.0	0.00005
2.10	8.1662	0.1225			
2.20	9.0250	0.1108			
2.30	9.9742	0.1003			
2.40	11.0232	0.0907			

TABLE 3 Natural Logarithms

n	0.0	0.1	0.2	0.3	0.4	0.5	0.6	0.7	0.8	0.9
0*		7.697	8.391	8.796	9.084	9.307	9.489	9.643	9.777	9.895
1	0.000	0.095	0.182	0.262	0.336	0.405	0.470	0.531	0.588	0.642
2	0.693	0.742	0.788	0.833	0.875	0.916	0.956	0.993	1.030	1.065
3	1.099	1.131	1.163	1.194	1.224	1.253	1.281	1.308	1.335	1.361
4	1.386	1.411	1.435	1.459	1.482	1.504	1.526	1.548	1.569	1.589
5	1.609	1.629	1.649	1.668	1.686	1.705	1.723	1.740	1.758	1.775
6	1.792	1.808	1.825	1.841	1.856	1.872	1.887	1.902	1.917	1.932
7	1.946	1.960	1.974	1.988	2.001	2.015	2.028	2.041	2.054	2.067
8	2.079	2.092	2.104	2.116	2.128	2.140	2.152	2.163	2.175	2.186
9	2.197	2.208	2.219	2.230	2.241	2.251	2.262	2.272	2.282	2.293
10	2.303	2.313	2.322	2.332	2.342	2.351	2.361	2.370	2.380	2.389

* Subtract 10 if $n < 1$; for example, $\ln 0.3 \approx 8.796 - 10 = -1.204$.

TABLE 4 Values of the Trigonometric Functions

t	t degrees	sin t	cos t	tan t	cot t	sec t	csc t		
.0000	**0°00′**	.0000	1.0000	.0000	—	1.000	—	**90°00′**	1.5708
.0029	10	.0029	1.0000	.0029	343.8	1.000	343.8	50	1.5679
.0058	20	.0058	1.0000	.0058	171.9	1.000	171.9	40	1.5650
.0087	30	.0087	1.0000	.0087	114.6	1.000	114.6	30	1.5621
.0116	40	.0116	.9999	.0116	85.94	1.000	85.95	20	1.5592
.0145	50	.0145	.9999	.0145	68.75	1.000	68.76	10	1.5563
.0175	**1°00′**	.0175	.9998	.0175	57.29	1.000	57.30	**89°00′**	1.5533
.0204	10	.0204	.9998	.0204	49.10	1.000	49.11	50	1.5504
.0233	20	.0233	.9997	.0233	42.96	1.000	42.98	40	1.5475
.0262	30	.0262	.9997	.0262	38.19	1.000	38.20	30	1.5446
.0291	40	.0291	.9996	.0291	34.37	1.000	34.38	20	1.5417
.0320	50	.0320	.9995	.0320	31.24	1.001	31.26	10	1.5388
.0349	**2°00′**	.0349	.9994	.0349	28.64	1.001	28.65	**88°00′**	1.5359
.0378	10	.0378	.9993	.0378	26.43	1.001	26.45	50	1.5330
.0407	20	.0407	.9992	.0407	24.54	1.001	24.56	40	1.5301
.0436	30	.0436	.9990	.0437	22.90	1.001	22.93	30	1.5272
.0465	40	.0465	.9989	.0466	21.47	1.001	21.49	20	1.5243
.0495	50	.0494	.9988	.0495	20.21	1.001	20.23	10	1.5213
.0524	**3°00′**	.0523	.9986	.0524	19.08	1.001	19.11	**87°00′**	1.5184
.0553	10	.0552	.9985	.0553	18.07	1.002	18.10	50	1.5155
.0582	20	.0581	.9983	.0582	17.17	1.002	17.20	40	1.5126
.0611	30	.0610	.9981	.0612	16.35	1.002	16.38	30	1.5097
.0640	40	.0640	.9980	.0641	15.60	1.002	15.64	20	1.5068
.0669	50	.0669	.9978	.0670	14.92	1.002	14.96	10	1.5039
.0698	**4°00′**	.0698	.9976	.0699	14.30	1.002	14.34	**86°00′**	1.5010
.0727	10	.0727	.9974	.0729	13.73	1.003	13.76	50	1.4981
.0756	20	.0756	.9971	.0758	13.20	1.003	13.23	40	1.4952
.0785	30	.0785	.9969	.0787	12.71	1.003	12.75	30	1.4923
.0814	40	.0814	.9967	.0816	12.25	1.003	12.29	20	1.4893
.0844	50	.0843	.9964	.0846	11.83	1.004	11.87	10	1.4864
.0873	**5°00′**	.0872	.9962	.0875	11.43	1.004	11.47	**85°00′**	1.4835
.0902	10	.0901	.9959	.0904	11.06	1.004	11.10	50	1.4806
.0931	20	.0929	.9957	.0934	10.71	1.004	10.76	40	1.4777
.0960	30	.0958	.9954	.0963	10.39	1.005	10.43	30	1.4748
.0989	40	.0987	.9951	.0992	10.08	1.005	10.13	20	1.4719
.1018	50	.1016	.9948	.1022	9.788	1.005	9.839	10	1.4690
.1047	**6°00′**	.1045	.9945	.1051	9.514	1.006	9.567	**84°00′**	1.4661
.1076	10	.1074	.9942	.1080	9.255	1.006	9.309	50	1.4632
.1105	20	.1103	.9939	.1110	9.010	1.006	9.065	40	1.4603
.1134	30	.1132	.9936	.1139	8.777	1.006	8.834	30	1.4573
.1164	40	.1161	.9932	.1169	8.556	1.007	8.614	20	1.4544
.1193	50	.1190	.9929	.1198	8.345	1.007	8.405	10	1.4515
.1222	**7°00′**	.1219	.9925	.1228	8.144	1.008	8.206	**83°00′**	1.4486
		cos t	sin t	cot t	tan t	csc t	sec t	t degrees	t

t	t degrees	sin t	cos t	tan t	cot t	sec t	csc t		
.1222	**7°00′**	.1219	.9925	.1228	8.144	1.008	8.206	**83°00′**	1.4486
.1251	10	.1248	.9922	.1257	7.953	1.008	8.016	50	1.4457
.1280	20	.1276	.9918	.1287	7.770	1.008	7.834	40	1.4428
.1309	30	.1305	.9914	.1317	7.596	1.009	7.661	30	1.4399
.1338	40	.1334	.9911	.1346	7.429	1.009	7.496	20	1.4370
.1367	50	.1363	.9907	.1376	7.269	1.009	7.337	10	1.4341
.1396	**8°00′**	.1392	.9903	.1405	7.115	1.010	7.185	**82°00′**	1.4312
.1425	10	.1421	.9899	.1435	6.968	1.010	7.040	50	1.4283
.1454	20	.1449	.9894	.1465	6.827	1.011	6.900	40	1.4254
.1484	30	.1478	.9890	.1495	6.691	1.011	6.765	30	1.4224
.1513	40	.1507	.9886	.1524	6.561	1.012	6.636	20	1.4195
.1542	50	.1536	.9881	.1554	6.435	1.012	6.512	10	1.4166
.1571	**9°00′**	.1564	.9877	.1584	6.314	1.012	6.392	**81°00′**	1.4137
.1600	10	.1593	.9872	.1614	6.197	1.013	6.277	50	1.4108
.1629	20	.1622	.9868	.1644	6.084	1.013	6.166	40	1.4079
.1658	30	.1650	.9863	.1673	5.976	1.014	6.059	30	1.4050
.1687	40	.1679	.9858	.1703	5.871	1.014	5.955	20	1.4021
.1716	50	.1708	.9853	.1733	5.769	1.015	5.855	10	1.3992
.1745	**10°00′**	.1736	.9848	.1763	5.671	1.015	5.759	**80°00′**	1.3963
.1774	10	.1765	.9843	.1793	5.576	1.016	5.665	50	1.3934
.1804	20	.1794	.9838	.1823	5.485	1.016	5.575	40	1.3904
.1833	30	.1822	.9833	.1853	5.396	1.017	5.487	50	1.3875
.1862	40	.1851	.9827	.1883	5.309	1.018	5.403	20	1.3846
.1891	50	.1880	.9822	.1914	5.226	1.018	5.320	10	1.3817
.1920	**11°00′**	.1908	.9816	.1944	5.145	1.019	5.241	**79°00′**	1.3788
.1949	10	.1937	.9811	.1974	5.066	1.019	5.164	50	1.3759
.1978	20	.1965	.9805	.2004	4.989	1.020	5.089	40	1.3730
.2007	30	.1994	.9799	.2035	4.915	1.020	5.016	30	1.3701
.2036	40	.2022	.9793	.2065	4.843	1.021	4.945	20	1.3672
.2065	50	.2051	.9787	.2095	4.773	1.022	4.876	10	1.3643
.2094	**12°00′**	.2079	.9781	.2126	4.705	1.022	4.810	**78°00′**	1.3614
.2123	10	.2108	.9775	.2156	4.638	1.023	4.745	50	1.3584
.2153	20	.2136	.9769	.2186	4.574	1.024	4.682	40	1.3555
.2182	30	.2164	.9763	.2217	4.511	1.024	4.620	30	1.3526
.2211	40	.2193	.9757	.2247	4.449	1.025	4.560	20	1.3497
.2240	50	.2221	.9750	.2278	4.390	1.026	4.502	10	1.3468
.2269	**13°00′**	.2250	.9744	.2309	4.331	1.026	4.445	**77°00′**	1.3439
.2298	10	.2278	.9737	.2339	4.275	1.027	4.390	50	1.3410
.2327	20	.2306	.9730	.2370	4.219	1.028	4.336	40	1.3381
.2356	30	.2334	.9724	.2401	4.165	1.028	4.284	30	1.3352
.2385	40	.2363	.9717	.2432	4.113	1.029	4.232	20	1.3323
.2414	50	.2391	.9710	.2462	4.061	1.030	4.182	10	1.3294
.2443	**14°00′**	.2419	.9703	.2493	4.011	1.031	4.134	**76°00′**	1.3265
		cos t	sin t	cot t	tan t	csc t	sec t	t degrees	t

TABLE 4 Values of the Trigonometric Functions (cont'd.)

t	t degrees	sin t	cos t	tan t	cot t	sec t	csc t		
.2443	**14°00′**	.2419	.9703	.2493	4.011	1.031	4.134	**76°00′**	1.3265
.2473	10	.2447	.9696	.2524	3.962	1.031	4.086	50	1.3235
.2502	20	.2476	.9689	.2555	3.914	1.032	4.039	40	1.3206
.2531	30	.2504	.9681	.2586	3.867	1.033	3.994	30	1.3177
.2560	40	.2532	.9674	.2617	3.821	1.034	3.950	20	1.3148
.2589	50	.2560	.9667	.2648	3.776	1.034	3.906	10	1.3119
.2618	**15°00′**	.2588	.9659	.2679	3.732	1.035	3.864	**75°00′**	1.3090
.2647	10	.2616	.9652	.2711	3.689	1.036	3.822	50	1.3061
.2676	20	.2644	.9644	.2742	3.647	1.037	3.782	40	1.3032
.2705	30	.2672	.9636	.2773	3.606	1.038	3.742	30	1.3003
.2734	40	.2700	.9628	.2805	3.566	1.039	3.703	20	1.2974
.2763	50	.2728	.9621	.2836	3.526	1.039	3.665	10	1.2945
.2793	**16°00′**	.2756	.9613	.2867	3.487	1.040	3.628	**74°00′**	1.2915
.2822	10	.2784	.9605	.2899	3.450	1.041	3.592	50	1.2886
.2851	20	.2812	.9596	.2931	3.412	1.042	3.556	40	1.2857
.2880	30	.2840	.9588	.2962	3.376	1.043	3.521	30	1.2828
.2909	40	.2868	.9580	.2994	3.340	1.044	3.487	20	1.2799
.2938	50	.2896	.9572	.3026	3.305	1.045	3.453	10	1.2770
.2967	**17°00′**	.2924	.9563	.3057	3.271	1.046	3.420	**73°00′**	1.2741
.2996	10	.2952	.9555	.3089	3.237	1.047	3.388	50	1.2712
.3025	20	.2979	.9546	.3121	3.204	1.048	3.356	40	1.2683
.3054	30	.3007	.9537	.3153	3.172	1.049	3.326	30	1.2654
.3083	40	.3035	.9528	.3185	3.140	1.049	3.295	20	1.2625
.3113	50	.3062	.9520	.3217	3.108	1.050	3.265	10	1.2595
.3142	**18°00′**	.3090	.9511	.3249	3.078	1.051	3.236	**72°00′**	1.2566
.3171	10	.3118	.9502	.3281	3.047	1.052	3.207	50	1.2537
.3200	20	.3145	.9492	.3314	3.018	1.053	3.179	40	1.2508
.3229	30	.3173	.9483	.3346	2.989	1.054	3.152	30	1.2479
.3258	40	.3201	.9474	.3378	2.960	1.056	3.124	20	1.2450
.3287	50	.3228	.9465	.3411	2.932	1.057	3.098	10	1.2421
.3316	**19°00′**	.3256	.9455	.3443	2.904	1.058	3.072	**71°00′**	1.2392
.3345	10	.3283	.9446	.3476	2.877	1.059	3.046	50	1.2363
.3374	20	.3311	.9436	.3508	2.850	1.060	3.021	40	1.2334
.3403	30	.3338	.9426	.3541	2.824	1.061	2.996	30	1.2305
.3432	40	.3365	.9417	.3574	2.798	1.062	2.971	20	1.2275
.3462	50	.3393	.9407	.3607	2.773	1.063	2.947	10	1.2246
.3491	**20°00′**	.3420	.9397	.3640	2.747	1.064	2.924	**70°00′**	1.2217
.3520	10	.3448	.9387	.3673	2.723	1.065	2.901	50	1.2188
.3549	20	.3475	.9377	.3706	2.699	1.066	2.878	40	1.2159
.3578	30	.3502	.9367	.3739	2.675	1.068	2.855	30	1.2130
.3607	40	.3529	.9356	.3772	2.651	1.069	2.833	20	1.2101
.3636	50	.3557	.9346	.3805	2.628	1.070	2.812	10	1.2072
.3665	**21°00′**	.3584	.9336	.3839	2.605	1.071	2.790	**69°00′**	1.2043
		cos t	sin t	cot t	tan t	csc t	sec t	t degrees	t

t	t degrees	sin t	cos t	tan t	cot t	sec t	csc t		
.3665	**21°00′**	.3584	.9336	.3839	2.605	1.071	2.790	**69°00′**	1.2043
.3694	10	.3611	.9325	.3872	2.583	1.072	2.769	50	1.2014
.3723	20	.3638	.9315	.3906	2.560	1.074	2.749	40	1.1985
.3752	30	.3665	.9304	.3939	2.539	1.075	2.729	30	1.1956
.3782	40	.3692	.9293	.3973	2.517	1.076	2.709	20	1.1926
.3811	50	.3719	.9283	.4006	2.496	1.077	2.689	10	1.1897
.3840	**22°00′**	.3746	.9272	.4040	2.475	1.079	2.669	**68°00′**	1.1868
.3869	10	.3773	.9261	.4074	2.455	1.080	2.650	50	1.1839
.3898	20	.3800	.9250	.4108	2.434	1.081	2.632	40	1.1810
.3927	30	.3827	.9239	.4142	2.414	1.082	2.613	30	1.1781
.3956	40	.3854	.9228	.4176	2.394	1.084	2.595	20	1.1752
.3985	50	.3881	.9216	.4210	2.375	1.085	2.577	10	1.1723
.4014	**23°00′**	.3907	.9205	.4245	2.356	1.086	2.559	**67°00′**	1.1694
.4043	10	.3934	.9194	.4279	2.337	1.088	2.542	50	1.1665
.4072	20	.3961	.9182	.4314	2.318	1.089	2.525	40	1.1636
.4102	30	.3987	.9771	.4348	2.300	1.090	2.508	30	1.1606
.4131	40	.4014	.9159	.4383	2.282	1.092	2.491	20	1.1577
.4160	50	.4041	.9147	.4417	2.264	1.093	2.475	10	1.1548
.4189	**24°00′**	.4067	.9135	.4452	2.246	1.095	2.459	**66°00′**	1.1519
.4218	10	.4094	.9124	.4487	2.229	1.096	2.443	50	1.1490
.4247	20	.4120	.9112	.4522	2.211	1.097	2.427	40	1.1461
.4276	30	.4147	.9100	.4557	2.194	1.099	2.411	30	1.1432
.4305	40	.4173	.9088	.4592	2.177	1.100	2.396	20	1.1403
.4334	50	.4200	.9075	.4628	2.161	1.102	2.381	10	1.1374
.4363	**25°00′**	.4226	.9063	.4663	2.145	1.103	2.366	**65°00′**	1.1345
.4392	10	.4253	.9051	.4699	2.128	1.105	2.352	50	1.1316
.4422	20	.4279	.9038	.4734	2.112	1.106	2.337	40	1.1286
.4451	30	.4305	.9026	.4770	2.097	1.108	2.323	30	1.1257
.4480	40	.4331	.9013	.4806	2.081	1.109	2.309	20	1.1228
.4509	50	.4358	.9001	.4841	2.066	1.111	2.295	10	1.1199
.4538	**26°00′**	.4384	.8988	.4877	2.050	1.113	2.281	**64°00′**	1.1170
.4567	10	.4410	.8975	.4913	2.035	1.114	2.268	50	1.1141
.4596	20	.4436	.8962	.4950	2.020	1.116	2.254	40	1.1112
.4625	30	.4462	.8949	.4986	2.006	1.117	2.241	30	1.1083
.4654	40	.4488	.8936	.5022	1.991	1.119	2.228	20	1.1054
.4683	50	.4514	.8923	.5059	1.977	1.121	2.215	10	1.1025
.4712	**27°00′**	.4540	.8910	.5095	1.963	1.122	2.203	**63°00′**	1.0996
.4741	10	.4566	.8897	.5132	1.949	1.124	2.190	50	1.0966
.4771	20	.4592	.8884	.5169	1.935	1.126	2.178	40	1.0937
.4800	30	.4617	.8870	.5206	1.921	1.127	2.166	30	1.0908
.4829	40	.4643	.8857	.5243	1.907	1.129	2.154	20	1.0879
.4858	50	.4669	.8843	.5280	1.894	1.131	2.142	10	1.0850
.4887	**28°00′**	.4695	.8829	.5317	1.881	1.133	2.130	**62°00′**	1.0821
		cos t	sin t	cot t	tan t	csc t	sec t	t degrees	t

TABLE 4 Values of the Trigonometric Functions (cont'd.)

t	t degrees	sin t	cos t	tan t	cot t	sec t	csc t		
.4887	**28°00′**	.4695	.8829	.5317	1.881	1.133	2.130	**62°00′**	1.0821
.4916	10	.4720	.8816	.5354	1.868	1.134	2.118	50	1.0792
.4945	20	.4746	.8802	.5392	1.855	1.136	2.107	40	1.0763
.4974	30	.4772	.8788	.5430	1.842	1.138	2.096	30	1.0734
.5003	40	.4797	.8774	.5467	1.829	1.140	2.085	20	1.0705
.5032	50	.4823	.8760	.5505	1.816	1.142	2.074	10	1.0676
.5061	**29°00′**	.4848	.8746	.5543	1.804	1.143	2.063	**61°00′**	1.0647
.5091	10	.4874	.8732	.5581	1.792	1.145	2.052	50	1.0617
.5120	20	.4899	.8718	.5619	1.780	1.147	2.041	40	1.0588
.5149	30	.4924	.8704	.5658	1.767	1.149	2.031	30	1.0559
.5178	40	.4950	.8689	.5696	1.756	1.151	2.020	20	1.0530
.5207	50	.4975	.8675	.5735	1.744	1.153	2.010	10	1.0501
.5236	**30°00′**	.5000	.8660	.5774	1.732	1.155	2.000	**60°00′**	1.0472
.5265	10	.5025	.8646	.5812	1.720	1.157	1.990	50	1.0443
.5294	20	.5050	.8631	.5851	1.709	1.159	1.980	40	1.0414
.5323	30	.5075	.8616	.5890	1.698	1.161	1.970	30	1.0385
.5352	40	.5100	.8601	.5930	1.686	1.163	1.961	20	1.0356
.5381	50	.5125	.8587	.5969	1.675	1.165	1.951	10	1.0327
.5411	**31°00′**	.5150	.8572	.6009	1.664	1.167	1.942	**59°00′**	1.0297
.5440	10	.5175	.8557	.6048	1.653	1.169	1.932	50	1.0268
.5469	20	.5200	.8542	.6088	1.643	1.171	1.923	40	1.0239
.5498	30	.5225	.8526	.6128	1.632	1.173	1.914	30	1.0210
.5527	40	.5250	.8511	.6168	1.621	1.175	1.905	20	1.0181
.5556	50	.5275	.8496	.6208	1.611	1.177	1.896	10	1.0152
.5585	**32°00′**	.5299	.8480	.6249	1.600	1.179	1.887	**58°00′**	1.0123
.5614	10	.5324	.8465	.6289	1.590	1.181	1.878	50	1.0094
.5643	20	.5348	.8450	.6330	1.580	1.184	1.870	40	1.0065
.5672	30	.5373	.8434	.6371	1.570	1.186	1.861	30	1.0036
.5701	40	.5398	.8418	.6412	1.560	1.188	1.853	20	1.0007
.5730	50	.5422	.8403	.6453	1.550	1.190	1.844	10	.9977
.5760	**33°00′**	.5446	.8387	.6494	1.540	1.192	1.836	**57°00′**	.9948
.5789	10	.5471	.8371	.6536	1.530	1.195	1.828	50	.9919
.5818	20	.5495	.8355	.6577	1.520	1.197	1.820	40	.9890
.5847	30	.5519	.8339	.6619	1.511	1.199	1.812	30	.9861
.5876	40	.5544	.8323	.6661	1.501	1.202	1.804	20	.9832
.5905	50	.5568	.8307	.6703	1.492	1.204	1.796	10	.9803
.5934	**34°00′**	.5592	.8290	.6745	1.483	1.206	1.788	**56°00′**	.9774
.5963	10	.5616	.8274	.6787	1.473	1.209	1.781	50	.9745
.5992	20	.5640	.8258	.6830	1.464	1.211	1.773	40	.9716
.6021	30	.5664	.8241	.6873	1.455	1.213	1.766	30	.9687
.6050	40	.5688	.8225	.6916	1.446	1.216	1.758	20	.9657
.6080	50	.5712	.8208	.6959	1.437	1.218	1.751	10	.9628
.6109	**35°00′**	.5736	.8192	.7002	1.428	1.221	1.743	**55°00′**	.9599
		cos t	sin t	cot t	tan t	csc t	sec t	t degrees	t

t	t degrees	sin t	cos t	tan t	cot t	sec t	csc t		
.6109	**35°00′**	.5736	.8192	.7002	1.428	1.221	1.743	**55°00′**	.9599
.6138	10	.5760	.8175	.7046	1.419	1.223	1.736	50	.9570
.6167	20	.5783	.8158	.7089	1.411	1.226	1.729	40	.9541
.6196	30	.5807	.8141	.7133	1.402	1.228	1.722	30	.9512
.6225	40	.5831	.8124	.7177	1.393	1.231	1.715	20	.9483
.6254	50	.5854	.8107	.7221	1.385	1.233	1.708	10	.9454
.6283	**36°00′**	.5878	.8090	.7265	1.376	1.236	1.701	**54°00′**	.9425
.6312	10	.5901	.8073	.7310	1.368	1.239	1.695	50	.9396
.6341	20	.5925	.8056	.7355	1.360	1.241	1.688	40	.9367
.6370	30	.5948	.8039	.7400	1.351	1.244	1.681	30	.9338
.6400	40	.5972	.8021	.7445	1.343	1.247	1.675	20	.9308
.6429	50	.5995	.8004	.7490	1.335	1.249	1.668	10	.9279
.6458	**37°00′**	.6018	.7986	.7536	1.327	1.252	1.662	**53°00′**	.9250
.6487	10	.6041	.7969	.7581	1.319	1.255	1.655	50	.9221
.6516	20	.6065	.7951	.7627	1.311	1.258	1.649	40	.9192
.6545	30	.6088	.7934	.7673	1.303	1.260	1.643	30	.9163
.6574	40	.6111	.7916	.7720	1.295	1.263	1.636	20	.9134
.6603	50	.6134	.7898	.7766	1.288	1.266	1.630	10	.9105
.6632	**38°00′**	.6157	.7880	.7813	1.280	1.269	1.624	**52°00′**	.9076
.6661	10	.6180	.7862	.7860	1.272	1.272	1.618	50	.9047
.6690	20	.6202	.7844	.7907	1.265	1.275	1.612	40	.0918
.6720	30	.6225	.7826	.7954	1.257	1.278	1.606	30	.8988
.6749	40	.6248	.7808	.8002	1.250	1.281	1.601	20	.8959
.6778	50	.6271	.7790	.8050	1.242	1.284	1.595	10	.8930
.6807	**39°00′**	.6293	.7771	.8098	1.235	1.287	1.589	**51°00′**	.8901
.6836	10	.6316	.7753	.8146	1.228	1.290	1.583	o0	.8872
.6865	20	.6338	.7735	.8195	1.220	1.293	1.578	40	.8843
.6894	30	.6361	.7716	.8243	1.213	1.296	1.572	30	.8814
.6923	40	.6383	.7698	.8292	1.206	1.299	1.567	20	.8785
.6952	50	.6406	.7679	.8342	1.199	1.302	1.561	10	.8756
.6981	**40°00′**	.6428	.7660	.8391	1.192	1.305	1.556	**50°00′**	.8727
.7010	10	.6450	.7642	.8441	1.185	1.309	1.550	50	.8698
.7039	20	.6472	.7623	.8491	1.178	1.312	1.545	40	.8668
.7069	30	.6494	.7604	.8541	1.171	1.315	1.540	30	.8639
.7098	40	.6417	.7585	.8591	1.164	1.318	1.535	20	.8610
.7127	50	.6539	.7566	.8642	1.157	1.322	1.529	10	.8581
.7156	**41°00′**	.6561	.7547	.8693	1.150	1.325	1.524	**49°00′**	.8552
.7185	10	.6583	.7528	.8744	1.144	1.328	1.519	50	.8523
.7214	20	.6604	.7509	.8796	1.137	1.332	1.514	40	.8494
.7243	30	.6626	.7490	.8847	1.130	1.335	1.509	30	.8465
.7272	40	.6648	.7470	.8899	1.124	1.339	1.504	20	.8436
.7301	50	.6670	.7451	.8952	1.117	1.342	1.499	10	.8407
.7330	**42°00′**	.6691	.7431	.9004	1.111	1.346	1.494	**48°00′**	.8378
		cos t	sin t	cot t	tan t	csc t	sec t	t degrees	t

TABLE 4 Values of the Trigonometric Functions (cont'd.)

t	t degrees	$\sin t$	$\cos t$	$\tan t$	$\cot t$	$\sec t$	$\csc t$		
.7330	**42°00′**	.6691	.7431	.9004	1.111	1.346	1.494	**48°00′**	.8378
.7359	10	.6713	.7412	.9057	1.104	1.349	1.490	50	.8348
.7389	20	.6734	.7392	.9110	1.098	1.353	1.485	40	.8319
.7418	30	.6756	.7373	.9163	1.091	1.356	1.480	30	.8290
.7447	40	.6777	.7353	.9217	1.085	1.360	1.476	20	.8261
.7476	50	.6799	.7333	.9271	1.079	1.364	1.471	10	.8232
.7505	**43°00′**	.6820	.7314	.9325	1.072	1.367	1.466	**47°00′**	.8203
.7534	10	.6841	.7294	.9380	1.066	1.371	1.462	50	.8174
.7563	20	.6862	.7274	.9435	1.060	1.375	1.457	40	.8145
.7592	30	.6884	.7254	.9490	1.054	1.379	1.453	30	.8116
.7621	40	.6905	.7234	.9545	1.048	1.382	1.448	20	.8087
.7650	50	.6926	.7214	.9601	1.042	1.386	1.444	10	.8058
.7679	**44°00′**	.6947	.7193	.9657	1.036	1.390	1.440	**46°00′**	.8029
.7709	10	.6967	.7173	.9713	1.030	1.394	1.435	50	.7999
.7738	20	.6988	.7153	.9770	1.024	1.398	1.431	40	.7970
.7767	30	.7009	.7133	.9827	1.018	1.402	1.427	30	.7941
.7796	40	.7030	.7112	.9884	1.012	1.406	1.423	20	.7912
.7825	50	.7050	.7092	.9942	1.006	1.410	1.418	10	.7883
.7854	**45°00′**	.7071	.7071	1.0000	1.0000	1.414	1.414	**45°00′**	.7854
		$\cos t$	$\sin t$	$\cot t$	$\tan t$	$\csc t$	$\sec t$	t degrees	t

TABLE 5 Trigonometric Functions of Radians and Real Numbers

t	$\sin t$	$\cos t$	$\tan t$	$\cot t$	$\sec t$	$\csc t$
.00	.0000	1.0000	.0000	—	1.000	—
.01	.0100	1.0000	.0100	99.997	1.000	100.00
.02	.0200	.9998	.0200	49.993	1.000	50.00
.03	.0300	.9996	.0300	33.323	1.000	33.34
.04	.0400	.9992	.0400	24.987	1.001	25.01
.05	.0500	.9988	.0500	19.983	1.001	20.01
.06	.0600	.9982	.0601	16.647	1.002	16.68
.07	.0699	.9976	.0701	14.262	1.002	14.30
.08	.0799	.9968	.0802	12.473	1.003	12.51
.09	.0899	.9960	.0902	11.081	1.004	11.13
.10	.0998	.9950	.1003	9.967	1.005	10.02
.11	.1098	.9940	.1104	9.054	1.006	9.109
.12	.1197	.9928	.1206	8.293	1.007	8.353
.13	.1296	.9916	.1307	7.649	1.009	7.714
.14	.1395	.9902	.1409	7.096	1.010	7.166
.15	.1494	.9888	.1511	6.617	1.011	6.692
.16	.1593	.9872	.1614	6.197	1.013	6.277
.17	.1692	.9856	.1717	5.826	1.015	5.911
.18	.1790	.9838	.1820	5.495	1.016	5.586
.19	.1889	.9820	.1923	5.200	1.018	5.295
.20	.1987	.9801	.2027	4.933	1.020	5.033
.21	.2085	.9780	.2131	4.692	1.022	4.797
.22	.2182	.9759	.2236	4.472	1.025	4.582
.23	.2280	.9737	.2341	4.271	1.027	4.386
.24	.2377	.9713	.2447	4.086	1.030	4.207
.25	.2474	.9689	.2553	3.916	1.032	4.042
.26	.2571	.9664	.2660	3.759	1.035	3.890
.27	.2667	.9638	.2768	3.613	1.038	3.749
.28	.2764	.9611	.2876	3.478	1.041	3.619
.29	.2860	.9582	.2984	3.351	1.044	3.497
.30	.2955	.9553	.3093	3.233	1.047	3.384
.31	.3051	.9523	.3203	3.122	1.050	3.278
.32	.3146	.9492	.3314	3.018	1.053	3.179
.33	.3240	.9460	.3425	2.920	1.057	3.086
.34	.3335	.9428	.3537	2.827	1.061	2.999
.35	.3429	.9394	.3650	2.740	1.065	2.916
.36	.3523	.9359	.3764	2.657	1.068	2.839
.37	.3616	.9323	.3879	2.578	4.073	2.765
.38	.3709	.9287	.3994	2.504	1.077	2.696
.39	.3802	.9249	.4111	2.433	1.081	2.630

t	$\sin t$	$\cos t$	$\tan t$	$\cot t$	$\sec t$	$\csc t$
.40	.3894	.9211	.4228	2.365	1.086	2.568
.41	.3986	.9171	.4346	2.301	1.090	2.509
.42	.4078	.9131	.4466	2.239	1.095	2.452
.43	.4169	.9090	.4586	2.180	1.100	2.399
.44	.4259	.9048	.4708	2.124	1.105	2.348
.45	.4350	.9004	.4831	2.070	1.111	2.299
.46	.4439	.8961	.4954	2.018	1.116	2.253
.47	.4529	.8916	.5080	1.969	1.122	2.208
.48	.4618	.8870	.5206	1.921	1.127	2.166
.49	.4706	.8823	.5334	1.875	1.133	2.125
.50	.4794	.8776	.5463	1.830	1.139	2.086
.51	.4882	.8727	.5594	1.788	1.146	2.048
.52	.4969	.8678	.5726	1.747	1.152	2.013
.53	.5055	.8628	.5859	1.707	1.159	1.978
.54	.5141	.8577	.5994	1.668	1.166	1.945
.55	.5227	.8525	.6131	1.631	1.173	1.913
.56	.5312	.8473	.6269	1.595	1.180	1.883
.57	.5396	.8419	.6410	1.560	1.188	1.853
.58	.5480	.8365	.6552	1.526	1.196	1.825
.59	.5564	.8309	.6696	1.494	1.203	1.797
.60	.5646	.8253	.6841	1.462	1.212	1.771
.61	.5729	.8196	.6989	1.431	1.220	1.746
.62	.5810	.8139	.7139	1.401	1.229	1.721
.63	.5891	.8080	.7291	1.372	1.238	1.697
.64	.5972	.8021	.7445	1.343	1.247	1.674
.65	.6052	.7961	.7602	1.315	1.256	1.652
.66	.6131	.7900	.7761	1.288	1.266	1.631
.67	.6210	.7838	.7923	1.262	1.276	1.610
.68	.6288	.7776	.8087	1.237	1.286	1.590
.69	.6365	.7712	.8253	1.212	1.297	1.571
.70	.6442	.7648	.8423	1.187	1.307	1.552
.71	.6518	.7584	.8595	1.163	1.319	1.534
.72	.6594	.7518	.8771	1.140	1.330	1.517
.73	.6669	.7452	.8949	1.117	1.342	1.500
.74	.6743	.7385	.9131	1.095	1.354	1.483
.75	.6816	.7317	.9316	1.073	1.367	1.467
.76	.6889	.7248	.9505	1.052	1.380	1.452
.77	.6961	.7179	.9697	1.031	1.393	1.437
.78	.7033	.7109	.9893	1.011	1.407	1.422
.79	.7104	.7038	1.009	.9908	1.421	1.408

TABLE 5 Trigonometric Functions of Radians and Real Numbers (cont'd.)

t	$\sin t$	$\cos t$	$\tan t$	$\cot t$	$\sec t$	$\csc t$
.80	.7174	.6967	1.030	.9712	1.435	1.394
.81	.7243	.6895	1.050	.9520	1.450	1.381
.82	.7311	.6822	1.072	.9331	1.466	1.368
.83	.7379	.6749	1.093	.9146	1.482	1.355
.84	.7446	.6675	1.116	.8964	1.498	1.343
.85	.7513	.6600	1.138	.8785	1.515	1.331
.86	.7578	.6524	1.162	.8609	1.533	1.320
.87	.7643	.6448	1.185	.8437	1.551	1.308
.88	.7707	.6372	1.210	.8267	1.569	1.297
.89	.7771	.6294	1.235	.8100	1.589	1.287
.90	.7833	.6216	1.260	.7936	1.609	1.277
.91	.7895	.6137	1.286	.7774	1.629	1.267
.92	.7956	.6058	1.313	.7615	1.651	1.257
.93	.8016	.5978	1.341	.7458	1.673	1.247
.94	.8076	.5898	1.369	.7303	1.696	1.238
.95	.8134	.5817	1.398	.7151	1.719	1.229
.96	.8192	.5735	1.428	.7001	1.744	1.221
.97	.8249	.5653	1.459	.6853	1.769	1.212
.98	.8305	.5570	1.491	.6707	1.795	1.204
.99	.8360	.5487	1.524	.6563	1.823	1.196
1.00	.8415	.5403	1.557	.6421	1.851	1.188
1.01	.8468	.5319	1.592	.6281	1.880	1.181
1.02	.8521	.5234	1.628	.6142	1.911	1.174
1.03	.8573	.5148	1.665	.6005	1.942	1.166
1.04	.8624	.5062	1.704	.5870	1.975	1.160
1.05	.8674	.4976	1.743	.5736	2.010	1.153
1.06	.8724	.4889	1.784	.5604	2.046	1.146
1.07	.8772	.4801	1.827	.5473	2.083	1.140
1.08	.8820	.4713	1.871	.5344	2.122	1.134
1.09	.8866	.4625	1.917	.5216	2.162	1.128
1.10	.8912	.4536	1.965	.5090	2.205	1.122
1.11	.8957	.4447	2.014	.4964	2.249	1.116
1.12	.9001	.4357	2.066	.4840	2.295	1.111
1.13	.9044	.4267	2.120	.4718	2.344	1.106
1.14	.9086	.4176	2.176	.4596	2.395	1.101
1.15	.9128	.4085	2.234	.4475	2.448	1.096
1.16	.9168	.3993	2.296	.4356	2.504	1.091
1.17	.9208	.3902	2.360	.4237	2.563	1.086
1.18	.9246	.3809	2.427	.4120	2.625	1.082
1.19	.9284	.3717	2.498	.4003	2.691	1.077

t	$\sin t$	$\cos t$	$\tan t$	$\cot t$	$\sec t$	$\csc t$
1.20	.9320	.3624	2.572	.3888	2.760	1.073
1.21	.9356	.3530	2.650	.3773	2.833	1.069
1.22	.9391	.3436	2.733	.3659	2.910	1.065
1.23	.9425	.3342	2.820	.3546	2.992	1.061
1.24	.9458	.3248	2.912	.3434	3.079	1.057
1.25	.9490	.3153	3.010	.3323	3.171	1.054
1.26	.9521	.3058	3.113	.3212	3.270	1.050
1.27	.9551	.2963	3.224	.3102	3.375	1.047
1.28	.9580	.2867	3.341	.2993	3.488	1.044
1.29	.9608	.2771	3.467	.2884	3.609	1.041
1.30	.9636	.2675	3.602	.2776	3.738	1.038
1.31	.9662	.2579	3.747	.2669	3.878	1.035
1.32	.9687	.2482	3.903	.2562	4.029	1.032
1.33	.9711	.2385	4.072	.2456	4.193	1.030
1.34	.9735	.2288	4.256	.2350	4.372	1.027
1.35	.9757	.2190	4.455	.2245	4.566	1.025
1.36	.9779	.2092	4.673	.2140	4.779	1.023
1.37	.9799	.1994	4.913	.2035	5.014	1.021
1.38	.9819	.1896	5.177	.1931	5.273	1.018
1.39	.9837	.1798	5.471	.1828	5.561	1.017
1.40	.9854	.1700	5.798	.1725	5.883	1.015
1.41	.9871	.1601	6.165	.1622	6.246	1.013
1.42	.9887	.1502	6.581	.1519	6.657	1.011
1.43	.9901	.1403	7.055	.1417	7.126	1.010
1.44	.9915	.1304	7.602	.1315	7.667	1.009
1.45	.9927	.1205	8.238	.1214	8.299	1.007
1.46	.9939	.1106	8.989	.1113	9.044	1.006
1.47	.9949	.1006	9.887	.1011	9.938	1.005
1.48	.9959	.0907	10.983	.0910	11.029	1.004
1.49	.9967	.0807	12.350	.0810	12.390	1.003
1.50	.9975	.0707	14.101	.0709	14.137	1.003
1.51	.9982	.0608	16.428	.0609	16.458	1.002
1.52	.9987	.0508	19.670	.0508	19.695	1.001
1.53	.9992	.0408	24.498	.0408	24.519	1.001
1.54	.9995	.0308	32.461	.0308	32.476	1.000
1.55	.9998	.0208	48.078	.0208	48.089	1.000
1.56	.9999	.0108	92.620	.0108	92.626	1.000
1.57	1.0000	.0008	1255.8	.0008	1255.8	1.000

Answers to Odd-Numbered Exercises

CHAPTER 1

Exercises 1.1, page 9

1 (a) $<$ (b) $>$ (c) $=$ **3** (a) $>$ (b) $<$ (c) $>$
5 $-8 < -5$ **7** $0 > -1$ **9** $x < 0$
11 $3 < a < 5$ **13** $-2 < x < 1$ **15** $3 < x \leq 5$
17 $0 \leq x \leq 2\pi$ **19** $x < 2$ **21** $x \geq 1$
23 $(2, 5)$ **25** $[1, 4]$

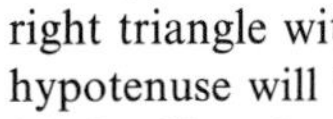

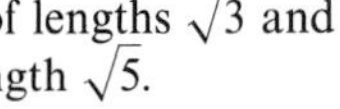

27 $(-1, \infty)$

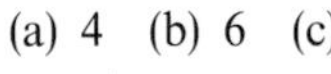

29 Construct a right triangle with legs of length $\sqrt{2}$ and 1. The hypotenuse will have length $\sqrt{3}$. Next construct a right triangle with legs of lengths $\sqrt{3}$ and $\sqrt{2}$. The hypotenuse will have length $\sqrt{5}$.
31 (a) 5 (b) -5 (c) 13
33 (a) 0 (b) $4 - \pi$ (c) -1
35 (a) 4 (b) 6 (c) 6 (d) 10 **37** $\frac{3}{2}, -1$
39 $2, -\frac{4}{3}$ **41** $(2 \pm \sqrt{14})/2$ **43** $\frac{5}{2}$ **45** 55 ft
47 (a) $d = \sqrt{(400t)^2 + (200t + 100)^2} = 100\sqrt{20t^2 + 4t + 1}$ (b) 3:30 P.M.
49 (a) 206.25 ft (b) 40 mph

Exercises 1.2, page 24

1 $12u^{11}/v^2$ **3** $4/xy$ **5** $81y^6/64$ **7** $9x^{10}y^{14}$
9 $1/(9a^4)$ **11** $4x^2y^4$ **13** $2a^2/b$ **15** $\sqrt{3uv}/3u^2v$
17 $x^{3/4}$ **19** $(a + b)^{2/3}$ **21** (a) $4x\sqrt{x}$ (b) $8x\sqrt[3]{x}$
23 $\neq$ **25** $\neq$ **27** $x^4 - 3x^3 + 4x^2 + 4x + 5; 4$
29 $-8x; 1$ **31** $2x^5 + 2x^4 + 3x^3 + 4x^2 + 3x + 10; 5$
33 $15x^4 - 29x^2y - 14y^2$ **35** $16x^4 - 40x^2y^2 + 25y^4$
37 $x^3 - 6x^2y + 12xy^2 - 8y^3$ **39** $(3x - 2)(x + 4)$
41 $x^2(3x + 2)(2x - 3)$ **43** $(5x - 3)(5x + 3)$
45 $(x^2 + 1)(x + 1)$
47 $(x^2 + 3)(x^2 - 3)(x^2 + 1)(x + 1)(x - 1)$
49 $(x - 1)(x^2 + x + 1)(x + 1)(x^2 - x + 1)$
51 $(x + 2)/(x + 3)$ **53** $3y/(5y + 1)$
55 $-(a + 2)/(2a + 1)$ **57** $(2x - 3)/x(5x - 2)$
59 $34r/(3r - 1)(2r + 5)$ **61** $6x/(2x - 1)(x + 1)$
63 $3/[2(4t^2 + 6t + 9)]$ **65** $a/(a^2 + 4)(5a + 2)$
67 $(6x - 7)/(3x + 1)^2$ **69** $(3 - 2c)/c^3$
71 $(5x^2 - 2x - 6)/(x - 1)^3$
73 $(18x^3 - 10x^2 - 66x + 63)/x^2(2x - 3)^2$
75 $11/(7x - 3)(2x + 1)$ **77** $a - b$
79 $(x^2 + xy + y^2)/(x + y)$
81 $(2x^2 + 7x + 15)/(x^2 + 10x + 7)$ **83** $2x + h - 7$
85 $-(3x^2 + 3xh + h^2)/x^3(x + h)^3$
87 $(3x + 2)^3(36x^2 - 37x + 6)$
89 $(2x + 1)^2(8x^2 + x - 24)/(x^2 - 4)^{1/2}$
91 $(27x^2 - 24x + 2)/(6x + 1)^4$

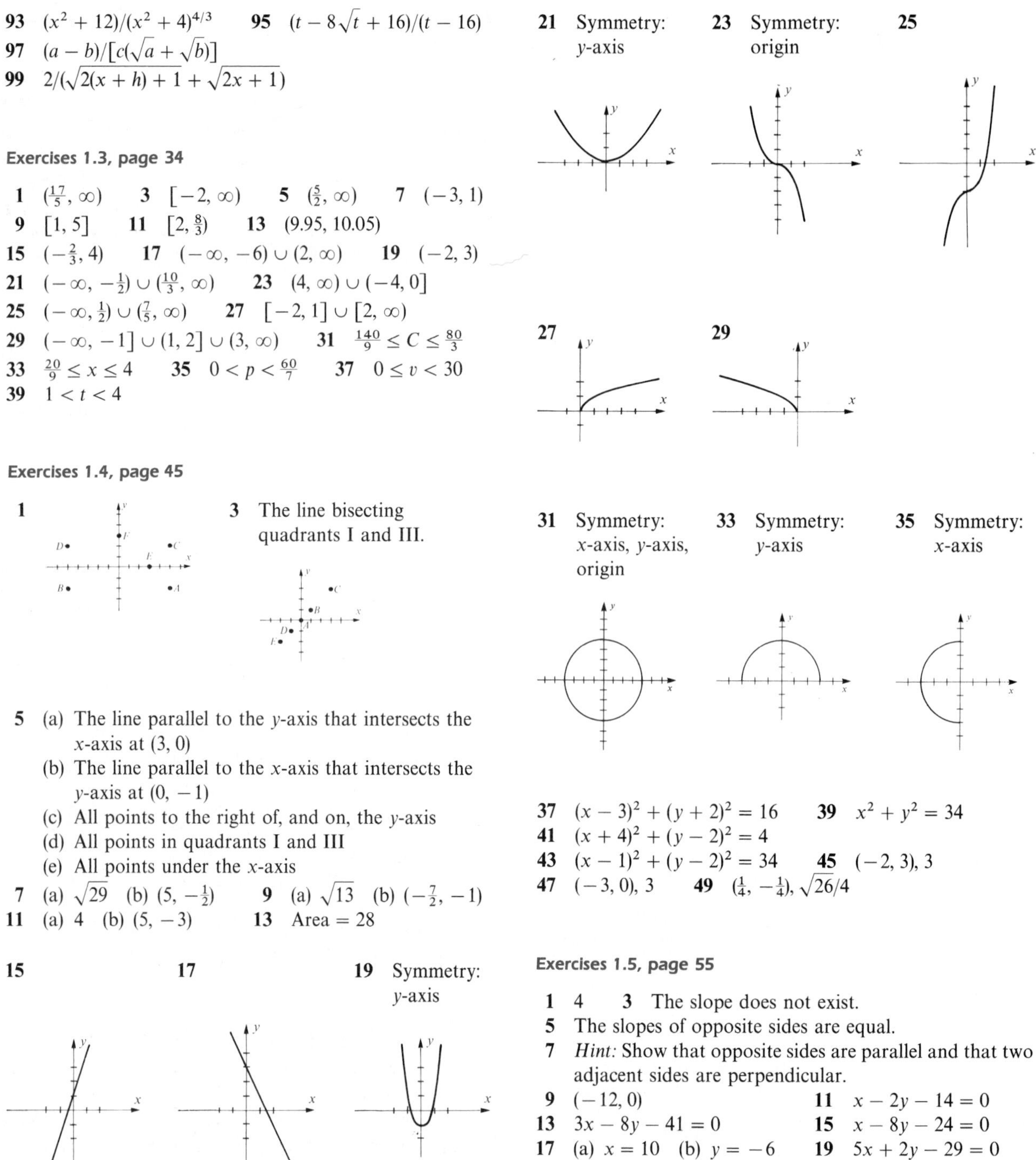

93 $(x^2+12)/(x^2+4)^{4/3}$ **95** $(t-8\sqrt{t}+16)/(t-16)$
97 $(a-b)/[c(\sqrt{a}+\sqrt{b})]$
99 $2/(\sqrt{2(x+h)+1}+\sqrt{2x+1})$

Exercises 1.3, page 34

1 $(\frac{17}{5}, \infty)$ **3** $[-2, \infty)$ **5** $(\frac{5}{2}, \infty)$ **7** $(-3, 1)$
9 $[1, 5]$ **11** $[2, \frac{8}{3})$ **13** $(9.95, 10.05)$
15 $(-\frac{2}{3}, 4)$ **17** $(-\infty, -6) \cup (2, \infty)$ **19** $(-2, 3)$
21 $(-\infty, -\frac{1}{2}) \cup (\frac{10}{3}, \infty)$ **23** $(4, \infty) \cup (-4, 0]$
25 $(-\infty, \frac{1}{2}) \cup (\frac{7}{5}, \infty)$ **27** $[-2, 1] \cup [2, \infty)$
29 $(-\infty, -1] \cup (1, 2] \cup (3, \infty)$ **31** $\frac{140}{9} \le C \le \frac{80}{3}$
33 $\frac{20}{9} \le x \le 4$ **35** $0 < p < \frac{60}{7}$ **37** $0 \le v < 30$
39 $1 < t < 4$

Exercises 1.4, page 45

1 (graph) **3** The line bisecting quadrants I and III.

5 (a) The line parallel to the y-axis that intersects the x-axis at $(3, 0)$
(b) The line parallel to the x-axis that intersects the y-axis at $(0, -1)$
(c) All points to the right of, and on, the y-axis
(d) All points in quadrants I and III
(e) All points under the x-axis

7 (a) $\sqrt{29}$ (b) $(5, -\frac{1}{2})$ **9** (a) $\sqrt{13}$ (b) $(-\frac{7}{2}, -1)$
11 (a) 4 (b) $(5, -3)$ **13** Area = 28

15 **17** **19** Symmetry: y-axis

21 Symmetry: y-axis **23** Symmetry: origin **25**

27 **29**

31 Symmetry: x-axis, y-axis, origin **33** Symmetry: y-axis **35** Symmetry: x-axis

37 $(x-3)^2+(y+2)^2=16$ **39** $x^2+y^2=34$
41 $(x+4)^2+(y-2)^2=4$
43 $(x-1)^2+(y-2)^2=34$ **45** $(-2, 3), 3$
47 $(-3, 0), 3$ **49** $(\frac{1}{4}, -\frac{1}{4}), \sqrt{26}/4$

Exercises 1.5, page 55

1 4 **3** The slope does not exist.
5 The slopes of opposite sides are equal.
7 *Hint:* Show that opposite sides are parallel and that two adjacent sides are perpendicular.
9 $(-12, 0)$ **11** $x-2y-14=0$
13 $3x-8y-41=0$ **15** $x-8y-24=0$
17 (a) $x=10$ (b) $y=-6$ **19** $5x+2y-29=0$
21 $5x-7y+15=0$

23 $m = \frac{3}{4}, b = 2$

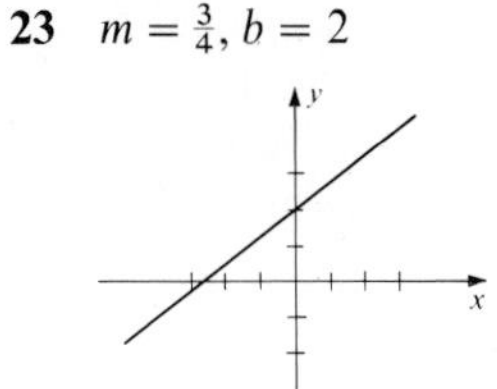

25 $m = -\frac{1}{2}, b = 0$

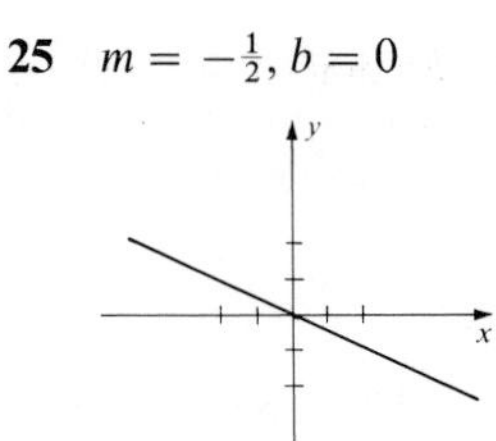

27 $m = 0, b = 4$

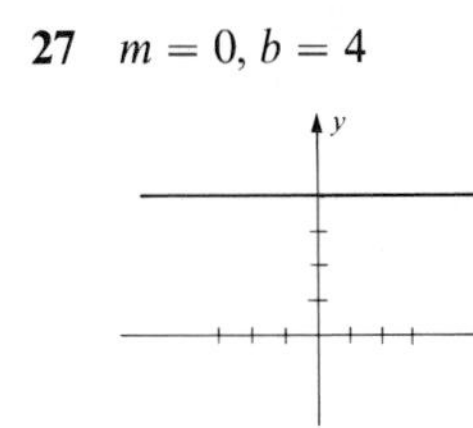

29 $m = -\frac{5}{4}, b = 5$

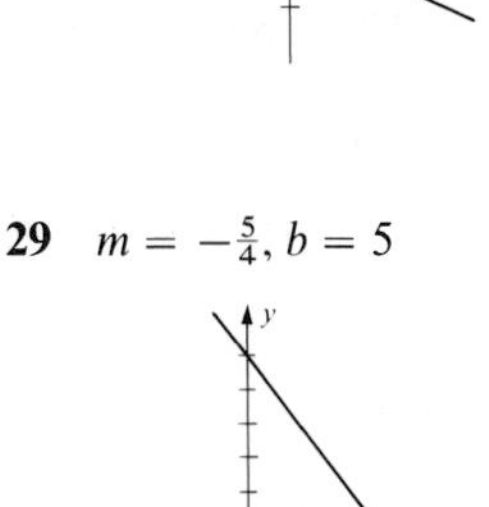

31 $y = 59{,}000 + 6000x$, where y is the value of the house and x is the number of years after the purchase date; 2 years and 4 months after the purchase date.
33 (a) R_0 is the resistance when $T = 0\,^\circ\text{C}$ (b) $\frac{1}{273}$ (c) $273\,^\circ\text{C}$
35 (a) 8.2 tons (b) as large as 3.4 tons
37 (a) $M = 0.45R - 900$ (b) \$2000 per month
39 If the player shoots when the plane is at P, the creature at $x = 3$ will be hit. All targets will be missed if the player shoots when the plane is at Q.

Exercises 1.6, page 58

1 (a) $<$ (b) $>$ (c) $>$ **3** (a) 7 (b) -1 (c) $\frac{1}{6}$
5 $(-5 \pm \sqrt{73})/2$ **7** 1 ft **9** $18a^5b^5$
11 $xy^5/9$ **13** x^8/y^2 **15** $(1 - 2\sqrt{x} + x)/(1 - x)$
17 $(14x + 9)(2x - 1)$ **19** $(2c^2 + 3)(c - 6)$
21 $8(x + 2y)(x^2 - 2xy + 4y^2)$
23 $x^4 + x^3 - x^2 + x - 2$
25 $3y^5 - 2y^4 - 8y^3 + 10y^2 - 3y - 12$
27 $(3x - 5)/(2x + 1)$ **29** $(3x + 2)/x(x - 2)$
31 $(5x^2 - 6x - 20)/x(x + 2)^2$ **33** $(-\frac{11}{4}, \frac{9}{4})$
35 $(-\infty, \frac{11}{3}] \cup [7, \infty)$ **37** $(-\infty, -\frac{3}{2}) \cup (\frac{2}{5}, \infty)$
39 $4 \le p \le 8$ **41** Area = 10
43 The points in quadrants II and IV
47 $(x + 5)^2 + (y + 1)^2 = 81$
49 (a) $18x + 6y - 7 = 0$ (b) $2x - 6y - 3 = 0$

51 **53** **55** **57** **59**

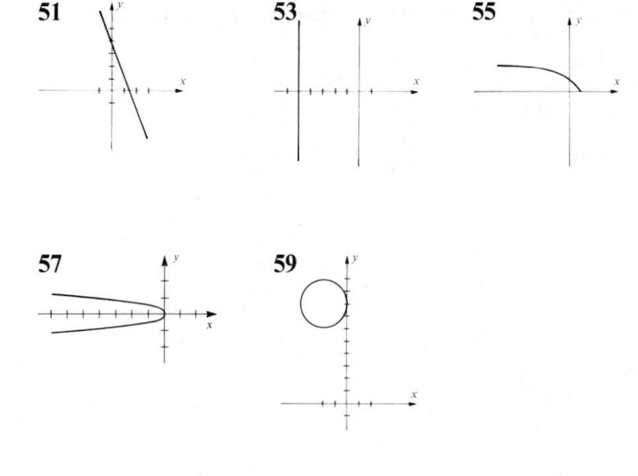

CHAPTER 2

Exercises 2.1, page 70

1 3, 9, 4, 6 **3** 2, $\sqrt{2} + 6$, 12, 23
5 (a) $5a - 2$ (b) $-5a - 2$ (c) $-5a + 2$ (d) $5a + 5h - 2$ (e) $5a + 5h - 4$ (f) 5
7 (a) $2a^2 - a + 3$ (b) $2a^2 + a + 3$ (c) $-2a^2 + a - 3$ (d) $2a^2 + 4ah + 2h^2 - a - h + 3$ (e) $2a^2 - a + 2h^2 - h + 6$ (f) $4a + 2h - 1$
9 (a) $3/a^2$ (b) $1/(3a^2)$ (c) $3a^4$ (d) $9a^4$ (e) $3a$ (f) $\sqrt{3a^2}$
11 (a) $2a/(a^2 + 1)$ (b) $(a^2 + 1)/2a$ (c) $2a^2/(a^4 + 1)$ (d) $4a^2/(a^4 + 2a^2 + 1)$ (e) $2\sqrt{a}/(a + 1)$ (f) $\sqrt{2a^3 + 2a}/(a^2 + 1)$
13 $[\frac{5}{3}, \infty)$ **15** $[-2, 2]$
17 All real numbers except 0, 3, and -3
19 All nonnegative real numbers except 4 and $\frac{3}{2}$
21 Domain and range: $\mathbb{R}$
23 Domain $[3, \infty)$; range $[0, \infty)$
25 Domain and range: $(-\infty, 0) \cup (0, \infty)$

21 **23** **25**

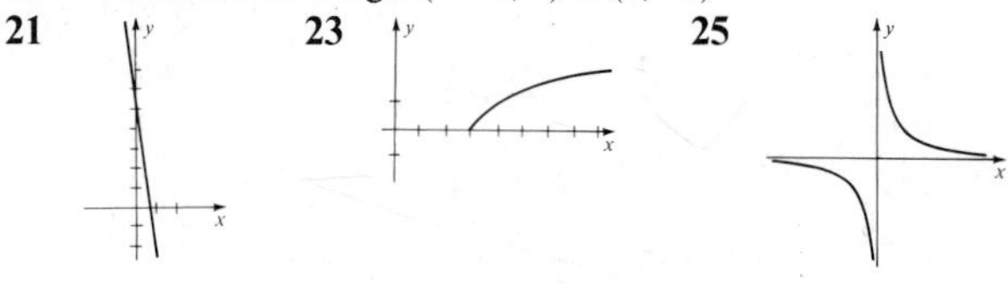

27 Odd **29** Even **31** Even **33** Neither
35 Neither **37** $V = 4x^3 - 100x^2 + 600x$

39 (a) $y = 500/x$ (b) $C = 100[3x + (1000/x) - 3]$
41 $d = 2\sqrt{t^2 + 2500}$
43 (a) $y = \sqrt{2rh + h^2}$ (b) 1280.6 mi
45 $S = 4\pi r(5 + r)$ **47** $d = \sqrt{90{,}400 + x^2}$
49 (a) $h = 3(4 - r)$ (b) $V = 3\pi r^2(4 - r)$ **51** Yes
53 No **55** Yes **57** No **59** No

Exercises 2.2, page 79

In Exercises 1–19, D denotes the domain of f and E denotes the range.

1 $D = \mathbb{R}, E = \mathbb{R}$; increasing on $\mathbb{R}$

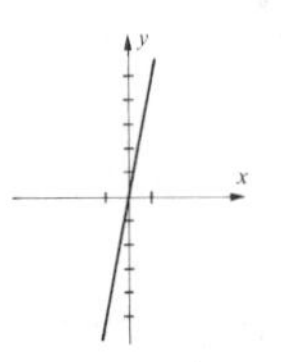

3 $D = \mathbb{R}, E = \mathbb{R}$; decreasing on $\mathbb{R}$

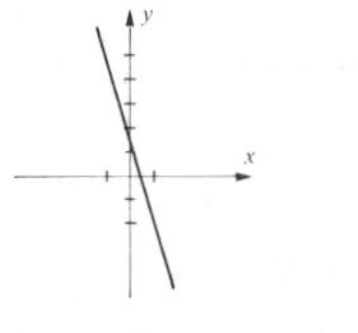

5 $D = \mathbb{R}, E = (-\infty, 3]$; increasing on $(-\infty, 0]$, decreasing on $[0, \infty)$

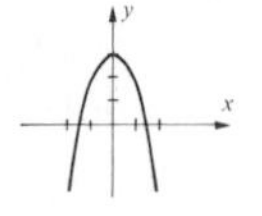

7 $D = \mathbb{R}, E = [-4, \infty)$; decreasing on $(-\infty, 0]$, increasing on $[0, \infty)$

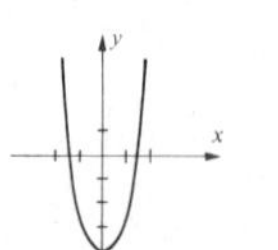

9 $D = [-4, \infty), E = [0, \infty)$; increasing on $[-4, \infty)$
11 $D = [0, \infty), E = [2, \infty)$; increasing on $[0, \infty)$

9 **11**

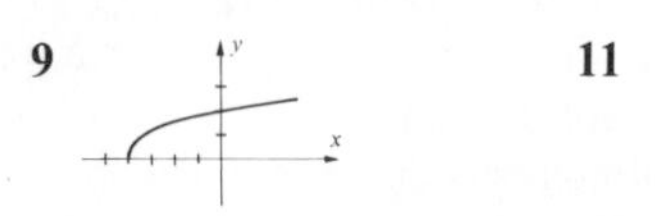

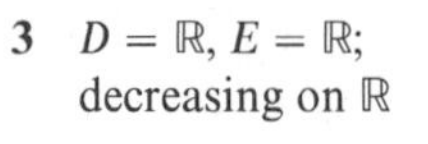

13 $D = (-\infty, 0) \cup (0, \infty) = E$; decreasing on $(-\infty, 0)$ and on $(0, \infty)$
15 $D = \mathbb{R}, E = [0, \infty)$; decreasing on $(-\infty, 2]$, increasing on $[2, \infty)$

13 **15**

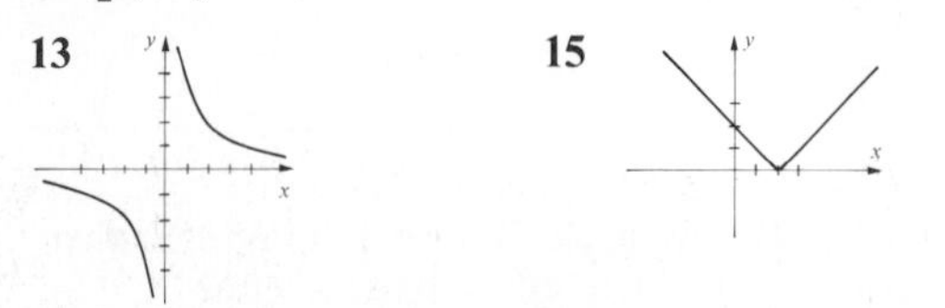

17 $D = \mathbb{R}, E = [-2, \infty)$; decreasing on $(-\infty, 0]$, increasing on $[0, \infty)$

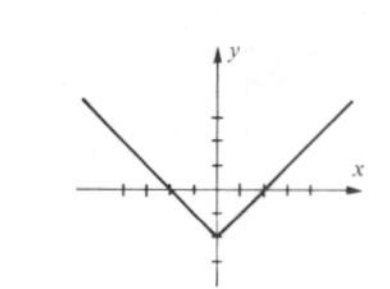

19 $D = (-\infty, 0) \cup (0, \infty)$, $E = \{-1, 1\}$; Neither increasing nor decreasing

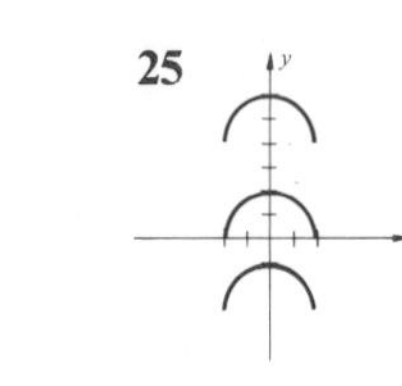

21

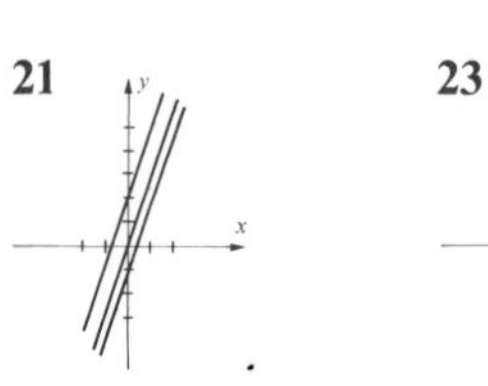

23

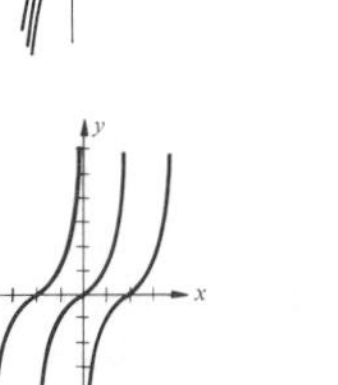

25

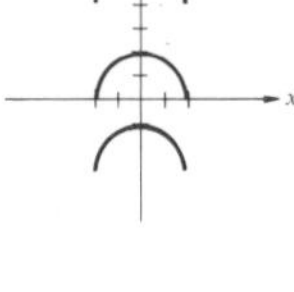

27

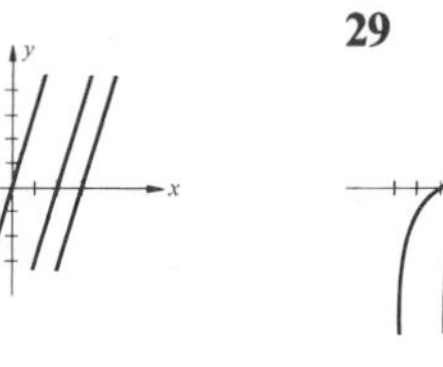

29

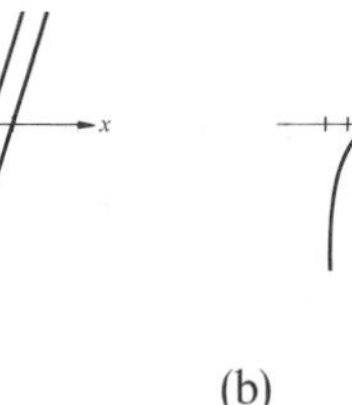

31 (a) (b) (c) (d)

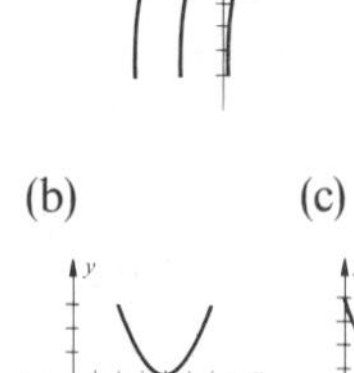

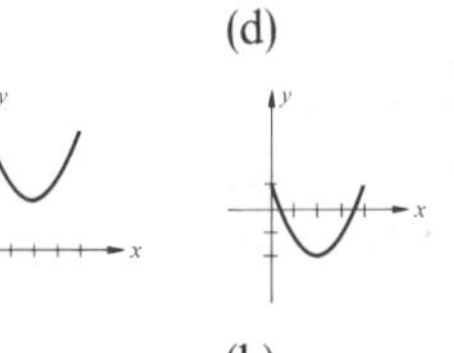

(e) (f) (g) (h)

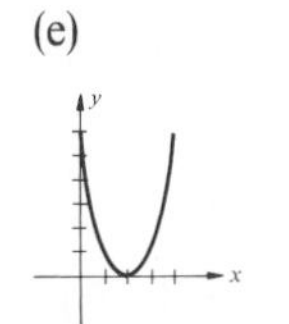

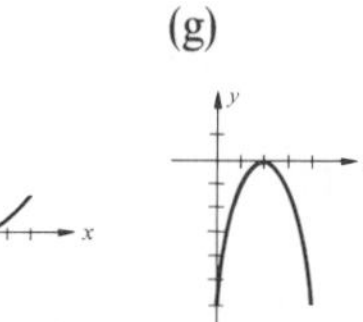

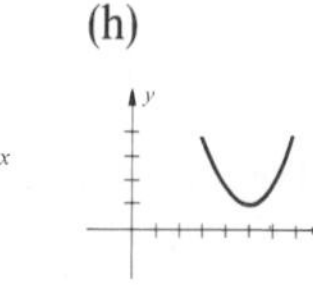

33

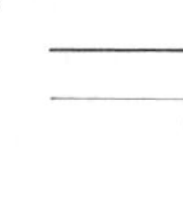

35

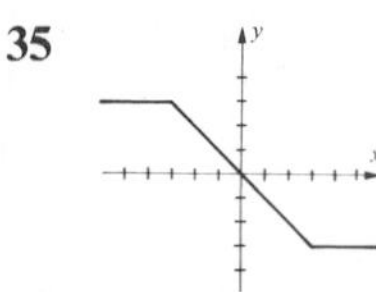

37 **39**

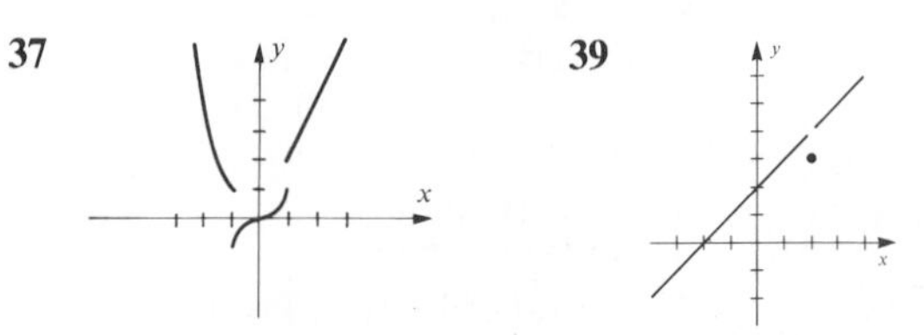

41 (a) (b) (c)

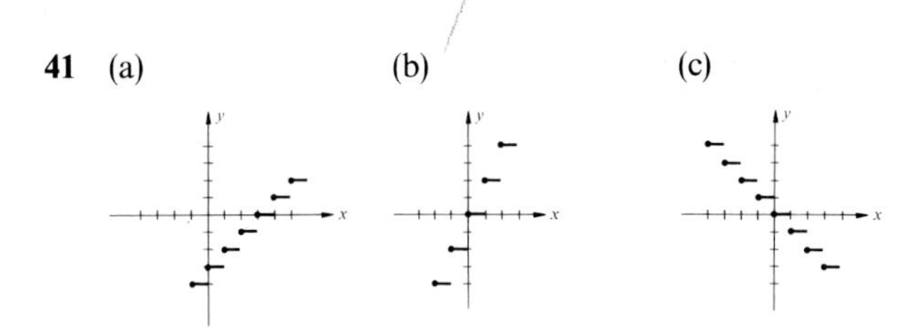

Exercises 2.3, page 86

1 (a) (b) (c) (d)

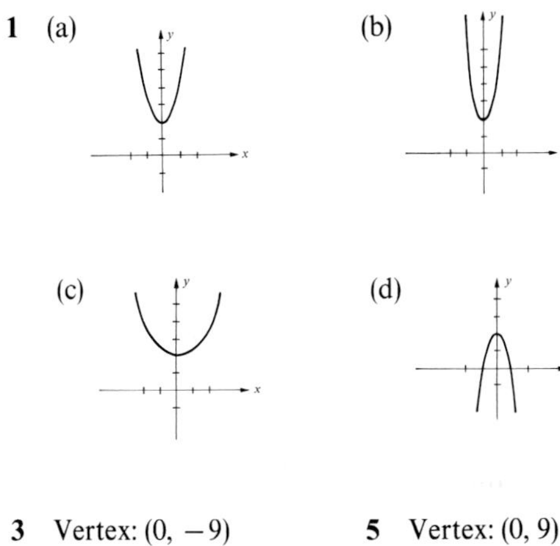

3 Vertex: $(0, -9)$ **5** Vertex: $(0, 9)$

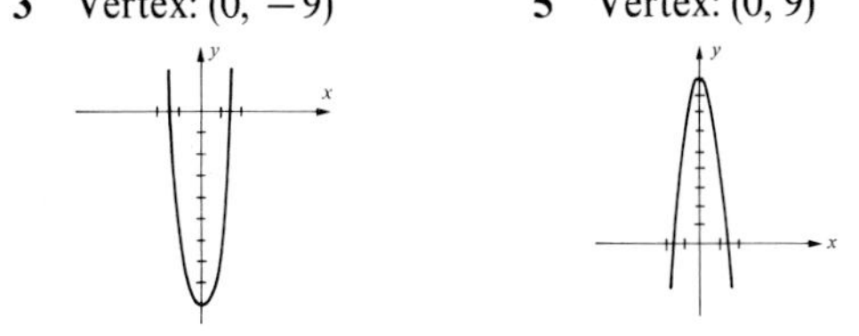

7 $3, -\frac{1}{4}$ **9** $\frac{2}{3}$
11 $2(x-4)^2 - 9$; min: $f(4) = -9$
13 $-5(x+1)^2 + 8$; max: $f(-1) = 8$
15 min: $f(-\frac{5}{2}) = -\frac{9}{4}$ **17** max: $f(4) = 4$

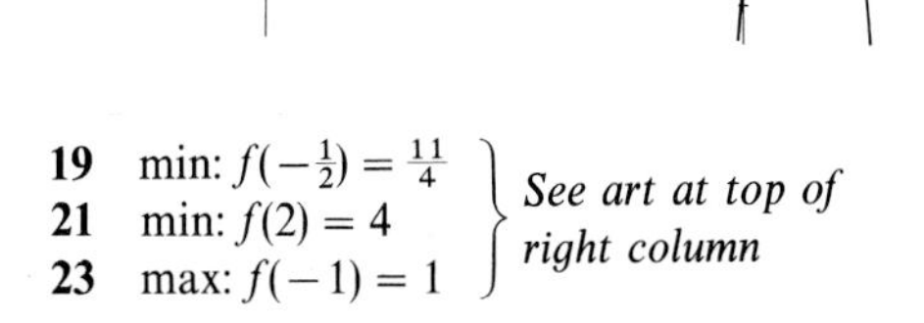

19 min: $f(-\frac{1}{2}) = \frac{11}{4}$
21 min: $f(2) = 4$
23 max: $f(-1) = 1$
See art at top of right column

19 **21** **23**

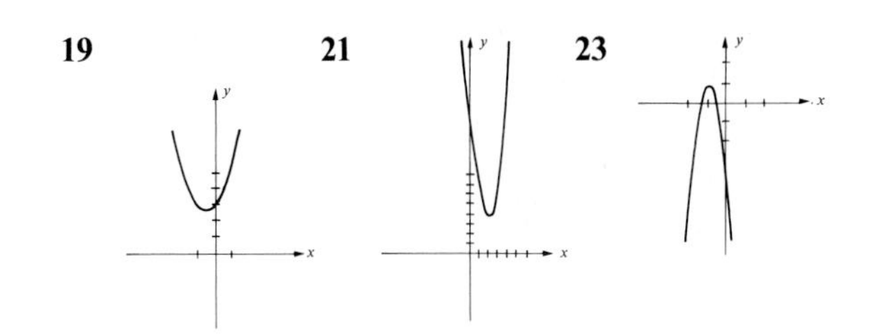

25 $y = \frac{4}{27}x(9-x)$
27 (a) $x^2 = 500(y-10)$ (b) 282 ft
29 $y = 2x^2 - 3x + 1$ **31** 424 ft, 100 ft **33** 10.5 lb
35 (a) $y = 250 - \frac{3}{4}x$ (b) $A = x(250 - \frac{3}{4}x)$
(c) $166\frac{2}{3}$ ft by 125 ft
37 (a) $y = 12 - x$ (b) $A = x(12 - x)$
39 (a) $y = -\frac{7}{160}x^2 + x$ (b) 17.5 ft

Exercises 2.4, page 93

1 $3x^2 + 1/(2x-3)$; $3x^2 - 1/(2x-3)$; $3x^2/(2x-3)$; $3x^2(2x-3)$
3 $2x$; $2/x$, $x^2 - (1/x^2)$, $(x^2+1)/(x^2-1)$
5 $2x^3 + x^2 + 7$; $2x^3 - x^2 - 2x + 3$; $2x^5 + 2x^4 + 3x^3 + 4x^2 + 3x + 10$; $(2x^3 - x + 5)/(x^2 + x + 2)$
7 $6x - 1, 6x + 3$ **9** $36x^2 - 5, 12x^2 - 15$
11 $12x^2 - 8x + 1, 6x^2 + 4x - 1$
13 $x^3 - 1, x^3 - 3x^2 + 3x - 1$
15 $x + 9 + 9\sqrt{x+9}, \sqrt{x^2 + 9x + 9}$
17 $x^2/(2 - 5x^2), 4x^2 - 20x + 25$ **19** $5, -5$
21 $1/x^4, 1/x^4$ **23** x, x **25** $A = 36\pi t^2$
27 $h = 5\sqrt{t^2 + 8t}$
29 $h = (14/\sqrt{821})t + 2 \approx 0.4886t + 2$

Exercises 2.5, page 99

1 Yes **3** No **5** Yes
7 No **9** No **11** Yes
13 and **15** Show that $f(g(x)) = x = g(f(x))$.

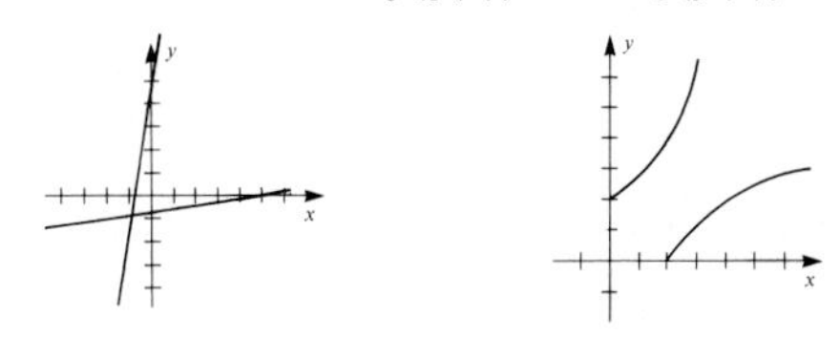

17 $f^{-1}(x) = (x+3)/4$ **19** $f^{-1}(x) = (1-5x)/2x, x > 0$
21 $f^{-1}(x) = \sqrt{9-x}, x \le 9$ **23** $f^{-1}(x) = \sqrt[3]{(x+2)/5}$
25 $f^{-1}(x) = (x^2+5)/3, x \ge 0$ **27** $f^{-1}(x) = (x-8)^3$

29 $f^{-1}(x) = x$

31 (a) $f^{-1}(x) = (x - b)/a$ (b) No (not one-to-one)

33 If g and h are both inverse functions of f, then $f(g(x)) = x = f(h(x))$ for all x. Since f is one-to-one, this implies that $g(x) = h(x)$ for all x, that is, $g = h$.

35 **37**

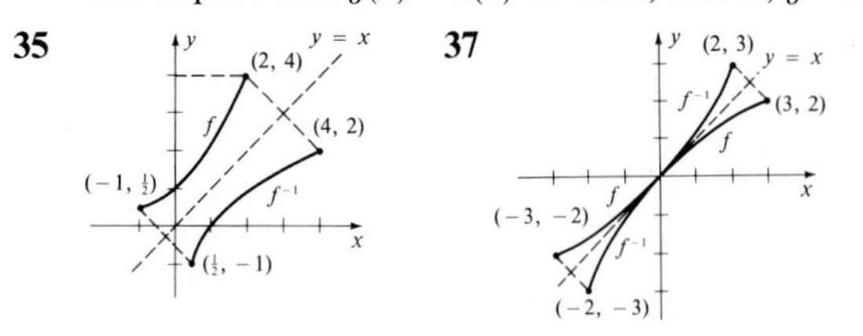

Exercises 2.6, page 100

1 (a) $\frac{1}{2}$ (b) $-\sqrt{2}/2$ (c) 0 (d) $-x/\sqrt{3 - x}$
(e) $-x/\sqrt{x + 3}$ (f) $x^2/\sqrt{x^2 + 3}$ (g) $x^2/(x + 3)$

3 $D = \mathbb{R}, E = \mathbb{R}$; decreasing on $\mathbb{R}$

5 $D = [-1, \infty), E = (-\infty, 1]$ decreasing on $[-1, \infty)$

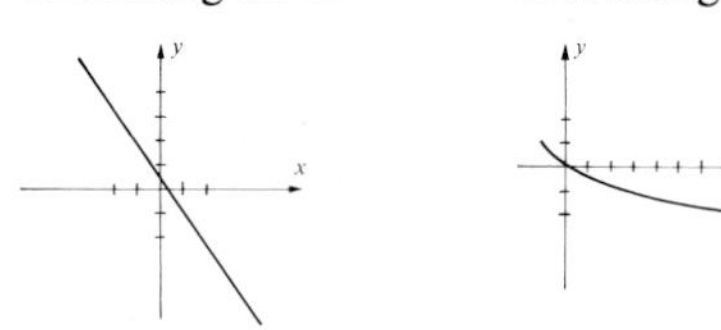

7 $D = \mathbb{R}, E = \{1000\}$

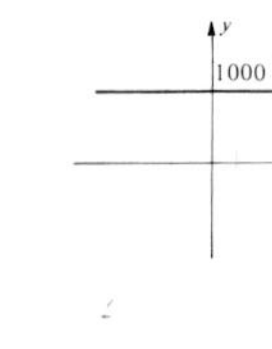

9 $D = \mathbb{R}, E = [7, \infty)$

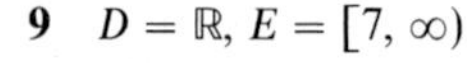

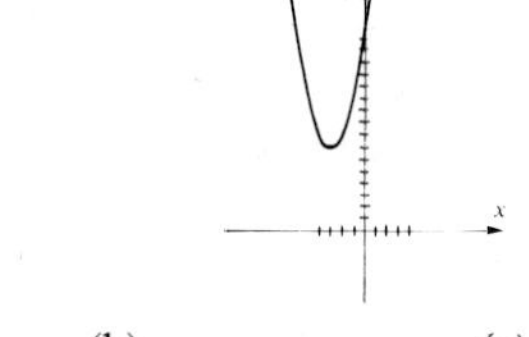

11 (a) (b) (c)

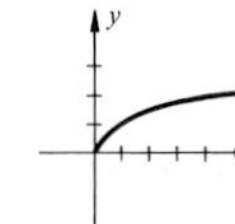

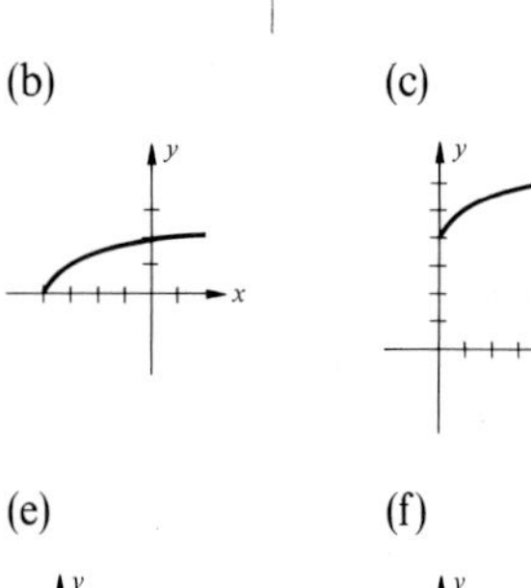

(d) (e) (f)

13 $f(x) = 5(x + 3)^2 + 4$; min: $f(-3) = 4$

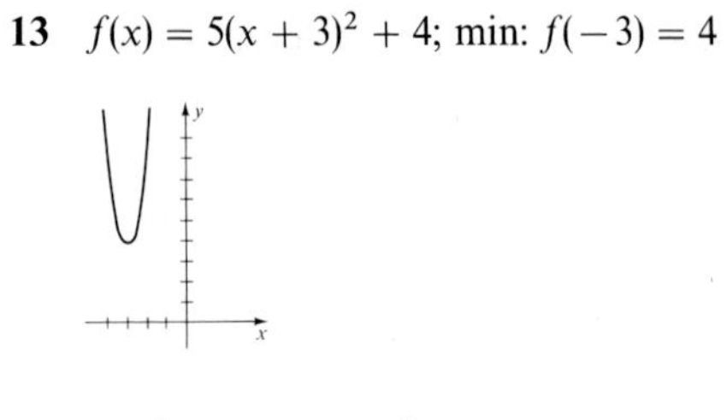

15 $18x^2 + 9x - 1$; $6x^2 - 15x + 5$

17 $f^{-1}(x) = (10 - x)/15$ **19** $V = C^3/8\pi^2$

21 (a) $y = 20 - \frac{4}{5}x$ (b) $V = 4x(20 - \frac{4}{5}x)$
(c) The shelter is 12.5 ft tall and 10 ft long.

23 (a) $V = 200h^2$ for $0 \le h \le 6$; $V = 7200 + 3200(h - 6)$ for $6 \le h \le 9$
(b) $V = 10t$; $h = \sqrt{t/20}$ for $0 \le t \le 720$; $h = 6 + (t - 720)/320$ for $720 \le t \le 1680$

25 Radius of semicircle is $1/(8\pi)$ mi; length of rectangle is $\frac{1}{8}$ mi.

27 (a) (87.5, 17.5) (b) 30.625 ft

CHAPTER 3

Exercises 3.1, page 109

1 (a) (b) (c) (d)

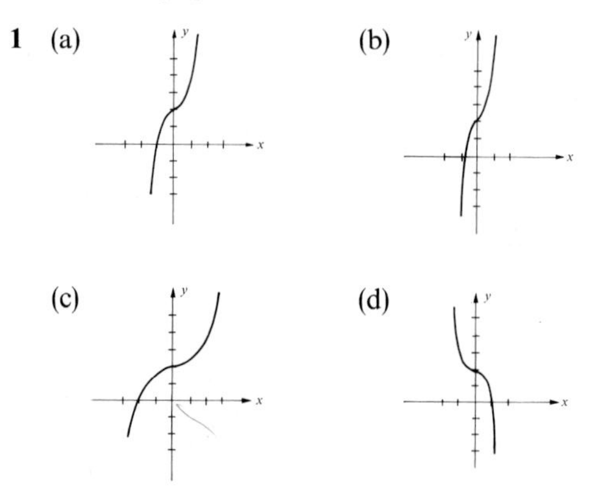

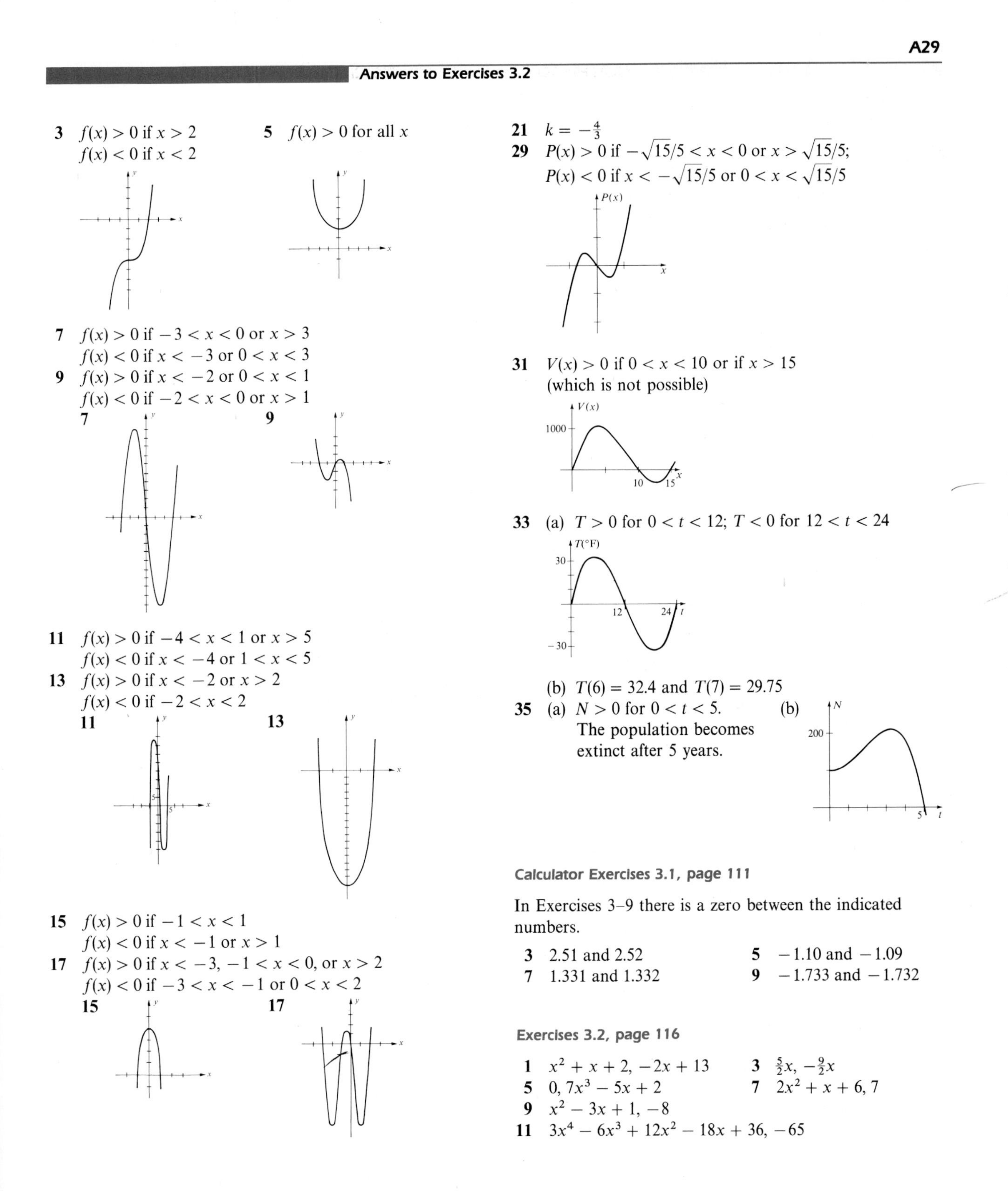

3 $f(x) > 0$ if $x > 2$
$f(x) < 0$ if $x < 2$

5 $f(x) > 0$ for all x

7 $f(x) > 0$ if $-3 < x < 0$ or $x > 3$
$f(x) < 0$ if $x < -3$ or $0 < x < 3$

9 $f(x) > 0$ if $x < -2$ or $0 < x < 1$
$f(x) < 0$ if $-2 < x < 0$ or $x > 1$

11 $f(x) > 0$ if $-4 < x < 1$ or $x > 5$
$f(x) < 0$ if $x < -4$ or $1 < x < 5$

13 $f(x) > 0$ if $x < -2$ or $x > 2$
$f(x) < 0$ if $-2 < x < 2$

15 $f(x) > 0$ if $-1 < x < 1$
$f(x) < 0$ if $x < -1$ or $x > 1$

17 $f(x) > 0$ if $x < -3$, $-1 < x < 0$, or $x > 2$
$f(x) < 0$ if $-3 < x < -1$ or $0 < x < 2$

21 $k = -\frac{4}{3}$

29 $P(x) > 0$ if $-\sqrt{15}/5 < x < 0$ or $x > \sqrt{15}/5$;
$P(x) < 0$ if $x < -\sqrt{15}/5$ or $0 < x < \sqrt{15}/5$

31 $V(x) > 0$ if $0 < x < 10$ or if $x > 15$
(which is not possible)

33 (a) $T > 0$ for $0 < t < 12$; $T < 0$ for $12 < t < 24$

(b) $T(6) = 32.4$ and $T(7) = 29.75$

35 (a) $N > 0$ for $0 < t < 5$.
The population becomes extinct after 5 years.
(b)

Calculator Exercises 3.1, page 111

In Exercises 3–9 there is a zero between the indicated numbers.

3 2.51 and 2.52
5 -1.10 and -1.09
7 1.331 and 1.332
9 -1.733 and -1.732

Exercises 3.2, page 116

1 $x^2 + x + 2, -2x + 13$
3 $\frac{5}{2}x, -\frac{9}{2}x$
5 $0, 7x^3 - 5x + 2$
7 $2x^2 + x + 6, 7$
9 $x^2 - 3x + 1, -8$
11 $3x^4 - 6x^3 + 12x^2 - 18x + 36, -65$

13 $4x^3 + 2x^2 - 4x - 2, 0$
15 $x^{n-1} + x^{n-2} + \cdots + x + 1, 0$
17 95 **19** -23 **21** 0
23 -23 **25** 3277 **27** $8 + 7\sqrt{3}$
33 $k = \frac{17}{12}$ **35** $f(2) = 0$ **37** $f(c) > 0$
39 If $f(x) = x^n - y^n$, then $f(y) = 0$. If n is even, then $f(-y) = 0$.

Exercises 3.3, page 123

1 $-2 + 6i$ **3** $-2 + 4i$
5 $7 - 5i$ **7** $4 + 7i$
9 $-6 + 3i$ **11** $-10 + 5i$
13 $20 - 10i$ **15** $-5 + 12i$
17 $-72 - 36i$ **19** 100
21 -1 **23** $\frac{3}{13} - \frac{2}{13}i$
25 $\frac{35}{61} + \frac{42}{61}i$ **27** $-\frac{1}{5} - \frac{11}{10}i$
29 $-\frac{7}{13} + \frac{17}{13}i$ **31** $-7 - 21i$
33 $x = 10, y = 3$ **35** $(3 \pm \sqrt{31}i)/2$
37 $-1 \pm 2i$ **39** $(-1 \pm \sqrt{47}i)/8$
41 $5, (-5 \pm 5\sqrt{3}i)/2$ **43** $2, -2, -1 \pm \sqrt{3}i, 1 \pm \sqrt{3}i$
45 $\pm 2i, \pm \frac{3}{2}i$ **47** $0, (-3 \pm \sqrt{7}i)/2$

Exercises 3.4, page 131

1 $-\frac{1}{15}x^3 + \frac{19}{15}x + 2$ **3** $-2x^3 + 12x^2 - 22x + 12$
5 $\frac{3}{10}x^3 - \frac{33}{10}x + 6$ **7** $x^4 - 2x^3 - 11x^2 + 12x + 36$
9 $\frac{2}{9}x^8 - \frac{4}{3}x^7 + \frac{8}{3}x^6 - \frac{16}{9}x^5$
11 -4 (multiplicity 3); $\frac{4}{3}$ (multiplicity 1)
13 0 (multiplicity 3); 5, -1 (each of multiplicity 1)
15 $\pm\frac{5}{3}$ (each of multiplicity 4); $\pm 4i$ (each of multiplicity 1)
17 -2 (multiplicity 3); 1 (multiplicity 2); 2 (multiplicity 1)
19 $(x + 3)^2(x + 2)(x - 1)$ **21** $(x - 1)^5(x + 1)$

In Exercises 23–29 the types of possible solutions are listed in the following order: positive, negative, nonreal complex.

23 Either 3, 0, 0, or 1, 0, 2 **25** 0, 1, 2
27 Either 2, 0, 2; 2, 2, 0; 0, 2, 2; or 0, 0, 4
29 Either 2, 3, 0; 2, 1, 2; 0, 3, 2; or 0, 1, 4
31 Upper 5, lower -2 **33** Upper 2, lower -2
35 Upper 3, lower -3

Exercises 3.5, page 139

1 $x^2 - 8x + 17$ **3** $x^3 - 10x^2 + 37x - 52$
5 $x^4 - 6x^3 + 30x^2 - 78x + 85$
7 $x^5 - 2x^4 + 27x^3 - 50x^2 + 50x$
9 No. If i is a root, then $-i$ is also a root. Hence, the polynomial would have factors $x - 1$, $x + 1$, $x - i$, $x + i$ and, therefore, would be of degree greater than 3.
11 $-1, -2, 4$ **13** $2, -3, \frac{5}{2}$
15 $4, -7, \pm\sqrt{2}$ **17** $4, -2, \frac{3}{2}$
19 $\frac{1}{2}, -\frac{2}{3}, -3$ **21** $-\frac{3}{4}, (-3 \pm 3\sqrt{7}i)/4$
23 $4, -3, \pm\sqrt{3}i$ **25** $1, -2, -\frac{4}{3}, \frac{2}{3}$
27 $3, -4, -2, \frac{1}{2}, -\frac{1}{2}$
37 Complex zeros occur in conjugate pairs.
41 The two boxes correspond to $x = 5$ and $x = 5(2 - \sqrt{2})$. The box corresponding to $x = 5$ has the smaller surface area.
43 $t = 4$ (10 A.M.) and 6.2020 (12:12 P.M.)
45 (c) The triangle has legs of length 5 and 12 ft.
47 (b) $x = 4$ ft

Exercises 3.6, page 151

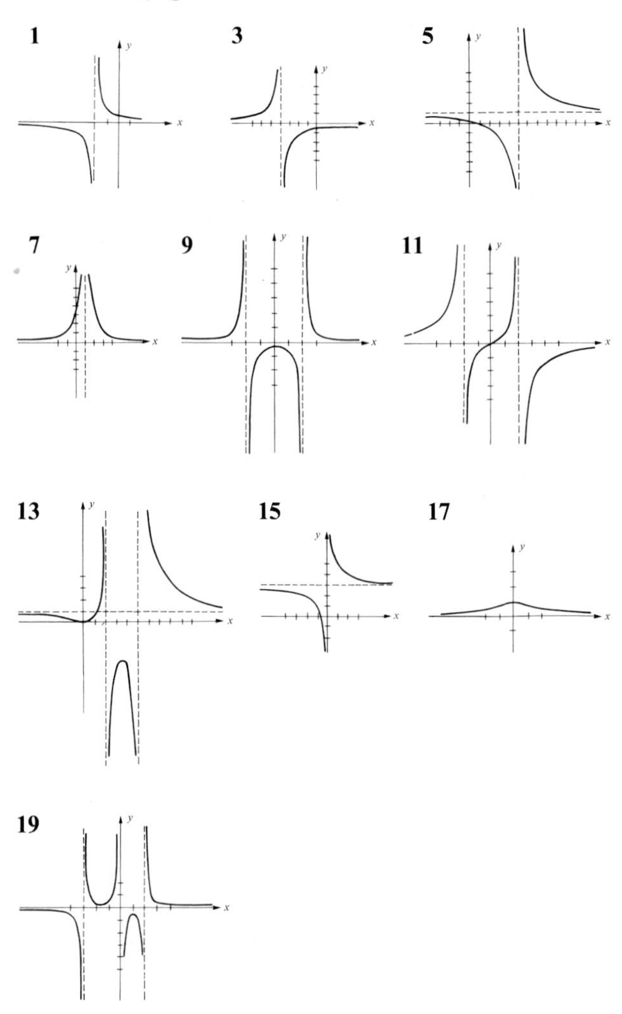

21 Vertical: $x = -1$
oblique: $y = x - 2$

23 Vertical: $x = 0$
oblique: $y = -\frac{1}{2}x$

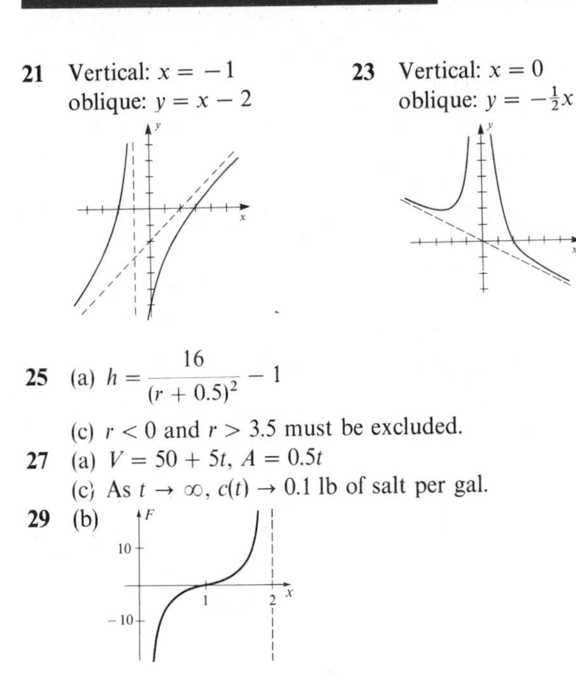

25 (a) $h = \dfrac{16}{(r + 0.5)^2} - 1$

(c) $r < 0$ and $r > 3.5$ must be excluded.

27 (a) $V = 50 + 5t$, $A = 0.5t$

(c) As $t \to \infty$, $c(t) \to 0.1$ lb of salt per gal.

29 (b)

Exercises 3.7, page 153

1 **3** **5**

7 **9**

11 Vertical: $x = -3$
oblique: $y = x - 1$

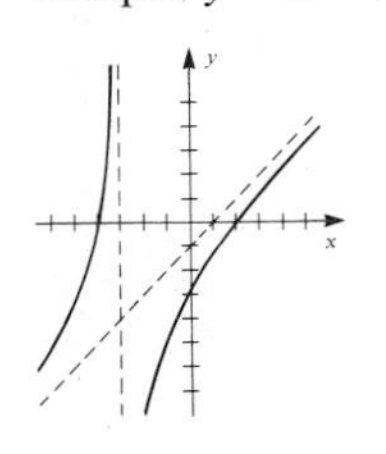

13 $3x^2 + 2, -21x^2 + 5x - 9$ **15** $\frac{9}{2}, \frac{53}{2}$ **17** -132

19 $6x^4 - 12x^3 + 24x^2 - 52x + 104, -200$

21 $\frac{2}{41}x^3 + \frac{14}{41}x^2 + \frac{80}{41}x + \frac{68}{41}$ **23** $x^7 + 6x^6 + 9x^5$

25 1 (multiplicity 5); -3 (multiplicity 1)

27 (a) Either 3 positive and 1 negative or 1 positive, 1 negative, and 2 nonreal complex

(b) Upper bound 3; lower bound -1

31 $-\frac{1}{2}, \frac{1}{4}, \frac{3}{2}$

33 (a) $C(100) = \$2{,}000{,}000$ and $C(90) = \$163{,}636.36$.

(b)

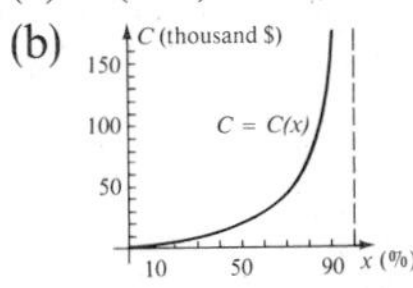

CHAPTER 4

Exercises 4.1, page 162

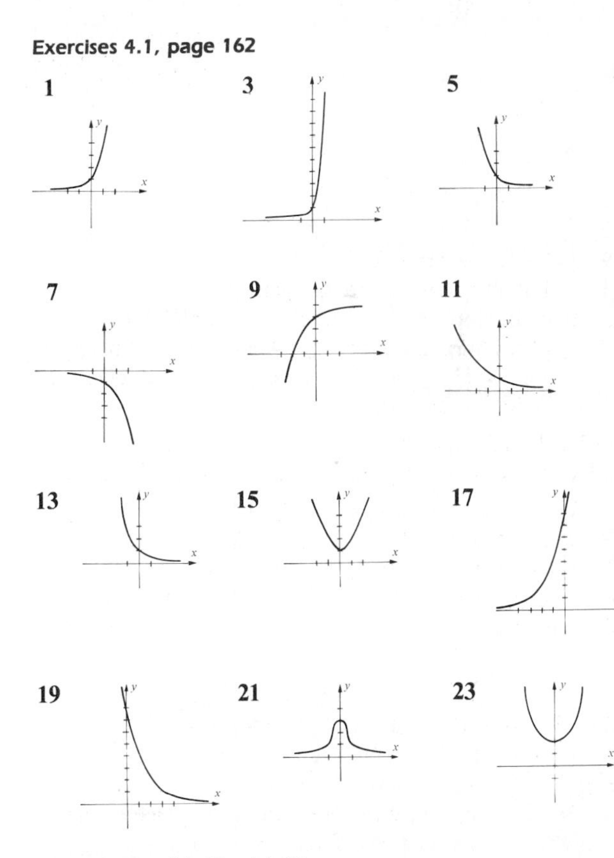

25 (a) 90 (b) 59 (c) 35

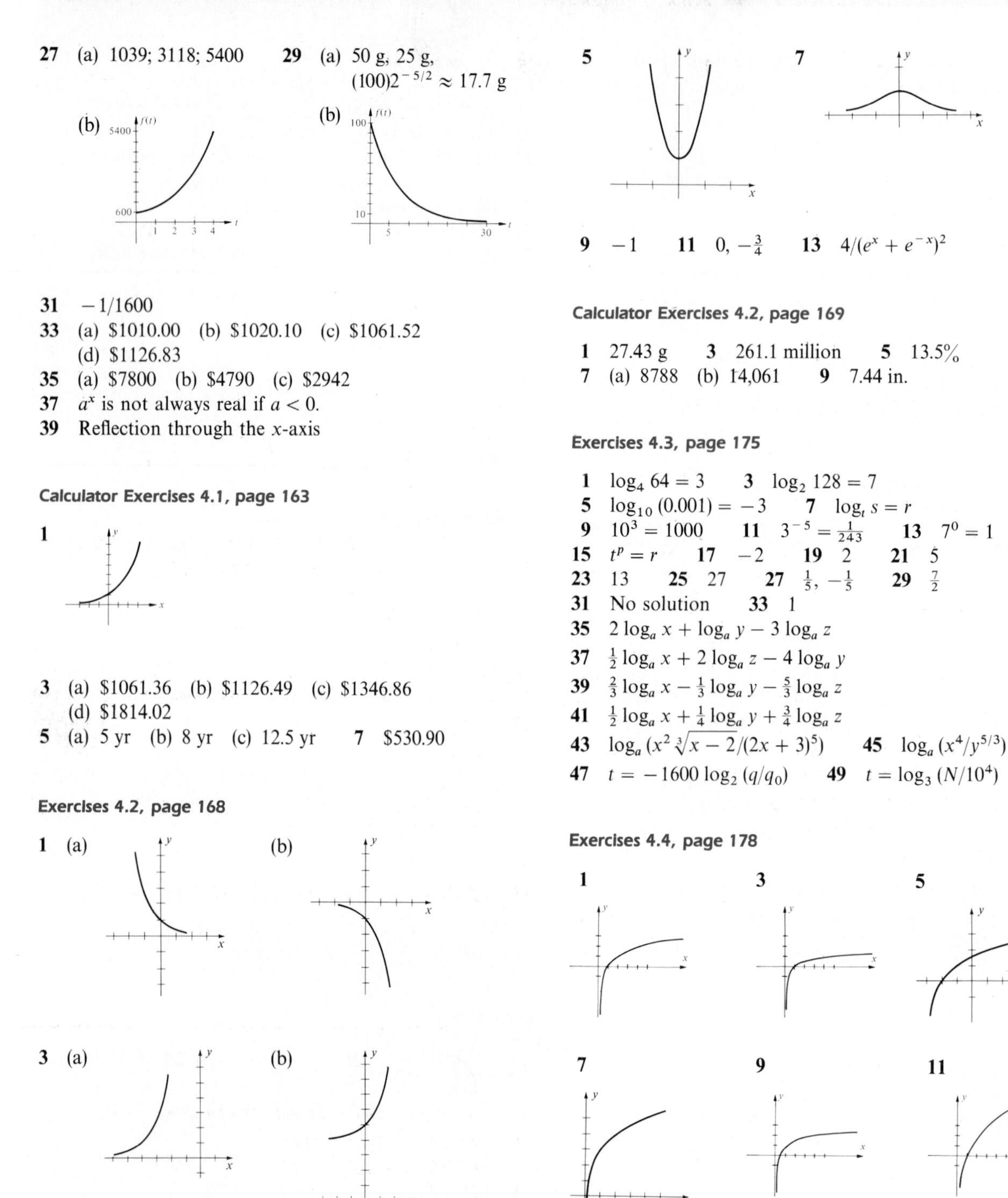

27 (a) 1039; 3118; 5400 (b)

29 (a) 50 g, 25 g, $(100)2^{-5/2} \approx 17.7$ g (b)

31 $-1/1600$

33 (a) \$1010.00 (b) \$1020.10 (c) \$1061.52 (d) \$1126.83

35 (a) \$7800 (b) \$4790 (c) \$2942

37 a^x is not always real if $a < 0$.

39 Reflection through the x-axis

Calculator Exercises 4.1, page 163

1

3 (a) \$1061.36 (b) \$1126.49 (c) \$1346.86 (d) \$1814.02

5 (a) 5 yr (b) 8 yr (c) 12.5 yr **7** \$530.90

Exercises 4.2, page 168

1 (a) (b)

3 (a) (b)

5

7

9 -1 **11** $0, -\frac{3}{4}$ **13** $4/(e^x + e^{-x})^2$

Calculator Exercises 4.2, page 169

1 27.43 g **3** 261.1 million **5** 13.5%

7 (a) 8788 (b) 14,061 **9** 7.44 in.

Exercises 4.3, page 175

1 $\log_4 64 = 3$ **3** $\log_2 128 = 7$

5 $\log_{10}(0.001) = -3$ **7** $\log_t s = r$

9 $10^3 = 1000$ **11** $3^{-5} = \frac{1}{243}$ **13** $7^0 = 1$

15 $t^p = r$ **17** -2 **19** 2 **21** 5

23 13 **25** 27 **27** $\frac{1}{5}, -\frac{1}{5}$ **29** $\frac{7}{2}$

31 No solution **33** 1

35 $2\log_a x + \log_a y - 3\log_a z$

37 $\frac{1}{2}\log_a x + 2\log_a z - 4\log_a y$

39 $\frac{2}{3}\log_a x - \frac{1}{3}\log_a y - \frac{5}{3}\log_a z$

41 $\frac{1}{2}\log_a x + \frac{1}{4}\log_a y + \frac{3}{4}\log_a z$

43 $\log_a (x^2 \sqrt[3]{x-2}/(2x+3)^5)$ **45** $\log_a (x^4/y^{5/3})$

47 $t = -1600 \log_2 (q/q_0)$ **49** $t = \log_3 (N/10^4)$

Exercises 4.4, page 178

1 **3** **5**

7 **9** **11**

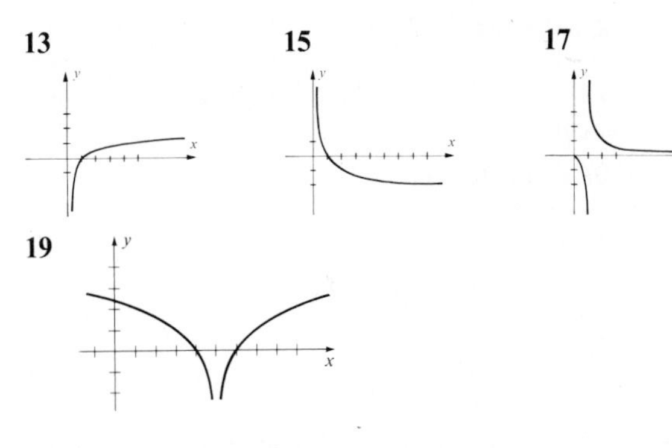

Exercises 4.5, page 183

1 4240 **3** 8.85 **5** 0.0237 **7** 9.97
9 1.05 **11** 0.202 **13** (a) 2 (b) 4 (c) 5
15 $A = 10^{(R+7.5)/2.3} - 34{,}000$
17 (a) 10 (b) 30 (c) 40 **19** (a) 2.2 (b) 5 (c) 8.3
21 Acidic if pH < 7, basic if pH > 7
23 $t = -(L/R)\ln(I/20)$
25 $h = (\ln 29 - \ln P)/0.000034$ **27** $W = 2.4e^{1.84h}$
29 (a) $n = 10^{7.7-0.9R}$ (b) 12,589; 1585; 200
31 110 days **33** 139 months **35** In the year 2079

Exercises 4.6, page 189

1 (a) log 7 or ln 7/ln 10 (b) 0.85
3 (a) log 3/log 4 or ln 3/ln 4 (b) 0.79
5 (a) $4 - (\ln 5/\ln 3)$ (b) 2.54
7 (a) $(\ln 2 - \ln 81)/\ln 24$ (b) -1.16
9 (a) -3 (b) -3 **11** (a) 5 (b) 5
13 (a) $\frac{2}{3}\sqrt{\frac{1001}{111}}$ (b) 2.00 **15** 100, 1
17 10^{100} **19** 10,000
21 $x = \log(y \pm \sqrt{y^2 - 1})$
23 $x = \frac{1}{2}\log[(1 + y)/(1 - y)]$
25 $x = \ln(y + \sqrt{y^2 + 1})$
27 $x = \frac{1}{2}\ln[(1 + y)/(1 - y)]$
29 $t = -(L/R)\ln(1 - RI/E)$
31 86.4 m **33** (a) 7.21 hr (b) 3.11 hr

Exercises 4.7, page 190

1 -4 **3** 4 **5** 6 **7** $\frac{1}{2}$ **9** 5

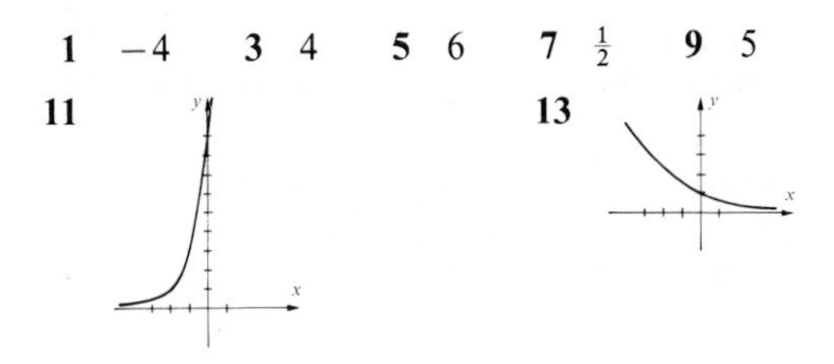

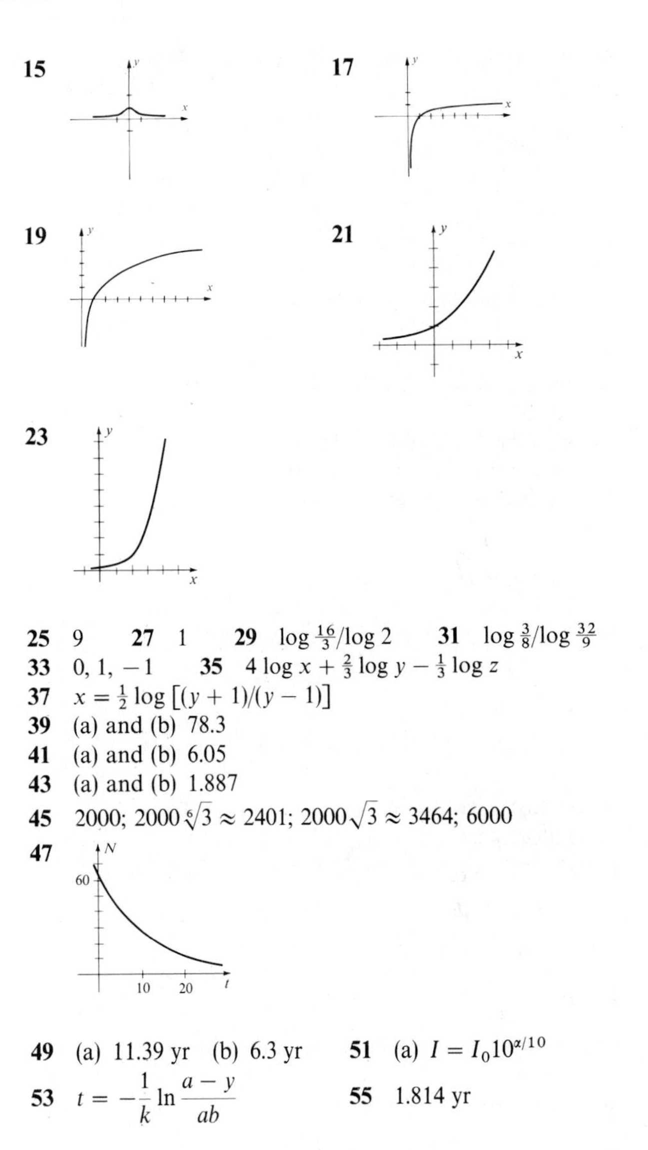

25 9 **27** 1 **29** $\log\frac{16}{3}/\log 2$ **31** $\log\frac{3}{8}/\log\frac{32}{9}$
33 0, 1, -1 **35** $4\log x + \frac{2}{3}\log y - \frac{1}{3}\log z$
37 $x = \frac{1}{2}\log[(y + 1)/(y - 1)]$
39 (a) and (b) 78.3
41 (a) and (b) 6.05
43 (a) and (b) 1.887
45 2000; $2000\sqrt[6]{3} \approx 2401$; $2000\sqrt{3} \approx 3464$; 6000
47

49 (a) 11.39 yr (b) 6.3 yr **51** (a) $I = I_0 10^{\alpha/10}$
53 $t = -\dfrac{1}{k}\ln\dfrac{a - y}{ab}$ **55** 1.814 yr

CHAPTER 5

Exercises 5.1, page 201

There are other possible answers for Exercises 1–11.

1 480°, 840°, $-240°$, $-600°$
3 495°, 855°, $-225°$, $-585°$
5 330°, 690°, $-390°$, $-750°$
7 260°, 980°, $-100°$, $-460°$

9 $17\pi/6, 29\pi/6, -7\pi/6, -19\pi/6$
11 $7\pi/4, 15\pi/4, -9\pi/4, -17\pi/4$
13 $5\pi/6$ **15** $-\pi/3$ **17** $5\pi/4$ **19** $5\pi/2$
21 $2\pi/5$ **23** $5\pi/9$ **25** 120° **27** 330°
29 135° **31** −630° **33** 1260° **35** 20°
37 114°35′30″ **39** 286°28′44″ **41** 37.6833°
43 115.4408° **45** 63°10′8″ **47** 310°37′17″
49 (a) 1.75 (b) 100.27° **51** 6.98 m **53** 8.59 km
55 Approximations (in miles):
(a) 4189 (b) 3142 (c) 2094 (d) 698 (e) 70
57 $\frac{1}{8}$ radian $= (45/2\pi)° \approx 7°10'$ **59** 7.162°
61 $\theta_2 = (r_1/r_2)\theta_1$

Exercises 5.2, page 215

Answers to Exercises 1–17 are in the order sin, cos, tan, csc, sec, cot.

1 $\frac{4}{5}, \frac{3}{5}, \frac{4}{3}, \frac{5}{4}, \frac{5}{3}, \frac{3}{4}$
3 $2\sqrt{13}/13, 3\sqrt{13}/13, \frac{2}{3}, \sqrt{13}/2, \sqrt{13}/3, \frac{3}{2}$
5 $\frac{2}{5}, \sqrt{21}/5, 2\sqrt{21}/21, \frac{5}{2}, 5\sqrt{21}/21, \sqrt{21}/2$
7 $a/\sqrt{a^2+b^2}, b/\sqrt{a^2+b^2}, a/b, \sqrt{a^2+b^2}/a,$
$\sqrt{a^2+b^2}/b, b/a$
9 $b/c, \sqrt{c^2-b^2}/c, b/\sqrt{c^2-b^2}, c/b, c/\sqrt{c^2-b^2},$
$\sqrt{c^2-b^2}/b$
11 $\frac{3}{5}, \frac{4}{5}, \frac{3}{4}, \frac{5}{3}, \frac{5}{4}, \frac{4}{3}$
13 $\sqrt{11}/6, \frac{5}{6}, \sqrt{11}/5, 6\sqrt{11}/11, \frac{6}{5}, 5\sqrt{11}/11$
15 $\sqrt{2}/2, \sqrt{2}/2, 1, \sqrt{2}, \sqrt{2}, 1$
17 0.7826 **19** 6.197 **21** 2.650
23 0.7776 **25** 0.4718 **27** 1.035
29 $\cot\theta = \sqrt{1-\sin^2\theta}/\sin\theta$ **31** $\sec\theta = 1/\sqrt{1-\sin^2\theta}$
33 $\tan\theta = \sqrt{\sec^2\theta - 1}$ **35** $\sin\theta = \sqrt{\sec^2\theta-1}/\sec\theta$

Answers for Exercises 39–51 are in the order sin, cos, tan, csc, sec, cot.

39 $\sqrt{3}/2, \frac{1}{2}, \sqrt{3}, 2\sqrt{3}/3, 2, \sqrt{3}/3$
41 $\frac{2}{3}, \sqrt{5}/3, 2\sqrt{5}/5, \frac{3}{2}, 3\sqrt{5}/5, \sqrt{5}/2$
43 $-\frac{3}{5}, \frac{4}{5}, -\frac{3}{4}, -\frac{5}{3}, \frac{5}{4}, -\frac{4}{3}$
45 $-5\sqrt{29}/29, -2\sqrt{29}/29, \frac{5}{2}, -\sqrt{29}/5, -\sqrt{29}/2, \frac{2}{5}$
47 $4\sqrt{17}/17, -\sqrt{17}/17, -4, \sqrt{17}/4, -\sqrt{17}, -\frac{1}{4}$
49 $-2\sqrt{53}/53, -7\sqrt{53}/53, \frac{2}{7}, -\sqrt{53}/2, -\sqrt{53}/7, \frac{7}{2}$
51 $\frac{4}{5}, \frac{3}{5}, \frac{4}{3}, \frac{5}{4}, \frac{5}{3}, \frac{3}{4}$
53 1, 0, —, 1, —, 0 **55** 0, −1, 0, —, −1, —
57 0, 1, 0, —, 1, — **59** −1, 0, —, −1, —, 0
61 IV **63** III **65** II **67** III **69** I

Exercises 5.3, page 224

In Exercises 1–11, answers for (b) are in the order sin, cos, tan, csc, sec, cot.

1 (a) (1, 0) (b) 0, 1, 0, —, 1, —
3 (a) (0, −1) (b) −1, 0, —, −1, —, 0
5 (a) (−1, 0) (b) 0, −1, 0, —, −1, —
7 (a) (0, 1) (b) 1, 0, —, 1, —, 0
9 (a) $(-\sqrt{2}/2, -\sqrt{2}/2)$
(b) $-\sqrt{2}/2, -\sqrt{2}/2, 1, -\sqrt{2}, -\sqrt{2}, 1$
11 (a) $(\sqrt{2}/2, -\sqrt{2}/2)$
(b) $-\sqrt{2}/2, \sqrt{2}/2, -1, -\sqrt{2}, \sqrt{2}, -1$
13 (a) $(-\frac{3}{5}, -\frac{4}{5})$ (b) $(-\frac{3}{5}, -\frac{4}{5})$ (c) $(\frac{3}{5}, -\frac{4}{5})$ (d) $(-\frac{3}{5}, \frac{4}{5})$
15 (a) $(\frac{8}{17}, -\frac{15}{17})$ (b) $(\frac{8}{17}, -\frac{15}{17})$
(c) $(-\frac{8}{17}, -\frac{15}{17})$ (d) $(\frac{8}{17}, \frac{15}{17})$
17 $\sin(-3\pi/4) = -\sin(3\pi/4) = -\sqrt{2}/2;$
$\cos(-3\pi/4) = \cos(3\pi/4) = -\sqrt{2}/2$
19 $\sin(-\pi/2) = -\sin(\pi/2) = -1;$
$\cos(-\pi/2) = \cos(\pi/2) = 0$
21 (a) $\csc(-t) = 1/\sin(-t) = 1/(-\sin t) = -(1/\sin t)$
$= -\csc t.$
23 Using the hint with $t = 0$ gives us $\sin k = \sin 0 = 0$. Since $0 < k < 2\pi$, it follows that $k = \pi$, and hence $\sin(t+\pi) = \sin t$ for all t. In particular, if $t = \pi/2$, then $\sin(3\pi/2) = \sin(\pi/2)$, or $-1 = 1$, an absurdity.
25 Use $\csc t = 1/\sin t$ and the result of Exercise 9.
27 If $y/x = a$, where $x^2 + y^2 = 1$, then $\pm\sqrt{1-x^2}/x = a$. Hence $(1-x^2)/x^2 = a^2$, and $1 - x^2 = a^2x^2$, or $1 = a^2x^2 + x^2 = (a^2+1)x^2$. Consequently, $x^2 = 1/(a^2+1)$, or $x = \pm 1/\sqrt{a^2+1}$. If $a > 0$, choose the point $P(x, y)$ on U where $x = 1/\sqrt{a^2+1}$. If $a < 0$, choose $P(x, y)$ such that $x = -1/\sqrt{a^2+1}$. Thus, there is always a point $P(x, y)$ on U such that $y/x = a$.
29 For n even, $P(n\pi) = (1, 0)$; for n odd, $P(n\pi) = (-1, 0)$.

Exercises 5.4, page 231

1 (a) $\pi/4$ (b) $\pi/3$ (c) $\pi/6$
3 (a) $\pi/4$ (b) $\pi/6$ (c) $\pi/3$
5 (a) 1.5 (b) $2\pi - 5$ **7** (a) 60° (b) 20° (c) 70°
9 (a) 49°20′ (b) 45° (c) 80°35′
11 (a) $\sqrt{3}/2$ (b) $-\sqrt{3}/2$ **13** (a) −1 (b) −1
15 (a) $-2\sqrt{3}/3$ (b) −2 **17** 0.7826
19 6.197 **21** −1.019 **23** −0.3035

Calculator Exercises 5.4, page 232

1 30°28′, 329°32′ **3** 74°53′, 254°53′
5 204°55′, 335°5′ **7** 124°29′, 304°29′
9 $P(\frac{1}{2}) = 0.479427\ldots$ and $\sin\frac{1}{2} = 0.4794255\ldots$
11 (a) $(-1.2624, 0.9650)$ (b) $(1.8415, -0.5403)$

Exercises 5.5, page 237

1 (a) $[-2\pi, -3\pi/2), (-3\pi/2, -\pi], [0, \pi/2), (\pi/2, \pi]$
(b) $[-\pi, -\pi/2), (-\pi/2, 0], [\pi, 3\pi/2), (3\pi/2, 2\pi]$
(c) $\sec x \to \infty$ as $x \to (\pi/2)^-$; $\sec x \to -\infty$ as $x \to (\pi/2)^+$
(d) Symmetric with respect to the y-axis

5

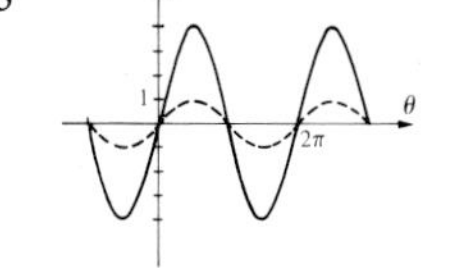

7

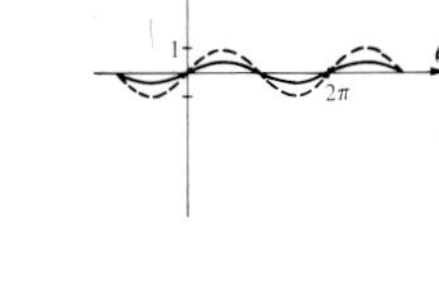

9

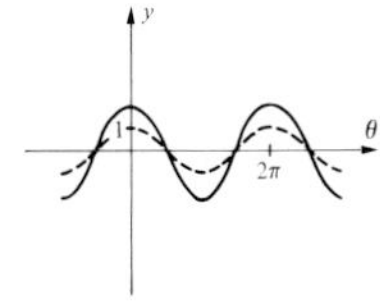

11

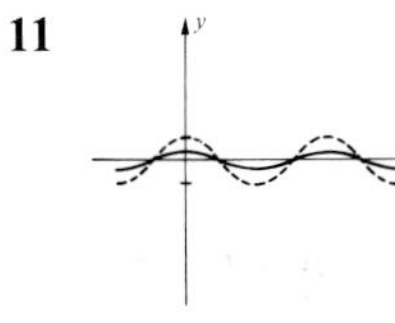

13

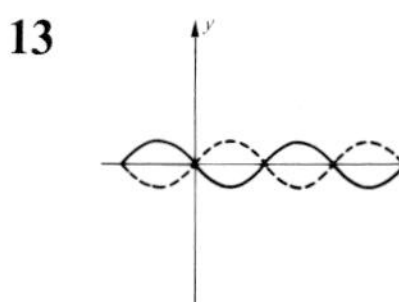

Exercises 5.6, page 243

1 The amplitudes and periods are:
(a) $4, 2\pi$ (b) $1, \pi/2$ (c) $\frac{1}{4}, 2\pi$ (d) $1, 8\pi$
(e) $2, 8\pi$ (f) $\frac{1}{2}, \pi/2$ (g) $4, 2\pi$ (h) $1, \pi/2$
3 The amplitudes and periods are:
(a) $3, 2\pi$ (b) $1, 2\pi/3$ (c) $\frac{1}{3}, 2\pi$ (d) $1, 6\pi$
(e) $2, 6\pi$ (f) $\frac{1}{3}, \pi$ (g) $3, 2\pi$ (h) $1, 2\pi/3$

5 **7** **9**

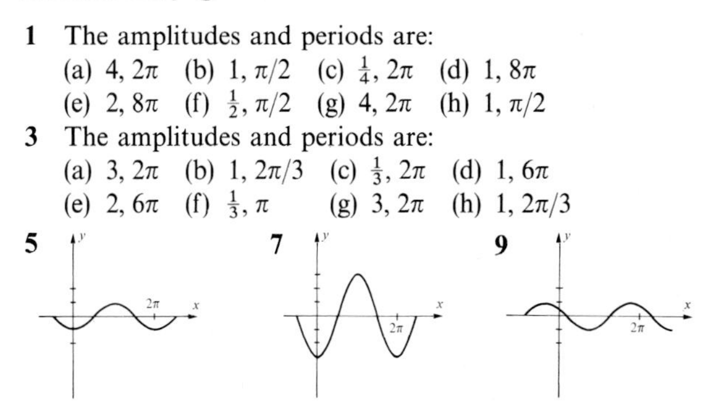

11 **13** **15**

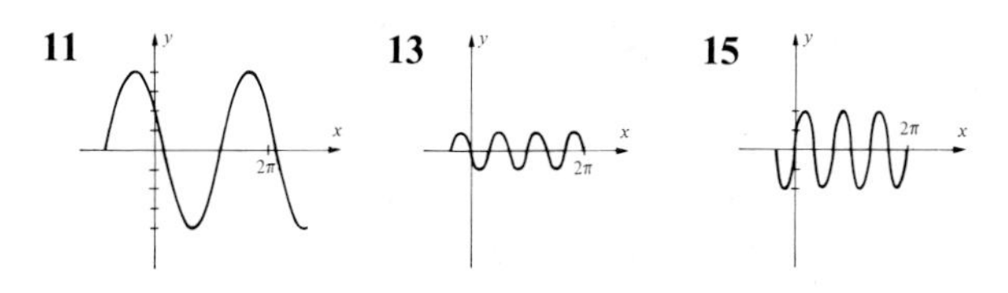

17

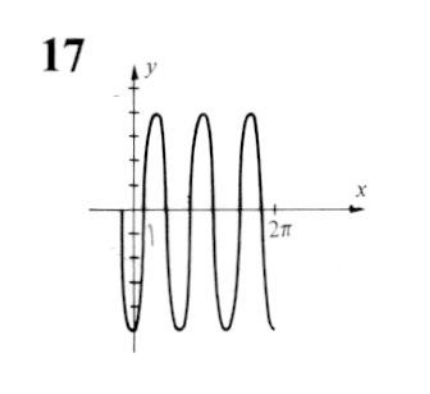

19 **21**

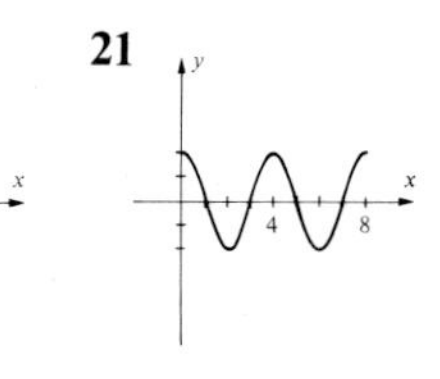

23

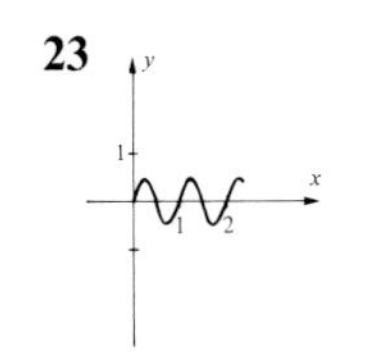

25 **27**

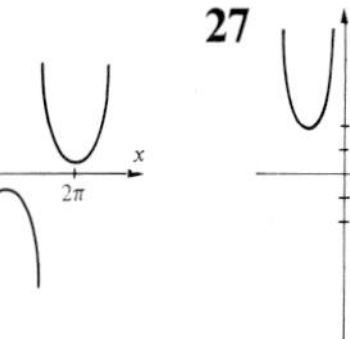

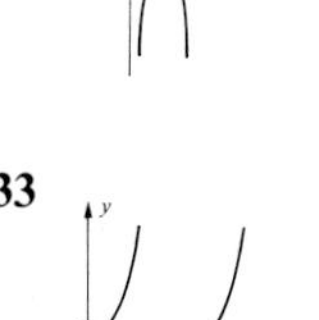

29 **31**

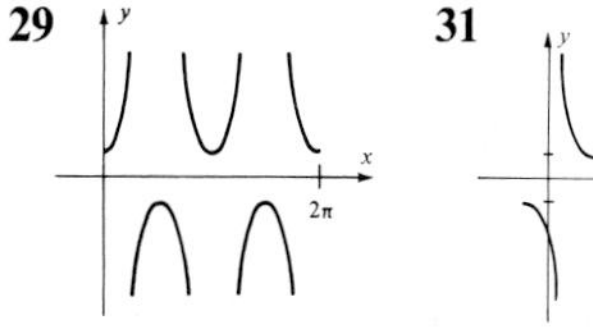

33

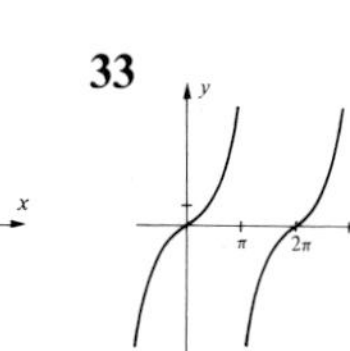

35 **37**

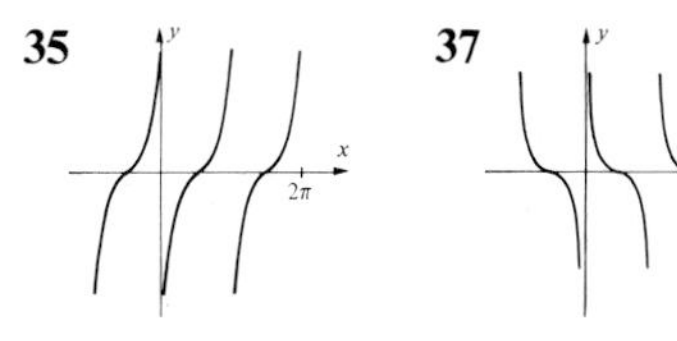

39

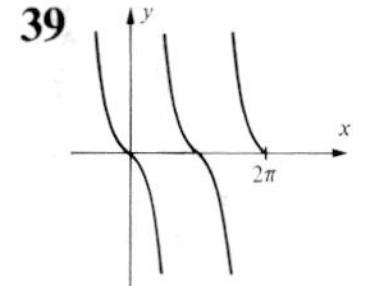

41 **43**

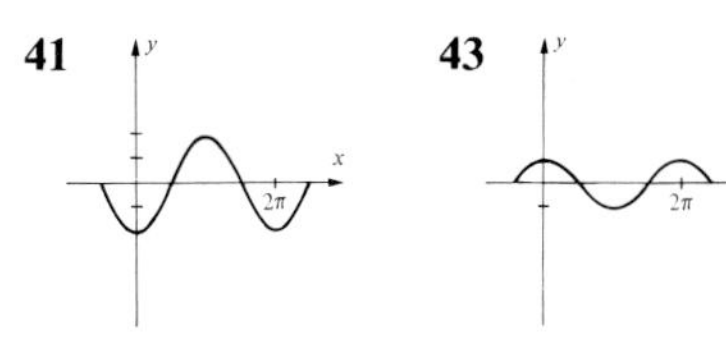

45 (a) Period, 2π; amplitude, 4 (b) $y = 4\sin(x - \pi)$
(c) π
47 (a) Period, π; amplitude, 3 (b) $y = 3\sin(2x - 3\pi/2)$
(c) $3\pi/4$
49 4π **51** $y = 8\sin 4\pi t$ during the systolic phase

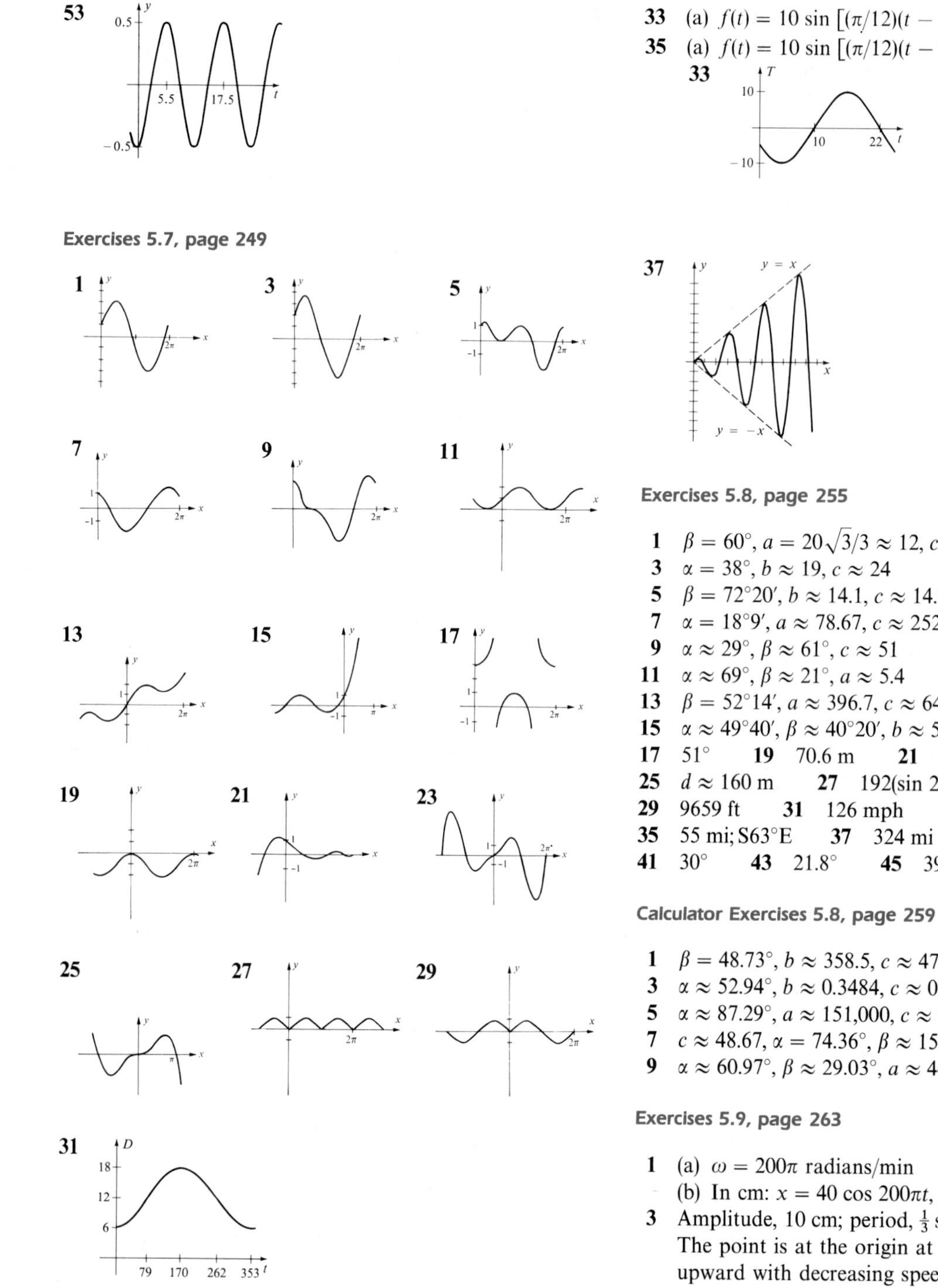

53

Exercises 5.7, page 249

1 3 5 7 9 11 13 15 17 19 21 23 25 27 29 31

33 (a) $f(t) = 10 \sin [(\pi/12)(t - 10)]$

35 (a) $f(t) = 10 \sin [(\pi/12)(t - 9)] + 20$

33 35

37

Exercises 5.8, page 255

1 $\beta = 60°$, $a = 20\sqrt{3}/3 \approx 12$, $c = 40\sqrt{3}/3 \approx 23$

3 $\alpha = 38°$, $b \approx 19$, $c \approx 24$

5 $\beta = 72°20'$, $b \approx 14.1$, $c \approx 14.8$

7 $\alpha = 18°9'$, $a \approx 78.67$, $c \approx 252.6$

9 $\alpha \approx 29°$, $\beta \approx 61°$, $c \approx 51$

11 $\alpha \approx 69°$, $\beta \approx 21°$, $a \approx 5.4$

13 $\beta = 52°14'$, $a \approx 396.7$, $c \approx 647.7$

15 $\alpha \approx 49°40'$, $\beta \approx 40°20'$, $b \approx 522$

17 51° **19** 70.6 m **21** 20.2 m **23** 29.7 km

25 $d \approx 160$ m **27** $192(\sin 22°30') \approx 73.5$ cm

29 9659 ft **31** 126 mph **33** 28,400 ft

35 55 mi; S63°E **37** 324 mi **39** (a) 58 ft (b) 27 ft

41 30° **43** 21.8° **45** 3944 mi

Calculator Exercises 5.8, page 259

1 $\beta = 48.73°$, $b \approx 358.5$, $c \approx 476.9$

3 $\alpha \approx 52.94°$, $b \approx 0.3484$, $c \approx 0.5781$

5 $\alpha \approx 87.29°$, $a \approx 151{,}000$, $c \approx 151{,}200$

7 $c \approx 48.67$, $\alpha = 74.36°$, $\beta \approx 15.64°$

9 $\alpha \approx 60.97°$, $\beta \approx 29.03°$, $a \approx 4437$

Exercises 5.9, page 263

1 (a) $\omega = 200\pi$ radians/min
(b) In cm: $x = 40 \cos 200\pi t$, $y = 40 \sin 200\pi t$

3 Amplitude, 10 cm; period, $\frac{1}{3}$ sec; frequency, 3 osc/sec. The point is at the origin at $t = 0$. It then moves upward with decreasing speed, reaching the point with

coordinate 10 at $t = \frac{1}{12}$. It then reverses direction and moves downward, gaining speed until it reaches the origin at $t = \frac{1}{6}$. It continues downward, with decreasing speed, reaching the point with coordinate -10 at $t = \frac{1}{4}$. It then reverses direction and moves upward with increasing speed, returning to the origin at $t = \frac{1}{3}$.

5 Amplitude, 4 cm; period, $\frac{4}{3}$ sec; frequency, $\frac{3}{4}$ osc/sec. The motion is similar to that in Exercise 3; however, the point starts 4 units above the origin and moves downward, reaching O at $t = \frac{1}{3}$, and the point with coordinate -4 at $t = \frac{2}{3}$. It then reverses direction and moves upward, reaching O at $t = 1$ and its initial point at $t = \frac{4}{3}$.

7 $d = 5 \cos (2\pi/3)t$ **9** Current lags emf by $\frac{1}{1440}$ sec.

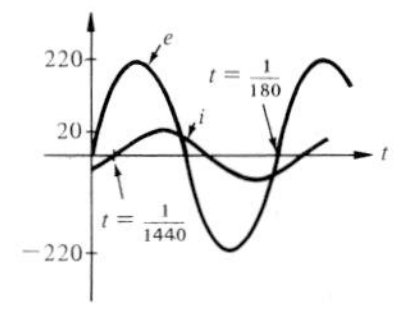

11 (a) $\pi/15$ radians/sec
(b) $h(t) = 60 - 50 \cos (\pi/15)t$
$= 60 + 50 \sin [(\pi/15)t - \pi/2]$

13 $y = (4.5) \sin [(\pi/6)(t - 10)] + 7.5$

Exercises 5.10, page 265

1 $11\pi/6, 9\pi/4, -5\pi/6, 4\pi/3, \pi/5$ **3** 0.1

5 $(-1, 0), (0, -1), (0, 1), (-\sqrt{2}/2, -\sqrt{2}/2), (1, 0), (\sqrt{3}/2, \frac{1}{2})$

7 (a) II (b) III (c) IV

9 Answers are in the order sin, cos, tan, csc, sec, cot:
(a) 1, 0, —, 1, —, 0
(b) $\sqrt{2}/2, -\sqrt{2}/2, -1, \sqrt{2}, -\sqrt{2}, -1$
(c) 0, 1, 0, —, 1, —
(d) $-1/2, \sqrt{3}/2, -\sqrt{3}/3, -2, 2\sqrt{3}/3, -\sqrt{3}$

11 65°, 42°50′, 8°; $\pi/4, \pi/6, \pi/8$

13 Answers are in the order sin, cos, tan, csc, sec, cot.
(a) $-\frac{4}{5}, \frac{3}{5}, -\frac{4}{3}, -\frac{5}{4}, \frac{5}{3}, -\frac{3}{4}$
(b) $2\sqrt{13}/13, -3\sqrt{13}/13, -\frac{2}{3}, \sqrt{13}/2, -\sqrt{13}/3, -\frac{3}{2}$
(c) $-1, 0,$ —, $-1,$ —, 0

15 **17** **19**

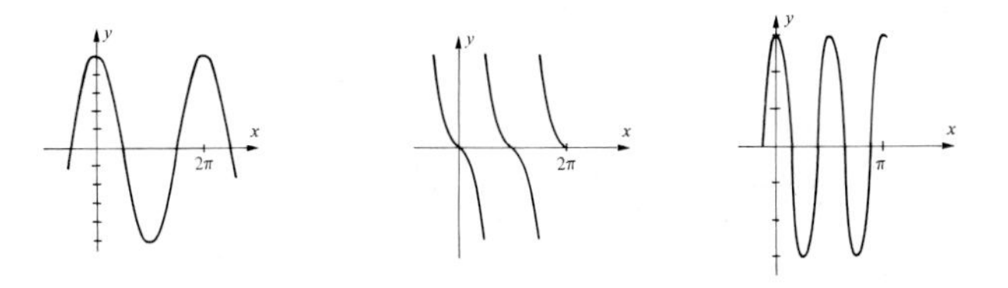

21 **23** **25** **27** **29**

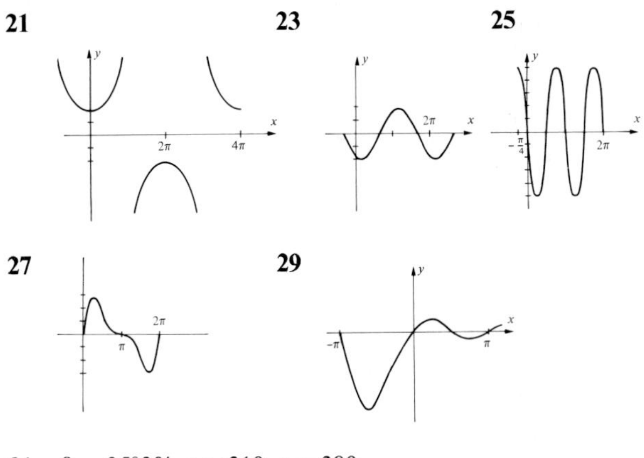

31 $\beta = 35°20', a \approx 310, c \approx 380$

33 $6\pi/5$ radians or 216°

35

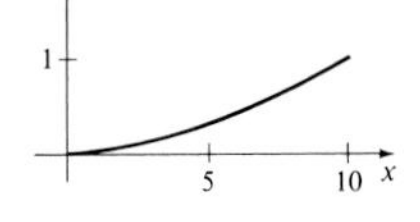

37 52° **39** 250 ft

41 Approximately 67,900,000 mi

CHAPTER 6

Exercises 6.1, page 273

101 $a^3 \cos^3 \theta$ **103** $a \tan \theta \sin \theta$
105 $a \sec \theta$ **107** $(1/a^2) \cos^2 \theta$
109 $a \tan \theta$ **111** $a^4 \sec^3 \theta \tan \theta$

Exercises 6.2, page 281

In the following, n denotes any integer.

1 $(2\pi/3) + 2\pi n, (4\pi/3) + 2\pi n$ **3** $(\pi/4) + (\pi/2)n$
5 $2\pi n, (3\pi/2) + 2\pi n$ **7** $(\pi/3) + \pi n, (2\pi/3) + \pi n$
9 $(4\pi/3) + 2\pi n, (5\pi/3) + 2\pi n$
11 $(\pi/6) + \pi n, (5\pi/6) + \pi n$
13 $(7\pi/6) + 2\pi n, (11\pi/6) + 2\pi n$
15 $(\pi/12) + n\pi, (5\pi/12) + n\pi$
17 $\pi/3, 2\pi/3, 4\pi/3, 5\pi/3$; 60°, 120°, 240°, 300°
19 $\pi/6, 5\pi/6, 3\pi/2$; 30°, 150°, 270°
21 $0, \pi, \pi/4, 3\pi/4, 5\pi/4, 7\pi/4$; 0°, 180°, 45°, 135°, 225°, 315°
23 $\pi/2, 3\pi/2, 2\pi/3, 4\pi/3$; 90°, 270°, 120°, 240°

25 No solutions **27** $11\pi/6, \pi/2$; $330°, 90°$
29 $0, \pi/2$; $0°, 90°$ **31** 0; $0°$
33 $7\pi/6, 11\pi/6, \pi/2, 3\pi/2$; $210°, 330°, 90°, 270°$
35 $3\pi/4, 7\pi/4$; $135°, 315°$ **37** $15°30', 164°30'$
39 $41°50', 138°10', 194°30', 345°30'$
41 6.145 min **43** $t \approx 0.29D$ and $t \approx 0.71D$
45 (a) 3.29 hr (b) 4 hr
47 $A = (-4\pi/3, \sqrt{3}/2 - 2\pi/3)$,
$B = (-2\pi/3, -\sqrt{3}/2 - \pi/3)$, $C = (2\pi/3, \sqrt{3}/2 + \pi/3)$,
$D = (4\pi/3, 2\pi/3 - \sqrt{3}/2)$

Exercises 6.3, page 290

1 (a) $\cos 43°23'$ (b) $\sin 16°48'$ (c) $\cot \pi/3$
(d) $\csc 72.62°$
3 (a) $\sin 3\pi/20$ (b) $\cos (2\pi - 1)/4$ (c) $\cot (\pi - 2)/2$
(d) $\sec (\frac{1}{2}\pi - 0.53)$
5 (a) $(\sqrt{2} + \sqrt{3})/2$ (b) $(\sqrt{6} - \sqrt{2})/4$
7 (a) $\sqrt{3} + 1$ (b) $-2 - \sqrt{3}$
9 (a) $(\sqrt{2} - 1)/2$ (b) $(\sqrt{6} + \sqrt{2})/4$
11 $\cos 25°$ **13** $\sin(-5°)$ **15** $-\sin 5$
17 $\frac{36}{85}, \frac{77}{85}$; I **19** $\frac{3}{5}, \frac{4}{5}, \frac{3}{4}, -\frac{117}{125}, \frac{44}{125}, -\frac{117}{44}$
41 $\sin u \cos v \cos w + \cos u \sin v \cos w + \cos u \cos v \sin w - \sin u \sin v \sin w$
49 $0, \pi/3, 2\pi/3, \pi, 4\pi/3, 5\pi/3$; $0°, 60°, 120°, 180°, 240°, 300°$
51 $f(x) = 2\cos(2x - \pi/6)$; $2, \pi, \pi/12$
53 $f(x) = 2\sqrt{2}\cos(3x + \pi/4)$; $2\sqrt{2}, 2\pi/3, -\pi/12$
55 (a) $y = \sqrt{13}\cos(t - C)$, where $\tan C = \frac{3}{2}$;
amplitude $= \sqrt{13}$, period $= 2\pi$
(b) $t = C + (\pi/2) + n\pi \approx 2.5536 + n\pi$, $n = 0, 1, 2, \ldots$
57 $y = 10\sqrt{41}\cos(60\pi t - C)$, where $\tan C = \frac{5}{4}$; or
$y \approx 10\sqrt{41}\cos(60\pi t - 0.8961)$
59 (b) $h \approx 4.69$ mi

Exercises 6.4, page 299

1 $\frac{24}{25}, -\frac{7}{25}, -\frac{24}{7}$ **3** $-4\sqrt{2}/9, -\frac{7}{9}, 4\sqrt{2}/7$
5 $\sqrt{10}/10, 3\sqrt{10}/10, \frac{1}{3}$
7 $-\sqrt{2 + \sqrt{2}}/2, \sqrt{2 - \sqrt{2}}/2, -\sqrt{2} - 1$
9 (a) $\sqrt{2 - \sqrt{2}}/2$ (b) $\sqrt{2 - \sqrt{3}}/2$ (c) $\sqrt{2} + 1$
29 $\frac{3}{8} + \frac{1}{2}\cos\theta + \frac{1}{8}\cos 2\theta$
31 $0, 2\pi/3, \pi, 4\pi/3$; $0°, 120°, 180°, 240°$
33 $\pi/3, \pi, 5\pi/3$; $60°, 180°, 300°$ **35** $0, \pi$; $0°, 180°$
37 $0, \pi/3, 5\pi/3$; $0°, 60°, 300°$
41 (a) 1.20, 5.09
(b) $P = (2\pi/3, -1.5)$, $Q = (\pi, -1)$, $R = (4\pi/3, -1.5)$
43 (a) $-3\pi/2, -\pi/2, \pi/2, 3\pi/2$
(b) $0, \pm\pi/4, \pm 3\pi/4, \pm\pi, \pm 5\pi/4, \pm 7\pi/4, \pm 2\pi$
45 (b) Yes, the branching point B is 25 miles from A.
47 $24.30°$ and $65.70°$

Exercises 6.5, page 305

1 $\sin 12\theta + \sin 6\theta$ **3** $\frac{1}{2}\cos 4t - \frac{1}{2}\cos 10t$
5 $\frac{1}{2}\cos 10u + \frac{1}{2}\cos 2u$ **7** $\frac{3}{2}\sin 3x + \frac{3}{2}\sin x$
9 $2\cos 2\theta \sin 4\theta$ **11** $-2\sin 4x \sin x$
13 $-2\cos 5t \sin 2t$ **15** $2\cos \frac{3}{2}x \cos \frac{1}{2}x$
25 $\frac{1}{2}\sin(a + b)x + \frac{1}{2}\sin(a - b)x$
27 $n\pi/4$ where n is any integer.
29 $n\pi/2$ where n is any integer.
37 $\pi/4, \pi/2, 3\pi/4, 5\pi/4, 3\pi/2, 7\pi/4$
39 $\pm\pi/4, \pm 3\pi/4, \pm\pi, \pm 5\pi/4, \pm 7\pi/4, \pm 2\pi$
41 $f(x) = \frac{1}{2}\sin[(n\pi/L)(x + kt)] + \frac{1}{2}\sin[(n\pi/L)(x - kt)]$

Exercises 6.6, page 315

1 (a) $\pi/6$ (b) $\pi/6$ **3** (a) 0 (b) 0
5 (a) $-\pi/2$ (b) π **7** (a) $\pi/3$ (b) $-\pi/3$
9 (a) 0.9745 (b) 0.974478
11 (a) -0.7069 (b) -0.7067952
13 (a) 1.1403 (b) 1.1402832
15 $\sqrt{3}/2$ **17** $\frac{3}{5}$ **19** $-\pi/4$ **21** 0 **23** $-\frac{77}{36}$
25 $-\frac{24}{25}$ **27** $x\sqrt{1 + x^2}/(1 + x^2)$ **29** $\sqrt{2 + 2x}/2$
39 $\cot^{-1} u = v$ if and only if $\cot v = u$.
41 $\csc^{-1} u = v$ if and only if $\csc v = u$.
43 **45** **47**
49 **51**

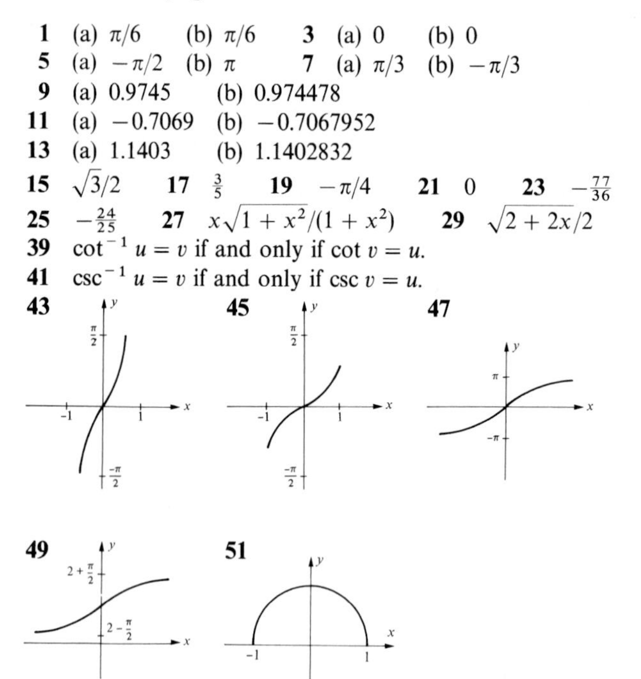

55 (a) $\arctan(-9 \pm \sqrt{57})/4$ (b) $-0.3478, -1.3336$

57 (a) $\arccos(\pm\sqrt{15}/5)$, $\arccos(\pm\sqrt{3}/3)$
(b) 0.6847, 2.4569, 0.9553, 2.1863

59 (a) $\arcsin(\pm\sqrt{30}/6)$ (b) ± 1.1503

Calculator Exercises 6.6, page 317

1 $\sin^{-1}(\sin 2)$ is between $-\pi/2$ and $\pi/2$ and 2 is not in $[-\pi/2, \pi/2]$. $\cos^{-1}(\cos 2)$ lies between 0 and π. Hence $\cos^{-1}(\cos 2) = 2$.

3 An error message will be avoided if $-\sin 1 \le x \le \sin 1$.

5 $\sin^{-1} x = \tan^{-1}(x/\sqrt{1 - x^2})$ for $-1 < x < 1$;
$\cos^{-1} x = \tan^{-1}(\sqrt{1 - x^2}/x)$ for $0 < x < 1$;
$\cot^{-1} x = \tan^{-1}(1/x)$ for $0 < x < \infty$

Exercises 6.7, page 317

17 $\pi/2, 3\pi/2, \pi/4, 3\pi/4, 5\pi/4, 7\pi/4$;
$90°, 270°, 45°, 135°, 225°, 315°$

19 $0, \pi$; $0°, 180°$

21 $0, \pi, 2\pi/3, 4\pi/3$; $0°, 180°, 120°, 240°$

23 $\pi/2, 7\pi/6, 11\pi/6$; $90°, 210°, 330°$

25 $\pi/6, 5\pi/6, \pi/3, 5\pi/3$; $30°, 150°, 60°, 300°$

27 $\pi/3, 5\pi/3$; $60°, 300°$

29 $\sqrt{2 - \sqrt{3}}/2$ or $(\sqrt{6} - \sqrt{2})/4$

31 $-\sqrt{2 - \sqrt{3}}/2$ or $(\sqrt{2} - \sqrt{6})/4$ **33** $\frac{84}{85}$

35 $-\frac{36}{77}$ **37** $\frac{240}{289}$ **39** $\frac{24}{7}$ **41** $\frac{1}{3}$

43 (a) $\frac{1}{2}\cos 3t - \frac{1}{2}\cos 11t$ (b) $\frac{1}{2}\cos\frac{1}{12}u + \frac{1}{2}\cos\frac{5}{12}u$
(c) $3\sin 8x - 3\sin 2x$

45 $5\pi/6$ **47** π **49** $\frac{1}{2}$ **51** $-\frac{7}{25}$ **53** $\pi/2$

55

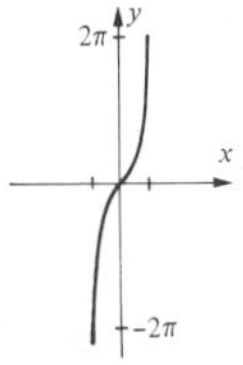

57 (b) $t = 0, \pm\pi/4b$ (c) $\frac{2}{3}\sqrt{2A}$

59 (a) $x = 2d(1 - \cos\theta)/\sin\theta$ (b) $d \le 1000$ ft

CHAPTER 7

Exercises 7.1, page 327

1 $\beta \approx 62°$, $b \approx 14.1$, $c \approx 15.6$

3 $\gamma \approx 100°10'$, $b \approx 55.1$, $c \approx 68.7$

5 $\alpha \approx 58°40'$, $a \approx 487$, $b \approx 442$

7 $\beta \approx 53°40'$, $\gamma \approx 61°10'$, $c \approx 20.6$

9 $\alpha \approx 77°30'$, $\beta \approx 49°10'$, $b \approx 108$; $\alpha \approx 102°30'$, $\beta = 24°10'$, $b \approx 58.7$

11 $\alpha \approx 20°30'$, $\gamma \approx 46°20'$, $a \approx 94.5$ **13** 219 yd

15 (a) 1.57 mi (b) 0.56 mi **17** 2.7 mi

19 3.7 mi from A and 5.4 mi from B **21** 628 m

23 350 ft **25** 18.7, 814

Calculator Exercises 7.1, page 330

1 $\alpha = 119.7°$, $a \approx 371.5$, $c \approx 243.5$

3 $\beta = 49.36°$, $a \approx 49.78$, $c \approx 23.39$

5 $\alpha \approx 25.993°$, $\gamma = 32.383°$, $a \approx 0.146$

Exercises 7.2, page 336

1 $a \approx 26$, $\beta \approx 41°$, $\gamma \approx 79°$

3 $b \approx 177$, $\alpha \approx 25°10'$, $\gamma \approx 4°50'$

5 $c \approx 2.75$, $\alpha \approx 21°10'$, $\beta \approx 43°40'$

7 $\alpha \approx 29°$, $\beta \approx 47°$, $\gamma = 104°$

9 $\alpha \approx 12°30'$, $\beta \approx 136°30'$, $\gamma = 31°$

11 196 ft **13** 24 mi **15** 39 mi

17 Approximately 2.3 mi

19 Approximately N55°31′E

21 63.7 ft from first and third base; 66.8 ft from second base

23 40,630 ft, or 7.7 mi

25 (a) 72°, 181.6 ft^2 (b) 37.6 ft

Answers to Exercises 27–33 are in square units.

27 260 **29** 1125 **31** 2.9 **33** 517

35 1.6 acres

Calculator Exercises 7.2, page 338

1 $a \approx 40.8$, $\beta \approx 26.9°$, $\gamma \approx 104.8°$

3 $c \approx 0.487$, $\beta \approx 5.11°$, $\alpha \approx 168.04°$

5 $\alpha \approx 157°16'$, $\beta \approx 7°49'$, $\gamma \approx 14°55'$

Exercises 7.3, page 343

1 5 **3** $\sqrt{85}$ **5** 8 **7** 1 **9** 0

The geometric representations in Exercises 11–19 are the following points:

11 $P(4, 2)$ **13** $P(3, -5)$ **15** $P(-3, 6)$

17 $P(-6, 4)$ **19** $P(0, 2)$

21 $\sqrt{2}(\cos 7\pi/4 + i \sin 7\pi/4)$

23 $8(\cos 5\pi/6 + i \sin 5\pi/6)$
25 $20(\cos 3\pi/2 + i \sin 3\pi/2)$
27 $10(\cos 4\pi/3 + i \sin 4\pi/3)$ **29** $7(\cos \pi + i \sin \pi)$
31 $\sqrt{5}(\cos \theta + i \sin \theta)$, where $\theta = \arctan \frac{1}{2}$.
33 $4\sqrt{2}(\cos 5\pi/4 + i \sin 5\pi/4)$
35 $6(\cos \pi/2 + i \sin \pi/2)$ **37** $2(\cos 7\pi/6 + i \sin 7\pi/6)$
39 $12(\cos 0 + i \sin 0)$ **41** $-2, i$
43 $10\sqrt{3} - 10i, (-2\sqrt{3}/5) + \frac{2}{5}i$
45 $40, \frac{5}{2}$ **47** $8 - 4i, \frac{8}{5} + \frac{4}{5}i$
49 $z_1z_2z_3 = r_1r_2r_3[\cos(\theta_1 + \theta_2 + \theta_3) + i \sin(\theta_1 + \theta_2 + \theta_3)]$, $z_1z_2 \cdots z_n = (r_1r_2 \cdots r_n)[\cos(\theta_1 + \theta_2 + \cdots + \theta_n) + i \sin(\theta_1 + \theta_2 + \cdots + \theta_n)]$

Exercises 7.4, page 348

1 $-972 - 972i$ **3** $-32i$ **5** -8
7 $-(\sqrt{2}/2) - (\sqrt{2}/2)i$ **9** $-\frac{1}{2} - (\sqrt{3}/2)i$
11 $-64\sqrt{3} - 64i$
13 $(\sqrt{6}/2) + (\sqrt{2}/2)i, -(\sqrt{6}/2) - (\sqrt{2}/2)i$
15 $(\sqrt[4]{2}/2) + (\sqrt[4]{18}/2)i, -(\sqrt[4]{18}/2) + (\sqrt[4]{4}/2)i$, $(\sqrt[4]{18}/2) - (\sqrt[4]{2}/2)i, -(\sqrt[4]{2}/2) - (\sqrt[4]{18}/2)i$
17 $3i, -(3\sqrt{3}/2) - \frac{3}{2}i, (3\sqrt{3}/2) - \frac{3}{2}i$
19 $\pm 1, \frac{1}{2} \pm (\sqrt{3}/2)i, -\frac{1}{2} \pm (\sqrt{3}/2)i$
21 $\sqrt[10]{2}(\cos \theta + i \sin \theta)$, where $\theta = 9°, 81°, 153°, 225°, 279°$
23 $\pm 2, \pm 2i$ **25** $\pm 2i, \pm\sqrt{3} + i, \pm\sqrt{3} - i$
27 $2i, -\sqrt{3} - i, \sqrt{3} - i$
29 $3 \cos \theta + (3 \sin \theta)i$, where $\theta = 0, 2\pi/5, 4\pi/5, 6\pi/5, 8\pi/5$

Exercises 7.5, page 356

1 $\langle 3, 1\rangle, \langle 1, -7\rangle, \langle 13, 8\rangle, \langle 3, -32\rangle$
3 $\langle -15, 6\rangle, \langle 1, -2\rangle, \langle -68, 28\rangle, \langle 12, -12\rangle$
5 $4\mathbf{i} - 3\mathbf{j}, -2\mathbf{i} + 7\mathbf{j}, 19\mathbf{i} - 17\mathbf{j}, -11\mathbf{i} + 33\mathbf{j}$
7 $-2\mathbf{i} - 5\mathbf{j}, -6\mathbf{i} + 7\mathbf{j}, -6\mathbf{i} - 26\mathbf{j}, -26\mathbf{i} + 34\mathbf{j}$
9 $-3\mathbf{i} + 2\mathbf{j}, 3\mathbf{i} + 2\mathbf{j}, -15\mathbf{i} + 8\mathbf{j}, 15\mathbf{i} + 8\mathbf{j}$

Answers for Exercises 11 and 13 are terminal points of the vectors.

11 $(3, 2), (-1, 5), (2, 7), (6, 4), (3, -15)$
13 $(-4, 6), (-2, 3), (-6, 9), (-8, 12), (6, -9)$
15 $\frac{1}{2}\mathbf{a}$ **17** $\mathbf{f}$ **19** $-\frac{1}{2}\mathbf{e}$
37 $3\sqrt{2}, 7\pi/4$ **39** $5, \pi$
41 $\sqrt{41}, \arccos(-4\sqrt{41}/41)$ **43** $18, 3\pi/2$
45 89 kg, S66°W **47** 5.8 lb, 129°
49 $\langle 40.96, 28.68\rangle$ **51** $\langle -61.80, 190.21\rangle$
53 (a) $\mathbf{F} = \langle 7, 2\rangle$ (b) $\mathbf{G} = -\mathbf{F} = \langle -7, -2\rangle$
55 (a) $\mathbf{F} \approx \langle -5.86, 1.13\rangle$ (b) $\mathbf{G} = -\mathbf{F} \approx \langle 5.86, -1.13\rangle$
57 23.58° **59** 56°, 232 mph
61 420 mph, 244° **63** N22°W

Exercises 7.6, page 359

1 $a = 50\sqrt{6}, c = 50(1 + \sqrt{3}), \gamma = 75°$
3 $a = \sqrt{43}, \beta = \arccos(4/\sqrt{43}), \gamma = \arccos(5/2\sqrt{43})$
5 $\alpha = 38°, a \approx 8, b \approx 13$
7 $b \approx 10.1, \gamma \approx 41°, \alpha \approx 24°$ **9** 289.8
11 $10\sqrt{2}(\cos 3\pi/4 + i \sin 3\pi/4)$ **13** $17(\cos \pi + i \sin \pi)$
15 $10(\cos 7\pi/6 + i \sin 7\pi/6)$ **17** $-512i$
19 $-972 + 972i$ **21** $-3, \frac{3}{2} \pm (3\sqrt{3}/2)i$
23 Terminal points are $(-2, -3), (-6, 13), (-8, 10), (-1, 4)$.
25 Circle with center (a_1, a_2) and radius C
27 $\langle 14 \cos 40°, -14 \sin 40°\rangle$ **29** (b) 2 mi
31 (a) 448.52 ft ≈ 449 ft (b) 434.45 ft ≈ 434 ft
33 (a) 102.6° (b) 37.4 ft (c) 290.3 ft

CHAPTER 8

Exercises 8.1, page 369

1 $(3, 5), (-1, -3)$ **3** $(1, 0), (-3, 2)$
5 $(0, 0), (\frac{1}{8}, \frac{1}{128})$ **7** $(3, -2)$ **9** No solutions
11 $(-4, 3), (5, 0)$ **13** $(-2, 2)$
15 $((-6 - \sqrt{86})/10, (2 - 3\sqrt{86})/10)$, $((-6 + \sqrt{86})/10, (2 + 3\sqrt{86})/10)$
17 $(-4, 0), (\frac{12}{5}, \frac{16}{5})$ **19** $(0, 1), (4, -3)$
21 $(\pm 2, 5), (\pm\sqrt{5}, 4)$ **23** $(\sqrt{2}, \pm 2\sqrt{3}), (-\sqrt{2}, \pm 2\sqrt{3})$
25 $(2\sqrt{2}, \pm 2), (-2\sqrt{2}, \pm 2)$ **27** $x = 3, y = -1, z = 2$
29 $(1, -1, 2), (-1, 3, -2)$ **31** 8 in., 12 in.
33 There is a single solution that occurs at $x \approx 0.6412$.

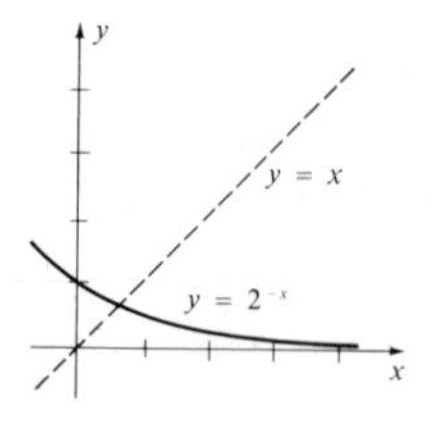

35 $r = 2$ in., $h = 50/\pi$ in.
37 (a) $a = 120{,}000$ and $b = 40{,}000$ (b) 77,143

39 (0, 0), (0, 100), and (50, 0). The fourth solution (−100, 150) is not meaningful.

Exercises 8.2, page 376

1 (4, −2) **3** (8, 0) **5** $(-1, \frac{3}{2})$
7 $(\frac{76}{53}, \frac{28}{53})$ **9** $(\frac{51}{13}, \frac{96}{13})$ **11** $(\frac{8}{7}, -3\sqrt{6/7})$
13 No solution
15 All ordered pairs (m, n) such that $3m - 4n = 2$
17 (0, 0) **19** $(-\frac{22}{7}, -\frac{11}{5})$
21 313 students and 137 nonstudents
23 $x = (30/\pi) - 4 \approx 5.55$ and $y = 12 - (30/\pi) \approx 2.45$
25 $x = 20/\pi$ ft and $y = 10$ ft
27 \$6,650 at 6.5% and \$3,325 at 8%
29 40 g of 35% alloy, 60 g of 60% alloy
31 540 mph, 60 mph **33** $v_0 = 10, a = 3$
35 20 sofas and 30 recliners
37 (a) $(c, \frac{4}{5}c)$ for an arbitrary $c > 0$ (b) \$16 per hour

Exercises 8.3, page 389

1 (2, 3, −1) **3** (−2, 4, 5)
5 No solution **7** $(\frac{2}{3}, \frac{31}{21}, \frac{1}{21})$

There are other forms for the answers in Exercises 9–15.

9 $(2c, -c, c)$ where c is any real number
11 $(0, -c, c)$ where c is any real number
13 $((12 - 9c)/7, (8c - 13)/14, c)$, where c is any real number
15 $((7c + 5)/10, (19c - 15)/10, c)$ where c is any real number
17 (1, 3, −1, 2) **19** $x = 2, y = -1, z = 3, s = 4, t = 1$
21 $(\frac{1}{11}, \frac{31}{11}, \frac{3}{11})$ **23** (−2, −3) **25** No solution
27 17 liters of 10%, 11 liters of 30%, 22 liters of 50%
29 4 hr for A, 2 hr for B, 5 hr for C
31 380 lb of G_1, 60 lb of G_2, 160 lb of G_3
33 (a) $I_1 = 0, I_2 = 2, I_3 = 2$ (b) $I_1 = \frac{3}{4}, I_2 = 3, I_3 = \frac{9}{4}$
35 $\frac{3}{8}$ lb Columbian, $\frac{1}{8}$ lb Brazilian, $\frac{1}{2}$ lb Kenyan
37 $f(x) = -\frac{1}{2}x^2 + x + \frac{5}{2}$
39 $x^2 + y^2 - x + 3y - 6 = 0$

Exercises 8.4, page 396

1 $3/(x - 2) + 5/(x + 3)$ **3** $5/(x - 6) - 4/(x + 2)$
5 $2/(x - 1) + 3/(x + 2) - 1/(x - 3)$
7 $(3/x) + 2/(x - 5) - 1/(x + 1)$
9 $2/(x - 1) + 5/(x - 1)^2$
11 $(-7/x) + (5/x^2) + 40/(3x - 5)$
13 $-\frac{23}{25}/(2x - 1) + \frac{24}{25}/(x + 2) + \frac{2}{5}/(x + 2)^2$
15 $(5/x) - 2/(x + 1) + 3/(x + 1)^3$
17 $-2/(x - 1) + (3x + 4)/(x^2 + 1)$
19 $(4/x) + (5x - 3)/(x^2 + 2)$
21 $(4x - 1)/(x^2 + 1) + 3/(x^2 + 1)^2$
23 $2x + 3x/(x^2 + 1) + 1/(x - 1)$
25 $(2x + 3) + 2/(x - 1) - 3/(2x + 1)$

Exercises 8.5, page 405

1 $\begin{bmatrix} 9 & -1 \\ -2 & 5 \end{bmatrix}, \begin{bmatrix} 1 & -3 \\ 4 & 1 \end{bmatrix}, \begin{bmatrix} 10 & -4 \\ 2 & 6 \end{bmatrix}, \begin{bmatrix} -12 & -3 \\ 9 & -6 \end{bmatrix}$

3 $\begin{bmatrix} 9 & 0 \\ 1 & 5 \\ 3 & 4 \end{bmatrix}, \begin{bmatrix} 3 & -2 \\ 3 & -5 \\ -9 & 4 \end{bmatrix}, \begin{bmatrix} 12 & -2 \\ 4 & 0 \\ -6 & 8 \end{bmatrix}, \begin{bmatrix} -9 & -3 \\ 3 & -15 \\ -18 & 0 \end{bmatrix}$

5 $[11 \ \ -3 \ \ -3], [-3 \ \ -3 \ \ 7], [8 \ \ -6 \ \ 4], [-21 \ \ 0 \ \ 15]$

7 $\begin{bmatrix} -3 & 4 & 1 & 6 \\ 3 & 2 & 7 & -7 \end{bmatrix}, \begin{bmatrix} 3 & 4 & -1 & 0 \\ -1 & 2 & -7 & -3 \end{bmatrix}, \begin{bmatrix} 0 & 8 & 0 & 6 \\ 2 & 4 & 0 & -10 \end{bmatrix}, \begin{bmatrix} 9 & 0 & -3 & -9 \\ -6 & 0 & -21 & 6 \end{bmatrix}$

9 $\begin{bmatrix} 16 & 38 \\ 11 & -34 \end{bmatrix}, \begin{bmatrix} 4 & 38 \\ 23 & -22 \end{bmatrix}$

11 $\begin{bmatrix} 3 & -14 & -3 \\ 16 & 2 & -2 \\ -7 & -29 & 9 \end{bmatrix}, \begin{bmatrix} 3 & -20 & -11 \\ 2 & 10 & -4 \\ 15 & -13 & 1 \end{bmatrix}$

13 $\begin{bmatrix} 4 & 8 \\ -18 & 11 \end{bmatrix}, \begin{bmatrix} 3 & -4 & 4 \\ -5 & 2 & 2 \\ -51 & 26 & 10 \end{bmatrix}$

15 $\begin{bmatrix} 1 & 2 & 3 \\ 4 & 5 & 6 \\ 7 & 8 & 9 \end{bmatrix}, \begin{bmatrix} 1 & 2 & 3 \\ 4 & 5 & 6 \\ 7 & 8 & 9 \end{bmatrix}$

17 $[15], \begin{bmatrix} -3 & 7 & 2 \\ -12 & 28 & 8 \\ 15 & -35 & -10 \end{bmatrix}$ **19** $\begin{bmatrix} 4 \\ 12 \\ -1 \end{bmatrix}$

21 $\begin{bmatrix} 18 & 0 & -2 \\ -40 & 10 & -10 \end{bmatrix}$ **31** $\frac{1}{10}\begin{bmatrix} 3 & 4 \\ -1 & 2 \end{bmatrix}$

33 Does not exist. **35** $\frac{1}{8}\begin{bmatrix} 2 & 1 & 0 \\ -2 & 3 & 0 \\ 0 & 0 & 2 \end{bmatrix}$

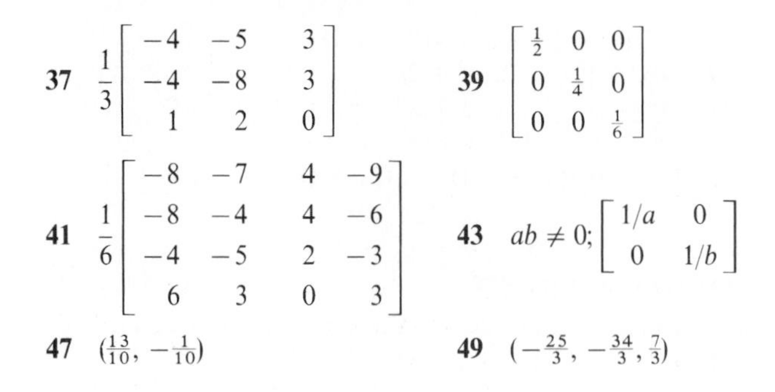

37 $\frac{1}{3}\begin{bmatrix} -4 & -5 & 3 \\ -4 & -8 & 3 \\ 1 & 2 & 0 \end{bmatrix}$ **39** $\begin{bmatrix} \frac{1}{2} & 0 & 0 \\ 0 & \frac{1}{4} & 0 \\ 0 & 0 & \frac{1}{6} \end{bmatrix}$

41 $\frac{1}{6}\begin{bmatrix} -8 & -7 & 4 & -9 \\ -8 & -4 & 4 & -6 \\ -4 & -5 & 2 & -3 \\ 6 & 3 & 0 & 3 \end{bmatrix}$ **43** $ab \neq 0; \begin{bmatrix} 1/a & 0 \\ 0 & 1/b \end{bmatrix}$

47 $(\frac{13}{10}, -\frac{1}{10})$ **49** $(-\frac{25}{3}, -\frac{34}{3}, \frac{7}{3})$

Exercises 8.6, page 412

1 $M_{11} = -14, M_{21} = 7, M_{31} = 11, M_{12} = 10,$
$M_{22} = -5, M_{32} = 4, M_{13} = 15, M_{23} = 34, M_{33} = 6;$
$A_{11} = -14, A_{21} = -7, A_{31} = 11, A_{12} = -10,$
$A_{22} = -5, A_{32} = -4, A_{13} = 15, A_{23} = -34, A_{33} = 6$
3 $M_{11} = 0, M_{12} = 5, M_{21} = -1, M_{22} = 7; A_{11} = 0,$
$A_{12} = -5, A_{21} = 1, A_{22} = 7$
5 -83 **7** 5 **9** 2 **11** 0
13 -125 **15** 48 **17** -216 **19** $abcd$
31 (a) $x^2 - 3x - 4$ (b) $-1, 4$
33 (a) $x^2 + x - 2$ (b) $1, -2$
35 (a) $-x^3 - 2x^2 + x + 2$ (b) $1, -1, -2$
37 (a) $-x^3 + 4x^2 + 4x - 16$ (b) $4, 2, -2$

39 $-31i - 20j + 7k$ **41** $-6i - 8j + 18k$

Exercises 8.7, page 417

1 R_{23} **3** $(-1)R_1 + R_3$
5 2 is a common factor of rows 1 and 3.
7 Two rows are identical. **9** $(-1)R_2$
11 Every number in column 2 is 0. **13** $2C_1 + C_3$
15 -10 **17** -142 **19** -183
21 44 **23** 359

Exercises 8.9, page 427

1 **3** **5**

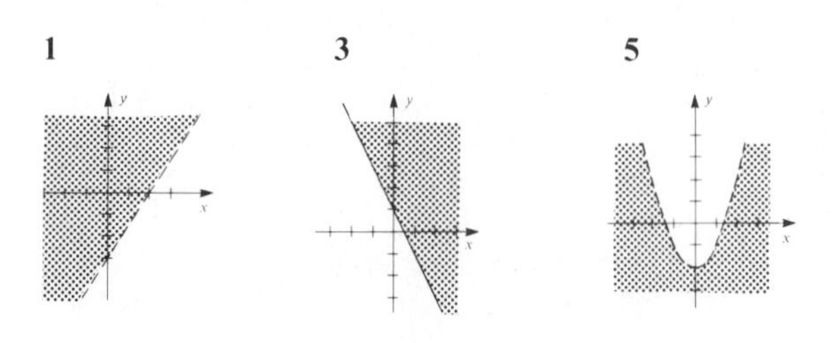

7 **9** **11**

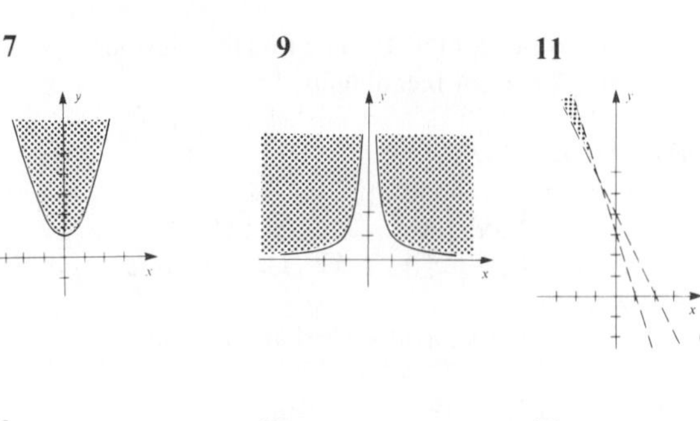

13 **15** **17**

19 **21** **23**

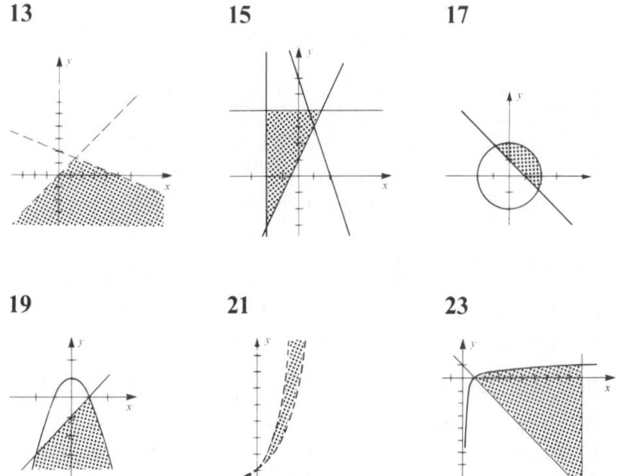

25 If x and y denote the number of brand A and brand B, respectively, then $x \geq 20$, $y \geq 10$, $x \geq 2y$, and $x + y \leq 100$. The graph is the region bounded by the triangle with vertices (20, 10), (90, 10), $(\frac{200}{3}, \frac{100}{3})$.
27 If x and y denote the amounts in the first and second accounts, respectively, then $x \geq 2000$, $y \geq 2000$, $y \geq 3x$, and $x + y \leq 15000$. The graph is the region bounded by the triangle with vertices (2000, 6000), (2000, 13000), and (3750, 11250).
29 $x + y \leq 9$, $y \geq x$, $x \geq 1$, where x = height of the cone and y = height of the cylinder.

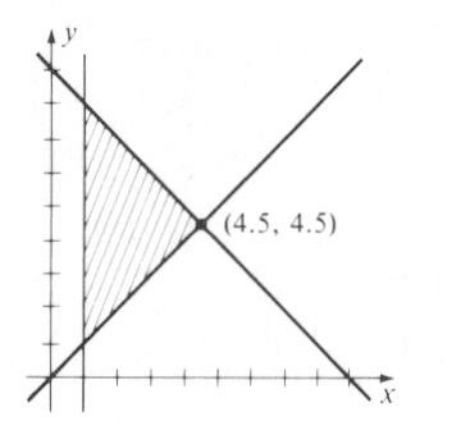

31 If the plant is located at (x, y), then
$3600 \le x^2 + y^2 \le 10{,}000$ and
$3600 \le (x - 100)^2 + y^2 \le 10{,}000$ and $y \ge 0$
The graph is the region in the first quadrant that lies between the two concentric circles with center (0, 0) and radii 60 and 100, and also between the two concentric circles with center (100, 0) and radii 60 and 100.

Exercises 8.10, page 433

1 50 Double Fault and 30 Set Point
3 3.51 lb of S and 1 lb of T
5 Send 25 from W_1 to A and 0 from W_1 to B. Send 10 from W_2 to A and 60 from W_2 to B.
7 0 acres of alfalfa and 80 acres of corn
9 Minimum cost: 16 oz X, 4 oz Y, 0 oz Z; maximum cost: 0 oz X, 8 oz Y, 12 oz Z
11 2 vans and 4 small buses
13 Two answers lead to a maximum of 12,000 lb: Either stock 4000 bass or stock 3000 trout and 2000 bass.
15 60 small units and 20 large units

Exercises 8.11, page 435

1 $(\frac{19}{23}, -\frac{18}{23})$ **3** $(-3, 5), (1, -3)$
5 $(2\sqrt{3}, \pm\sqrt{2}), (-2\sqrt{3}, \pm\sqrt{2})$ **7** $(\frac{14}{17}, \frac{14}{27})$
9 $(\frac{6}{11}, -\frac{7}{11}, 1)$
11 $(-2c, -3c, c)$ where c is any real number.
13 $(5c - 1, (-19c + 5)/2, c)$ where c is any real number
15 $(-1, \frac{1}{2}, \frac{1}{3})$
17

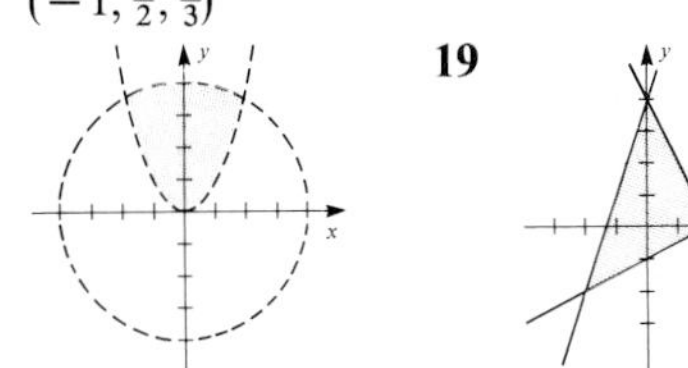

19

21 -6 **23** 48 **25** -84 **27** 120 **29** 0

31 $a_{11}a_{22}a_{33}\cdots a_{nn}$ **33** $\left(-\frac{1}{2}\right)\begin{bmatrix} 2 & 4 \\ 3 & 5 \end{bmatrix}$

35 $\frac{1}{7}\begin{bmatrix} 2 & 1 & 0 & 0 \\ -1 & 3 & 0 & 0 \\ 0 & 0 & 3 & 2 \\ 0 & 0 & -5 & -1 \end{bmatrix}$ **37** $\begin{bmatrix} 4 & -5 & 6 \\ 4 & -11 & 5 \end{bmatrix}$

39 $\begin{bmatrix} 0 & 4 & -6 \\ 16 & 22 & 1 \\ 12 & 11 & 9 \end{bmatrix}$ **41** $\begin{bmatrix} -12 & 4 & -11 \\ 6 & -11 & 5 \end{bmatrix}$

43 $\begin{bmatrix} a & 3a \\ 2b & 4b \end{bmatrix}$ **45** $\begin{bmatrix} 5 & 9 \\ 13 & 19 \end{bmatrix}$ **49** $-1 \pm 2\sqrt{3}$
53 $8/(x - 1) - 3/(x + 5) - 1/(x + 3)$
55 The field is $40\sqrt{5}$ ft by $20\sqrt{5}$ ft.
57 Tax = \$18,750; Bonus = \$3125
59 Pipe A: 30 ft^3 per hr; pipe B: 20 ft^3 per hr; pipe C: 50 ft^3 per hr
61 80 mowers and 30 edgers

CHAPTER 9

Exercises 9.2, page 453

1 720 **3** 12 **5** 336 **7** 35
9 56 **11** 126
13 $a^6 + 6a^5b + 15a^4b^2 + 20a^3b^3 + 15a^2b^4 + 6ab^5 + b^6$
15 $a^8 - 8a^7b + 28a^6b^2 - 56a^5b^3 + 70a^4b^4 - 56a^3b^5 + 28a^2b^6 - 8ab^7 + b^8$
17 $81x^4 - 540x^3y + 1350x^2y^2 - 1500xy^3 + 625y^4$
19 $u^{10} + 20u^8v + 160u^6v^2 + 640u^4v^3 + 1280u^2v^4 + 1024v^5$
21 $r^{-12} - 12r^{-9} + 60r^{-6} - 160r^{-3} + 240 - 192r^3 + 64r^6$
23 $1 + 10x + 45x^2 + 120x^3 + 210x^4 + 252x^5 + 210x^6 + 120x^7 + 45x^8 + 10x^9 + x^{10}$
25 $(3^{25})c^{10} + 25(3^{24})c^{52/5} + 300(3^{23})c^{54/5} + 2300(3^{22})c^{56/5}$
27 $60(3^{14})b^{13} - (3^{15})b^{15}$ **29** $30{,}618a^{10}b^2$
31 $840u^4v^6$ **33** $924x^2y^2$ **35** $-61{,}236$
37 $24x^2y^6$ **39** $210c^3d^2$
41 $1 + 0.2 + 0.018 + 0.00096 = 1.21896$

Exercises 9.3, page 461

1 $9, 6, 3, 0, -3; -12$ **3** $\frac{1}{2}, \frac{4}{5}, \frac{7}{10}, \frac{10}{17}, \frac{13}{26}; \frac{22}{65}$
5 9, 9, 9, 9, 9; 9
7 1.9, 2.01, 1.999, 2.0001, 1.99999; 2.00000001
9 $4, -\frac{9}{4}, \frac{5}{3}, -\frac{11}{8}, \frac{6}{5}; -\frac{15}{16}$ **11** 2, 0, 2, 0, 2; 0
13 $\frac{2}{3}, \frac{2}{3}, \frac{8}{11}, \frac{8}{9}, \frac{32}{27}; \frac{128}{33}$ **15** 1, 2, 3, 4, 5; 8
17 $2, 1, -2, -11, -38$ **19** $-3, 3^2, 3^4, 3^8, 3^{16}$
21 5, 5, 10, 30, 120 **23** $2, 2, 4, 4^3, 4^{12}$
25 $a_n = 2n + \frac{1}{24}(n - 1)(n - 2)(n - 3)(n - 4)(a - 10)$
(There are many other answers.)
27 -5 **29** 10 **31** 25

33 $-\frac{17}{15}$ **35** 61 **37** 10,000
39 $(n^3 + 6n^2 + 20n)/3$ **41** $(4n^3 - 12n^2 + 11n)/3$
43 $\sum_{k=1}^{5} (4k - 3)$ **45** $\sum_{k=1}^{4} k/(3k - 1)$
47 $1 + \sum_{k=1}^{n} (-1)^k x^{2k}/(2k)$ **49** $\sum_{k=1}^{7} (-1)^{k-1}/k$
51 $\sum_{n=1}^{99} 1/n(n + 1)$

Calculator Exercises 9.3, page 462

1 The terms of the sequence approach 1.
3 3.142546543, 3.141592653, 3.141592654, 3.141592654, 3.141592654; when $x_1 = 6$, the terms of the sequence approach 2π.
5 0.4, 0.7, 1, 1.6, 2.8
7 (a) $y_n = K$ for all n
(b) 400, 560, 425.6, 552.3, 436.8, 547.2, 443.9, 543.5, 448.9, 540.7; The terms appear to be oscillating about 500.

Exercises 9.4, page 466

1 18, 38, $4n - 2$ **3** 1.8, 0.3, $3.3 - 0.3n$
5 5.4, 20.9, $(3.1)n - (10.1)$ **7** $\ln 3^5, \ln 3^{10}, \ln 3^n$
9 -8.5 **11** -9.8 **13** -105 **15** 30
17 530 **19** $\frac{423}{2}$ **21** 60; 12,780 **23** 24
25 $\frac{10}{3}, \frac{14}{3}, 6, \frac{22}{3}, \frac{26}{3}$ **27** 255
29 23.25, 22.5, 21.75, 21, 20.25, 19.5, 18.75
31 \$1200 **33** Four sides

Exercises 9.5, page 473

1 $\frac{1}{2}, \frac{1}{16}; 8(\frac{1}{2})^{n-1} = 2^{4-n}$
3 $0.03, -0.00003; 300(-0.1)^{n-1}$ **5** $3125, 5^8; 5^n$
7 $\frac{81}{4}, -3^7/2^5; 4(-\frac{3}{2})^{n-1}$ **9** $x^8, -x^{14}; (-1)^{n-1}x^{2n-2}$
11 $2^{4x+1}, 2^{7x+1}; 2^{nx-x+1}$ **13** $\frac{243}{8}$
15 $a_1 = \frac{1}{81}, S_5 = \frac{211}{1296}$ **17** $-\frac{3}{2}(1 - 3^{10}) = 88{,}572$
19 $-\frac{1}{3}(1 - 2^{-10})$ **21** $\frac{25}{256}\%$
23 $10{,}000(\frac{6}{5})^t, 10{,}000(\frac{6}{5})^{10}$ **25** $\frac{2}{3}$ **27** $\frac{50}{33}$
29 Sum does not exist ($|r| = \sqrt{2} > 1$). **31** $\frac{23}{99}$
33 $\frac{2393}{990}$ **35** $\frac{5141}{999}$ **37** $\frac{16123}{9999}$ **39** 30 m
41 \$3,000,000 **43** (b) 375 mg
45 (a) $a_{k+1} = \frac{1}{4}\sqrt{10}a_k$
(b) $a_n = (\frac{1}{4}\sqrt{10})^{n-1}a_1, A_n = (\frac{5}{8})^{n-1}A_1, P_n = (\frac{1}{4}\sqrt{10})^{n-1}P_1$
(c) $[16/(4 - \sqrt{10})]a_1$

Exercises 9.6, page 475

7 $x^{12} - 18x^{10}y + 135x^8y^2 - 540x^6y^3 + 1215x^4y^4 - 1458x^2y^5 + 729y^6$
9 $-\frac{63}{16}b^{12}c^{10}$ **11** $5, -2, -1, -\frac{20}{29}, -\frac{7}{19}$
13 $2, \frac{1}{2}, \frac{5}{4}, \frac{7}{8}, \frac{65}{64}$ **15** $10, \frac{11}{10}, \frac{21}{11}, \frac{32}{21}, \frac{53}{32}$
17 $9, 3, \sqrt{3}, \sqrt[4]{3}, \sqrt[8]{3}$ **19** 75 **21** 1000
23 $\sum_{k=1}^{5} 3k$ **25** $\sum_{k=0}^{4} (-1)^k 5(20 - k)$
27 $-5 - 8\sqrt{3}, -5 - 35\sqrt{3}$ **29** $-31, 50$ **31** 64
33 $4\sqrt{2}$ **35** 570 **37** 2041 **39** $\frac{5}{7}$
41 (b) The other four pieces have lengths $1\frac{1}{4}$ ft, 2 ft, $2\frac{3}{4}$ ft, and $3\frac{1}{2}$ ft.

CHAPTER 10

Exercises 10.2, page 485

1 $V(0, 0); F(0, -3); y = 3$

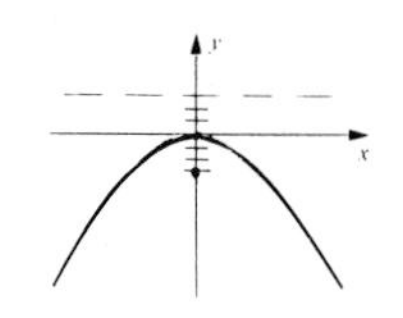

3 $V(0, 0); F(-\frac{3}{8}, 0); x = \frac{3}{8}$

5 $V(0, 0); F(0, \frac{1}{32}); y = -\frac{1}{32}$

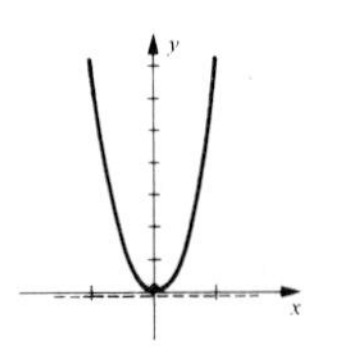

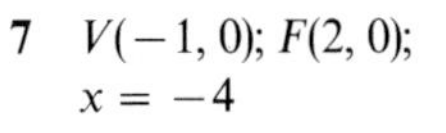

7 $V(-1, 0); F(2, 0); x = -4$

9 $V(2, -2); F(2, -\frac{7}{4}); y = -\frac{9}{4}$

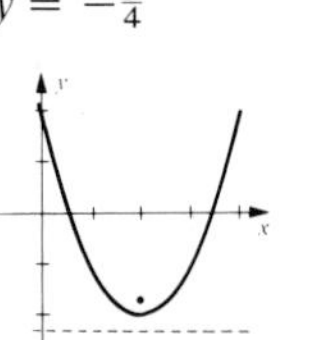

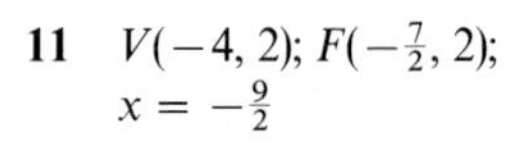

11 $V(-4, 2); F(-\frac{7}{2}, 2); x = -\frac{9}{2}$

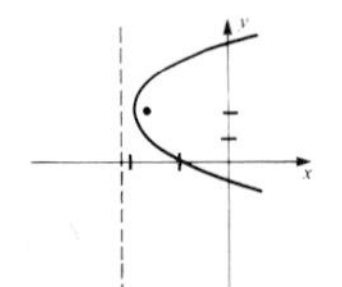

13 $V(-5, -6); F(-5, -\frac{97}{16}); y = -\frac{95}{16}$

15 $V(0, \frac{1}{2}); F(0, -\frac{9}{2}); y = \frac{11}{2}$

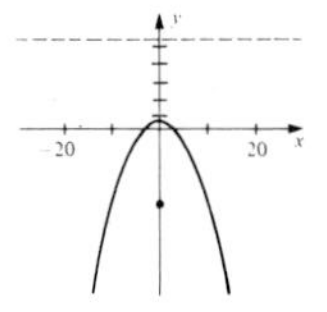

17 $y^2 = 8x$ **19** $(x-6)^2 = 12(y-1)$

21 $3x^2 = -4y$ **23** $\frac{9}{16}$ ft from the vertex

Exercises 10.3, page 491

1 $V(\pm 3, 0)$; $F(\pm\sqrt{5}, 0)$ **3** $V(0, \pm 4)$; $F(0, \pm 2\sqrt{3})$

5 $V(0, \pm\sqrt{5})$; $F(0, \pm\sqrt{3})$ **7** $V(\pm\frac{1}{2}, 0)$; $F(\pm\sqrt{21}/10, 0)$

9 Center (4, 2), vertices (1, 2) and (7, 2), $F(4 \pm \sqrt{5}, 2)$; endpoints of minor axis (4, 4) and (4, 0)

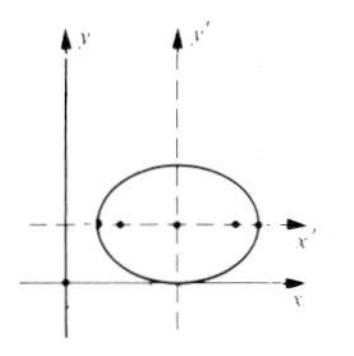

11 Center $(-3, 1)$, vertices $(-7, 1)$ and $(1, 1)$, $F(-3 \pm \sqrt{7}, 1)$; endpoints of minor axis $(-3, 4)$ and $(-3, -2)$

13 Center (5, 2), vertices (5, 7) and $(5, -3)$, $F(5, 2 \pm \sqrt{21})$; endpoints of minor axis (3, 2) and (7, 2)

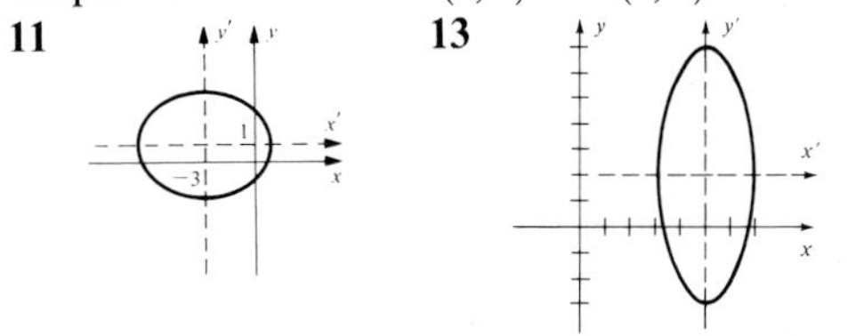

15 $x^2/64 + y^2/39 = 1$ **17** $4x^2/9 + y^2/25 = 1$

19 $8x^2/81 + y^2/36 = 1$

21 $\{(2, 2), (4, 1)\}$

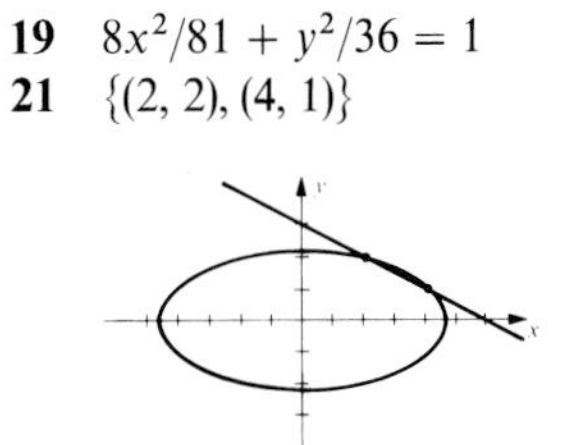

23 $2\sqrt{21}$ ft

27 $A = 4a^2b^2/(a^2 + b^2)$

Exercises 10.4, page 498

1 $V(\pm 3, 0)$; $F(\pm\sqrt{13}, 0)$; $y = \pm\frac{2}{3}x$

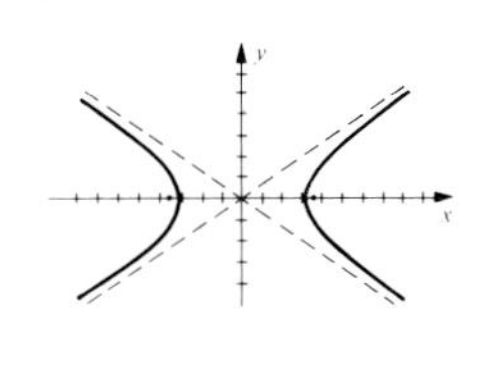

3 $V(0, \pm 3)$; $F(0, \pm\sqrt{13})$; $y = \pm\frac{3}{2}x$

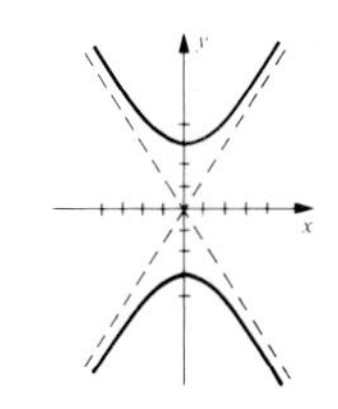

5 $V(0, \pm 4)$; $F(0, \pm 2\sqrt{5})$; $y = \pm 2x$

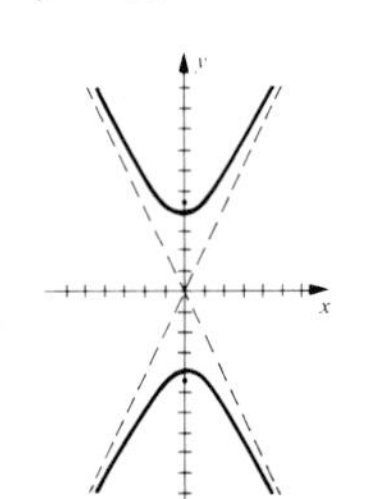

7 $V(\pm 1, 0)$; $F(\pm\sqrt{2}, 0)$; $y = \pm x$

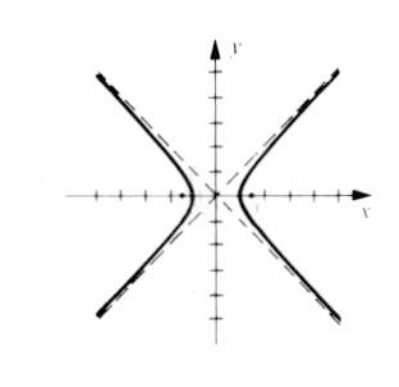

9 $V(\pm 5, 0)$; $F(\pm\sqrt{30}, 0)$; $y = \pm(\sqrt{5}/5)x$

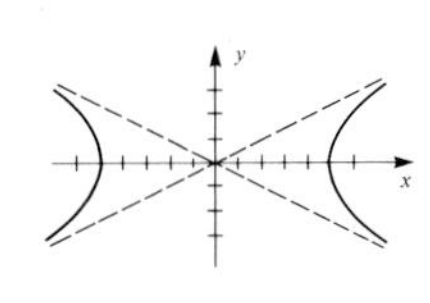

11 $V(0, \pm\sqrt{3})$; $F(0, \pm 2)$; $y = \pm\sqrt{3}x$

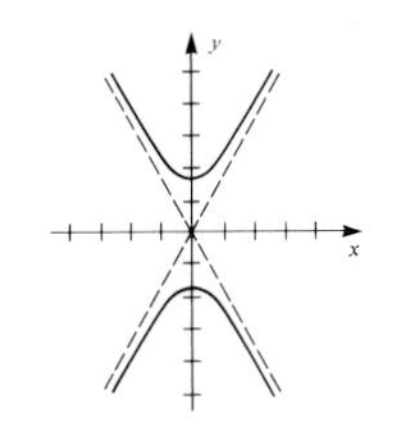

13 Center $(-5, 1)$; vertices $(-5 \pm 2\sqrt{5}, 1)$; $F(-5 \pm \sqrt{205}/2, 1)$; $y - 1 = \pm\frac{5}{4}(x + 5)$

15 Center $(-2, -5)$; vertices $(-2, -2)$ and $(-2, -8)$; $F(-2, -5 \pm 3\sqrt{5})$; $y + 5 = \pm\frac{1}{2}(x + 2)$

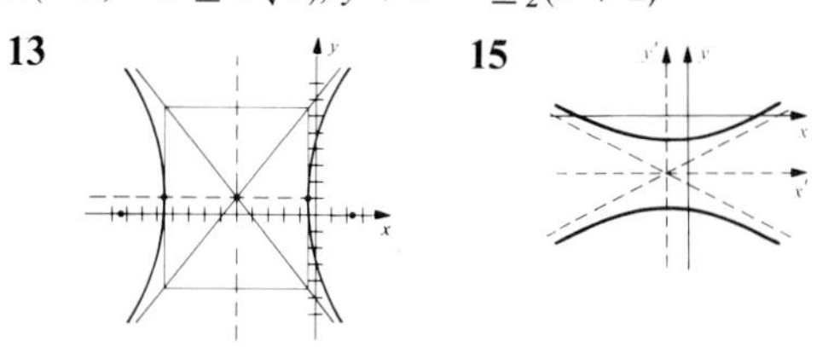

17 Center $(6, 2)$; vertices $(6, 4)$ and $(6, 0)$; $F(6, 2 \pm 2\sqrt{10})$ $y - 2 = \pm\frac{1}{3}(x - 6)$

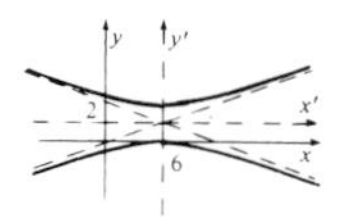

19 $15y^2 - x^2 = 15$

21 $x^2/9 - y^2/16 = 1$

23 $y^2/21 - x^2/4 = 1$

25 $x^2/9 - y^2/36 = 1$

27 $\{(0, 4), (\frac{8}{3}, \frac{20}{3})\}$

29 Conjugate hyperbolas have the same asymptotes.

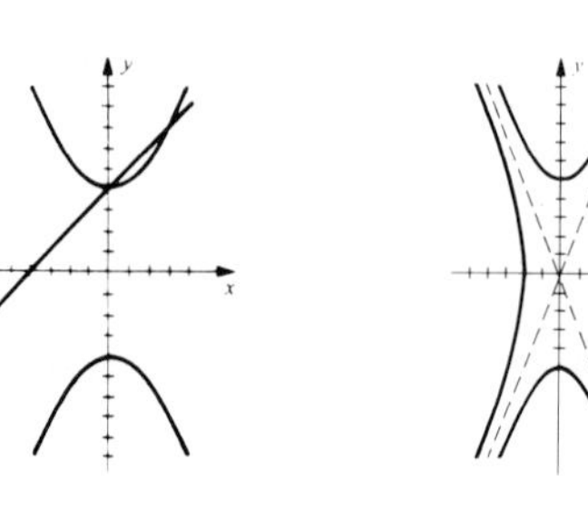

31 $x^2 - 4y^2 = 9$

Exercises 10.5, page 503

The following answers contain equations in x' and y' resulting from a rotation of axis.

1 Ellipse, $(x')^2 + 16(y')^2 = 16$

3 Hyperbola, $4(x')^2 - (y')^2 = 1$

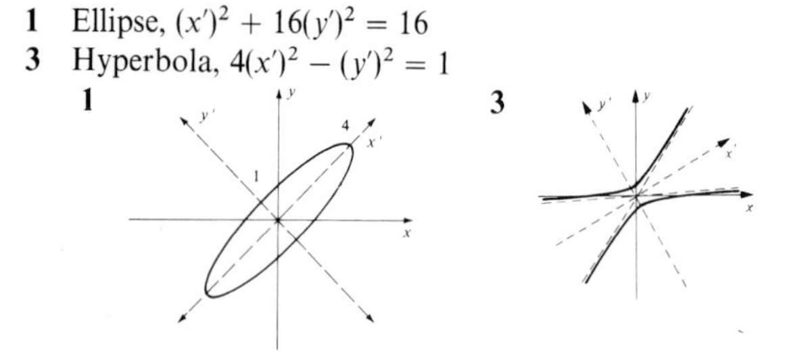

5 Ellipse, $(x')^2 + 9(y')^2 = 9$

7 Parabola, $(y')^2 = 4(x' - 1)$

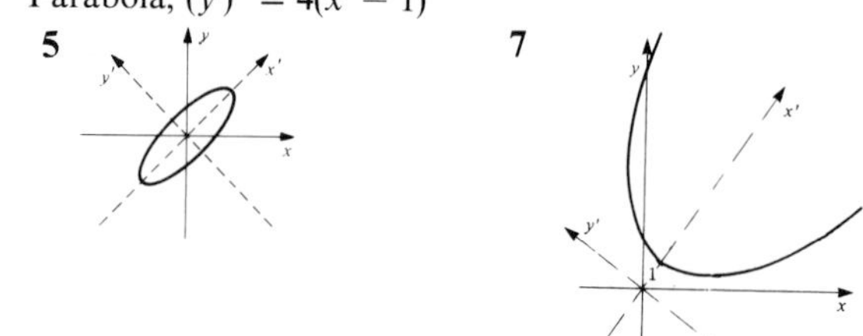

9 Hyperbola, $2(x')^2 - (y')^2 - 4y' - 3 = 0$

11 Parabola, $(x')^2 - 6x' - 6y' + 9 = 0$

13 Ellipse, $(x')^2 + 4(y')^2 - 4x' = 0$

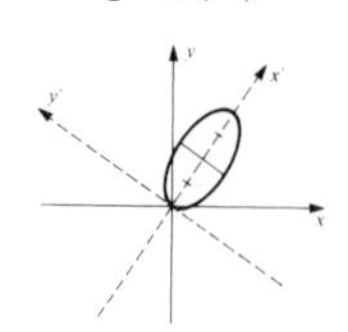

15 Sketch of proof: It can be shown (see page 501) that $B^2 - 4AC = B'^2 - 4A'C'$. For a suitable rotation of axes we obtain $B' = 0$ and the transformed equation has the form $A'x'^2 + C'y'^2 + D'x' + E'y' + F' = 0$. Except for degenerate cases, the graph of the last equation is an ellipse, hyperbola, or parabola if $A'C' > 0$, $A'C' < 0$, or $A'C' = 0$, respectively. However, if $B' = 0$, then $B^2 - 4AC = -4A'C'$ and hence the graph is an ellipse, hyperbola, or parabola if $B^2 - 4AC < 0$, $B^2 - 4AC > 0$, or $B^2 - 4AC = 0$, respectively.

Exercises 10.6, page 510

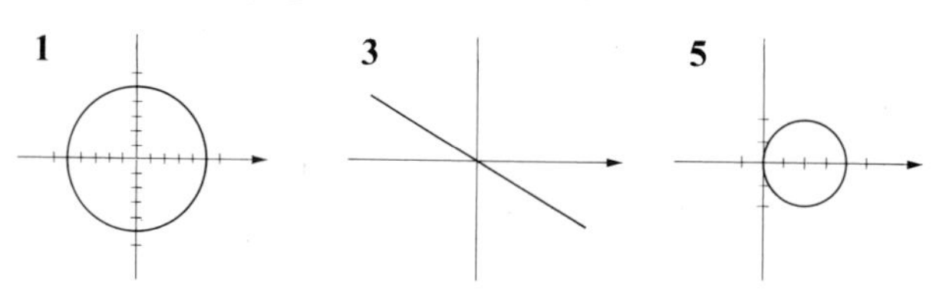

7 **9** **11**

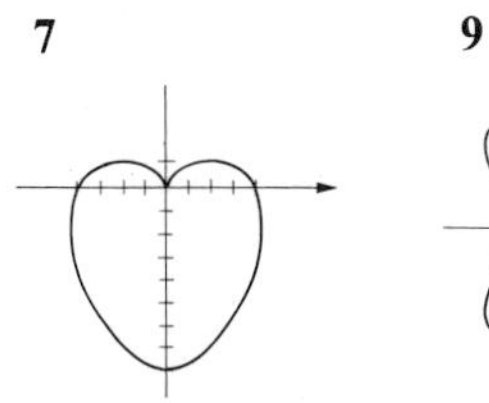

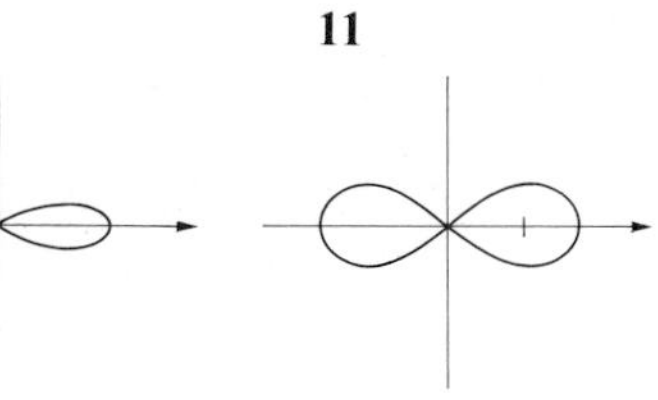

13 **15** **17**

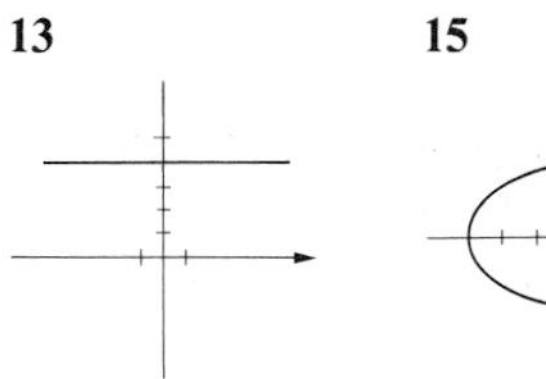

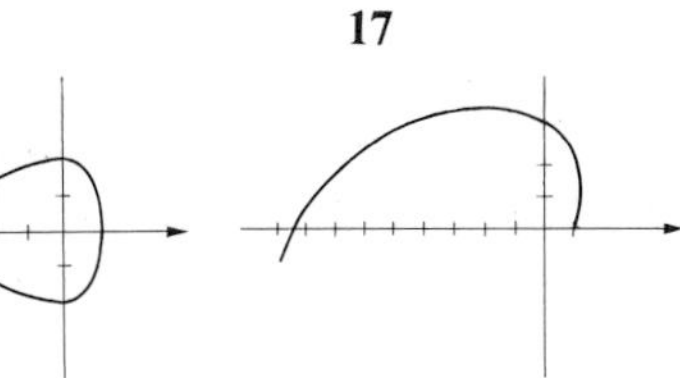

19 **21** **23**

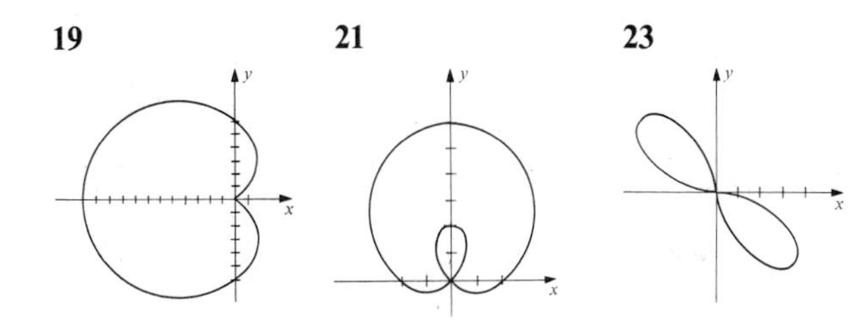

25 $r = -3 \sec \theta$ **27** $r = 4$ **29** $r = 6 \csc \theta$

31 $r^2 = 16 \sec 2\theta$

33 $x = 5$ **35** $x^2 + y^2 - 6y = 0$

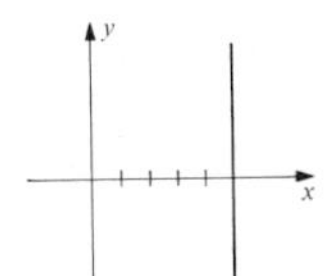

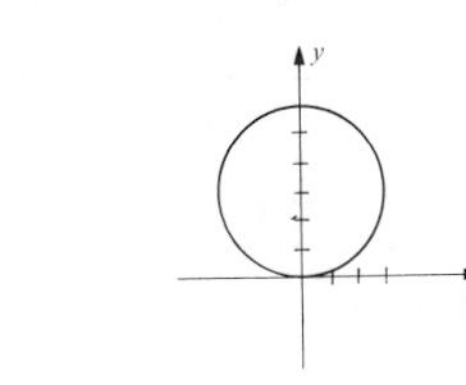

37 $x^2 + y^2 = 4$ **39** $y^2 = x^4/(1 - x^2)$

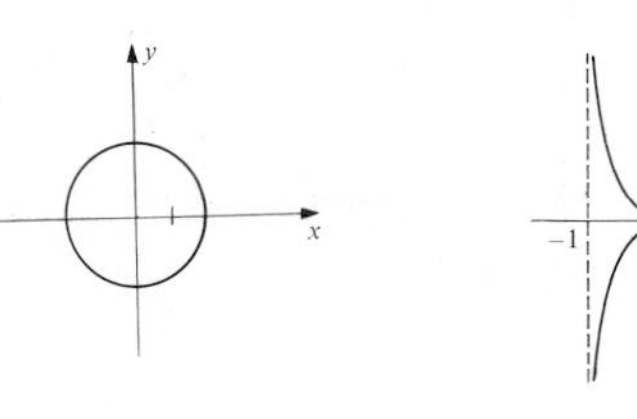

41 $y^2/9 - x^2/4 = 1$ **43** $x^2 - y^2 = 1$

45 $y - 2x = 6$ **47** $y^2 = 1 - 2x$

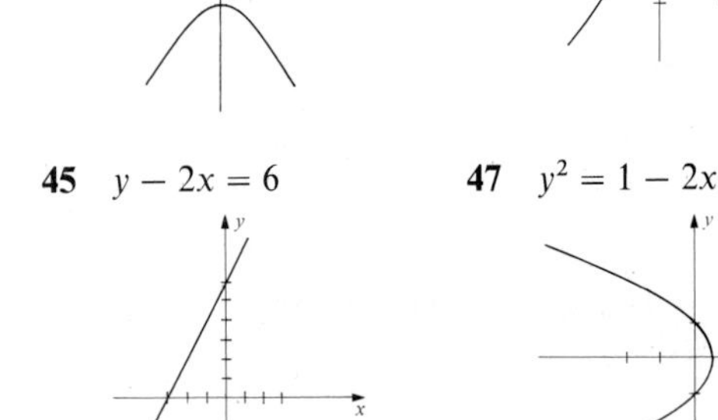

Exercises 10.7, page 515

1 Ellipse: vertices $(\frac{3}{2}, \pi/2)$ and $(3, 3\pi/2)$; foci $(0, 0)$ and $(\frac{3}{2}, 3\pi/2)$
3 Hyperbola; vertices $(-3, 0)$ and $(\frac{3}{2}, \pi)$; foci $(0, 0)$ and $(-\frac{9}{2}, 0)$
5 Parabola; $V(\frac{3}{4}, 0)$; $F(0, 0)$
7 Ellipse; vertices $(-4, 0)$ and $(-\frac{4}{3}, \pi)$; foci $(0, 0)$ and $(-\frac{8}{3}, 0)$
9 Hyperbola (except for the points $(\pm 3, 0)$); vertices $(\frac{6}{5}, \pi/2)$ and $(-6, 3\pi/2)$; foci $(0, 0)$ and $(-\frac{36}{5}, 3\pi/2)$
11 $9x^2 + 8y^2 + 12y - 36 = 0$
13 $y^2 - 8x^2 - 36x - 36 = 0$
15 $4y^2 = 9 - 12x$ **17** $3x^2 + 4y^2 + 8x - 16 = 0$
19 $4x^2 - 5y^2 + 36y - 36 = 0$
21 $r = 2/(1 + \cos \theta)$; $e = 1$; $r = 2 \sec \theta$
23 $r = 4/(1 + 2 \sin \theta)$; $e = 2$; $r = 2 \csc \theta$
25 $r = 2/(3 + \cos \theta)$; $e = \frac{1}{3}$; $r = 2 \sec \theta$
27 $r = 2/(3 + \cos \theta)$ **29** $r = 12/(1 - 4 \sin \theta)$
31 $r = 5/(1 + \cos \theta)$ **33** $r = 8/(1 + \sin \theta)$

Exercises 10.8, page 522

1 $y = 2x + 7$ **3** $y = x - 2$

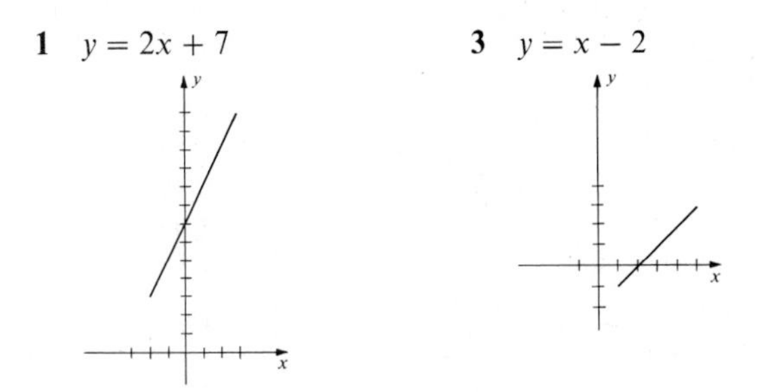

5 $x = y^2 - 6y + 4$

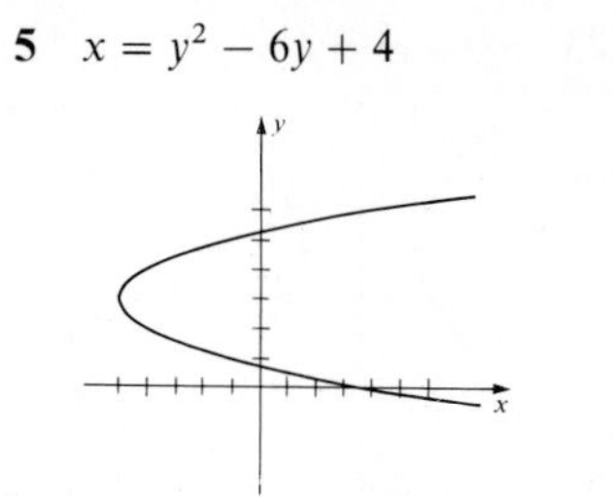

7 $y = 1/x^2$

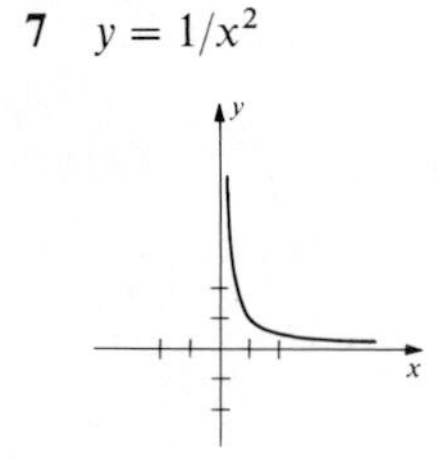

9 $x^2/4 + y^2/9 = 1$

11 $x^2 - y^2 = 1$

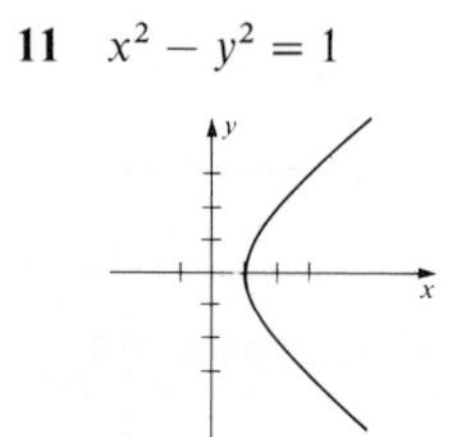

13 $y = \ln x$

15 $y = 1/x$

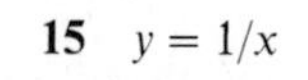

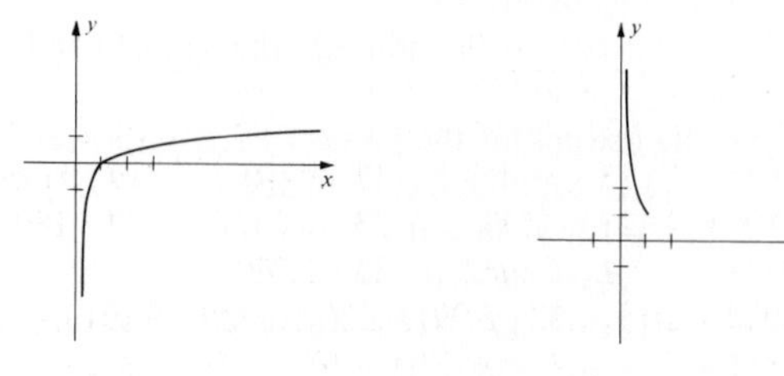

17 $y = \sqrt{x^2 - 1}$

19

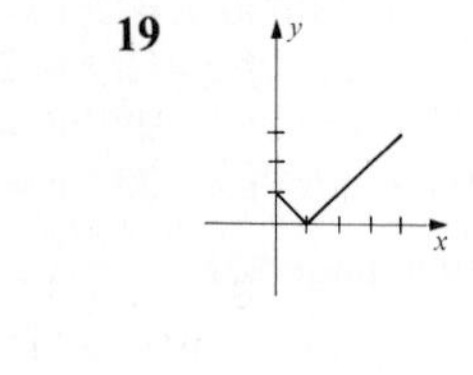

21

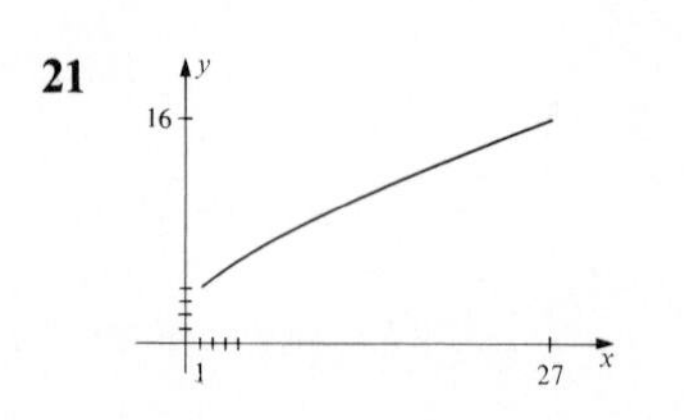

23

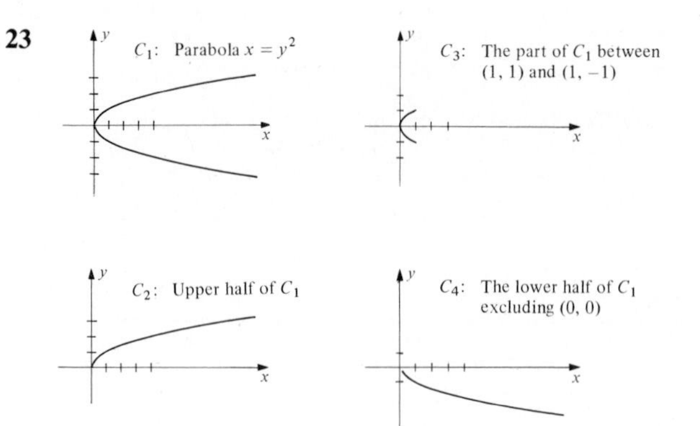

25 $x = (x_2 - x_1)t^n + x_1$, $y = (y_2 - y_1)t^n + y_1$ are parametric equations for l if n is any odd positive integer.

27

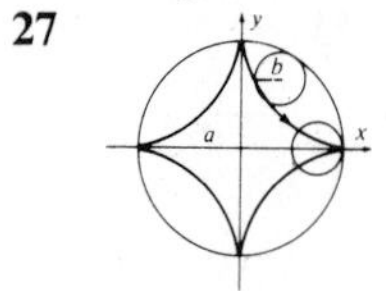

29 $x = 4b \cos t - b \cos 4t$,
$y = 4b \sin t - b \sin 4t$

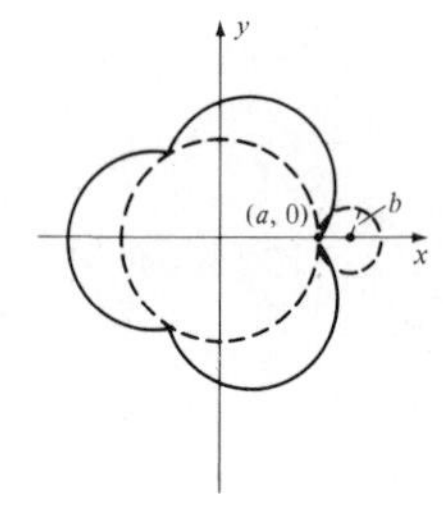

Exercises 10.9, page 524

1 Parabola; $F(16, 0)$; $V(0, 0)$

3 Ellipse; $F(0, \pm\sqrt{7})$; $V(0, \pm 4)$

5 Hyperbola; $F(\pm 2\sqrt{2}, 0)$; $V(\pm 2, 0)$

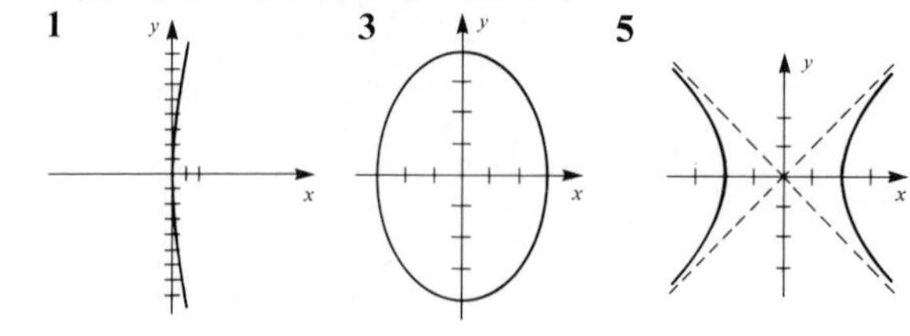

7 Parabola; $F(0, -\frac{9}{4})$; $V(0, 4)$
9 Hyperbola: vertices $(-5, 5)$ and $(-3, 5)$; foci $(-4 \pm \sqrt{10}/3, 5)$

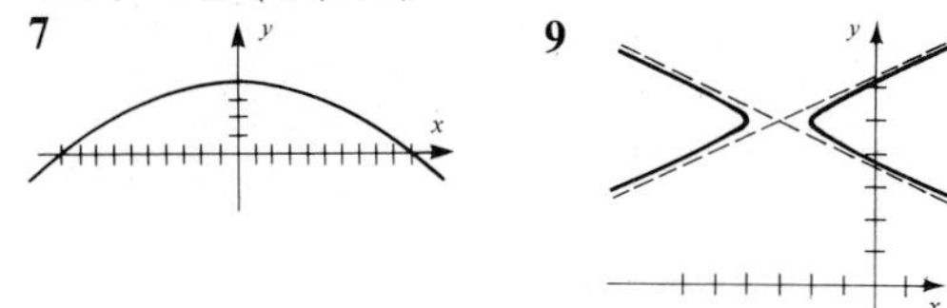

11 $9y^2 - 49x^2 = 441$ **13** $x^2 = -40y$
15 $4x^2 + 3y^2 = 300$ **17** $y^2 - 81x^2 = 36$
19 Ellipse; center $(-3, 2)$, vertices $(-6, 2)$ and $(0, 2)$, endpoints of minor axis $(-3, 0)$ and $(-3, 4)$
21 Parabola; $V(2, -4)$, $F(4, -4)$
23 Hyperbola; center $(2, -3)$, vertices $(2, -1)$ and $(2, -5)$, endpoints of conjugate axis $(2 \pm \sqrt{2}, -3)$

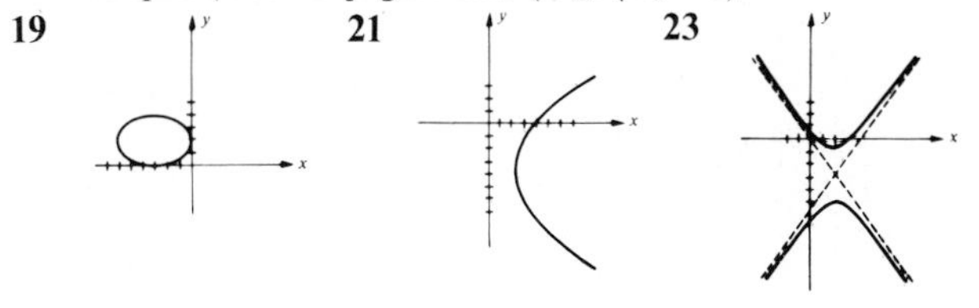

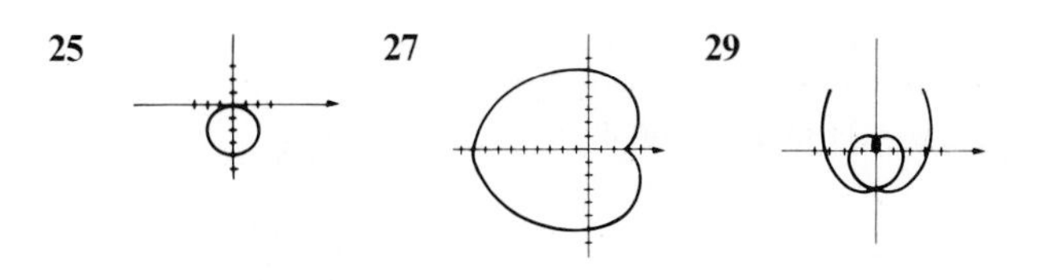

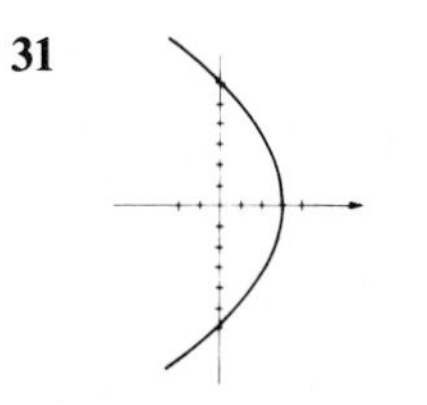

33 $r \sin^2 \theta = 4 \cos \theta$ **35** $r(2 \cos \theta - 3 \sin \theta) = 8$
37 $x^3 + xy^2 = y$ **39** $(x^2 + y^2)^2 = 8xy$

41 $y = 2(x - 1) - 1/(x - 1)$ **43** $y = 2^{-x^2}$

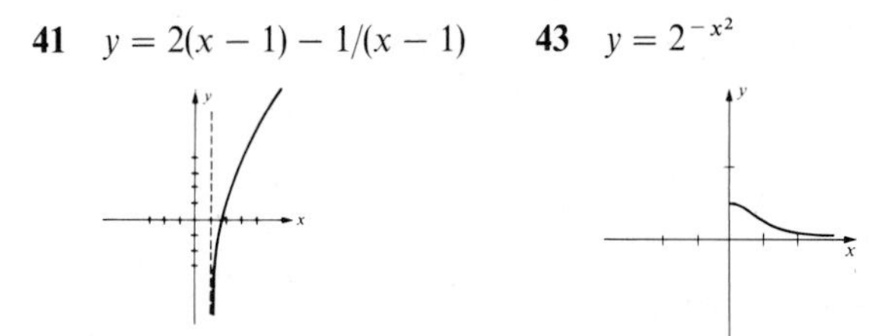

45 Parabola: $(y')^2 - 3x' = 0$ is an equation after a rotation of axes.

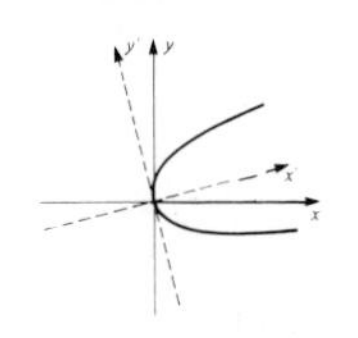

47 $\dfrac{x^2}{10000} + \dfrac{y^2}{1200} = 1$; $20\sqrt{3} \approx 34.64$ ft

Exercises A.1, page A11

1 2.5403, 7.5403 − 10, 0.5403
3 9.7324 − 10, 2.7324, 5.7324
5 1.7796, 5.7796 − 10, 2.7796
7 3.3044, 0.8261, 6.6956 − 10
9 9.9235 − 10, 0.0765, 9.9915 − 10 **11** 4.0428
13 2.0878 **15** 8.1462 **17** 4240 **19** 8.85
21 162,000 **23** 0.543 **25** 0.0677 **27** 189
29 0.0237 **31** 1.4062 **33** 3.7294
35 7.1000 − 10 **37** 5.0913 **39** 9.8913 − 10
41 2.5851 **43** 9.9760 − 10 **45** 4.8234
47 8.6356 − 10 **49** 0.4776 **51** 27.78
53 49,330 **55** 0.1467 **57** 0.006536
59 234.7 **61** 1.367 **63** 0.09445
65 109,900 **67** 0.001345 **69** 0.2442
71 0.4440 **73** −0.1426 **75** 1.032
77 0.7911 **79** 0.5217 **81** 3.436
83 0.5315 **85** 1.3521 **87** 0.7703
89 21°33′, 158°27′ **91** 26°45′, 206°45′
93 69°44′, 290°16′ **95** 43°46′, 316°14′

Index

THE TRIGONOMETRIC FUNCTIONS

I. OF ACUTE ANGLES

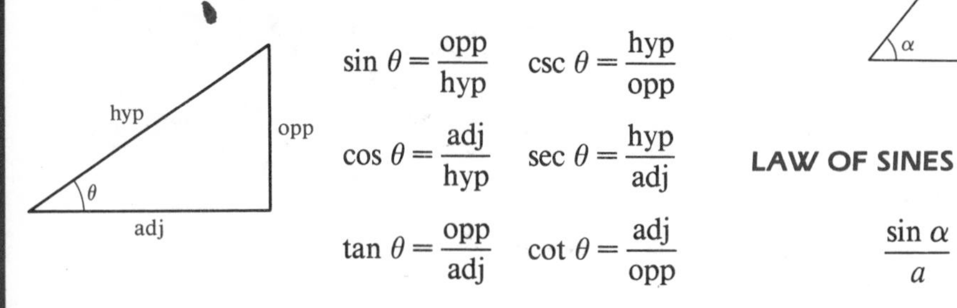

$$\sin\theta = \frac{\text{opp}}{\text{hyp}} \qquad \csc\theta = \frac{\text{hyp}}{\text{opp}}$$

$$\cos\theta = \frac{\text{adj}}{\text{hyp}} \qquad \sec\theta = \frac{\text{hyp}}{\text{adj}}$$

$$\tan\theta = \frac{\text{opp}}{\text{adj}} \qquad \cot\theta = \frac{\text{adj}}{\text{opp}}$$

II. OF ARBITRARY ANGLES

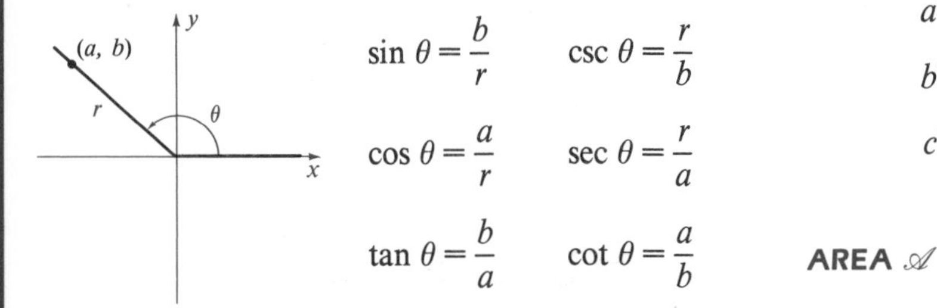

$$\sin\theta = \frac{b}{r} \qquad \csc\theta = \frac{r}{b}$$

$$\cos\theta = \frac{a}{r} \qquad \sec\theta = \frac{r}{a}$$

$$\tan\theta = \frac{b}{a} \qquad \cot\theta = \frac{a}{b}$$

III. OF REAL NUMBERS

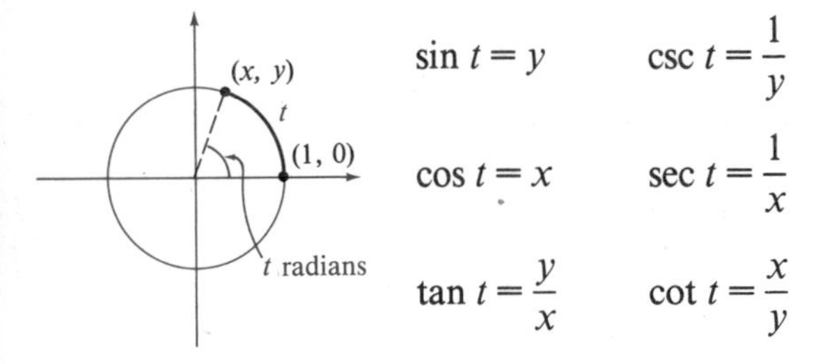

$$\sin t = y \qquad \csc t = \frac{1}{y}$$

$$\cos t = x \qquad \sec t = \frac{1}{x}$$

$$\tan t = \frac{y}{x} \qquad \cot t = \frac{x}{y}$$

OBLIQUE TRIANGLES

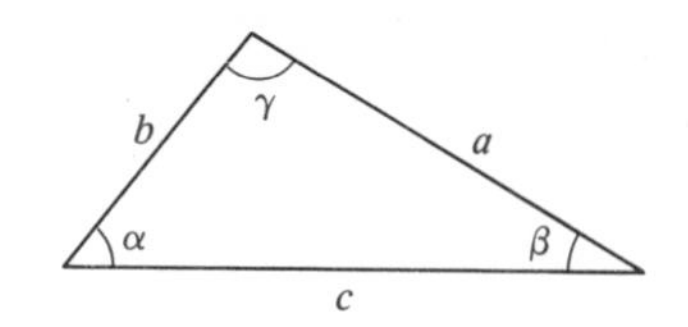

LAW OF SINES

$$\frac{\sin\alpha}{a} = \frac{\sin\beta}{b} = \frac{\sin\gamma}{c}$$

LAW OF COSINES

$$a^2 = b^2 + c^2 - 2bc\cos\alpha$$

$$b^2 = a^2 + c^2 - 2ac\cos\beta$$

$$c^2 = a^2 + b^2 - 2ab\cos\gamma$$

AREA $\mathscr{A}$

$$\mathscr{A} = \tfrac{1}{2}ab\sin\gamma = \tfrac{1}{2}bc\sin\alpha = \tfrac{1}{2}ac\sin\beta$$

$$\mathscr{A} = s\sqrt{(s-a)(s-b)(s-c)}$$

where $s = \frac{1}{2}(a+b+c)$